HARVARD ORIENTAL SERIES

Edited by MICHAEL WITZEL

VOLUME SEVENTY-FOUR

ETHNOGRAPHIC NOTES

ON THE

MRU

AND KHUMI

of the Chittagong and Arakan Hill Tracts.
A Contribution to our Knowledge of South and Southeast Asian Indigenous Peoples
mainly based on field research in the Southern Chittagong Hill Tracts

by

Lorenz G. Löffler

PUBLISHED BY THE DEPARTMENT
OF SOUTH ASIAN STUDIES,
HARVARD UNIVERSITY

DISTRIBUTED BY
HARVARD UNIVERSITY PRESS
CAMBRIDGE, MASSACHUSETTS
AND LONDON, ENGLAND

2012

For information write to Editor, Harvard Oriental Series,
Department of South Asian Studies,
1 Bow Street, Cambridge MA 02138, USA
617-495 3295; email: witzel@fas.harvard.edu

Library of Congress Cataloguing in Publication Data
Ethnographic Notes on the Mru and Khumi of the Chittagong and Arakan Hill Tracts.
A Contribution to our Knowledge of South and Southeast Asian Indigenous Peoples
mainly based on field research in the Southern Chittagong Hill Tracts
by Lorenz G. Löffler

(Harvard Oriental Series; v. 74)
ISBN 978-0-674-06686-1
I. Lorenz G. Löffler, 1930-
II. Title
III. Series: Harvard Oriental Series; 74

CIP

Contents

A The research

A1 Previous literature

When I wrote the first (German) version of this chapter in 1960, the most important ethnographic details on the Chittagong Hill Tracts in general were to be found in Lewin (1869, 1870), Hutchinson (1906, 1909) and Mills (1931). By and large this is still true today. Though the bibliography lists a number of further titles, none of these works includes more detailed data of general importance: up to the latest time they consist of unsystematic journey notes, in which occasional interesting details can be found, or just statements offering nothing new but quotations from older works.

What can be found in the three major old works about the Mru and Khumi lacks exhaustiveness in every regard. A major trait of the Mru culture, the cattle sacrifice, is not mentioned at all. Lewin's details are generally exact, although there are two or three remarks which I could not confirm. Also Mills contains a couple of errors. Hutchinson is only reliable insofar as he takes (without acknowledgement) his details from Lewin; his own additions are frequently inaccurate and occasionally wrong.

Apart from the few data in the works just mentioned, little can be found on the other ethnic groups of the Chittagong Hill Tracts, and what exists about the Mru and Khumi of this area till now is still far more meagre and more inadequate. Apart from some notes on them in the three "major works" mentioned at the beginning (the Khumi only in Lewin and Hutchinson), they are barely mentioned at all (only Riebeck 1885 gave some information). Notes on the Mru in Dalton (1872) and in the Linguistic Survey (1904) are second-hand and, moreover, refer mainly to the Mru of Arakan. Kauffmann was able to copy some exact and hitherto unpublished data on house forms and equipment of the Mru and Khumi, in Kohima in 1936, which were collected by A. S. Hands, who later became Deputy Commissioner of the Chittagong Hill Tracts, in my opinion during a tour in the year 1928. Though newer reports on the Mru and Khumi of Arakan are completely missing, a large number of older notes dealing mainly with the Khumi (or Khami) can be found in Phayre (1841), Latter (1846), Tickell (1854), St. Andrew St. John (1872), and Hughes (1881). They do not always distinguish exactly between Khami and Mru; according to Dalton (1872:113) "Mru" was, besides "Chin" a general term for mountain dwellers (O'Donel 1864:405 wrote Mroo-Khyeng), a usage that occasionally led to mistakes (e.g., Webb 1901 "Mro" instead of "Khami"). Details on different topics of the culture of the Khami, Mru and Khumi of Arakan were first given in the Census reports of U Ohn Pe (1931), U Ba Thin (1931) and primarily U Ba Myaing (1934); Mru vocabularies were published by Phayre (1841), Lewin (1870), Campbell 1874 (probably collected by Lewin as well), Konow (Linguistic Survey 1904) and U Ba Myaing (1934). The more recent the date of these glossaries the poorer their quality, and for practical purposes they are all useless; what can be gleaned from them for comparative linguistics has been collected by Shafer (1943). More numerous are the word lists published on the dialects of the Khumi and Khami, which were critically evaluated and completed by me in a separate publication (Löffler 1960). I could not check the ethnographic details made by the sources for Arakan, but I can find little reason to call them seriously into question by comparison with my materials. The main value of the data is for the time before the British took possession of the area. A detailed critical review of every single publication is not necessary here, since in the following I will quote and, whenever necessary, critically scrutinise all notes accessible to me on the culture of the Mru and Khumi in their context, provided that they tell us something essential.

The first scientific enterprise devoted to ethnographic investigation in the Chittagong Hill Tracts was by Denise and Lucien Bernot, who in 1950/51 concentrated mainly on a study of the Marma. The main part of their results was not yet published when I started to write my Mru manuscript, but in 1967 Lucien Bernot published "Les paysans arakanais du Pakistan oriental. L'histoire, le monde végétal et l'organisation sociale des réfugiés Marma (Mog)"; 793 pp. in two volumes, followed in the same year by "Les Cak, contributions à l'étude ethnographique d'une population de langue *loi*"; 385 pp.

I did my first research on the Mru and Khumi in 1955–1957; in 1964 I wanted to complete some of my data and took some of my students with me to make a study of the Bawm; our time was cut short, however, since we were refused any Government permit to pursue our studies, and we had to return to Chittagong: the Hill Tracts were closed to foreigners for the next 30 years. I managed to receive a special permit at the end of 1990, but was only allowed to proceed up to Banderban and to make a short visit to Rangamati, the main town of the Chittagong Hill Tracts. (My report on this short stay will be added at the end of this book.) However, in 1964 one of my students, Hans-Jürgen Spielmann, managed to evade the local police by walking from the Bawm country north to Rangamati and to collect data, which he finally published in 1968 in his dissertation: "Die Bawm-Zo. Eine Chin-Gruppe in den Chittagong Hill Tracts (Ostpakistan)"; 342 pp.

Another of my students, Almut Holtheuer, who later married Wolfgang Mey, with whom she made a short (illegal) visit to the Chittagong Hills, published, in 1979, "Untersuchungen zur Wirtschaft in den Chittagong Hill Tracts (Bangladesh)"; 368 pp. Despite its general title, this dissertation contains mainly data on the Bawm, collected after the field work by means of a survey conducted by an educated Bawm. Almut's husband, Wolfgang Mey, in 1980 published a dissertation entitled: "Politische Systeme in den Chittagong Hill Tracts"; 275 + 16 pp.; mainly dealing with the Chakma, and in part based on historical data collected in London. In the subsequent years W. Mey became active on behalf of the indigenous (mainly Chakma) resistance force of the Chittagong Hills and published several papers and booklets on the situation there. The Chakma themselves also started to publish and one of the most easily available sources was published by IWGIA (Copenhagen) and written by Rajkumari (the daughter of the Chakma Raja Tridiv Roy) Chandra Roy: "Land rights of the indigenous peoples of the Chittagong Hill Tracts, Bangladesh"; 231 pp. Other books on the situation in the Chittagong Hill Tracts followed, I shall mention here only the book edited by Subir Baumik et al. 1997. "Living on the Edge, essays on the Chittagong Hill Tracts"; 289 pp.; it also contains a contribution by the present Chakma Raja, Devasish Roy.

Returning to the Mru, I must also mention a German photographer, Claus-Dieter Brauns, who, with the help of a Bengali "tourist guide", managed to make several visits to a Mru hamlet near the Matamuri during the time when the Chittagong Hills were officially closed. He had published several papers with his photographs in three international magazines, but when he wanted to publish a book, for which he already had found a publisher (Birkhäuser, Basel), the publisher wanted a text as well, and so Brauns turned to me and asked me whether I couldn't write 100 pages on the Mru. I accepted, wrote my text in two months and helped with the layout of the book. In 1986 the book was published in a German version: "Mru, Bergbewohner im Grenzgebiet von Bangladesh", but this "coffee-table book" (as one of the reviewers called it) sold so badly, that Brauns decided to spend a further 100,000 CHF to publish an English version as well. This appeared in 1990 under the title "Mru, hill people on the border of Bangladesh", but sales did not go up significantly and Birkhäuser decided to sell off the remaining stock of this rather high-priced book, now at a cheap sales rate.

As for photographs, only one picture of a Mru group was published before 1957: it is to be found in Hutchinson 1906 and can hardly satisfy today's requirements. Six photos of later date were published in the "Tribesmen of the Chittagong Hill Tracts", compiled in 1958 by Pierre Bessaignet. The quality of the reproduction, however, is completely inadequate. Several photos taken in 1926 by Mills were not published; the author kindly provided me with them and they were published (posthumously), together with those of C.-D. Brauns and some of my own, in the editions of my first book on the Mru. For this edition I have replaced most of them by my own photos.

A2 The German Chittagong Hills expedition

A2a) The preparations

The main burden of the preparations was carried by the leader of the expedition, Professor Dr. Hans E. Kauffmann, who in the beginning acted as my tutor. Credit for the planning and commitment for the materialisation of the expedition, as also the naming, are due to him alone. I therefore wanted to leave the writing of this sub-chapter to him. This plan could not be realised, however, Kauffmann died in the meantime, and since I lack the requisite knowledge, now the corresponding details are missing too.

A2b) The course of events

In the following I give only a short summary; a detailed timetable with a short summary of my activities can be found at the end of this chapter.

The first month in the Hill Tracts began favourably, insofar as our first visits both in Rangamati and in Banderban coincided with the Punyah of the Rajas (the "lucky day" of the princes on which they received the collected taxes). Since representatives from all areas of the administration "circles" meet for these celebrations, possibilities offered themselves for initial contacts. Accompanied by a Marma tax collector, a relative of the local Raja, who had offered himself for this job, we spent Christmas on a tour to the Boga lake, situated without outlet high up on a mountain. Following an invitation, we took part in a celebration of Hrui Pyo Ong, the Marma headman of Polika-Mouza at the end of December.

At the beginning of the year 1956 we were in Ruma (a police station with a small bazaar), waiting in vain for the promised return of our first companion (the tax collector) and staying put because we lacked sufficient knowledge of the country, had no linguistic proficiency and were having difficulties with the money supply. The leader of the expedition had to make new arrangements for our supply, to find a new boy (the first was a Bengali, who had offered himself when we arrived in the port of Chittagong, he was later succeeded by a Marma who had been raised in the house of the Raja as an orphan). In addition we lacked a map of the region, which preferably would also show the tribal distribution but this did not exist at all. Finally the leader also had to settle the difficulties with the custom office for sending back the exposed film material. Therefore in the middle of the month he went back to Banderban and Chittagong and at the end of January also to Dhaka. By the end of January I myself set out, in the company of my new boy, Kyo Thwân Ong (to be pronounced "kyothwàe🕒õ") on my first reconnaissance tour to the Mru area, followed from the end of February to the middle of March by a tour to the Twipra (formerly written Tipperah or Tipra) and the Khumi near the upper Songu river. By the end of March I was in Banderban (with an intermediate visit to a spring festival of the Marma), where I met the leader of the expedition (who had to break off a journey of his own to the Chüngma Mru on the Songu river, due to illness).

At the beginning of April I went to the Mru hamlet of Tapwüa-Kua where I looked for "Mastor" Menkröi Ngarua', who was said to possess a book with all Mru words in it. This book turned out to be a translation of the gospel of St. John, prepared by Menkröi when, in the '40s, he had worked as missionary for the Baptist mission of the Lushai pastor Dala. Menkröi had had no success in converting the Mru, but was willing to work with me. In the middle of April I was back in Banderban again, where I took part in the Buddhist New Year's celebrations of the Marma and said goodbye to the leader of our expedition, who had to return to Germany in order to be treated for his illness, an elephantiasis infection, which he had probably contracted during our stay in Madras.

At the end of the month I moved with all my luggage to Tapwüa-Kua, where at first I took residence in Menkröi's house. This man was to become my main informant. Before the beginning of the rainy season I wanted still to do a short supply tour to Chittagong. As it was the end of May, I was, however, only able to return to Banderban, where I was told that the road leading in the direction of Tapwüa-Kua had been washed away, tried to reach Galengya by boat (from where I could climb uphill to Tapwüa-Kua), but due to the rising waters of the Songu, the boat could not be poled so far and finally, after a week, I allowed the tired boatman to return to Banderban. We reached this town in a few hours, I stayed in the house of Kyo Thwân Ong, made several new attempts to reach Tapwüa-Kua, but in the end did not manage to reach it until the end of June. A period of intensive ethnographic work followed until the beginning of October. Then, until the middle of November, I went again to Chittagong (to find somebody to repair my tape recorder, but in the end the trouble was in vain) and to Rangamati (to visit the authorities and to acquire the material for a tribal map of the central Hill Tracts). I spent from the end of November until December in Tapwüa-Kua again, from where I made two tours into the area of the Dopreng Mru.

At the beginning of January 1957, after another short visit to Chittagong, I travelled up the Songu, where I purchased a boat of my own and visited the local Mru and Twipra. By the end of the month, I was in Tapwüa-Kua again. February was a month in which I was not very successful (except for a short visit to the Betchora-Khumi), not only because almost all plans failed due to unexpected delays, but also because I myself was tied down by repeated illness. In March I did an extensive tour from Tapwüa-Kua to the south and visited the Mru groups of the Rümma, Dömrong, Rengmitca and Chüngma as well as the Sak and Khumi. I spent April in Tapwüa-Kua and on different occasions went to surrounding hamlets. In May I was again permanently away: to Chittagong and Rangamati, to Manikchori (to collect material for a tribal map of the northern Hill Tracts), to Artha' (the main place of the Christian Bawm), to Chittagong again and finally back to Tapwüa-Kua where until the beginning of August I tried to complete my ethnographic notes, permanently troubled by amoebic dysentery, which weakened me so much that I decided to return home. Before embarkation in the harbour of Chittagong at the end of August, I still had to make my last visits to the authorities in Banderban, Rangamati and Dhaka.

A2c) Limitations of the work

Good training and freedom from prejudice are ideal demands for field research. They are, however, in this respect in a certain contradiction with each other as hand in hand with the training goes the influence of "schools", which is the more influential, the less the researcher is conscious of it. But even when tendentious questions are largely avoided, there remain certain preferences of the researcher (caused not least by previous knowledge) which, relatively independent of the material at hand, determine the course of his activities and the main emphases of his interest. I therefore regard it as important to briefly explain the preconditions given for me in this regard.

In my case I had no previous field experience or training in the methodology of field research (before 1964 no such training was offered in courses for social and cultural anthropology.) My first *in situ* introduction I owe to the leader of the expedition. My interests, and with them the research direction, were determined by the course of my studies, namely (briefly summarised): the connection of culture and speech (Hestermann), history of civilization, legal systems and social economy (Lips), economic systems and conceptions of the world (Friedrich), group dynamics and social systems (Mühlmann). I was most familiar with the connected questions through my thesis on the religious significance of cattle in "Further India" (done in a culture morphological view under Friedrich).

In my private studies I had become more thoroughly acquainted with the clan and sib structures in northwestern Southeast Asia. Concerning the cultivation of swidden fields I received special support from the leader of the expedition, who in 1957 sent me the excellent work of J. D. Freeman (on Iban Agriculture) according to whose example I tried to complete my material and was able to improve it. In all other fields my previous knowledge was poor or insufficient, especially in the so-called bordering areas such as medicine, botany and zoology, anthropogeography, education, art, etc., so that in these directions some questions were taken into account only inadequately or not at all. Deliberately, however, I did not do any anthropological measuring, though I had received an appropriate training.

Beyond the personal prerequisites, limitations are always also set by the possibilities of contact. Therefore they shall also be mentioned briefly to show the conditions of my ethnographic work in the Hill Tracts and the ways taken by me (with their advantages and faults), in order to provide a basis for the assessment and critical evaluation of the data offered here.

Communication by language should be regarded as the basic means of field research. It is not only important for bare observation in cases in which one sees without being able to understand what is happening, but also always as soon as human partners are present. The languages spoken in the Hill Tracts cannot be learned in Europe. In the '50s only some officials and more senior personalities understood English, the knowledge of the Bengali standard language (Bangla), at least in the southern area of Bangladesh, is restricted to a hardly larger group of people. A knowledge of English is indispensable for dealing with officials and Bangla can be used as a starting point for learning the *lingua franca*, the Chittagong Bangla. Many people have an elementary knowledge of the Bengali script and also understand some expressions of the standard language, which deviates considerably from the local dialect both in grammar and vocabulary.

the Bengali, Romjan Ali from Comilla, who accompanied us for the first months, and the Marma, Kyo Thwân Ong from Banderban, who accompanied me for the duration of my independent activity, had this elementary knowledge. After a certain period of getting acquainted with the mutual vocabulary, we could communicate. In order to reach the level of the local language, I used every opportunity in which Kyo Thwân Ong talked with others in local Bangla, to have him translate words and passages incomprehensible for me into the vocabulary known by me. In the beginning I also tried to get an understanding of the local Arakanese Burmese (Marma), but I soon realised that I could progress more easily and express myself better when I used Bangla. In the course of time, however, I could also understand a bit of Marma.

In selecting Tapwüa-Kua as my base, I made a good choice insofar as most Mru men of this hamlet also preferred local Bangla to Marma when they had to speak a foreign language, and so Kyo Thwân Ong also had to use it and I could profit more from his conversations. Although I continued to enlarge my vocabulary until the end of my stay, by the summer of 1956 it already sufficed to talk with the Mru, who for their part also had only a limited vocabulary in local Bangla. The difficulties I faced in the beginning were caused less by the limitations of mutual vocabulary than by the fact that I first had to learn to adapt the complicated ideas I wanted to express to my available means of expression.

By the beginning of 1956 I had already started to note down some Mru words. An intensification was, however, possible only in Tapwüa-Kua after I could work without an interpreter (standard Bangla > local Bangla or Marma). By autumn 1956 I understood enough Mru to be able to follow the course of a conversation, but I, for my part, was understood only to a very limited extent. The reason was undoubtedly lack of training, since I preferred to speak local Bangla (using Mru expressions only for special cases) and to meet the interlocutor at roughly the same level of knowledge, instead of in Mru, where I was obviously inferior in every regard since Mru is a tonal language! On the other hand I used my knowledge of Mru in contact with the Khumi, since as an unknown person speaking Bangla I was received with the greatest unease (particularly with the Betchora-Khumi who spoke Mru adequately); after switching over to Mru a basis of friendship and confidence was immediately created to me as a strange "hill dweller".

By working nearly daily together with Menkröi Ngarua' in Tapwüa-Kua we were able to completely overcome our difficulties of understanding, since we came to know not only the limitations of our vocabulary, but also the peculiarities of our ways of expressing ourselves. I therefore appreciated having Menkröi present during all important inquiries. He frequently functioned as my interpreter: to the persons with whom I talked, he explained my questions, which they could largely understand themselves in their own language (in which I could check the correctness of his explanation), while it was left to my informants to answer me in Bangla or in Mru, in which I in turn often had a translation which I could check myself and correct it when necessary. Since Menkröi was himself a Mru, he could at the same time judge the probability of the statements in the context of his knowledge of his own culture, particularly since he took very seriously his task of giving me correct information and ensuring the correctness of the information which I received from others. Due to his help, the reliability of my data was much improved compared with those which I received in the time when Kyo Thwân Ong had to interpret for me. My lack of understanding of the language, combined with his inability to render data from a culture foreign to him as a Marma without adding his own interpretation, caused such unreliable results that I have not taken any important point of this material without discussing it again with Menkröi. It still will have to be explained why this relationship with Menkröi also had some disadvantages (see A2f).

A2e) First contacts

An analysis of the position of the field researcher requires an observation of the attribution of roles and the expectations evinced on both sides based on speculations. Attitudes based on prejudices change in the course of the contacts, a corrected picture arises and in retrospect this leads to further reinterpretations of the initial attitudes. However, a scientific analysis can result only from the overcome position, from a distance. Differences in prejudice (pre-judgements) arise from previous knowledge, varies depending on contact experience; a restriction to my experiences among the hill dwellers (Mru and Khumi) reduces the variation only to a certain degree; with persons of greater sophistication a more subtly differentiated position scale will also result. In first position I could generally be identified me as a Sahb (Bangla, mostly pronounced “shab” from Arabian sahib) or Phalong (Burmese). Sahbs (sirs) are people who have authority and the necessary means for exercising it; that is primarily access to political power. For most people it remained difficult to understand that this did not apply to me. However, everybody who can keep others in his service due to special abilities (knowledge or money) or to whom one wants (even if only out of courtesy) to grant a corresponding status is also classified as a Sahb. This applied to me, and in this respect I could not object when I was called Sahb. Kyo Thwân Ong and Menkröi knew both my first and my family name, but they found these names too complicated and difficult to pronounce and preferred to continue to call me “Sahb”.

A variation of the Sahb is the Babu. A Babu has superior knowledge which one can make use of. Someone who applies to a Babu, thereby becomes dependent for a certain time, but does not need to submit to his command. Babus are people who know to write and office assistants, but one can also address doctors and even simple businessmen as Babu to please them. In a certain sense, therefore, I also belonged to the category of the Babus while at the same time the odium of presumption and corruptibility that stuck to them resulted in my companions' directly forbidding such a salutation to people who wanted to call me so: I was a Sahb. In one case a man rebuked for calling me Babu wanted to used "Kaphameng" (Government) as a salutation thenceforth, which makes obvious the connection of the concept Sahb with the authority of office. Within the category "Sahb" I belonged by my skin colour and as a non-Bengali to a special category: to the upper circles (Huzur Sahb), which are characterised by a (revocable) assignment of authority over the lower orders.

In the beginning of my stay the topmost person in authority in the Hill Tracts was a European (as had been the case since the beginning of the colonial administration), although he was hardly responsive to a "normal mortal"; so it was quite normal that in local eyes I must be one of his relatives, either personally (for instance one of his sons) or as a jati ("caste") member. This kind of Sahb position gave rise to conflicting expectations of my role: first one of discomfort and fear of disaster, and then one of hope for support against the Bengali Sahbs. If flight was not possible, both expectations resulted in trying to keep me in a good mood and to leave an impression of kindness. My indication that I could do nothing in the area of "troubles with the administration" often remained difficult to understand and had to be interpreted as lack of readiness to help; I appeared as both harmless and incompetent at best, but this did not render my intentions less suspect.

I reached these insights only with time. At first I felt quite worried by the situation when the usual effusive kindness during my first visits to a hamlet turned into an uncomfortable restraint during the second visit. Only in Tapwüa-Kua did I finally learn that the initial kindness was dictated by fear and the openly shown uneasiness was already progress, when the people themselves talked to me about their own first reaction. Thus I learned that the man who had in reality most opposed my presence on the first day typically asked me to stay in his house.

A2f) My position in Tapwüa-Kua

For an exact evaluation of the development of my relations with the residents of this hamlet I lack some clues – firstly because I could not follow the consultations of the villagers at the beginning (and moreover came back only some weeks after I had proclaimed my decision to take up my residence there), and secondly, because in the following period I cooperated very closely with Menkröi Ngarua', whose situation, which until then had already been a little exposed (see A3b), was strengthened by his cooperation with me. My thereby conditioned partiality of course also influenced the relations of the other villagers with me and forms the reason that I do not feel able to give an unbiased description of the "domestic" life of Tapwüa-Kua.

My interest in their culture did not form any advantage for them, except in the form of my tape recordings, which enjoyed great popularity and were a great help to me in finding contacts. But even this little joy was not free from troubles, because the many visitors who came to me to listen to my recordings had to be fed by their relatives or acquaintances in the village and thereby reduced their stocks even more, which were meagre anyway in summer 1956. The financial help I granted to some needy villagers formed actual advantages, which to a certain degree also relieved their relatives. On the whole, this financial assistance (which was given nominally as a loan only and in any case hardly exceeded one hundred rupees in total) testified more to my good will than as something that could be understood as a decisive advantage for everyone, since it did no more than to compensate them for the additional burden due to my presence.

My medicine box, however, was obviously regarded as a benefit, though there were also difficulties after the first successes, since I and with me the whole hamlet was increasingly overrun by people from the neighbouring area. For the villagers, my first system of free distribution to all who needed medicine and whom I could help with my resources meant a loss of "their" stock of medicine and once more a loss of food for all who had to cater for those coming from remote villages. Since the permanently growing stream of patients obviously not only reduced my limited stock of medicines, which was far from sufficient to satisfy such requirements (and could hardly be replenished because of the difficult traffic situation during the rainy season), but also in increasing measure absorbed my working time. I therefore without hesitation followed the decision of the villagers that I should no longer distribute medicine to people from outside. In the following healthier seasons and after a better stockpiling, these restrictions could be lifted to some extent and I could distribute more generously – when I was in the hamlet.

Some further consequences of my presence did not immediately become visible to the residents from Tapwüa-Kua, since they were based in a change of their own attitude. There were mainly two phenomena which (contrarily) determined themselves by their relations to me: firstly those of the social behaviour in the community and secondly those of the resistance to requirements from outside (not immediately equivalent to a greater consolidation). The turning up in the hamletvillage of an unusual stranger who did not make any secret of his intentions to observe them at first led to an amplified control of their own behaviour, based on the already mentioned desire to make a good impression. Petty disputes were replaced by the impression of the new unknown, which had priority for the whole community. This new feeling of solidarity was expressed openly by the end of summer 1956: since I was in the village, everything became settled much more amicably and the standard of general good conduct had been raised noticeably.

In the measure, however, in which the villagers accepted me as a usual component of their social life and brought their behaviour towards me back to normal, i.e., in the measure in which the reduction of my Sahb role made progress, my influence on their everyday behaviour also faded, and the relations of the villagers to each other returned to normal. As a consequence their fear of foreign authorities was also reduced. By regarding me as component of their village and furthermore so to say as their prerogative, they made use of my presence and my functions of Sahb to strengthen their own position opposite outsiders. With me the Mru had more "rights" than the Bengali (they were, for instance, allowed to enter my house any time, while Bengali peddlers had to stay in front of my door). While the residents of Tapwüa-Kua (like most Mru) were accustomed largely to yield to the desires of the Bengali who were regarding themselves as representatives of the ruling *jati*, they now increasingly permitted themselves to insist on their own rights, partly by reference to my "protection". Hence, by and large my presence was beneficial to the community, and Karbari Kangku Catumma did not hesitate to give expression to this feeling: he asked me to stay with them for at least five years and said he would take care of my maintenance. When I pointed out to him, that the hoped-for advantages of my presence (for instance the supply of medicines) would still be dependent on my own supply of money, which was limited, the idea arose of putting my existence on a broader basis, namely that the Mru of the four *mouzas* (areas under a headman), the borderlines of which crossed each other directly in front of Tapwüa-Kua, should consider erecting a new house for me on the crossing point, so that all inhabitants of the four areas would have equal rights and corresponding duties to care for my livelihood and maintenance – which does not mean that some people nevertheless surely had nothing against my definitely leaving their hamlet.

But when I came back in 1964, many people from the neighbouring hamlets also came to welcome me, and those who had left Tapwüa-Kua also came back and told me that they were happy to see me again, expressing their idea to also move back into the hamlet, hoping that everything would be fine again in the years to come. They could not understand that the Government gave me no permit to stay.

A3 Employees and other remarkable persons

A3a) Kyo Thwân Ong

The two persons on whose help my work decisively depended, have been mentioned several times already: on the one hand Kyo Thwân Ong, on the other hand Menkröi Ngarua'. But let me first mention Romjan Ali, a Bengali from Comilla, who just after our disembarkation in Chittagong asked whether we might not employ him as a boy and who accompanied Dr Kauffmann up to the end of his stay. As a docker, Romjan had learned some strange chunks of English and now tried his best to improve our modest knowledge of Bangla. With indestructibly good mood and a trained eye for of his neighbours' weak sides, he knew splendidly how to defend our interests and to keep his own. By comparison with this vivacious, sanguine person, Kyo Thwân Ong seemed apathetically tolerant, but because of this very passiveness was rather suited to accompany me in the Mru country, where people are afraid of nothing more than the superior craftiness of the Bengali.

Born in 1930 approximately, Kyo Thwân Ong had been raised as an orphan and had become used to service in the house of the Bohmong, the Marma Raja of Banderban. After his marriage (he meanwhile had two daughters and a son) he tried to become an independent man by doing some small trading and preparing a swidden; his time with me helped him to improve his situation a little. An early dismissal would have meant not only a material loss for him but also a moral disgrace, so that in the end he felt compelled to submit to my rules of conduct, even if he had some difficulties with them. It was his main task to care for the physical well-being of his "Sahb", i.e., to look after kitchen and chamber. Besides this, especially in the initial period, he had to play the role of an interpreter and to organise the technical side of my tours. To manage this organisation successfully (e.g., the hiring of porters and boatmen) he had to count during his negotiations on my position and therefore to assume bigger power than he had at his command (I should add that I allowed him to conduct most negotiations, because I was afraid of being overcharged and cheated). The role of a mediator which I left to him, someone who could refer to my wishes, did not always work to my advantage, however, namely when he used it for the fulfilment of his own wishes which were not in my sense, for instance when he maintained that he was acting as a mediator of my goodwill when he himself was entertained with arak (rice-spirits), which in the end was harmful to his efficiency as a servant (when drunk, he fell asleep), or when, in his effort to protect the integrity of my role as Sahb and to protect me from annoyances, he tried to put the very people in their place in whose nonchalance I was interested, especially the Mru. Although oftentimes rebuked and reduced to absurdity, his self-esteem could not really deny itself the maintenance of the ethnic prejudice of the superiority of the Marma over the hill people. In vain he fought against the putative slanders with which the Mru tried to oust him from his position: I preferred it unalterably to leave Kyo Thwân Ong behind on my tours from Tapwüa-Kua to the Mru area and to take only Mru's company (as the most effective contact persons). His prerogatives in the valleys and the plains, however, remained undisputed.

Apart from being surrounded by "sly Mru" (as he saw them) who endangered his position and worry about the welfare of his family which he had to leave behind without being able to support them in critical times (of which there were many), the unconventional views of his Sahb often caused him inner conflicts. Finally he came to the conclusion, as far as I think to be able to gather from his actions and remarks, that I was a person who was very advanced in the field of ethics and that it would be useful to his progress to trust me and to follow me in this respect. He formulated this approximately so: "If you only wanted, you could pass through a completely overgrown jungle, and I could come along." One of my habits that really astonished him was that compared with him I ate very little. He explained this to the people by saying that I had descended from God's country and only partook of their food out of sheer courtesy.

The following examples may show how real this jungle sometimes looked for him: On the morning of 19.02.1956 when (on one of my first tours with Kyo Thwân Ong) I wanted to start from Pantola, a heated quarrel started between my boy and the boatman so that at the most dangerous moments I felt it necessary to intervene in order to prevent a fight. Kyo Thwân Ong was so enraged that in compensation he struck against baskets and the walls of the roof overarching the boat. I stood by and with the best will could not find out what this was all about, since Kyo Thwân Ong was too irritated to provide an understandable explanation to me even after a final armistice. Only in the afternoon, when the battle of words threatened to get lively in the boat again, I finally learned the cause of the quarrel: the boatman had refused to carry all the luggage from the boat to the village and back again unaided while Kyo Thwân Ong accompanied him more or less empty-handed. In his view the boatman was also paid for this, the latter, however, maintained to have got the worst of the bargain. Well, I thought that it could not do any harm to Kyo Thwân Ong to carry something too.

What a mistake! Only in case the boatman got ill or if we were to change the boat from village to village – otherwise, however, it was impossible and he could not do it even for five hundred rupees. The reason: he must keep face, both his and thereby also that of the Sahb. And as he, the Sahb's boy, had employed the boatman, it would be impossible for the reputation of the Sahb if his boy now did coolie work. The Sahb, however, declared, if this was asking too much of the boy, the Sahb himself would carry his things and the boy could play the Sahb. Embarrassed silence. Impossible: what can one do with such a mad Sahb? When landing in Wapran, I was just wondering whether I should defy all face-saving and order Kyo Thwân to carry the tape box (the heaviest but also the most distinguished piece of my luggage), amazingly he carried it of his own accord.

A3e) The inhabitants of "my" hamlet

Ethnographers have taken to the habit to speak of "my" village or even of "my" people, whenever they have stayed for some longer time in a certain place. I just follow suit, not because I think this designation justified, but just because I lack a more appropriate term to characterise the place where I stayed. I call it hamlet, because Tapwüa-Kua was just a small hamlet consisting (if I leave out my own house) of only five houses, but the settlement (in 1956/57) consisted of ten household, five of which, however, would not consider themselves to be inhabitants of Tapwüa-Kua but of Atwangwüa-Kua, because most of them belonged to the Atwang sib. The Atwang were also to be found in the next hamlet and the "founding father" of the sib, called Empu, had long since moved there (Empu was said to have reached his 90th year in the meantime, but had become deaf and blind). Also "my" hamlet was still known as "Empu-Para", though in reality this name accrued only to the houses of the Atuangwüs sib,

Nearly exactly a year later to the day, on 17.02.1957. I meanwhile had a boat of my own, which saved the expense of having Kyo Thwân Ong hire a boatman. In the early afternoon we arrived in Banderban. I climbed up the shore and sat down in Kyo Thwân Ong's house in the expectation that my luggage would soon be brought by a daily labourer, as usual from the neighbouring house. Here in Banderban one must pay something for one's status, especially if at the same time one can help a poor devil. Instead of the poor man, however, Kyo Thwân Ong turned up repeatedly and at the end uttered his opinion: "Now we have saved another rupee!" I did not give him any reward, but the new regulation remained in force from then on.

Supported by my method of checking the whereabouts of the last penny, Kyo Thwân Ong developed the virtue of refraining from using my means to promote his own interests and instead developed a commendable trustworthiness. That he nevertheless did not manage to apply my strict bookkeeping to his own household as well may only be noted in passing. The inconveniences he caused me were mainly due to my own faults: while I could rely on his honest care, I pursued my aims largely without consideration for his situation. Nevertheless, I think that I was also of some help to him, and that neither of us would hesitate to entrust himself to the other once more.

It remains to be stated that Kyo Thwân Ong still helped me to settle some smaller questions in writing after my return to Germany. I had bequeathed my boat (without engine) to him as remuneration for the work he had done for me and as an "advance" for further help when I left the Hill Tracts. He used it to build up a new existence for himself and his family as a boatman on the Kornofuli river.

A3b) Menkröi Khatpo-Ngarua'

As the "Mastor", Menkröi had already been made known to me a long time before I came to know him personally: he was the man of whom it was said that he had a book in which all Mru words were recorded. Already two months before I came to his hamlet, I had stayed in a neighbouring hamlet (Yöngtu-KP, 15.02.1956), but then turned north again in the search of a stone with an inscription (which turned out to be a trigonometric point). In Yöngtu-KP I learned that the book was far from there, in the south. In answer to my question "how far?" one man said "approximately four miles". But I already knew by this time that the Mru did not have a particularly good idea either of miles or of hours, and Kyo Thwân Ong commented: "If they say 'far' here, it is near, and 'near' means far." I learned only too late how right he was; in the meantime I was told, since the book was too far away, the hamlet could be reached better by starting from the Songu river from the mouza of Headman Sa Bai U. Although Sa Bai U was no longer the headman of this mouza, I met him five days later (on 20.02.1956) in Galengya and when I asked him I learned that the master I was looking for consisted of two Christian missionaries living in the Horinjhiri-Mouza, one of whom understood English (he had studied almost up to an M. A.), the other one, however, did not. And one of their hamlets was eight and the other ten miles removed. To reach them one would have to climb the whole mountain range, it would be easier to come from the Dokhinhangor-Mouza (which I had left five days ago). On 4.04.1956 I met in the bungalow of Chinbok (from where it was in Mru opinion another four miles to reach the hamlet of Menkröi) the Road Suboverseer K. A. Rahman, who supervised repair works and should have known by his profession the actual distance, and he explained to me it would be still fifteen miles, that is, still rather far. I set out nevertheless, passed Yöngtu-KP, once more and after one and a half hours I arrived. I doubted that I might already be where I wanted to go, but I was led into a house in which, although the head of the household was absent, several small booklets were to be found, among other things a "Maru-Primer" and a translation of St John's Gospel, the last copies of which were being slowly eaten up by humidity and mice or as long as the paper was still intact were now being used by the lady of the house (as I saw later) as cigarette paper. The master's house was situated in the Galengya-Mouza (of former Headman Sa Bai U) and a hundred metres further, in the same hamlet but in the Horinjhiri-Mouza, stood the house of the other (younger) "master" who allegedly could speak English. Of him, Cönglök Atwang, I shall talk later. The old master, who returned the next day, agreed to take me in his teaching, and thus started my stay in Tapwüa-Kua.

Menkröi of the sib Khatpo-Ngarua', was born around 1913 in Chinbok-Para. When he was approximately five years old, his father died. Two years later his ailing mother with her five children moved to Tapwüa-Kua close to the house of her brother (Elai Atwang) and died there in the following year. The two eldest sons, Khamlai (ten years older than Menkröi) and Klönglai managed the household. Then the Lushai missionaries came into the hamlet, among them Ro Hming Lian, earlier headman in Lungleh. He worked in the

Mru area until after 1934. In the company of the missionaries and their children, Menkröi learned Lushai, stayed in Lungleh for three months and in 1928 accompanied Ro Hming Lian as interpreter and evangelist into the areas of the Dopreng and Dömrong Mru. In 1929 he went to Manipur, visited the school there for five years and studied up to the sixth class. After coming home, he settled in Mitca-Para (in the Dokhinhangor-Mouza) where the missionaries had established their headquarters. Among them was also pastor Dala (formerly in Demagri, since 1928 in Mitca, he died 1962 in Artha'). Together with him, Menkröi wrote the Primer and translated (from the Lushai version) St. John's Gospel and a Christian songbook (which appeared in 1939, '40, and '41). After a year Menkröi married Thanvel, the sister of Dala's wife. Up until 1940 Menkröi stayed as a teacher in Mitca, then (during the war) moved back to Tapwüa-Kua and after a further five years gave up the teaching profession and returned to conventional village life. His eldest son, Dingte, he sent (1950) for two years to Artha', the then headquarters of the Lushai mission in the Hill Tracts; – if Dingte had learned something, he had largely forgotten it again by my time. (He told me that he had hated the school and the teachers, who were in the habit of hitting their pupils on the fingers with a stick.) He was a proper Mru and also his younger brothers and sisters were no different from their contemporaries, they were not educated as "Christians". Menkröi himself could read still and write, however, he did not use it any more, since he had become short sighted and did not own any glasses. His knowledge of English had to be called fragmentary at best and was insufficient for any conversation. He continued to be fluent in Lushai (although his wife Thanvel meanwhile spoke only Mru), knew sufficient Marma and (even up to a certain degree elevated) Bangla. His journeys to the other Mru groups had given him an adequate knowledge of his country (his old acquaintances still proved useful for me), and his stay abroad had helped him to get an open-minded understanding of cultural differences, so that he regarded his own culture interestedly, but with relative impartiality. I would doubt that he was ever very dedicated to the field of religion; his principal interest was in the teaching profession, until he came to the conclusion that the effort was not worth while: after he had greeted me, he at once told me that if I were a missionary, I had better just return home, since his people were not interested in Christian teaching. But he agreed to become my teacher. I came to know him as a sceptic who, like many others, simply followed the usual customs. For questions going beyond that, he put me into contact with the corresponding experts.

I am sure that without Menkröi's help I would not have managed to collect even a part of the material presented here. Toward the end of my stay he himself tried to think over what I still might have forgotten to ask and to note down. During the time when I stayed sedentarily in Tapwüa-Kua, I usually bade him for half a day to talk with me and to answer my questions (I spent the rest of the day trying to formulate a text from my notes). In the summer of 1957, when I recorded texts and songs on tape, he often helped me the whole day long (with short pauses in between) to listen to, understand and translate the texts – none of the more highly educated men, with whom I had worked in Ruma and Banderban, would have been able to bear this unfamiliar mental effort.

During this time Menkröi's everyday work was taken over mainly by Thanvel and Dingte, who moreover did their best to lend me their hand whenever necessary. A large part of my meal was cooked (by Kyo Thwân Ong) on wood which Thanvel had collected. Thanvel was called Thanni by all villagers, and I followed suit. Although she still wore the Lushai skirt, she did not regard herself as a Christian any more. On the first Sunday of my stay in Menkröi's house she told me a lesson by surprising me with the communication, that today I would not get any food to eat because as a Christian she would have to abstain from work on Sunday, but afterwards she invited me for a meal: "You see, we must eat anyway. The Christian teachings are just not good for our life." Thanni often coughed, I could not help her, and she died of TB soon after my departure in 1957.

I became closely acquainted with her son Dingte who was fifteen years old at that time. Among other things he saw to it that I was accepted in the circle of young people. (On one tour I also engaged him as a boy, but he did not like it.)

Although the residents of Tapwüa-Kua may have found some faults with their "Mastor" Menkröi, I did not find any reason for this. If this book should have any merits, it is largely his; for faults in the material my lack of understanding must take responsibility.

A3c) Information

The strong tie with an informant which will also clearly come to expression in the following text (see A5b) raises a methodical problem. It forms a clear violation of the requirement of achieving a high degree of general validity by a comparative evaluation of a variety of statements. Having taken an opposite stance, I distinguish 1) between statements of one determined informant at a particular time, and 2) generalising details. By attaching higher value to the former, I negate the ideal of objective observations, which would have to be abstracted from the statements of as many sources as possible. On the other hand I found in none of the previous ethnographic works after every paragraph any details of how many statements from which informants it was secured or whether it corresponded only to the author's opinion. I also did not want to do any opinion research, but, whenever possible, to state rules. At first these rules exist in the idea. When they are realised in actions, it should be stated whether the ideas of a single person or those of a community (depending on number of participants) are decisive. So actions appropriate to the ideas of a rule of every particular participant are possible. I have correspondingly noted them down. In all probability there are other possibilities as well; if by chance I was able to notice remarkable differences, I also noted them down. As a further argument I adduce that informants are not equal; the greater their number, the less one can secure their reliability. Ideas about a certain process or belief are certainly not the same among all members of a group: adults know more than children, believers more than sceptics, some follow other principles than those they profess, and finally it is human to err. And as soon as we have to do with specialist knowledge, statement of general common knowledge can never bring the "rule" to light.

The restriction to a main informant is not only a rational and efficient means to reduce misunderstandings thanks to the mutual acclimatisation, it is at the same time a means (provided that the statements are made recognisable as such), to differentiate more general knowledge from that of a more special nature, for which special informants had to be consulted.

As has been said, at first it is to be assumed that the information gained in this way has only a restricted general validity. But even this position is based on the prerequisite that the statements are "regular". Just this reliability is to be gained by the general checking of parallel information. The aim of this check now, however, is not to state the general validity, but the rule validity of a statement. The possibility for this is always given when the ethnographer himself can observe the execution of such a rule. I have therefore given the greatest value to checking the information I had received in the past, besides by confirmation of the details in other conversations, primarily by observations of my own (or by seeking afterwards to confirm earlier observations by new information). An informant can only be regarded as trustworthy if he gives the researcher opportunity again and again to find the information confirmed by an observation of his own. And in this sense I can say: Menkröi Ngarua' was a reliable informant.

But with practical performances one also may and indeed must expect variants: these are examples. Who here does not ask for examples, but for the type of the events, I should like to ask how he wants to state the typical language when, for instance, there are countless related dialects without the existence of standard language, fixed in writing between yesterday and tomorrow. It also would result in something like statistical semantics, if one wanted to have limiting values, where I tried to describe the object chosen as an example by its measurements ("How small may a house be, and still be valid as such?") since, seen culturally, the "measurement" is not pre-defined in metres but conditioned by the material in the not measurable opinion of the manufacturers.

With this perspective, I should like to distinguish theoretically three approximations to a perceived reality: 1) the researcher's subjective description, 2) the subjective description by one member (or several members) of the culture in question, and 3) the combination of both. The greater the trustworthiness of the informant, the more views 2) to 3) come closer to each other. Whenever possible I have replaced the doubtful general validity of my opinion by the detail of a certain informant at a particular time. The informant mentioned, however, is by no means alone responsible for the description I have given here: on the one hand frequently further persons participating in the conversation were available at the time of the information, on the other hand I may occasionally have misunderstood the answers or re-interpreted them a little in the résumé.

A3d) Friends

In reports on field research it has become customary to thank all the countless "others" without whose support the enterprise could not have succeeded. I regard this sweeping statement as unjustified; not everybody one meets is ready to welcome the uninvited guest. Not everybody knows how to offer the visitor what he would like to receive, and some who would be entitled to be remembered will drop unnoticed into oblivion, but after some years only a few that one remembers and to whom perhaps one does not even owe some debt of gratitude – they would not even know what to do with it, since such phrases are not common in the Hill Tracts – but whose memory one would like to preserve because they were of some importance.

Without wanting to connect their mentioning with an order of evaluation, at first there was Romol Krishna, a Chakma, who provided medical services in Ruma and kindly helped me with much advice and information, not only in the first difficult time, but whenever I visited him. Near Ruma, in Polika-Para, the Marma Headman Hrui Pyo Ong has to be mentioned, who invited us to his appointment celebration (which we filmed), and later gave me some of the clothes he wore for the celebration, and with whom I spent many a pleasant hour, until in melancholy about a personal stroke of fate and the sad situation of the people of his mouza he surrendered himself to opium. There also is the Karbari Thwân Pho Ong of Galengya and his son Shwe Hla, with whom I always found a welcome whenever I used the river on the way to or from Tapwüa-Kua, and was catered for in a friendly fashion despite their economic crisis, which revealed itself to me only after some time. With them I never needed to wait for porters, since if necessary the karbari and his son themselves carried my luggage to the next hamlet. I also knew that here my boat was in good hands when I stayed on the hill. There is also the Karbari Kyo Co Khöng, who was honoured with a medal because of his brave stand during the last war, and now made a new honour to himself to see me as his guest; the Roaja Ong Sa Phru of Kyokcarong who sometimes still had to take care of me late at night, the Karbari Thwi Phru whose house offered me a kind service station when I went by foot from Banderban to Tapwüa-Kua; and finally also the Headman Kyo Mong of Gozaliya, who at first, due to a misunderstanding, ran the risk of being judged by me as an inhospitable person, although the opposite was the case. Among the Marma particularly great services to our expedition were rendered by Ong Shwe Phru, a well-educated man, closely related to the Bohmong (the Marma-Raja of the southern Hill Tracts) in Banderban. His support reached from accommodation in his house over important information to temporary financial help. I do not know how many difficulties (also of semi-official type) were removed from our way without our knowledge by his influence. He later became Bohmong himself, had to endure imprisonment several times by the changing Governmental powers of East Pakistan, since 1971 Bangladesh, until he finally had been restored when I saw him the last time during my short visit in 1990.

The residents of Tapwüa-Kua will be spoken of in the following subchapter; for my tours in the Mru area I primarily remember the Karbari Thümru Chimlung-Ngarua', in whose house I stayed as a guest once in a while, and who himself prepared my meal and bed, when one evening I reached his house without company and being ill; the Karbari Kröilö Ngaplo, who always received me hospitably and all porters (even if they were indebted to him) on our way between Galengya and Tapwüa-Kua; the Karbari Thonru Nang who tried his best to talk to me in Mru; Menyem Rüm, whom I learned to appreciate as a particularly kind person; and the Headman Haiwön Macar, an intelligent and intrepid man, such as I would have liked to find more frequently.

Among the Khumi, I should like to mention the Karbari Rennoi who invited me to his house in order to help me when I was temporarily short of food, and who as "Raja of the hills" also consciously upheld his tradition in times of economic decline; and the Karbari Müilün Tamchaa who not only housed me for some time but also informed me with pleasure and brought forward a festival for my sake. Finally three kind Twipra men shall be mentioned here: Bijonmoni, my first Twipra singer and player, Roaja Noliram, who welcomed me with a gun salute when I visited his buffalo sacrifice and Karbari Chigohan who knew how to talk to me unrestrictedly. Some other remarkable men will still have to be mentioned in other places, as for instance Menyom Tang, the great song master, and Rengthon Changkan, the superior villain. It may be, as I said, that I have unjustifiably left out some names of persons who should have deserved mention as well; but which man always counts the right persons as those who were important for and near to him. Since the opposite can also be the case I permit myself to do without mentioning the handful of those who gave me serious trouble.

Nevertheless, during the first months of my stay I always thought (and told the people) that I was staying in "Empu-Para", until Menkröi told me the correct old name: Tapwüa-Kua, as it had been called already in the time before Empu. Tapwüa is not a sib name, but means, literally translated, "those of the fortified place" – though nobody remembered the time when (and why) this settlement had been fortified. The Atwangwüa had moved in later and had their own reason to call the place differently, because their hamlet belonged to a different *mouza,* an artificial unit created by the English administration, when it wanted to establish something like the villages in the plains, even if their borderlines, following the ridge of the hills, divided a settlement in two different parts. Therefore "Atwangwüa-Kua" had a *karbari* ("manager" of the hamlet) of its own and belonged to the Horinjhiri-Mouza (under a headman of its own), while modern Tapwüa-Kua also had a karbari of its own and belonged to the Galengya-Mouza (equally with a headman of its own). Nevertheless, the two hamlets formed one settlement, a twin-hamlet, which for convenience sake, I called "my" hamlet. The inhabitants had had to accept this artificial division and with it the necessity to pay their "jhum-tax" (see J1b) twice if they made a swidden on the other side of the ridge. In the beginning (by the end of the 19th century) this double taxation may have proved a burden, in the meantime, however, it had remained fixed at Rs. 6 per year, which they normally could pay.

After this somewhat lengthy explanation, and since I have already described my relations with the inhabitants in general under A2f, I shall now list the names of all of them as noted down by me in 1956–1957 (developments until 1964 will be noted afterwards). I start with the houses East of the ridge (that is in Tapwüa-Kua proper) and proceed West to the houses of Atwangwüa-Kua, listing first the name of the (male) household head followed in the same line by that of his wife, followed in the next lines (preceded by a sign for their sex) by that of their children in the sequence of their birth. In round brackets I have noted their age, which, however, especially for older people, is only approximate, since nobody really remembers the year in which he or she was born. For the children the age is more reliable, since it could be compared. In square brackets the names are given that were used as a kind of nickname (see N1e). The sib name of the wife was rarely stated. Children who died (but were still remembered) were marked by †.

I Yenpau Khongtör (46) – Chongpoi [Chingkong] Ngarua' (35)
♀ Tönching [Kaikring] (17)
♂ Menlong (15)

♂ Nia-ok (13)
♀ Pumpau (11)
♂ Klangwai (6)
♂ Menpio (2)

II Klingtui [Thengkia] Khongtör (59) – Changlong (47)
(2 daughters, †)
♂ Bulum (20) – Kaiche (29)
♂ Menpoi (0)
(1 daughter, †)
♂ Khaiyuk (16)
♀Nganrui (5)
♀ Nganli (* May 1956, † February 1957)
(adopted): ♂ Niahut (39, became mad, † 1957)

III Menkröi Ngarua' (42) – Thanvel [Thanni] (36, † 1957)
♂ Dingte (16)
♀ Tharpau (14)
♀ Ruirong [Kuidong] (11)
♂ Ilong (Menkong) [Kangding] (8)
♂ Menpaa (6)
♀ Roupau (4)
(adopted): ♀ Kaiche (19)

IV Netkhai Atwang (56) – Tumcong (52)
♀ Runring (23)
♂ Chingkrat (20) – Kaitön (19)
♀ Taibia (18)

V (Karbari) Kangku Catumma (60) – Nguiyom (50)
♂ Mowai (son of Nguiyom, 33) – Komlap (36)
♂ Menrau (11)
♂ Langrung (9)
♀ Ruirau (7)
♀ Wurwai (4)
♂ Chingrau (2)
♂ Menleng (* June 1957)
(3 elder ♀ married)
♀ Thithop (19)
♀ Namplong (17)
♀ Camrau (12)

VI Menching Atwang (26) – Camru (24)
(Mc.s children):
♂ Longngan (6)
♂ Ngaucang (†)
♂ Langpung (0)
(Mc's father):
Niadöng [Rengyüa] (46) – Premprong (†)
(M's siblings):
♂ Rümtui (18) – Komnang (18)

♀ Löli Thüryam (*, † 1957)
♂ Niachöm (15)
♂ Langrau (13)
♀ Rümpau (10)
♂ Menpung (7)

VII Netnöng Atwang (37) – Kaichüa (30)
♀ Paurung (5)
♂ Menlong (1)

VIII (N's brother): Cönglök (32) – Thidon (29)
♀ Paingpau (1)
(born after 1957):
♀ Paingpyo
♀ Pyo-chamaa

VIII Khamcöng Khongör (33) – Kraapau (30)
♂ Citcöi (5)
♂ Limwai (3)
(K's brother's son): ♂ Chakching (15)

IX Karbari Elai Atwang (66) – (†)
Reng-et [Ret-yaa] (31) – Chekkre (38)
♀ Ruitum (14)
♀ Hirwai (11)
♂ Langngi (9)
♀ Tumdi (6)
♂ Yongring (4)
♀ Campung (2)
(E's 2nd son): Klangwai (28) – Kaichong (24)
(E's daughters): 2 †, 2 married

X Khamlai Ngarua' (53, never married)

In 1964, Dingte in the meantime had married Kangku's daughter, Thithop, they had one daughter (Cinrau) and one son (Rengtai); Menkröi had lost Thanni and his elder brother Klönglai had died, his two sons, Mangcüng and Mangling [Mangbür] had moved into the house which now was headed by Dingte; his sister Tharpau had married and was pregnant, but wanted to leave her husband and return to her father, who, however, was not inclined to let her come back. Khamlai, since he no longer felt strong enough to earn his own living, had settled down in the Kyong (Buddhist monastery) of Dokhinhangor.

Thithop's father, Karbari Kangku, was still alive, but Mowai's eldest son had died. For his second son, Langrung, he had given a big marriage feast (*machi-pok-poi*), killing at least 120 chickens, but his wife did not care for the household and after one year returned home. Now Langrung had married Retyaa's daughter Hirwai and they had a daughter (Kaichong). Thithop's sisters, Namplong and Camrau, had married and left the hamlet. Also Hirwai's sister Ruitum had married and left the hamlet.

Rengyüa's sons Niachöm and Langrau had married too. Niachöm's wife was Yarpau, they had one daughter (Tumpoi) and one son (Menyang). Langrau had first married Komnang, the wife of his elder brother Rümtui, who had died, but then Komnang had withdrawn and now he had married Ruichong, but they had not yet had any children. Their eldest brother, Menching, had by now married his fourth wife, Tönching [Kaikring], daughter of Yenpau, who had to care for four children (two sons: Ngancang and Lengpang, and two daughters: Kaico and Tharpau); I do not know who was the mother of these children, but Ngancang was the son of Menching's first wife (Camru), who had never been happy with him, since her husband was a philanderer. She had then given birth to a *yang-caa* (extra-marital child), taken it away with her and left him. He had certainly soon found another wife.

Karbari Elai had died after becoming mad in the last year of his life and having to be tied up. Elai's eldest son, Retyaa, after his daughters Ruitum and Hirwai had married, still stayed with his wife Chekkre, his eldest son Langngi and his three other children (see above) in his house, Klangwai (Elai's second son) and Kaichong had a son (Kaingwöi). Netkhai, who had never been on good terms with his neighbours, had moved out and founded a new hamlet of his own, but three new Atwang families had moved in: 1) Angkaa and his wife Haili; they had two daughters (Chüapau and Ruichi) and one son (Menwai); 2) Krongtön and his wife Donram, with two sons (Nganyan and Langyang) and one daughter (Chongpoi). Nganyan had married Tumpang, but she had returned to her parents, leaving her daughter (Pletdöi [literally: "does not thrive"] with him; 3) widow Khönwan (from Menlai-KP), with her son Chongpria married to Kaikök, who had five children. Despite the fact that the two Khongtör families, Yenpau and Klingtui (with his son Bulum), had also left the hamlet and moved to the South, beyond Thanchi (in search of better swidden areas), there still remained ten houses in the twin hamlet, but only two houses on the Galengya side. These were to be reduced to one house only after 1964, when Dingte also moved with his family to the South.

If one thinks that this might have meant the end of Tapwüa-Kua, I must add that it still existed in 1990, as I was told by Klangwai Atwang, the grandchild of old Klangwai, when he came to see me on my short visit to Banderban. The hamlet had even become bigger under its new karbari, Pastor Cönglök Atwang, who had made it the centre of his newly created Christian missionary station. (During my stay, I never had built up good relations with him, he once wanted me to help him with some money, but I only reproached him for addressing me on the card he wrote to me as "my god and helper"). He, in the meantime, had made a new translation of the whole bible. The new church and the school, however, like most houses, were not built on stilts any longer, since trees and stronger bamboo canes had become too short in supply. Some people still made swidden fields, but most lived from large-scale orange plantations, as Menkröi had seen in Lungleh and started to plant, against the strong opposition of his conservative brother-in-law Yenpau, who feared that this practice would lead to a gradual privatisation of the community soil. But the scarcity of water persisted, despite Cönglök's efforts to have a well dug.

Most people I had known had died, but after 2000 Céline Mouchet conveyed me greetings from a comparatively very old Mru lady, whom she had met by chance and who had asked her whether she knew me: Taibia, Netkhai's daughter, who still remembered me; when young she still was very shy, but in my view the nicest of all the girls of the twin hamlet.

A4 Equipment and cost of living

A4a) Household objects and clothes

According to the experience of the leader of our expedition during his Naga Hills expedition (1936/37) the number of the objects bought in Europe as equipment was largely limited. His experience in 1936/37 had told him that most equipment objects for the tropics could be better and more cheaply bought in India. In Calcutta and especially in Chittagong, however, we had to notice that in many cases unfortunately this was not the situation any more; the Indian market in 1955 was no longer prepared for European consumers.

But even equipment like Petromax lamps, locally called "pump-lights", could be bought with only difficulty and for higher prices than in Germany. Simple kerosene lanterns might do as well, but if more light is needed for writing and reading the pump-light is more serviceable – as long as it functions well. My Petromax, however, after more than one year stopped working, probably because the kerosene available in the local market contained too many impurities. I had to pump in more kerosene before I could light it, and one evening when it had already become rather dark and I had put the Petromax on the floor behind the door opening, I pumped too much, the kerosene ran over and when I tried to light the lamp with a match, a big flame flared up from the floor, and I had to fear that my house would burn down and with it all my notes and results, which I had collected and which were lying on the bamboo board over my (i.e., Menkröi's) table. It came to me as a shock, and I felt that I would be ruined. But then, for the first time, I came to see the advantages of a bamboo floor: the big flame burned down, the roof had not caught fire, no new flames came up from below (as I had surmised would be the case), and no harm was done to my house.

There were no country bedsteads from light metal available in the market, the wood constructions in stock formed a load of my luggage, which I kept in my house, but never took with me when travelling. It is quite possible to sleep on the bamboo floors of the houses on stilts but it offers less safety against vermin (ants, mice, snakes and leeches) and is even less to be recommended during the rainy season because of the rising damp (many older local people suffer from rheumatism). Perhaps hammocks would be more practical, but they would have to be brought along from Europe.

However, the usual travel equipment was easily and cheaply obtainable: "bedding", for example, an easily transportable roll, into which blankets can be wrapped up, and also smaller utensils such as toilet things, laundry, writing paper, film material and the mosquito net, which is very useful to ensure undisturbed rest at night. Kitchen and dinner equipment can also be bought locally but it is best to bring one's own cutlery, because in Bangladesh people are used to eating with their fingers. Plastic plates and cups are less durable and moreover difficult to obtain, while aluminium bowls and drinking glasses are more practical and also available in the country bazaars. Soap, toothpaste, etc. from the international companies can be bought in Chittagong, but not flannels or sponges. For shaving I used a clockwork dry shaver (from the now dissolved company O. H. Meyer KG.) which did a good and reliable job and, during hot season, was gentler on the skin than blades or a knife.

Clothes were mainly brought from Europe. Clothes in the European style are made only to order and the material is either inferior or disproportionately expensive. Local clothing can be worn temporarily (see below) but is regarded as unseemly for Europeans when travelling in the plains, and on tours in the hilly country they are also less practical than European clothes, since the local clothes have no pockets. Two types of clothing are necessary: 1) correct European street clothes, which also suffice for official occasions. During the cold season warm clothes are definitely useful, although a coat is not necessary (raincoats cannot be recommended either, since in the rainy season it is so warm that any protecting coat is not warranted); a waterproof cloth-hat fulfils all requirements and pith helmets are out of fashion; 2) bush clothes, i.e., bush shirts and short trousers in sufficient numbers, which can be washed every day if necessary. Although during the rainy season clothes should be as light as possible, short sleeves and legs have the great disadvantage that they afford the mosquitoes free access to arms and legs; long trousers hardly permit the necessary ventilation; the customary sarong (lunggi), however, proved to be very advantageous when sitting, since it covers the legs down to the ground without fitting tightly. For the cold season, long trousers made from a solid cloth are recommended; in the cooler times of day a roll neck jumper (next to the jacket) did a good job for me. If necessary, a shawl or wrap can offer additional protection.

Footwear poses a special problem. After some experiments I used 1) when touring in the hills, a pair of heavy mountain-shoes with studs and firm heels (to be brought along) and 2) a pair of (locally made) canvas shoes when walking in brooks. If the journey calls for frequent changes, this is very uncomfortable, however, as wet climbing boots are a burden, while it is difficult to get a proper grip with the gym shoes on the mountain slopes. Half-high canvas shoes permeable to water would be advisable, if they had a well protected toe and heel, firm studs and well-fastened (!) heels (which are also effective on slippery paths or screes). But these were not, and are probably still not, available in the market. As for stockings, woollen socks proved to be useless for me in two different ways: I had to mend them constantly (when wet they wore through too easily) and they kept shrinking, because my boy would not desist from trying to iron them after washing them (an iron for charcoal can be obtained locally). I was very content, however, with my nylon-crepe socks which did not wear through and in addition offered a valuable advantage during the rainy season: leeches could almost never bite through them to gorge themselves with blood, which was easy for them when I wore woollen socks.

The bathing shorts should not be forgotten; not because the watercourses offered many possibilities for bathing (the water is either too shallow or too dirty and sometimes raging, depending on the season), but the daily ablutions can be taken care of most easily on the house platform, where some spectators may always come in order to see how the Sahb washes his body.

A4b) Apparatus

The following equipment and apparatus were carried: per participant two 35-mm cameras, two exposure meters, a flashlight with bulbs, one pair of field glasses, one portable typewriter, several ring files and diaries, furthermore one cinematographic camera, one tape-recorder, one thermometer, one hygrometer, one tape measure and some small tools. The cameras used by the leader of the expedition were not exposed to an endurance test for the rainy season and the shutters of the two cameras (Agfa Super Silette) used by me started to get rusty after one and a half years; microbes settled in one of the objectives, but both remained in working order until the end, except for occasional failures (also after one of them had fallen into the water). It is questionable whether more expensive apparatus would have shown better powers of endurance. The light meters (Ikophot) had already started to indicate wrong values shortly after the beginning of the journey, and later I did not use them anymore.

The following films were used): 1) “Agfacolor”, though only used before the rainy season: material influences due to the dry heat were not noticeable, but after some years the colours started to change. 2) “Kodachrome”: the stocks were in sealed packs, however the containers were too large, with the result that after they had been opened, more films than could be used within the next months had to be exposed to the atmospheric humidity. However, in the rainy season as well (in which a film occasionally had to stay longer in the camera for lack of suitable weather and the despatch could be somewhat delayed because the paths to the next post-office were closed or could not be used for some time) in general no material defects occurred, except that a few films showed lighter or stronger discolouration, occasional microbe spots and in one case pink colour spots. Kodak, however, generally proved resistant, despite (involuntary) violations of the storage and transport rules. Even for the rainy season, minimal failures would have to be expected if the cartridges were opened one by one. 3) “Agfa Isopan F” (black and white), which also survived the rainy season without protection. Damage only occurred when the film was left too long in the camera; in one case (after a period of continuous rain) the film stuck together, in another the whole coating dissolved.

My experiences with the tape recorder (“Nagra II b”) were less favourable because although it was especially protected against heat and cold, it proved not to be robust enough to withstand the humidity. Although simpler repairs could be executed by a specialist (of “tropics company”) in Chittagong, spare parts obtained locally were not guaranteed, so that after several vain (and very time-consuming!) trials, a spare apparatus had to be obtained for tests from Europe, which, however, did not hold out either till the end of my stay. Fortunately the faults of both recorders did not influence the recording quality, but I was able to check this only when I listened to my recordings in Europe. The tapes used initially (“Scotch Tape”) also proved unsatisfactory, since they started to become mildewy and slimy. The BASF tapes I used later did not show these disadvantages; except that the sound quality deteriorated noticeably when recording for a longer time under conditions of high atmospheric humidity.

The portable typewriter (Olivetti Lettera 22), on the other hand, always stayed intact and was stable enough to withstand being dropped occasionally.

Kauffmann, the leader of the expedition, had bought himself a double-barrel hunting rifle in Chittagong (he had to apply for the licence with the Deputy Commissioner). Ammunition was available in the larger bazaars, although not always. A good chance for a profitable bag, however, was to be had in a few places only, and when he left the country, he left the rifle to me. However, I never went hunting and left it to my boy, who once managed to shoot a hornbill.

Kauffmann also bought an outboard motor (Johnson Sea-Horse, 3½ hp) in Chittagong. When returning to Germany, he had left it with the bank. In January 1957 I fetched it from storage, but it had suffered damage and did not even prove very useful when it was finally working again (repaired mainly by myself): petrol and oil were available only in Dohazari (in the plains) on the Songu river and even then not always and not in sufficient quantities. Although I could reach Banderban from there in a few hours, when further upstream and the petrol was exhausted my boy had to use the punting pole or the paddle again, and in the dry season the shallowness of the river can be especially dangerous for the screw in rapids. If a serious fault developed, the engine would be useless, since spare parts were not available – which the man who bought the outboard motor from me unfortunately had to learn even before my departure. When the waterway is used frequently, possession of a little boat (dug-out) of one's own is of advantage provided that one's boy is also capable of taking over the role of a boatman. Though I had to pay the (at that time) enormous sum of 200 Rs. for the dug-out, the expenditure proved reasonable, since in this way in the second year I spared myself many otherwise unavoidable troubles and losses of time and in the end could bequeath it to Kyo Thwâng Ong as remuneration for his services.

A4c) Consumer goods

A complete renunciation of European cooking is a condition of a longer stay in the Chittagong Hill Tracts. People who can cook (or pretend to be able to do so) in the English style offer their services in the centres of civilisation, but in the hilly area such knowledge is completely useless, since here one has to live from what is obtainable. Since in general all men know how to cook, practically any local man can be hired as a cook, provided that he can handle the money entrusted to him for the purchases appropriately. On the other hand one certainly does not need to eat as frugally as the hill people, whose meal ingredients are proportionate to their fortune and therefore usually largely missing. Oil, spices and tea as well as vegetables and eggs are often to be had even in the small bazaars and available to everybody who has the money. In 1956 this was different for sugar: provided that one did not want to buy it "black" at excessive prices, an allocation certificate from the Subdivisional Officer was necessary. I always received the certificate in Banderban without difficulties, but could not always use it for lack of available stocks. For vegetables one depends on the seasonally common country fruits in the hilly area; concerning animal food I was able to obtain eggs (however rarely in larger quantities); meat is rare since bigger animals are slaughtered only for festivities and although chickens may be slaughtered for the visitor, during a longer stay in the same area they can be bought only occasionally (essentially from peddlers from the plains). Hunted game (predominating jungle fowl and wild pigeons) can sporadically provide an extra, but bigger game must be shared to prevent the meat getting spoiled. Fresh fish is a rarity in the hilly area and the local style of preparation is a little strange (see K1b).

Bread is only available in town and therefore not used for daily meals. Rusk is available in some bazaars, however, and can be bought in bulk, as can flour, from which the customary chapati can be made and baked in oil. While I ate rice at noon and in the evening, rusk or chapati formed my breakfast, although occasionally (when the stocks were exhausted or when on a tour) I had to content myself with eating nothing but rice. As a spread for rusk and chapati only jam or cheese could be used, from tins imported from Australia, which were available in town and had to be brought from there (or could also be ordered from the "Paramount" department stores and sent from Chittagong to the nearest post office). Kerosene tins also had to be bought in bulk for the Petromax or stable lamp, since the hill people generally use only unrefined fuel ("red kerosene") for their small dim lights, which cannot be used for the lamps because of excessive soot.

Tobacco is available locally in the form of dried leaves, which the hill people produce themselves, to be smoked in cigar form or in a pipe, or to be chewed in the form of a quid. Cigarettes of better quality, which can also be used as gifts, could be bought only in the larger bazaars, and even there they were occasionally scarce. As a precaution, it was therefore advisable to buy them in larger quantities. (One reason for the scarcity of better cigarettes is said to be that the main customers, the civil service and the armed forces, insist on the official price, which does not afford the dealers the necessary income to covering the costs of procuring them: higher prices are paid only for scarce articles.) Matches are also best bought in bulk, if a longer stay in the hills is planned; during the rainy season they must be dried over the fireplace before they can be used.

All goods with an unregulated price are subject to seasonal price fluctuations. However, great differences also arise depending on the place of purchase: the prices increase with the distance from the main bazaar or purchase place of the dealer or peddler. The following table is intended not only to give a summary of the main food and consumer items but at the same time an insight into the local price structure, which (unlike calculations from the official exchange rate) is decisive for the evaluation of the economic situation of the residents of the Hill Tracts. The prices are for 1 Sher, i.e., roughly 1 kg, unless indicated differently. (Prices are indicated in local rupees of 1956/57, 1 Re. = 16 annas.)

rice	-/5 – 1/8
onions and garlic	-/4 – -/12
aubergines (eggplants)	-/3 – -/6
potatoes	-/3 – -/14
taro	-/3 – -/5
dal (Indian lentils)	-/12 – 1/8
flour	-/12 – 2/-
rusk (10 pieces)	-/7 – 1/-
sugar	1/- – 4/-
salt	-/4
pepper pods (chilli pepper)	3/-
caraway	3/- – 4/-

cardamom (12 gr)	-/11 – 1/-
cloves (12 gr)	-/8
cinnamon	6/8
tea	5/8 – 8/-
oil	2/- – 3/8
clarified butter (ghee)	8/- – 11/-
eggs (piece)	-/1 – -/4
chicken (1, according to size)	1/- – 4/-
pigeon (1)	-/12 – 1/-
coconut (1)	-/6
matches (12 dozen)	8/2 (1 dozen: -/11)
cigarettes (10 Capstan)	-/10 – -/12
torch battery (each)	-/12 – 1/4
"mantle" for Petromax	1/-
kerosene (tin) (for Petromax)	7/8 – 8/8
soap (1 kg)	2/- – 2/8

A4d) Medicine

A basic stock of medicine was brought from Germany, arranged as tropical equipment. This would not have been necessary, since apart from privately preferred or needed medicaments, as long as one knows what kind of medicine one wants to buy, the pharmacies in Chittagong sell all major medicines (primarily those produced by Burroughs, Wellcome & Co, but also by Bayer, Ciba, etc.) for the usual illnesses. Snakebite serum, however, was not to be had, (two ampoules were brought from Calcutta). Apart from the remedies needed for private malaria prevention, for the hill people the main remedies needed are against malaria and amoebiasis, since the supply of medicine in the hills is completely insufficient. This situation also means that anyone who wants to do research in this area is dependent on his own knowledge and remedies. In enlarging my elementary medical knowledge I received the help of Romol Krishna Chakma, who in Ruma had to care for the local population with the remedies provided from state funds (with allocations that were far from sufficient for the size of the area), and also primarily of the fathers of the Roman Catholic mission in Chittagong. Only in Chittagong is it possible to find medical examination and support in the European style, and only here was a dentist to be found. Since the journey from the hills to Chittagong requires several days, remedies for first aid and medicine for emergencies had to be kept close at hand.

Before being allowed to enter the country, a vaccination is required against smallpox, typhoid fever, paratyphoid and cholera. In the country, especially during the rainy season, malarial infections cannot be avoided, but have only a barely imperceptible effect on general condition if Resochin is used regularly. However, over a period of time, amoebiasis infections prove a strain, weakening the body; they are hardly avoidable if one stays a whole year, since the whole area is contaminated: the most dangerous time is in the first weeks after the rains have set in. Inflammations of the eyes and infections with ringworm (tinea) are possible but can easily be cured. Coughing is problematic, however, which develops in chilly and damp nights and can become chronic with time (as in the case of the local population), but fades away after the return to our climate. However (as I learned only afterwards), one could also be infected with TB, since people habitually expectorate everywhere.

To my surprise I found out that today, fifty years after my first stay, basically the same medicines are still in use and no fundamental progress seems to have been made in this field. I shall therefore list the remedies most frequently taken by me, given to patients and felt useful (manufacturer's name in brackets and, provided that it was available in Chittagong, price in rupees at 16 annas):

Resochin (Bayer, 100 tablets Rs. 13/8, sold on remote bazaars for up to 4 annas per tablet), a malaria remedy, for personal use (for prevention) and for the treatment of local people with malarial fever (4–6 tablets) and enlarged spleen (20–40 tablets, to be taken over a period of time). Eked out by:

Daraprim (Burroughs, Wellcome & Co, 8 tablets Rs. -/7), a bit more expensive, but better suited for children, since it does not taste bitter.

Entero-Vioform (Ciba, 100 tablets 13/4), against dysentery and for regulation of the intestines (except in cases of constipation) for some time (also preventively), but does not cure amoebiasis; for the treatment of local people in acute cases of dysentery: 1–2-day doses; in greatest demand at the beginning of the rainy season.

Chrysophanic acid (1 oz 12/4) against tinea (ringworm) of the local people, depending on strength of the infection (to be applied mixed with kerosene); for personal protection on the appearance of a minor infection Tineafax (BW&Co, tube Rs. 2/1) is sufficient; it was also effective against ringworm and against sore toes of the local people. (Mitigal, Bayer, tube Rs. 1/5, was less effective.)

Dressing material: plasters (e.g., Elastoplast) were only available in small strips but not in rolls; surgical cotton wool, clips and gauze bandages (4", Rs. -/8), iodine (4 oz, Rs. 1/4), Marfanil-Prontalbin powder (Bayer, not available), Vaseline (chemically pure hardly available), Azulon ointment (Homburg, not available), penicillin ointment: Crystapen (Glaxo, tube Rs. -/14); Novocain (Hoechst) for local operations, for this surgical cutlery and hypodermic syringe.

Irgapyrin (Geigy, tablets not available, ampoules of 5 pieces Rs. 10, not used, 2–3 tablets were enough for noticeable successes with local people) against rheumatic pains.

Acedicon (Ingelheim, not available) for the alleviation of irritations of the throat.

Scheroson-Ophthalmicum (Schering, not available) against inflammations of the eyes, very much praised by the local people.

Antepar (piperazine, BW&Co, 4 oz, Rs. 5/14) against pinworms (*Enterobius vermicularis*): local children can be strongly infected and personal infection is possible; the remedy was thought not to be harmless and therefore employed with care.

Aspirin Phenacetin (Boots, 100 tablets, Rs. 2/9), cheap fever and pain alleviation medicament, predominantly for personal use. Caligesic (Sharp & Dohne, per tube Rs. 2/1) against skin irritations from poisonous leaves etc.

Potassium permanganate (against scorpion bites). Also used were Novalgin quinine, Supronal, Gelonida, Yatren, weak laxatives, medical charcoal and magnesia.

Finally I shall mention two other important means for killing insects and protecting against them that I had to use frequently: Dichlorodiphenyl trichlorethane (DDT), spray cans and solution were available in Chittagong; and Kita (a mosquito repellent, which when rubbed in, had a restricted but welcome temporary effect, especially during the rainy season; not available in Chittagong).

A4e) Maintenance and accommodation

For the regulation of our financial supply we opened an account with the National Bank of India in Chittagong. Here arbitrary amounts could be held ready for disposal. A transfer from here to the post offices in the three main places of the Subdivisions of the Hill Tracts was possible, but did not prove to be advantageous in the long run: apart from the delay (from mailing the instruction until receipt of the money) these little post offices often did not have (or pretended not to have) the necessary cash at their disposal for an immediate payment and moreover postal orders (as also insured transfers) could not be kept longer than a week. To police outposts (with a weekly postal delivery) only amounts up to one hundred rupees could be remitted, and one could not even rely on regular delivery. After a few not very successful trials of receiving smaller sums regularly, I therefore finally changed the procurement procedure completely and went to Chittagong every few months to collect larger amounts (e.g., one thousand rupees) and brought them back to my location. Thus, at my own discretion I could take the money I required, for instance for tours in the country, from my own reserves, which I left behind in Tapwüa-Kua. This method was only possible thanks to the honesty of the villagers, with whom I occasionally left over one thousand rupees and who in my absence guarded my personal belongings in a bamboo house, which for certain could not be locked. This honesty was a considerably better guaranty for the safety of my money than permanently carrying it with me would have been. (My only loss happened when, on my way back from Chittagong, in Kyo Thwân Ong's house in Banderban, a ten- and a five-rupee-note were stolen from under my pillow. My boy later confessed to me that his wife had probably been the thief as she had become an opium addict.) Since one-hundred-rupee-notes could be changed only in the main bazaars, my reserve had to consist mainly of ten- and five-rupee-notes, which, however, could already cause difficulties, when I wanted to change them in some hamlets), so that in the main I had to use one-rupee-notes (which I could receive in the bank in bundles of hundred notes). Coins would have been even more convenient, since the villagers preferred them to the notes, but they would have been heavy to transport and not even available in the bank in the desired amount.

On a journey I could always find accommodation in the houses of the local people, mostly in the house of the karbari, headman or some other richer man. With very few exceptions, overnight stays are granted willingly and everywhere with great hospitality. In Tapwüa-Kua where I stayed for months I had a house of my own set up for me; this house cost me 80/- Rs. in all, and Menkröi lent me his table and chair.

Places with a police station usually have a bungalow which can be provided to the traveller. Early in our stay, we "camped" here in Ruma. For accommodation in the bigger places the following were used: in Chittagong the Hotel "Mishka" near the station (Anglo-Indian cooking, daily rate approximately 20/- Rs., visitors mainly better-off Pakistani) or the Europeans' Club, situated a little remotely (approximately the same prices, but very snobbish attitude and low standard of comfort); in Rangamati: Circuit House (only with permission of the Deputy Commissioner) in cultivated bungalow style with catering; during our first stays in Banderban, Ong Shwe Phru was kind enough to provide us with accommodation in a house that was empty at the time; on later visits I always stayed in Banderban in the house of my boy, Kyo Thwân Ong.

A5 The structure of the book

A5a) Division into chapters

In the present work the data collected in the Chittagong Hill Tracts is intended to be represented in an organized way. Such an order, however, has its limitations since not all information and impressions can be explained. Not only must many data that are based only on assumptions be left out, but also an ordering frame has to be provided for the totality of data described. As a central topic for my collected materials I chose a monographic representation of a part of the culture of the Mru. Whenever data that I came to know about the culture of the other residents of the Hill Tracts offer themselves for comparative purposes, they are included provided that they do not blow up the chosen frame. Fragments which cannot be united to any uniform picture must be left aside. To these belong, for instance, some knowledge of the vocabulary and grammar of "hill Bangla", which I could never subject to any conscious systematisation, although they were fundamental for my field research and though I constantly tried to improve them, since the longer I stayed, the more I interspersed my own "hill Bangla" with Mru elements and words.

But even in the realm of the topic not everything can be said. The tiresome question arises here: how to draw an overall picture of a culture? How to project space and time into a line, to systematise what one has experienced, to take to pieces an organism and line up its components so that everything joins itself together again into a clear whole? This "clear whole" is undoubtedly only an ideal; but perhaps it is not even this. Who will read a monograph chapter by chapter? Does the reader not rather seek details on single topics? Can such topics be represented for themselves at all without hiding something in another unexpected place? And where would such a topic catalogue be that would do justice to later claims? Does a questionnaire monograph really make sense? Would it not be better to write a detailed diary, in which not only the data, but also the circumstances of their collection could be found, by looking up the topics in the index? But what researcher could note down hour for hour a detailed interim report of his impressions and results? Where would the insights be mentioned if one discovers them one day and is unable to give any information about how they were acquired? Is the problem of falsification not rather to be found in the memory?

After some trials and considerations, which shall not be listed here in detail, I finally decided to follow the conventional form and breakdown of the presentation despite all its defects. Some problems which form the topics of special examinations today, are inevitably taken to pieces or will be seen only from one side, but if I had chosen them as a principle of dividing up my data, the conventional chapters would have met the same fate. Moreover, in these modern times of specialised research, a complete monograph and with it also its inner coherence is something that can no longer be handled by a single researcher. A general overview may, therefore, safely use conventional methods, and to its advantage also contain materials that are of no immediate interest, and I will bear the full responsibility for the fact that the inevitable biases are neither completely adequate to the data nor correspond to those required by possibly interested people, to whom the Mru themselves may also belong in the future.

A5b) Presentation

Although I have subjected my material to a conventional division scheme, I do not feel able to follow the conventional mode of presentation. I do not want to claim that I am authorised to maintain that "the Mru" would do or permit this and that etc., and I am not convinced that my opinion, which was formed under their influence in the course of my stay, and the generalizations I reached are right in every case or could have any more weight than a statement by a member of the culture I studied. Unfortunately, my text still contains very many of these "unverified" generalisations; on the one hand because I did not register my informants' statements as a matter of principle, on the other hand because I did not recognise the information received as such in time, so that the question only arose when I tried to write down my results and order them more systematically than I had done in my note-book.

It would also undoubtedly have made little sense, if I had wanted to make even more evident the events and facts which I had daily before my eyes by using a statement from an informant. However, whenever I regarded a question as useful for the confirmation of my observations, I have noted down substantial answers with person, place and date and will quote them here accordingly. In special points, i.e., for statements that cannot be confirmed by frequent observation, especially for coherent processes on special occasions, I similarly added place and date to my own observation. I have already explained that these details should also be understood only as examples (see A3c). If such special details are not dated twice, they are based on a simple statement from an informant. When I received such information one or several other persons were almost always present who could give or refuse their consent. This, however, does not apply to data I received from Menkröi Ngarua', which as such already have a high degree of validity. Menkröi's explanations, whether general, specially controlled, or unchecked, form the basis of my data. If another informant is indicated from Tapwüa-Kua or the surroundings, then it can be assumed that the topic dealt with is specialist knowledge which is not present with the precision necessary for execution by other people. To be distinguished from these data are details from other Mru groups which mainly refer to differences in rules.

Since I did not normally use tapes for recording the information I collected, I shall refrain from quoting statements verbatim. Any mistakes in interpreting the information received are therefore mine, especially since, as already indicated, I used mainly "hill Bangla" mixed with Mru terms in my conversations. This even applies to some extent to my first translations of texts and songs that I recorded on tape in Mru. These texts were first listened to, several times, passage by passage, by me and Menkröi (not by the story-teller or singer, since they constantly tended to tell me a somewhat different text), noted down by me line by line or sentence by sentence in

Mru, for which I had to develop a transcription of my own. Menkröi helped me to translate words, expressions and passages which I could not understand, so that I could later try to translate them into German or English by myself.

All references to informants (which I give in brackets) only refer, unless separated from the previous sentence by a full stop, to the immediately preceding statement. If, however, it is separated by a full stop, it refers to the whole previous part of the paragraph in question. With many informants only a date but no place is stated. When the informant is called Karbari or Headman, however, this at the same time implies a place name, since it is customary to call the hamlets after their karbari or headman. When even such a detail is missing, I refer to some inhabitant of Tapwüa-Kua. As the name of the informant I use the personal name, followed (as far as known to me) by his sib name. The name of my main informant, Menkröi Ngarua', I have abbreviated to MK. In place names the abbreviations mean:

KP = Karbari-Para

RP = Roaja-Para

HP = Headman-Para

Para is a Bangla word, also used by the hill people, and means "hamlet".

A5c) References and transcription

Dividing the overall picture of a culture up into chapters necessarily means that topics and data that seem necessary for understanding are frequently touched upon, although they are not wanted in detail in the actual connection. In all cases where it seemed important to me I have added cross-references, not only to spare the reader the trouble of looking up an index, but also to make it clear that more details can be found on the topic in question. In order to make these references easy to handle, a correspondingly simple chapter numbering was necessary, while simultaneously a far-reaching subdivision was required. The system used now is the result of several numbering trials and has the purpose of reconciling the dissimilar size of the single topics with a subdivision into sections of as nearly as possible equal length.

When beginning to write up my data, I was of the opinion (based on my own experiences) that it is unreasonable to ask of every reader not fully familiar with the material to deal with an accumulation of words from the vocabulary of the treated culture. Every attempt to write my text without using these special terms, however, regularly ended when I wanted to be precise without having at hand an equivalent expression in my own language, which would at the same time be convenient. I have therefore in many cases given non-English terms not only in brackets (where they are legitimate), but also whenever I felt I should use them as "technical terms". In the beginning I followed the rule of using them only when they re-appeared without explanation within the chapter in which they were explained, while in all other cases the explanation was added anew, or, when this would have required a lengthy text, by giving a reference to the subchapter where they were explained. In the end, however, I felt that even this would be troublesome for the reader, and therefore decided to omit these additional references.

I have transcribed the Mru words without marking their syllable tone, since I could not have guaranteed correctness in all cases. The Mru themselves do not use tone marks, neither in the old transcription as introduced by Menkröi, when he translated the Christian hymn book and St. John's gospel, nor in the new script invented in the 1980s by Menlè [Menlei. Ssee Chapter T] for spreading his new religion. Both forms of transcriptions (those using the Latin alphabet and that of Menlè) therefore have the disadvantage that words are written identically, even though they have a different tone. As a rule, only the main syllable of a word carries a full tone; prefixes and suffixes, which without exception end in "-a" (as used by Menkröi) or "-ö" (as used by Menlè) remain unstressed, and are pronounced like an unstressed "a" in English,. To distinguish these –"a" from the full vowel "a" (like in English "father") I use –"aa" for the latter, which exist in open syllables only. Whenever not pronounced long, (but when short), as in closed syllables, I use "a". In all names I have written the two full syllables of which they are mostly made up, or unhyphenated together; in other cases I have used hyphens in the interest of better syntactic clarity.

More details on the pronunciation of the Mru and Khumi words are not necessary. Major deviations from the pronunciation according to a Latin version must be noted only for kh-, ch-, th- and ph- (all to be pronounced aspirated, "th" and "ph" are never as in English); c- is approaching ts-, never to be pronounced as k-, and ch- is aspirated c-, but approaching s, with a weak t- in front; e and o are always open ("e" as in "pet" or "pair", "o" as in "pot" or "awe"), ö and ü are central vowels ("ö" roughly represented by open German open "ö" or in English by the first vowel in "clergy", and "ü" like Saxonian "u"). The ' indicates a glottal stop which, however, is not obligatory. (For "a" and "aa" see the preceding paragraph.)

Also in Bangla words "e" and "o" are open, "ô", however, stands for a closed "o". Thus my transcription follows the pronunciation, not the official transcription. Correspondingly, dentals which in official standard Bangla are written and spoken as retroflexes (but spoken as normal dentals in the Chittagong-Bangla) are written as normal dentals. "j" stands for a sound which can vary from "dj" to "z".

Major deviations from this notation are used only for Marma words, since my transliteration represents a transcription of the written words and does not correspond to the pronunciation, which cannot be represented phonetically without overloading it with diacritic marks. The following rules are generally valid: initial "s" is to be pronounced like unvoiced English "th", "c" as Italian "ci" with a trend to "s", "ch" like aspirated "c", "j" between "dj" and "dz", "hr" and "sh" like German "ch" (in "ich"). The aspiration of voiced consonants appears only in the script. All initial consonants can become voiced in sandhi, provided that the preceding syllab-le does not end in an occlusive, "ph" tends to become "f" in sandhi. Final occlusives are not spoken as such: they are, when standing alone, pronounced like a hard glottal stop, in sandhi they assimilate to the following consonant. Nasals nasalize the previous vowel and become blurred, final "-ö" is closed "-o", "-ai" is to be spoken as open "-e", "-o" is always open, "-a" in the short tone becomes noticeably more fronted. – Medially "-a-" in front of a guttural tends to become closed to "o", before all other final consonants it becomes diphthongized and fronted to -ae- (in front of -t and -n almost to -ei-), "-o-" (which appears only in front of gutturals) becomes diphthongised to "-au-", "-u-" (only in front of labials and dentals) to “-ou-”, “-ö-” (only in front of gutturals) to an open “-oi-”, “-i-” usually remains unchanged, tends, however, in the falling tone to be preceded by an “u”-sound. I have marked the light (rising) tone (which in an open syllable implies a light glottal closure) by ’ after the syllable, the heavy (falling) tone with an ^ over the vowel.

A6 The course of the expedition the Chittagong aeea

The following overview is reproduced, with insignificant cuts and new formulations, from our/my activity reports to the German Research Council (Deutsche Forschungsgemeinschaft) of March 22nd, 1956, April 7th, 1957 and September 22nd, 1957. Two maps have been added: one showing eastern Bangladesh, the Bangali plains in the west, the Indian States of Tippera in the North and of the Mizo-Hills in the East, the other map centers on the southern Chittagong Hill Tracts and shows my tours.

30-XII-1955 Arival Chittagong port

30-XI/2-XII-1955 Chittagong. Visiting authorities and purchases.

3/5-XII-1955 Rangamati. Visit to the Deputy Commissioner and the Chakma prince at the time of his Punyah; first meeting with different ethnic groups.

6-XII-1955 Chittagong. Copying part of the Census (1951) of the Chittagong Hill Tracts, lent by the Dy. Commissioner; shopping.

7-XII-1955 Dhaka. Visiting the Govt. main office to receive maps of the area; dentist.

8-XII-1955 Chittagong. Remaining purchases.

9/10-XII-1955 Travelling from Chittagong via Dohazari to Banderban.

11/19-XII-1955 Banderban. Visiting authorities; participation in the Punyah of the Bohmong (Marma prince); recording music and dances for the celebration and Bengali songs; conversations with educated Marma on general conditions in our field of research; trying to regulate our money supplies.

20/22-XII-1955 Travelling from Banderban to Ruma in the company of the Subdivisional Publicity Officer (Bohmong’s son); first acquaintance with Marma hamlets on the Songu river.

23-XII-1955 Ruma. Accommodation in the bungalow; visit to the bazaar.

24/26-XII-1955 tour Ruma > Boga Lake > Polika. Visit to a mythically significant lake, experiences about difficulties of mountain tours, visit to Khumi hamlets.

26/28-XII-1955 Polika. Participation in the great celebration of a Marma headman; shooting cine-film.

29-XII-1955/25-I-1956 Ruma. Loss of time because of false promises (tour companion), lack of knowledge and suitable people to go on tours independently, difficulties with food and money supply and official negotiations. Listing previous expenditures, evaluation of previous knowledge, memorandum to the German Research Council. Trying to clarify the variety of the ethnic names (who is who?). Vocabulary comparisons, esp. the kinship terms of the Chakma in comparison with those of the Bengali in Hindu and Muslim form; tape recording of Bengali and Marma songs, (long) attempt to have the Bengali texts noted down and translated; (partly by comparison with older reports) notes about recent cultural change; notes on school systems and medical welfare. Leader of the expedition goes to Banderban and Chittagong 13/25-I.

25/26-I-1956 Chitpup. Participation in a Thakur celebration of the Marma; tape recordings.
27-I-1956 Ruma. Post handling and tour preparation. Leader of the expedition has to visit Dhaka.
28/29-I-1956 Ruma > Polika > Pantola > Yöngtu-HP.
29-I/2-II-1956 Yöngtu-HP. Participation in a buffalo sacrifice of the Mru; notes on the course and the paraphernalia of the festival; tape recording of Mru songs and music; small vocabulary.
2/4-II-1956 Menring-KP. Another buffalo sacrifice of the Mru; comparisons.
4/6-II-1956 Menring-KP > Ruma. Participation in a small festival of the Marma in Polika.
7/10-II-1956 Ruma. Ordering the Mru notes; making arrangements for the food supply.
11/12-II-1956 Ruma > Yöngtu-HP.
13-II-1956 Yöngtu-HP > Menring-KP. Supplements to and corrections of the notes on the first Mru tour.
14/17-II-1956 tour Menring-KP > Yöngtu-KP > Chinbok > Sengum > Ruma. Various notes; inquiries about a suitable place for a longer stay.
18/20-II-1956 boat tour Ruma > Galengya.
20/2l-II-1956 Inglai-KP. Visit to an ear-piercing festival of the Mru.
21/24-II-1956 tour Inglai-KP > Galengya > Bolipara > Hamajon-KP > Jogendro-KP > Bolipara. Visit to Twipra hamlets, recording of music, notes about house and faith forms.
25/26-II-1956 boat tour Bolipara > Thanchi. Visit to Tongpüi-HP (Chüngma Mru).
27/28-II-1956 tour Tongpüi-HP > Kwamcwe > Ongpha-HP.
29-II-1956 Ongpha-HP. Recording of Khumi songs; notes on faith forms.
1-III-1956 tour Ongpha-HP > Onghruimo-KP.
2/3-III-1956 visit to Krawu-HP in the company of the Officer-in-charge of Thanchi. Notes about the relations of the southern Hill Tracts to Arakan and on the local jurisdiction.
4-III-1956 tour Onghruimo-KP > Tindu.
5-III-1956 Tindu. Notes on the acculturation Khumi > Marma; recording of Khumi songs; trying to make a Khumi vocabulary.
6/7-III-1956 visit to Royno-KP; notes on house forms and prestige symbols of the Khumi.
8/10-III-1956 Tindu. Participation in a cattle sacrifice of the Khumi; notes on the course of the festival, adornments, faith ideas; music recordings.
10/13-III-1956 return Tindu > Ruma.
14-III-1956 Ruma. Updating of my diary; post.
15/17-III-1956 tour to Artha' (Calkhup-HP). Studies on the acculturation Bawm < Lushai. Recording of old tribal and modern Christian songs of the Bawm.
18/l9-III-1956 Ruma > Banderban.
20/24-III-1956 Banderban. Updating of my diary, post, activity report; participation in the celebration of the Islamic Republic Day; tour preparations.
25/28-III-1956 tour to Rajbila. Participation in the spring celebration of the Marma. On the way back in Mrokhyong recording of old Marma songs and Khyang songs.

29-III/l-IV-1956 Banderban. Meeting the leader of the expedition, mutual information on experiences; work schedule planning; packing up of the previous tape recordings; learning to write the Burmese script.
2/7-IV-1956 tour Banderban > Chinbok > Tapwüa-Kua; passing Mru hamlets in the Rankhyong area. On 5/6-IV participation in a cattle feast in Rengyong-KP (Alekhyong area).
8/9-IV-1956 Tapwüa-Kua. Preparations for a base.
10/11-IV-1956 way back Tapwüa-Kua > Banderban.
12/13-IV-1956 farewell to the expedition leader; post; participation in the New Year's Day of the Marma.
14/20-IV-1956 Banderban. Updating of my diary; preparations for moving to my new base; visiting authorities; participation in celebrations of the Marma. Very hot days.
21/23-IV-1956 boat tour Banderban > Ruma > Galengya.
24-IV-1956 ascent Galengya > Tapwüa-Kua.
25-IV/2-V-1956 Tapwüa-Kua. Observing the building of a house; participation in a cattle feast; enlarging acquaintances in the hamlet; studies on the relationship and sib system; medical care.
3/5-V-1956 Tapwüa-Kua > Tankaboti valley > Satkania and back. Studies in the Mru-Bengali contact zone; bazaar visit.
6/13-V-1956 Tapwüa-Kua. Ethnographic notes; getting acquainted with visitors from the surrounding area; distribution of medicine, tape demonstrations and new recordings; trying to learn and to improve my knowledge of languages (hill Bangla and Mru).
14/19-V-1956 Tapwüa-Kua. Updating of my diary, post, report to the DFG; moving into my own house, visits in the neighbouring hamlets.
20/2l-V-1956 Tapwüa-Kua > Banderban > Chittagong.
22-V-1956 Chittagong: Bank, purchases, doctor visits. Return to Dohazari.
23/24-V-1956 Dohazari > Banderban. In B. participation in the procession on the day of Lord Buddha.
25/28-V-1956 Banderban. Visiting authorities; delay by post officer who cannot pay out money; studies in Marma [?] grammar as preparation for Mru questions.
29-V/4-VI-1956 boat tour Banderban > Ruma > Polika and back. Vain attempt to reach Galengya by boat: monsoon start with inundations; notes on Marma folklore.
5/6-VI-1956 Banderban. Notes on the Santal of the Hill Tracts, on lending systems and interest, on the Marma medicine-men.
7-VI-1956 Banderban > Cökkhyong and back. Vain attempt to reach Tapwüa-Kua via the land route.
8/9-VI-1956 Banderban > Dohazari > Chittagong.
9/11-VI-1956 Chittagong. Dentist, purchases, post.
12/13-VI-1956 return Chittagong > Banderban.
14/15-VI-1956 Banderban. Visiting authorities; post; tour preparation.
16/19-VI-1956 boat tour Banderban > Galengya.
20-VI-1956 journey up Galengya > Tapwüa-Kua.
21-VI/4-X-1956 Tapwüa-Kua. Systematic studies on the culture and language of the Anok Mru according to the following weekly overview:
21/23-VI (after listing of expenditures and updating my diary) weeding;
24/30-VI cattle and buffalo sacrifice, song texts;
1/7-VII text interpretation, grammar; pig sacrifice;

8/14-VII pig festivals, texts and vocabulary, heir, divorce; participation in a cattle sacrifice;
15/21-VII colour production and use; attitude of the Mru regarding Buddhism, house parts and furnishings; grammar and words; own pig festival;
22/28-VII language studies (pronouns, conjunctions, particles), text interpretations; paddy harvest;
29-VII/4-VIII fairytales and nursery rhymes, settlement and hamlet history, sacred house stones, birth and naming; participation in pig festival and field sacrifice;
5/11-VIII field sacrifices, pacts between sibs, conception and abortion, veneral diseases, vocabulary (esp. loan and foreign words from Marma);
12/18-VIII dreams, days, months, possession, celestial phenomena, further field sacrifices (with participation);
19/25-VIII old weapons, oaths, sib relations, river spirit and sacrifices, spirit stones, sacrifice to the border spirit (with participation), marriage; word construction;
26-VIII/1-IX closing the hamlet, religious prohibitions, traps, song record and text interpretation;
2/8-IX text interpretation, treading out paddy and field sacrifice (with participation), further pig sacrifices, ceremonies for a dead body, children's drawings, hunting ceremonies;
9/15-IX death ceremonies, sacrifices for sick persons (especially lunatics), magic, further pig sacrifices (with participation), rules for swiddening and building a house;
16/22-IX further sacrifices, hunting songs (text recording and interpretation);
23/29-IX new song texts, incorporation of tribal strangers; harvest yield and consumption, house measures, baskets, goat sacrifice (participation);
30-IX/4-X clothes and ornaments, bamboo toys; swidden visits (cutting and treading out), participation in a wedding feast.

5/6-X-1956 tour Tapwüa-Kua > Galengya > Ruma.
7-X-1956 Ruma. Updating of my diary, post.
8/9-X-1956 boat trip Ruma > Banderban.
10/11-X-1956 Banderban. Noting the text of old Marma songs.
11/12-X-1956 Banderban > Chittagong.
13/16-X-1956 Chittagong. Trying to have the tape recorder repaired; bank, purchases; delay through holidays, contacts with Germans in Chittagong; post.
17-X-1956 Chittagong > Rangamati.
18/22-X-1956 Rangamati. Visiting authorities; studying the swidden tax lists for the preparation of an ethnographic map of the Chakma-Circle.
23-X-1956 Rangamati > Chittagong.
24/28-X-1956 Chittagong > Kaptai > Chandraghona > Chittagong. Visit to Tongzengya settlements; breaking off the tour before reaching the Kaingsa Mru because of a renewed defect of the tape recorder.
29-X/3-XI Chittagong. Another attempt to have the tape recorder repaired, bank, purchase; delayed by political events (strike to show sympathy for Egypt).
4/5-XI-1956 Chittagong > Banderban.
6-XI-1956 Banderban. Post office.

7/11-XI-1956 boat trip and tour Banderban > Ruma > Galengya > Tapwüa-Kua.
12/28-XI-1956 Tapwüa-Kua. Smaller ethnographic notes, updating of my diary; participation in ceremonies when bringing in a bag and fetching a bride; notes on sib relations, "spinning-room" customs, bamboo work; hill Bangla vocabulary; working on a mouza map; tour preparation; post.
29-XI/8-XII first Dopreng tour (area of the rivers Bommu, Lama and Prumma). Observation of cultural differences between Anok and Dopreng; extension of the sib list.
9/20-XII-1956 Tapwüa-Kua. Updating of my diary, participation in a swidden festival, smaller ethnographic notes.
21/27-XII-1956 second Dopreng tour (Bommu and Prumma area). Participation in a great buffalo feast.
28/3l-XII-1956 Tapwüa-Kua. Writing down tour results; participation in a great swidden festival.
1/2-I-1957 tour to Borpara for first aid.
3/5-I-1957 Tapwüa-Kua > Banderban > Chittagong
6/8-I-1957 Chittagong. Repairing of tape recorder, bank, medicine; contact with new missionaries for the Hill Tracts; fetching the outboard motor from bank.
9/14-I-1957 Chittagong > Banderban (leaving outboard motor) > Ruma > Thanchi.
15/16-I-1957 Thanchi (Üikük-HP, Tuktöng-P, Ongkacan-Mohajon-P). Inquiries about festivals and ceremonies of the Chüngma Mru.
17/18-I-1957 tour to Noliram-RP. Participation in a buffalo feast of the Twipra; on 18-I-1957 boat bought in Ongkacan-Mohajon-Para.
18/19-I-1957 Thanchi > Galengya > Tapwüa-Kua.
20/25-I-1957 Tapwüa-Kua. Participation in the spring prohibition time; noting down the results of the Chüngma tour.
26/27-I-1957 tour to Chamklaa-KP. Participation in a death ceremonial.
28/30-I-1957 Tapwüa-Kua > Galengya > Banderban.
31-I/1-II-1957 Banderban. Authorities, post; ill.
2/3-II-1957 Banderban > Chinbok > Tapwüa-Kua.
4/11-II-1957 Tapwüa-Kua. Curing myself; taking notes and making additions to the death ceremonies, *Khang* and swidden festivals; smaller ethnographic notes; problems of soil requirements.
12/14-II-1957 Tapwüa-Kua > Galengya > Banderban.
15/18-II-1957 tour via Boido-P to Lenten-KP, visiting the Betchora-Khumi (with strong Mru influence), comparisons, small vocabulary.
19/20-II-1957 Banderban. Vocabulary comparisons; post; expecting news.
21/22-II-1957 Banderban > Kuwalong > Dohazari.
23/25-II-1957 Dohazari. Repairing the outboard motor of my boat, buying petrol and oil; ill.
25/26-II-1957 Dohazari > Banderban.
27/28-II-1957 Banderban. Ill; procuring money.
1/3-III-1957 Banderban > Galengya > Tapwüa-Kua.
4/6-III-1957 Tapwüa-Kua. Listing of expenditures, updating of my diary, tour preparation.
7/30-III-1957 tour to the Mru groups of the Rümma, Dömrong, Rengmitca and Chüngma as well as the Sak and Khumi.
7/9 Tapwüa-Kua > Lama > Yongsa;

10/15 visit to Rümma hamlets, participation in cattle feast and the swidden sacrifices, notes on sib migrations;
15/16 visit to a Dömrong cattle feast;
16/17 visit to two Sak hamlets, notes on language and material culture;
18 crossing over the Maranca chain to the Matamuri valley;
19/20 participation in a buffalo feast in the Alikodong area; notes on the migrations of the Tongzengya;
20/21 visit to a Rengmitca hamlet, linguistic studies;
22/26 visit to Khumi hamlets, notes on cattle feasts and swidden sacrifices, divorce and inheritance, wedding and death ceremonies, kinship and sib relations, acculturation Mru > Khumi, language; participation in a small cattle feast;
27/28 visit to a Chüngma hamlet, notes on swidden ceremonies, marriage, feasts of merit;
29/30 return Thanchi > Galengya and climbing up to Tapwüa-Kua.
31-III-1957 Tapwüa-Kua. Notes on the tour.
1/3-IV-1957 tour to Chikcaron-P. Participation in a buffalo feast in the Anok-Chüngma mixed zone.
4/7-IV-1957 Tapwüa-Kua. Notes on the tour, activity report.
8/9-IV-1957 tour to Cong'uk-KP. Medical help; notes on sib relations, bad death.
10/18-IV-1957 Tapwüa-Kua. Notes on the tour, scientific report, post; plant collection; check of language tone rules.
19/21-IV-1957 visit in Rengnok-KP. Participation in a great wedding celebration.
22/25-IV-1957 Tapwüa-Kua. Updating of diary; swidden visits (sacrifice and seed).
26/30-IV-1957 Tapwüa-Kua > Galengya > Banderban > Dohazari > Chittagong.
1-V-1957 Chittagong. Purchases.
2-V-1957 Chittagong > Rangamati.
3/6-V-1957 Rangamati. Authorities, visiting the new Deputy Commissioner and the Chakma Raja (questions on the history of the Chakma).
7-V-1957 Rangamati > Chandraghona > Chittagong.
8-V-1957 Chittagong. Bank; purchases.
9-V-1957 bus Chittagong > Fatikchori, tour Fatikchori > Manikchori.
10/11-V-1957 Manikchori (court of the Mong-Chief). Studies of the swidden tax lists for the preparation of an ethnographic map of the northern Circle.
12/14-V-1957 walking back to Fatikchori; journey via Chittagong > Banderban.
15/17-V-1957 Banderban. Visits; continuation of the journey to Ruma.
18-V-1957 tour Ruma > Artha' (Calkhup-HP).
19/20-V-1957 Artha. Notes on the Lushai mission and on the linguistic position of the Bawm.
21/22-V-1957 tour Artha' > Ruma, boat Ruma > Banderban.
23/25-V-1957 Banderban. Notes on the tour, listing of expenditures; vocabulary studies.
26/27-V-1957 Banderban > Dohazari > Chittagong.
28/31-V-1957 Chittagong. Visits; waiting for a new tape recorder; updating of notes.

1/7-VI-1957 Chittagong via Banderban (purchases) > Ruma > Galengya. Calculations from population lists. Way up to Tapwüa-Kua.
8-VI/5-VIII-1957 Tapwüa-Kua. Completion of the previous studies.
8/16-VI listing of expenditures, updating of notes, post; notes on the household budget of the Mru;
17/23-VI village plan, fishing, tobacco cultivation, noting song text;
24/30-VI sib lists, swidden cultivation (yield calculation), stories, *Tu*-songs;
1/7-VII further text recordings and translations, supplements on material culture;
8/14-VII rotation, yield, erosion (comparison); new song recording and interpretation;
15/21-VII song interpretation, vocabulary of the texts; games;
22/28-VII song recording and interpretation; smaller notes; ill;
29-VII/5-VIII swidden measuring and calculations; supplements on traps; intermediate listing of expenditures; vocabulary revision and checking; text recording and interpretation, noting further song contents; farewell.
6/7-VIII-1957 Tapwüa-Kua > Galengya, boat to Banderban.
8/11-VIII-1957 Banderban > Galengya > Banderban (in search of lost collected plants and keys for cases and trunk)
12/16-VIII-1957 Banderban. Noting text of and translating Marma songs recorded earlier; listing of expenditures, up-dating; farewell.
16/18-VIII-1957 Banderban > Chittagong > Rangamati.
19/20-VIII-1957 Rangamati. Cancellation of my registration, farewell; ill.
21/22-VIII-1957 Rangamati > Chittagong > Dhaka.
22/23-VIII-1957 Dhaka (authorities), return to Chittagong.
23/26-VIII-1957 Chittagong. Preparing for departure; embarkation.

B The Chittagog Hill Tracts and their inhabitants

The Chittagong Hill Tracts (henceforth abbreviated CHTs) are bounded on the west by the narrow Chittagong Plains running along the Bay of Bengal, on the north and east by the Indian states of Tripura and Mizoram respectively, and, finally, in the south and south-east by the Burmese region of Arakan; they lie between 91° 45' and 92° 50' East longitude and 21° 35' and 23° 45' North latitude. At the partition of British India, the CHTs became part of the eastern wing of Pakistan; since Bangladesh gained its independence, they are part of that country's territory.

B1 Landscape and climate

The chains of large mountains which reach the sea in Arakan are part of the extension of the Himalayas which bends southward, cutting through Assam and Arakan. From a very modest height in the backcountry of Chittagong, the mountains rise gradually toward the east in three or four ranges. Running approximately north to south, the hill and mountain ranges are occasionally traversed in a westerly direction by the rivers which flow down between them. The Feni river is found in the north (it also forms the border with the Indian state of Tripura); the Kornofuli, fed by many tributaries, forms the largest river and cuts across the central area; the Songu and the Matamuri rivers respectively are found in the central and extreme south. Unless one wishes to climb the mountains, these few rivers offer the only access to the Hill Tracts. At the mouth of the Kornofuli lies Chittagong town, the largest port in Bangladesh.

The monsoon rains have set in

Whoever chooses the route over the mountains must master considerably steeper slopes in the south than in the north. In the south-east, the highest point – Kyokra-Tong (Burmese for "stony mountain"), which peaks at about 1,300 meters (40,346 ft.) – marks the place where three countries come together: India (Assam), Bangladesh and Burma (Arakan). Generally, however, the ranges rising from a low elevation climb to an altitude of only 500 to 700 meters (1,500 to 2,000 ft.). The hills are formed of poorly consolidated sandstone and shale which conglomerate into a harder stone only in certain places; where the slopes were exposed; they were quickly eroded by the monsoon rains and converted into scree. Cultivation on the hillsides is therefore possible only as long as the soil itself is not broken up or moved. How such cultivation is done will later be described (see J).

Between October and April the sky is almost always bright; from May to September the monsoon rains, which are occasionally accompanied by typhoons, sweep across the area, with the heaviest rains occurring in June and July. With an average yearly rainfall of 2.5 meters (100 in.) – an amount which, however, has declined due to progressive deforestation – the Hill Tracts belong to those areas receiving an extremely high amount of precipitation. Cherrapunji, which receives the highest annual amount of precipitation in the world (more than 10 meters [400 in.]), lies somewhat north of the Hill Tracts in the mountainous area of west Assam. Not only during the rainy season does the relative humidity of the Hill Tracts lie almost always between 90 and 100 %; during the dry season the valleys are also covered by a fog so thick that it is normally penetrated only briefly by the noonday sun. At night the fog is so heavy that water drips from the trees. Over the mountain ridges, in contrast, the dry northeast monsoon wind blows; nights on the mountains are generally clear and the temperatures five to seven degrees *warmer* than those on the plains and in the valleys. (During the cold months of January and February, temperatures in the valleys and on the plains can sink below 10 °C.) During the hot season, though, from the end of March to the beginning of the rains, the temperatures on the ridges are a few degrees *colder* than below. In April, around the middle of the day, temperatures in the valleys can go above 40 °C. Since the humidity decreases at the same time, however, these midday periods are comparatively pleasant when compared with the afternoons and evenings; for even during this season fog reappears in the afternoon and evening. Far from reminding a European of November weather back home, this fog seems more like a sauna – but with one small difference: in the valleys there is no door through which to escape; and in the civilised steam of the sauna there are no frolicking clouds of sand flies and malaria mosquitoes.

The oncoming monsoon rains bring some relief with respect to the heat. During the months of the monsoon, the temperature generally hovers around 28 ° to 30 °C. day and night, though continual rainfall may push it down even a little lower. At the same time, however, the rains wash up all kinds of refuse which during the dry months was hardly able to penetrate the earth's surface. Everyone has intestinal troubles, mostly in the form of amoebic dysentery; and small children die during this season due to the resulting dehydration. It should therefore be obvious, not only that everyone in the Hill Tracts suffers from malaria, but also that, during the cold months, colds and respiratory diseases are likewise universal (at night one can even hear the monkeys in the trees coughing). This is not to mention the fact that tuberculosis is fairly common; that until a few years ago cholera and smallpox epidemics belonged to the annually-recurring plagues; and that persons suffering from leprosy are also occasionally found. Chronic malaria leads periodically to a painful enlargement of the spleen; yet these periods, like mosquito and leech bites, are tolerable. Amoebas, on the other hand, can in the course of time eat through the walls of the intestines and, by way of the spinal cord, enter the brain, causing insanity. The most commonly used remedy against dysentery (at times officially prescribed by doctors) is opium, well-known to be addictive and to ruin a person's physical health. In addition, the Mru claim that it will destroy a person morally; and they complained bitterly that the government withheld from them the healing remedies, so that it could instead sell them expensive poison. To this political tendency we shall return shortly.

In view of the climatic conditions and related health problems just described, it is not surprising that an earlier author described the climate of the Hill Tracts as "deadly". Yet the population increased yearly by some two percent. This rate of increase is, to be sure, substantially lower than that of the Bangali plains; but it has been high enough over the last hundred years to have effected a fivefold increase in population, in spite of some emigration during past decades. If the valleys were in places almost free of people one hundred and thirty years ago, today, under the present conditions of land utilisation, they are overpopulated. Except for a few places in the governmental forest reservations, what was once virgin forest has long disappeared; and in many places an infertile grassland is already spreading – a fact which, however, did not deter the government from moving hundreds of thousands of landless Bangalis from the plains into the hills where, under military protection, they compete with the indigenous people for land.

B2 Major ethnic groups

The fundamental opposition just hinted at – namely, between the plains people and those of the hills – has a long history. To be sure, the Hill Tracts can be geographically considered the hinterland of Chittagong; yet still during the first half of the 20th century one crossed from South to Southeast Asia when one moved from the plains into the hills. There was a striking difference in culture between the plains people of Caucasoid race (speaking an Indo-European language) and the hill people of Mongoloid race (speaking Sino-Tibetan languages). The disparity was further accentuated by the fact that the inhabitants of the plains had for a long time been subjects of a state, whereas the hill dwellers traditionally understood themselves to be their own masters. The ethnic boundary, however, was not always so sharp. Prior to 1666, the kings of Arakan ruled several times also over Chittagong. Those occasionally challenging such rule were the kings of Tripura, whose subjects were no less of the "yellow race" and made no less use of a Sino-Tibetan language than the Arakanese themselves.

During the 16th century Chittagong was the headquarters of Portuguese soldiers who, in the service of Arakan, plundered the neighbouring Muslim provinces (as far north as Dhaka) and carried off the indigenous people to Arakan into slavery – unless the latter had already fled farther northwards. To what extent Bangalis already at that time populated the Chittagong Plains remains an unanswered question, as does the question of who – if anyone –settled at that time in the hills. For some years during the 16th century, the kings of Tripura had their residence in the northern part of the Chittagong Hill Tracts, an area characterised by wide valleys. The so-called **Tippera** (originally Twipra, sanscriticised Tripura) people live not only in the contemporary Indian state of Tripura, but for quite some time they also constituted the majority of the population in the northern part of the Hill Tracts bordering on Tripura state. Only during the last decades have the Tippera been forcibly pushed further and further out of the Hill Tracts by Bangali Muslim settlers – not least of all because they profess Hinduism.

In the southern parts of the Hill Tracts, the names of rivers refer to an earlier people whose language no longer exists. Whether this people was absorbed into the **Chakma** remains just as open to question as Chakma history itself. On the one hand, with regard to many cultural characteristics, the Chakma are quite similar to their northern neighbours, the Tippera. On the other hand, many things indicate that they could have had something to do with the **Sak** (or Chak), a small group living in the far south of the Hill Tracts, whose closest linguistic relatives are found in central Burma. According to Burmese chronicles, the Sak once formed a large kingdom in that area. The Chakma themselves, however, do not speak Sak, but rather a dialect of Bangla – although, with respect to their physical features, they have little in common with the Bangalis. (They are rather clearly of "eastern" origin.) This Bangla dialect is to a large extent unintelligible for the Bangalis of the Chittagong Plains and was traditionally not written in Bangali characters but rather in a script peculiar to the Chakma alone (and unknown to the Sak). This script, at times wrongly classed with the Khmer script of Cambodia, is actually closely related to the Burmese script. In those cases where Hindus and Muslims use different words to denote the same things, the Chakma follow the Muslim usage; however, they, like the Tippera, make use of Hindu first names. Other Hindu influence may also be found, although the Chakma, like the Marma who came from Arakan, officially confess Buddhism.

According to their own traditions, in the 17th century several Chakma clans lived on the upper Matamuri, that is, in the southern Hill Tracts. During the two preceding centuries, however, Chak or Sak probably constituted the majority of the inhabitants of the southern plains. There they became involved at various times in the struggle for control over Chittagong – a struggle carried on between Arakanese and Muslim rulers. Chittagong was definitively incorporated into the Mogul empire in 1666. Yet until the beginning of the 18th century Chakma chiefs still sought to have their position confirmed by the Arakanese king; and only at that time did an ancestor of the present chiefly line, who was coming from exile in Arakan, move his residence as far north as Rongunia on the Kornofuli.

While the history of the Chakma still poses many unsolved questions, there can be little doubt about the origin of the **Marma.** Their name for themselves is already suggestive of their connection to Arakan and Burma: "Marma", like "Burma", can be traced back to the old designation "Mranma", which the Burmese used for themselves. In older sources the Marma are referred to as "Mogh" (also written "Magh"), a pejorative term still used today by Bangalis. In the most ancient sources on Chittagong and Arakan, however, the use appears to be mixed with a certain amount of fear and respect. "Mogh" is probably to be derived from the same Persian source as our word "magician" and, in our context, denoted the "infidels" of Buddhist creed. To state that all Marma came from Arakan, however, is not to say that all contemporary Marma are indeed descendants of the erstwhile rulers of the Chittagong Plains – even if the *raja* of the Marma in the southern Hill Tracts, the so-called "Bohmong" (field marshal), may claim to have descended from the brother-in-law of the former king of Arakan. (This brother-in-law, incidentally, was not a Marma, but a descendant of the Mon dynasty of Pegu in lower Burma; and he was appointed governor of Chittagong by the king.) On the contrary, most of the forefathers of the Marma moved out of Arakan as that kingdom declined and was in 1784 incorporated into the Burmese empire, that is, during a time when the East Indian Company had already taken over the administration of the Chittagong region (1760). Repeated, but futile, attempts to throw off the yoke of the new Burmese rule brought more and more refugees into the English-controlled area. For the year 1798 alone a report speaks of 10,000 refugees; and in addition to the Marma, the Chakma and other hill tribes are also mentioned. One Marma group migrated from a tributary of the Koladan in Arakan; after intermediate stops on the Matamuri and in the low hills of the plains north of Chittagong, it reached its present place of residence in the north-western part of the Hill Tracts around 1830. Marma, however, also lived on the southern Chittagong Plains: in Cox's Bazar and its backcountry (see Bernot 1967: Chapter 1, II)

There the Bangalis are the new-comers; and since the end of the colonial period, more and more Marma have relinquished their possessions to these new-comers and sought protection and security once again in Arakan and Burma. A witness to this exodus is the deterioration of the once magnificent Buddhist temples (made of teakwood), occurring simultaneously with the erection of new mosques. Not that Buddhists, as such, are persecuted. On the Chittagong Plains one finds a not insignificant minority of Buddhist Bangalis, who in the older sources are related at times to the Marma and are referred to as Marma-gri. These Bangalis refer to themselves, however, as **Borua;** and under this designation they made a name for themselves in Calcutta – as cooks. In other places, though, they tend to belong to the more educated and highly respected strata of the Chittagong Plains. A Borua even appears in the chronicle of the royal house of the Chakma, as one of the *rajas.*

B3 Colonial administration

The three major ethnic groups living in the valleys of the Chittagong Hill Tracts are 1) the Tippera in the north (without a *raja* of their own, as long as one does not want to relate them to the *raja* of the Indian state of Tripura), 2) the Chakma (with their own *raja*) in the central and north-eastern regions, and 3) the Marma with a *raja* both in the south and the north-west. The order in which they have just been mentioned corresponds approximately to the order of their movement into the Chittagong Hill Tracts. With respect to the history of the royal houses of the *rajas,* prior to 1750 it fades into darkness. In the case of the Chakma, it seems that the position of the *raja* was not consolidated before the Mogul rule; in the case of the Marma, as late as the English administration. Prior to that, each of the regional groups of both peoples had its own representatives.

For both Muslim and English administrations, the Hill Tracts produced an interesting raw material: cotton. The inhabitants of the Hill Tracts were, at the same time, dependent on the plains for iron and salt. By throttling the free flow of these goods from the plains, permitting such trade only against payment of a tribute of cotton (and even leasing a monopoly of trade to the highest bidder), the rulers of the plains contributed to the gradual subordination, or elimination, of the minor group leaders who had acted as middlemen. Those coming out on top were those chiefs who had built their places of residence at the main gates of entry to the Hill Tracts, that is, at those places where the Kornofuli and the Songu flow onto the plains. After the British had set up their own administration over the Hill Tracts in 1860, it was these chiefs – namely, the Marma Bohmong and the Chakma Chief – who were held ultimately responsible by the British for the collection of taxes from all inhabitants of the southern and central regions. Only a few years earlier, the northern Marma had established themselves in such a way as to select their own *raja;* and because the British did not want to concede too much to the reigning *rani* of the Chakma, who was less than subservient, they eventually turned over the northern region to this "Mong Raja" (from the Burmese word *mang,* meaning "ruler"), although the area was peopled mainly by Tippera.

The Bohmong (Marma-Raja) collecting the Government tribute on 12.12.1955
(to his right: the then still English Deputy Commissioner of the CHTs, to his left: the Commissioner of Chittagong; photo: H.-E. Kauffmann)

Thus was created the Mong Circle in the north (with the centre of administration in Ramgor and the rajbari in Manikchori); the Chakma Circle, in the central region (with administrative centre and rajbari in Rangamati, the principal city of the Chittagong Hill Tracts); and the Bohmong Circle, in the south (with administrative centre and rajbari in Banderban). In addition to the portion of tax and revenue to which these rajas were entitled – by that time these leaders were officially called "chiefs" – the jurisdiction over minor legal matters and the selection of "headmen" for the newly-created smallest administrative units (the "mouzas") were also entrusted to them by the colonial administration.

The fact that in the course of these duties the *rajas* feathered their own nest did not escape the attention of the administration; yet in spite of suggestions for fundamental modifications, the system survived the colonial period. The Marma Chiefs particularly were looked upon by their subjects (including the Marma themselves) more as exploiters than leaders. One of the Chakma groups, the Tongcengya, does not traditionally recognise the Chakma *raja* as its representative. Yet a Chakma Chief was even able to legitimate democratically his position as representative for the entire Hill Tracts. In 1971, when all of Bengal voted for the Awami Party of Mujib-ur-Rahman, thereby signifying the commencement of the struggle for independence, the voters of the Hill Tracts chose the Chakma Chief, Tridiv Roy, as their representative to the Pakistani parliament. With his election, the Chakma Chief accepted his exile into Pakistan, while those who had elected him were, against their wills, incorporated into the new state of Bangladesh.

The Chakma Raja (photo: Wolfgang Mey)

But we rushed ahead of events. Let us return again to the beginning of the colonial era. Surely the *rajas* owed their rise to power to their role as middlemen between the inhabitants of the Hill Tracts and the representatives of what for them was a foreign state in Chittagong. Yet they were anything but zealous civil servants. They handed over hesitantly and irregularly the tributes they were required to pay. East of their region lived hill tribes, generally referred to as "Kuki" in the old sources, who made a name for themselves by raids reaching even into the district under British administration. The *rajas* received money and weapons in order to defend themselves against such attacks, but occasionally they made a pact with the raiders. This gave the British an excuse to extend their own system of control over the mountains, which they did in 1860. The area was placed under the jurisdiction of a Superintendent (from 1900 onward Deputy Commissioner) and was henceforth known as the "District of the Hill Tracts of Chittagong". In 1880–83 one-fifth of the region was declared forest reserve – most probably at that time there was almost no one living in these lands bordering on Lushai and "Shendu" country. Around 1900 the remainder was divided up into more than three hundred administrative units, called "*mouza"*. These units were meant to correspond to the village administrative bodies of the plains; in the mountains, however, they incorporated rather arbitrarily into one unit a varying number of hamlets, or villages, irrespective of ethnic affiliation. (Since official usage tends to regard the mouza as the equivalent of a "village", the several small villages included within the borders of a mouza will be called hamlets or settlements in order to avoid confusion.) It was the task of the "headmen" who oversaw these mouza to collect taxes – part of which they were entitled to keep – and settle minor disputes. In the hamlets of the mouzas, which otherwise were independent, so-called *karbari* (managers) served without pay; these men were named to this post by the villagers themselves. The control was effective, and the inhabitants of the Hill Tracts proved to be peaceful people.

B4 Minor ethnic groups

The free hill tribes of the east, however, continued to cause trouble. These tribes included above all the **Lushai** and the so-called "Shendu" (Lakher or Maraa, Poi, and others). The Lushai were at the time expanding toward the north-west, superimposing themselves on the so-called "Old Kuki" group of tribes; the Shendu peoples were located south of the Lushai. The raids of these tribes were feared not only in the Chittagong region, but also in Arakan, Tripura, Kachar and Manipur. The weakness of the Lushai and other related groups was their practice of fighting each other in shifting alliances, as well as their willingness to include the British in this game. After fifty years of what were partially painful experiences in guerrilla warfare, the British eventually dispatched large units of troops from the west, north, and east and burned down every village whose chief they considered untrustworthy. Since the hill people could harvest only once a year, this policy helped to starve them out and thereby force the insubordinate Lushai to their knees. In 1898 the land of the Lushai was added to the province of Assam. Twenty years later the "Shendu" settling south of Lushai land – namely, the Lakher – also had to surrender; and for the sake of control, in 1924 their land was divided between the Chin Hills, the Arakan Hills and the Lushai Hills. This meant that at the 1947 partition of British India the Lakher, too, were apportioned to two countries: India and Burma.

After the "pacification" of the hill region by the British, the Lushai turned very quickly to Christianity and set out, with the Bible rather than armed violence, to bring their new culture to their neighbours. The fellowship of believers grew quickly and eventually had to face the consequences. The Lushai had once been nothing but a tribe; now it was made up of hundreds of thousands of Christians. Their old name seemed, therefore, inappropriate; so a new and more comprehensive designation was introduced. They became "Mizo" – in English something like "highlanders”. As Lushai, they had been forced to accept the sovereignty of the British; as Mizo Christians, they reasserted their equality with the British. When they did not also receive *their* independence at the time the British pulled out and British India was partitioned, but were rather subordinated to India, the Mizo eventually took up arms again. In doing this they followed the lead of their northern brothers, the Naga, who also resorted to arms in their struggle for independence. In their armed fight the Mizo were supported by the other partner of the partition, namely, Pakistan.

Those people of the Chittagong Hill Tracts standing culturally closest to the Lushai are the **Paangkhua** (in older sources written: Pankhua) and the **Bawm.** Linguistically, the Paangkhua are probably to be grouped with the so-called Old Kuki; ethnographically, however, very little is known about them. By absorbing parts of other tribes, the Bawm came into being during the 18th century under the leadership of a “Shendu” aristocracy. (“Banjogi”, the name by which they are called in the older literature, would mean “forest yogi” and is nothing but a Bangali corruption of “Bawm-Zo”, the name these people use for themselves.) The aristocratic dynasty which united the Bawm had already died out by the beginning of this century, and the social hierarchy weakened. This was due not least of all to the missionary work of the Lushai. If the border had not been drawn between India and East Pakistan (now Bangladesh), the Bawm would surely in the long run also have been absorbed into the Mizo. Meanwhile, however, between 1960 and 1970, that is, during the Mizo struggle for independence, this small ethnic group doubled its some 3,500 people as a result of newcomers fleeing the war-torn neighbouring country. The mission work which the Lushai introduced into the Hill Tracts during the 1920s is now being carried on independently by the Bawm and the church they founded, the Christian Church of the Chittagong Hill Tracts. Those most likely to be attracted to the work of the mission are the other small ethnic groups of the Hill Tracts: the Khumi, the Tongcengya, the Hill Tippera, the Khyang, and, finally, also the Mru.

These small ethnic groups are generally found in the transitional areas – that is, in small river valleys and on the lower crests of hills – between those unmistakable hill dwellers of eastern origin and those who are clearly valley dwellers. The hill dwellers include the Lushai, the Paangkhua and the Bawm. They are never attracted to the valleys and their villages are therefore nearly always found on hill tops and the spurs of hills. The Chakma, the Marma and the Tippera are, on the other hand, valley dwellers who will settle in higher regions only when pressed for lack of land. The largest of the smaller ethnic groups mentioned above is the **Mru,** called “Murong” or “Murung” by the Bangalis, and “Mro” by the Marma. “Mru”, as the people call themselves, denotes “men” in general; in order to set themselves apart from others, “Mru-ca” (“children of man”) may also be used. They inhabit a relatively closed area in the southern part of the Hill Tracts, an area to which, according to their own tradition, they immigrated from Arakan several hundred years ago. Perhaps more than half of the Mru still live there today. The Arakan chronicles mention them as early as the first millennium and speak even of a Mru ruler of Arakan.

The language of the Mru can be assigned neither to the group of Burmese languages nor to the Naga-Chin languages. (In addition to Lushai, the Naga-Chin languages include Bawm and Paangkhua, Khumi and

Khyang.) It rather claims an independent status within the framework of the Tibeto-Burman family. The Mru of the Chittagong region divide themselves into five linguistically and culturally distinctive groups: the Anok, the Chüngma, the Dömrong, the Dopreng, and the Rümma. Anok means "West", although today this is the most northerly group south of Banderban; the northern section of the Chüngma ("Mountain People") has joined the Anok, while its southern part in the upper Songu valley has adopted Khumi practices of house construction and festival celebration and calls itself "Longhu", the Khumi word for the Mru; the Dömrong, or "Lowlands" group, is at home north of the Matamuri; and, finally, the Dopreng and the Rümma ("Forest People") live in the far south and on into Arakan. The ethnic group which is culturally most closely related to the Mru is the **Khumi.** In the Hill Tracts this people is represented by a relatively weak contingent in the far south-east; in Arakan, however, the Khumi comprise several 10,000 people. We know little about them, however. A small group of Khami, a tribe closely related to the Khumi, must also have come out of Arakan; these people have been absorbed today, as Rengmittsa, into the Mru living on the upper Matamuri. Though they still know some of their old language, they know nothing of the Khami in Arakan.

As far as the small group of **Khyang** is concerned, they claim, according to one source, to have always lived in the Hill Tracts; yet their language relates them to the most southerly Chin group in Arakan. ("Chin" [in Bawm pronounced Ciin] is the anglicised form of the High Burmese pronunciation of "Khyang".) The Khyang live surrounded by Marma in the area between the Bohmong and Chakma Circles; and they have more or less taken over the culture of the Marma. It is also quite possible that the two groups came into the area together at the time the Burmese empire was expanding.

Brung (Hill Tippera) mother with small child (photo: H.-E. Kauffmann)

Finally, there are two groups of hill people who in the census enumerations are normally subsumed under groups whose language they share.

On the one hand, there are those "Hill Tippera", who live spatially separated from the Twipra, widely scattered over the southern part of the Hill Tracts. As they came from the north, they possibly settled down in that area wherever they could still find unclaimed land. They are called by the Bangali "Uchoi", but call themselves **Brung,** and are closest related to those "Hill Tippera" in the Northern hills, known as **Riang.** I was also given the form "Brong", a little bit nearer to the "standard" form, that is the general Twipra term "Borok" which in its turn gave rise to (the Assamese form) "Bodo", a term used by linguists for the even larger subgroup of Tibeto-Burman languages in Assam. "Brung" was transformed in Marma to "Mrung", and as such these people made their first appearance in the census reports. In consequence they sometimes were confused with the Mru, who are called "Murung" by the Bangalis. The valley "Tippera", as members of the Hindu caste system, do not recognise the Brung and Riang as their equals, who nevertheless claim to believe in Hinduism as well.

On the other hand, we have the **Tongcengya** (also written "Tanchangya", etc.), who are often called "Doi(n)gnak" in the older sources. ("Tong-cen" is a Marma word for "mountain clans"; "Doing-na(k)" is the Marma pronunciation of the Mru designation for the Tongcengya, namely, "Dengnak"). Like their neighbours, the Chakma, the Tongcengya speak a local dialect of Bangla; but there are differences in culture, for instance in their house form and style of dress. For a long time the Tongcengya attached much importance to their *not* being subsumed under the Chakma. They are found both in the Kornofuli area, on the hills between Chakma and Marma, and, along with the Marma, in the southernmost part of the Hill Tracts on the border of Arakan. (An old source tells of Dengnak who came out of the mountainous area of Arakan, but who then returned.) It is improbable that these two groups were originally Tippera or Chakma who over time "degenerated" into hill people. It is more likely that they have largely assimilated themselves to the more civilised valley dwellers, but actually represent the "aboriginal people" of the Hill Tracts – with the Brung/Riang being formerly (that is, before the Marma entered the hills) located in the northern, and the Tongcengya/Dengnak in the southern, part of the Tracts.

This distribution corresponds to the spheres of influence of the Tippera and Chakma (or Sak?) prior to the massive influx of Marma, that is, roughly in the seventeenth century. Because by that time, however, Mru may well have already moved into the southern Chittagong hills, while Dengnak may have migrated into Arakan, such early relationships between the two ethnic groups would help to explain the existence of an isolated Mru village which, though located in the middle of the Dengnak in the Chakma Circle, still preserved an exceedingly large number of old traditions. It is not at all impossible, however, that Mru were also absorbed into the Dengnak, though today the two cultures have little in common.

B5 Population figures and development

After this short, historical overview of the ethnic groups of the Chittagong Hill Tracts, let us enumerate them again. There are 1) the

Chakma and 2) the Tongcengya/Dengnak, both of whom speak a dialect of Bangla; 3) the Tippera and 4) the Brung/Riang, both speaking a Bodo language; 5) the Marma, who speak a Burmese dialect; 6) the Bawm, 7) the Paangkhua, and 8) the Lushai, whose languages are classed as Central Chin; 9) the Khumi and 10) the Khyang, whose languages are classed with the Southern Chin; 11) the Mru and, finally, 12) the Sak, with comparatively isolated languages, but languages which, nevertheless, like Burmese, Chin, and Bodo, are to be placed in the large Tibeto-Burman language family. To these twelve indigenous ethnic groups have been added in more recent times a few Santal, some Assamese, and, above all, Bangalis. (The Santal are representatives of a people whose home is in West Bengal and whose language belongs neither to the Indo-European, nor the Tibetan, language family.) Most of the Bangalis are Muslims, but some are Hindus or Buddhists, that is, Borua. All of these peoples are gathered in an area, including the forest reserves, of only 13,000 km^2, which is approximately the size of Connecticut.

If there is anywhere on earth where one can find within an area of a few square miles several different ethnic groups exhibiting distinctly different cultures, then it is in certain regions of the southern Chittagong Hill Tracts. Here, within one and the same mouza, one may find four groups speaking completely different languages, building different types of houses, wearing different clothing, and following different customs and different religions (Buddhism, Hinduism, Christianity and Animism). In order to be able to communicate with each other in spite of these differences, one must master sufficiently at least one additional language. This is particularly true for members of the smaller ethnic groups. Some Bangla, in the form of a reduced Chittagong dialect, is understood by most. In the Bohmong Circle, however, Marma is often preferred. English, the language of their erstwhile colonial masters, is common only to those who have a higher education, that is, above all the better-off social strata of the Chakma and Marma. English, however, is also mastered by a growing number of representatives of the Christianised ethnic groups, particularly the Bawm.

The diversity which is evident in a mere listing of ethnic groups and languages is, however, deceptive if it does not take the size of the groups into account. In order to compare group sizes, we would have to have accurate figures at our disposal. The official census, however, which was begun by the British in 1871 and which has been repeated every ten years since that time, has more and more neglected ethnic identification the closer one comes to the present. (In the first census reports the small figure of Sak most probably was enumerated as Marma.) While the census of 1931 contained interesting material, those of 1941 and 1951, under the circumstances of world war and newly gained independence, were not particularly informative; in 1961 notes on ethnic affiliation seem still to have been taken but are not to be found in the published report. The census of 1971 fell at the time of the Bangladesh war for independence. The official figures published in 1981 may not be fully reliable; but, in any case, they come much closer to my own extrapolations than privately published, imaginary figures.

During my first period of field research in the Hill Tracts (1955–1957), the leader of the expedition, Dr. H.-E. Kauffmann, secured the field-tax lists from the office of the Bohmong in Banderban. These lists recorded, by name and village, the household heads charged with the tax. From the names one can with some certainty determine the ethnic affiliation. Summing up the number of ethnically-identified households and multiplying the sum by the average number of persons per household, one gets the approximate strength of the ethnic groups (although in each case a few households which were not obliged to pay the tax will be missing). I myself was able to view the lists for the central and northern Circles. Difficulties, however, arose in relation to name identification, since both the Chakma and Tongcengya, as well as the Tippera, have Hindu names. The ethnic affiliations were eventually established by means of additional information and were finally checked against the census data, inasmuch as the religious affiliation – though not the ethnic affiliation – was included; and most of those persons classified as members of Hindu castes are probably Tippera. The figures given here for 1981 are rough estimates, which need not be accurate.

Ethnic group	1901	1956	1981
Chakma	44,500	140,000	230,000
Tongcengya		15,000	20,000
Marma	35,000	80,000	120,000
Sak		2,000	1,500
Khyang	500	1,000	1,500
Tippera	23,000	30,000	40,000
Brung/Riang		7,000	10,000
Mru	10,500	17,000	20,000
Khumi	1,500	2,500	1,000
Bawm	1,500	3,500	8,000
Paangkhua	200	1,500	2,000
Lushai	?	500	1,000
Tribal total	118,000	300,000	455,000
Bangali	7,000	30,000	290,000

Taking all uncertainties regarding the estimates into account, the table still shows that the Chakma, who once made up one-third of the indigenous population, now account for at least one-half of it. The expansion is due primarily to the constantly growing number of children over the past decades. This growth, however, is no sign of wealth, but rather of increasing poverty. The comparatively small growth rate of the Tippera, which contrasts the high one of the Chakma, is to be explained by the fact that the Tippera, as Hindu, have increasingly been driven from their homeland. The Marma and the Tippera seem to be the smaller ethnic groups when compared to the Chakma, but only because their areas of settlement were cut up by the political boundaries established at the partition of British India. If the Bangalis today want to force these ethnic groups back into their areas of origin, then it should be pointed out that when the Chittagong Hill Tracts were first consolidated as a political territory by the British, the one and only people who were positively not yet settled in the Tracts were the Bangalis themselves. The present government would like to transform the

Hill Tracts into a home for Bangalis, in spite of the unpleasant fact that the one and only ethnic group already speaking a Bangla dialect is the selfsame group offering the most obstinate resistance to the government's policy.

The British pursued completely different goals. When they began from 1789 onward to require the representatives of the Hill Tracts to pay their tribute in money rather than in cotton, as had formerly been the case, they could not have known what the consequences would be. Surely the money economy would in time, and without such measures, have displaced the barter economy. The forced use of money, however, resulted immediately in an increasing indebtedness to Bangali moneylenders, since money was obtainable only through trade with the plains dwellers. The goods imported from the plains increased in number and kind, and the volume of produce obtained by the hill dwellers from their traditional agriculture was substantial enough to enable them to buy luxury goods. But the hill dwellers did not understand much about market trading. As far as the traditional agriculturists are concerned, this is still true today, even though, in the meantime, successful traders and entrepreneurs also appeared among the Chakma and Marma. Lending businesses were not part of the traditional culture. "Debts" were in the form of long-term mutual obligations; only the aristocrats among the valley dwellers and eastern ethnic groups grew rich by turning debtors into slaves. Price fluctuations and usury, which enabled clever moneylenders, for example, during the 1950s to squeeze annually up to 500 % of a loan out of their debtors, was – and still is today – something far beyond that with which a simple farmer can come to grips. Since such lending and borrowing relationships are understandably loaded with tension, the common man generally tries to avoid giving a loan to a person of his own ethnic group. This means, however, that those suddenly finding themselves in need almost inevitably fall into the hands of foreign moneylenders. Against the exploitative manipulations of such persons, the debtors are again unable to defend themselves, since all legal means are in the hands of those who also control the market, namely, the Bangalis.

As early as 1870 Lewin described the numerous tricks and types of chicanery used by Bangali moneylenders who attempted in court to extract all they could from some inexperienced hillman – a hillman who had become deeply indebted due to a poor harvest or a wedding. In such cases, it was not rare for a debtor to suddenly find himself in a situation of lifelong dependency. Lewin, therefore, fixed the annual interest rate at 12 %; and whenever a court had validated a claim, the outstanding debts had to be collected immediately or they were written off. The efforts of the British, however, appear in the long run to have changed little. When in the 1950s, the courts were for a time supervised by the Pakistani military government, a certain Bangali moneylender was, against all expectations, condemned to pay back the debtor. The moneylender, however, had already wrangled an exorbitant payment out of the Marma involved. Over a period of three years, the moneylender had collected ten times the amount of rice originally lent, and as settlement for the remaining debt was now demanding more than the man's entire harvest.

When a few Bangali policemen were finally disciplined for blackmail and rape, the inhabitants of the Hill Tracts recognised the Pakistani government as their new friend; and it is no wonder that Bangali endeavours to attain independence elicited little positive response from the hill people. The simple hill dwellers would have much preferred to have again been placed under the protection of the British, even if the British had demanded multiple taxes. Such statements, voiced by the people themselves, throw some light on the close connection between economy and politics.

When the British administration recognised that it could not solve the region's problems through the courts alone, it drew up special regulations for the Hill Tracts. The initial Rules of 1892 were followed by the Chittagong Hill Tracts Regulations of 1900; in 1921 the Hill Tracts were declared a "backward tract", and in 1935, by receiving the status of a "totally excluded area", they were withdrawn from the central and provincial legislature. Besides laying down special functions for the Chiefs and headmen, these regulations, by means of special rules regarding land acquisition and residence rights, provided for an effective check on a further influx of Bangalis. The markets which had in the meantime grown up were, likewise, placed under special supervision, in order to prevent the intrusion of "dishonest" traders and merchants. During the Pakistani era, however, in spite of the resistance of the Chiefs, the special status was partially revoked and immigration became possible again. The Bangali government completely eliminated the discrimination against its own people, abrogated in 1981 all special privileges of the Chiefs, and in 1989 finally tried to abolish the Regulation of 1900 altogether.

Prior to 1900, though, the British administration had already taken measures which inadvertently furthered the influx of Bangali settlers. The traditional method of tilling a field, slash-and-burn cultivation (which will be discussed in Chapter 3), had seemed to the British a harmful practice – all the more so since the easy mobility it afforded the farmers was thought to be politically risky. In 1870 slash-and-burn cultivation was therefore forbidden on the plains; however, sedentary plowing, which in the meantime had become the ideal and generally-practised method on the plains, could not be introduced in the mountains. Only in the wide valleys of the central and northern Hill Tracts were there a few places where plowing appeared possible. The first "development-aid" money was used to help purchase the tools and animals needed for plowing, and what the *rajas* already practised on their estates on the plains now gained entry into the valleys: plowing done by Bangali share-croppers. Even though the hill farmers at first seemed rather reluctant to set their own hand to the plow, after 1900, a growing number of Chakma, Tippera, and Marma turned to plow cultivation; and this changeover indicates that land was already becoming scarce. The traditional slash-and-burn method of cultivation brings higher yields as long as ample fallow periods are possible. Fields which are plowed but not regularly irrigated yield less; yet because they may be cultivated every year, plowed fields allow for a substantially higher density of population.

During the first half of the 20th century, land in the valley regions was brought under the plow. The largest part of this land was located in the Chakma Circle (the rapid growth of the Chakma has already been mentioned); and precisely this land in the valleys of the Kornofuli and its tributaries had to fall victim to a new project, namely, an artificial lake. This "Kaptai Hydro-electric Project" was to produce the badly needed energy for the electrification of Chittagong. Of the some 85,000 people whose land disappeared in 1960 under the new man-made lake – among them approximately 70,000 Chakma – 46,000 lived primarily from plow cultivation.

To replace the lost 585 km^2, the government allowed 40 km^2 of forest-reserve land to be taken over by farmers and turned into plowed fields, and relocated Bangalis received preference. Following the resettlement of the Bangalis, the remaining “available” land was mostly land already under slash-and-burn cultivation. The shortage of suitable land resulted therefore in massive unrest and overexploitation of existing resources. In search of new land, some Chakma migrated all the way to north Assam and the foothills of the Himalayas. The Pakistani military government, however, under whose protection this whole process had taken place, knew very well how to blame all mismanagement on subordinate Bangali officers; the blame placed on these officers only contributed to the traditional Chakma mistrust of Bangalis, a mistrust which was subsequently transformed into bitter hostility.

Living in an area already strongly influenced by the economic system of the plains, the concerned Chakma and Marma were surely not people with whom other hill people would have felt any particular solidarity. Similarly, the valley dwellers, with their *raja,* had always considered the less civilised hill people to be quite different from themselves and not in the least tied to them by a common origin. The fact that they now had to contend with these people for land only added to their feeling of separateness and a lack of common interest; thus the building of the Kaptai dam had no further political repercussions. It is true that in the European embassies in Dhaka the rumour went around that Christian Chakma were being persecuted; yet if D. E. Sopher, an American geographer interested in the transition from slash-and-burn to plow agriculture, had not almost accidentally been eyewitness to the 1960 flooding, we would not know even today what really happened there. The government prevented further research by declaring Sopher a *persona non grata.* A few years later, in 1964, the entire Hill Tracts were closed to foreigners; and up until 1980 they have not been reopened. What has happened in the area since that time was not supposed to be known to the outside world. The information which in spite of these measures has managed to make its way out of the area will be briefly discussed in the following.

B6 The development during the last decades of the 20th century

Not only is the fertility of the land in the Chittagong Hills decreasing – land upon which too many people’s existence was and still is dependent – but the hill people’s right to the land itself has been disputed.

As a result of the building of the Kaptai dam in the Chakma region, the hill people had already been forced to squeeze more closely together; yet because there were still fifteen times more people per square mile in the plains than in the Hill Tracts the government concluded – without taking into consideration the ecological differences of the two areas – that there should still be ample opportunity for landless Bangalis to settle in the hill region. As early as 1960, therefore, the government began to settle additional people in the outer edges of the Tracts. Still, the immigration policy became legalised only in 1964, when the restrictions imposed by the British on non-indigenous peoples were lifted, while at the same time the Hill Tracts were declared off-limits for foreigners. The fact that the population of the Hill Tracts has doubled since that time might appear to support the government’s view of ample land reserves. Yet part of the immigrant Bangalis live in the main towns and market areas, being themselves not active agriculturists, but rather additional consumers of local resources. The indigenous population, then, has had to tighten its belt even tighter in order to feed themselves and their "uninvited guests". Wherever Bangali farmers settled, indigenous people had to be expelled. The flat valleys of the central and northern Hill Tracts offer the best example of this process of settlement: Hindu Tippera and Buddhist Chakma by the thousands left their homeland and moved into India; Marma moved to Arakan and lower Burma.

The first large stream of refugees left the northern Hill Tracts in 1971. Along with these people there also fled Hindu Bangalis from the plains area bordering the Tracts. During the Bangladesh struggle for independence, these people had to fear not only the Pakistani army, but also – and above all – their Muslim neighbours. When Pakistani rule in Bangladesh was brought to an end by the intervention of Indian troops, the Bangalis who had fled during the war were invited to return home to what was then to become a secular state; and they were promised the return of their properties. The tribal inhabitants of the Hill Tracts, on the other hand, whose chosen representative, the Chakma Chief, had taken sides with Pakistan, were suspected of collaboration. They were powerless against land-grabbing Muslim Bangalis who, during the turmoil of war, had settled in their villages. The search for hidden collaborators offered paramilitary troops a good excuse to maraud in the hills. A delegation of the hill people, which wanted to negotiate with the new government the re-establishment of the Hill Tracts' protected status (abrogated in 1964) in order to stop further immigration, were simply told that ethnic identities must disappear. Shortly thereafter, the Bangali army, supported by the air force, invaded the Hill Tracts for the first time. The army wanted to crush the enemies of the new regime; however, by massacring villagers and ransacking and burning down villages, it succeeded in doing just the opposite. The villagers became convinced that they could expect no help from the new government, and that they would have to take care of themselves. In order to defend the interests of the indigenous population, an armed resistance movement, the Shanti Bahini ("fighters for peace") was formed, primarily among the Chakma.

The general decay of public order in Bangladesh led to a take-over of power by a new military government – a government which reverted to a decidedly pro-Muslim and anti-Indian stance. Although the government's policy vis-à-vis the wishes of the political representatives of the Hill Tracts remained in fact the same, the resistance movement in the Hill Tracts profited from the anti-Indian stance of the new military government: it began to receive support from India. Yet the Shanti Bahini did not present a united front; they rather consisted of various groups having different ethnic ties and varying political alignments. Some were prepared to make concessions; others were determined to resist, making no concessions whatsoever. Occasionally, therefore, they even fought among themselves. In addition to these Shanti Bahini, there were Bangali-Muslim groups, also resisting the military government, which had retreated into the Hill Tracts. So even in the Bangali Chittagong area one could denounce unpopular neighbours as "fighters for peace" and count on their leaving the remand prison only as cripples. Finally, the Hill Tracts also became a refuge for members of the Mizo liberation front, a movement which fought against the

Indian central government in Assam. As time went on, the Indian government therefore changed its attitude to the Hill Tracts and, as a result, tightened its borders with the region. Officially, the Indians wanted to block the way of guerrilla troops into their country; in fact, however, they primarily blocked the influx of refugees.

Yet time and again whole tribal villages tried to flee; for wherever the guerrillas struck, the Bangali army devastated everything in the surrounding villages. They plundered the houses, raped the women, and tortured, mutilated, and killed all the men they could get their hands on. In 1981 alone, 10,000 members of the hill tribes are believed to have been killed. The purpose of this reign of terror was not only to frighten the inhabitants out of lending any support to the guerrillas, but also to threaten, in a most basic way, their very existence in Bangladesh. As far as the new Bangali settlers were concerned, they had to fear that the Shanti Bahini would attack their villages and attempt to drive them back to their original homes. As a response to this threat, the government adopted a measure which had proven successful in many a counter-insurgency program: the erection of so-called "strategic villages". Inhabitants of these villages, which were under military control and which were denounced by the guerrillas as concentration camps, could leave the village to do prescribed jobs only after a thorough search and only without rations. Anyone walking about freely, therefore, had to reckon with the fact that all provisions for the road, as well as all paddy, salt, medicine, and money, would be confiscated. In spite of this, the guerrillas knew how to survive. A large proportion of the population may well have been in agreement with the guerrillas' goals, namely, the reestablishment of a society where the indigenous inhabitants may live in freedom and security, the establishment of an indigenous administration, and the winning of autonomy for the Hill Tracts. The situation of the Shanti Bahini, however, might have appeared hopeless; for over and against their some 5,000 men the Bangladesh military government could and did deploy what experience in other countries had shown to be effective in crushing a guerrilla movement, namely, an army and police force of at least ten times the number of the guerrillas.

In the long run, however, this strategy has cost the state a disproportionately large amount of money; and were it not for international development aid, the government would perhaps have been forced to consent to a cheaper solution, that is, to a peaceful arrangement. Development aid money has been used to finance the constantly growing army and to train and equip it for anti-guerrilla warfare; it has also been used to construct strategic all-weather roads through the impassable Hill Tracts, making them (and their forest reserves) more easily accessible. For the donor countries it amounted to sheer hypocrisy to deny support to anything obviously serving the purposes of genocide, while allocating at the same time funds for such useful purposes as the development of communication facilities which not only benefited military operations, but also facilitated the harassment, displacement, expropriation and expulsion of the indigenous population. When internationally pressured to comment on the reported violations of human rights in the Chittagong Hill Tracts, the representatives of the government vacillated between a flat denial of the mere existence of an indigenous tribal population, and a formal pledge to protect the lives and interests of these tribal people. The only credible pledge, however, would have been the re-opening of the Hill Tracts to international humanitarian organisations, to social scientists from other countries, or journalists independent of any government. For more than twenty years, however, only economic and technical development experts under official government contract were permitted to visit the Hill Tracts, for example, representatives of Shell Oil Company in order to search for oil.

As early as 1964 the government (at that time still Pakistan) requested that an international team of experts study how one could make best use of the land. These experts rightly discovered that the traditional system of slash-and-burn cultivation was perfectly adapted to local conditions; they also found, however, that with this type of cultivation the population would soon no longer be able to maintain itself. The experts therefore made the following proposal: the traditional swidden cultivation should be done away with altogether; the land should subsequently be used primarily for plantations and forest production; and the inhabitants should work in forestry or "other development industries". In one of the first areas to be tested, the people were forbidden to cultivate their traditional fields. Under the instructions of forestry officials, they were instead forced to plant quick-growing timber, bananas, pineapple, cashew nuts, guava, papaya, and citrus fruits. According to official calculations, after a few years the people should have received a good income from these products; but following the first harvest there were in fact no markets for the products. Having grown no paddy and receiving no money to buy it, the people began to starve; so whoever could manage left the controlled area and went in search of a new area in which to make his living. In subsequent years, however, due to the massive immigration of Bangalis, the displaced indigenous inhabitants had practically no other choice than to allow themselves to be "rehabilitated". "Rehabilitation" meant placing together in one village up to sixty families of various ethnic backgrounds (in such a way, cultural peculiarities can most easily be eliminated); each family was given a maximum of five acres of land and placed under obligation to use the new methods of cultivation. With the help of billions of dollars worth of credit from the Asian Development Bank, the new order should be realised over the whole of the Hill Tracts within the next few years. One could foresee the results: those who had nothing left to eat would pawn their land to moneylenders and would eventually become landless day labourers.

The wide valleys in the north and the area around the artificial lake were more agreeable to the Bangali settlers than the steep mountains of the south; however, since the Shanti Bahini were active primarily in the north and since the government was interested in Banga-lising all of the Hill Tracts, the southern valleys also became a settlement area. The Mru could not evade the developments accompanying this settlement: not only were two strategic roads constructed through their region, but substantial army units were stationed in the valleys of the southern rivers, so that life in the surrounding villages was no longer safe. More importantly, government policy resulted in the forced eviction of large groups of Mru. In January of 1983, for example, the army forced the Anok Mru in three mouza bordering the Chittagong Plains to leave their villages without any compensation whatsoever.

One might suppose that under these pressures also some Mru joined the resistance forces, the "fighters for peace". But they were peaceful people since long, and neither the Christian teaching nor the Kramma movement preached violence. They remained long-suffering and also accepted Shanti Bahini camps trying to hide in their area for some time. However, one day these Chakma fighters exasperated even the Mru. The Chakma had brought a girl of the neighbouring Mru village into their camp (in the Matamuri region). This they still tolerated, but when she died there, they took out their old guns which were outlawed since long (that's why I did not mention them before – though they still knew to manufacture them in the 1950s), surrounded the camp and overwhelmed the Chakma soldiers in such a way that they preferred to withdraw. To be sure, this event could not remain unnoticed by the government army which henceforth thought that "the Mru" were their allies in their fight against the Shanti Bahini, subsequently legalised the old and promised the supply of new guns. They even tried to set up and train a "Mru Bahini", paid by the government. But these new forces never saw a payment (nor a uniform except a white band around the upper arm) and never became active. Nevertheless the antagonism of "the Chakma" and "the Mru" henceforth remained an established fact, while the government pursued its nocuous policies.

"We want the land and not the people", a high-ranking military official reportedly said. However, the main concerns of the government were neither the natural resources (such as natural gas and oil) nor the forestry products, since these things could be exploited even *with* "the people". It had to do with land for more and more Bangalis. By means of terrorism, resistance to Bangali expansion could be repressed; but it could not be completely eliminated until the entire indigenous population had been either exterminated or driven off. Surely no one could believe that this "Final Solution" for the non-Muslim inhabitants of the Hill Tracts would in fact at the same time solve the problems of Bangladesh. One thing is certain, however: without foreign aid the government of Bangladesh could not have flourished. Would it have been asking too much to require the Bangladesh government – as a small remuneration for the aid – to allow the inhabitants of the Hill Tracts to survive? Put more blatantly: would it have been asking too much to require the government to discontinue its military and police terrorism and, in place of the terrorism, to grant the ethnic groups their own administration and allow them to use their own land and their own reserves? Instead of carrying out "development projects" to destroy traditional cultures and ethnic groups, should the government not have refrained from hindering – if not supported – those projects which were in the interests of the indigenous population and open up for them an acceptable path to the future?

In 1987, despite several previous efforts at repatriation, more than 10 % of the population of the Chittagong Hill Tracts were living in Indian refugee camps under miserable conditions. They had been beaten, robbed, tortured, raped, mutilated, or nearly hacked to death by Bangladeshi military forces or armed settlers and now insisted that it would be more humane just to shoot them dead on the spot, rather than send them back to what had once been their homeland. The tribal resistance movement demanded the withdrawal of the settlers and the repatriation of the refugees under the supervision of the UN. The government of Bangladesh then finally ordered the armed forces to stop persecution of the hill people, even if provoked by the Shanti Bahini; it also appointed a committee to negotiate with representatives of the hill people, in order to come up with a political solution.

In the spring of 1989, just in time for the UN conference on human rights, the parliament of Bangladesh adopted three new bills providing for the constitution of local government councils in the districts of Rangamati, Khagrachori and Banderban. A majority of seats on the councils, as well as the chairmanship of the same, was reserved for the local tribesmen. Each council could draw up its own regulations for running the affairs of the district; each would be responsible for the maintenance of law and order and should, for this purpose, recruit a district police force; each was meant to assess the social and economic needs of the people, levy taxes, and initiate development projects financially supported by the central government. In all of these activities, however, the councils would have to come in for the approval, or to secure the consent, of the government, which reserved for itself the right to dissolve the councils.

The councils would also have to settle land questions. However, they had no say over the forest reserves nor the area of the Kaptai lake. Nor would they be able to give the land back to the tribesmen who once cultivated it, since under state law swidden cultivators were never recognised as owners of their land. Their land was "unclassified state forest", they themselves were nothing but squatters. Bangali businessmen and local officials, on the other hand, managed to buy thousands of acres of swidden land for conversion into rubber plantations; their newly acquired rights to tribal lands were no less officially registered than those of the poor Bangali settlers who, upon government initiative, were provided with five acres.

The district councils also were entitled to appoint minor officials, although these remained under the command of the higher officials appointed by the government. And, last but not least, the local councils had no authority to order the military forces out of their districts – they might well prove to not even be interested in doing this, since the new "political solution" fell far short of the demands of the "fighters for peace". As a result, the war continued, with some of the tribal people having been bought over by the government. Hill people killed hill people without realising that they needed each other. There would be no local councils, and the councils would have no claims to press for, without the resistance movement; likewise, the resistance movement would achieve nothing but bloodshed were it not for the legalised representatives of the hill people, who, as intermediaries, were willing to collaborate with the government.

In the meantime, however, a peace treaty between the Shanti Bahini and the government finally has been signed. I will not go into the details. Much has been promised; little has been realised. Real democracy could not be achieved, nor did the continued presence of all Bangali "security forces" really improve the security situation. In the renewed local government bodies, due to Chakma influence, the Mru are under-represented. Their traditional culture as described later on cannot survive, but deserves to be remembered.

An increasing part of the impoverished Mru hoped – in vain – for better chances of survival by adopting new religions which have at least one feature in common: abolishment of the feasts (to be described in detail below) as well as the many other so-called "sacrifices", which in reality were nothing but opportunities to eat some meat in common. Christian teaching had not been very successful until the 1960s, but got new impetus under the influence of the Bawm, who combined practical advice in horticulture with their teachings. However, they still made but slow progress. Moreover, in the 1980s these "old Christians" had to face a competing tendency due to a renewed interest of the US Baptist mission. More successful was the Kramma religion invented by the end of the 1980s by a Mru prophet together with a new script. Much more could be said about him. But I am at present not going to write a new chapter. Suffice it to say that today religiously there are three factions among the Mru: the Kramma adherents, the Christians, and the "boudho" (Buddhists), who tend to become a minority, since they still follow the old rules outlined in this book. Still, all these new creeds, despite all their promises, changed nothing regarding the general circumstances.

C Hamlets and houses

C1 Hamlets

C1a) General remarks

The most important consideration when founding a village has to do with the available water supply: there must be a water source near the hamlet which does not run dry even in the dry season. This may be a small brook or stream. That a settlement is located above a stream is unavoidable; that it, however, is also always found above a spring – so that water must be carried *up* to the hamlet – surely indicates that higher residential locations are preferred whenever available. If the source runs dry too often, one has a reason to move; illnesses which occur too frequently may also prompt the relocation of a settlement.

In general, however, hamlets remain in the same place for decades, although the inhabitants may well change. Whenever official statements all too gladly characterise the hill dwellers of the Hill Tracts as semi-nomads, they paint a false picture, but one which is supported by the official government practice of recognising a hamlet under the name of its karbari or of the headman. This practice results in the constant "disappearance" of old settlements and the "appearance" of new ones, although it is the settlement *names* only which actually disappear. Some hamlets may not have their own proper names; in such a case the settlement is known by the name of its founding sib or the name of the stream on which it is located. "My" hamlet, which was situated on the crest of a hill, was bisected by the boundary of a mouza; and each of its two parts was registered in the tax lists under the name of its respective karbari. The people living in the other mouza, the majority of whom were members of the Atwang sib, called the entire hamlet "Atwangwüa-Kua"; in my mouza, however, the hamlet was known as "Tapwüa-Kua" ("Fortress-people-hamlet"). This latter name recalls a previous time, more than a hundred years ago, when the settlement was still fortified. The last traces of even earlier inhabitants were found in the form of an old tombstone located just outside the hamlet. (Incised on this stone was a head with feather decorations.) These inhabitants were no doubt the Bawm, who for a short time pushed into the mountain chain west of the Songu valley. According to their own traditions, however, the Bawm retreated again to the eastern chain of mountains because of too many diseases (see Spielmann 1968:54).

It is generally difficult to identify a Dengnak hamlet, since their houses stand isolated in the middle of the fields. On the other hand, one may find hundreds of Lushai houses grouped into one village; and the paths to their fields are correspondingly long. The Mru hit a happy medium between the two extremes: five to twenty houses generally make up a hamlet. Among the Marma, who have larger settlements and therefore longer paths to the fields, entire families often move into the field houses during the time of cultivation. The Mru walk each day to and from their fields; and often two or three families will join together, working the field of one family one day and the field of another family the next. Among the Marma such forms of co-operation are rare. The correlation between co-operation and settlement size is, however, not so simple. The Lushai and Bawm also prefer to co-operate in the fields rather than live there as single families, even though they must walk longer distances. In the past when warring was common, large villages could better defend themselves against attack; isolated families in the fields had to fear for their own safety. Today, on the other hand, people would like to have churches and schools in the villages; and such can be afforded only when settlements are large.

C1b) Closing the hamlet

Today normally anyone may enter a hamlet, whether Mru or not, whether especially welcome or not. This, however, was not and is not always the case. In earlier times Mru settlements were fortified and normally had but one gate (*plon*) reinforced with bamboo spikes. Even the paths leading to a hamlet could be "mined" with caltrops in case of danger (see C2). Similar fortifications have been used even in modern times, for instance by Bengali soldiers in a camp, but also by hamlets near the border where the inhabitants had to fear the invasion of robbers. No full fortifications, but spiked gates were also constructed in the 1950s, in order to signify the symbolic closing (*khang*) of a hamlet, for instance in the case of a "bad death" (*char*, see Q2a). These gates might be circumvented easily, but as a rule would be respected, since the intrudor would be fined by having to pay all the costs incurred by the inhabitants for repeating the *khang* ceremonies.

Guardian monkey (photo: H.-E. Kauffmann)

But hamlets might also be closed to foreigners when no bad death has occurred, especially in case of a threatening epidemic of cholera which in the 1950s was still quite common, spreading from the plains into the hills. Against cholera no medecines were available, and besides humans also spirits were believed to be responsible for the spreading. In order to ward them off it might be doubted that a mere gate would do, and for supernatural help a monkey was shot and placed, armed with bow and arrow, on a bamboo pole or in a tree near the entrance of the hamlet.

The closing of a hamlet will need the consent of all inhabitants, who make a meeting to decide it. Especially old men may make the proposal, but all heads of the households will have to consent, old men, married men, and in case there is no adult male in a house, also unmarried men, or even women (MK, 26.06.1956). One reason to close the hamlet might be simply the appearance of unwanted spirits. In order to drive them away, a *chüng-nam-khang* (see P3a) would be held.

C1c) Hamlets and *mouzas*

Hamlets are living settlements, mouzas are nothing but administrative units. Small hamlets do not mean isolated hamlets. Unmarried Mru rarely find proper marriage partners in their home hamlet; during the dry season, therefore, young men may occasionally leave their hamlet in order to pay an evening visit to some neighbouring hamlet. If a celebration (for instance a cattle feast) should be taking place in one of the hamlets, young girls also go along. Married persons may also be especially invited for one reason or another. It can happen, however, that two neighbouring hamlets are not on particularly good terms.

Neither hamlet nor mouza boundaries are officially marked, local people are just expected to know them. The mouza boundaries were defined by the English administration at the end of the 19th century, and people have to know them, since it is the duty of the headmen of the mouzas to collect the field taxes of its inhabitants, including those who just come to their mouza in order to work on their swidden which they made there. People who have their swiddens in two mouzas, have to pay the tax twice. Though this was primarily meant as a means to settle the "semi-nomads", it probably never really served this purpose, but just imposed the plight to pay the field tax twice, in case someone made his field in the neighbouring mouza. Since the field tax has not been raised substantially while the currency constantly lost in value, the burden of being taxed twice, which may have been felt heavily hundred years ago, however, is not really felt anymore.

But as land becomes more and more scarce, people more frequently tend to transgress the boundary claimed by the hamlet and plant their crops on a plot which is also claimed by a neighbouring hamlet. In such a case it is the duty of the headman (or in case the two villages belong to different mouzas, the headmen) to settle the problem, call at least the two karbari and the responsible village members together and try to mediate. Should his (or their) decision not be accepted by the two parties, should, for instance, someone from the neighbouring hamlet answer the "encroachment" by a similar invasion, the relationship between the two hamlets can be permanently damaged. In such a case, the footpath connecting the two settlements may even be allowed to become grown over. In order to avoid such a qualm, it might be better first to ask the karbari of the neighbouring village whether someone of his village claims the plot on which one wants to make his swidden. Since, however, there will always be someone claiming it for future use, there remain only two ways: to desist or to venture the encroachment.

One might assume that the growing scarcity of land available for cultivation has made encroachments more frequent, but this is not the case. The Mru are a peaceful people, and they prefer to live in peace also with the inhabitants of the neighbouring villages. Therefore, in case there is no more land available, one knows that the best way to survive is to move (alone or in case one is married together with one's family) to another village, that is, growing scarcity of land increases "semi-nomadism". And this may have been already the case 100 years ago, when more land was available but not cultivated, since it did not "rent" (according to the then prevailing standards). I'll return to the problem in chapter J.

C2 Paths

It takes only one rainy season for a path to grow over, and the only footpaths always kept cleared are those leading to the hamlet of the mouza headman. Wherever relationships between neighbouring hamlets are good, however, villagers agree upon the boundary up to which each hamlet is obliged to keep the connecting path cleared; most of the time this boundary corresponds to the dividing line between the fields being worked by the inhabitants of the two hamlets. Once a year, after the paddy has been cut and the rainy season has almost come to an end (in October), every household supplies one man; and together these men clear the connecting paths (*thak tia*).

The footpath to one's own field is cleared by the person himself, and it is cleaned up to four times a year: before the fieldwork begins; in July, during the weeding period; before the rice is harvested; and, if necessary, again before the cotton harvest. Footpaths used frequently by many people doubtless grow over less easily or quickly than private ones. If an obstacle, such as a fallen tree or a landslide, blocks a public path, however, no one takes the trouble to remove it if he or she can go around it in less time or with less trouble. Should a really lovely path leading along the crest of a hill suddenly take a sharp, unpleasant turn downward and continue so for some distance before turning up again, one can be sure that there was once a cultivated field located on that spot below the ridge. (Why should people take the trouble to keep a comfortable or pleasant public path cleared, when someone is going to keep the footpath to his own field cleared anyway?) If a person cuts bamboo close to a footpath, without any hesitation he leaves the cut-off branches and chips on the path, in spite of the fact that these bamboo chips, which are as sharp as arrows, can be dangerous for even the most thickly-callused Mru feet.

During the earlier days of warfare, two bamboo sticks sharpened on both ends were used to "mine" a footpath. The sticks were twisted tightly together in such a way that, no matter how they fell, one point always stuck upward over the other three. Dozens of these caltrops were hidden on a path, if possible under leaves; and whoever stepped on one of these points could count on its going straight through his foot.

Caltrop (drawing: C.-D. Brauns)

In spite of all the discomforts they offer, Mru footpaths are laid out in a truly masterful way. First of all, one makes sure that during the rainy season the paths are not washed out by the rain and thereby transformed into slopes of scree. Wherever it is possible, therefore, the paths run along the top of a watershed, even when this leads down the slope; and at places where this is not feasible, paths are moved to the right or the left for short stretches. One does not by any means always choose the shortest route. Smaller valleys and depressions, for example, are often circumvented by walking around the rim. Steepness, however, is no obstacle; if necessary, one uses a notched tree trunk as a staircase – like the one used at home.

Dried up bed of a brook, used as path

The most popular paths, though, are those which do not have to be cleared, those which nature herself, so to speak, makes freely available. These are the waterways. Path and riverbed are often more or less identical. Most of the time during the dry season, only a small trickle of water flows over these paths made of scree; but it does not in any way disturb a Mru to have to wade through water up to his waist. The Mru, however, normally cannot swim. For this reason they prefer to walk along the edge of a river – even a large river – rather than use the dugout of one of the river dwellers. Walking is, first of all, cheaper, since one would have to pay to be ferried across. Secondly, it is safer, since a Mru would constantly fear that the boat might capsize. Under such circumstances, riverbed paths are much preferred by the Mru. For me they were horrible: slipping from the slick stones and scrapping my ankles on the scree, with shoes full of water and, often enough, with wet trousers, I could never keep pace with the Mru.

Until I had learned to use the Mru whistle technique of breathing, I also had some trouble keeping pace when climbing uphill – whistling enables one to better control the rhythm of breathing, thus preventing the mucous in the mouth from drying out. Steep descents, however, remained problematic. The footholds which offered Mru toes a secure foothold did not give my boots much help; and the two times I attempted in desperation to go unshod, thinking it might work better, I failed due to the relative instability of the soles of my feet. The fact that Bangali police and military personnel all wear shoes and, in addition, know nothing about mountain climbing has proved an advantage for the hill dwellers: they have been protected from too many unwanted visitors. (This will also continue to protect them until the roads necessary for "the opening up of the Hill Tracts" have been constructed with development-aid money.)

During the last world war, the attack of the Japanese on India, which was launched from Burma, got stuck in these "hills" – and already on the Arakan side. I have walked hundreds of miles on Mru footpaths and have heard even the Mru complain that their footpaths are "not fit for a dog", but the Mru love their land and can with justification become enthusiastic about its beauty. Why should they construct paths which would offer easier access to the foreigners who want to snatch their land from them?

C3 Houses in the hamlet

C3a) The building material

When one approaches a Mru hamlet after climbing up the slope, the hamlet may not look particularly inviting, since one is confronted with the solid backs of houses. They tower on posts far above the head and are broken only occasionally by narrow windows.

Backside of a row of houses (photo: C.-D. Brauns)

Normally, however, one will enter the hamlet by one of its entrances (*plon*), and see the houses only from the side of the village street or square. When the ground is uneven, as it normally is, one may get the impression that the houses are built in complete disarray. There is nothing to suggest the clear order which can be found among the Chakma – the Chakma turn all of their houses toward the east because they love to have the morning sun on their platforms. Mru houses, however, are much larger than those of the Chakma.

Compared with Mru houses, for example, the dwellings of the simple Bangali plains farmers are nothing but huts. In contrast to the Bangalis, though, the Mru and the Chakma have something in common: their houses stand on stilts. (This is a practice common to much of Southeast Asia where traditional cultures still prevail.) In a country of monsoon rains, the Mru would consider it quite unwise to build their houses directly on the ground as the Bangalis do. Due to the natural unevenness of their land, they would, in addition, be obliged to move a considerable amount of soil.

As a general rule, trunks of hardwood trees, from which the bark has been removed, serve as the main posts and major cross-beams of Mru houses. (The better the wood, the more durable and long-lasting the house.) Other posts may, if necessary, be replaced by bundles of bamboo. The choice of material is dependent upon availability, since increasing deforestation has reduced the constant supply of large hardwood trees; and in some places even bamboo has become scarce. Whereas in former times one needed only go to the edge of the village to cut whatever one wanted, today one is often obliged to seek materials in neighbouring villages. Before one is allowed to chop down what one needs, however, one must attain the consent of the villages concerned; otherwise, they might file a complaint with the headman, and the transgressor may be fined.

Moreover, one has the inconvenience of carrying the supplies home over long distances; and if villagers assist in the cutting and carrying of materials, they must be paid for their labour. Since the floor and walls of a Mru house are made of bamboo, a shortage of bamboo means that houses can be refurbished less often. After a few rainy seasons, the original yellow-brown colour of the bamboo walls is transformed into a speckled grey; and the village begins to take on a dowdy and poor appearance – no mere deception when the material for renovation is no longer available. The growing depletion of wood and bamboo in modern times has indeed induced some Mru to contend themselves with very short stilts and floors near to the ground.

Houses situated on exposed slopes are more subject to wind and rain than those lying in protected valleys. Generally speaking, however, bamboo and softwood pieces should be renewed every 5 to 10 years, though this depends much on location: on exposed places one reckons 5–6 years, in protected places and in the plains 10–12 years. In the past, when building materials were plentiful and when one was allowed to cut wood anywhere except in official forest reserves, purely hardwood constructions would surely have made for more durable houses. One could, for example have used teakwood boards. Yet even the Marma, whose Buddhist temples were made of teakwood, preferred bamboo for their homes. This was not without reason. A floor made of boards is hard, whereas one of woven bamboo strips is flexible. (With a simple blanket, one sleeps on a bamboo floor as on a mattress.) Similarly, a board wall means a dark room, whereas a bamboo wall can be penetrated by hundreds of narrow beams of light. Bamboo also allows fresh air to enter; so even an inside room of a bamboo house having small windows is never dark or stuffy.

C3b) The major parts

A Mru house consists of three basic parts. First, there is a large room, the *kim-tom* (ca. 7 x 7 m), with an outside entrance. This room is a kind of living room in which people cook and eat, receive visitors, and accommodate guests. Children and the unmarried members of the family also sleep in the *kim-tom.* Secondly, there is a narrower and (approx. 30 cm) lower room, the *kimma* (length: 6 m, breadth: 3.70 m, maximal room heighth: 3 m) with a separate roof.

Though being situated on the side of the village square and in front of the *kim-tom,* this room is accessible only from the *kim-tom.* At night the *kimma* serves as a bedroom for the married couple and their smaller children; in the front-most part of the room, where there is a second fireplace (the main hearth is always in the *kim-tom*), children are born. Strangers are not permitted to enter the *kimma* (in case they would do it nevertheless they might be fined 30 Rs.); and if the door to the *kimma* is open, they should not even look inside. The view is blocked anyway, though, by the large, round storage bin placed directly behind the door. (This large basket holds the year's paddy harvest.) Spears, cloth, and other things of value, as well as cotton and vegetables, are also kept in the *kimma.* If a stranger would still like to look into the *kimma,* in spite of the prohibition, he must be prepared to get an inflammation of his eyes (see P1g). It is therefore better to wait until the wall of the *kimma* must be renewed; for at that time one has a full view of the room for an entire day – and a view which embarrasses no one. During the period of repair, the married couple sleeps in the *kim-tom;* and no one would renovate a house at the time when a birth is expected. The prohibition then is really intended only as a defence against the "evil eye": it will return to its sender.

Finally, there is the open-air platform, the *car* (4.50 m x 4.50 m), which is built onto the right or left side of the *kim-tom* and is accessible only from that room. If one wishes to dry the rice – which one must do in order to avoid mould – one spreads it out on a large bamboo mat on the *car.* If there are vegetables to be cleaned and washed, animals to be carved up, vessels and dishes to be washed, or other work to be done which would dirty the inside of the house, this is also done on the *car,* since water runs immediately through the bamboo slats and, thereby, off the floor. On the *car* one rinses one's mouth out after eating; on this platform one could also wash clothing or oneself. Since water for washing or bathing must first be fetched in gourds from the spring or brook, however, one generally prefers to wash or bathe at the water-source itself. (Using warm water for washing is consequently rare.) Finally, if a corner of the *car* is particularly isolated – or even if not – it serves as a toilet. Even if this area of the *car* is not specifically marked, one recognises it immediately by the bamboo strips placed there to be used as toilet paper, as well as the opening left in the slat grid-work. Generally, the designated corner serves only as a urinal. (The women do their business standing, the men squatting – just the opposite, therefore, from our practice). In an emergency, however, it may also be used for defecation, though one normally prefers to do the latter in a place on the backside of the village. At night, when the path to this outside toilet is dark, one could unexpectedly step on a poisonous snake or, apart from that, meet an evil spirit. Children particularly, then, cannot be expected to go out at night.

One might think that this practice would create a stench in the neighbourhood of every house, but that is not the case. The pigs, when allowed to run loose, demolish all kitchen trash and human faeces; and the slanting incline which is always found under the *car* sees to it that all water runs off quickly. The only disadvantage of the *car* is the fact that its thin bamboo slats rot quickly, since they are often washed over with water. Unless the slating is annually repaired, one is occasionally in danger of stepping down into open space. This disadvantage, however, means once again that the *car* never becomes very dirty, for its renovation sooner or later becomes imperative.

Wealthy people occasionally build themselves an additional storage room, a *kim-ca,* on the other side of the *car.* This room is shaped like a *kimma* and, when a son of the family marries, can also be used as such. Houses with two or three *kimma* are unusual. (The *kimma* of the household head is always the one built directly onto the *kim-tom.)* A French sociologist, P. Bessaignet, had the bad luck of landing in just such a house and therefore came to the conclusion that the Mru live in extended families. This is, however, completely false, since, generally speaking, every married Mru male values greatly being the head of his own household. A widow or a widower who lives with a married son must again, like the unmarried, sleep in the *kim-tom;* in such a case, a room is often partitioned off by a special wall. A private sleeping compartment may also be constructed for the oldest unmarried daughter still living at home. The *kim-tom,* with its floor-space of approximately 50 m^2, is always large enough to take care of the everyday needs of any Mru family – as well as those special needs on festive occasions.

Since the central village square is often the highest point of the locality, the front wall of the *kimma,* which faces the square, usually stands on lower stilts (or poles) than the back wall of the *kim-tom.* This means that the distance of open space under the two walls can vary from one-half to five meters. The smaller space (under the *kimma)* is panelled laterally with small bamboo canes; at night the front part of this basement serves as a pigsty and the back part as a chicken coop. Firewood is stacked in the large, open space beneath the *kim-tom* and testifies to the industriousness – or lack of such – of the housewife and (possibly) her marriageable daughter.

Houses should whenever possible be built side by side (with the *kimma* in front), and nobody should build his house in such a way that the front side of his house stands against the side wall of his neighbour. In case he did so, either he himself or his neighbour would die. This is generally known; less known (and therefore not always complied with) is the traditional saying that the same danger would arise in case one would build his house (and be it only his open-air platform) in such a way that the oblique supporting posts would cross those of one's neighbour (MK, 13.09.1956).

In the following presentation I shall first of all describe the house as it normally presents itself among the Anok Mru and especially in Tapwüa-Kua. At the end I shall give a summary of the major deviant features to be found with the other Mru groups. No informants will be quoted, since everyone may see the buildings with his own eyes; the Mru terms, however, mostly were elicited from Menkröi.

C3c) The house construction in general

The building of a house is a communal undertaking involving all the village men and lasts several days. First of all, trees and bamboo must be selected, cut, and carried to the village. Long tree trunks which have a diameter of more than 20 cm are so heavy that no one person alone could transport them or lift them up into place. Neither cars nor machines are available; and given the circumstances, such conveniences could not be utilised anyway. The axe and the hewing knife are the only two tools suitable for the work at hand.

First, the ten main posts of the *kim-tom* are set into position. The holes for these posts, dug with hewing knives converted into digging sticks, are dug in two parallel rows about 4.20 m apart. Each row has four posts ([*kim-*]*cüng*, 15 cm diameter, 2.10 m high [room heighth, the total height will vary according to the distance of the floor from the earth]) spaced about 2.10 m apart; then two posts (*klong-cüng*, 20 cm diameter, 3.30 m high [room heighth]) are placed between the rows, at the front and back. These two posts tower above the others and will later support the ridge pole (*klong*) of the roof. All of these posts will stand within the interior of the house, only 20 cm from the back wall (*kim-döng*) but 1.40 m from the side walls (*kim-reng*). Slightly inside the line of the wall, somewhat thinner and shorter posts (*cüng-ca*, 1.60 m [room height]) are set up. These are notched on the top end; and a pole which will later serve as the purlin (*pang*) is tied lengthwise in the notch.

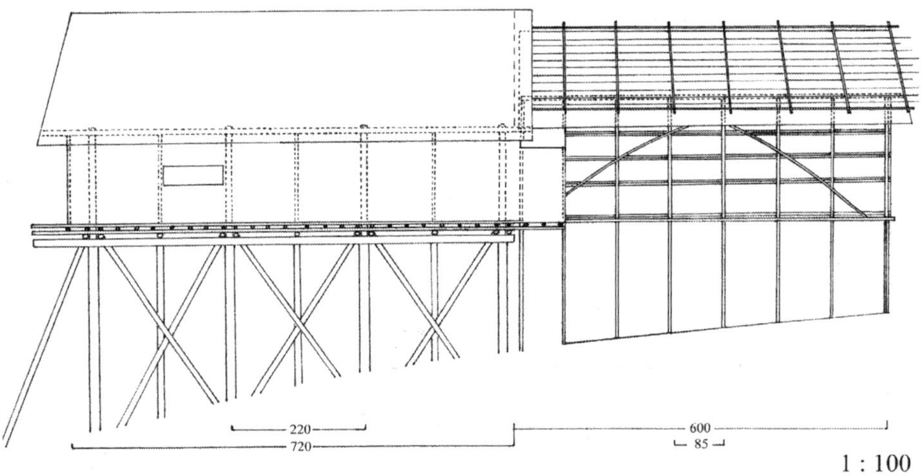

House (Anok region): side-view

Additional rows of posts are set up still closer to the edges; these posts (*tüa*), which are notched at the top, reach up only to the bottom of the floor and support the lowest beams (*hong*) of the floor, which are approximately 10 cm in diameter. Upon this lowest layer, which runs lengthwise, eleven pieces of wood (equally called *hong*) are placed crosswise: in the middle between, and on either side of the main posts. They are supported by perpendicular posts (*tüa*) staggered with the main posts. The perpendicular posts, moreover, are propped up from left and right – and the crosswise, horizontal posts, from front and back – by still other wooden pieces (*tok*) some 10 cm thick. Then follow a layer of bamboo poles lengthwise (*hong klang*), a layer crosswise (*kan*), and then a second layer lengthwise (*prop*).

The poles laid out crosswise are placed about 30 cm apart, and those of the second lengthwise layer, some 5 cm apart. The floor itself (*phai*) is finally laid out on the third and last layer of poles (*prop*) in the inside of the room, that is, inside the walls. (At the back side the lower floor construction juts out 1 m beyond the wall.) It consists of interwoven bamboo strips (*bang*, 15–20 cm wide), which horizontally (*phai klang*) are placed close together, while lengthwise (*phai ku*) they may lie 8 cm apart. This spacing is necessary, in order to allow the interweaving. Since the strips are not always equally wide, from time to time there appear small slits or holes, which are welcome, since through them dirt and water can fall or drip down. (In some houses there are no *phai ku*, that is, the floor matting is not interwoven and the *phai klang* are only held in place by slats fastened over their ends. This makes sleeping on the *phai* more comfortable, but this construction is much less durable.) In order to prepare the bamboo strips, one side of a bamboo cane is split open so that the inner side of the knotted joints (nodes) can be reached; these joints are then notched in such a way that the canes bend completely open and can be freed of all remainders of the nodes.

In the construction of the foundation (*kim-do*), all points of intersection (where perpendicular posts come together with props or bamboo or wooden pieces running lengthwise or crosswise) are secured by thongs or lashings (*mia*) cut from bamboo bark. No use is made at all of nails, screws, or wire. Props and thong bindings offer ample support for a Mru house. Not only can dozens of people sit on the floor at the same time; they can also dance on it and jump on it in rhythm. During the monsoon season these houses hold up repeatedly to storms. In some villages, however, special typhoon houses (*li-ku kim*) are also built. Consisting only of a roof made of tree trunks, these shelters serve as protection in case of an extremely heavy storm.

C3d) Details of house construction

Each of the two rows of four main posts (*kim-cüng*) is notched at the side of the top, and into this notch a pole is tied out lengthwise; upon these poles (*klöng pang*) four thick beams (*klöng*), which have been notched at the ends, are placed crosswise. These beams are also fixed from above by another pole running lengthwise (*klöng-nüm*). Other *pang* (which serve as purlins) are tied into notches at the upper outside end of 4 smaller vertical poles (*cüng-ca*) on both lengthsides of the *kim-tom*. The *cüng-ca* are interspaced with other vertical bamboo rods (*kim-cök*), which reach from the purlins down to the *kan* of the floor (but not down to the earth). An arrangement with 2 wooden *cüng-ca* and 6 *cök* is possible. The latter are cut in such a way, that at their ends a long "thorn" of the outer bamboo bark is left standing, which then can be bent backward around the wooden poles of the *pang* and the *kan* and is secured by a lashing.

In the twin hamlet of Tapwüa-Kua (Atwangwüa-Kua in the Horinjhiri-Mouza) there were, however, also two houses where rather thick *kim-cüng* took the place of the *cüngca* of the side walls, so that the 2nd and the 3rd *klöng* (beam crossing the room) had to transverse the room without some additional support, since only the first and the last *klöng* could be tied to one of the *klong-cüng*.

Above the purlins come the rafters (*krang*) of the roof; they run (30 cm apart) from the ridge pole (*klong*), which is supported by the two central posts (*klong cüng*) and (in the middle of the room) by stilts placed in the middle of the *klöng*, to about half a meter past the side purlins (*pang*). To be more exact, the rafters (*krang*) do not start from the ridge pole, but are incised above it, so that they can be bend down parallel to the rafters coming up on the other side of the roof. Below these rafters one ties (20 cm apart) bamboo rails (*bai-kop*), while on the rafters bundles of thatching grass (*ai-daa* or *ci-daa*) or bamboo leaves (*kau-daa*) are placed; the thatching material is then secured against wind damage by an outer layer of rails (*kau-praa*) which is tied with bamboo thongs to the lower parallel-running layer (*bai-kop*). The thatching grass (*imperata spec.*) is cut in the jungle and carried home in thick bundles, tied together at their foot and in the middle. These bundles are placed (head-down) on the roof starting with a row on the lower end, fixing them with a *kau-praa*, adding the next row overlapping that previously fixed, and so on until one reaches the ridge, which will receive an extra thick crosswise layer. Due to this process, the *kau-praa* of the *kim-tom* will in the end become invisible. A layer of grass (or leaves) which is roughly 20 cm thick can keep even a heavy monsoon rain out, though it will have to be renewed every year. This is especially true of the somewhat thinner roof of the *kimma*.

Here the *kau-praa* normally are fixed only after the bundles have been laid. To keep these *kau-praa* in place they in turn are overlaid by *mong-küa*, bamboo canes (90 cm apart) which cross their upper ends over the ridge pole. Into the crossings another pole (*klong-nüm*) is laid lengthwise and tied together with the *klong* running below the roof. Other bamboo poles (*don-höng*) running lengthwise over the lower ends of the *mong-küa* on both sides of the roof are tied to a second *don-höng* below the thatch and the rafters (*krang*), outside of the *kimma*. The *ai-daa* (thatch) jutting out 10–20 cm below the *don-höng* is called *daa-yör* (height above the floor of the house: 1.30 m). Where the *kim-tom* and the *kimma* meet, the roof of the *kim-tom* overarches that of the *kimma* by approximately 50 cm, that is, the last *mong-küa* lies under the last *krang* of the *kim-tom*.

In the case of Menkröi's house the right roof side of the *kimma* was longer than the left side: it stretched down over the front door to the same breadth as the roof of the *kim-tom* and was supported by a second wall (*wang-khör bang*) tied to the outside of 5 posts. By this construction a kind of channel (*wang-khör*) was created, through which one had to pass before one could climb up the log staircase. I have not seen a similar *li-tung* (wind-"screen") in other houses, and I just mention it to show that variations are possible. (Menköi's neighbour's *kimma*, for instance, had a similar roof, but without *wang-khör-bang*.)

Finally, the walls (*bang*) are added. They normally consist of vertical bamboo strips (20 cm broad), placed narrowly together, interwoven with more widely spaced horizontal strips (*bang-lang*, 5 cm broad). In some cases, however, all strips (of *bang-lang* size) may also be plaited diagonally. The walls of the *kim-tom* are secured from without to double horizontal rails (*bang-yang*) which have been attached to the outside of the wall posts (*cüng-ca* and *kim-cök*). At the lowest end of the wall three or more *bang-yang* are fixed.

The back wall (*kim-döng*) of the *kim-tom* may be tied also to the two *kim-heng* inside of the room, which in turn may be crossed by another pair of *heng*. The first pair of *heng* (the *klöng-nüm-heng*) rests on the *hong* of the lowest layer of the floor on both sides of the *klong-cüng* and reaches up to the *klöng nüm*, which here must help to carry the last *klöng* (the beam crossing the room) before the back wall, since the *klöng-pang* end before they reach the back wall and are tied from below to the *klöng*, which here directly rests on the *kim-cüng*. The second pair of *heng* (the *klong-heng*) rests on the *hong* at the inner side of the *cüng-ca* and reach up to *klong* (the ridge pole).

The side walls (*kim-reng*) of the *kimma* are put up in precisely the opposite manner to those of the *kim-tom*, that is, the woven wall is tied to the inside of the 3 x 2 *bang-yang* (in the midle) and 2 x 5 horizontal bamboo canes (below the roof and above the floor), which in turn are tied to 8 posts (normally all bamboo, some 80–90 cm apart), reaching down to the earth. Finally, the whole may be secured from outside by two oblique bamboo canes (*kim-heng*) resting with their lower ends on the *hong* (at the inside the last post) and reaching up to the *pang* on the opposite side.

The front wall (*kim-döng*) of the *kimma*, however, is, once more, tied from outside to the *bang-yang*, which in turn are tied to 7 *kim-cüng* (made of bamboo), and also the two *klong-heng*, here serving to replace the *klong-cüng*, are to be found inside the wall. Above the floor and in heighth of the side walls the front wall is transversed by two halved bamboo canes. As a general rule wooden poles and beams are avoided in the interior of the *kimma*. This has the advantage of keeping out wood-boring insects which might produce wood dust which may fall down into the eyes of those resting or sleeping below.

In case the *kimma* should become shackle, instead of rebuilding it, one may try to stabilize it by additional wooden posts (*kim-ngüa*) which are erected at the outside, one each before and tied to every second *cüngca* of the side walls and one below the ridgepole (*klong*) in front.

The *kimma* may be provided with a window (80 x 80 cm) of its own. It is cut out of the side wall and can be closed by a shutter which hangs at the outside fixed by thongs to the lowest of the transversal bamboo canes of the wall. The shutter consists of horizontal small strips (alternately 2 thin and 1 broad) interwoven by vertical strips, and resembles the door to the *car*, except that it has no vertical framing. The window itself has a frame (lining) of halved bamboo slats.

The space below the *kimma* serves in its front part (up to the 5th *cüngca*) as a pig-sty and in its back part as a room where the fowls are kept overnight. It is surrounded by a kind of wall (*tung-ping*), consisting of horizontally laid bamboo poles, which are held in place by outer vertical halved bamboo canes tied to the interior *cüngca* by oblique bamboo lashings which each time embrace two *tung-ping* canes. The front and the back part are also separated by a *tung-ping* wall. In order to let the animals out – the pigs to feed them at the trough and to let them cleanse the village, the fowls the whole day long, so that they might fend for themselves – there are (right and left of the 5th *cüngca*) two openings in the *tung-ping*, 30 cm wide, lined by 2 halved bamboos and bordered from above by 4 ping bamboos. In front of these and the *cüngca* additional wooden posts are erected which are bored through above the earth and 20 cm from above. Through these holes bamboo transversals are inserted by which the doors (consisting of wooden shutters or vertical bamboo canes) can be kept shut. The doors are closed at night, so that the pigs and the chickens may not fall prey to a wild-cat (see F2b and F5a).

Windows (*tang-pok*) can also be cut out of the walls of the *kim-tom*, in the space left between two (horizontal) *bang-yang* and two vertical bamboos (*cök)*. The upper and the lower rim of the opening may be protected against splintering by a halved bamboo (*mom*, "lining"). Generally, only one window is constructed (on the back side or one of the length sides, depending on the weather front) of the house. The window can, if necessary, be closed with a piece of bamboo matting (*tang-pok-büng*), plaited from thin (8 mm wide) bamboo strips interwoven like those of the floor, i.e., the transversal strips are placed narrow together, the vertical strips 12 cm apart. The outer vertical strips lie 2 cm before the end of the horizontal strips which stick out freely to the sides. The vertical strips may have their ends inserted into an upper and a lower bamboo, the lower one, resting on the *bang-yang*, sticks out 5–8 cm over the breadth of the window and runs behind the *cök*. With the help of this lower bamboo the window matting can be moved to the side in order to be opened or closed.

Doors (*kim-büng*) equally consist of woven pieces of bamboo matting. The matting closes the door on the inside of the room and can be moved to one side either by hanging on rails (*tang-pok klia*) on the upper part of the door frame or resting on a bamboo at its lower end. The door opening being roughly 90 cm broad (and 1.50 m high), the lower bamboo will measure at least 1.40 m. Into it 5 vertical strips (1 cm broad, 20 cm apart) are inserted, transversed by broader strips, which often are interwoven in such a way as to show a pattern: one 6 cm broad strip followed by two 2 cm broad strips, all of alternatingly green or white colour (outer or inner side of the bamboo). The whole matting can be strengthened by inserting also diagonal strips. There are also doors in which the smaller transversals are replaced by two smaller strips (only 1 cm broad) which on one side end in a node and (by reversing the outer and inner side) may "embrace" the outer vertical strip. The next vertical strip is then placed only 10 cm apart and at a convenient height the transversal strips are shorter by these 10 cm so as to embrace the second vertical strip only. Thereby a hole is formed through which the hand can reach and open the door – both from the outside and the inside.

Among southern Mru groups (see C3f) one may enter the house from the *car;* this platform is reached by climbing up a log staircase (*dong*, a tree trunk with notched steps). Among the Anok Mru, the *car* is accessible only from the *kim-tom,* which in turn is accessible directly from the outside – again by way of a log staircase. The staircase (*dong*) is normally leaned to a projection of the floor of the *kim-tom*, called *pröng-prüa*, and covered by *chek-rek* (bamboo slats). The climbing up the ladder may be facilitated by erecting a *rai-cün* (banister) at its side. According to early sources, in the 19th century it was still customary to pull up the log staircase at night; today one leaves it down. One could not possibly enter the house unnoticed anyway, since the door can be locked from the inside; in addition to that, it would be impossible to open the door without making some noise.

Some houses of the Anok region have a kind of small annex to the *kim-tom* on the opposite side of the entrance. It serves as *tui-krök* (water-shelf), on which the *tui-yia* (water-bottles) are stored. Outwardly the side wall of the *kim-tom* is prolonged by 80 cm, but it has a slanting roof of its own below that of the *kimma* as well as that of the overarching *kim-tom*. It can be accessed from the *kim-tom*, the opening being 1.50 m wide. The water-shelves are erected left and right (80 cm high) and consist, like the floor, of *chek-rek* (bamboo slats), so that, in case one bottle should be spilled, the water may run off freely. In order to prevent tumbling down of the bottles, the borders of the shelves are secured by two tied-on bamboo lengths. The side wall may be be provided with a small window (20 x 20 cm) of its own. In houses without such a special annex, the *tui-krök* can be erected also below the *kim-tom* roof of the *car*.

The *car* (open-air platform, 4.50 m x 4.50 m) has a less stable floor than the *kim-tom*, that is, one cannot dance on it and, especially when the slats covering it have become rotten, one should beware not to brake through. To prevent this the slats should be renewed every year. These slats (*chek-rek*) consisting of halved bamboo canes, are the third layer which rests on two lower layers of spaced full bamboo canes. The lowest layer is just a frame (2.20 cm spaced), the outer bamboo being tied to the side posts of the the *car*, the middle one supported by vertical bamboo poles (stilts). The next one (50 cm spaced) is also carried by vertical stilts, which in part may be dug in at first, but most of them will be added after the layer has been placed. This has the advantage that the stilts can be cut as required: their upper end carries a "thorn" of the outer bark of the bamboo, which can be bent round the longitudinal canes and fixed to the vertical stilt by a lashing. The upper layer of halved bamboos (*chek-rek*) is narrowly spaced (some 5–10 cm apart) and kept in place by small lengthy strips (*chaa*) carrying short bamboo strips (*wa-naa*), the ends of which are cramped below the next layer of bamboo canes (spaced 50 cm apart).

Finally the *car* is surrounded by a low (1 m high) wall of 6 broad bamboo strips (which alternatingly show their green and white side), interwoven by small vertical strips (*chaa*), the ends of which are inserted in holes cut into the horizontal bamboo canes (*wa-kaa-pang*) at the foot and the top of the wall and tied to the outside of some longer bamboo poles (*car-cüng*). The *car-cüng* at their upper end (1.80 m above the floor) may carry (once more tied by means of a "thorn" left standing) another horizontal pole (*yang-pang*), used for hanging up the washing or yarn to be dried after dying. One corner of the *car* is used as a toilet and may be partitioned off by a wall of its own. The *car* is used to perform various tasks (see C3b), for instance to prepare vegetables or to cut up killed animals, even weaving can be done outside, or to spread mats on it (to dry paddy), but there also are to be found old vessels, basins, water-bottles and baskets. One of the basins is filled with water and used to empty in husk and bran (called, after being soaked in water, *pak-di*), to be cooked during the day by a woman (wife or daughter) for the pigs (see F2b). To this pig's food left-over bits from daily meals will be added.

C3e) Hearths and shelves

After the house has been erected, the fireplaces (hearths, *taping*) are built. They consist of 4 logs carved in at their ends, so that they can be placed in a square (1.50 m x 1.50 m) close to the wall, where they will be fixed by lashings. The square will then be filled up (30 cm high) with humid earth and have three stones placed in it. These stones (projecting approx. 15 cm) are a little bit slanting towards the centre (*chi-kua*) and serve to carry the cooking pots. After the earth has dried, its surface is plastered by mud mixed with cowdung. The main hearth used for daily cooking is that in the *kim-tom*. It is usually placed near the door leading to the *car*, but protected against draught by an additional wall reaching up to the roof and to the front to the next *kim-cüng*. At the side of the hearth this wall is also plastered over with clay. A second hearth is built in the *kimma*, near the front wall. It mainly will be used after child-birth only (see N1b).

Lightening the first fire on a hearth can be done without any ceremonies; for the lightening matches can be used or (since the new place is still a little bit damp) some embers may be fetched from a neighbouring house. Before, however, the family may move into a new house, the performance of a *cari-yöng* (see P5b) is required. Generally one sees to it that the fire never extinguishes: a thick log (*mai-phum*) is kept glowing day and night. When the fire is needed for cooking, fresh firewood (*chop*) is fetched from under the *kim-tom*, where it is stored, some embers are beaten off from the *mai-phum* and the new fire is kindled by blowing, if necessary with the help of a bamboo internode used as a blowpipe (*ria-ca*). To regulate the strength of the fire needs some skill in pushing the pieces of firewood under the pot or withdrawing them.

80 cm over the fireplace a shelf (*krök*, 1.20 m x 1.50 m) is fastened. It consists of two layers of framing bamboo, the lower one fastened to the house wall where it rests on a bamboo tied to the *cüng-ca*, the upper one to an additional small wall at one side of the fireplace, and at its front side it is tied by a thong to the *kri* (see below). On this frame lies an open wickerwork mat (with strips ca. 10 cm apart), tied to it by bamboo halves which serve as lining (*mom*). On the walls the mat is fixed by an upper and a lower halved bamboo which are tied together. Whatever needs to be dried or to be kept dry (and cannot be tucked in below the roof or must be spread on a larger mat) may be placed on this shelf.

An especially large shelf or rafter, called *kri*, is hanging head-high in the middle (between the second and the third main post) of the *kim-tom*. It consists of two widely spaced layers of bamboo canes, the lower one resting on the wall side on the *pang*, is tied to the *kim-cüng* and hung up by means of long thongs (*kri-tük*) running over the *klöng* (house-beams) and the *klöng-pang* (see C3d). On this frame large rolled-up mats (used for drying paddy on the *car*) are deposited, and (where the upper layer is strengthened by two additional canes, 50 cm apart) an openly plaited mat (made of strips 2 cm broad and spaced by 5 cm) is placed, which serves as a pad for smaller things and baskets (like *kruk*, see D2a), while larger carrying baskets may be stored together with the mats.

Larger standing baskets (see D2), which keep the clothing, are mostly kept in the *kimma*, where also the ceremonial weapons are preserved. Rich people may possess a cupboard (*long-pan tök-raa*) to store empty bottles (*phalang*), cups and mugs (*bia*, needed for drinking *arak*) and perhaps also plates (*long-pan*) which, however, like cutlery, are rarely needed, since people use their fingers and eat from banana leaves or (more modernly) flat metal bowls. For drinking water they take a water-bottle. Larger pots will be kept on the *car*. Instead of a *long-pan-tök-raa* one may also erect another *krök* (shelf) between the 2nd and the 3rd *cüngca* opposite to the *taping*: a rack (30 cm deep, 1.20 cm long), reaching up to the roof, with two layers plaited in open wicker-work to deposit pots, bowls, cups, mugs and plates. The lower layer is framed with full bamboo canes, the upper layer with halved bamboo slats.

There remain some utensils, fabricated from wood, which most of the time are used outside the house: the wooden trough for feeding the pigs (*pak-ko*) and the mortar (*chum*), used for hulling grain with the help of one or two pestles (*li*). The mortar is 30–50 cm high, cut out of a thick trunk (up to 40 cm in diameter), hollowed out at one side. The pestle is 1.80 m long, 10 cm thick (somewhat narrower in the middle) and rounded at the ends (one end broader than the other). Though rather heavy, the mortar, in case of continuous rain, may also be taken inside the house. The pig trough, on the other hand, always will stay outside near the pig sty. It consists of an approximately 1 m long trunk, 25 cm in diameter, flattened on the long side which rests on the earth (so that it cannot be toppled over) and hollowed out at the upper long side (see F2b).

C3f) Variations with the southern Mru groups

Though I was told that houses should whenever possible be built in such a way that the houses are aligned side by side with the kimma pointing to the street, on the 26th of December 1956 I saw an arrangement in such a way that all houses stood with their lengthside to the street, that is, the *kimma* of one house stood opposite the *kim-tom* of the neighbouring house. This deviating arrangement was the case in the Anok hamlet Rekca-Kua (Menpung-KP) in the Lemupalong-Mouza. This hamlet was situated quite close to the Palong river and built on a low, but very narrow ridge standing at a right angle to the river. The ridge formed the ascending village street, the open-air platforms stood on high stilts on the flanks of that narrow ridge, all houses were built with their *kim-tom* towards the (lower) side of the river and could be entered by their lengthside right and left of the street.

The most obvious variation among the southern Mru groups (like Dömrong and Rümma) is that one normally enters the house not directly by the *kim-tom* (respectively the preceding *pröng-prüa*) but via the *car*. For the Dopreng no general rule can be given: North of the Prumma river (up to Rünghim-KP, Lama-Mouza) the Anok way is used, starting from Taa-uk-Kua the entrance is via the *car*, but already in Rünghim-KP the rule is valid that the hearth is placed on the wall opposite to the entrance from the *car*, not as with the Anok (and still in Rüm-putma-Kua, Luleng-Mouza) on the same side as the *car*-entrance. A different arrangement I found in the house of Karbari Rengdon Changpröca (Dopreng) of Chingkhung-Kua (Dordori-Mouza): one hearth was to be found in the corner of the length wall and the front wall with the *kimma*-entrance, another one at the opposite length wall, but framed in by a separate wall: this separate room served as the sleeping place of the adult girls, who used the hearth for distilling. The distance from the last row of posts to the back wall measured 1 m, was not covered by the *phai* (floor matting) but by *chek-rek* (bamboo slats) and served as a storeroom for baskets (as is the rule in all houses of the Dopreng).

In Chingkhung-Kua the *kri* connected all posts; longitudinal canes formed the lower layer, transversal canes the upper layer. Here 3–4 m long mats, secured against sagging by 4 tied-on bamboo canes, could be stored. For storing smaller things a wickerwork (110 x 70 cm) was tied 60 cm below the *kri* and in addition fastened to the wall. The shelf above the hearth was stabilized in one corner by a special bamboo pole reaching from the floor up to the roof. In all Dopreng houses there is a little window (20 x 20 cm) near the corner of the length wall behind the hearth, and maybe a look-out in the opposite side wall, but no window in the back wall. No annex is built to contain the *tui-krök* (water-shelves); the *tui-yia* (water-bottles) are just placed on the floor near the door inside the house, but a *tui-krök* can be built outside on the *car*. Doors consist of narrowly placed vertical bamboo slats interwoven by spaced horizontal slats and are hanging on a slide outside the room.

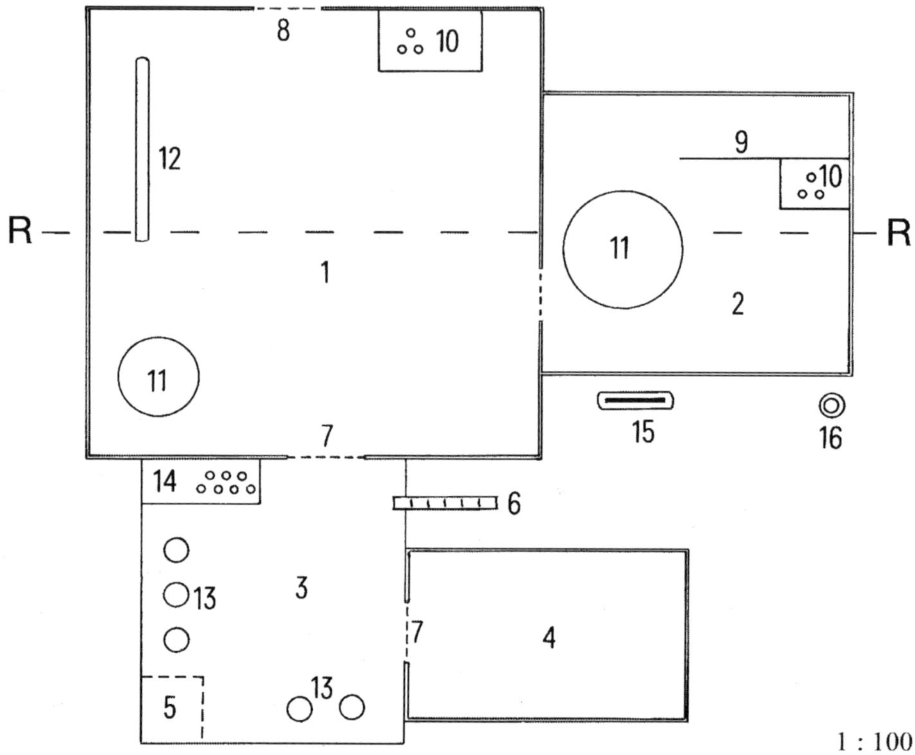

Ground plan of a house in the Matamuri region

1. common living room (*kim-tom*), 2. private quarters (*kimma*), 3. open-air platform (*car*), 4. (optional) annex (*kimca*), 5. toilet (*ek-raa*), 6. log staircase (*dong*), 7. doors (*kim-büng*), 8. small window, 9. partition, 10. cooking area (*taping*), 11 paddy bin (*ca-pam*), 12. pillow beam (*lu-ngua*), 13. (optional) pots and stationary baskets, 14. rack for water-bottles (*tui-krök*), 15 feeding trough (*pak ko*), 16. mortar (*chum*), R. ridge pole (*klong*).

In the Bommu-, Luleng- and Songkö-Mouzas there are some *cen* (sibs), the groups of which are differently identified as Dopreng, Dömrong, or Rümma, depending on the informant (see L2b). Let us take the Phen (whose group was identified as originally Anok). Their houses (like also some in the Rümma area) show a specialty: the normally plaited *bang* (see C3d) of the back wall ends ca. 80 cm above the floor with three *bang-yang*. The lower part is made by interweaving (2 up and 2 down) small bamboo strips with 3 double strips on top, in the middle and above the floor. In the house of headman Chamklaa Tou (who identified his *cen* as Dömrong, while others identified the Tou as Rümma) the back wall even up to a height of 1.20 m consisted of a horizontal layer of full bamboo canes kept in place by vertical bamboo poles inside and halved bamboos outside.

Dömrong and Rümma houses show yet another specialty: the lower *kimma*-wall (behind which the fowls are kept overnight) is closed like the upper room. It is accessible only from inside, that is from a second door in the *kim-tom* (in the place where with the Anok the entrance is to be found) which leads to a special room of the *kimma*, completely separated from the sleeping place. Here vegetables may be stored, but one may also climb down to the earth to fetch eggs from the chicken room, and at night the chicken must be caught to be thrown down into it from above. (With the Anok and the Dopreng the "chicken room" is accessible only from outside.)

C3g) The houses of the Khumi and the Longhu

In distinction to the houses of all other Mru groups, the Longhu (Southern Chüngma) build their houses in the Khumi style, that is, instead of a *kim-tom* and a *kimma* the houses have one room only, which (at least by its roof) looks like a big *kimma*. Rich people may, however, have two of these houses, connected by a common front platform. Many of the houses have two open-air platforms, one in front (*kim-pur-car*, leading to the main entrance) and possibly one at the back side, but never at the length side. The platforms often have no special walls and no railing. (Both must be added, however, in case the house owner wants to give a feast of merit, see R5 and R6.) The *kim-pur-car* may have one smaller side producing a corner into which the log staircase is leading.

In the arrangement of the house posts I found no rule except that the ridgepole is supported by two *klong-cüng*, standing outside of the front and the back wall. In the interior only one or two beams are crossing the room, resting on the *pang* of the length walls and supported by posts in the middle line of the house, but those in the middle may consist only of a bamboo *cüng*. The left side of the interior may be more or less empty (on the back wall one hearth is situated), in the middle of the right side the *ca-pam* (paddy bin) and some baskets bar off the back side. In the corner (behind more baskets) is the sleeping place of the parents. As an additional borderline in the middle of the backside serves the stand to which the skulls of cows, buffaloes and pigs sacrificed during feasts are tied. In the right front corner the main hearth is to be found, and in the space between hearth and *ca-pam* the children find a resting place. Visitors may sleep on the left side of the house.

*

Differently as the Longhu- and Khumi-houses may appear, slight resemblances could also be found with the houses of the Rengmitca (Twainfa-Mouza) and even the Dömrong in Kamichora-Mouza. Here the prominence of the *kim-tom* is reduced by bringing it under one roof with the *kimma*. But the difference of the two parts can still be seen by that only the part of the *kimma* is closed from outside by a *tung-ping*, while the open-air platform (*car*), over which one may enter the house, is built on the side of the *kim-tom*.

C4 The field house

C4a) Anok

For building the field house (*bok*) only bamboo is used. The house always stands on stilts and basically has just one room and an open-air platform before it. It may be (and often is) enlarged by a covered platform built at the side of the walled-in room. This room has but one door, which always openes southward, so that in the following description I can take recourse to the four points of the compass. In the case of Menkröi's field house the open-air platform was in the South, the covered platform in the East. The room was broader in the EW (East-West) direction (5.50 m), smaller in the NS (North-South) direction (3.50 m); the covered platform added 1.40 m to the breath. The open-air platform measured 4 x 4 m. These measures, however, may vary with the surface of the earth, as will the heighth of the stilts bearing the platform. In the present case it varied fom 30 cm in the southwestern corner to 1.40 m in the northeastern corner. Generally, however, the open-air platform (*car*) shows the longest stilts (they may well reach 5 m down-hill).

The walled-in room (*kim*) always has the ridgepole (carried by two *klong-cüng*) in the broader extension, i.e., mostly in EW direction. The lengthsides show 5 *cüng-ca*, standing ca. 1.40 m apart and reaching down to the earth, which carry the *pang*. They are interpaced by 2–3 bamboos which help to carry the platform. To each of these *cüng-ca* are tied (right and left) two bamboo canes which run NS and form the lowest layer of the platform. These in turn are supported by oblique bamboo canes, one each East and West, while another oblique bamboo supports the *cüngca* from the South or the North. On this lower frame of the floor there rests (EW) a second layer of bamboo canes (5–10 cm apart) which carry a last layer (NS) of *chek-rek* (halved bamboo slats). These are held in position by interweaving them three times (two up and two down) with two long thongs. Finally the whole platform inside the *kim* is covered with 2 or 3 large bamboo mats.

The house walls at the lengthsides (1.80 m high) consist of longitudinally placed strips vertically interwoven with small strips which are tied (from the inside) to the *cüngca* and in each intermediate space to another bamboo pole (reaching down only to the first layer of the floor). The side walls show two additional vertical bamboos right and left of the *klong-cüng* and the last *cüngca* at the end of the lengthwalls. On one or two sides the two uppermost broad longitudinal strips may be replaced by two thin strips letting in more light and allowing the circulation of fresh air. The front and the back wall may be further strengthened by two *kim-heng*, resting on the last NS frame bamboo of the platform border and reaching up to the *pang*. Only on the northern wall the *kim-heng* are placed fully symmetrically, on the southern side, however, they are placed somewhat asymmetrically, so that they do not cross the door. On the side walls one *kim-heng* (resting in the West on the last EW bamboo of the platform, in the East on an EW bamboo in front of the southern wall), reaching up to the *pang* of the opposite house side, will do.

Over the ends of the *pang* (which rest on the *cüngca*) a longitudinal bamboo transverses (below the *kim-heng*) the side walls. On the eastern side this transversal bamboo serves as the upper rest of the (at a right angle) slanting roof of the covered open platform. The *pang* moreover serve as a support for three thick bamboo canes transversing the interior of the *kim* (so to say as *klöng*), which in turn carry in their middle a pole supporting the ridgepole in addition to the *klong-cüng* of the side walls. These supporting poles are fixed at their ends by tying a lashing around the bent back long thorns of the outer skin of the poles.

The roof resembles that of the *kimma*. The rafters (*krang*) (30 cm apart) are bent over the ridgepole, the *bai-kop* are placed over (not under) them,

20 cm apart and the thatch-grass is added. Follow the *kau-praa* (also 30 cm apart), with which the thatch is tied to the *bai-kop*. The *kau-praa* are kept in place by *mong-küa* (ca. 1 m apart, but one each near the ends and one in the middle may be regarded as sufficient), which in turn are secured by a *klong-nüm* on top and a *don-höng* over their lower ends (for the Mru terms see C3d). For fixing *klong-nüm* and *don-höng* it is considered sufficient to tie them (cross-wise) to some rafters below the thatch. In case the house is built while it is still raining and the roof becomes leaky, additional thatch may be added by just placing it on top of the leaky places. This means that the field house is not expected to withstand a real storm and to last more than a few months (until the cotton harvest in November). (On the harvest see chapter H.)

The door opening (80 cm wide) reaching up to the *pang* may be closed by a simple *büng* (shutter), which consists of vertically placed strips interwoven with very small strips (30–40 cm apart). The *büng* has no frame and can be just placed before the opening. The two sides of the door opening are framed by vertical bamboo poles. The 40 cm of the front wall to right (East) side are plaited like the *büng*, that is, the wall strips are standing upright and are interwoven by horizontal small strips, the ends of which are fixed in holes cut into the right door post.

The additional platform (1.50 m broad and long like the side wall of the house) with a pent roof of its own (slanting from 1.80 to 1.40 m) in front of the East wall of the house can be dispensed with, since it mainly serves as an afternoon rest place for those harvesting, but during this time, when the interior of the house is used for treading out and storing the harvested paddy, it may also be used to temporarily keep in a shady place things brought along from home.

Its floor is the continuation of that of the house, supported by additional vertical bamboo poles with a bent-back thorn by which four eastern NS bamboo canes of the floor are fixed. The EW canes of the floor carry a layer of broad bamboo strips (placed NS) which are fixed by 3 *chekrek* (halved bamboo slats, EW), one each at both ends and one in the middle, tied to the canes below. To carry the slanting roof at its East side a NS bamboo is tied to the upper ends of three (rather thin) bamboo poles, the lower ends of which do not reach the earth but rest on the easternmost of the NS frame of the floor. In order to support this construction two outer *kim-heng* (crossing each other in front of the thin middle pole) may be added. The slanting roof shows 5 *krang*, over which the thatch held by (NS) *kau-praa* is placed. The *kau-praa* in turn are fixed by 5 bamboo canes equally slanting EW and tied through the thatch to the *krang* below. Moreover the East end is secured by a NS cane on top tied to a halved bamboo below the *krang*.

The open-air platform, though somewhat smaller (4 m instead of 7 m), in principle continues that of the house and the roofed platform. It shows 5 EW canes supported by vertical bamboo poles with a bent-over "thorn" at their top. Over this lowest layer is placed the NS layer (5–10 cm apart), covered by an EW layer of halved bamboo (*chek-rek*), which are held in place by three NS *chaa* (see C3d), one in the midle and one each at both ends, carrying short bamboo strips (*wa-naa*), the ends of which are cramped below the canes of the NS layer. This *car*, which never has a wall of its own, on its South end is marked by three (one in the middle and one each at both ends) bamboo poles, 1 m high, which jut out obliquely towards the

South and are decorated with a tassel, serving as a warning that transgressing this borderline one would fall down. Access is provided by a log staircase leaned in the corner formed by the smaller open-air platform and the larger one of the pent-roofed part.

C4b) Rümma

The most remarkable difference of the *üa-kim* of the Rümma in Rengtan-KP to that of the Anok in Tapwüa-Kua is the use of bamboo canes instead of strips for the wall of the main room and the existence of a double annex. The main room measures 5 x 5 m, the inner part of the annex 2 x 5 m (the outer part 2 x 5.50 m), so that the total length of the house is 9 m. The length of the open-air platform is 7 m – that is, it does not extend in front of the second (outer) annex –, its breadth is 5 m.

The canes of the 5 m long length wall of the main room are placed horizontally and held by 5 vertical poles (including the corner poles) in the interior tied to 5 halved bamboo slats outside. (At its left side the canes are shorter by 80 cm leaving a door-opening.) Those of the side walls, however, are placed vertically and held by 3 horizontal canes inside tied to halved slats outside. This is valid, however, for the right side only, since the left-side wall exists only in part and is held only by two horizontal canes. The right side (seen from within) of this wall is open towards the annex building. The vertically placed poles of the two side walls reach up only to the heighth of the length walls (1.70 m), the remaining space up to the ridgepole is again filled with horizontal canes, the lowest of which rests on the *pang* (purlins) of the length walls (which in turn rests on the vertical posts in the interior). The upper horizontally placed canes are held in place in the middle of the side walls by a vertical pole inside (tied to a halved slat outside) reaching up to the ridgepole.

The part of the field house which I called "annex" has a wall plaited from horizontally placed bamboo strips. The annex has two rooms, the first (inner) part has a window (created by leaving open the upper part of the wall), to be closed by a shutter (made from strips like the wall), which can be moved sidewards to the front of the second (outer) annex. The first (inner) annex has (below the window) a hearth (*taping*) with a shelf (*krök*) of its own and a rafter (*kri*) in wall heighth, that is, the field house resembles a house in the hamlet insofar as the annex forms a small *kim-tom*, while the main room (used for treading out the paddy) takes the place of the *kimma*. The impression is strengthened by the fact that the roof of the "annex" is a little bit higher than the main part (walled-in by canes). The *üa-kim* is even provided with 3 *klöng* (beams crossing the room), one between the first and the second annex, the second between the inner annex and the main room, and a third one in the main room crossing it before its last third part. Each of the *klöng* is resting on the purlins (*pang*) and fixed to a central pole which also serves to carry the ridgepole.

The floor consists (as usual) of three layers, the first one of canes (50 cm apart) in length direction, the second one of canes (10 cm apart) at a right angle with the first layer, the third one (on the open-air platform and in the first [inner] annex) of narrowly spaced halved slats (*chek-rek*), again in length direction. The upper layer in the main-room, however, consists of bamboo strips.

The second (outer) annex is ca. 50 cm broader than the first one (the 50 cm of the corner have a wall with vertical strips). It has a floor of its own raised against the rest by 50 cm and covered with broad bamboo strips, interwoven (1 m apart) by small bamboo strips. It serves as a sleeping room. Except for the fact that the roof over the annex is a bit higher than that over the main room, the roof corresponds to that of the Anok *üa-kim*. This short description, however, may be sufficient to show that the *üa-kim* of the Rümma is more fit than that of the Anok to serve as a temporary dwelling house during the time when the paddy is harvested.

D Basketry and other utensils

D1 General remarks

There are baskets for seeds, others for harvesting; there are baskets for women, others for men; there are portable baskets and stationary ones; baskets for clothing or for chickens. There are very small and very large baskets. They range from delicate, lidded baskets in which women keep their jewellery to the simply woven baskets which serve to transport cotton to market – unless, of course, the cotton is used at home. The baskets used to transport cotton are crude, however, only because they go with the cotton. Baskets made for personal use are always of high quality. They are both stable and flexible and, therefore, capable of adapting themselves to a person's back even when heavy loads are being carried. A very poor Mru may weave mats and baskets to sell to Bengali peddlers, but Mru do not buy and sell basketware among themselves. Similarly, no trading takes place between the ethnic groups. Every man weaves his own baskets – weaving is a man's job; and the baskets a man weaves are peculiar to his own ethnic group. The Marma may well recognise that the Mru have better baskets than they, but they stick to their own forms even if these basically fulfil the same kinds of functions as among the Mru.

The Longhu Mru, as well as the Khumi and the Bawm, use an additional form of basket in which they tread out paddy with their feet. The other Mru groups use one of their many types of mats for this activity. The Bengalis have still another form of basket – a basket with handles which is used together with a carrying pole to transport goods. In the Hill Tracts a carrying pole is never used unless a large game animal must be carried by two people – or unless a pig must be transported during a wedding ceremony or a corpse to a place of cremation. If possible, hill people always carry their load in a basket on their back. The big baskets, however, are not suspended from a shoulder strap, but rather from a headband which is grasped by the hands between forehead and basket. This, of course, helps to lighten the load. The headband is fashioned from the inner bark of a tree (*Sterculia villosa*), or it may be woven out of fine bamboo thongs.

Animals are not used for transport. Horses and donkeys may have been known to the plains-people (the tribal languages have words for them), but in the plains, too, normal people went on foot – in the hills only the *rajas* kept elephants; so the only mode of overland transport left for the hill tribes (before modern streets were built and the car arrived) was man and his baskets. In baskets, harvests were (and still are) carried home from the swidden-field, and in baskets women fetch their daily water, which is poured into gourd bottles beforehand. Children learn to carry baskets at an early age, although when young girls must begin helping their mothers carry water, their necks hurt until their muscles have become strong. As grown-up women, they, like the men, are able to carry very heavy loads up the steep slopes.

D2 Carrying and standing baskets

D2a) *em*

The *em* is a small carrying basket mostly (but not exclusively) used by men. Of its many uses testify the many times it is mentioned in this book (cf., for instance, F1d, H8b, J4c, J6c6, J6d2, N1a, P2d, P2h, P2i, P3d, Q1a, R1g). The first basket which I will describe here was 27 cm high, its bottom measured 12 x 18 cm, its upper opening had a length of 29 cm and a breadth of 25 cm (total circumference 85 cm) and was provided with two small handles of 3 cm circumference below the brim at a distance of 23 cm, through which the carrying strip (1.70 m long and 1.5 cm broad) was passed.

The bottom was woven with 16 x 24 (0.4 cm broad) bamboo strips, crossing each other in a simple two down two up (*nüm pre hau pre*) pattern, bent at the end of the base and leading upward into the sides for about 15 cm. To strengthen the bottom two small wooden sticks (*akar*, like the bamboo strips 0.4 cm broad, but 21 cm long) were added diagonally in the interior of the *em*, crossing each other at the outer side between the 5th and the 6th length strip and the 7th and the 8th breadth strip. Also the sides were not too tightly woven, that is the length strips (*acöng*) were 0.4 cm broad and about equally distant from each other. A smaller but much longer thong (*apai*) was plaited through them (3 up, 1 down, 1 up, 1 down, 3 up, 1 down, 1 up, etc.), i.e., in a kind of special twill form, winding round and round at a distance of 1.5–2 mm. Since the long thong (leading from below left to up right) in its next round will repeat the sequence (3 up, 1 down, 1 up, 1 down, 3 up, etc.) as if shifted two lengths to the right, the outside of the basket shows a pattern of 3 ups climbing from below right to up left. Where the long thong comes to an end and has to be lengthened, it will be doubled over ca. 6 lengths of vertical strips.

Twill binding of *em*

The upper rim is formed by bending over ca. 3 cm of the vertical strips (at a right angle) to the left, overlaying these ends with strips of 0.4 cm breadth on the inside and 0.2 cm breadth on the outside, and wrapping round them 0.2 cm strips, which pass through the space between the vertical strips. Over the 0.2 cm strip on the outside another strip of 0.6 cm is placed and wrapped round similarly (at the left side of the inner wrapping). The wrapping strips on the inside of the basket are led vertically, on the outside diagonally, so that they return to the inside to the left of the next vertical strip. The ends (and beginnings) of the wrapping strips are bent over and led 10 cm downward between the vertical strips.

Also the bottom is provided with an additional rim consisting of two strips (of 0.4 cm) each on the outer side of the smaller sides and of the broader sides. The one end of those side strips is bent over and doubles the last bottom strip, the other end is bent over and leads ca. 15 cm upward. A further strip of 0.4 cm is lead around the whole bottom, lying at the corners over those of the longer sides but below those of the smaller sides. These additional strips are wrapped round 3 times (on one long side 5 times) with a 0.2 cm (rattan) strip, the ends of which are bent over and run upwards between the vertical strips, passing over or under 3–7 rounds of the long thong. This wrapping strip is (as at the upper rim) led vertically on the inside of the *em*, and diagonally on its outside.

The basket is carried with the help of a double string of bast of the *Sterculia* tree, knotted together after having passed the two handles and the bottom of the *em*. The handles are made of a double ring of 0.2 cm strips closely wrapped round with an equally broad strip, the ends of which are plaited horizontally into the wall of the basket.

A second *em*, on which I took notes, was somewhat smaller; its bottom measured 9 x 14 cm, its height was 20 cm, and the upper opening had a minimum breadth of 19 cm, and a maximum breadth of 21 cm. The bottom was plaited 2 down, 2 up (*nüm pre, hau pre*), and this form was continued for the first 3 cm of the wall, when it changed for the horizontal strips into 3 down 3 up, a form called *nüm krong* (low lifting), because the next row moves 2 verticals to the right, thereby producing a pattern climbing up to the right.

A third *em*, which I was able to measure, belonged to Menkröi. It was 37 cm high, its bottom measured 24 x 26 cm, the upper circumference was 102 cm, (minimum diameter 29 cm, maximum diameter 36 cm), that is, it was considerably larger than the first one, and the wall was plaited in *klikca* (see D2d).

D2b) *emca*

This basket is just a smaller edition of the *em* and often called like it, but can be distinguished from it not only by its size but also by its being woven so closely that no grain can drop out, since it is used mainly as a basket for sowing paddy (see J3c). Its bottom measures 6.5 x 11.5 cm, at its upper end it is 15 x 18 cm broad. The strips have an average breadth of 0.3 cm, but those of the long thong are only 1 mm broad. The bottom is plaited in 1 up, 1 down pattern (to be correct, I should say 2 up, 2 down, since always two strips are taken together). Also for the first four rounds of the wall the long thong continues the pattern 2 up, 2 down, but the remainder is plaited in the twill form described under D2a, with the difference that no space is left. Also, the bottom rim resembles that of a normal *em*, while the upper rim resembles that of a *klai-puk* (see D2c), but differs from it by an additional strip wrapped round the outer and the inner ring in every space left by the verticals bent left. To be sure this means that the *emca* becomes wider at its upper end, there appears a space of ca. 0.2 cm between the upper ends of the vertical strips (*acöng*), only partially filled up by the ends of the wrapping strips of the upper rim leading downward, but this does not produce holes of equal height, since this distance is kept at a minimum by the fact that the long thong (which acts as a horizontal, *apai*) is woven as closely as possible.

D2c) *klai-puk*

Though the *klai-puk* is not a carrying basket, I treat it next, since it is a basket of medium size, just a little larger than an *em*. I'll describe a rather small basket (there are bigger ones) made by Kyo Thwân Ong under the guidance of Mowai Catumma (28.09.1956). Its bottom measured 22 x 22 cm, its upper opening measured nearly 37 cm in diameter and 116 cm in circumference. Its height was 30 cm. What really distinguished it from an *em* was the breadth of its strips, which were 0.8 cm broad. The pattern in which it was woven is based on a simple 2 up, 2 down pattern, but made more complicated by starting in the middle with a pattern called *chak-ak* (parting heart) taking a middle vertical strip (*acöng*) passing over 3 horizontal strips (*apai*), its neighbours right and left, however, passing below the central horizontal strip only. The next vertical strips will then pass under the three horizontals, the next verticals right and left will pass over the central horizontal strip only, while the next vertical strips will repeat the role of the central vertical strip, and so on. The centre of the resulting pattern (which can be varied, as will be shown under D2f and g) can appear like this:

This pattern is continued in the wall, but with a difference: the vertical middle strip is now woven (seen from the inside) 4 up, 2 down, 2 up, 2 down. The strips to the left and the right of it are plaited uniformly 2 up, 2 down, but in such a manner that the 2 up to the left lie just in the middle of the preceding strip, those to the right at the beginning and the end of it, while the middle is followed by 2 down. Then follow once more 4 up, 2 down, 2 up, 2 down, so that to the right the 4 up have the 2 down of the preceding strip in their middle, while to the left it once more has the 2 up in its middle. As a result the 4 up at the inside (repeated every second strip) seem to climb up to the left.

The drawing to the right shows the wall seen from outside; the 4 up *cöng* have disappeared, their place is taken by 3 up *pai* (alternating with 1 down *pai*) every second round. The picture in the middle shows a *klai-puk* (belonging to Kangku Catumma) turned upside down (the right-left orientation is not changed by this), the *akar* is clearly visible (see also left picture of D2e). As can be seen the uniform pattern ascending to the left is obliquely crossed 3 times by a spiral pattern (*pia*) ascending to the right, which is produced by a reversal of the primary pattern. (The details will be shown in D2f).

The construction of the bottom corresponds (apart from the *chak-ak*-pattern) to that of an *em* (with sticks crossing each other in the middle of the bottom). But the rim, on which the *klai-puk* normally stands, is thicker than that of an *em*. It consists of 4 strips, the ends of which are bent upward in the corners, joining the wall, and one thicker strip leading 2 times round the four corners and tied together with the strips just mentioned by a small rattan strip, which is wrapped around them and repeatedly tied into double knots. These knots are made when the basket is lying on its side, the rattan strip is two times inserted and brought up again, then plucked through these two "loops", once more inserted and brought up again, then pulled tight, so that the "loops" are pulled together to a knot. The same procedure is repeated besides the first and all further knots until the whole strengthening of the rim is fastened.

Also the upper rim is done in approximately the same form as with an *em*: the first ascending strip (*cöng*) is bent over to the left, then the next strip (to the right of the first) is bent over it, and so on. On the outside a 0.3 cm broad strip is laid over these bent-over ends, on the inside a 1.2 cm broad strip is placed, and all are wrapped round by a (0.15–0.2 cm broad) rattan strip. Actually this is wrapped around (proceeding to the right) while the next *cöng* (vertical strip) is bent over to the left, so that its end can be fixed. When the circle of the rim comes to an end, the bent-over ends are tucked under those of the beginning, but for security's sake wrapped around a second time.

Another *klai-puk* I was able to study in detail was made by Menkröi. The bottom measured 25 x 25 cm, its height was 35 cm, the upper opening had a diameter of 38 cm, the upper rim was wrapped round twice and therefore was ca. 0.5 cm thick and the lower rim consisted of 7 double strips on each side leading up into the wall. These made the wall become rather straight after 13 cm, while at first it was bulging out by 4 cm. (In the corners themselves it bulged only 2 cm and up to 7 cm.) The vertical strips of the wall (*cöng*) were 0.7 cm broad, the transverse strips (*pai*), however, only 0.25 cm. The whole basket was very closely woven and as a result kept a square form until 10 cm below the upper rim.

D2d) *man*

There are small and big *man*. I measured 2 small ones and 2 big ones. The small ones had approximately the size of a *klai-puk* (bottom (1) 19 x 19 cm, (2) 21 x 21 cm, height (1) 50 cm, (2) 53 cm, diameter of upper opening (1) 53 cm, (2) 1957 cm), the big ones had bottom dimensions of (1) 20 x 20, (2) 32 x 32 cm, and were (1) 100, (2) 117 cm high, diameter of upper opening (1) 63 cm, (2) 74 cm. Since the walls are woven in the same way, I shall restrict my description to the smallest *man*. The pattern of the lower half is different from that of the upper half; the horizontal strips (*apai*) of the latter are plaited 1 down 2 up, moved one length strip to the left in the next row. (In the second small *man* the last 5 rounds were plaited 1 down 1 up.) The pattern of the lower half starts with a spiral (*pia*) of four strips on each side climbing up to the right (2 up 2 down). After 1¼ round this is changed to a pattern where the first and third horizontally ascending strips cross 3 vertical strips, the second and the fourth cross 2 verticals, but the first up starts above the preceding fourth, the second and third ups two verticals to the right, etc. The following diagram shows the two patterns of the lower half to the left and the simple pattern of the upper half to the right.

One may be inclined to think that these are two completely different patterns, but on a closer look this impression will disappear. Putting the right part over (instead of beside) the left one, a common element will appear: the 2 ups, in the lowest (left) part followed by 2 downs, in the second (upper left) part modified to 3 up 1 down every second round and ending (in the uniform right part) in 2 up 1 down for the upper half. – The upper rim is formed by bending the vertical strips to the left and wrapping them twice.

There exists also a special small kind of *man*, called *caa-ram-pok manca* (small *man* for taking paddy leaves) and mainly used for this special purpose only. Girls may also use it when they go to another hamlet to participate in a festival – the boys would use their *em* instead. The bottom measures 14 x 14 cm, the diameter of its upper opening is 38 cm (circumference 120 cm), its standing height is 34 cm (wall height 37 cm), up to 20 cm the wall is rather straight (circumference 70 cm). The bottom is woven like the *klai-puk*, the first 20 cm of the wall in 2 down 2 up (*nüm-pre hau-pre*) moving up to the right, the next 25 cm in *klikca* (2 down 1 up), moving up to the left (*cöng*-breadth 0.4 cm, *pai*-breadth 0.2 cm); the last 5 rounds of the *apai* (horizontal strips of only 0.1 cm breadth) are done in 1 down 1 up. The upper rim (*töm-wer*) consists of an outer 0.15 cm narrow rattan strip and an inner bamboo strip of 0.5 cm breadth, wrapped round with a similarly narrow rattan strip.

A basket woven similarly to a big *man*, but with a larger bottom and a wide upper opening, is called *pung-töm*. I saw it in a field hut. Its bottom measured 37 x 37 cm, the standing height was 90 cm, the upper opening had a diameter of 96 cm (circumference ca. 3 m). Its walls started to bulge at a height of 30–40 cm. The *cöng*-strips were 0.8 cm broad, the *pai*-strips 0.4 cm. The first 50 cm were plaited in *pia* pattern, the rest in *klikca* (1 up 2 down) rising to the left. The strengthening of the bottom rim seemed to be drawn in to the middle.

D2e) *yo* and *par*

I unite under this heading a basket (called *yo*) for carrying water-bottles (*tui-yia*) and another one called *par*, because in all songs the basket used for carrying water-bottles is called *par*. Let me start with the *yo*. Its bottom measures 20 x 20 cm (diameter 28 cm), the wall is 57 cm high, the diameter of the upper opening is 58 cm (circumference under the upper rim: 182 cm, at a height of 22 cm: 100 cm, at 30 cm: 112 cm, from there onwards widening nearly straight). The bottom is woven like in open *klikca* (1 down 1 up), but what can be counted as 1 actually consists of a pair of strips (each 0.7 cm broad) placed narrowly together, the distance to the next pair of strips being also 0.7 cm. These pairs of strips at the lower rim are bent upwards (they form the *acöng* of the wall) and (1 pair down 1 pair up) narrowly interwoven with (0.3 cm broad) double *pai* strips for the first 5 rounds, then the pairs are split up and 1 up 2 down interwoven with the pairs of *pai* strips, keeping a mutual distance of 0.5 cm, the "2 down" moving after each row 1 to the left. In the last 4 rounds below the upper rim, the *apai* (horizontal strips) once more are closely woven (1 down, 1 up), while the distance between the *acöng* (vertical strips) here reaches 1.2 cm. The 2.5 cm broad upper rim is called *par-wer* or *pan-kom-wer* (raft strip rim) in distinction to the normal basket rim (*töm-wer*). The *acöng* (vertical strips) pass beyond the last *pai* (horizontal strip) by roughly 1.5 cm before they are as usual bent to the left side, but then twisted around themselves and after the next three bent-over *acöng* led obliquely downward on the interior side, brought to the outside of the 4th *acöng* and back again behind the 5th, where (at a distance of 3 cm from the upper rim) they are bent once more obliquely upwards, after 7 cm reaching the rim again after crossing the next *acöng* to the left 2 up 2 down. At the rim they are again bent back to the interior side and plaited 2 up 2 down into the first plait, thus forming an interwoven double layer, for 2 cm lying over the 1 down 1 up interwoven part of the outside of the upper wall. In this way the whole texture is not really stable (with old baskets it may start to move somewhat), but very pliable, it can bulge from the weight of the water bottles carried in it. At last the bottom is stabilised by adding not only an *akar* in the middle but also at each side a thick framing bamboo strip bent up at the corners and protruding into the lower third of the wall and finally with another frame running around the whole rim of the bottom.

The *par* is a special basket used by women for catching small fish (see H8b). The bottom measures 14 x 14 cm, the upper opening has a diameter of 40 cm, its standing height is 39 cm. Neither the bottom nor the walls are closely woven, and the pattern of the bottom and the first 6 rounds of the wall is very simple: 2 down 2 up, *pre-nüm pre-hau,* crosswise. The rest of the wall is plaited in 2 down 1 up, *pre-nüm pa-hau*, forming (like with the *man*) steps rising to the left. The bottom is woven and framed as with the *em*, and strengthened with an *akar* (see D2a), the upper rim, however, shows the pattern special to the *yo.* (The fishing basket photographed by Brauns, see K1b, has a rim in the normal form.)

The *yo* (the picture to the left shows a father putting in the frame of the bottom) are carried by means of a carrying strap (*nam*) (as shown by the picture to the right). Both photographs were taken among the Dopreng Mru by C.-D. Brauns, who also made the drawing of a *nam*. This type of *nam* is also used to carry other large baskets as those used for harvesting paddy (*te*).

D2f) *tê*

This basket is used for harvesting and carrying home paddy (see J4e). The bottom measures 23 x 23 cm, the standing height is 64 cm. The upper rim has a diameter of 69 cm, the circumference is 216 cm. At a height of 12 cm the wall starts to become fully round, and at a height of 35 cm – where the basket starts to reach its maximum width – the circumference is 130 cm. Below the upper rim the vertical strips (*cöng*) reach a same distance of 1.3 cm. The *töm-wer* consists of two rattan strips, an inner one of 0.8 cm breadth, and an outer one of 0.3 cm, wrapped round the bent-over *cöng* (8 cm) by a further rattan strip of 0.4 cm.

Left: A father weaving a *tê* ; right: woman using it when collecting cotton.

While the upper part of the wall is woven 1 down 1 up, the lower part (see photo) is "embroidered" with a spiral pattern (*pia*) climbing up to the right, similar to that with the *klai-puk*.

The bottom shows the following pattern:

```
| – – | | – – | | – – | | | – – | | – – | | – –
| | – – | | – – | | – – | – – | | – – | | – – |
– | | – – | | – – | | – – – | | – – | | – – | |
– – | | – – | | – – | | – | | – – | | – – | | –
| – – | | – – | | – – | | | – – | | – – | | – –
| | – – | | – – | | – – | – – | | – – | | – – |
– | | – – | | – – | | – – – | | – – | | – – | |
– – | | – – | | – – | | – | | – – | | – – | | –
| – – | | – – | | – – | | | – – | | – – | | – –
| | – – | | – – | | – – | – – | | – – | | – – |
– | | – – | | – – | | – – – | | – – | | – – | |
– – | | – – | | – – | | – | | – – | | – – | | –
– | | – – | | – – | | – | | – – | | – – | | – –
| | – – | | – – | | – – – | | – – | | – – | | –
| – – | | – – | | – – | – – | | – – | | – – | |
– – | |   – | | – – | | | – – | | – – | | – – |
– | | – – | | – – | | – | | – – | | – – | | – –
| | – – | | – – | | – – – | | – – | | – – | | –
| – – | | – – | | – – | – – | | – – | | – – | |
– – | | – – | | – – | | | – – | | – – | | – – |
- | | – – | | – – | | – | | – – | | – – | | – –
| | – – | | – – | | – – – | | – – | | – – | | –
| – – | | – – | | – – | – – | | – – | | – – | |
– – | | – – | | – – | | | – – | | – – | | – – |
```

I shall call the fields marked by "|" u (= up) and those marked by "–" d (down). Though new horizontal strips will have to be plaited in for the wall, the pattern could well be continued, but it is changed in such a way that there is no "dividing line" in the middle from which the pattern rises left and right, instead a spiral (*pia*) rising to the right crosses a basic continuous pattern of 2u 2d. To show this I'll try to describe the path of the 2 *pai*-strips which are inserted at the same time in one corner and pass through the 24 *cöng*-strips of every side. I restrict myself to the full first 7 rounds only, the rest would be a repetition. The strip of the 1st round is continued in the 3rd round, that of the 2nd in the 4th, etc. (u = up, d = down, *italics* = forming the spiral (*pia*) ascending to the right, / = new side, underlined = continued on the next side). Starting in the upper left corner of the bottom scheme and going (1) to the right, then (2) down, then (3) back to the front side and finally (4) up again on the left side, the pattern on sides 3 and 4 will repeat that of sides 1 and 2. An exact continuation of the pattern initialised with the bottom would produce the following scheme:

7th : *2u 2d 2u* 2d 2u 2d 2u 2d 2u *1d 2u 2d 1u* / *2d 2u 2d* 2u 2d 2u 2d 2u 2d *1u 2d 2u 1d* /

6th : *1d 2u 2d* 2u 2d 2u 2d 2u 2d *3u 2d 2u* / *1u 2d 2u* 2d 2u 2d 2u 2d 2u *3d 2u 2d* /

5th : *2d 2u* 2d 2u 2d 2u 2d 2u *1d 2u 2d 2u 1d* / *2u 2d* 2u 2d 2u 2d 2u 2d *1u 2d 2u 2d 1u* /

4th : *1u 2d* 2u 2d 2u 2d 2u 2d *3u 2d 2u 2d* / *1d 2u* 2d 2u 2d 2u 2d 2u *3d 2u 2d 2u* /

3rd : *2u* 2d 2u 2d 2u 2d 2u *1d 2u 2d 2u 2d 1u* / *2d* 2u 2d 2u 2d 2u 2d *1u 2d 2u 2d 2u 1d* /

2nd : *1d* 2u 2d 2u 2d 2u 2d *3u 2d 2u 2d 2u* / *1u* 2d 2u 2d 2u 2d 2u *3d 2u 2d 2u 2d* /

1st : 2d 2u 2d 2u 2d 2u *1d 2u 2d 2u 2d 2u 1d* / 2u 2d 2u 2d 2u 2d *1u 2d 2u 2d 2u 2d 1u* /

This scheme, however, is not fully realised, since it would require an "irregular" insertion of 3d and 3u at each corner. Moreover the 3d in a corner might produce a small hole in the wall, in which an ear of paddy might remain stuck. Therefore the scheme is simplified to correspond to the following pattern (in the first rounds – before the wall becomes really round – "errors" may be made which retard the onset of the correct pattern). Due to this changed sequence the danger of having 3d in a corner is delayed till (at least) the 20th round when the wall has become nearly round.

1d 2u 2d 2u 2d 2u 2d *1u 2d 2u 2d 2u 2d* / *2u 3d* 2u 2d 2u 2d *3u 2d 2u 2d 2u* /

2u 2d 2u 2d 2u 2d *3u 2d 2u 2d 2u 1d* / *1d 2u 1d* 2u 2d 2u 2d *1u 2d 2u 2d 2u 2d 1u* /

1u 2d 2u 2d 2u 2d *1u 2d 2u 2d 2u 2d 2u* / *3d* 2u 2d 2u 2d *3u 2d 2u 2d 2u 2d* /

2d 2u 2d 2u 2d *3u 2d 2u 2d 2u 2d 1u* / *1u 1d* 2u 2d 2u 2d *1u 2d 2u 2d 2u 2d 1d* /

D2g) (*caa-liu*) *yöng* and *lung*

The *caa-liu yöng* is used to carry the seed to the future field. Its bottom measures 23 x 23 cm (diagonally 32.5 cm), the wall height (nearly equal to the standing height) is 65 cm, the upper rim has a diameter of 45 cm (circumference 142 cm). It becomes fully round at a height of 12 cm, at a height of 15 cm the circumference is 113 cm. The breadth of the bottom and vertical strips (*acöng*) is 0.8 cm, and that of the horizontal strips (*apai*) 0.6 cm. Though the centre seems to be different, the effect is the same as with the weaving of the bottom of a *te* (the graph shows the 12 central of a total of 24 strips, seen from outside) – a Khumi *klai-puk* shows the same pattern –, that of the wall and the upper rim resemble that of a Mru *klai-puk*. In distinction to a *te*, the wall is tightly woven all over (so that no paddy kernels may drop out during carrying the *yöng* to the field).

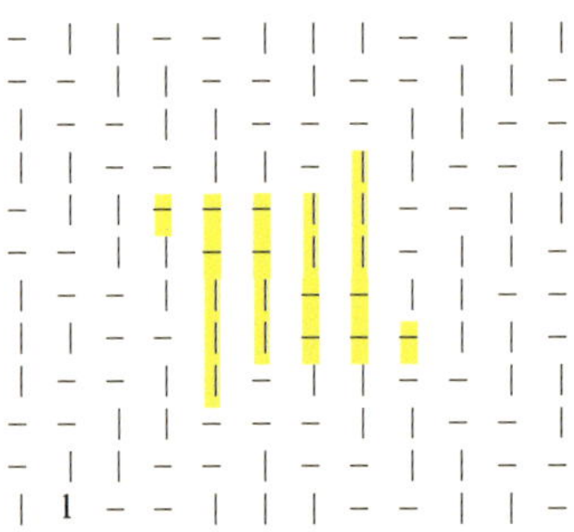

A *lung* is a standing basket used for keeping paddy or cotton seeds. It is formed like a *yöng*, but is bigger in size. The bottom measures 44 x 44 cm, it is 102 cm high and the diameter of the upper rim is 86 cm (circumference 270 cm). The bottom resembles that of a *klai-puk*, the breadth of the strips (*acöng*) is 1.2 cm. The *apai*-breadth of the wall is 0.8 cm. The wall is woven in a spiral (*pia*) like that of a *klai-puk* in *nüm-krong* (crossing three repeated every second round but moved two *cöng* to the right, followed in the next round by crossing one – in the middle over the preceding three) alternating with *nüm-pre hau-pre* (2 down 2 up).

The lower rim is also done as with a *klai-puk*, but stronger: wrapped five times. The upper rim shows a stripe of 1 cm on the inside and a rattan strip of 0.3 cm on the outside, wrapped around at first like a *klai-puk*, but then followed by a second wrapping with a rattan strip which comes to lie in the nicks left by the first wrapping. Finally the whole *lung* (if it is to be used for storing paddy seed) is coated with mud which, after drying, will seal all gaps.

D2h) *khong-töm*

This is a large standing basket, kept in the *kimma*, which serves mainly to store women's clothes. When a girl marries she will take a *khong-töm* with her to her husband (see M4e), whose side has to pay her father for it the traditional (relatively high) price of Rs. 10. It is the only one of the Mru baskets that has a special lid. The form can be seen in a drawing by C.-D. Brauns.

What cannot be seen from the drawing is that the basket has a double wall and bottom, that is, it consists of an outer and an inner basket, one placed into the other. The inner bottom measures 25 x 25 cm (the outer one with legs 30 x 30 cm), the standing height (without the upper lid) is 71 cm (of which 4 cm legs). When the cover is added the total height is 93 cm. The upper opening, after reaching a maximum diameter of 63 cm, narrows down (at an angle of 120 °) to 58 cm at the upper rim. The lid is placed over this narrowed part. The *acöng* have a breadth of 0.7 cm, the *apai* at first are also 0.7 cm broad, but starting from the middle narrow towards the top where they measure only 0.2 cm. The two baskets are not closely woven, the *acöng* below the upper rim have a common distance of 2 cm. The outer bottom shows an open (distance of one strip) crossed plaiting, the wall is woven *nüm-lök hau-lök* (1 down 1 up). The inner bottom is woven like that of a *klai-puk* (see D2c), the wall shows up to a height of 37 cm the spiral pattern (*pia*) of a *te*, it ends with one row which is plaited (seen from inside) 2 down 1 up and the next row with 2 down 3 up. Then follow rows with 2 up 1 down, and the last (slowly inward bending) part is plaited 1 down 1 up.

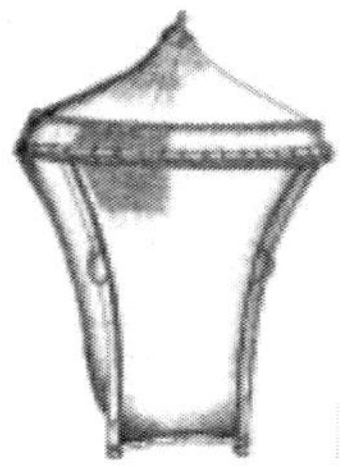

From the four corners (feet) of the bottom to the top run bamboo "linings" which strengthen the basket. These "linings" near the bottom are 3 cm thick, near the top 1 cm. The bamboo sticks used for them are cut from the lowest part of canes with a very thick wall, their lower parts are cut slightly square and form the feet of the basket. After 4 cm their inner half (seen from the basket) is cut away and the remainder narrowed down upwards. Beyond the outer rim it is so small that it can be bent inward again and finally fixed together with the upper rim. These "linings" are bored through every 5–8 cm to allow the insertion of a small rattan strip, which also passes through the double wall of the basket, is then led upward in the interior, passed back outside and through the next whole of the "lining", back inside again and so on. The ends of this rattan strip are knotted in the interior (shortly above the bottom and the upper rim). Also the bottom is strengthened by two strips of 1 cm, the first leading below the "linings", the second above them. Both together are fixed by a rattan strip which passes to the interior and back outside after every second *cöng* (vertical strip) of the outer basket. To two of these "linings" (at the same side) a rattan handle is fixed, 16–22 cm distant from the outer rim. Moreover a little rattan handle (of 3 cm) is fixed to each upper end of the "linings". They can be used to fix the lid, the lower rim of which is also provided with four little handles, by passing a strip or thread through the handles of both the basket and the cover. The outer rim is lined with one inner and one outer strip of 0.8 cm, the upper rim with an outer strip of 0.4 cm. Both are wrapped with a rattan strip, passing to and fro through the double wall and knotted at the interior side.

The lid has a pinnacle and a lower rim which fits in with the outer rim of the basket. It is woven in a double layer 1 down 1 up. Between the layers there is a layer of *ku-ram*-leaves which makes the cover rainproof. Above the rim the lid raises at an angle of 70 °, bends over after 7 cm and runs on with about 35 ° towards the middle, where it raises again forming a knob of 6 cm. The length of the plaiting (from rim to centre) is 39 cm, the height of the lid is 27 cm (the lower part of which covers the upper part of the basket wall). The number of *cöng* is halved twice in that every other one of them is led downward and broken off, the first reduction 14 cm and the second one 24 cm from the first bend. In the centre the strips run over each other, thereby forming the knob. The last strip running round on the interior side forms a ring of 4 cm diameter, the rest is woven through with threads. On the outer side a kind of cobweb of rattan strips (0.2 cm broad) is added above the knob in such a way that it ends in another handle with which the cover can be lifted. As already mentioned, above the lower rim the lid is provided with four little rattan handles by which it can be fastened to the basket.

As well as being used to keep women's clothes, the *khong-töm* may also be used to keep the *khong-rau*. These are little pebbles that are considered to be sacred stones and are never shown to any foreigner, since the guardian spirits of the family are supposed to reside in them. Whenever *köm-pot* (morsels of spirit-food, wrapped up in a little piece of banana leaf) are distributed in the house, one of them will also be placed on or in the *khong-töm*.

D2i) *mi-chau-tum*

This is a basket for washing (*chau*) husked paddy, that is rice before cooking (*mi*). Unlike all other baskets of that shape in this case we cannot make the distinction of *acöng* (vertical) and *apai* (horizontal) strips, since its strips are plaited (2 down 2 up) diagonally. The bottom measures 19 x 19 cm, the upper opening has a diameter of 24 cm and the height is 35 cm.

The general pattern of 2 down 2 up need not be valid for the centre of the bottom which may look like what we have seen with the *te*, with the difference that it is soon changed so that the rim is reached with 2 down 2 up, where the strips are bent upward in such a way that every two strips (whether down or up), which until this point have run parallel, now are crossed over, one leading upward 45 ° to the left, the other 45 ° to the right and can be plaited 2 down 2 up with the next strips, without any further strengthening of the bottom rim or the addition of any *akar* (see D2a). Two cm below the upper rim all strips which point left are cut off, those pointing to the right continue until they reach the (future) rim where they are bent outward and led backward over themselves, thus running obliquely back to the right, crossing (at a right angle) and interlacing those strips that have been cut off. After 9 cm they are bent back again and (running to the left) led diagonally upward for 4 cm, then bent back once more downward where they end behind the strips that lead upward. Thus the rim is plaited double for the last 5 cm, while for the next 4 cm downward it is even plaited fourfold so that a fairly stable rim is produced.

D2j) *caa-pam*

This basket is found in every house. It is a stationary basket, built directly behind the door opening in the *kimma* (into which foreigners are not allowed to look, see C3b), and in it the paddy for daily use is stored. There are two types: the Anok type and the Chüngma type. In 1957, however, the latter type had become predominant among the Anok as well; in the twin hamlet of Tapwüa-Kua the Anok type was still found in only one house in the Atwang-wüa hamlet, viz. the house of Elai Atwang. I detected it when visiting his house on 4.02.1957. The *kimma* door was open, I showed interest, and Elai was kind enough to allow me to step into the *kimma* and to have a closer look at it. (The restrictions mentioned are the reason that I do not have any notes on the *caa-pam* of the other Mru groups.)

The Anok type consists of broad vertical bamboo strips (with their originally outer side turned to the interior) interwoven (15 cm apart) with horizontal strips of 2 cm breadth, except for the lower rim and another strip in 1 cm height (*pam khen*). Here the horizontal strip (also with the bamboo bark to the interior) is double, the lower one plaited 3 up 1 down, the upper one the same way but 2 vertical strips further on, so that the vertical strips are crossed over alternately by 1 and by 2 horizontal strips. Over these double rings the 1.80 m high wall of the *pam* bulges somewhat, even though when the *pam* is filled the upper rim, too, is (and some places in between may be) surrounded with a double ring. In order to prevent the paddy dropping out through the slits in the wall, at least the upper part must be laid out with mats in the interior. In Elai's case, however, the mats had been removed and three of the vertical strips were broken off at a height of 1.30 m (while the others stood freely), and through this "hole" the paddy could be taken out easily.

With the Chüngma type the wall is made from horizontal strips of 6–8 cm breadth, interwoven (1 down 1 up) vertically with strips of equal breadth 10 cm apart, i.e., it is plaited similarly to the floor of a house. The bark side of the vertical strips faces the interior, and of the horizontal strips the exterior, with the exception of the middle strip (called *pam-khen*), which serves as a mark: when reached, it shows that the *pam* is only half filled. This type is easier to plait, but the wall shows even more slots than the Anok type. That is also why it needs to be lined with mats, while, on the other hand, the wall (normally also 1.80 m high) cannot be broken down easily, so that the paddy must be taken out from above, unless the upper part is dismantled. The diameter of the *pam* varies from 1.20–2 m, according to the total amount of paddy needed for one year.

D2k) *laa-par*

This is a type of simple basket, prepared in November/December, for carrying cotton for sale to the market. The only similarity with a (Dopreng) *par* used for carrying water-bottles (see D2e) is the weaving of the wall in simple *klikca* pattern (1 down 1 up). The bottom, measuring ca. 40 x 40 cm, is woven in rather open *klikca* (breadth of strips 0.8 cm) without a special framing of its rim. At the rim the strips are bent upward and in two corners one *pai* (horizontal strip) is inserted which is led upwards anticlockwise. When after 120 cm the upper rim is reached, the vertical strips are bent over to the right, passed over the last *pai* back outward and after the next *cöng* back again to the interior. This will do as the upper rim. Since the *apai* are woven rather tightly there is no danger that the cotton may drop out through the wall, but in case the bottom is woven too openly a small simple *klikca*-mat can be laid into it, and a similar mat may be used to cover the basket from above; but in Kröilöi-KP I saw a *laa-par* covered with a mat woven in *lüm-chi* pattern (for a drawing see Q1b). One basket can be filled with 20 seer, (around 1957 the price to be fetched for 1 md. of cotton would vary from 15–20 to 35 Rs.).

Weaving a basket for selling cotton (photo: C.-D. Brauns)

D3 Jewellery baskets

All of these baskets are special small baskets, in which girls and women keep their jewellery, glass beads, arm-rings, ear plugs and (if possible silver) bracelets. Despite the fact that these baskets are owned by females only, they are (as all basketry work) the products of men. One of the things a lover will make as a present for his beloved will be a *kruk* (without lid) or a *tum-pia* (with lid). He will use all his skill to make it look attractive, and correspondingly there is great variation, that is, the specimens described here (mostly belonging to the girls or wives in Menkröi's and Menching's house) are not typical but just examples.

D3a) *kruk*

(Owner: Thithop Catumma, later wife of Dingte Ngarua)

The basket measures 23 x 23 cm in breadth and is 3.5 cm high, the breadth of the bottom strips is 0.3 cm, and that of the *apai* strips of the wall 0.1 cm only. The woven pattern of the bottom was an open (with 0.4 cm distance) simple *klikca* (1 down 1 up), continued into the wall, though here the very small *apai* strips were narrowly woven for 2.5 cm, the remaining 1 cm was plaited like a *pan-kom-wer* (maybe meant to show that the manufacturer could also weave a *par*, a woman's basket for carrying water-bottles, see D2e). The bottom was strengthened by a crossed *akar* (see D2a) of 0.4 cm breadth, which passed the bottom 6 cm from the middle. The rim of the bottom was strengthened on all sides by 0.3 cm broad bamboo sticks fully wrapped with 0.3 cm broad strips and bound three times on every side onto the lower rim. Above all the *kruk* also had little legs (probably an allusion to a *khong-töm*, see D2h), 4 cm long sticks, sharpened at their top, plugged in under the 0.1 broad *apai* of the wall and jutting out 1 cm beyond the lower rim. The whole *kruk* was coloured red.

Another *kruk*, also made for Thithop, resembled a *mi-chau-tum* (rice-washing-basket, see D2h) insofar as its bottom was diagonally woven. The middle of the bottom was formed by a rhombus: 4 x 5 strips, the innermost strip of each quadrant crossing 5, each following strip one less, just the outermost one of the strips running at a right angle to them. The graph shows the rhombus turned 45 °

```
         –|
        ––||
       –––|||
      ––––||||
     –––––|||||
    |||||–––––
     ||||––––
      |||––-
       ||––
        |-
```

Thereafter the five strips are crossing 2 of those running at a right angle to them. At the corners of the rhombus, the *akar* are led from the outside to the inside. At the upper rim those pointing anticlockwise end on the inside 3 strips below those crossing them, which at the rim are bent over towards the outside, plaited back (anticlockwise) and tucked under the next crossing so that their ends cannot be seen. The breadth of the strips is 0.4 cm, the height of the basket 7 cm, diagonally the bottom measures 15 cm. The basket is not high enough for the wall to become round and so its upper rim keeps a square format.

Menching showed me a *kruk*, which also was woven diagonally. Its bottom was not square but measured 13 x 28 cm (that is, it was roughly twice as long as it was broad), its height was 8 cm. The *akar* consisted of 3 lengthwise (right, middle, left) sticks and two others crossing them (front and back), coming to the inside at 9 and 20 cm (length) and at 4 and 9 cm (breadth). Its upper rim was done as in *kruk* No. 2, but broader (3 instead of 2 backward). The middle of the bottom showed a uniform pattern: lb, 3p, 3b, 3p, 1b, 3p, 3b, 3p, etc. (b = the strips diagonally to the broad side, p = those diagonally to the long side) with the p-strips always moved one step to the right, see the following graph (p = |, b = _, turned 45 °):

```
– | | | – – – | | | – | | |   – – | | |
| | | – – – | | | – | | | – – – | | | –
| | – – – | | | – | | | – – – | | | – |
| – – – | | | – | | | – – – | | | – | |
– – – | | | – | | | – – – | | | – | | |
```

At the rim the sequence 3 3 3 is shortened to 3 2 1 (as indicated at the right end of the diagram). This *kruk* had a cover plaited in the same way.

D3b) *tum-pia*

This special jewellery basket belonged to Kaichüa, the wife of Netnöng Atwang (and daughter-in-law of Elai). It was shown to me by Menching, the elder brother's son of Netnöng. It was plaited from strips tinted blue(-green) and (lilac-)red. It had a bottom part (12 x 12 cm, 7 cm high) and a cover (12.5 x 12.5 x 8 cm). The bottom part resembled that of Thithop's kruk (with a rhombus in the middle of the bottom, continued with 3 [instead of 2] up, strip breadth 0.4 cm). The special feature was the pattern on the lid. Its top showed a rudimentary lüm-chi-pattern (see Q1b and graph), 14 x 14 strips broad and high, framed with 2 further red strips, not forming a square, however, but a rectangle (7.2 x 5.2 cm), since the strips were not plaited at a right angle. Then followed blue, red, blue, and after the rim, once more red, blue, red, blue and another rudimentary lüm-chi-pattern 16 strips broad and 8 high, equally in red, and finally, after another blue ring, a last red ring.At the rim the sequence 3 3 3 is shortened to 3 2 1 (as indicated at the right end of the diagram). This kruk had a cover plaited in the same way.

D4 Other basketry

D4a) *per-cing* (winnow)

This winnowing tray has the form of a shovel. It is mainly used for fanning the paddy after treading it out (see J4e), but it is also used at home for winnowing the husked paddy and as a tray on which things used in sacrifices are placed. The breadth of the lower end and the greatest length are 54 cm, the breadth at the beginning of the upper U-shaped part is 36 cm, the length up to its point is 48 cm. This upper U-shaped part has a rim rising to 9 cm, the circumference of the total U-shaped part measures 137 cm. The breadth of the plaiting strips is 0.7 cm, but that of those of the rim of the U is 1.5 cm and that of those at the lower end 1 cm..

The *per* is woven 2 down 2 up, for the first 22 cm in such a way that the pattern shows steps to the left, then follows one transverse strip which takes up all "downs" and traverses all "ups" of the preceding strip, whereupon the pattern continues in steps leading in steps up to the right. This pattern is continued in the middle to the upper rim. At the corners, however, where the upper U-vault starts, the transverse strips are not continued in full breadth, while the longitudinal strips on the sides for 9 cm are bent over towards the outside and plaited back inside over themselves, so as to form an isosceles right-angled triangle with a tip of 45 ° at the outside. Correspondingly the transverse strips are bent over at a right angle towards the upper rim and interwoven with themselves. The future tray consequently until now shows "indentations" at its upper corners up to 9 cm from the upper side and the length sides. These triangles are now drawn together and over each other, so that the upper triangles lie on those of the sides, until the upper and the lengthwise rims form one line and point up (into the third dimension). The U-shaped raised rim thus formed is strengthened at its outer and inner side by two strong (up to 3 cm broad) strips and overlaid on its border with a smaller strip, all wrapped by a rattan strip forming a knotted double loop every 5 cm. The lower (not rising) rim of the *per* is strengthened in the same way. The U-shaped (rising) rim is further strengthened by two thicker (but equally broad) transverse strips, 2 and 6 cm above the corners, interwoven in such a way that each strip crosses 6 up 2 down the longitudinal strips. The corners themselves are strengthened by 3 strips, one running immediately above the corner and one each starting from the long and the broad side, crossing each other and then leading (over or under the first strip) parallel upward, finally tucked in under the strengthening strip of the upper rim.

D4b) *per-lum*

This *per* is a nearly round (greatest diameter 63 cm, smallest diameter 57 cm) tray in daily use by the women for winnowing the husked paddy. (Especially among the Dopreng, but also among the Anok a *per-cing* may be used instead.) The pattern in which it is very narrowly plaited is called *beng*: The transverse strips are plaited 5 up and 2 down, the next strip repeats the same pattern, moved 4 longitudinal strips to the right (respectively 3 to the left). The rim is bent up and at its outside and inside strengthened with 2.5 cm broad strips, topped with a 0.5 broad strip, and wrapped with a rattan strip with a double loop every 4 cm. The rattan strip runs at the outer lower side, from where it is passed through the bamboo strips to the inside, up to the rim and down again on the outside, then passed inside again, up, over the rim and down again, knotted, passed inside, once more up and down, knotted and passed on to the next loop.

D4c) *bai-lum* ("round mat")

This is a round flat basin-like basket used for drying paddy on the *krök* (see C3e) over the fireplace and also for collecting the surplus paddy when measuring it with baskets (in the form of a *klai-puk*) holding exactly (when filled to the brim) 16 *seri* (10 kg). Its middle is woven like a rhombus (see the bottom of the *kruk* described under D3a), the remainder up to the rim in rather narrowly plaited *nüm-pre hau-pre* (2 down 2 up). The radius measures 37 cm. There is no lower rim but the strips turn slightly upwards until they reach a height of 14 cm. At the (upper) rim the strips that point clockwise are bent over downward to the outside, then tucked back inside (first over one, then below the next two strips leading upwards), then outwards again, first over two strips, then below the next two, finally broken off when they reappear. The strips that point anticlockwise are already broken off on the inside before they reach the rim, when they start to come up (after having passed 2 down) for the last time and their place is taken by the – formerly clockwise – strips now coming down again anticlockwise.

D4d) *yua* (rain cover)

Not everyone possesses a rain cover. Most people just put up with the rain or use an umbrella (if they have the money to buy one). A *yua* is shaped like a big, but rather flat *per-cing*. It is loosely woven (distance between strips 2.5 cm) and in duplicate, since it consists of two layers, the intermediate space is filled with *ku-ram*-leaves. For the sake of simplicity the measures given are valid for the outer layer which, however, fits exactly the lower (inner) layer. Up to the corners its length is 93 cm, its total length 108 cm, the breadth of its lower rim is 68 cm, the distance from the corner to the place of the maximum length is 31 cm, the whole U-shaped part measures 256 cm. This rim consists of a strong (0.8 cm broad) bamboo strip (which serves as the frame and which one might even call a stick if it could not be bent round). At the lower end of it is fixed a smaller (easily pliable) strip which then is led to the other end of the frame, twice wrapped around it so that it slightly climbs up the frame, then led back to the other end (from where it started), again wrapped twice around it and led back to the other end, and so on, to and fro, until it finally reaches the upper corner. Then vertical strips are interwoven (1 down, 1 up) with the transverse strips.

Of these vertical strips the second from the right (the directions "right" and "left" can be exchanged) is led in such a way that it reaches the upper U-shaped rim above the last transversal, is then wrapped twice around and downward (towards the corner of) the frame, led (3 cm above and parallel to the last transverse strip) towards the left side of the frame, slung twice around it and then led downward as the first vertical strip to the left. The third vertical strip from the right is treated accordingly, crossing the frame 3 cm above the first, and leading down again as the second strip to the left, and so on, until the right middle strip is led up, slung twice around the centre of the frame and led downward again as the left middle strip. The last strip to be drawn in is the first strip to the right, its upper end is not led transversally over the frame, but interwoven with the other vertical strips in a bow just below the upper rim of the frame. In such way also the upper part of the frame is filled, while at the lower end the strips hang free.

After the duplicate of the cover has been completed, the first is laid on the ground, filled with leaves and the second pressed in, a new frame is laid around them and all three bound together by strips. The lower ends for a length of 6 cm have no leaves between them; here four new transverse strips are narrowly plaited through both covers, fixing them firmly together. Finally a belt made from the bark of the *Sterculia villosa* tree is fixed ca. 15 cm below the upper corners to the sides of transverse strips of the inner frame, so that it can be carried with the belt on and over the head, the lower part of the shield hanging over the back.

D4e) *be* (cradle)

The Anok Mru use two types of cradle, one being the old Mru form, the other the Marma form. The first one is outdated, I only found it (on

30.09.1955) on a baby's grave. It had the form of a *yua*, but was only woven once (not in duplicate and not filled with leaves). It measured 82 x 67 cm. The rim consisted of 5 strong strips placed side by side and wrapped with rattan strips. To prevent the baby from falling out it was most probably necessary to fix it with a towel.

The Marma form has practically replaced the original Mru form. It has a bottom of 40 x 63 cm, with an obliquely raising rim of 12 cm. The strips are 0.8 cm broad and woven openly (common distance 1.5 cm) 1 down 1 up. The rim of the bottom is strengthened as with the *em* (see D2a). The simple *klikca* pattern of the bottom is continued with the wall, but the *acöng* are now plaited narrowly. The upper rim resembles the form described for the *par* (see D2e: *pan-kom-wer*), a double layer, but without open space. For hanging the cradle up near the ends of the two broad sides 2 x 2 handles are fixed round the four last *apai*. The handles have the form also used for the *em*, but are larger: distance over the rim 7 cm, full length 9 cm (of these above the rim 2 cm, wrapped with rattan strips). Through these handles a long rope is passed and knotted together. With this rope the cradle can be hung up on any hook or bamboo tube, at night in the *kimma*, during day in the *kim-tom*, whenever the baby is not (as usual) carried around in a sling by its mother, father or elder sister.

D4f) *pom*

This is an openly woven basket (see drawing by C.-D. Brauns), in the form of a *waa-pom* used for carrying fowls or piglets (usually before they are sacrificed on the fields), or carrying them into another hamlet as a present to a relative during a festival (for which they will be sacrificed as well). In the form of a *long-pan-pom* (woven more carefully and durably), seen with the Dopreng Mru, it is used in the kitchen for storing plates and bowls. A *waa-pom* is 30 high and 42 cm long. When starting weaving 6 (or 8) strips will be used: first 2 (or 4) strips are laid out, onto these 2 times 2 further strips crossing each preceding pair at an angle of 30 ° are laid and interwoven with each other. Additional obliquely running strips will be added and interwoven to form the side walls, the first 2 (or 4) strips may form the "bottom" (which without rim is continued into the wall), and one (or 3) of them raising vertically at both (smaller) sides. In any case only one is continued horizontally in the wall and at the upper rim led three times around, then one round downward, led around once more two times and finally tucked in under the rim, while all obliquely rising strips are here bent over the rim anticlockwise to the outside, led back at 60 °, tucked inside again and broken off. With an additional strip the two (longer) sides of the *pom* may be bound together, or a longer thong (fastened at the smaller sides) may be used to hang it over the shoulder (for carrying a fowl) or to the wall near the fireplace (when it is used for storing plates).

D4g) *waa-dör* (chicken-baskets)

These baskets are made for hens when they sit on their eggs. They are not absolutely necessary: the hens can also be allowed to brood in the place below the *kimma* reserved for the fowls (see chicken coop, C3b), but using *waa-dör* has the advantage that one can control the breeding and take out eggs without difficulty. There are two types in use:

1) real baskets with a square (approx. 30 x 30 cm) bottom, 20 cm high, plaited in a simple open *klikca* (1 down 1 up). At the upper rim the *acöng* (vertical) strips are bent over anticlockwise to the interior and plugged out again to the outside. These baskets are hung up or placed on the open platform (with the Longhu and Khumi on the *kim-pur-car*, see C3g), or even besides the entrance on the small ledge of the *kimma*.

2) baskets produced by splitting up the upper part (over a node) of a bamboo tube. The strips into which this part is split up are first bent (nearly horizontally) to the sides and interwoven with additional strips, then bent upward again and once more interwoven with a thong. The upper diameter of these "baskets" measures 25–30 cm. The lower end of the tube may be used to set the basket up; on the platform of the house of Phungkri Patlaica (Longhu Mru) several of these baskets were tucked in (slightly obliquely) at the house wall.

D4h) *ling-kö cua*

This is a special basket with an inset for sowing pepper (*ling-kö*) and letting it grow until it is planted out. The bottom of the outer basket measures 30 x 30 cm and is woven with 4.5 cm broad strips in an open (2 cm distance) simple *klikca* (1 down 1 up). Its height is 42 cm, the wall is plaited with a distance of 2 cm between the *acöng*, but with narrowly inserted *apai*. The inset is plaited narrowly and diagonally, those strips of the wall which (anticlockwise) point to the right 10 cm below the upper rim pass over a first thin (0.2 cm) strip leading horizontally around, then under the next but one strip of those running clockwise, crossing the next strip of these, and then leading over a second horizontal strip (1.5 cm below the upper rim) to be finally bent inside and interwoven with more horizontal strips in order to form a kind of cover with a hole (of 9 cm diameter) in the middle into which their ends jut out anticlockwise. Grasping it the inset can be taken out.

D4i) fish trap (*cong*)

I reproduce here a drawing by C.-D. Brauns. The plaiting of a fish trap

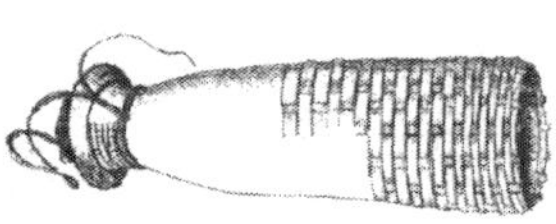

will be described under H8a. [A specimen was bought and brought back for the Linden-Museum in Stuttgart in 1965.]

D5 Khumi baskets

The following short notes were taken in Müilün-KP on 26.03.1957. They represent only a small part of the baskets used by the Khumi.

D5a) *klai-puk* (K: *teom*)

The bottom measured 22 x 22 cm, the height was 26 cm, the diameter of the upper opening 32 cm. The middle of the bottom was woven like this:

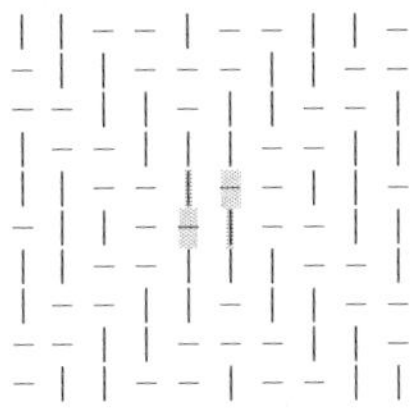

That is, exactly in the same way as described for the Mru *yöng* (D2g). Its rim is strengthened on each side by a thicker bamboo strip which is bent over and led up to the upper rim on the outside of the wall after having passed 3 x 2 horizontal strips of it, but there is no strengthening strip encircling the whole lower rim. The horizontal strips of the wall are inserted at two opposite corners and plaited 2 down 2 up, but the pattern of the bottom is continued by that in the middle the horizontal strips are plaited once 1 up, once 3 up, then once 1 down, once 3 down, followed by a reversal, so that each wall side shows a small rhomboid pattern, before it is continued with broader stripes as with the Mru *klai-puk*. At the upper rim the vertical strips are bent over towards the outside to the right, strengthened with an 8 mm broad counter-strip and wrapped with rattan, as with the normal Mru *töm-wer*.

D5b) *em* (K: *hai-tlem*)

Its height was 31 cm, the bottom was not square but rectangular (23 x 10 cm), woven (with 0.8 cm broad strips) 2 down 2 up, the pattern moving to the right:

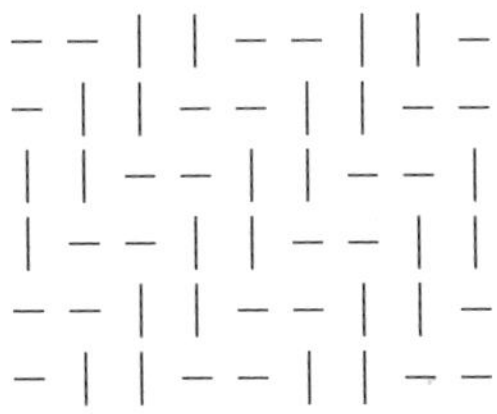

The *pai* are 0.2 cm broad, start in diagonally opposite corners and are led (first on the broader sides), 1 down 1 up, to the right. The bottom is strengthened by additional strips, overlapping in the corners and fastened by rattan strips, the ends of the strengthening strips are drawn up to the upper rim, on their way tugged 4 times under 2–5 *pai*. The upper opening is oval, greatest diameter 35 cm, smallest diameter 29 cm. The upper rim is done as with the *klai-puk*, i.e., the *acöng* are bent over to the outside and to the right, wrapped with a rattan strip which at the same time enwraps a 0.1 cm broad strip at the outside and a 1 cm broad strip at the inside. Finally the *em* is provided with two handles, fastened to two of the bamboo strips strengthening one of the broader sides of the bottom leading up (on the smaller sides) to the upper rim. The handles are made by wrapping two loops of a small strip with another rattan strip, starting from below right, leading it to the left behind the two loops, then back right over them, again back left downward behind it and the first round and finally pulling it up right to the front again, and so on, making one "knot" after the other. Through these two handles a longer string or band can be passed with which the *em* can be carried slung over the right shoulder and hanging down on the left backside.

[Rectangular *em* are also to be found among the Mru, I saw some in Cong'uk-KP.]

D5c) *yo* (here called *par*) (K: *co-wang*)

I shall describe a small specimen, there are others which have double the size. The bottom measured 21 x 21 cm, its standing height was 41 cm, its upper diameter 48 cm. The bottom was woven openly (distance approx. equal to strip breadth 0.8 cm) 1 down 1 up. The breadth of the *apai* (horizontal strips) was 0.3 cm. The lowest 12 cm of the wall were plaited like that of an *em*; the strengthening of the bottom was drawn up to the same height. The next 12.5 cm were woven more openly: each *cöng* (vertical strip) was split up in two parts (of 0.4 cm breadth), for 5 rounds interwoven with a double horizontal strip (together 0.6 cm broad), the first round near the end of the lower part after the splitting of the vertical strips, the next three rounds confirming the splitting by alternately passing over and under the split parts. The fifth round marks the beginning of the upper 18 cm, which start with one strip (0.2/0.3 cm broad) 1 down 1 up (alternating to the 5th round of double *pai*), the next 5 rows are plaited 2 down 1 up, in a pattern raising to the right, thereafter raising to the left, then follow again 5 rows 1 down 1 up, ending with two strips which are plaited front-back front-back over each other, thus forming a twisted ring. Lastly follows the upper rim plaited with the vertical strips like with an *em*, but enveloped in a second rattan strip, which by running in the opposite direction fills the rills left by the vertical strips turned over to the right.

D5d) *yöng* (K: *ding-khom*)

Its bottom is woven like that of a *klai-puk* (K: *teom*), 26 x 26 cm, its standing height: 53 cm. The wall has a straight lower part of 23 cm (with the strengthening of the bottom reaching to the same height) and an oblique upper part of 36 cm. The upper diameter measures 58 cm. Its lower part is plaited as with the Mru with a spiral pattern rising to the right, its upper part is woven 2 down 1 up raising to the left, the last part of it (15 rows), however, 1 down 1 up. The upper rim is woven like that of the *cô-wang* (see D5c). In the same way a Khumi *pung-töm* (K: *caleong ding-khom*) is woven, the difference is its size: it is roughly 1 m high, and the bottom measures 40 x 40 cm.

D5e) *ngun*

This is a special basket used by the Khumi and Longhu for treading out the paddy (see J4e). Its bottom measures roughly 1 m x 1 m, its walls are 60 cm high, all plaited 2 down 2 up. The lower and upper rims are strengthened by 2 cm broad strong strips which also reach up on the sides of the corners from the lower to the upper rim.

D6 Other utensils

D6a) Gourd bottles and pots

Among the Marma, women prefer to fetch (and preserve) water in the round earthen pots made by Bengalis. The great disadvantage of these pots is the fact that they break much more easily than the gourd bottles which, unlike the pots, can simply be stoppered with a piece of banana leaf. These gourds (*tui-yia*, *Lagenaria vulgaris* see J3d) have the form of a thick-bellied bottle, and are grown in the fields for no other purpose. Before the gourds can be used as water containers, however, one must allow the soft inside to rot out until only the hard shell remains. If the soft tissue inside the gourd is not completely removed, the water kept in it will soon taste foul. After each use the gourds must again be allowed to dry out as completely as possible. In order to dry them, they are placed upside down on a rack over the hearth; this means that in time the gourds turn black on the outside. If the fire gives off too much smoke, the water which is afterwards fetched in the gourds will also take on a smoky taste.

Men cooking meat under open-air in iron "wok"-pots

The big round clay pots will keep water cool; for this reason, the Mru occasionally store water in such pots after it has been brought into the house. In general, however, the Mru use clay pots for one purpose only: the brewing of beer (see K2b). These beer pots are called *yu-khong*. Since rice-beer plays an important role in all ceremonies and since the clay used in making the pots is not found in the hills – and neither the Mru nor any other ethnic group of the Hill Tracts know how to make pottery – these pots may well be among those items which the Mru have always been obliged to acquire from the plains people.

As well as the big clay pots the Bengali market also offers small clay pots of less than 20 cm diameter, which are called *ci-ö-ca* in Mru (the word probably being an old loan from Marma *jhe:-ö:* = "market-pot" + Mru suffix *ca* = "small") and needed for certain ceremonies, mostly to be filled with *hom-noi* (rice soaked in water) to replace beer (see, for instance, J6b-d; P2d, Q1g, Q1j, R4a, R4e, R8a).

No less common than clay pots are modern metal (mostly aluminium) pots, which the Mru purchase in the market. They have a round belly and a neck; with such shape they rest firmly on the stones and can be taken from the fire with bamboo tongs. For cooking bigger portions of meat, such as pork or beef (on an open fire, for which a channel can easily be dug in the open air), men use big iron pots in the form of a Chinese wok.

D6b) Iron, wood, and glass

Another long-imported commodity important to everyday life has just been mentioned: iron (*long-haa*, a word borrowed from Bangla *loha*, though the Mru formerly had their own "caste" of smiths, called *plamphai-kila*, who, however, may also have been imported Bengalis). Without hewing knives (choppers, *charai*) a Mru cannot cut and split bamboo; without bamboo he would have no house and no baskets. The hewing knife, which Mru men always carry with them as regularly as any of us would carry a purse, is at the same time their universal tool. Because of its size (with the bamboo handle, provided by the Mru, as long as the lower arm) it cannot be carried in a pocket, but must be put into a basket (*em*, see D2a); men mostly carry it in the belt of their loincloth. With the hewing knife the Mru cut everything, from the largest to the smallest object. For felling big trees, however, an axe (*m'rek*, word probably loaned from Marma) may also be used (when available). For clipping nails and cutting hair, scissors (*kaing-kep*, word also probably loaned from Marma) may be used, if available. If one does not have these additional instruments, the hewing knife must do.

(However, cutting hair is not a traditional custom and it is the pride of every Mru, whether female or male, to have long hair. The women wear their knot at the back of the head, the men in front over the forehead. In order to make this knot bigger, a young man may add a bunch of additional strands of hair, which are either cut off from his own hair or bought from a poor young man who would have his hair cut in order to sell it to make some money. If these strands are strong he might receive for them as much as 5 Rs. [the price of a very big fowl or a small pig]. Only some elderly modern men have started to cut their hair short in Bengali fashion.)

When the cutting edge of the hewing knife becomes worn down by repeated sharpening on a hone, the blade is given the form of a chisel. Fitted with a longer handle, it then serves as a dibbling, or digging, stick. Spear points are also made from iron, and there are arrows and spears made completely of iron. Except for the spear (*re*) with an iron point, which is occasionally used to kill a cow, these weapons are no longer used in everyday life. Pigs are normally killed by piercing them with a pointed bamboo (*chau*) – though an iron arrow or dart (*chia*) could be used –, dogs are clubbed with a piece of wood (for instance a pestle), but cattle are normally killed by piercing them with a spear between their ribs, though one may also kill them by cutting their throat with a hewing knife. The killing of a cow is a ritual act and is done only as a sacrifice (see chapter R) connected with celebrations – the very context in which the erstwhile weapons still play a strategic role in the ceremonial exchange of gifts. Formerly iron swords (longer than hewing knifes) were also in use, but they mostly survive in memory (and songs) only. If an old sword should still survive as an inherited item, it would have become rusty and disintegrated.

Two other imported iron tools are the hoe (*tim*, a proper Mru word), used for weeding, and the sickle (*tang-cen*, a Marma loan word, *ta~cañ'*), used for cutting paddy. Both are (like the hewing knife) provided by the Mru with a bamboo handle. The hoe is a rather short instrument and cannot be a "recent" import, since it plays a role of its own in certain ceremonies (see J4d, J6c5, J6d2, P3e).

Other iron instruments used for ceremonies are the *ner* (flat gongs, see E5c) and, sometimes, the cymbals (*chokca*). Both can be bought from the Bengalis, but the big buckled bronze gongs (*mong*) are rarely to be found in the market. These are said to be of Burmese (Arakanese) origin and are treated as precious heirlooms.

One of the visitors takes rest by using the *lungua*

In the 1950s it was rare to find tables, stools, or cupboards in normal households. The only large wooden furniture – if one wishes to call it that – is a tree-trunk "pillow" (*lu-ngua*), that is, a halved tree trunk which has been smoothed off on the rounded side and is used like a pillow. Also made of wood are the pig trough and the mortar (*chum*) and pestle (*li*) used daily in every household. For husking paddy all other tribes of the Hill Tracts, except the Mru and Khumi, by the 1950s had already started to use the Bengali husking machine, which is worked by treading on a wooden lever. Fortunately the big mortars can stand rough treatment, but if one should split, it is difficult to find a replacement, since hardwood trees of that size (½ m in diameter) have become extremely rare and hollowing them out with the hewing knife is hard work requiring much skill. The only wooden machine the Mru use is the cotton gin, which is normally produced by a Bengali carpenter, though some clever Mru have learned to make it themselves. It will be described in chapter F1c.

A large wooden stem is also required for making drums (*tömma*, see E5a), and even more skill is needed for making a proper Mru wooden comb (*charüt*). It would be quite easy to cut one out of a larger piece of wood with a saw, but though there is a word for saw (*chok-chak*), the Mru rarely have one. Therefore the young man who wants to craft a comb has to take the hewing knife in order first to cut a piece of wood in the form of a wedge with a bow like half a moon, and then to indent it to make the teeth as fine as possible without breaking any of them. When the young man has finally finished this work and if he is lucky enough to find a piece of tin foil he will wrap this round the frame after staining his comb with red colour. Under these conditions it is no wonder that the art of comb making soon fell into oblivion once plastic combs appeared in the market that were not only cheap, but also coloured. In memory of Dingte Ngarua who presented a *charüt* to me when I left his hamlet in 1957, I reproduce here a photograph of his ornaments (comb, earrings and glass pearl chain).

Dingte's ornaments

Another good imported from the market is glass. Every family has a few glass bottles, which are utilised for preserving *arak* (distilled spirit, see K2b). For drinking *arak* one uses water glasses, which, however, are rarely cleaned. (For drinking water a bottle gourd is used.) More commonly bought are yellow, red and green glass beads which the women, especially grown-up girls, need to make their colourful necklaces. They like other jewellery (formerly in silver, but nowadays mostly in aluminium) which will be described in chapter F2.

E Musical instruments

E1 Gourd-pipes (*plung*)

E1a) *plai-plung* (dance-pipes)

There exist several kinds of *plung*: *plai-plung* (instruments played by a whole ensemble), *rina-plung* (used as a solo instrument), *rek-rüa-plung* (played only when the bride is brought from her parents' house to her new home, see M4b). All *plung* consist of a hollow bottle-gourd (*plung-no*) into which two kinds of small bamboo canes are inserted: 1) one *pho-müa* (pipe to blow air into the gourd) and 2) two rows of *plung-long* (pipes by which tones can be produced). In order to be able to sound, these pipes are equipped with a vibrating reed (*plung-bür*) and a finger hole. Their lower ends are closed by a node.

In the foreground: boy with a very small *plung*

The *plai-plung* show the greatest variety in the size of their pipes; some of them have very short pipes (barely 20 cm long) and produce high tones, others have very long pipes (more than 6 m long) and produce low tones. In distinction to other *plung*, the *plai-plung* normally have a bourdon placed like a phial on top of the pipes, which serve to make the tones more resounding. For the long pipes these bourdons consist of two internodes of a very large bamboo species ("giant bamboo"), for smaller pipes part of such an internode will do. In order to keep the bourdons in place (and to prevent them from falling down), they are fixed from outside with the help of another thin bamboo cane tied to the pipe. If two or three pipes are used in one row, they may be tied together in such a way that the additional cane can be dispensed with. The largest *plung* when provided with bourdons may be so heavy that they, though having only three pipes (1 in the upper, 2 in the lower row), are best carried in a towel slung over the right shoulder of the player. The large bourdons are often decorated with drawings in red or green colour showing flowers or trees with drooping branches. Whatever the size of the pipes, the *plung* when played are always held grasped with both hands, the right hand holding the upper (and longer) pipes, the left hand the lower (and shorter) pipes.

Dance accompanied by 3 very long *plung* with bourdons painted red

The size of the gourd varies according to that of the pipes. For the small pipes the calabash may be 5 cm in diameter, for the long pipes it may be 20 cm in diameter. The bigger the calabash, the more air one has to blow in by mouth through the blowpipe. The length of the blowpipe varies from 30 to 40 cm, depending on the length of the elongated snout of the calabash, into which its lower part is inserted, and the convenience of the player. Small gourds are carried before the breast, larger gourds at stomach height.

The pipes are inserted in two horizontal rows into the calabash, the upper row sticking out at an angle of approximately 40 °, the lower row of 80 ° to the blowpipe. Each row of pipes is then tied together by a lashing.

So far I have only mentioned the gourds and the bamboo canes, but a *plung* could not be made with these alone. Another material is absolutely necessary: wax of the earth wasp (*kwai ut-naa*, see G9). Without it the canes could not be glued and fixed – air-tightly – into the gourds. This blackish-brown wax has the consistency of plasticine and remains pliable for years, but, especially when dancing with the large *plai-plung,* care has to be taken that the wax does not develop fissures: otherwise air will leak out and no tone can be produced anymore.

Boy trying to mend a fissure in the wax of his *plung*

That's the main reason why players prefer to carry the heavy *plung* in a towel. The player will have to leave the group and try to close the leakage by rubbing the wax with his finger over the fissure. The more wax is available, the less is the danger that it will develop fissures. Wax is also needed whenever the pipes contain more than one node at their lower end. To cut out the additional nodes, a small piece of the wall of the tube (0.4 x 2 cm will be sufficient) is removed and the rest of the node scraped out. The cut-out piece is reinserted and fixed in its original place by *kwai-ut-naa* wax.

Another two other materials are also necessary: *chang-ku* (an unidentified tree, which is also mentioned in a *tu-long*, see R3g) and *cim* (chalk). The reeds are made from little tiles of its wood (2.5 cm long and 0.7 cm broad), and their vibrating tongues (1.8 cm long and 0.3 cm broad) are smeared with chalk, produced by grinding the shell of the *chaling* mussel. This chalking serves to intensify the sound and prevents the tongue of the reed from sticking to its frame. This frame is stuck by *kwai-ut-naa* wax to the opening cut into the pipe shortly above its lower end. But a little lump of wax (3 mm long, 2 mm high and 0.6 mm broad) must also be stuck to the tongue of the reed itself: the tone may be tuned by moving it upward or downward. And the tuning is necessary whenever the length of the different pipes is not quite exact and especially when the tones of several *plung* have to be harmonised. To do so the young men of the hamlet will assemble every night until they are fully in tune with each other and can play together without producing a disharmonious concert.

The measurements of two small *plai-plung* with 4 pipes are summarised in the following table::

pitch	row	side	larger instrument	bourdon length	bourdon width	smaller instrument
(a)	upper	left	28 (+13) = 41	9.5	4	4 (+13) = 17
(g)	upper	right	33 (+13) = 46*	9.6	4.5	7 (+13) = 20*
(f)	lower	left	41 (+11) = 52*	9.5	3.5^{+}	15 (+11) = 26*
(d)	lower	right	52 (+11) = 63	19.5	6	13 (+11) =24

The values given in this table, are not meant to represent the absolute values (the Mru have no names for them), but only serve to indicate the relative distance. Moreover they are valid for the larger instrument only. Pipe lengths (in cm) were measured outside the gourd (first value), the value added for calculating the whole length was assumed (on the basis of the gourd diameter). The bourdons were bevelled at their outer side, the longer sides clamped together by the pipes and in addition fixed by some wax between them. Both pairs of pipes were tied together 5–10 cm below the bourdons by a lashing.

Since the upper end of the pipes would be closed by the bourdons put over them, they are opened here by cutting a hole into their wall. If the hole was made too big (so that the tone would be raised), it can be shortened again by partly closing it with wax, but the main tuning is always done by changing the size and the position of the little lump of wax stuck to the tip of the tongue of the reed. The larger the lump and the closer it is placed to the end of the tongue, the slower the vibration and the lower the tone. The tones of the resounding bourdons can be changed only by shortening them, if they are cut too short, however, they will not resound anymore.

The *plai-plung* are used only at feasts of merit. During the dances of the young people of both sexes the instruments are blown by the young men. All Mru groups (see R2) use it. The Khumi (see R5b) have taken over the custom from the Mru, but they do not use the very big and the very small *plai-plung* (Khumi term: *alum*). All *plung* are fabricated by young (unmarried) men, provided only that they have the necessary skill. Those lacking the skill are not excluded from blowing a *plung* during a dance, they may receive one from a friend. To blow one of the very large *plai-plung* does not require more skill than blowing a tiny one, but more breath is needed. That's why only strong men (they may even be married) can blow such an instrument.

On December 9, 2005, Céline Mouchet provided me with the following additional information: Among the Dopreng Mru the *plai-plung* are named according to the number of pipes they have:

name	size*	number of pipes in the	
		upper row	lower row
kak-ma	very big	1	1
plung dang	big	1	2
don-ma	"	2	2
dong tro rut	"	3	2
plung dang cöi	medium	1	2
ting tong plung	"	2	3
ting tong plung cöi	small	2	3
win ci cöi	"	1	2
don don cöi	"	2	2

* Size (estimates): "small" = below 40 cm, "medium" = 70–100 cm, "big" = more than 1.50 m, "very big" more than 2 m.

The smallest *plung* is said to set the rhythm to the bigger *plung*. There were not only unmarried young men playing, but also married middle-aged men. The players often take hours to adjust their *plung* with the *plung-charaa* before the ceremonies.

E1b) r*i-naa-plung*

Ri-naa is a loan-word from Marma ("dirty water") and means "cholera". Though this translation makes little sense with regard to this type of *plung*, nobody could give me a better translation. The *ri-naa-plung* is said to have been invented by the Mru in Arakan, around 1920 it reached the Matamuri region and in the following 30 years it spread among all Mru groups in the Chittagong Hill Tracts (MK and Khamlai Ngarua' 13.05.1956). This type of *plung* distinguishes itself from the *plai-plung* in that it has no bourdons, but 8, 9, or 10 pipes, and is used as a solo instrument only, not for accompanying the dance but for playing it when young men visit girls to court them.

For the rest it consists of exactly the same pieces and the same materials as the *plai-plung*. While for the latter I hesitated to take the pipes out (which would have been necessary in order to measure them exactly), since I would have to break the wax open (though every owner will have to do this, whenever he wants to tune his reeds), Dingte, the son of my main informant, freely allowed me to measure his *rinaa-plung* and even presented me with one for my own use when I left his country. The following table therefore will summarise my measurements of two instruments, a bigger one with 8 pipes and a smaller one with 10 pipes. The pitches, once more, do not represent the absolute values, but only serve to indicate the relative distance.

pitch	row	full length (in cm)		length up to finger hole	
		big	small	big	small
(a’)	lower		13		6.5
(e’)	lower	32	18.5	14.5	7.5
(c’)	upper		22		10
(a)	lower	44 (48)	26	17	9.5
(g)	lower	54	30	18.5*	10.5*
(e)	lower	65 (64)	37	19.5	12*
(d)	upper	72	38.5 (41)	23*	13*
(c)	upper	85 (81)	45 (44)	22.5	11
(A)	upper	102 (96)	53 (52)	23*	13*
(F)	upper		68		14
(E)	upper	128	(74)	24.5	

* = Finger hole on the side opposite to the reed, to be closed by the thumb.

The diameter of the tubes varies from 0.8 to 1 cm. Regarding the length, it would seem that by halving the length the higher octave can be reached, but the actual lengths do not exactly comply with this expectation. (In brackets I have added the length to be expected). Moreover, some of the pipes of the smaller instrument are incised at their upper end and in part closed again by wax. I assume that they formerly served as pipes of a *plai-plung*. Nevertheless, the fifths, fourths and octaves can be blown correctly, because they are tuned by manipulating the tiny lump of wax stuck to the interior of the tongue of the reed (see E1a). The columns containing the measurements from the bottom up to the finger hole have been added just to give an idea of the position of the finger holes, but they are nevertheless misleading in so far as only the longer pipes reach down to the bottom of the gourd, while the shortest pipes of each row do not, and since the bottom of the gourd is round even the remaining finger holes do not exactly have the position shown in the table. A better picture therefore may be given by mentioning the arrangement of the pipes.

The 2 x 4 pipes are not arranged according to the gamut, but in the order c E d A in the upper row and g e a e’ in the lower row. If 10 pipes are used the order is: A d F c c’ and g e a a’ e’ (or in the case of a third instrument with 9 pipes) c E d A and g G e a e’, that is each row will comprise at least one octave, the remainder consisting of fourths and fifths. With the arrangement of 8 pipes, the fourth A-d can always be blown by putting the thumb of the right hand over the finger holes. With the thumb of the left hand the upper tone of the next fourth (g) can be reached. In the case of the *rinaa-plung* with 10 pipes the thumb of the right hand will cover c and c’, that of the left hand e and g, that is, by pressing both thumbs the tonic can be blown. (I cannot exclude the possibility that this nicety was just built in because this *plung* was presented to me.) In the table the lowest tone is given once as E (8 pipes) and once as F (10 pipes). At first I assumed that my notation of F was a mistake until it came to my mind that it added just another fifth (F-c) to the gamut, the others being c-g, d-a, a-e’, A-e, but only F-c and a-e’ are available in the same row.

From the information furnished to me by Céline Mouchet (see above) I add the following: In the Anok area there is the centre of the new Kramma religion, introduced (together with a newly invented script) by the Mru prophet Menlè around 1983/84. The centre is called Malau-O-Kua (in Bangla: Pura-Para), and when Céline visited it in 2000, the village had not had a *plung* for a few years because it was too closely linked with the *cia-chot-poi* (cattle sacrifice, see chapter R), which had been abolished by Menlè. But in 2002 the *up* (the leading priests) from Malau-O-Kua, thinking they had perhaps made a mistake, decided to bring back the *plung*, since Menlè (who disappeared, most probably died, in Arakan, shortly after 1985) never gave any instruction regarding the *plung* and never said that it should be forbidden. (That's how the *up* justified to Céline their change of mind). Some middle-aged men of the village now used to assemble in one house in the evening in order to play the *plai-plung* for long hours, but without dancing, and the girls did not participate: that's probably why younger men (unless the *up* change the new rules and women are again allowed to dance) were not really interested in joining. However, in December 2000 Céline participated in a large Kramma meeting, where some boys also brought along their *rinaa-plung* and played them in the evening. Despite the ban, the *up* tolerated the playing as well as the flirting of the younger generation. The meeting was so big (maybe more than 1000 people joined) that it became more of a fair than a religious event, the *up* just lost control and some participants even drank spirits and smoked opium, although it was forbidden by the Kramma rules.

In 2002 this re-introduction seemed still limited to Malau-O-Kua. In the other Kramma villages visited by Céline it was almost impossible to find any *plung*, and apparently the first prohibition of the *up* was still in force. Also in hamlets converted to Christianity *rinaa-plung* were seldom played. The missionaries (Bawm and other Christian Mru) discourage the newly converted from playing *plung* or *rinaa-plung*. So it is difficult to find it in the converted villages.

On a few occasions Céline was also told that the Dopreng women could play *rinaa-plung*, but she never saw it. The Dopreng are known to be very good singers, better than the other groups. In many buddho (new term for Mru who still stick to their old creed) villages Céline had the opportunity to hear some *rinaa-plung* and songs, played by men of every age. The people often like to drink at that time, and the player appreciates having some glasses of *arak* (distilled rice-beer) in exchange for his performance – except for the younger men who do not dare to drink much in the presence of the elders. Most of the songs she heard were love songs, sometimes just improvised at the time of playing.

E1c) Other types of *plung*

The *rek-rüa-plung* looks like a *rinaa-plung*, but has only 6 pipes. It is used only at a big marriage feast, which because of its expensiveness has become rare. I had the chance of observing the first two days of such a feast on the 19th to 21st of April 1957 in Rengnok-KP, Horinjhiri-Mouza (see M4b). The *rek-rüa-plung* was made by a young man of the bride's sib and blown on the whole way accompanying the bride on her journey from her own hamlet to the bridegroom's hamlet. Only one instrument was used, though I was told in advance that there should be two, and it had 7 instead of the predicted 6 pipes. Due to the extreme rareness of such a feast not all rules were correctly remembered, and it may be doubted whether a genuine *rek-rüa-plung* was made or whether it was just replaced by a normal *rinaa-plung*.

Nevertheless I give here my notations: upper row (longer pipes) f a d, lower row (shorter pipes) f' c' a d'. The finger holes of the underlined pipes are to be closed by the thumb. Due to the fact that an "a"-pipe appears in both rows, 6 pipes would have been sufficient indeed. The gamut is made up of three minor thirds and two fourths, while by pressing the thumbs two fifths (d-a and c'-f') will be blown. A newly introduced type of *plung* is the *nom-um-plung*. A *nom* is a bottle-gourd with cut-off neck. This new type of *plung* is said to have been invented in Arakan, but in the meantime it was also used in some places in the Southern Chittagong Hill Tracts. In Mangpung-KP (Kamichora-Mouza) I was told that it could be found among the Mru in the Dochori-Mouza and that they called it "Dömrong-long", while the *rinaa-plung* here was called *tingtong-plung*. A group of players with the new *nom-um-plung* was expected to participate in a feast of merit in Cöngpat-KP (Alikodong-Mouza), but did not turn up. Thus, I had no chance to see such a *plung* myself, but I was told that it is a *plai-plung* in which the bourdons are not made of giant bamboo, but are replaced by *nom* for *plung* with 2 to 4 or with 6 and more pipes.

rek-rüa-plung played (on the left side) while leading the bride away from home

Another type of *plung* is the *le-wö-plung*, so called because the "bottom" (*le*) of the gourd is "opened" (*wö*), that is, pierced by the pipes. The pipes stick out approximately 20 cm on the upper side and 2 cm on the lower side. With the Anok this instrument is used when going to visit girls in a neighbouring hamlet and can be blown also in the swidden, but it is rare and was not in use in Tapwüa-Kua, where no one even knew how to fabricate it. I came to know about it with the Rümma (in the hamlet of Karbari Nongneng Krimchang, 12.03.1957), where it is used only in the two months when a *khang* is to be observed. At other times it is strictly forbidden to blow it, and its openings are plugged up. If anyone should violate this rule he would be bitten by the tiger, and the karbari even hesitated to show me a *le-wö-plung* and had to be persuaded by my two Anok companions, who found this restriction ridiculous. But when Khamcöng showed his intention to try to blow it, the instrument was immediately withdrawn, and I could not even take a closer look at it.

A note on another type of *plung* should be added here: the *tang-plung*. I never saw real instruments, only imitations (see P5b), a set of four, which could be blown into, and looked like the *tu* (see E1d), in that they had one pipe only, but this pipe did not traverse the gourd. They are to be played in triple sets (together with the *tu*) at the big *reng-tang* cattle sacrifice (see especially chapter R6j).

Three *tang-plung* imitations

A *plung* of which I got hardly any more than the name is the *plung kok-ma*. It is said to have two long pipes only and a short tube. I now assume that it corresponds to the very big *plai-plung* mentioned by Céline Mouchet (see above) for the Dopreng Mru under the name *kak-ma*.

From Céline's notes I add the following remark: "I have noticed that more and more villages are now gradually stopping keeping a special place in their fields to plant some gourds which will be used to make water-bottles or *plung* (*rinaa plung* and *tu*) [see E1d]. Indeed these villagers find it more practical (though unfortunately it is much less hygienic) to store their water in plastic bottles (recycled soda bottles) or aluminium jars bought in the bazaar. I saw this mainly in some places close to the road or the bazaar. People living far away still have enough space to plant these. We may wonder then if the *plung* will still exist much longer, considering as well that the younger Mru sometimes prefer to listen to modern Indian or Bengali music on their transistors. But anyway there should still be some *plung* as long as there are some Buddho Mru, willing to perpetuate their traditions (and having enough land to plant the gourds!)."

E1d) *tu*

Another kind of gourd-pipes are the *tu*. They also have a gourd (*tu no*), but in distinction to the *plung* only one pipe traversing the gourd (see R4a). They also should have a *bür,* a reed, so that with a set of three *tu*, varying sequences of three tones could be blown – but most instruments called *tu* cannot really be used for this purpose, they just look like a *tu*, are needed in different offerings (as for instance a *bong-kom* ceremony, see M4 and P2e), for which, however, one *tu* will do, is never blown and therefore need not even have a reed. After all, it would seem that the *tu* are outdated instruments, reduced to a mere ceremonial function.

Blowing three *tu* and beating the *taröng*-instruments on the occasion of an ear-piercing ceremony (shortly after day-break)

But that's not quite true, since there are ceremonies in which three *tu* have to be used and blown, and there are many *tu-long* (tu-verses), known by special men, called *tu-charaa* (*tu* masters), who should be invited for directing these ceremonies. The *tu* are to be blown at death ceremonies (for the accompanying 60 *tu-long* see Q1d) and on the occasion of big feasts of merit (see R3g with 18 *tu-long*). On the one hand the *tu* are said (see R3f) to represent the *plung* of Cöng-cöi (see P2c), an evil spirit who when he takes possession of a human being will drive him mad and can only be placated by offering him a head of cattle. On the other hand the use of the *tu* in death ceremonies is said to have been introduced by the sib of Tamtuca, who also invented the accompanying death rites (the *tamtuca ri* and its 60 *tu-long*) – and perhaps even the story of how death came to be the fate of man. These two versions do not necessarily contradict each other, though Cöng-cöi does not play any role in the story of how death came to be introduced in the human world due to the imitation of death rites as formerly performed in the heavenly country (see Q1e).

Furthermore the two versions do not tell us anything about the age of the *tu*, though in the case of the use for death ceremonies it is said that the rules were introduced only "a short while ago". At first I thought that the use of *tu* might be restricted to the Anok, since they are not used by Chüngma Mru and I had not seen them with the other Mru groups, but there is a report by Céline Mouchet (2001:64–65), who saw the *tu* used in a small *cia-chot-poi* (cattle sacrifice), which she observed in a Dopreng hamlet (Khailu-KP = Dopru-Para, Thwain-Mouza). These *tu* were decorated with *tür-ram* (leaves of the fish-tail palm), "rubbed with a mixture of rice, ginger, water and chicken", blown "a bit" and deposited near the beer pots, but never carried around in the dance (cf. R3f). With the Khumi, on the other hand, I heard some kind of *tu* blown (however, I did not see the instruments) during the opening ceremonies of a cow festival (Tindu-Para, 8.03.1956).

The Tamtuca (see L5b) are living in the Horinjhiri-Mouza and are regarded as belonging to the Prencü, which are to be found in the mouzas of Takerpanchori, Galengya, and Lemupalong (see L2b). Though I visited a hamlet in Lemupalong-Mouza on the occasion of a marriage, where by chance I met a Tamtuca karbari, who, however, unfortunately was rather intoxicated, so I had to forego the chance to ask him about his version of the invention of the *tu*.

E2 Flute (*prui*)

While the *plung* (gourd-pipes) are played by males only, the *prui* (bamboo flutes) are said to be played by females only. However, I never heard or saw a Mru girl playing one – maybe they were just too shy to play it in my presence. All I heard is their assertion that they would play when they had an instrument, but at the same time they maintained that they had none.

Therefore I can only give a very brief description of a flute that I saw (lying around unguarded) during a short rest in the karbari's house of the Marma hamlet of Zainragro (Poli-Mouza, 15.03.1956). The owner of the house not being present, I could not ask him to whom it belonged. It had 6 finger holes and to play the basic tone (lets call it c) one had to keep them all closed. By opening the 1st hole, the next full tone (d) could be played, opening the 2nd hole a half-tone (d sharp) followed, the 3rd hole again a half-tone (e), the 4th a full tone (f sharp), the 5th a half-tone (g), and the 6th a major third (c), that is, the fifth (c-g) was divided up in 6 tones. As mentioned, I do not know whether the Mru flutes are tuned in a similar way. In view of the Mru sequence of 2.5, 1.5, 1, 1, 1.5, 1, 3.5 used for the *rinaa-plung* with 8 pipes (so that a fifth never comprises more than 4 tones), however, it would seem improbable.

E3 *taro* (two-stringed fiddle)

This instrument consists of two main parts: the instrument proper (*taro*) and, the bow (*küng-müa*). It is rarely to be found and most probably was taken over from the Marma (though I could not trace the word in Judson's Burmese-English dictionary). For manufacturing a *taro* several materials are used: wood (for the main rod and the pegs), iron wire (for the strings), coconut shell (for the sounding-bowl), goat's leather (for the skin of the latter), bamboo (for the lashing of the latter and especially for the bow and its string), wax (applied to the bow string). On the instrument I measured, the main rod was 81 cm long (of which 5 cm were below the sounding-bowl). The sounding-bowl was 10 cm long, 11 cm broad and 6 cm high, its wall was 2–3 mm thick. The distance between the sounding-bowl and the bridge was 36 cm, the distance between the bridge and the first peg was 4 cm, the second peg was inserted 4 cm higher, from the second peg to the upper end of the main rod remained 22 cm.

The main rod at its lowest end had a diameter of 1 cm only, below the sounding bowl its diameter was 2.5 cm, above it was flattened and 2.3 cm broad dwindling to 2 cm at the bridge. Its upper end (above the pegs) was bent somewhat to the back, was four-sided and became smaller. The lower peg was 9 cm long (with a thickened head of 2.5 cm), the upper peg 8 cm (head 1.5 cm). The "needles" had a diameter of 3 mm at their one end, 5 mm at the other (head) end. The strings were fixed on the lower peg at the right side, on the upper peg at the left side (seen from the front). The distance between the strings measured 1.3 cm.

For the bridge the main rod was carved in by 0.5 mm, to receive a small piece of bamboo (1 mm thick). This was fixed by the strings, which ran above it at the sides of the main rod. The strings were tied immediately above the bridge to the main rod by a slightly twisted string of two thin bamboo thongs.

The sounding-bowl was pierced by the main rod approximately 7 mm below its rim. Just after leaving the bowl the main rod was thickened, thus preventing the bowl from sliding down. This was further prevented by the strings, led over the bowl and tied immediately below it to the main rod. The lower side of the bowl had one larger sound hole (2.3 cm diameter) and its sides had ten smaller sound holes (0.7 cm diameter). The skin was tied to the bowl by small (1.5–2 mm broad) bamboo lashings, leading through 8 + 9 (right and left) holes bored through the skin and kept in place by a ring of 3 bamboo (slightly twisted) thongs of similar quality. Another ring of 4 thongs was tied round the bowl at a distance of 6 cm from the upper side, and between the two rings a thong running up and down was used to tighten the skin (similar to the tightening of a drum-head, see E4). Since one thong only would not have been long enough, it was actually made up of three thongs tied together by simple knots, while the free ends were just tucked under.

The strings at the brim of the bowl rested immediately upon (and slightly cut into) the skin, but in its middle two small pieces of wood were placed under the strings. They had the form of a hipped roof (1 cm long, on their upper side 4 cm broad and 5 mm thick, on their lower side 2 broad and 1 mm thick), with the broad side of the "roof" resting on the skin. When, however, the skin was not fully tightened, they would drop out rather easily.

The bow was 58 cm long. Where it was to be taken into the hand it was 1 cm broad, 3 mm thick, at the other end it was 3 mm broad and 2 mm thick. It was bent mainly in the last 12 cm. The bow string was 50 cm long, 3 cm broad and 3 mm thick. The largest distance of the string from the bow was 4.5 cm, approximately 9 cm from its upper end. It was (partially) coated with wax, tied to the bow in a notch 5 cm from the lower end and (without a notch) 0.5 cm from the upper end.

The strings of the *taro* were tuned at a fifth (a-e') by the Mru when they played it together with a *rinaa-plung*. After being played for a recording I made on May 10, 1956, the instrument was left in my house, so that I could measure it. Before it was taken back by the owner, my Marma boy Kyo Thwân Ong tried to play it and tuned it in his own way, lowering the fifth to a fourth.

E4 *ting-teng* (bamboo zither)

Like the *taro* the *ting-teng* normally is used only as a solo instrument and played by young men (who also make it) when they go to court girls in the evening. It is possible for *rinaa-plung*, *taro* and *ting-teng* to play together, but it is rarely done, probably because of the different acoustic qualities of these different instruments. The *ting-teng* would be difficult to discern while the other instruments are played, though it is not small (the instrument measured by me was 45 cm long), but its diameter is only 4 cm, since it consists of one bamboo internode only.

The two nodes are kept, but one is bored through and therefore has a small hole. The length of the internode, which serves as resonating body, is 42 cm between the nodes. In the middle of its upper side a larger sound hole (4 cm long and 1.8 cm broad) is cut out. At first 8 mm and then 6 mm to the side of this hole the outer skin of the internode is lifted, resulting in 4 strips on the upper side, to which are added 3 strips on the lower side. These strips, though running over the whole length of the internode, are secured against tearing off and coming loose by being tied down at the ends by a bamboo thong (3 mm broad) wound four times round the instrument, leaving a of 34 cm long remainder on which the middle strips (2 on the upper and 1 on the lower side) are underlaid (and lifted) at both ends by small bridges of wood. The strips are less than 1 mm broad and serve as strings, of which, however, only those running over bridges emit a full sound. The remaining 4 outer strips when clicked with the finger emit only a buzzing sound. (There are, however, also instruments on which all 7 strings can be tuned.) The tunes of the three-stringed instrument were recorded at the end of tape 24, which has since been lost.

E5 Drums and other percussion instruments

E5a) Mru drum (*tömma*)

The drum is a very common instrument, to be found in roughly the same form with all Mru groups, with the Khumi, the Bawm, and the Marma. The Mru beat it on the occasion of several ceremonies, for instance *cari-yöng* (see P5b), on the death of a person (see Q1c), and especially at all *cia-chot-poi* (cattle sacrifices, see R1h). In the latter case the drum may (but need not) be beaten during all dances; it is indispensible, however, for the *taröng*, the circumambulations of the cattle to be sacrificed (necessary also in case of a small festival without a dance of the young people). For the *taröng* not only the drum is beaten, but a total of 5 to 6 players is necessary: 3 beating flat gongs (*ner*), 1 a cymbal (*chokca*) and 1 beating, whenever available, a big buckled gong (*mong*). The drum is normally carried by a strap fixed on both ends of the instrument, hung over the shoulder and beaten with the flat of the hands. For performing the death ceremonies in the form usual with the Anok Mru even 9 players should be present: 3 for beating the flat gongs (*ner*), 3 for blowing the *tu*, and 3 for beating 2 (a bigger and a smaller one) drums (*tömma*). (On the occasion of the death ceremonies, which I observed in Chamklaa-KP, the two drums were hung up besides the coffin below the roof and were not beaten by the hand as usual, but by simple sticks.) Only for a *cari-yöng*, when one pig is to be sacrificed, a single drum will do.

In ceremonies performed for the swidden cycle, however, no drum is used. But this may be true only for the Anok, since with the Rümma a drum and three *ner* were beaten while the pig was killed (see J6c2).

Beating a drum (*tömma*) and three flat gongs (*ner*) for the *cari-yöng* in front of my finished house in Tapwüa-Kua

Despite the rather common use of drums, I never saw one made. My description therefore will be restricted to its outer form. The instrument is approximately 50 cm long, slightly convex to rather straight, with a diameter of 25–30 cm, and covered by a cow hide at both ends (which rarely have the same diameter). The hides are drawn over the rim of the drum body and here fixed, 2–3 cm under the rim, by a bamboo ring, the remaining margin of the hide being tucked away under the ring. The two rings are connected by a long string of hide, which pierces the skin and approximately every 5 cm runs up and down between the two rings in a narrow zigzag. The ends of this string are tied below one of the rings by a simple knot, done in such a way that the string passes below the ring and back over it, passing again below the string before it passes below the ring and back again below the end just drawn back over the ring.

E5b) Khumi drums

Among the Khumi I found, in addition to the Mru form of drum just described, two variations: a more slender drum (Rengnok-KP, 7.03.1956) and a shorter drum (Kangnong-KP [=Tindu-Para], 10.03.1956). The first one had (like the Mru drums) a length of approximately 50 cm, but the diameter was only 15 cm at the one side and 10 cm at the other. The skin was drawn over the rim by two strings piercing it in an interval of approximately 2 cm in such a way that alternately one of the strings hung down 2 cm, while the other stayed put running parallel to the rim, thus forming a sequence of triangles, alternately of the one and the other string. Through the lower ends of these triangles a third string ran, hanging down in similar triangles itself. Only between this second row of triangles and its counterpart on the other side of the drum a long string ran up and down in zigzag form.

The shorter (but also slender) drum showed an even more complicated system. The body of the drum was surrounded by 3 threefold thongs (1 in the upper quarter, 1 in its middle and 1 in the lower quarter) and covered by a net of strings, meandering up and down in such a way that a string leading downward to the left changed its direction in the middle to continue its way downward to the right. On the lower rim it turned left, followed the rim for about 2 cm and then led backward, at first to the right and then, after crossing the middle, turning to the right until it reached the upper rim again, following it for about 4 cm to the left, and started a new way down, as described before. While on its way down it passed below the rings, on its way up it passed over the rings, but on the same way down it passed over any string it met (leading upward), while on its way up it passed under any string it met (leading onward). Below the rim the strings crossed each another once, in the middle the upward and downward leading strings crossed each other twice, once before and once after they passed the ring (the downward leading string passing below the ring, the upward leading string passing over it), so that they ensured the turning of the strings to the right or to the left.

In a more formalised way the downward and upward path of the string can be described as follows (d: downward, u: upward, p: parallel to the rim, l: left, r: right, b: below, o: over, in brackets: string crossing, without brackets: ring crossed): dl, (o), b, (o) b, dr, (o), b, (o), pl (2 cm); ur, (b), o, (b), o, ul, (b), o, (b), pl (4 cm). In order to prevent the hide being drawn together at the rim, a second string is drawn in, running completely parallel to the first, the sole difference being that on the rim it is led to the right instead of to the left.

The drum could be carried by a strap tied to two "eyes" (one each at the upper and the lower end of the drum), consisting of a wooden ring (5 cm in diameter) with a hole in the middle.

I must admit that I failed to note on which occasions these two special types of drum are used. Most probably they are beaten like the "normal" drums at festivals, at least I saw the shorter type used, together with a buckled gong, a big cymbal and seven gourd-pipes (*alum*) at a cattle sacrifice in Tindu-Para on the 9th of March 1956.

E5c) Gongs and cymbals

Gongs are among the metal objects imported by the Mru. The gongs consist of two types: flat (plate-shaped) gongs (*ner*), always (i.e., whenever available) to be played in a set of three instruments, and round buckled gongs (*mong*), of Arakanese origin, available only to wealthy persons. A *ner* looks like a plate, 30–40 cm in diameter, with a rim of 3–4 cm raised obliquely outward. For beating the plate a stick with a bulb at its lower end is used, while the left hand holds the *ner* by means of a short string stretching across it between the raised rim. A *mong* is a rather heavy instrument, 50–60 cm in diameter, with a buckle in the middle, and a straightly raised rim of 5–6 cm. It can be carried around, but mostly a special stand is built for it, to which it is hung, to be beaten (on the buckle) by a stick with a bulb.

Two different sizes of cymbal are used, a bigger one (*chokca*) of more than 20 cm in diameter and a smaller one (*klingca*), 10–15 cm in diameter. No difference in function was noted. The two buckled plates of the cymbals are clapped together with both hands, holding them on the little eye that crowns the buckle and that serves moreover to tie them together with a string of approximately 1 m length, so that one part cannot get lost during storage.

A small boy with a big cymbal (*chokca*)

Flat gongs and cymbals are made by Bengali craftsmen and can be bought in the market; big buckled gongs, however, are rarely to be had and therefore treated as precious heirlooms.

Khumi dance (in Müilün) with gong stand and a group with *taröng* instruments

Among the Khumi the large buckled and smaller flat gongs are, like the cymbals, the same as among the Mru. Flat gongs, however, seem to be rarely used: of more common occurrence (even with small festivals without *alum*-players) are the large gongs. They are hung on a stand (Müilün-KP, 26.03.1957) or (as in Tindu-Para 9.03.1956) carried around on a bamboo pole by two men, with one walking behind beating it with a stick with a large “ball” of cloth on its head.

Quite uncommon was the use of a bugle, but blown at the festival dance in Tindu-Para. I was told that the instrument had been left behind in the village by a soldier during the last world war. A Khumi man had learned to master it perfectly.

F Textiles and jewellery

F1 Textiles

F1a) General remarks

In the 1950s and '60s textiles were already available in the market, not only in town but also in the countryside, and some Bengali peddlers even visited bigger Mru festivals like cattle sacrifices, in order to sell them some textiles. Men might buy a new loin-cloth or turban, and girls might persuade their lovers to buy them a new shawl or cape (*pahaa*), coloured red or green. It seems to have been an old custom to buy coloured shawls of finely woven cotton for women, since when a woman died, it was usual to tear one into streamers which were hung up as "flags" on 3 m high bamboo poles at the corners of their grave-huts (*tur-kim*, see Q11). In Chamklaa-KP, this *pahaa* was printed with flowers on a black background, a form no longer available in the market. Whether it had been used much, I could not ascertain. However, these *pahaa* were the only type of cloth which the women got from the market, all other types of cloth (skirts, blankets) they were still weaving themselves, while the men had started to buy their daily costume, called *dong* (a turban which could also be used as a loin-cloth). Women did not weave this cloth anymore (though they could easily have done so).

As a matter of fact, more modern-minded men had already abandoned their old costume of a loin-cloth and anyone who could afford it bought a *lungi* of machine manufacture, taking care not to buy one that was normally bought by the Bengali, but preferably one as worn by the Marma. All *lungi* consist of just a piece of coloured fabric sewn together at one end, so that they form a large tube into which one can slip from above or from below and fasten it over the stomach by pulling it together and tucking in the open end so that it sticks firmly behind the fabric drawn together – it is possible to make a knot, but it is not approved unless one constantly has to bend down while working. In addition, the end of the side that hangs low behind the body can be pulled between the legs and tucked in front: it would then form a kind of short pants. Care must be taken when buying a Marma *lungi*, one could buy the wrong pattern – as I myself found out when I tried to buy one that afterwards I could not wear in public. I had one of green silken fabric which was presented to me by the Marma headman Hrui Pyo Ong during the first days of my stay, and when I wanted to replace it for everyday use by a cotton *lungi*, I had just made sure that it was long enough for me, but it was later identified as a woman's *lungi*! A *lungi* can easily be washed, but it soon wears out and becomes torn from constant wear.

Men who had the means to buy a strong pair of trousers with long legs preferred this. Menkröi, for instance, had taken the chance to acquire a second-hand pair of military trousers, which he always wore when going out. (I did not ask him whether he still also used a loin-cloth as a kind of underwear.) But long trousers are difficult to wash, and by constantly wearing them during the cold and hot season the natural ability of the body to acclimatise itself gets lost. Therefore it would be better to wear a pair of short pants, reaching (English style) down to the knees, during the hot season, but as soon as the rains set in, the bare legs will offer ample opportunity for the mosquitoes to bite – although this they can do anyhow and even better when one is wearing only a loin-cloth. (I therefore preferred the long hanging *lungi*, which one just needed to lift for defecating, but when touring through the country I used my (German style) short pants, though these (because of their shortness) were considered slightly shameful and I had to answer many questions in this regard – but I declared it to be German national custom, comparing it with the different length of Mru and Khumi skirts.)

A modern Mru (with short trousers) and a Marma (with *lungi*), sowing

Men who wear trousers and pants normally also wear shirts, which also have become common with the Bengali and hence are also available in the market. Most Mru, however, have not enough money to buy a long-sleeved shirt. Polo shirts were available, but for women only, and the only ones who bought them instead of women's blouses were Khumi girls. Marma women had accustomed themselves to wear just a bra to cover their breasts, but Mru and most Khumi women still went topless. When dancing, Khumi girls wore a self-made small shawl (hanging down over the breasts) over their shoulder, some of the Southern Mru girls still danced topless, while the Anok girls tended to wear a *pahaa*, irrespective of whether the weather was cold or hot. Especially when strangers were present they would rather sweat than discard their *pahaa*. Even elderly women, who always would go topless when at home, would wear a *pahaa* when they visited another hamlet (which they rarely did) and when they (even more rarely) visited a bazaar.

I should have surmised that the *pahaa* is worn exclusively by women, but C.-D. Brauns on the Lama bazaar took a photograph of a man wearing one. Apparently this poor chap did not like to be seen by the Bengali, clothed in nothing but a ragged loin-cloth. Formerly Mru women also wove loin-clothes for their men and among the Khumi there still exist such clothes embroidered with black design in *Chamtu* (= Maraa) style, which a great feast-giver may still wear and show proudly on festive occasions. It can be assumed that similarly embroidered clothes were once common among the Mru as well, since at least the word *Chamtu dong* is still known among them. The women once wove this material and decorated the ends with red or black embroidery, like they still embroider the *wan-li*. But this material apparently was thought to be too coarse, and hence lost out against the smoother cloths which could be bought. Whether Mru women formerly wove some cloaks that could be used instead of the modern *pahaa* remains doubtful. The blankets which are still produced cannot be used as cloaks, since they would be far too heavy. These blankets are more than two yards long and, like the women's skirts, of double thickness (i.e., the strip of material has a total length of nearly 2 m). Where the ends of the strip meet, the warp threads are twisted together into cords. Since the blankets, which are made only of cotton, must be warm even on cold nights, they are woven of thick threads.

Women visiting the bazaar looking for pots (photo: C.-D. Brauns)

Yet technique also plays a role. Marma blankets are considerably thinner and it may therefore be necessary to use several blankets in order to keep warm. The fabrics the Marma women weave for their ankle-length, wraparound skirts (unless they prefer to buy a *lungi* in the market) are also thinner than those woven by the Mru. If Mru women wanted to produce such thin threads and fabrics, they would be obliged to spend much more time on their spinning and weaving, or they would have to take over the spinning wheel and loom of the Marma. Since daughters, however, learn by observing and imitating their mothers, rather than by verbal instruction as such, the dissemination of such knowledge is extremely limited – not to mention the fact that every ethnic group has its own style of which it is quite proud. (The Marma cannot spin and weave like the Mru either.) Under these circumstances, therefore, it makes more sense for Mru women to sell raw cotton and buy finer cloth, rather than to try with disproportionately more energy to produce such cloth themselves. However, the Mru cannot buy thick blankets like their own type (*wanma*) or women's skirts (*wan-klai*).

F1b) Preparing cotton

In October/November the cotton is harvested (see J4g). The white bolls of cotton fibre are picked with the right hand, then collected in the left hand, and when this is full thrown over the shoulder into a *te* (see D2f) – the same baskets that were used for the paddy are now used to gather and carry home the cotton. In contrast to the paddy harvest, no special ceremonies are necessary this time. Cotton, therefore, is not given the same meaning as paddy, though during one of the swidden sacrifices the same spirit (*caa-kom-laa-kom-nam-ma*) is invoked to protect the growth of both, paddy and cotton (see J4c). Cotton is an old cultigen (called in songs *ting-thi*) which is still valued today by the Mru. They need it in order to produce much of their clothing, especially their skirts and blankets and also men's jackets, which, as far as the traditional types are concerned, are not available in the market. For this reason, even the Mru villages situated at higher elevations cultivate cotton, although the harvest there is generally scant. This is due to the fact that the climate on the ridges is a few degrees cooler than in the valleys; and in order to grow more profusely the cotton would need just those few degrees of warmer weather. There are also other factors, likewise outside the control of the Mru, which may render the harvest less than that which was sown; under very favourable conditions, however, 1 kg of cottonseed will yield 200 kg of cotton fibre. The best bolls are sorted and worked separately, and the seeds are preserved for future planting. In spite of this selection process, the quality and quantity of the fibre, which at one time served to produce the famous Dacca muslin material, lags behind the modern varieties. If the government had really been interested in bettering the income of the swidden farmers, the cultivation of newer varieties could long have been promoted; until now Bangladesh has had to import cotton from Pakistan.

Mru blanket (*wanma*)

F1c) Ginning, spinning and dyeing

The cotton gin (*rahat*, old loan-word from Arakanese, in Burmese [Judson p. 928] designating "a reel from which thread is wound from the spindle", while the Marma use this word for the spinning wheel [Bernot p.395]) is the only machine utilised by the Mru. The type they use, normally produced by Bengali carpenters, is common all over the Hill Tracts. This contraption consists of a wooden frame with two rollers geared to revolve in opposite directions, the lower part of which is set into motion by a hand crank. The rollers allow only the fibre to pass; the seeds are kept behind and caught in a basket placed below the machine. Ginning, which is done by the women, is rather time-consuming. It is, however, worth the trouble because clean fibre fetches twice the price of unginned cotton; and the seeds can be sold separately. When fibre is to be sold, it is packed into large, cylindrical baskets, *laa-par* (see D2k) – each holding 19 kg (41 lb) –, and carried to the closest market or to any place on the main rivers where it is taken over by Bengali peddlers.

Ginning cotton (photo: C.-D. Brauns)

On a sunny day the fibre which has been kept back for personal use is spread out on a mat on the *car*. Here it is teased with the vibrating string of a special bow (*laa-thop-baa*). This string is set into motion by scraping it with a notched piece of wood. This scutching removes debris from the fibre and fluffs it up. The downy soft fibre is then shaped into small sausages (*laa-pe*) and put aside until the women have time to spin (*rui laa*). Women prefer to spin during the evenings when there is an opportunity for mutual visits. If several young girls are gathered together in one house, young men usually show up in order to make some music and chat with the girls by the light of the fire. While Chakma, Marma, Twipra, and Bawm women make use of the spinning wheel, which they acquire from the Bengalis, Mru women use only the manual spindle (*laa-rui-müa*).

This instrument consists of a rod, some 25 cm long, with a small, round disk (the whorl) attached just above its lower end. The women sit down on their buttocks, placing the right foot over the left lower shank. The spindle carries a remnant of thread over the whorl and this thread is spirally wound upward to the top of the rod where new fibres are attached to its end from a *laa-pe* which the woman holds in her left hand. She then takes the spindle in her right hand and sets it in motion by twisting with the first three fingers, while those of the left hand slowly draw out more fibres, which are twisted by the twirling spindle into a thread. Then the woman holds the twirling spindle so that it hangs over the right side of her right leg, and with the right hand rolls it upward on her right lower shank, which is drawn up and stretched again, so that the spindle is set in quicker spinning motion.

While it is spinning freely, it hangs on the thread developing from the *laa-pe*, which the woman slowly draws up by rising her left arm, while the right hand, starting from the top of the spindle and moving slowly upwards, is smoothening the developing thread, by holding and pulling it in its thicker places, so that only the thread below a thinner place is twirling. When the twist becomes slower, it is again speeded by rolling it over the right lower shank. When the thread becomes so long that the arm cannot be raised any higher and the spindle would not be dangling freely any more, the thread is wound up on the lower part of the rod over the whorl by twisting the upper end of the rod with the right hand, and when finally too much thread has accumulated on the spindle, the bigger part of the newly spun thread (a smaller part always rests on the spindle) is wrapped up into a ball (*laa-lum*). The ball is held with the left hand, the thread wrapped onto it by the right hand, then put into a basket and kept for weaving, unless it is to be dyed before.

If the yarn is to be dyed, it is first wound around a simple reel (*laa-wang*). This consists of a 4-cornered piece of wood (2.5 x 2.5 cm thick and 74 cm long), with a bamboo stick (0.5 cm thick and 52 cm long) fixed to it at each end. The thread is wound up and down between the bamboo sticks in the following way: from the right front above to the right back below, back to the opposite – left – side (front below to front above, on from back above to front below, and on from back below left to back above right, and then, repeating the first round, from front above right to back below right and so on). Being wound up like this it can easily be taken off from the *laa-wang* in approximately 3 m long strands. To prevent the strands from becoming tangled, a small thread string is normally tied around them.

Today dyes are generally bought in the market, likewise the fine, coloured yarn used in weaving patterns. The Mru, however, still know how to dye red and blue-black (indigo) colours with home-made dyes. When one wishes to dye some yarn red (*pring*), the root (*hing*) of a certain tree (*long-ching*, not identified) is dug up, cleaned, cut with the hewing knife into small pieces, and crushed in a mortar with a bit of water; the resulting solution is then filtered through a piece of cloth and boiled together with the yarn. For blue-black colour (*ing*), indigo (*charam*) is planted in the field, its seeds being mixed in with the paddy seeds. The small leaves are pulled off the indigo plants by hand and soaked overnight in a round earthen pot used only for this purpose (*charam-pu*); bamboo ashes, which contain alkali, are added to the water as a mordant, producing a lye (*ung-cer*). For preparing the ashes stems of *kau-ting* (*Bambusa tulda*) are cut by the women in February, brought home and dried (three loads will be sufficient per house and year). After drying they will be burned outside the house, the ashes (called *oi*) can be preserved in a big lump (bigger than a football) at least until after the *chia-dong-khang* (see P3c), since before it has been performed no dyeing is allowed, normally till November after the cotton has been harvested and yarn has been spun. The yarn is punched down into the solution and boiled together with the *charam*-leaves and the *oi*-ashes. The ashes are necessary to make the colour durable. After the yarn has been dyed in this manner, the strands are hung up for drying between the necks of two gourd bottles, put up on the *car* at a distance of 1.50 cm. This entire procedure must be repeated three or four times and in between the strands when still wet may be beaten with a bamboo hammer, before they are cooked again and finally become completely black. When they have been dried, the strands can again be wound into balls – after the thread has been spun a second time in case it should have become loose and uneven by the dyeing procedure.

Those who steep often the yarn into the lye with their hands, afterwards will have to live for a few days with blackened hands, but the skin will not be affected otherwise and will become clean again after the hands have been washed several times. Likewise, material made from yarn dyed with natural indigo fades after several washes and takes on the bluish tinge (*ing*) of jeans; nevertheless for their skirts (*wan-klai*) the Mru women still prefer their indigo to the aniline colour which may be bought in the market. The same is true for the red (more or less yellow) stripes in their blankets. Also whenever coloured strings are necessary for sacrifices (*ing* or *pring* for *bong-kom* threads, see P2e) home-made colours will be used, though it is not forbidden to use aniline colours.

F1d) Weaving

Weaving, like spinning and dyeing, is exclusively women's work; however, the men make the looms (*chin-pöng*). While still a marriageable girl, a Mru woman receives her first and prettiest loom (stained red) from her future husband. It is considered to be a kind of wedding present, though it is given to the girl before the wedding. It is not a large machine, but consists of a minimum of six separate bamboo rods (*chin-cia*), all ca. 50 cm long, yet of different diameter, plus a wooden sword (64 cm long, largest breadth on its long side 5.2 cm, on the small side 1 cm). The ends for 11–12 cm of the larger loom rods, for 9 cm of the smaller rods, (the first internodes on both sides) are decorated on the outermost sides by incisions (little crosses) and on the following length by scraping off the outer skin. Two of the larger rods (2.5 cm in diameter) form the bar of the loom; by means of two strings these rods are attached to the wall in such a way that the warp, when tightened, does not shift or slip. On the opposite end, the breast beam is provided with two deep rills, 3.5 cm from each end, into which a sling can be fastened. The slings in their turn are fastened to a leather strap (50 x 6 cm) around the back of the weaver, so that she can regulate the tension of the warp by the movements of her body. In order to ensure a smooth sliding of the warp threads over the rods the latter are repeatedly waxed in with beeswax (*yön*, see F2a).

The thickest stick (3 cm in diameter) is the lease rod (*nam-phai*); the warp threads (*chin-cöng*) run alternately over and under this stick. A thinner shedding rod (0.8 cm in diameter) with an alternate threading pattern is placed just behind it, so that the lease rod is kept straight and does not slip back in the direction of the bar. Within the space between weaver and lease rod an exchange between the shed and countershed is made possible in that the warp threads running under the lease rod are caught in a continuous heddle thread. This passes over the heddle stick (*nat*), runs down to catch a warp thread, and up again, from one side to the other. If the weaver pulls the heddle stick up, the warp threads passing under the lease rod are brought up over the upper warp threads. The countershed is thus formed. By placing the sword (*pöng-pai*) edgewise, the weaver keeps the countershed open, in order to shoot through the shuttle. Then she loosens the sword and uses it to beat down the weft thread (*chin-pai*) firmly against the previously woven material. The weaver next pulls the lease rod close to the heddle; this pushes the lower threads, which were pulled up, back into their original position. The shed is now formed again, and the weft can be threaded back through and beaten down again. When the material becomes too long for the weaver to ply the shuttle, the weaver loosens the bar of the warp, passes the woven piece under the breast-beam, readjusts the other rods of the loom, and continues her work. By using threads of different colours for the weft, she can weave horizontal stripes. Longitudinal stripes are more common, however; these result when threads of various colours are used for the warp.

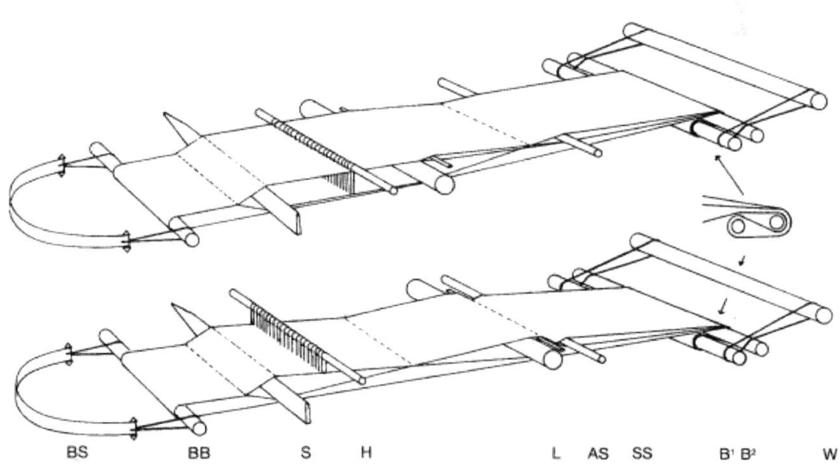

Tension loom: shed (above) and countershed (below)
BS: backstrap, BB: breast beam, S: sword, H: heddle, L: lease rod, AS: additional shedding stick, SS: shedding stick, B^1, B^2: bars of the loom, W: wall mounting

The width of a piece of woven material is restricted by the fact that generally the rods are only half a meter long. In order to make wider pieces like blankets (*wanma*), three strips of the same length are sewn together. For the narrower coverlets in which children are carried (*wan-li*), one strip is sufficient. A measured piece was 33.5 cm broad and 169 cm long, ending in a knotted fringe of 9 cm. The *wan-li* is basically white, but decorated with interwoven red and black threads forming a kind of checkerboard (see part of the weft near the body in the photograph). However, there are also *wan-li* with a kind of *lüm-chi* pattern woven in at the small sides.

Thithop Catumma sitting on the *car* and weaving a *wan-li*

A single strip is also wide enough for a work jacket (*körma*, reaching just down to the navel, used by both sexes when working in the swidden). Its main part is woven in one piece from the end of the back side. About 5 cm below the neck the strip divides into two parts, which are then continued to the lower end of the front side and sewn to the back part on the sides under the armholes. The two side parts of the front can also be sewn together. Alternately, the back and front parts may be woven separately and sewn together at the sides and over the shoulders. The three-quarter-length sleeves are woven as separate strips and then sewn on. The word *körma* is also used for modern shirts, which is understandable since the old form also has a similar shape and when sewn together in front, it must be pulled over the head in order to wear it.

Sewing is a little bit complicated, since the Mru do not have and cannot use sewing needles, but have to prick in the thread by means of a spike, whether (today) the long hairpin (*lui-cuk*) or (formerly) the quill of a porcupine (*chi-cuk*). Consequently it is nearly impossible to mend a torn piece of material, as one would have to ask a Bengali tailor to do that. This, however, is beyond the means of the Mru, and the thick material woven by Mru women normally will not rend, only the thinner material, like that of a *lungi* bought in the market will tear – and when this is so badly torn that it cannot be used any more it is thrown away.

Considerably more skill than for a *körma* is needed when a woman weaves herself a skirt (*wan-klai*). A skirt is made of an indigo-coloured strip (normally some coloured warp threads are added at its borders) slightly more than 25 cm wide and 150 (or a little bit more) cm long. The length of the primary warp threads normally is 2 m, but the weft will shorten the total length, and the rest will have to be cut off when the ends of the strip are sewn together so that the skirt actually consists of two layers of cloth. This double-layered skirt is wrapped around the hip (the two ends meeting on the left side) and secured by a chain (see F2). A multicoloured pattern (*bong*), 10–20 cm wide, decorates the outer layer of the skirt at the back; however, the pattern is visible only on festival days when a woman wears her skirt right side out. For everyday attire, women wear their skirts wrong side out, so that one sees only the colourful weft threads on the reverse side of the material, but not the embroidery itself. The pattern design is left up to each weaver, though it often contains rhomboid motifs and meanders, which in a similar form are also woven by the men into many *kruk*-baskets (see D3a) and especially into the mats used for a coffin (see Q1b, *lüm-chi* and *kan-kok*).

The *bong* is marked by coloured weft threads (not woven 1 down 1 up, but 2 down 2 up), for instance (in the *wan-klai* of Kaichüa, wife of Netnöng Atwang, renowned for her skill) with the following sequence: 1 yellow, 1 green, 1 yellow, 2 black, followed by 2 times 1 yellow, 1 green, 1 yellow, 3 black, and then 1 green, 1 red, 1 green, 3 black, 1 green, 1 red, 1 green, 19 black, then the same sequence backward. Into the black strips threads of other colours are pricked in (3 up 1 down), for instance 2 orange threads alternating with 1 green thread in the middle, or 1st line: 4 up, 2 down, 2nd line: 3 down, 2 up, 1 down, 2 up, 3 down, 3rd line: like 1st, then back again to the 2nd line and the 1st line, all with the same red thread, to produce the following pattern: _/‾_. This pattern can be reversed, enlarged (and additionally inserted with white, yellow and green threads), and repeated in yellow. The last black strip carries blocks of three lines in white, yellow, green, red and orange, while the broad middle strip carries a red *lüm-chi* pattern with blue threads on the black fond.

One of the *wan-klai* of Thithop Catumma (later wife of Dingte) had a breadth of 27 cm, but the bong was only 10 cm broad. The strips woven in were mainly red and yellow; the middle of the *bong* showed red rhomboids pricked in and separated by yellow and green zigzag lines. Another one, equally 10 cm broad, showed a narrow *lüm-chi* pattern.

Embroidering is done with the help of the eyeless needle which customarily serves as a hairpin. Modern pieces take advantage of the great variety of coloured threads which can be bought in the market. Old pieces which were dyed with home-made dyes are easily recognisable: their brownish-red hue clearly contrasts with the bright modern red. The selvages of the skirts are trimmed with tiny white or red ringlets or beads, depending on the region; formerly, the halved shells of small, pearl-like fruits (*leu-ca*, sown in the swidden) served this purpose. The beads are strung on a thread, passed through them by hand (see F2a).

The next photographs show two old *wan-klai*.

Another piece of material which women know to weave, is the *im*, a kind of shoulder bag, which must have been invented a considerable time ago, since it is known not only to all tribes in the CHT, but also to all people in Bangladesh, Burma and Thailand, and in the meantime has also made its way to Europe. (The Lai of Burma, former neighbours of the Mru, even designate it by a related word: *ip*). A Mru *im* is 17 cm high and 24 cm broad (consisting of one folded piece of 34 x 24 cm). The sides are closed by two equally folded straps (126 cm long and 16 cm broad) which overlap the sides by 8 cm and serve to hang the bag over the shoulder, their lower ends carrying 5 cm long fringe threads which are knotted together at their ends. 9 cm above the upper rim of the bag the carrying straps are sewn together, so that their breadth is reduced to 8 cm. A bag in typical Anok style is recognisable in that both the straps and the front side of the bag are decorated with black and red threads, the straps by 1 black, 1 red, 1 black warp thread, the bag (and the straps where they are sown over the bag) in that every 7th weft thread is red plus a pattern of black zigzag lines vertically pricked into it: every time three lines, then black rhomboids, after which the zigzag is turned, black rhomboids again, the zigzag turned once more, and so on. The Southern Mru do not take so much trouble: they just introduce broader strips of red and black threads into the warp. But, as a matter of fact, such *im* are rarely used, the men still prefer to use their own *em*-baskets, even when "going shopping" at a bazaar.

Some special bags are only used to make a show when dancing. Their lower rim is decorated with *klema-phak* (wing covers of the emerald coloured *buprestis* beetle). These *im* are never sold, because their preparation involves too much work (MK, 6.08.1956).

F2 Women's jewellery

F2a) Chains and belts

Tiny glass-beads (*keng*), approximately one millimetre in diameter, are available in the market. These beads can be purchased in various colours – primarily red (*keng-wi*), but also green, white, yellow and black – and fashioned into elaborate chains and belts. The holes of the beads are so small that even the tiniest needles cannot be used for stringing. Mru women must therefore string them by hand; if necessary they use their hairpin to push the thread through the hole. In order to keep the thread from unravelling during this stringing, it is rubbed with beeswax. This treatment also prolongs the life of the chains in general. All thread used to sew blankets and jackets should likewise be rubbed with wax. Beeswax is also necessary for weaving, since all loom rods are waxed with it, so as to slide more smoothly; and the sword is repeatedly waxed even during the weaving itself.

The securing and preparation of beeswax (*yon*, see H9) is the responsibility of the men. They smoke out the hives of wild bees, cut up the honeycomb, and press out the honey. The remaining comb is then boiled, and the liquid wax that results is poured into water to eliminate impurities. Finally, the wax is reheated and poured into bamboo tubes, where it hardens into round sticks. This beeswax is used exclusively for weaving and sewing. For a different type of wax (*kwai-ut-naa*), see F9.

In addition to the small beads, the market also offers Mru women larger beads (*kengma*) which are easier to thread. These white and iridescent green beads, however, have the disadvantage of being made of very thin glass; so they break quite easily. Likewise available at the market are old coins to which a small ring has been soldered. Alternated with red beads or hard, red fruits, these coins – formerly made of silver – are fashioned into decorative chains (*tangkaa-leng*). The longer chains, which are worn only for feasts, are hung from the right shoulder and over the left hip; shorter types may be wrapped, together with chains of beads, around the hair-knot or slung about the hips. It is only poorer or elderly women who fasten their *wan-klai* with a mere string (or in more modern fashion with a safety pin at the left side where the double length of the skirt must be fixed somehow.) Marriageable girls will prefer a *keng-ko* (the word suggesting a "white" chain), which may consist of a broad band of yellow (or red) beads strung on threads knotted together, or (if available) of small round silver plates strung together.

Taa'uk-para; girls dancing wearing combs and orchids in their hair; their *wan-klai* are fastened by yellow *keng-ko* chains

F2b) Bangles

Different types of bangles are also available in the market. Narrow silver bangles and aluminium bracelets (*kwai-che*) are worn around the wrist – the latter are more common; broader silver bangles are worn about the upper arm and are equipped with a hinge. Young girls begin wearing the upper-arm bangles as soon as they reach puberty; and since the bangles do not grow along with the girls, they cut considerably into the lower part of the biceps. However, only older women who are no longer so concerned with beauty refrain from wearing them. These bracelets are made expressly for Mru customers by Bengali silversmiths, who also fashion the special type of hourglass-shaped ear plugs (*param-cheng*), worn exclusively by Mru women. Young girls may further adorn these ear plugs with flowers and fragrant herbs. On festive occasions the young Mru ladies also wear silver anklets (*khok-khen*) and brass foot-rattles (*röu*) while dancing.

In contrast to Mru women, Mru men wear less jewellery. Though young men do in fact possess bangles and flat, disk-shaped earrings (*ram-cheng*) which can be decorated with beaded pendants, they wear them almost exclusively for festive occasions. A red wooden comb (*charüt*, see D6b), which is decorated on the rounded top with a piece of aluminium belongs, on the other hand, to traditional everyday attire, especially of young men, but young women also sometimes use it when dancing. Yet self-made combs are becoming rarer, since the colourful plastic ones available at the market represent cheap competition. The hair-knots in which the combs are worn are becoming less fashionable among older men who prefer to cut their hair short. Young men, in contrast, take great pains with the grooming of their hair-knot. The knot is worn on the upper, left side of the head and is often reinforced with strands of the person's own hair which were earlier cut for that purpose. (These strands may also be bought, one bunch for the very high price of Rs. 5/- [MK, 16.09.1956].) Women never cut their hair short, but rather knot it on the back of the head.

In contrast to the men, women always adorn their hair-knot with a hairpin (*lui-cuk*) – which is also needed for their needlework. In former times this hairpin was probably a porcupine quill (*chi-cuk*); porcupines (*chi-pakma*), however, have become rare, and the metal needles bought in the market are more durable. For festive occasions, colourful paper rosettes may also be bought in the market; they are used to adorn the hair and are particularly popular among young girls.

F2c) Other decorations

The body, too, may be decorated. Tattooing (of the arms) is practised by the Marma but the Mru do not use it. The young people (mostly girls) often decorate their faces (forehead, cheeks and lips) red, by applying the coloured spittle made by dissolving in the mouth the lilac-green crystals of potassium permanganate, which are cheaply available in the market and which many young men carry around wrapped in a little piece of paper and hidden behind the loin-cloth tied around their waist. The young men bring it with them when courting girls, and when a group of young people sits together in a house on an evening during the cold season, the girls spinning, the boys singing or playing an instrument, the young men may try to spit a whole mouthful of coloured spittle on the bare back or the shoulders of the women. These "decorations" are considered as ugly, but fortunately can be easily washed off with water.

Young people of both sexes often try to decorate their faces by tainting their teeth shiningly black by *shuli*, a word used in hill-Bangla for an old hewing knife, in Mru [*chuli*] for the soot prepared on it in the following way: the hewing knife is scraped clean and a piece of burning bamboo pressed on it. The soot thus prepared in the evening is applied on the teeth. This process is called *ti chuli* (*ti*: the move with the finger over the teeth) and will be repeated every night. After one or two weeks the teeth will look black, but the colouring will disappear after a short time. If one wishes it to remain black, the *ti-chuli* has to be repeated every night (MK, 19.07.1956). However, no elderly people will be seen with blackened teeth; their teeth, on the other hand, tend to look brown. The brown colour does not result from decay, but is the result of often chewing betel quids which serve as a stimulant and are thought to preserve the teeth. In fact most people live with teeth intact, until they start to fall out at an old age (which, however, people rarely reach).

G1 Cattle

G1a) Types

The Mru distinguish three types of cattle (*cia*): 1. *cia-nöm* (Bangla: goyal), gayal, 2. *mang-chang* (Bangla: tong-goru), mountain-cattle, 3. *ratca* (Bangla: goru), zebu. Buffaloes (*naa*) are not cattle (*cia*) but the Mru use them for the same purposes and for this reason they are counted here together with them. Probably the connotation *cia* originally meant only the gaur and the gayal: in the Kuki-Chin languages *sial*, cognate with the Mru *cia* (<**cial*), is used predominantly for the gayal and not for the mountain-cattle or the zebu, and Mru *cia-nöm* (goyal) means literally "true/real *cia*", and the gaur is called *cia-röng*, the "wild *cia*". The gaur is not domesticated, but is still to be found in the Eastern hills (see H1a).

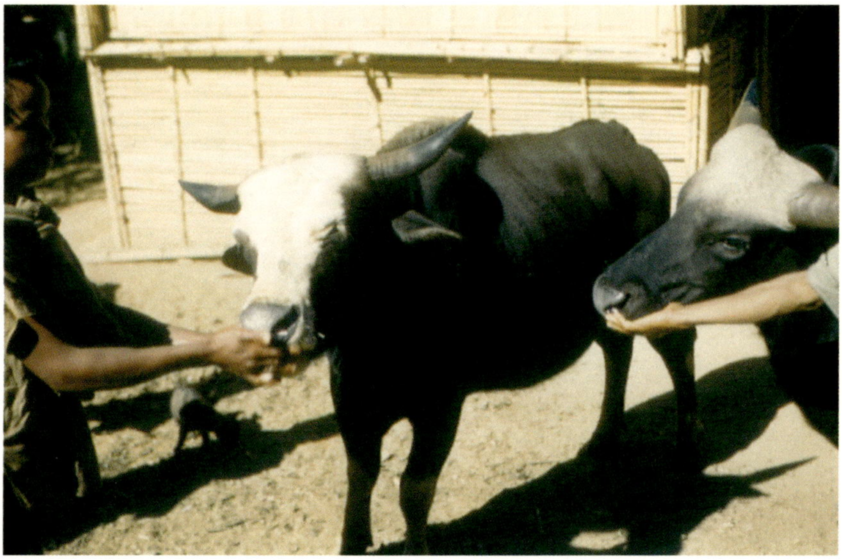

Young (2–3 years old) *cia nöm* in Moloi-Para, Lemupalong-Mouza

Of all types of cattle, the gayal is the most highly esteemed; however, it has become very rare. I myself encountered gayal only in one location (Moloi-Para, Lemupalong-Mouza), however, gayal skulls stored (particularly in Khumi houses) prove that these animals must have been still widespread two to three generations previously to the west of the middle Songu. The difficulties of keeping them (see G1d) may have been the main contribution to the decline in gayal ownership; the price of the animals (200–500 Rs), i.e., approximately five times that of an ordinary cow, is beyond the means of an average feast-giver: they are used as sacrificial animals for feasts of merit only.[447]

[447] In the meantime prices for goyal have gone up. Céline Mouchet reports: "I myself have seen some gayals in few hamlets, mainly in the south (and not close to the main communication ways). People told me that they can now be sold at the average (huge) price of 40,000 taka (roughly 500 euros). I heard once of a gayal being sacrificed for a feast in 2004, however, I couldn't check the information."

Buffaloes, which deliver no less meat, cost at most half of that price (up to 200 Rs); but also they are rarely kept by the Mru. They are to be found predominantly on the western slopes of the hill ranges in the valley areas adjoining the lowlands of the coast; in the middle Songu valley they are rather rare. Furthermore, where they are found in the Mru area they are usually in the possession of Bengali, who use them for field work and from whom the Mru buy them when required (for slaughtering them at festivals).[448]

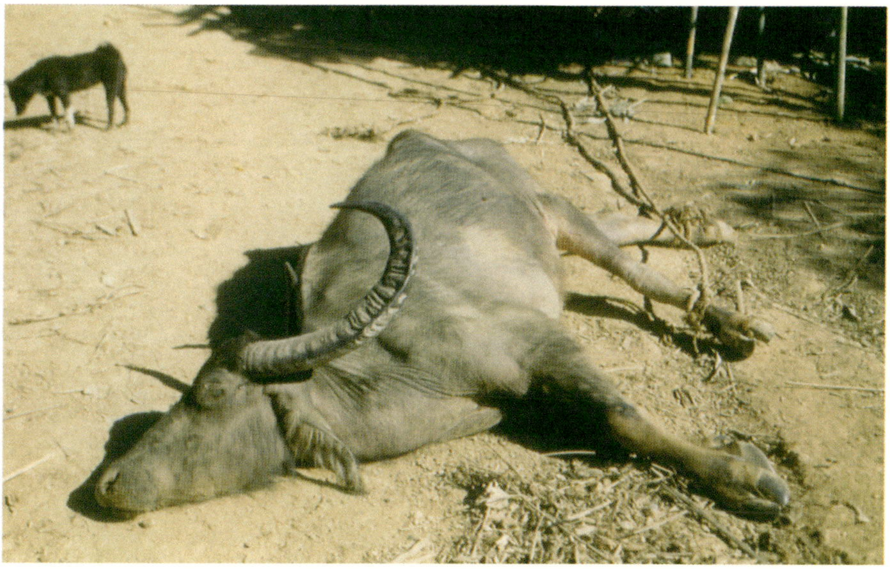

Sacrificed buffalo, Rümpöng-Karbari-Para

The ownership of cattle of a prosperous Mru almost always consists only of *mang-chang* (hill cattle) and *ratca* (zebu) or their crossbreed, called *ratca-mang-chang*. Further crossings with *mang-chang* or *ratca* are simply designated again as *mang-chang* or *ratca*, a differentiation to which the Mru attach little importance, in everyday language all are *cia*, and when one wants to differentiate, all cattle with a hump are *ratca* and all without a hump *mang-chang*. Both types are reddish brown (the zebu occasionally also grey) and approximately equally big, occasionally, however, the mountain-cattle are built somewhat more strongly and have shorter, straighter but stronger horns. Thus their shape somewhat resembles that of the gayal, with which they also cross-breed. According to the parents these

[448] To quote Céline Mouchet again. "Once, I was in a Mru hamlet on Chimbuk, and my informant was happy to have some dried buffalo meat to eat for dinner. Our host, the Karbari, told us that his buffalo had recently died as a result of having fallen from a cliff. On the following day, as we were leaving the hamlet, the Karbari got the bad news that another of his buffalo had died, strangled by the rope which attached him to a fence. The son of the Karbari told us that they had to attach them with ropes as the buffalos were causing a lot of damage to the surrounding fields and that they often had to pay fines to the owners. He concluded, saying that it was not worth owning buffaloes as they gave far too much trouble (and my Mru informant agreed)."

cross-breeds are called *cia-nöm-mang-chang* (gayal-cattle), further crossings with *mang-chang* are once more called only as such. Mountain-cattle and zebu are also used like gayal and buffaloes essentially only as sacrificial animals for feasts. Since they are needed for other working purposes, they are not castrated by the Mru; oxen can be bought from the Bengalis when required. A Mru cattle-owner almost never sells his animals, they are his pride, from which he parts only when he himself must give a feast or when he has run into difficulties and is in urgent need of money. Therefore whenever possible sacrificial animals are always bought from the Bengalis; their price varies according to size between 40 rupees for young animals, and 100 rupees or more for full-grown animals.

G1b) Keeping

The cattle are neither looked after nor accommodated in stables. They stay in and around the village; during the day usually in the jungle in search for food, they come back into the village in the late afternoon and lie down, wherever they like, usually in groups, when the sun shines in the shade of trees or next to the houses, or under the houses when it rains. The owners take little notice of the animals, every now and then they give them some salt; if they don't get enough, they try to help themselves by licking the bamboo supports under the privy corners of the open-air platform (*car*) or by catching fresh urine with their tongues. For the gayal, the need for salt is the only reason for their visits to the village: they often stay away for days without letting themselves be seen, but they come back again and again to their owners, even if one managed steering them into another village. Accordingly the owners have no need to worry about their whereabouts; it is often some days before the owner finds out that a cow has calved. The gayal are sufficiently strong not to fear the most dangerous beast of prey of the mountains, the leopard, and they are sufficiently surefooted not to fall anywhere and to hurt themselves.

This is not so for ordinary cows, and the owner looks every evening to see if all of his animals have returned home. He never has a whole herd; many people have not even one head of cattle and most of them only one or two. If the animals don't appear in the evening, at first nobody worries much because almost always they are there again in the next morning or at the latest next evening. They never move far from the village and have certain places, where which one can look for them, and if necessary, everyone in the hamlet will keep an eye out for the missing animal, in case it has got caught in a thicket, from which it cannot free itself, or has gone round a fence and couldn't find the way back. The danger from leopards is minimal; if a leopard's tracks are spotted during the day, people make sure that the cattle are all in the village at night. Should a leopard creep up at night, the cattle band together and the unrest among all the animals soon wakes up the villagers who drive the robber away with noise (by beating on the floor) and with lights. Moreover, the leopard is less keen to prey on cattle than on pigs. Actual danger for the cattle in the jungle exists only during the rainy season, when every day flocks of leeches attach themselves to them, so that the animals often return home in the evening with blood dripping from their bodies. The smell of blood also attracts beasts of prey during the day. But during the rainy season the roaming range of the cattle is very much restricted, fences obstruct their way to the fields around the village, so that they always have to stay close together and thus remain unassailable in the group.

The wounds caused by the leech plague constantly bleed the animals, so that they look emaciated and miserable in the rainy season even though ample forage can be found. Other vermin do not spare the animals either, above all worms, which sneak into the body cavities, and larvae, which hollow out finger-deep burrows in the soft parts. The Mru have no idea about how to treat such illnesses; occasionally a remedy is bought from the Bengalis and tried. If there is no hope of helping the animal, it is slaughtered as soon as possible. For the treatment the beast is caught with a rope (which is not especially difficult, if something is held out for it to eat), and tethered strongly to a house post by the head with its legs tied up. A jerk of the cord draws the hind legs aside so that the animal falls down and then on the back, so that it cannot kick any more. The animal is not handled particularly sensitively, but the financial expenditure is already a sign of quite special affection for an animal.

Cattle are the only animals which can be given a human name. Pigs and dogs can also be given names as a sign of affection, but not names that are also used for people (MK, 26.05.1956). Such a naming is not necessary and not connected with any ceremony, but the Mru appreciate prize cattle the most of all the domestic animals. Occasionally a young man buys a beast with his own money which is then his personal property, to which he particularly attends, feeds and looks after. Such private purchases are usual only with cattle; if a father of a family buys a head of cattle, it becomes family property (MK, 12.07.1956). The cattle are called by their names, they are driven away by a loud brrr! (rolled R) and a stick. Driving them away is particularly necessary when they are in the rut and if the animals, while chasing each other through the hamlet, can trample down fences and, if they come under the house, break or knock over supports. Taken over from the Bengali is the reassuring *bo*! *bo*! (shortened form of boyo < bosho "sit down"), with which one wants to calm down animals which are restless or turned stubborn.

G1c) Supervision

Not all cattle which can be seen in the Mru hamlets belong to the villagers. Some of the animals are only under their supervision. The Mru themselves usually have enough room in their hamlets for all animals which belong to the villagers so that giving them away is hardly ever necessary. In the Bengali plains villages every fertile ground is used as arable land and so the little grass, growing on the narrow dikes between the fields and at the waysides does not suffice to provide all animals of the village with food. Moreover somebody must keep permanent watch, so that the hungry animals do not run into the fields. The Bengali therefore like to give their animals to the Mru hamlets where the jungle offers them sufficient food. The nearer the cattle are to their native village, the more often they feel the urge to run back home, and therefore the Bengali often drive them up to the ridges of the mountains. The animals are not accustomed to long tours and mountaineering and are not invited friendly at all, they may take time neither for feeding nor ruminating. Ordinarily only one animal is driven per tour; the driver keeps at its tail which he twists: depending on the strength of the pain the animal accelerates its slow trot and occasionally it happens that the cattle arrives at the destination with its tail twisted off. If necessary they can also be driven with the help of a fire brand. The cattle driven by the Mru even to other hamlets (for festivals) do not undergo any better treatment.

Buffaloes fare still worse when they climb up the mountain someplace to end up above as sacrificial animals. Some days before the food is already withdrawn from them and if they then are still too strong and too stubborn, a joist is bound around the neck, which drags between their legs, so that they cannot break out. The main thing is that they arrive still with a remnant of life at their destination; but I have come to know of a case in which the animal broke down on the way so that it had to be shot and cut up into pieces, so that one could at least still eat the meat of the buffalo in the festive hamlet (Noliram-RP, 17.01.1957).

The animals brought to the pasture into another hamlet must be supervised there. Whoever takes over the task must pay fully for their loss or any damage caused by them. That is why he is entitled to get a remuneration from the owner. This payment for the supervision depends on the sex of the animals: for bulls and oxen, the keeper gets 8 annas up to 1 Re. per month as *corani* (Bangla: pasturage). For cows the supervisor gets a *bak-ka* (from Bangla hhaga: share), namely half of the calves born during the supervision time or a corresponding compensation (MK, 10.07.1956).

G1d) Field damage

The damage that cattle may cause and for which the owner or supervisor must pay, concerns primarily the fields. Minor damage caused by the cattle in the hamlet, at garden fences and house supports during the rutting season, can be repaired easily without compensation. However, if they invade a field the owner of the animal has to pay at least a little pig. Every owner of a field in his own interest tries to bar access to the field by a stable wooden fence especially if paddy grows there. Most of these barriers are not situated directly in front of the field, but, since the animals don't generally break through the dense jungle, and do not use steep slopes, it suffices that the animals still have sufficient grazing area within the restricted area. The difficulty with this confinement is greater for those who keep gayals. The gayal don't trot docilely on the paths or in the thin grove in a narrow circle around the village, they want much more space and to make themselves independent. They need to be treated with more consideration than the cattle and they also need more room where uncultivated jungle areas extend towards a mountain crest, so that the animals can roam freely while the country area located in another direction can be barricaded against them. The more areas are required for cultivation, the more the gayals must disappear.

If in spite of the obstacles a cow or a gayal enters a field and causes damage there, the owner of the field can demand compensation from the owner or supervisor of the animal in the form of paddy to the estimated amount of that destroyed. This demand doesn't need to be raised, but in any case the owner of the cattle must even sacrifice a pig for the damaged field, even if it is his own field, either by killing his own pig, or by paying the corresponding price to the field owner, who has to sacrifice a pig from his own stock. The pig serves to reconciliate the angered rice spirit, it is a *dim-üa-pak* to repeat the *üa dim* which is sacrificed after the clearing fire (see J2e). When driving it off, the field-owner cannot seriously hurt or even kill the foreign cow which invaded his paddy field, otherwise he would have to refund the value of the animal to the owner or supervisor. Reaped fields are open for pasture to all animals (MK, 10./12.07.1956).

G1e) Utilisation

To possess cattle is a sign of affluence for the Mru, and their utilisation is restricted almost exclusively to the sacrifice and the consumption of their meat. Slaughter for other than sacrificial purposes, i.e., without ceremonies, is possible, but very rare, since a reason for a feast can almost always be found. There are two ways of killing the animals: ordinarily by stabbing them to death with the spear, in rare cases by cutting their throat through with the chopper. The animals used for sacrifices do not need any special qualification, but a beast with a broken horn cannot be used for the *pram-rui-poi* (ear-piercing festival, see N3a). There is no difference between female, male and castrated animals with regard to the sacrificial value.

For other forms of use there is hardly any need, although they are not forbidden. Whenever a Mru has a small parcel of level field for tobacco cultivation or a water paddy-field, he can harness his animals in order to plough the field with them. Anyone who has no skill in this employs a Bengali, if he can afford it. The plough used is also produced by the Bengali. In the Bommu valley I saw a Mru ploughing a tobacco field with a zebu, in the upper Matamuri valley where the Mru have some water paddy-fields in the area of Alikodong buffaloes are also harnessed. But these are exceptional cases; when arable fields are found by smaller streamlets or on the hills, they are either made utilisable by and in possession of Bengali or not used at all and overgrown by the jungle.

Like ploughing, milking is little practised among the Mru. Though many Mru would get stomach ache if they drank too much milk, there is no general aversion to drinking it; but the cows rarely give milk, which seems understandable when one sees their permanently emaciated condition and the rare efforts to milk them. So I met on the 20.05.1956 in Thümru-KP, a man who searched in all surrounding villages for milk for a little child whose mother had died; he had no success. Neither he nor anyone else knew that small babies could not digest cow's milk. In Tapwüa-Kua I got milk from the hamlet once only during the whole time of my stay: shortly after a cow had calved. This is the only opportunity to milk a cow, sometimes, when it has enough milk. As with the Marma and Bengali, milking is the business of the men. For catching the milk a pot will be used, for transport a bamboo quiver. The milk is drunk only scalded, other types of use are unknown. Only the Bengali know how to produce ghee (clarified butter). If available they prefer milk from a buffalo cow.

The dung is also little used; normally it is left lying, unless the children take pleasure to poke in it and thereby to distribute it, anyway it is dried out quickly by the sun, crushed to dust or (during the rainy season) washed away by the rain. Fresh dung is ritually used for throwing it at a feast-giver's house during a feast of merit[449], practically it is used for the

---

[449] Céline Mouchet remarks on this: "I've seen only once some cow dung thrown on the wall of the house of a *cia chot poi* giver, in a Dopreng hamlet (situated between Lama and Alikodong). The dung was coming from the intestines of the cow which just had been sacrificed and cut into pieces. The

production of the loam necessitated for the hearthside: it is the only thing that binds the sandy soil. It is also often picked up before the preparation of the garden beds and used as a fertilizer; other excrement is never used for it; only cow dung is considered to be clean, because "cattle do not eat cooked rice" (MK, 11.05.1956). On the use of the skin for drumheads see E5a.

*

Goats are occasionally needed for sacrifices, they are bought for this end from the Bengali. In Mru hamlets goat keeping is not usual and trials of single people in this regard meet with the resistance of the other villagers, since the animals would cause too much damage. Céline Mouchet reports (without stating whether the goats had been raised in the hamlet or bought): "The only goats I saw among the Mru were in a Dopreng hamlet [...]. Two goats were going to be sacrificed in the bed of a stream for a *khang* ritual." (See P3b–d.)

G2 Pigs

G2a) Esteem

The pigs (*pak*) of the Mru, like those of all other residents of the CHTs, belong to the breed of the Indian black pigs, have long snouts, bent backs and hanging bellies, "rat's tails", and thick, black and (especially at the belly) also white bristly hairs. They can reach a weight of more than two metric hundredweights, most of them, however, are killed much earlier. Slaughtering can be carried out without special ceremonies, but the animals are hardly ever killed just for their meat and even in cases when one has to put them down, this is always used as a special occasion for giving a feast or performing a sacrifice. The uncastrated boars are an exception (MK, 25.06.1956) since they are useless as sacrificial animals and their meat may not be given to *pen*-relatives (see L3b) either. Whether pigs are sacrificed or not, they are normally killed by stabbing them to death with a sharpened bamboo (*cau*) or a spear (*re*), very rarely (see P5d) they are killed by cutting their throat with a hewing knife or chopper (as is the custom of the Marma).

Even the poorest Mru has one, two or three pigs, unless an unavoidable sacrifice has just fetched the last pig out of the stable before one could buy a substitute. Pigs are needed for all possible occasions, the rules of the annual fieldwork and *the rites de passage* cannot be carried out without pigs and besides these occasions, at any arbitrary time a man can all of a sudden come into the situation to have to sacrifice a pig, so that it is appropriate to have one always ready. The best guarantee for this is a sow in the stable which produces offspring. Since the permanent need for pigs reduces the stock again and again, there is a general aversion to selling pigs. Correspondingly their price is relatively high, it is determined by the weight of the animals and the kilo cost approximately 8 annas so that for a little pig 3–4 Rs. and for a pig of medium size 10–15 Rs. have to be paid. Bigger animals are not sold any more, but only fattened for special opportunities, like wedding celebrations. If, as seldom happens, pigs are sold or given away for rearing, they are kept locked up for some time in their new sty, since otherwise they run back to their old home and must be captured repeatedly, especially if the

---

charaa of the hamlet explained to me that the dung was thrown in order to protect the house from the *chüngnam*; in seeing and smelling the dung, the *chüngnam* would see that the house had recently performed a ritual."

former owner lives in the same hamlet. When it is time for them to be fed, they sometimes do push to strange troughs, where they are chased away with sticks. It is mainly the duty of the women to care for the pigs, though not to the same extent among the Mru as among the Marma, where the pigs really belong to the woman so that she can do with them what she wants, and even sell them against the will of her husband. Among the Mru the pigs are family property and the father of a family negotiates and sacrifices them. The castration of the animals is also incumbent on him.

Sow with her piglets

G2b) Sty and food

During the night the pigs are shut in their sty completely closed all around under the front part of the front-facing house (*kimma,* see C3b and d). This place is never cleaned (except when rebuilding *the kimma*), but it remains dry, since it is slightly raised above the flat earth. Admittedly the smell penetrates through the thin bamboo floor from the pig-sty into the rear part of the room lying above this, but the sleeping place of the residents lies in the front part, over the chicken-coop and gets enough fresh air through the walls, and moreover the parents sleeping in the *kimma* have got used to these smells since childhood, so that they no longer smell them. The pigs in the sty allegedly do not contribute to an increase of vermin in the house, since the walls of bamboo poles of the sty prevent insects flying in from outside and allow free ventilation (MK, 25.06.1956). The door of the sty is locked overnight so that the pigs are protected from beasts of prey. Occasionally, however, a pig may remain locked out (one can count them only with difficulty once the pigs are in the sty), and small or strong pigs also succeed in breaking out when the bamboo tubes become brittle, or a leopard may try to break in, so that negligence with regard to the security of the sty occasionally may have to be paid for with a tribute to the leopard.

In the early morning the door of the sty is opened, the pigs are allowed to roam freely and look for their food in the village primarily under the *car*, the open-air platform, where all garbage is thrown down and water poured off so that the pigs here also use to make their wallows. In the rainy season they also dig holes for themselves in the village paths. They never move far from these places and when they do then only for brief surveys in the bushes which are used as latrines. By their permanent and immediate removal of all human excrement the pigs are of greatest importance for the hygiene of the hamlet. Some pig is always paying attention to the access to these places and if anybody goes there, the pig also comes running at a gallop. Its grunting in the bushes also excites the attention of other pigs and a person going to defecate in the bushes is soon surrounded by animals and is well advised to take a stick along and to be on guard against their intrusiveness by striking them on their muzzles; the backs of the pigs are insensitive against such a treatment. The defence gets particularly unpleasant with several pigs of the same age which attack from all sides at the same time; if pigs of different size have come, however, then the strongest animal tries to drive away all others by biting them. Old sows in particular vigorously defend their position and then happily wait at their leisure: if they did not, even an adult person would have a heavy stand against it. Therefore dangerous old animals are not released any more, and the owner on request of the villagers must kill even violent younger animals, against the attacks of which children could not defend themselves.

Apart from excrement and refuse, which they search out for themselves and thus clean the village, the pigs are fed with rice waste. Old metal pots, pails or buckets into which the meal waste is poured stand on the open-air platform (*car*). These vessels contain mainly water, in which husks and broken rice, collected after winnowing into a basket of its own, are soaked. This mixture is called *pak di* (pig bran). This food is soaked during the day and other leftovers are added to it. In the afternoon it is cooked and given to the pigs in a trough (or if there are only a few animals also from the bucket). These troughs are hollowed out with the axe from a tree-trunk and flattened below; they are approximately 1 m long, 25 cm wide and 20 cm high. Between four and five in the afternoon the women stand on the log staircase of their houses and call through the hamlet: *pakca wâng*, *pakca wâng* (piglets come!!) and when the animals come running, a call in a high voice of *dük dük, dük*! still lures them. During the feeding the woman, or a child with a bamboo stick, makes sure that no strange pig joins them and that the piglets and smaller animals can come to the trough, too, without being pushed away by the rapacious bigger ones.

G2c) Rearing

Heavily pregnant sows are kept in the sty whenever possible so that they litter there. With the Anok no taboos for the occupants of the house are connected with it as no ceremonies at all are usual for young animals of any type. The Dopreng, however, hold *khang* for one day when piglets are born, i.e., foreigners are forbidden to enter the house on this day (MK, 15.11.1956).

Male piglets are castrated at the age of 1–2 months. Nobody deliberately lets an animal uncastrated, since the small boars do not get fat, their meat has a bad taste and moreover they cannot be used for sacrifices. Now and then, however, it may happen that an animal escapes the procedure. With 4–5 months it is thought too old for a castration, and slaughtered on the next occasion. The general castration is no reason to worry that a sow may not get pregnant, since the male piglets are sexually mature shortly after their birth and do not only drink at the lying mother sow but also occupy themselves with copulation. The castration is carried out on the *car*. One man keeps the piglet stretched on its fore and hind legs, and another man cuts its scrotum open with a sharpened bamboo, pushes the testicles out and cuts them off. Earth or ash serves to stop the bleeding. The piglets usually squeak only a short time during the operation, the mother sow from which the male piglets are taken away is more furious, but has to remain downstairs. The best opportunity for capturing offers itself at feeding time. To avoid the risk of being bitten by the mother sow when holding the piglets, one must try to shut the sow into the sty by moving a bowl with food to and fro while the piglets are still outside and can be grabbed easily. A man who has a net can simplify the work by driving the crowd of piglets into the net, held up by two people, and so capturing them.

Bigger animals can be captured only while they are being fed; skillful women can manage this even with an empty feeding bucket. If pigs are approached otherwise, they dash away at the last moment; one can grab them by their hind legs, however, while they are feeding and pull them up so that they hang with the dangerous muzzle downward. Before they are killed or transported, one has to tie up the muzzle and bind their front and hind legs together. Openly woven bags (*pom*, see D4f) which one can close above and carry by a strap over the shoulder are used to transport piglets; bigger animals are wrapped up in a layer of broad bamboo strips in which they are carried to the offering place. Heavy animals are killed before they are transported, their legs are tied together, a long bamboo pole is inserted between the legs and the pig can be carried by two men walking one ahead of the other (*pak-laa kot*).

Pigs which have got too big and strong to be captured by anybody are rarely allowed to leave the sty. They are fed so much that they grow as fat as possible in their enforced peace. Even if it is the secret hope of everyone to possess a very fat pig, only few people have such a de luxe animal in the sty which one day can be slaughtered at a wedding feast. Since it becomes too strong to be held and tied up by anybody, it must be shot.

G3 Dogs

G3a) Keeping

It can be said that dogs are not kept, they just keep the company of men, like mice and other vermin, over which though they have the advantage that people are not always trying to kill them. Instead they live in permanent fear of hard objects which people beat them with or throw at them. They seem to be less impressed by the constant insults hurled at them with the help of their family name: *kui!* (dog!). This *kui!* is besides *le!* (away) and *döm!* (down!, i.e., down the stairs) nearly all what a dog gets to hear; if it is nervous, it winces, if not, it waits for more violent reprimands which completely unsystematically attack it until it, depending on the pain inflicted, withdraws for a few moments yelping or silently a few metres, to come back again after few moments. The dogs' presence is most annoying when it is most urgent for him: when the family members are taking their meal. He never feeds from the banana leaf or the plate, but only tries to catch the chunks that fall on the floor, but even these are refused most of the time and what lies around is swept away in front of its muzzle at the end of the meal. What remains on the plates is tipped over into the pig trough; the dog gets nothing. It is left to the dog to care for its survival; sometimes it even has to contest with the pigs for human excrements; a real dog's life; *kui-yong* ("pig-like", in English one would say "lousy") is the most pejorative title which a Mru can append to a thing or event. If one asks somebody on what his dogs live, he says that every two days or so he would give them something to eat, and to the objection that one has never seen this, one receives the answer: "Now and then, we may just forget it." Fact is that the dogs are hardly ever fed and most of them die before they have grown up. But dogs are always used and even called for their services when babies defecate in the house, especially when they have dysentery, as also has been noted by Céline Mouchet: "I was in a Mru hamlet close to Thanchi; in a house, an old lady was taking care of a baby when suddenly the poor child, having diarrhoea, defecated everywhere. The old woman then loudly called '*kui, kui, kui*' and in less than two seconds, the dogs were licking not only the floor but also the buttocks of the baby that the woman was presenting meticulously to them!" But the dog may also refuse to do this service, as I observed in Rümputma-Kua, when a drunken man could not help it and vomited the contents of his stomach filled with beer and *arak* all over the floor of the house. The dog was called, but its nose apparently could not stand the stench and it refused to lick up the vomit. The daughter of the house had to fetch bottle after bottle of water and to pour it out over the floor until finally all traces had been washed down through the slits of the floor.

A dog sneaking around while people are eating

A crowd of harmless yapping curs greets the stranger entering a hamlet, and at first one has the impression of dealing with a breed which does not grow much bigger than the Pomeranian. In reality these Indian village dogs can achieve the size of dog foxes, which they also resemble in shape. Their hair is short and fallow or black. I saw also white and spotted dogs with longer fur in some Mru hamlets.

One seldom meets big fully-grown animals; all bigger ones which I saw belonged to the brown kind. Better fed they make a quite stately but peaceful impression. The biggest dog I saw belonged to Ong Shwe Phru, at that time the cousin of the Bohmong in Banderban. I thought that it belonged to a special breed, perhaps an offspring of a German shepherd dog, but was told that it was nothing but a well-fed local country-dog.

Ong Shwe Phru's well-fed dog with a monkey on its back

Children can safely play with them. In Ongpha-HP (1.03.1956) I saw the animal ambled and pulled at its ears and tail by a two-year-old child whose head was at the same height as that of the dog. The dog licked the face of the child and the mother sitting close by let it happen quietly. Big dogs can get a name of their own, as in Tapwüa-Kua Kaichüa's bitch: Kaichau (*kai* = vagina, *chau* = fat!).

G3b) Use

Even if one cares little for them, the dogs remain faithful to their owners, i.e., they stay as long as they live in the house to which they are used since birth or into which they were brought when young. They occasionally accompany their "masters" on the paths to the field or in neighbouring hamlets; they, however, usually stay near the house, for which they allegedly also provide guardian services. I say "allegedly" because they do not bark regularly if a stranger approaches the house or enters it. Barking is more a pastime of pups, big animals hardly ever bark. Sometimes they stay in the house and sleep there at night, sometimes they are locked out and look for a place under the house; they then pay little attention to visitors. If the dogs bark, this is usually because a beast of prey is nearby or because evil spirits have showed up. Dogs can see these spirits and take on the task of protecting the house and its residents from them, whether by driving the spirits away by their barking, or by being sacrificed to them.

For killing them, a short cord or a long bamboo lash is put around the neck of the dog, so that it cannot run away, and then they are clubbed to death, that is beaten on the head with a strong stick until they are dead. The meat is prepared and eaten exactly like that of the pigs. With the Anok, however, dogs are only sacrificed at rare events and never slaughtered just for meat consumption (though this is not forbidden). In their eyes, most probably under the influence of their more "civilised" neighbours, it is regarded as "barbarous" to eat dog meat and the young generation shows little inclination to taste it.

Elai cutting up the dog sacrificed on the roof (see P5a)

This aversion against the consumption of dog meat is also felt among the Khumi, in perhaps even stronger form, though dog sacrifices belong to every greater feast of merit and dogs were also slaughtered for field ceremonies (not as sacrificial animals, but only for consumption, see J7b). However, as late as 2004 Céline Mouchet noted a special trend of the Khumi to eat dog meat: "I heard several times from the Mru that the Khumi were fond of dog. Once, I was in a Twipra hamlet and some Khumi men came to buy a dog (in order to eat it)."

Most probably the aversion was not as great in the past as is today, since Lewin wrote (1870:230): "A favourite festival dish among them is a dog stuffed with rice. The young cur is plentifully fed with cooked rice about half an hour before cooking time, and when stuffed to repletion is knocked on the head, skinned and roasted. The rice is left in the stomach and eaten with the dog's flesh as a concomitant relish. Surely men's appetites and their wits are sharpened on the same grindstone." I never watched this way of preparation, also attested for other areas of north-western India. In the CHTs of the 1950s also the dogs were prepared in the way generally used according to which all meat (apart from poultry) is cooked to goulash (roasting of undivided animals is unknown); stomach and intestines are also cut up and cooked together with the other meat, though not before the intestines have been thoroughly cleaned with water.

Systematic use of dogs for hunting is not usual, although some dogs are said to be suitable for deer hunting. But since nobody trains them and game ordinarily is caught with traps, it is hardly more than by chance that a dog helps to hunt an animal. So Dingte Ngarua succeeded to catch a wildcat with the hand which had been located and driven by two dogs on to an empty branch (19.06.1957). On the way to the swidden field the dogs occasionally find turtles and lizards: if nobody is near enough the lizards soon end up in the hungry dog's stomach; turtles, however, the dog reports by barking and running to and fro.

To conclude let me quote Céline Mouchet once more: "A Mru told me one day that the dog was used to look after the outside of the house, and the cat the inside. In this same house, they don't usually feed the cat as they think otherwise he will not try to catch mice."

G4 Cats

While the behaviour of the Mru and Khumi towards dogs is, with rare exceptions, alike everywhere, their relations to cats *(min)* seem different from house to house. Cats are rarely seen, and in many houses there is no cat at all. Cats are very independent and frequently abscond for good into the jungle. New cats can be bought from Bengali dealers. How little is known of how to handle cats can be shown by the example of Kyaw Thwân Ong, who one evening tried to fetch Menkröi's cat, since the mouse in the roof of my house could not be caught by traps of any kind. He wanted to use it by holding it tight long enough until the mouse appeared and he could let loose the cat to catch it. But first he had to have the cat, and fearing that it would bite or scratch him, he took a big stick along when he went to catch the cat. Menkröi reported that it had come to his house, Kyaw Thwân Ong went to him, the animal showed itself and also could be lured nearer with rice thrown for it on the floor, however, whenever he tried to grab it, it soon escaped; even when he finally tried without the stick. Such shy cats never appear at mealtime; they come down from the roof only after darkness begins to fall or creep through the window of the room to take up their night accommodation on the rack over the fire place on cold nights.

I found the cats a little more trusting in the house of Karbari Menring (13.02. 1956). They came while the people ate and waited aside, now and then mewling, so that one might throw something down to them. Unlike the dogs, they also got something, as well as rice they also received chicken bones, which they chewed to pieces and ate. From the bowls standing on the ground they themselves took nothing, even if nobody paid attention. Rarely people scared them away by clicking the tongue on the middle palate; but they gave the impression of being on permanent watch. Petting or fondling them was unknown.

I got yet another impression in the house of Karbari Yeopo (25.03.1957). The two cats of the house, an old one and a young one, were little intimidated and rubbed against the legs of the residents to have somebody petting them; when trying to fondle them, they immediately started to play and to bite quite strongly. But the youngest children were not deterred by that and did not stop playing with the little cats, stroked their fur and (without torturing them) let them run off again. At the meal both animals came and begged by permanent mewling. They were not chased away, but they got nothing either. Generally cats would not lack food, unless they were bought so young that they were not yet strong enough to hunt mice and rats for themselves.

The house mice usually build their nests in the grass of the ridge of the roof, from where they have an easy way to the rice barns, without someone being able to catch them with the customary traps. If one cannot manage to stab them to death in their nest or kill them when they run, they can lead an undisturbed life unless a snake finds the way to their hiding-place. Since snakes are not welcome visitors to a house either, the cat remains the only reliable helper. Cats are not eaten – as far as I learned.

What I have reported here about the different attitudes to cats may be easily explained by the different ways men and cats come to learn and get accustomed to each other. This, however, cannot be said for the following note which I owe to Céline Mouchet. Here the cat, on the one hand, is treated as fully responsible for its acts as a human being, on the other hand, it is punished for its acts (probably a "theft") in a quite inhuman form: "In Hapru-para (Chüngma hamlet south of Thanchi), one morning, a cat was killed and roasted (to be eaten). The reason for it, according to the owner, is that the cat was guilty of disturbing the household!"

G5 Poultry

G5a) Chickens

Ownership of chickens is as important to the Mru as possession of pigs. Chickens are also needed for sacrifices throughout the whole year. An individual's stock of chickens is often not sufficient for the annual requirement, so that further chickens must be bought (or exchanged for paddy) from the Bengalis. The Mru could keep more chickens, but they deliberately restrict their stock for reasons of risk and lack of profitability. The risk is the possible loss by the frequent fowl pest, which can empty the whole henhouse in a few days, the lack of profitability is due to the paddy needed as food: while for the Bengalis chaff and bran are available daily so to speak free of charge when husking and cleaning their rice, the Mru must feed the chickens with precious paddy since they need the bran for the pigs which are not kept by the Muhammadan Bengali. The Bengali can therefore convert their rice waste into chickens and exchange these with the Mru for rice (paddy); but the Mru also find the exchange profitable since they would have to buy even more paddy to raise the same number of chickens. Depending on the size of the chicken, the exchange value amounts to 1 to 3, or even to 5 Rs. Chicken is therefore by far the most expensive meat, but also enjoys the highest esteem and plays a special role in showing honour to the parents-in-law and the members of their sib (*tutma*, see L3b).

The Mru keep their chickens under the back part of the front-facing house (*kimma).* Like the pig-sty in the front part, the chicken-coop is also fenced all around with bamboo tubes; poles fixed crossways inside these serve as perches for the chickens. Only for hens laying eggs special baskets are woven, round or pointed, in both cases without any upholstery (*waa-dör*, see D4g). When the *kimma* is rebuilt, the chickens are accommodated on the open platform (*car*). This is done during the dry time so that the animals can stay in the open. Also among those Longhu and Khumi whose houses have no stable, the chickens spend the night on the *car* in chicken baskets attached to the house wall. Since these baskets cannot be closed (unless they are woven in the Marma way as a cage) there is a danger that they will be taken by a wildcat at night. An enemy exists during the day too: the goshawk. It waits patiently for its opportunity on a high tree in the neighbourhood of the hamlet. It may not be successful every time, but oftentimes a chicken is hurt so badly that it must be killed. It is difficult for the villagers to scare the goshawk away from its observation place, and a gun, the only effective defence, is not available in every village, so that as long as the robber is nearby, one must either lock up the chickens or must stay permanently watching it.

The chickens are fed in the morning; they are called by *ai! ai! ai!* The quantity of paddy fed to them depends on the stock and the opportunities the chickens have for finding food: particularly in the rainy season they find many worms and vermin which, however, slightly taints the taste of the meat. In every hamlet there is a cock, all richer people keep at least one. Since, however, many chickens are killed before they lay eggs (for a sacrifice, a chicken is a chicken, but a small animal costs less) the production of eggs is rarely sufficient to allow the consumption of bigger quantities; the eggs are collected (if necessary bartering with a neighbour) in a basket, in order to let a broody hen (a household often has but one for this purpose) sit on them.

Cockfights are watched with interest; but the animals are not trained for this and no sports or bets are connected with these fights.

G5b) Pigeons, ducks, etc.

In general pigeons are kept by rich people only. Only rarely is a pigeon needed for a sacrifice (see P2f), they mostly serve as a welcome enrichment of the menu, but they also help themselves to many a grain from the rice barn (*ca-pam*) or disappear into the jungle, never to be seen again. But otherwise they are not difficult to care for. An old cotton basket (*laa-par*, see D2k) sometimes serves as a dovecote, as I could see with Karbari Menring, who had hung up three such baskets horizontally under the overhanging roof of his house on the wall to the *car*; into the open side of the baskets at a depth of approx. 20 cm he had built in an open wickerwork, in the lower part of which an opening was left. If proper pigeon baskets are made, they are also woven like cotton baskets, but closed on both sides, an opening is cut out in the long wall, which can be closed by a little stick placed crossways. The hole is big enough to put one's arm through it: if a pigeon is in the basket, the way out remains blocked to it and it can be caught.

Like pigeons, geese are also needed for only a few sacrifices. They are kept relatively seldom, unless there is a large watercourse nearby. They are housed in the chicken-coop and are also called like the chickens: in the morning and in the afternoon, they are fed with bran which has been soaked (not cooked) in water and with old cooked rice. Their eggs are eaten with pleasure; they themselves can be slaughtered at arbitrary times. It is the prerogative of karbari and headmen to keep geese, "since the animals are too dangerous to allow everyone to keep them".

***

Finally I should mention a very valuable domestic animal, the identity of which, however, remains doubtful. It is said that in the past, as well as the present people also kept animals called *me-ma*, from which pieces of meat could be cut off without killing them. These pieces grew again directly afterwards, so that no harm was caused to the animals. Unfortunately, they have now died out (MK, 19.09.1956).

H Hunting, fishing, and use of bees

H1 Importance of hunting and hunting weapons

H1a) Hunting enthusiasm and hunting with rifles

The Mru do not hunt just to obtain food, as their domestic animals supply most of the meat they require. However, the number of animals they keep is barely enough to provide a dish for a festive occasion, and accordingly any kind of game is always welcome as an addition to their otherwise mainly vegetarian menu. This desire for variety in diet is supported by a marked enthusiasm for hunting, which goes beyond the mere prospect of getting some meat for the cooking pot. Only a passion for the experience for hunting itself can explain their readiness to bear the financial sacrifice of obtaining a shotgun, even an ancient muzzle-loader, in disregard of all other needs.

I never went hunting myself, but Kauffmann had bought a hunting rifle, which he left for me when he returned to Germany. Kyo Thwân Ong, my "boy", used it and once shot a *rang-pang*, a hornbill, with which he proudly posed for this photograph.

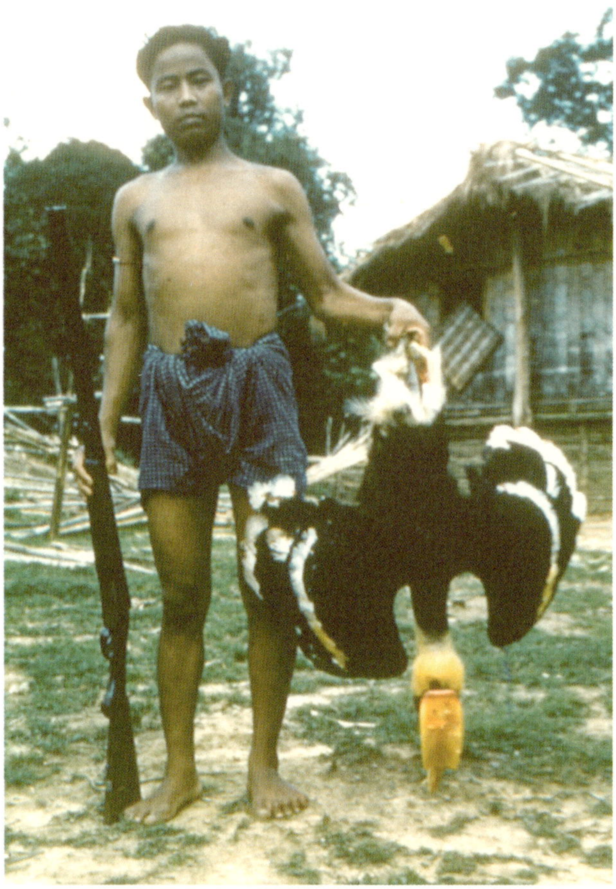

There would be many more owners of hunting rifles if everyone who applied for a licence could get one. However, the administration issues licences only in special cases, and if somebody wants to be one of these special cases, it will cost him, unofficially, a good sum of money in addition to the cost of the licence. There are whole mouza areas in which there is not a single licence holder, although this does not mean that there are really no shotguns at all in the area, even if only a home-made muzzle-loader with a barrel made from an old water pipe (bought in the plains). Such secrets are carefully guarded and not shown to any stranger; but they cost a constant tribute of good will to the other villagers – and sometimes a quite severe punishment of imprisonment or a heavy fine that can be paid only with the help of all relatives. But all these troubles are accepted: surely not for the occasional meat (parts of which, moreover, will be distributed to all villagers in accordance with tradition), but because of the undeniable passion for hunting.

Admittedly not everyone is an enthusiastic hunter and only youthful enthusiasts spend every free minute roaming in the jungle as soon as they get their hands on a shotgun. Of course, first it is necessary to find rewarding quarry to track – and big game has become quite rare in many areas of the Hill Tracts. Since the Chittagong hills are generally so densely populated that every square kilometre is subject to daily scrutiny, there remain few possibilities for game to escape the sharp senses of the inhabitants. What still saves the game from the hunter's gun is the impenetrability of the jungle. The inadequacy of most guns forces the hunter to creep up very close to the game, and even the hill men do not always succeed in moving noiselessly through the undergrowth, which a European would find impassable. Jungle fowl undoubtedly suffer most from the gun, but the red deer, too, which at night stands helplessly dazzled when confronted with a hunter with a pocket torch, is increasingly becoming a victim of technology.

Times when hunting is forbidden are prescribed by the administration: however, someone who cannot read has little idea what is on his licence and regards it as an injustice if by chance he is caught and has to pay a fine. This was the case for instance with a Mru from the Thanchi area who had shot a wild gaur cow with a muzzle-loader after having pursued it for days. The head with the horns alone weighed six metric hundredweight, and, proud of his performance (as the most powerful animal of the mountains the gaur is thought to be more dangerous than an elephant), the good man brought his trophy to the police station, where he received the opposite of the hoped-for bonus. Helplessly he came to me, but I only could confirm to him that he had violated the law.

I never saw a gaur myself (by now the species has almost died out), but Céline Mouchet encountered one in 2004. She reports: "I had once the chance to see a wild gayal (gaur?) not far from Ruma, near the Ruma river. A young boy ran towards us to alert us of the coming danger. Indeed, the huge animal was very nervous and agitated, staring at us and showing us that it would not accept being annoyed! The Bawm who was accompanying us deeply regretted not having his gun with him."

H1b) Conventional hunting methods

Hunting with the gun is undoubtedly the most popular way of hunting these days because of the immediacy of the hunting experience; but since, as said, only a fortunate few can follow this sport, the main hunting weapon is still the trap. Anyone may set traps but not everyone does so, because setting traps successfully is an art: not everybody knows how to find suitable places for them and not everybody has the necessary practised feel for a certain kind of trap. Some only specialise in only one or two types.

What Kauffmann (1939:18) reported about the knowledge of details among the Thadou-Kuki also applies to the Mru: “The necessary strength of these components is exactly fixed by experience, just like also the often very difficult balancing and adjusting to the easiest touch can be managed reliably and promisingly only by people who have grown up in the hunting way of setting traps.”

Any animal caught in a trap becomes the property of the person who set it (alone or with the help of family members). A co-operation of possibly the whole village and an appropriate division of the spoils as reported for the Thadou by Kauffmann (1939:2) is unknown among the Mru; but it is usual for them to distribute part of a larger kill, like a red deer or a wild boar, to all houses of the hamlet. Setting traps is the only important conventional hunting method; bow and arrow are unknown as hunting weapons (and likewise as children’s toys). There are, however, two other kinds of hunting weapons; but I have never seen adult men hunting with them. They could be described as children’s toys, if they were not used as real weapons for shooting birds, though only by adolescent boys. These are the pellet bow and the blow-pipe. Among the Anok blow-pipes are very rare, I never saw them used. They could be found more often among the Longhu, and used not only by young boys to shoot little birds (from the ground or from a seat in the tree), but also occasionally by the head of the household to shoot a chicken which had evaded attempts to catch it by hand.

H1c) Pellet bow and blow-pipe

The pellet bow (*yai-baa*) described here was manufactured by Mowai Catumma (of Tapwüa-Kua). The bow stave and string are both made of the same material: bamboo. The bow stave is cut out of a firm tube: it is 105 cm long, in the middle 2.2 cm broad and 0.8 cm thick. The breadth reduces to 1.1 cm toward the ends. At one end it is tapered to a 3 cm long cone, over which the loop of the string can be put on and removed, and 4 cm from the other end a sloping notch runs around the bow stave, into which the string is tied. The bow is drawn in such a way, that the former inside of the bamboo tube is now bent to the outside; the exact middle is marked by a node that is left protruding to form a 1.5 cm ridge, which rests between the fingertips when the bow is drawn. The string consists of a bamboo thong (4 x 15 mm strong) which is twisted around itself at both ends and made into a loop with a knot. The length of the string between the loops is 95 cm. The middle section of the string is divided for 12 cm lengthwise. Both parts of the string are held apart by little sticks (2 x 2 mm strong), the length of which is adjusted to the breadth of the opening (in the middle 15 mm). These sticks are held in position by bast wrapped around them and serving also to fix the ends of the opening. This widened part, like that of the whole string (regulated by its twist in front of the loops) is set at a slight angle to the bow stave, so that little stones laid on it do not strike the bow stave (and thus the hand), but pass a little to the side if the bow is held at an angle. Nevertheless it is advisable to practise at first only with a light pull on the string, because a fully drawn bow is powerful enough to send little stones up to the treetops with enough force to hit smaller birds there so hard that they fall dead to the ground. The pellet bow therefore corresponds approximately in strength and effect to a good slingshot as used by European schoolboys.

I saw the blowpipe (*pho-bük*) described here used in the house of Karbari Wöila (Tuktöng-Para 15.01.1957). The pipe, 220 cm long and approximately 2 cm in diameter (narrowing slightly towards the front) consists of a bamboo tube with six internodes, the nodes of which were removed by cutting the wall of the pipe open in these places, taking the partitions out and smoothing the inside of the pipe. The resulting holes in the pipe wall were sealed with wax. The arrows, rounded and sharpened thin bamboo sticks, are about 70 cm long, with the last 45 cm wrapped with little chicken feathers (quill length 6–7 cm). Starting from the end of the arrow, the feathers are tied to it using a thin bamboo strap moving spirally to the right, with the ends of the quills pointing to the tip of the arrow. This imparts a spin to the arrows blown through the pipe and over short distances experienced hunters can aim with the blowpipe so reliably that their success rate is much better than with the pellet bow. While I have never seen a considerable bag shot with the pellet bow, I once watched a young man in Uikük-HP shooting half a dozen birds from a tree with the blow-pipe in a short time. Poisoned arrows are not used.

H2 Noose traps/snares

H2a) *yot*

Yot are noose traps with bait, but without a triggering mechanism, which serve to catch jungle fowl (particularly at the mating time). A domestic cock is tied up in the jungle and with his crowing attracts on the one hand the infuriated jungle cocks for a fight, but on the other hand also the jungle hens. Around the cock a ring of 20 to 25 *yot-kua* (noose-trap loops) is set up, while other entries to the surrounded place are barred by bushes and bamboo thongs.

To make the loops, pineapple leaves are soaked in water for three to four days; thereafter the ribs can be separated. They resemble white hair and can be twisted into approximately 30 cm long threads (*leng*). About 25 cm long bamboo thongs (*chön*) of best quality, i.e., soft, flexible and strong, which at their lower end are cut somewhat broader (approx. 4 mm), while their upper end, only about 1 mm broad, is split. Into this split end of the *chön* one end of the *leng* is firmly twilled in so that the two ends pass into each other without a knot. The free end of the thread is tied into an open noose which is passed over the *chön,* which must be thin and soft enough in its upper half to be bent over by the weight of the thread and to form a loop together with the thread around it, but strong enough in its lower half to keep the loop upright. To this lower end of the *chön* a sharpened firm bamboo stake (*cung*), approximately 5 cm long, is tied at its lower end which is notched slightly at the side where it is tied to the *chön*. With the *cung* the loop is put into the ground. A cotton thread (*rui*) leads on the right and on the left from the place where the *chön* and *cung* are tied together to the next loops (*kua*). If there is a tree or stronger branch between the loops, the cotton thread is tied to it, too: the whole construction is thus secured against the possible tearing of an individual loop.

If a jungle cock wants to rush at the domestic rooster, it will run, holding its head forward, and its neck will enter a noose, which, however, is too narrow to let its body pass, and when the cock tries to fly away the noose tightens around its neck. Any jungle hens that try to approach the cock will also be caught this way. At the end of the mating season the domestic rooster is released and brought, unharmed, back into the hamlet (MK, 5.07.1957).

H2b) *dong*

A *dong* is a spring-pole trap, a small version of which (with a triangular frame) is used mainly to catch mice and other small rodents. A larger version (with a ring-like frame) can also be used to catch monkeys. The triangular frame of a *dong* is tied to a branch or bamboo stake (*dong-klöng* [6][450]), so that it stands vertically.

The frame [5] is made of an evenly 2 mm thick bamboo thong (or piece of rattan), the ends of which are brought together at the base and crossed over each other. The frame of the *dong* is approximately 10 cm high and about 7 cm wide at its base. From the middle of the base a thin fibre (*prün* [8]) leads to the pusher (*prün-tüa müa* or *atüama* [7]), which is put crossways in the upper third of the frame, held there and pressed tightly against the frame by a peg (*chit-pot* [4]). The fibre used for the *prün,* which resembles horsehair, grows on the bark of the *ramshupari* (Bangla) palm and can withstand the weather for years, i.e., it does not rot. Pusher and peg consist of little bamboo sticks. The peg is a short cone, the tip of which just touches the pusher, while its base, thickened by a node, rests against the topmost point of the frame, against which it is pulled upward under the topmost frame by a cord (*leng* [1]) that leads to the spring-pole (*püt* [3]).

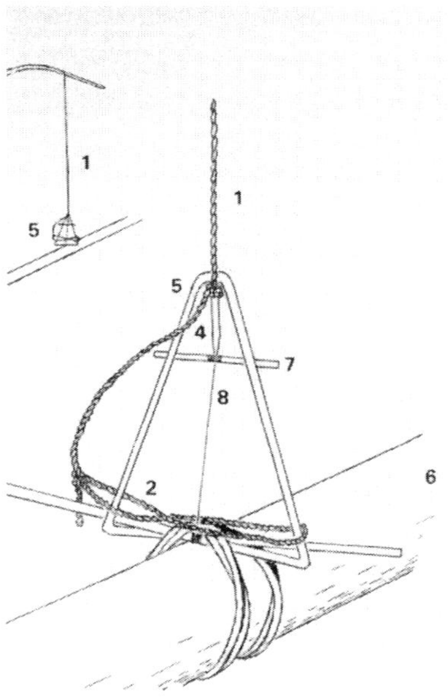

The spring-pole is a bamboo cane, the top of which is pulled downward and thereby stretched by the cord (*leng*). The cord leads from the node of the peg (*chit-pot*), without touching the fibre of the pusher (*prün*), to the base of the frame in such a way that it is looped around the feet of both legs of the triangle [2]. This cord, like those of the noose traps (see H2a), is made of pineapple fibres. The pull transferred by the cord of the spring-pole to the peg on the one hand pulls on the upper hanger of the frame that serves as a holder and on the other hand on the top of the peg pressing the pusher, which therefore cannot slip down on the legs that act as a carrier for the frame. Only a pull on the *prün*-thread [8] will draw the pusher [7] downward, so that the peg loses its hold and turns over, and while the cord is pulled up by the strength of the spring-pole, it also draws up the noose [2] over the legs of the frame and jams the animal in between noose rope (*leng*) and the upper frame.

The trap will work only if an animal (that is smaller than the opening of the frame) pushes the *prün*-thread aside; the *dong* therefore is always attached in such a way that it cannot be evaded, for instance by putting a bamboo as *dong-klöng* (on to which the frame is tied) at some height as a

[450] The numbers (in square brackets) refer to those used in the drawing.

"bridge" over a brook, or by setting a *dong* up in a hole in the underground corridor of a bamboo rat. The *dong* can also be used against house mice, tied in larger number to the sloping supporting bamboo tubes of a house, which mice make use of to climb up and down to their nests in the roof. Since no bait is used, the success rate is usually poor, and so steel torsion traps bought for this purpose in the market are also used.

As for the big *dong* for monkeys, mentioned at the beginning, I saw only a single specimen tied to a bamboo tube high up between the trees crossing the path at Chinglök-KP (29.08.1956). A ring twisted from roots served as a frame, and instead of one barring thread (*prün*) two threads ran from a common base in the form of a V to the pusher. The rest of the construction corresponded completely to that of a small *dong*.

H2c) *cak-leng*

This spring-pole trap for jungle fowl and pheasants is used only in connection with a fence. With this fence (*ban*) occasionally whole strips of an area are enclosed over distances of more than 50 m, normally close to a field, but also sometimes in the jungle. This fence consists of bamboo slats approximately 30 cm high, put into the ground roughly 5 cm apart and whenever thought necessary strengthened by a bamboo strip woven in above and below. Every 3 to 4 m an opening of about 20 cm is left. The side slats which limit the opening are set at an angle so that the opening widens upwards. On one side a bamboo thong bent into a bow is put into the ground crossways to the fence. The height of this so-called *kung-kaa* corresponds approximately to that of the bow that can be stretched between thumb and forefinger. A cord (*leng*) approximately 1 m long made from tree bast (*nam*) is tied to a young bamboo cane standing nearby which serves as the spring-pole (*püt*). Approximately into the middle of the cord a peg (*chit-pot*) is tied and the rest of the cord is made into a loop (*kua*). The peg, a little bamboo stick 5 cm long, is (on the side facing the spring-pole) jammed with its thicker upper end behind the upper edge of the bow (*kung-kaa*) serving as a holder, so that its tip presses against the pusher (*atüama*) which, 1 to 2 cm above the ground, is inserted between the top of the peg and the bow of the holder (*kung-kaa*). On the pusher, crossways over the breadth of the opening in the fence, little thin flat bamboo slats are laid which together form a sort of a running board (*capher*), on which the loop is laid. On the side opposite to the *kung-kaa* in front of and behind the opening in the fence and crossways to it, little bamboo sticks are put into the ground at a very flat angle of inclination, i.e., in the same direction as the sticks of the *capher* so that they form a little barrier in front of and behind the *capher*. If a running bird wants to pass through the opening in the fence, it must lift its legs over the barring sticks and then steps exactly on the *capher*. Since one end of the sticks lies on the pusher, they pull it off from the tip of the peg: the peg shoots out, the cord is pulled up and the loop contracts around the leg of the captured bird.

With another form of the *cak-leng*, a taller *kung-kaa* is used, which does not stand crossways to, but lengthways in the fence direction and reaches over the whole opening. Two *capher* lead to the pusher, one from the front and one from behind, and the loop hangs in the *kung-kaa*, so that the head of the animal will be caught (MK, 5.07.1957).

H2d) *nuk-ring leng*

The *nuk ring (=* seed red) is a red bean which is not used for food purposes but readily consumed by jungle fowl. Apart from the spring-pole (*püt*) and the cord (*leng*) the trap consists of two parts only: *cung* and *kap*. The *cung* is a piece of wood approximately 15 cm long sharpened at the lower end to be stuck into the ground, with a bit of a former branch sticking out horizontally from the upper end, and the upper end of the peg rests under this projection. The *kap* is a double loop, 10 cm long, made from a 7 mm broad bamboo thong. A little bamboo stick is pushed through the narrow end of the loop, which is just strong enough to fix the *kap*, so that it does not slip. With this little stick the *kap* is put into the earth at such a distance from the *cung*, that the end of the outer bamboo loop lies under the projection of the *cung* so that the top of the peg (*chit-pot*) reaches just to the end of the loop, which thereby takes over the task of the pusher. The red bean is fixed into the inner loop, reaching approximately up to the middle of the *kap*. The capturing rope from the spring-pole, after it is tightened with the peg, is laid as a loop in a ring around the *kap* (leaving the *cung* outside), not directly on the ground, but raised by three little bamboo sticks a bit higher than the loop of the *kap*. When a chicken or a pigeon pecks at the bean, this makes the *kap* slide a little lower, triggers the peg and lets the cord jump up, so that the noose tightens around the neck of the animal. As far as jungle fowl is concerned this trap is very reliable and it can also be set up without great effort, particularly since a fence is not necessary; but it has one disadvantage: it can be used only in the hot dry weather, since the bait decays too fast during the rainy and foggy seasons.

H3 Deadfalls

H3a) *köm*

The *köm* is a baited deadfall set up particularly for shore running birds and jungle fowl. Using bamboo strips, a network is made of approximately 80 x 100 cm (the type of weave corresponds to that used for house doors etc.) and on this two bamboo tubes are tied, which cross each other approximately 10 cm in front of one of the narrow sides. About 30 cm distant from the other narrow side, a *dong* is set up (see H2b). The cord of the *dong*, however, does not lead to a spring-pole but to a *khia*, a balance beam (see below). A 30 cm high *cüng* (post) is rammed in and the bamboo placed so that the post rises through the triangle formed by the front edge and the bamboo tubes crossing each other in front. While the back edge of the plate rests on the ground, the crossing of the bamboo tubes rests on the very front end of the horizontal bamboo tube (*khia*) which lies on the *cüng* and juts out over the plate (*köm*) with its other side, where the cord (*leng*) of the triggering mechanism (*dong*) is fastened to it. The *khia* therefore is a lever with two arms: the front (shorter) one carries the weight of the trap, the cord of the *dong* pulls on the back (longer) end. From the pusher of the *dong* a thread (*prün-leng*) runs through the wickerwork of the plate to the ground where the bait is tied to it. A pull at the bait pulls the pusher off from the peg, so that the pull on the back lever arm of the *khia* is released and it tips over, the crossed bamboo tubes slip from its end and the *köm* falls down (Klingtui Khongtör, 30.07.1957).

A larger and heavier form of the *köm* on the same principle can even be used as a tiger (or leopard) trap. In this case the plate of the *köm* is made from tree-trunks tied together (in a single or double layer) and in addition weighed down with stones. The "bamboo tubes" crossing each other in such cases are stout trunks, which are tied under or between the layers of the plate. An already injured animal is used as the bait. Since according to the principles of leverage the main pressure of the plate rests primarily on the vertical post and the remaining burden can be reduced by moving the balance beam minimally at the *dong* end, it is possible to put several metric hundredweights on the thin tip of the peg and the pusher, so that even leopards or tigers can be crushed under the heavy plate of the *köm*.

H3b) *klong*

The type of deadfall called *klong* is not often used singly (on tree-trunks as natural or artificial bridges) but usually built in a whole system connected by fences. Little creatures of all kinds, primarily lizards and rodents, but also birds, can be caught with a *klong*. The fence described here was erected near the field of Roaja Rengyua Atwang (Tapwüa-Kua), but extended downward over the whole mountain slope, and thus protected more than just the swidden field. Altogether 21 traps were installed at 5 to 6 m intervals in the fence. The fence (*ban*) was approximately 90 cm high and its main support was strong bamboo poles about 2 m apart. The intermediate space was enclosed by closely-spaced bamboo slats (*kwai-chek*) very near to each other which were woven over three crossways bamboo bars: the undermost (very near the ground) and the middle one (at a height of approximately 30 cm) were strong bamboo slats, the topmost (in a height of approximately 60 cm) consisted of a thin full bamboo tube [9] which was also led over the trap openings. These bamboo tubes passed through the vertical slats of the fence and jutted out approximately 10 cm at their upper ends.

The openings in the fence for the traps were approximately 12 cm wide and bounded by strong bamboo rods on the right and on the left, which at their upper end, i.e., immediately under the bamboo passing through, were tied together by a loop strap and thereby secured against being drawn apart and becoming loose.

The deadfall beams (*klong-müa*) were tree-trunks [2], approximately 3 m long and 10 cm thick. Each beam was suspended in the fence opening [1], so that it jutted out about 40 cm at the front. The counter-end (at the back of the fence) did not lie on the ground, but was suspended approximately 25 cm over the ground in a loop under two crossed bamboo rods (*king-kai*). The front end of the beam was also suspended in a loop (*apüng-müa nam* = hanging up carrying-strap from bark bast) so that there remained an opening of about 30 cm to the ground.[451]

The loop of the *apüng-müa nam* hung over the balancing beam (*khia* [8]), which consisted of a bamboo tube, which in its middle lied on a bamboo post (*cüng*). This post stood on the left besides the front end of the deadfall beam, approximately 20 cm in front of the fence. On the same side but directly by the fence, another approximately 50 cm long bamboo cane was rammed in (rising at an angle to the left front), its top was split open and spread to a fork by a little wooden stick (*cakung* [6]). The little piece of wood could not slip out, because it was inserted under the inner thickening of the node at the end of the tube. This split end of the bamboo (*wa-kaa*)

[451] The numbers (in square brackets) refer to those used in the drawing.

served as a carrier for the trigger of the trap: the *cakung* formed the holder for the peg (*chit-pot* [6]), which (on the side pointing towards the fence and the trap opening) with its tip rested on the pusher (*prün-tüa müa*).

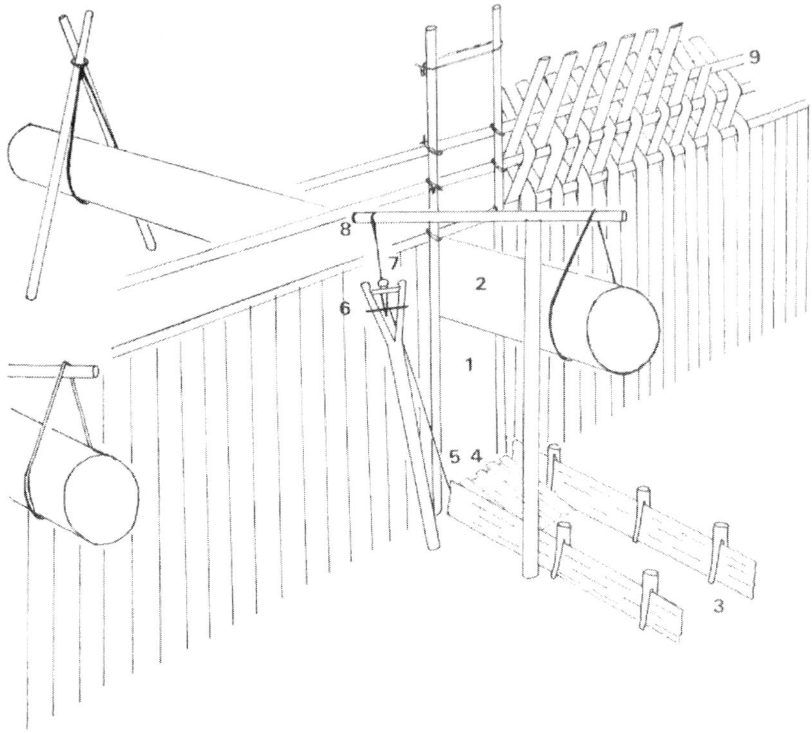

The peg was cut from a thin bamboo rod so that its form was like a children's shovel bent round at its front end: though it was only about 3 cm long and the bend at its broader end was the remainder of a node. From this broader (upper) end of the peg a cord (*cakung-leng* [7]) led (after passing under the holder) up to the balance beam (*khia* [8]). A thinner thread (*prün-leng* [5]) led down from the pusher to the ground and over the trap opening to the right frame bamboo of the opening, to which it was fastened at a height of approximately 5 cm. Approximately 20 cm long thin slats (*capher* [4]) led, alternately from the front and from behind, up to the *prün* thread hanging (crossways to the fence, i.e., exactly under the deadfall beam) a little bit over the ground.

To make it more difficult to run in at an angle into the opening of the trap, on the right and the left side of the opening, a small wall (*akhia-müa bang*), approximately 5 cm high, was set up crossways to the fence under the deadfall. For this purpose strips of flattened bamboo were jammed in between small bamboo sticks split up to the node [3]. The stronger part of these standing "staples" was located on the outside; the frame bamboo poles at the side of the fence opening were also split up to the same point. This *akhia müa bang* formed a kind of channel [3] with the running board of the *capher* in its middle. An animal that wants to pass from one side of the fence to the other must use the channel and hence will step on the slats of the *capher* [4]. The cord of the *prün* lying under this is pressed down, thus removing the pusher from the *wa-kaa* carrier, the peg comes loose and the *cakung* cord [7] is released, the balance beam [8] loses the counterbalance and tipping over lets the deadfall beam drop [2]. With such a *klong* fence something can be caught almost daily; the owner checks the traps every day when walking to his swidden.

H4 Spear traps

Spear throwing traps called *chot* are the most popular but also most dangerous trap of the Mru. In order to warn the people, *lak wa-kaa* (term taken from Marma *lak* = hand, *wa-kaa* = spread) are set up in front of a hidden *chot* trap as a warning sign on every path passing in its proximity. This sign consists of a vertical bamboo pole, a little bit higher than 1 m, provided with two bamboo sticks, about 30 cm long, crossing each other and plugged into a split-off part at its upper side, plus (near to it) a horizontally placed bamboo pointing in the direction of the trap, into which bamboo crosses (or at least a crossing piece) have also been set in.

The trap itself is set up on a deer path. The trap described here was set up by Klingtui Khongtör (of Tapwüa-Kua) about 2 m above a brook on a slope where the trail was narrowed down to approximately 60 cm by grown bamboo canes. Between these lateral bamboo canes two bamboo tubes (*ban*) were tied crossways over the trail at a height of 40 or 60 cm, so that the muntjac, for which this trap was meant, could still slip through under these bars, but could no longer jump freely past the dangerous place.

Approximately 10 cm over the ground ran a thin liana (*prün-leng* [5][452]), which at the right side was fastened to a side bamboo post of the gateway and to the left led to the triggering mechanism (*chot-kim*) of the trap. This "spear house" (*chot-kim*) lies only insignificantly uphill from the crossways bamboo poles of the *ban*. Once more two bamboo poles are tied crossways between two large bamboo canes (about 2 m distant from each other), this time approximately 20 and 40 cm above the ground. Into this frame lead from the upper slope two more than 5-m-long bamboo bars (*püt* [6]) tied together, which at their upper thicker end are jammed between two large bamboo canes, so that in relaxed position they push with their lower thinner end from the left against the right bamboo cane of the *chot-kim*. Into this thinner lower end of the *püt,* which juts out (about 30 cm) over the *chot-kim*, the end of the spear (*chot* [2]) is jammed so that it can neither slip forwards nor backwards, but within certain limits can still move to the right and to the left. The spear itself is made of bamboo and is approximately 3 m long with a set-in bamboo tip (*cau*) shaped like a spear blade and hardened in the fire. While the back end of the spear is jammed in the *püt*, the front part rests on two approximately l-m-long crossed bamboo poles (*king kai*) put into the ground on the left side of the trail; the height of the *king kai* (approx. 25 cm) determines the position of the spear.

To the right bamboo cane (or in a hole cut into the right half of the lower transverse bamboo) of the *chot-kim* a cord (*cakung-leng* [5]) of bark bast (*nam*) is fastened which reaches approximately into the middle of the *chot-kim* and is fastened there to the peg (*chit-pot* [4]). This peg is a thin bamboo wedge, approximately 50 cm long, and the mentioned cord (*cakung-leng*) is tied into a notch shortly before its thicker end. With this end the peg (with its top pointing to the left) is drawn over the upper bamboo of the *püt* [6] reaching into the *chot-kim*, where from the right side it comes to rest beside

[452] The numbers (in square brackets) refer to those used in the drawing.

the spear between the two bamboo of the *püt*, which for this is bent back about 60 cm to the left side, so that it is tightened. The tip of the peg is jammed below a ring (*long kua* [3]) which lies over the shaft of the spear [2] and at which also the liana (*prün-leng* [1]) leading over the trail is fastened. The ring is put at an angle when tightening so that any pull at the liana to the right removes it from the tip of the peg [4]. The bow (*püt* [6]) loses its hold, shoots back to the right and thereby pushes the spear [2] the tip of which lay immediately behind the left frame bamboo of the trail over the whole breadth of the trail. The shooting height is adjusted by the crossed bamboo rods of the *king-kai* so that the spear strikes the animal in its shoulder. Since the end of the spear remains jammed in the *püt*, it cannot be dragged away. In the case described, the barrier was laid out so that the ground below it steeply descended for 1 m. On the level under this steep place, six bamboo poles, approximately 60 cm long, sharpened above were plugged in, so

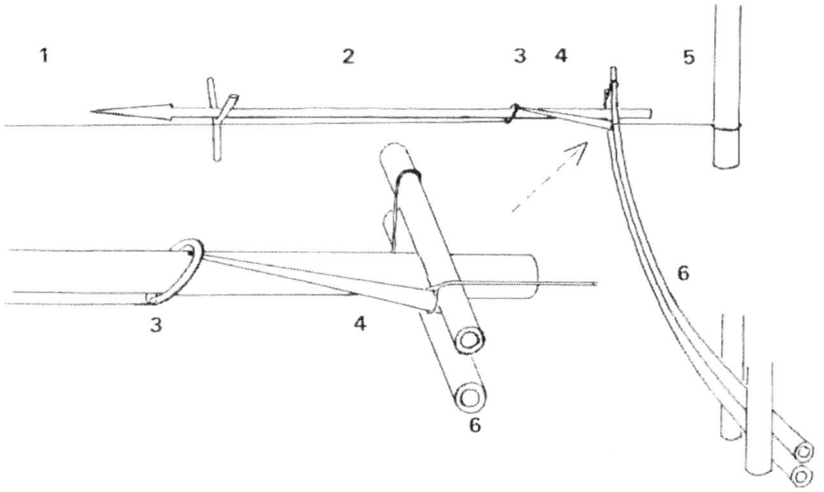

hatif the animal tried to escape nevertheless and jumped over the hurdle (*ban*), it would fall into this "spear pit" (*cau-ö*).

Gun traps are nearly unknown. They are only used when a head of cattle has been attacked by a tiger. The place around the injured animal is barred off so that only one entrance remains. When the tiger comes back and tries to pass through the entrance, it triggers the shot. I could not get a more detailed report of the construction, since no such cases were known.

The spear pit described is the only form of pit prepared today. Formerly covered pits (*wam*) were dug for people and elephants. Since neither may be caught and killed today anymore, pits are no longer needed and the details of their construction have fallen into oblivion. However, I heard that among the Khumi on the Arakanese border, pits were still used during the last World War.

H5 Lime twigs

Lime twigs are primarily used to catch little birds. The twigs (split bamboo sticks) are set up at an angle on fruit-bearing tree branches or also put into the earth at an angle at water places. The lime (*nai*) is drawn to threads and wrapped around the twig for about 30 to 50 cm. For the production of the lime the juice of the *pongma ching* (banyan; *Ficus bengalensis*) is used. For drawing the juice off in the top of the tree in several places at the same time, slanting notches are cut into the tree trunk. Small bamboo tubes (*röng*) sharpened in front at an angle are jammed into the lower end of the notches, which are filled in less than an hour. After taking off of the tapping tubes the juice is cooked at home in an old pot. Some juice of the *raima ching* (gorjon tree, *Dipterocarpus turbinatus*) is added to the banyan juice. This is necessary for keeping it smooth: the banyan juice alone would soon become hard. Also cooked together with the juice are *ching-dan* and *kau-dan*, joined wood and joined bamboo; they have no practical value but are used as sympathetic magic. It is forbidden to speak or eat while preparing the lime. When the juice begins to grow dark, it is removed from the fire, allowed to cool and water is poured over it: the water prevents the lime sticking to the hands, and therefore the hands are also moistened when wrapping the lime around the twigs.

Lime twig with caught bird (photo: C.-D. Brauns)

It is not necessary to use the lime immediately; it can be kept for months and is therefore poured into some vessel, for instance a halved calabash (MK, 5.07.1957).

H6 Hunting ceremonial

H6a) Restrictions for the hunter

Among the Mru most ceremonies have become frozen in an external stereotyped form and ordinarily no one is able to attribute to them any other meaning than that it is just custom, but nevertheless there are some rather clear ideas connected with the hunting ceremonies which are connected both with a live belief in hunting spirits and with a legend. The relative intensity of these ideas is astonishing in view of the minor significance that hunting has today. But there also are traces of decay recognisable as concerns the hunting taboos: the night before he intends to go hunting, the hunter should for instance abstain from sexual intercourse, a rule which is no longer unconditionally observed today. However, this idea of the incompatibility of the relations to animals and women, which in its primary form was originally widespread throughout the world, suddenly surfaces on other occasions: thus on 7.12.1956, Ma'ung Catumma in Moloi-Para wanted to show me his gayals. After he had been searching in vain for these self-willed animals, he thought it over for a moment and his eyes fell on the necklaces which some girls had hung around my neck during my tour after they had performed a *kham eng*, a drinking ceremony, with me. His face brightened and he added that now he knew why his animals were refusing to show themselves to me, and then with a grin explained the reason to me: they had probably caught sight of my jewellery; a man standing nearby intervened and forbade Ma'ung such "indecent" talk.

But the hunter should avoid not only women, but also a life of luxury. On the day when he sets up a spear trap he must keep a *caa-reng-khang* (hunter dish taboo), i.e., he may not eat any side dish with his rice. As mentioned already, there is an eating and speaking taboo connected with the making of birdlime. A particular rule applies to making hunting fences (*klong*): until completion, work must be done on the fence every day. It is thought that any interruption of the work would affect the success of the hunt, and if anyone of those employed with the fence falls ill, a decision will be made to stop the work completely, since it is not possible to continue. A connection to the cultivation of the fields is established by the rule that nobody may eat game together with the rice of the new harvest before the ceremony of the *caa-moi* had been held. Today, this rule, too, is only partly observed (MK, 10.09.1956). After a day when game has been caught with a spear trap for the first time, the next day no residents of the hamlet may work in the field, but must observe the *cak-kak khang* (game getting taboo).

H6b) Bringing in a kill

There are no special rules for bringing in small hunted animals. For bigger game, however, whether caught in a trap or shot with a gun, a special ceremony must be held. This applies for sambar (*nga-hiu*), serow (*nga-yaa*), muntjac (*nga-ki*) and wild boar (*pak-tia*). The villagers already learn that one of these animals has been shot, as those carrying the game call out as they approach the hamlet *kot-caa kot-caa üü*! (Carrying game, carrying game, ho!) The game is carried by two men on their shoulders on a strong bamboo tube which is put through the legs of the animal tied together. At the entrance of the hamlet (*plon*) two bamboo tubes are set up on the left side of the path (seen from outside), approximately 1 m high, the upper ends of which are scraped to tassles. Both ends are split, into one a leaf of *wia-*

chok is plugged (Tiliaceae: *Grewia microcos* L.) or *ram hoi* (not identified); it is not permitted to use other leaves; a bit of the left ear of the game is put on the other end, the left ear because in Mru *tawai praa* (= left side) at the same time denotes the "wrong way" which for the animal is the way to the trap.

When the carriers climb up the house stairs, some water is poured down from above. The head of the animal is cut off and put into the middle of the main room (*kim-tom*) on a banana leaf so that it faces the private room (*kimma*), then its body is cut up on the open platform (*car*). If available a jug of rice-beer is set up behind the head. When the game is a muntjac (*nga-ki*), an egg is placed in the mouth and someone spits *hom-noi* (water with cooked rice as a beer substitute), which is prepared in a little ceremonial earthen pot, *ci-ö-ca.* After the invocation (*tamma*) connected with it, the egg is taken out again. The head can now be chopped to pieces and in principle could be prepared like the other meat, but the hunter or trapper may not eat any of the meat from the head: this rule also applies to other game. If the kill is a sambar, a serow or a wild boar, instead of an egg a chicken is required for the ceremonies in the house. For one to two minutes, the chicken is tied to one of the animal's ears by a cord fastened to its leg. The *hom-noi* (rice-beer water) is prepared in a pumpkin bowl. The invocation, which starts with *phyoi!*, is addressed to Lima-Nam, the master of the wild animals. The head is left the whole night in the middle of the house and people dance around. A ring with tassel sticks (*chit-pap*, like those used for the *dong-lu-kau* at cattle feasts [see R1c]) is fastened to the back main post of the house (*klong-cüng*). Then seven little pipes are made: little bamboo tubes over which one blows. These little pipes are called *tum hut*; *tum* is said to designate the small pipes, *hut* was translated as "to blow over something", but *tum* normally means "to blow or sound (an instrument)" and *hut* means "bone", so that the thought seems reasonable that the original meaning was "whistling bone". On these small pipes special "songs", called *tum long,* are played, but generally not sung, unless one of the players does not know the tunes, so that a man must first sing the *long* in question (song for ceremonial occasions played on instruments). Anyone who wants to can blow the pipes and, if not enough men are present, one man may blow two *tum hut* at the same time. Whistling pleases Lima-Nam, the master of the wild animals; and the people young and old are also happy at being able to eat some of the meat, drink beer and *arak* and dance around the head until the early hours. At the end of the evening the *tum hut* pipes are put into the tassel ring (*chit-pap*) at the house post and the next morning the meat of the head is also prepared; the skull is kept for some time in the house (MK, 8.09.1956, observations with Reng'et Atwang, 13.11.1956).

Among the Khumi no ceremonies take place when a muntjac is brought home. For wild boar, sambar and serow two bamboo poles are set up at the entrance of the hamlet: something from the left ear of the animal is plugged into one of them, on the other any kind of leaves. When carrying the game up into the house some water is emptied out; the animal is cut into pieces on the open platform, its head set in the middle of the house and ceremonially spit at with rice-beer. But neither a chicken nor an egg is needed for it. In the night a dance is performed to the music of the little *tum-hut*-pipes and the head of the household rewards the players by presenting them with a turban. The next morning the head is skinned and pinned to the outer wall of the house. It can be thrown away after some time, but the Khumi set much more store on a "skull collection" than the Mru. If a leopard (or tiger, both are called *pri*) is killed, no ceremonies take place but care has to be taken that it is cut to pieces outside the house, and its head is on no account to be brought into the hamlet and therefore has to be thrown away beforehand (in an arbitrary place) (Müilün Tamchaa, 25.03.1957). I could learn nothing about special ceremonies with the Mru (including the Longhu) when a leopard had been killed; but in Moloi-Para I was shown a little leopard skull kept in the house, so that the ban of the Khumi might not apply with the Mru.

H6c) Animal spirits

It is said that the master of the animals, *lima-nam*, sees to it that not too many animals are killed. In general he is well-disposed towards people and is pleased by the blowing of the *tum-hut* pipes (see above). If, however, a hunter shoots too much game, Lima-Nam will cause him to die, which is why a particularly successful hunter is well advised to offer a cattle feast from time to time. Lima-Nam lives in *hua-puk* (Marma: *wan-tong*), a tower-like rock in which there is a cave (MK, 8.09.1956). I saw two such rocks in the southern Chittagong Hill Tracts: one in the upper Songu valley (Mibokya-Mouza) and the other near the Bawm village of Artha' (Komangkhyong-Mouza). At Artha' the rock or tower stands like a fortress of the European Middle Ages in front of a witness mountain, with differently coloured rock giving the impression that it is covered with a conical roof, and the Bawm say that a cavity in the shape of a door means that in the past the rock was hollow and people lived in it. The Marma who live in this region maintain that in the past the sounds of chickens and pigs could be heard coming from the rock and also human-like voices, whose language, however, could not be understood, although no living beings could be seen: so they inferred that the rock must be the dwelling place of a spirit prince (Kyaw Thwân Ong, 15.03.1956).

According to the Mru, the *mat* live in the *hua-puk* together with *lima-nam*. The *mat* are little spirits that accompany wild animals, particularly wild boars: when a wild boar has gone into a trap, their footprints (*mat-kung*) can be seen behind the animal like minute children's feet. The *mat* can lead people astray: they show them a way, like the one leading to their hamlet, but if they follow it, they must wander around for up to four or five days in the jungle without finding the right way. The *mat* also can leave a man up a tree or keep him as a prisoner in the *puk*. In such cases, i.e., if somebody is missing in the hamlet, the women go to the village entrance (*plon*) and pound in empty rice mortars and turn cotton ginning machines backwards: the *mat* then will release the captured person (MK, 8.09.1956).

H7 Hunting songs

H7a) The legend

The *tum-long* (songs played on little bamboo pipes), blown for sambar, serow or wild boar the night after bringing in the game, refer to a legend. The single verses (*long*) are not completely intelligible even to the Mru (the meaning of many words of the text is unknown), and the songs do not provide enough material to reconstruct the legend in its entirety. I therefore assume that these hunting songs are of considerable age and that the verses known today are merely fragments of a once bigger cycle.

It is said that there was once a man who had neither father nor mother, and his name was Klang-maica. In his field the paddy grew particularly well and his wife had a lot of work to bring in all the harvest. Now it happened that a big wild boar broke into the field at night and devoured the paddy. Klang-maica set traps, but he could not catch it, and every day he saw the exceptionally large footprints of the pig. Finally one night the pig was hit by a spear from a trap but it escaped. Thus Klang-maica set out, cut a piece of bamboo that just fitted into the track made by the pig, followed the animal's trail and arrived at the village of Cöng-cöi-ma, the daughter of the lord of the animals, who provided for the pigs. Cöng-cöi-ma invited Klang-maica for a meal, but forbade him to eat certain vegetables (which are mentioned in the song) and ordered him not to eat with his hand but to take the food with sticks. After the meal she called her pigs, but none was the right one; she called and called and finally a very big boar came, into whose footprint the bamboo cut by Klang-maica fitted. Cöng-cöi-ma scolded her boar and told him it was very wicked of him to devour this man's paddy. Then she sent her visitor home, telling him he would find the pig in his trap the next day. That night at home, Klang-maica dreamt that he was carrying lemons and tearing an old cloth, and next morning when he went to the field, in his trap at the brook, he found the gigantic boar, on whose back the dirt lay so thickly that plants grew on it. Since he could not pull the animal away unaided, he returned to the village to fetch help. But his wife scolded him: first he blamed her for not being able to carry home the paddy without help, and now he himself had to fetch help to carry a pig. When Klang-maica came back with his helpers to the pig, they were met by a tiger which had come to contest the kill. They told the tiger that, if he could drag the pig away, it would belong to him; but the tiger could not drag it from the spot and so the men carried it home. At home Klang-maica organised a feast and thanked Cöng-cöi-ma; they blew a tune on the piece of bamboo with whose help the pig had been found and sang for the first time the very *tum-long* that tells this story (Khamcöng Khongtör, 21.09.1956).

H7b) *tum long*

The text reproduced here was first recorded on tape and then written down and translated with the help of the singer, Klingtui Khongtör, and my main collaborator, Menkröi Ngarua (21.09.1956).

mang khi long
tabai mang kur ro-ui khai yua-lö,
tabai mang yang wan-tu khai yua-lö,
taruk war mum, war mang rong köi baa yua-lö.

Dream see verse
Tabai dreams he carries lemons, yualö,
Tabai dreams he tears an old cloth, yualö,
in the night, when one dreams, yualö.

waaca lai ngak long
rau-cin rau-nai, rau-cin töm-wang,
reng-ce-khö-kwak-, khö-raa-net-nöng,
chur-plon-tom köi,

ching-töm-klang-ri-, klang-bong-tawai-wa ö.
palai-chong ngak khai, yua-lö hon-ang.

Bird scolding verse
The dawn approaches,
at the place where one throws the little sacrificial baskets away,
at the thorn-armoured village entrance,
the Chingtöm bird cock scolds, yualö hon ang.

(Almost the whole verse consists of stereotyped poetic phrases; line 1 means "in the morning", line 2 and 3 "at the village entrance", line 4 contains the poetic name of the bird, the call of which announces the death of a man or an animal.)

tum chot long
praa le yua-1ö,
pai praa mang po, ting-che ban wöi yua-lö,
yu praa mang po, teu-che ban wöi yua-lö,
praa tawai, klöng-chong chan töi,
kaa ting chur chat, kaa tari,
kaa teu chur chat, kaa tari,
yua-1ö hon-ang panö.

Verse of looking after the trap
The sides, yualö.
Even if one goes to this side,
there is a *ting* bamboo bar (a trap),
even if one goes to that side, there is a *teu* bamboo bar,
the "wrong way", put the foot carefully:
if one gets a *ting* stab wound, one gets it properly,
if one gets a *teu* stab wound, one gets it properly, yualö, ...

(i.e., on the right and on the left of the path traps are set. Caution: traps! They are set well and hit for certain.)

kröi mü-mi long
yua-lö, yu-mi da, o-ce nang krang küi,
nga wön rua-ce, dük-dai ö, dük-dai ö,
dük nöm le, yua-lö, hon-ang pa.

Verse of discovering
Yualö, there: above at the bend of the brook,
the blue wild boar, fallen oh, fallen oh,
really fallen, yualö ...

wing-üng prum mi long
wing-üng kuai kot, lai ang khai,
plan chik, plan chik, chak wang döi, yua-lö,
wing-üng chur-cher, lai ang khai, yua-lö, hon-ang.

Verse of the bluebottle swarm
The bluebottles tell me: put it on the carrying bar,
let us go back, the heart does not come, yualö,
the bluebottles tell me: put it on the point, yualö

(i.e., the bluebottles buzz: load the animal on the carrying bar, turn around, it is too heavy, come back and impale it on the carrying bar).

pri hai takhün cak wak mi long
chüng pri chui-chük rak dö-ba,
pacöng krongma prün, yua-lö,
klangca chui-chük rak dö-ba,
panöm ploi ho ba, yua-lö, hon-ang pa

Verse in which they dispute the tiger's right to eat the dead animal.
The mountain tiger's prey,
like a root firm in the earth, yualö,
man's prey, as light as ashes, yualö ...

kot long
lu pre-pre yö, le pre-pre yö,
adong dapong, klangca reng arü tari, yua-lö, hon-ang pa

Carrying verse
At the head two men, behind two men,
in the middle the prey, so the people climb upwards, yualö ...,

kim kö wang ting mi long
yu-mi da, mia kua cö? Lai-hou kua;
khang ba? khang, khang döi, yua-1ö, hon-ang pe

Verse of arriving home
Which village is this there? Laihou village.
Do they keep *khang* (ceremonial shutting of the village)? Yes, no!, yualö

kup kua mi long
kup tan-dong lak, kup tari, yua-lö,
kup wang-bong lak, kup tari, yua-lö.

Verse of entering the village
Enter the village, enter the village properly, yualö (2 x)

hot mü mi long
rum-reng lang yap, rum-reng lang mü,
mü pong-ui min, khöt cam mün tu toi, yua-lö,
rum-reng lang mü, mü chai-ui min,
khöt cam mün khai, yua-lö, hon-ang pa

Verse of being on the look-out
The hornbill looks, the hornbill sees
a mature Ficus fruit, quite similar on the look-out, yualö,
the hornbill sees, sees a mature Banyan fruit,
so (the villagers) look, yualö ...

tui tut long / wang cak long
chong-ce kaa caa, kaa tui döi.
dam-ce kaa caa, kaa tui döi,
wang-cak carö, carö cak, cak krom-krom.

Verse of water shedding (rain verse)
The crab gets young ones, does not get water,
the fish gets young ones, does not get water,
the rain murmurs, murmurs, *krom-krom*.

(This verse refers to the shedding of water, when bringing in the prey. No one knows how to explain this custom or how it is to be connected with the mentioned animals.)

li khöng long
cin-dong li-khöng, khöng, lai khöng lai, khök-khök, yua-lö, hon-ang pa.

Likhöng verse
The Li-khöng sings its call at night: *khök-khök*, yualö ...
(The call of the bird announces that there will be dancing at night, i.e., a feast will be celebrated.)

caa reng khang long
le yua-lö mi, le yua-lö, cang caa-reng khang-mi da,
hang caa chong-chin chong-wöng kan,
hang caa. chong-chin chong-nat kan, yua-lö, hon-ang pa

Hunter-taboo verse
He yualö ... you who keep the hunter dish taboo, do not eat (*chong-chin*, etc.) vegetables, yualö, ...

pak-mi long
achek-arek akom ayong, ram che pak-la, le yua-lö,
döm-ko lau lam, lau bong chüm-tu,
achek-arek akom ayong, duk-duk klöng-klöng, duk klöng-klöng.

Verse of the pigs
… cleft ear boar, he yualö, ...,
similar to the end of the bamboo cane, to the bamboo sample,
... *duk-duk klöng-klöng* ...
(According to Klingtui the first line refers to the names of the daughters of Cöng-cöi, the owners of the pigs. But perhaps it also means "the form of one hundred claws". The cleft ear boars are the wild boars. The pigs are called to their food with the call of the last line [*duk-duk*]. The meaning of the verse would therefore be: The daughters of the Cöng-cöi call the herd of wild boars (so that Tabai can look and see if their hoof imprints are like the one that matches the bamboo cane).

pak toi long
ting bok ting ro, ting khöm do-dak, kha-e ka,
ting bok mau-ram, khöt nak döi, yua-lö,
lang-ang krüng-rüng khai, yua-lö, hon-ang pa.

Pig-feeding verse
Broken bamboo, rotten bamboo, bamboo lying down; when
the bamboo is broken down, like the *mau-ram* tree, yualö,
it juts out visibly, yualö ...
(i.e., just like the *mau-ram* [= bamboo leave] tree towers above the rotten, broken down bamboo, the wild boar surpasses in height the other pigs called to be fed.)

The two last verses together are also known as *Cöng-cöi pak lai long,* Cöng-cöi's pig calling song. Cöng-cöi is also an important figure in other Mru beliefs concerning spirits (cf. P1c), but only in the *tum long* does he (or she) play a role as a lord of the wild animals. That nowadays the lord of the wild animals is called *lima-nam* disturbs the Mru just as little as that the hero is called Klang-maica in the legend and Tabai in the verses.

H8 Fishing

H8a) Fish traps

Fish of every kind is eaten with great pleasure. In the hamlets on the top of the hills, far from the greater watercourses, however, there are not many possibilities of catching fish, therefore dried fish is bought from the Bengali pedlars visiting the Mru hamlets on the hills, or from the market.

What the trap is for hunting, the fish trap (*cong*) is for fishing, which may be set with or without bait and with or without blocking part of the watercourse. If no bait is used, as is usually the case, ordinarily a dam is erected. The smallest of these dams are to be found in narrow stony mountain brooks. Little fish, barely 10 cm long, take shelter in such waters under flat stones and boulders: the sides of such cavities are blocked with branches and leaves and a little fish trap is set in front of the opening that remains. These small, approximately 30 cm long, fish traps (oftentimes bought from the Bengali) are also used in flat sandy brooks. There shallow places can be blocked with little stones and the fish traps set into the openings of these miniature dykes. The blockage causes the embanked places to dry up slowly and the fishes get stuck in the fish trap as they try to withdraw. The process can be repeated by opening the dykes from time to time. Setting up such shallow water dykes is the work of the women and takes place mainly in December and January.

With brooks that divide into two branches (an effect which can also be induced artificially by building a dyke lengthways in the bed of the brook), a fish trap is placed in the deepest place of the lower end of the water branch (it is necessary to block the places to the right and left of the trap only if the branch is too broad) and when the upper part of the branch is barred with stones the water will carry the fishes in the branch into the trap as it flows out. In small watercourses one man alone can cope with this work; when the flow is stronger and where it is necessary to dam simultaneously in several places the work is done by a collective, as for instance in the rapids of the Songu above Thanchi, which is called by the Mru *tang-bung-hor*: as many inhabitants of a hamlet as possible, up to 30 people, come together here to create branches in the river bed many metres long that can be dammed alternately with the help of big stone blocks. The work is worthwhile because as much as 20 to 30 kg can be caught in a single day. The size of the fish traps is adapted to the availability of fish; particularly big fish traps, up to 2 m long, are used after the rainy season (in September and October).

Where dams are not possible, baited fish traps are set facing at an angle into the current on the edge of the watercourse. The fishes are usually caught at night when they swim into the quieter shore waters searching for food. Burned chicken bones or old yeast wrapped in a cloth serve as bait. Crabs are also caught with fish traps, baited with the thick skins of jackfruit, cucumbers, roasted maize or sword beans, depending on what is available at the time.

A fish trap has been mentioned already in chapter D4i), where I reproduced a drawing by C.-D. Brauns. Here follows a description of the manufacture of a fish-trap. All fish traps are woven from long bamboo thongs. As an example I shall describe here a fish trap of medium size, 50 cm in length, with a diameter of 24 cm. The thongs are 1 cm broad and spaced about 3 cm apart in the lengthways wickerwork. When weaving one starts by putting ten (approximately one-and-a-half-metre long) thongs on top of each other so that they form a star. Around the place where they cross each other a narrow double thong (about 3 mm broad) is first woven, one up one down, in a circle of about 12 cm diameter. After the middle is determined and fixed in this way, a new thong of normal breadth (1 cm) and as long as possible is folded together and put (about 8 cm away from the middle) around one of the thongs of the star so that the ends of this new thong point to the left. These ends are now woven while they are twisted together in the circle around the star thongs so that they cross each other after every thong of the star, and are raised a little and pulled closer together. After the first round the two ring thongs are not twisted any more but are now led on side by side, instead of behind and over each other, while they pull up the thongs of the star vertically, that is to say along the length of the fish trap. The spiral remains as narrow as possible and the type of weave is always the same: one up one down. When the ring thongs have to be prolonged they are led doubled over more than three perpendicular thongs, at the end they run out freely. At this upper end every vertical thong is bent inside counter-clockwise at a right angle, then led outside again over the next but one vertical thong, twisted in itself, pulled up and back down again over the bulging rim, plugged in again and led to the outside once more, to be finally plugged in with its end behind the next vertical thong,

This open end of the fish trap can be closed by a removable inset (*chür*). The length of the inset is 20 cm, and at its widest, like that of the fish trap opening, 24 cm. The lower 12 cm of the lengthways thongs (of 1 cm breadth) are not woven and form a passage that narrows down to about 4 cm diameter, after which comes a doubly woven thong twisted around itself. After a further five spiral rounds of the horizontal thongs, led flatly and slowly widening, the lengthways thongs are divided in two and are now much further apart from each other, while the ring thong continues in a spiral for the last 8 cm of another four circles. The rim is formed like that woven for the tube of the fish trap, but the bulge is thinner since the vertical thongs have been halved. The inset is tied into the tube opening with two little strips before the fish trap is set.

H8b) Other ways of fishing

Fish poisoning, an old method of fishing, is now forbidden by the government. The roots of a plant called in Mru *lu-ri* (not identified) were used to make the poison. The roots were crushed in the water with a piece of wood and for this they were put into an open bamboo basket (*bang krang*), so that they could not be washed away. The watercourse was not dammed, but a place was chosen for crushing the roots above a narrowing of the course where the fish floating on the water could be gripped by hand. Shallow water was needed for this method; wherever the water is deeper, fish-nets (*lok*) are used. Although the Mru do not know how to produce these nets themselves, and have to buy them from the Bengali, nets have probably been used for a long time (see *büi ko chang-cia*, R8b). Fishing hooks (*kwai*) are also bought from the Bengali – today mostly imported from Italy. They are tied to a line, fastened to a bamboo rod. Earthworms are used as bait, but floats are not usual. Since one usually has something more important to do than wait for a fish to bite, the fishing line is tied to a *dong* (see H2b) and the rod used as a spring-pole: when a fish pulls at the bait, it triggers the trap mechanism and is pulled into the air (MK, 3.11.1956 and 19.06.1957).

Often one sees groups of women wading in shallow brooks, each with two baskets, a small *par* (see D2e) in the hand and an *em* (see D2a) over the shoulder (for a photo, taken by C.-D. Brauns, see K1b). They lift up stones and catch little creatures, which try to swim away (small fishes, fresh-water crabs, crayfish, and snails), by holding the *par* in the water, and from time to time tipping the catch into the *em*. Crabs are also caught by hand, mainly in the darkness with torches. Fishing with the hand, as practised with the Marma, however, is not usual among the Mru. The Mru do not use the spear to catch fish while the Marma use it in two different ways: either with a short handle for stabbing the fish – the fisherman goes and stands on either a stone or on a block rising above the water –, or with a long shaft for throwing at the fish. For this spear, the two-pronged iron, which is approximately 20 cm long and bought from the Bengali, is mounted on a thin bamboo tube about 5 m long; the water supports the shaft, but the tip with the iron sinks a little below the surface. Standing in the water singly, or by twos from a boat, the fishermen hurl the spear. While the fish may notice the net thrown after it and can escape, the spear arrives without warning, and skilled fishermen are more successful with the spear (the catch being primarily a fish called in local Bangla *thanda mas*), than with the net.

H9 Use of bees

Since bees are neither captured nor artificial hives made for them, people depend on finding wild bees' hives. The main purpose is the extraction of honey and the gathering time falls mainly in the months before the rainy season, i.e., April and May. The right to the honey is claimed by the one who discovers the hive first; it occasionally happens, however, that two men claim the right, they quarrel, and if the one who has to withdraw is annoyed and bears his neighbour ill-will, he beats the tree on which the bees are to be found with garlic: the smell provokes the bees and they will sting anyone who tries to approach them so fiercely that for some time the idea of exploiting the hive has to be given up.

However, not all bees sting similarly and in some cases the hive can even be cleared without smoking it out beforehand. Since no protection is worn against the bees it is advisable to go at night when the bees are less inclined to sting. Normally the gatherer climbs the tree and first smokes out the hive using torches made of bamboo slats wrapped with leaves. Then the lower part of the hanging honeycomb containing the larvae is cut off with the hewing knife; a clean petrol tin is held under it and the rest is cut off. If a tin is not available, a round basket with a handle may be used (like the baskets the Bengali hang on their carrying bars), which has been sealed up beforehand with loam. The tin or basket is tied to a rope and let down from the tree. The honey to be gained from a honeycomb may weigh 5 kg, and in particularly good cases as much as 15 kg. The honeycombs are pressed out by hand, and the honey that flows out is sifted through a cloth so that any adhering dirt stays behind. The part of the honeycomb containing the larvae is also pressed out; the larvae drip out like milk, are caught in a pot, cooked together with spices until they coagulate, and consumed. The honey is used mainly as a medicine; it is taken against stomach ache, for instance, mixed with lizard gall, and is painted into the oral cavity of babies, if they develop a sore mouth a week or so after their birth or later.

The pressed-out honeycombs are cooked in bamboo tubes; the resulting liquid is poured onto water so that the dirt is eliminated; the remaining wax is liquified again by heating and poured into small bamboo tubes, where it cools again and is shaped into candle-sticks. This beeswax (*yon*) is indispensable for textile work (see F2a).

Another kind of wax (*kwai-ut naa*) is produced from the honeycombs of earthwasps. Their hives are discovered mainly when weeding the fields: the entrance to the comb is a small hard cone of earth. Since these wasps do not sting, the honeycomb can be dug up straight away with the hewing knife. The honey can be eaten, but is barely sweet; the part with the larvae is not used. The honeycombs are very important for the people: they are pressed out by hand and cleaned in water to remove any remaining earth. The wax is not cooked but only kneaded together after it has been washed. It is used as an adhesive, particularly for the gourd-pipes (*plung,* see E1a), which could not be produced without this wax. It is dark brown in colour and, to a European, its consistency is reminiscent of plasticine (MK, 15.11.1956).

A species of little bees is kept in round clay pots: These bees are only as big as flies and do not sting. A little hole is knocked into the wall of the pot and the opening of the pot is closed with a calabash which is stuck to the pot with wax. Preferably in April the pot is crushed, the bees are driven away with smoke and the honeycombs emptied. The wax is taken out and what is left of the hive is placed into a new pot and hung up near the door entrance (MK, 31.10.1964).

J Swidden cultivation

J1 Land acquisition

J1a) The quest for land

The text below was written 50 years ago. The situation described here therefore corresponds to what I found during my fieldwork. Due to increasing scarcity of land and Government influence it has changed during the last 50 years, at least in part.

Swidden cultivation provides the main source of sustenance for the Mru. Corresponding to the oikotic conditions, in the hilly areas, it is almost impossible to use the land except by swiddening. Terracing of slopes has never been practised anywhere in the Chittagong Hill Tracts, since the means and knowledge for building terraces are absent. That terracing would be completely impossible is doubtful; in most places, however, the stones that would be needed to reinforce the ground are simply non-existent. In order to prevent the monsoon rains from washing away the thin layer of humus, it is not possible to clear the ground by removing the roots of the old growth. Therefore only the vegetation above ground is cut and burned and the root system is left intact. In contrast to the type of cultivation where old growth is uprooted, however, this so-called swidden farming puts a double burden on the soil, for in addition to the old root system new plants are also put into the ground. The available soil nutrients are quickly depleted, and the yields decrease accordingly if a field is cultivated in this manner two years in a row. In order to restore the former fertility at least partly, fallow periods of many years are necessary, and during such periods a secondary forest grows up under which the nutrients can replenish themselves.

Relatively flat lands are rare in the mountains, and should there be some occasional small flat patches as natural formations, swidden cultivators confront them with a certain helplessness. In areas with broad valley bottoms, as for instance in some places near the Matamuri river, the Mru also followed the Bengali example and got used to embankments and irrigation of the fields, and even in the hills some people believe that the possession of a level field might offer them the chance of a sure sustenance. Basically, however, the Mru are not plough cultivators, and nobody has a mind to spend his money on a piece of land in the plains, and so most flat lands to be found in the Mru area are in the hands of Bengalis, Marma or Chakma. The Mru were and have remained hill dwellers who engage mainly in swidden cultivation on the slopes surrounding their hamlets.

The main crop is dry rice, and the yield of paddy is decisive for the prosperity or the need of the peasants. Not every plot of the soil produces the same yields, and the choice of the land and the size of the cultivated area are therefore of crucial importance for ensuring sustenance. To be sure, the favour of the weather, primarily during seed time, also plays a decisive role for the success of the harvest, but the weather is not among the calculable yield determining factors. The experienced peasant, however, has a good eye for the quality of the soil: the forms and strength of the previous vegetation yield the main clues. To be sure, everyone endeavours to select the best possible place for the seed and to keep the necessary work to the minimum. An ideal place would be easy to access, simple to clear, with scanty weeds and a rich yield. In areas newly opened up for cultivation, the best places will be swiddened first. But by the next year or the next but one, new plots will have to be cultivated, because once used, soils are not fertile every year, so one must move on and cultivate new plots, give rest to those that have been used and let the jungle grow back. In this way, new pieces of land are cultivated year after year in the surroundings of a hamlet, which means that if a man does not want to move his dwelling, soils of inferior quality must also be accepted.

Where continually decreasing yield must be expected, an appropriately bigger area must be cultivated. As long as the country was sparsely populated and there was a wealth of choice, no particular problems arose; a densely populated country, however, makes it necessary to take the interests of others into consideration also: since everybody wants to have his share of good soil, everybody must accept some soil of lesser quality. Today the whole area is densely populated and the times when virgin jungle could be swiddened are past. In former times (at the end of the 19th century) a patch could be cleared and if the first harvest turned out well it might be used for a second year; today one must be glad if the first harvest turns out adequate. Formerly the swiddened places could be allowed to rest until they were used again after twelve to fifteen years, by this time the soil could recover and be replenished with nutrients; today such a long fallow time is granted at best to soils of such inferior quality that they would produce too poor a harvest if left fallow for a shorter time. Good soils will be swiddened more frequently, since otherwise neither the need for soil nor the food supply could be met, even if it meant damaging the fertility of the soil in the process. The related problems will have to be examined in greater detail.

J1b) Legal rights

In the 1950s it was still theoretically possible for everyone within the whole area of the Chittagong Hill Tracts (with the exception of the forest reservations) to choose a patch for swiddening in any arbitrary place. Only a few areas with level fields and mountain places with orchards were registered as private property. Due to Government pressure, this situation has changed fundamentally during the last 50 years. In the following text, however, I will describe the situation as it was in the 'fifties', even though today there are only a few places left where the old rules still apply and land is available to everyone, provided that a "jhum-tax" is paid for its use. The tax (6 Rs. per annum) is not dependent on the size of the area brought under cultivation, but it is limited to the mouza in which the taxpayer lives. If he cuts a swidden in a neighbouring mouza, he must pay a second tax. Therefore there is a certain restriction to the area of his mouza; the boundaries between hamlets of one and the same mouza, however, are not fixed. Understandably, everyone wants to have his fields as near as possible to his own hamlet; some hamlets, however, lie rather close together and only habit and custom fences off from each other the areas used by the individual hamlets. It can happen that a man has to walk more than half an hour, or even more than an hour, in order to reach his swidden, while the neighbouring hamlet can be reached within five minutes. There are almost no proper legal measures against encroachments on the area of another hamlet, but as a rule these are avoided for the sake of decency and good agreement.

Just as the hamlets have little claim to a certain area, the single villager also has few guaranteed rights to certain areas. There is merely a customary rule that, after the necessary fallow time, a man may re-use an area which he once cultivated, but frequent exceptions must be made to this rule, since the households do not remain constant in size and therefore their annual need for land varies. In cutting a swidden some considerations of a practical and religious nature also have to be taken into account (see below), which make a redistribution necessary. Therefore every year before selecting the swidden, the family heads hold meetings to discuss the distribution. Everyone informs his co-villagers about the plots he is claiming and the assembly co-ordinates the plans. Food needs, expected yields, available workers and distribution of the soils are discussed, claims are accepted or rejected, and it may be that someone gives way or several combine forces to help someone in difficulties. Every larger household does not cut only one swidden; the reasons for that are for one thing the size of the available areas, for another soil qualities and finally opportunities can be distributed better in two places and unforeseen crop failures in one place may find a certain compensation in the other. The decisions taken by the assembly are not absolutely binding, and it may still become necessary to change the arrangements during the cutting of the swiddens (for instance if a reason is detected which forbids the continuation of work in this place or if there is a conflict with a neighbouring hamlet), but everyone who attaches importance to friendly relationship with his neighbours will hold on to them as far as possible in his own interest.

J1c) Ecological restrictions on land utilisation

Places covered with bamboo are considered ideal swidden areas; but precisely this bamboo is permanently required for house building and the production of baskets and equipment, etc. Therefore, in the general interest, a certain area of bamboo should remain uncut near the village: on good soil and with correspondingly good growth; an area of 1–1.5 hectares will suffice for ten houses. However, the increasing demand for soil has also contested the bamboo's soil: in many hamlets nowadays the bamboo grove has disappeared and the villagers are often forced to search for bamboo far away from the hamlet, in places unfit for cultivation and with correspondingly poor vegetation. Given enough fallow time, the bamboo will grow again on swiddened areas, too. However, if a place is burned five or six times with less than five years fallow in between, the bamboo will die out completely there. With short fallow periods there is also a danger for the growth of trees; both big and small trees are needed to build a house, although not very frequently, but good timber does not grow so fast, and so bigger jungle spots with utilizable timber must be kept in reserve. Furthermore, areas with at least rudimentary jungle are needed (although these places can change from year to year) to allow the cattle to graze during the day time, and these areas should be adequately connected and easily reached from the hamlet, so that the cattle can be prevented from encroaching on the fields.

Areas overgrown with *Imperata,* which increase every year, are not suitable for swiddening, since this grass grows too fast, too luxuriantly and obstinately to be weeded with success. Moreover, to avoid the formation of further and ever growing *Imperata* grass areas, it is necessary to take care that new swiddens are not cut too close to a fallow field in successive years: when burning the new swidden the fire may spread into the field lying fallow, and may find special food in the dry grass and destroy the new vegetation, while after the first rain the grass fertilized by the ash will grow up even more luxuriantly and will suffocate all other slower growing plants. An area that is burned every year for three to four successive years will be completely overgrown with *Imperata* and become useless for cultivation for at least the next decade. The Mru experience fewer problems with areas overgrown by *wöi-la-ram* (Bangla: Assam-Lota; *Eupatorium odoratum*). Although this creeping brush makes cutting and cleaning the swidden area more difficult, its ash is a good fertiliser, weeding is not especially difficult, and the yield is not less than in other places.

J1d) Religious restrictions

Every area has its spirits which must be treated with due consideration. Near particularly dangerous spirits no swidden can be cut, the individual cases and spots being precisely determined. Thus, for instance, no swidden is cut on a slope below which a *nat-hua* (spirit-stone, see P2f) is to be found, and whenever a *rui-yong-kar* (liana-like-crossing) lies over the watercourse: a fallen tree which has put down roots on the opposite side. Other places that are not suited for a swidden are sites where *tui-ching*, *rua-lim-kua*, *kau-pir* or *kwar-yua* are to be found, since all of them are considered to be dwelling places of evil spirits: *tui-ching* (water-green) are moist places where water seeps from the ground, without forming a spring; *rua-lim-kua* (python-hole) are holes mostly filled with dead water, considered to form the entrance to the dwelling of some Naga, a snake-like water-spirit (compare the legend of the Boga lake); *kau-pir* (bamboo-double) are bamboo canes with forked shafts; and *kwar-yua* ("Bengali-umbrella") are mounds in the form of a grave. (Understandably enough, places formerly or still used as cemeteries are not thought suitable for preparing a swidden either.) If such places are discovered only during the cutting of the swidden, the work immediately stops.

Regarding the placement of a swidden, it is forbidden to cut it between two swidden fields belonging to another man (*üa takot* = swidden across). It is also forbidden to cut three swiddens, but every other number is allowed. One and the same man may not cut one of two opposite slopes (separated by a brook) in one year and the opposite side the next year, while some other man may use the counter-slope (*üa tang*) without danger. A swidden is not allowed to have two upper ends, and when it is used again after lying fallow it may have the same size only once. When the place is cultivated for a third time, the swidden must be cut either in a larger or in a smaller size. In all these cases if the man who cuts and cultivates the swidden fails to observe the rules, he must expect the wrath of the spirits: either he or someone of his family may fall ill or even die.

Two persons will not normally cut two swiddens side by side, though it is not forbidden to do so. But if something edible grows on the boundary between the swiddens, it must not be used: otherwise *ra-hi-nam* (a border spirit, also called *chüng-hi-nam*) would send eye disease. If somebody cuts a swidden in an area of another hamlet, though he may have the approval of the villagers, he must avoid passing this hamlet on his way to it and back from the field; when carrying home the harvest he may be allowed to pass it, otherwise he will incur the anger of the villagers, particularly if he should carry some game or rodents through their hamlet, since this would diminish the harvest prospects of the villagers (MK, 13.9.1956, 24.06.1957).

J1e) Selecting a swidden (*kir üa*)

The discussions in the hamlet about the land distribution for the next year start soon after the harvest (provided that one can talk of a beginning for arrangements which will last for years): everyone can then clearly see his conditions arising from the last harvest and, when the heads of the families come together for a celebration in a house, there is also something to drink so that the problem of mutual co-ordination can be tackled more easily. When, after consideration of the ecological conditions (the distribution of old and new plots in relation to each other must be carefully considered; for under no circumstance, even accidentally, should old plots be burned off twice) and observing the rules of a religious nature, everybody has announced his plans and claims for the new swidden period and an (at least provisional) agreement has been reached, one day in November or December the men proceed to the *kham üa* (promise of the field). On the edge of the future field a short bamboo cane is cut, a small tassel is scraped at its upper end, the tube is beaten into the ground on the side of the path near the field, the upper end is split up and a short little stick put into the opening. By this sign (*plang kap* = sign-split) the field is marked: nobody can come afterwards and claim the place for this year. The following night the dream is of decisive importance; if the dream turns out bad (if a hewing knife gets broken, a tooth falls out, a chicken is sacrificed or fire breaks out (cf. the significance of dreams, P2f), the swidden place is abandoned or, nowadays, the chosen site is at least slightly altered. Since the sign (*plang-kap*) put up on the edge of the field is not removed when giving up the place (although it is up to everybody to use this place in his turn), its function as a property sign seems to be secondary: its original meaning (although no sacrifice is made and no spirit invoked) must have been to inform the local spirits of one's intention to prepare a swidden in this place. Seeking their consent for the further work in the swidden is no longer necessary: all other work, including burning, sowing and harvesting can be started without previous observation of one's dreams (MK, 24.6.1957).

J2 Preparing the swidden

J2a) *cia-üa* (slashing the swidden)

At the end of January work starts on cutting the swiddens. Anyone who is still busy with other activities (building or repairing a house, celebrating a feast) can start later, there is no particular or generally binding date for the beginning of the work. February is the main working month, and everybody should have finished cutting his swidden by the middle of March, so that enough time is left for drying out, before the burning starts. In the past, cutting the fields was exclusively men's work, but today in many places, when brushwood has to be cut, where the work is not hard but a long drudgery, more often than not the women and even the children must help, too. The main burden of work falls, however, on the father of a family and the adult sons, if there are any. For the work to go on better, it is usual for several men from different houses to co-operate in a swidden, whose owner then for his part provides them with a working time corresponding to their help. This does not decrease the quota incumbent on the single worker, but working together makes the work more pleasant and increases the "fitness" for work. If someone prefers to work alone, he may do so, and nobody will impose himself. This system, bound to no firm rules of teamwork, also applies to all further work in the swidden. It is of special value when there is no man available in a house, so that the woman must take on all work alone: her neighbours may well take her position into consideration when the land is distributed and allot her a place that is not too difficult to cultivate. However, if bigger trees have to be felled, she will need the help of a man, whom she for her part then helps to clean his field. Young women are able to do all swidden work by themselves if necessary, as can be seen in cases in which (contrary to the rule that the fields are cultivated by the individual families) young girls combine forces and cut a little swidden place of their own, the yield of which they will sell (mainly burned to spirits) in order to buy jewellery for themselves, which then will become their private property. The young girls, who make a swidden together, need not be related by kin; friendship forms a sufficient basis (MK, 9.04.1957).

One starts with slashing the swidden, like with all work on the field, at the lower end of the slope and then proceeds upwards without rule, depending on the ability of the single worker. The only tool used in the first operation is the hewing knife; everything that can be chopped down with it is cut off in this same go: brushwood, small trees and bamboo. Whenever possible cutting with the hewing knife is done from the right above to the left below, and everything is chopped off about 10 cm above the ground. With bamboo this method produces very dangerous (dagger sharp) stumps, but a horizontal blow would bounce off at the shaft of the bamboo. A difficult and lengthy task is the cleaning of areas overgrown with "Assam lota" (see J1c): creeping on all fours in order to cut off the stems below the brushwood from the ground and chopping from above in order to separate the interwoven lianas from each other, piece after piece is pulled off from the thicket.

New swidden cut recently (in 1957)

In a second operation, trees that were left standing when using just the hewing knife can be either felled or trimmed. The more common and older method is to fell everything and thus make the field completely clean because living trees constantly suck nutrients from the ground and little can grow within the radius of their roots. When the trees are felled, one usually does not take the pain to lop off all branches, in the hope that the fire will finish the work later. In recent times, however, bigger trees are also left standing (because they are too thick and the wood is too hard – they may be needed later for house building purposes – or because by leaving single trees one wants to support the regrowth of the jungle during the fallow time, because the tree growth has continued to deteriorate, or just because – even though it spoils one's reputation – one wants to spare the work. With smaller trees all the thinner branches are lopped off starting from below, so that all that is left is the stump, which often dies and chars during the burning. Bigger branches are chopped off with the axe; till now saws are rarely used for felling trees.

The time necessary for cutting the swidden cannot be determined exactly. Apart from the fact that not all places are equally densely covered with vegetation, the speed with which one can clean a swidden depends completely on the number of trees. The least time is needed with a good growth of bamboo. Brushwood and particularly hardwood require considerably more time. It can take a man a whole day to fell a hardwood giant; whereas a smaller swidden area is needed on a place where there are many old trees growing, since the soil still contains more nutrients and promises a better yield. Bamboo soils promise the best yields and provide the optimal conditions, because at the same time the work is relatively easy: a man can cut about 10 ares per day and by doing so clean a field big enough to maintain a family of five within two weeks (but only if he has the necessary area of such good soils at his disposal!). Under the current soil conditions, however, almost double this area of swidden must be cut so that with mixed vegetation one must reckon with four times as long and in unfavourable cases with an even longer time.

J2b) *tök üa* (burning the swidden)

From the beginning of March until the beginning of April the vegetation cut on the swidden will dry out. During this time the days are very sunny and warm, with the noon temperature always reaching 30 ° and oftentimes as much as 35–39 °. Rain falls very rarely; between 10.03 and 10.04.1956, there was only one thunderstorm (13.03, 18–20°°) and a short shower (1.04, 23°°) and in 1957 only one shower (1.04., 6–7°°). The atmospheric humidity usually sinks to less than 40 % during the day time, but it can reach 100 % again and occasionally fog will still form in the valleys overnight. But even in the valleys it takes less than a month for the cut brushwood and jungle vegetation to dry, and fire is set to it in the middle of the first half of April. April is the hottest month of the year. The preceding days are not passed in complete idleness (although there is now the opportunity to celebrate feasts that had to be postponed, because the preparations for them could not be finished before cutting the swidden). In all hamlets where a swidden is to be ignited near the habitations, breaks must be cleared in order to prevent the fire spreading. Corridors of 1½–2 m breadth are cleared of everything combustible that grows or lies there.

Whenever possible these breaks follow a path that will reduce the not insignificant work in which all young men, including the older boys of the hamlet, participate, whenever it is necessary to protect the hamlet.

Anyone who has an isolated field can set fire to it whenever he wishes; but if there is a possibility of the fire from his swidden spreading to another person's swidden, the owners of the places in question must reach an agreement on when to start the fire. These agreements must also be made between hamlets (by arrangement of the karbari of the individual hamlets with the headman of the mouza): firstly swiddens connected by proximity must be set on fire at the same time, secondly it is desirable not to set fire to too big areas at the same time, in order not to bring the jungle areas in between into the vortex of a fire circle. Before setting fire to a swidden, paths crossing the field are blocked by bamboo thongs bent to a bow and put into the earth at the two ends (*kung-kaa-kar*): Strangers who might use this path are thereby warned not to enter the field and put themselves in serious danger. To be burned on the swidden is a form of the "evil death" (see Q2a). Before actually igniting, a little fire is lit next to the pathway on the edge of the field. In this fire long bamboo slats (cut beforehand) are ignited; with these torches one then runs round the swidden setting fire to particularly combustible places at shorter and longer distances. It is necessary to work quickly for reasons of one's own safety and the uniform distribution of the fire. The fire quickly blazes up and, getting stronger and stronger, burns from all sides to the middle of the field. Thus the vortex of the fire is steered away from the edges, but the fire also creeps along the ground into the surrounding jungle. If, in spite of the surrounding cleared corridors, flying sparks and smouldering ashes cause the fire to spread across the jungle floor, no one is particularly concerned. The smoulder spreads only among dried leaves and grass and goes no further than the next path. Live bushes and trees remain basically undisturbed by a fire of this type, for in spite of the preceding six months of dry season, the nightly fog still prevails and the humidity is high enough to keep green wood from catching fire. Even in the field, some trunks and branches may be merely scorched by the fire, but not reduced to ashes.

The fire is usually started in the morning or evening, because in the dry season winds often arise after noon and can drive the fire in undesired directions. Because of possible gusts of wind and the flying sparks connected with these, particular care must be taken if a swidden is burned very near to a hamlet: all inhabitants are informed beforehand and everyone stays apprehensively in his house to save what can still be saved if necessary. All the roofs of the houses are covered with grass (*Imperata arundinacea*) or bamboo leaves, materials which burn like tinder in the hot season, so if a roof catches fire the house is lost and often the whole village with it. In order to protect oneself against the consequences of possible flying sparks a man sits on the roof of every house, provided with a water vessel, half a calabash as a scoop and green banana leaves as beaters (Menring-KP, 3.04.1956). But despite all precautions it happens from time to time (as also in 1956), that a hamlet burns down during the firing of the swidden. Since this means that the paddy stored in the houses is also burned, at the same time not only the basic food supply for the next half year is lost, but also the seed paddy for the coming season, with the result that until the next harvest the people must go begging and will probably also have to starve for the next years, a situation which turns out bad enough under the interest and working-off system, but might lead to lifelong ruin as a consequence of the rice price speculations prevalent in the 1950s.

Burning new swidden fields

J2c) *rat üa* (cleaning the swidden)

During the days when the swiddens are burned little can be seen of the blue sky and the sun as the whole sky is covered with a grey veil of smoke. On the day after the fire thin smoke trails still climb up from the swiddens and big tree trunks and stumps that the fire did not consume completely continue to smoulder. And now the dirtiest work of the year starts for the men: like black pixies they climb around in the field looking for unburned branches and collecting them. Everything that can be picked up and that can be got rid of is taken to the edge of the swidden and stacked there on a heap (*mung*). This wood will later serve as firewood whenever a meal is being prepared while working the field. If bigger trees have been left (hard wood only chars), all branch stumps are cut off and the trees themselves divided if possible, so that every piece can be rolled or dragged away. Frequently the remaining embers must be extinguished before the wood can be touched. If there are fallen trees that cannot be hauled from the field by collective effort, even after all the remaining branches have been chopped off, these are usually placed diagonally across the slope. They then serve two purposes: firstly they act as barriers against possible erosion; secondly they serve as paths through the field once the seeds have come up. On well-burned patches one man needs about three days to clear one hectare.

No attempt is made to get a special distribution of the ashes, although the value of the ashes is appreciated as they are the only form of manure; the more ashes, the better the harvest prospects. But the work of distributing the ashes uniformly does not pay, since the wind will blow them away from prominent places and collect them again in hollows anyway, and they cannot penetrate the hard ground before the rain. The only hope is that the first rains are not long in coming and that they are not too heavy when they do come. If the first rains are too heavy, the ashes are simply washed away. On steep slopes, moreover, the danger of erosion is always present.

Burning a swidden near Banderban

For this reason, additional burning material cut in other places is not brought on to a swidden either (the stronger the fire, the better the opening up of the soil by the heat). The edges adjacent to the field are left uncared for, too. Where the fire has penetrated strongly enough into the bush, weaker bamboo and rods are charred and broken down by the wind vortex. They often lie and hang down over the paths, making them largely impassable. Anyone who wants to pass must try to slip through them below or stride over them, or he must cut his way through. After some days, frequently used paths are re-opened to some extent and others will be overgrown anyway during the following rainy season, so the work of cutting them free is left until then.

J2d) Ceremonies preceding the *dim üa*

Despite the dangers connected with the cutting and especially the burning of the swidden, no ceremonies take place until the field is cleared. One of the ceremonial closings of the hamlet, the *chatma-khang* (see P3b) is observed before or after this time, but this *khang* is not closely connected with the actual work in the swidden.

Eroded steep hill side of a swidden

(This is different among the Marma as I learned in Aw-Phaw-KP, where on 11.03.1956 I was called to a serious casualty. While cutting a swidden a tree had fallen on the man's back and since he fell on a pointed bamboo stump his thigh had been ripped to the bone. Since I had no suitable medicine and instruments and his leg was already swollen and deep blue, I recommended an immediate removal down river to the doctor [a two day's journey], but they refused: this would be impossible, since they had missed the sacrifice before the beginning of the cutting, this must be made up first and this would not be possible before the next Monday [the accident had happened on Thursday] and only then, the following Wednesday could the sick man be taken down river. Thus I had to leave him there. For this sacrifice paraphernalia are used which largely resemble those of the *khang* of the Mru; but there is no rule among the Mru that the *khang* has to be observed before beginning to cut the swidden.) Burning the swidden also takes place without ceremonies. However, when igniting one exclaims: *kam ba, kam ba, hü* (burn, burn!) and the final *hü* is a strange howling sound, otherwise only pronounced by the leader of ceremonies at the harvest thanksgiving festival before inviting the spirits to the meal.

The first sacrifice is made in the new swidden only after clearing it and is called *dim üa*. If somebody has cut several swidden places, each place has so to speak an independent existence, and all sacrifices to be performed in a swidden should be carried out in every place. With pig sacrifices it has become usual, however, to perform the sacrifice in only one of the places, or to repeat the ceremonies in the other place briefly with parts of the animal. No restrictions, however, are allowed for the most important chicken sacrifices of the *chia-dong-khang* (see P3c) when the necessary chickens must be sacrificed on every swidden place.

J2e) *dim üa*

I cannot provide a translation for *dim.* The word apparently is a verb in the connection *dim üa*, but is used only in connection with a pig sacrifice of this kind. With the *üa-dim-pak* (swidden-appeasement?-pig) or its sacrifice, the *üa-dim,* the *tur-tut* place is determined. I cannot provide a translation for *tur-tut* either; the place is also called *liu-klak-raa* (seed-throwing place). The *tur-tut* will always be fixed at the stump of a tree. Four types of tree can be selected, either *ram-nget, mrut-ma*, *muk-ma* or *lo-hat* (none identified). A stone need not to be found at that place. The *tur-tut* is the sacred place of the swidden, here sowing is started, the work finished, and here the sacrifices are offered. But the place does not stand in connection with any particular spirit. Trying to translate *tur-tut* word by word, *tut* means "below", "at the foot of", while *tur* might be the same word as in *tur-kim* (house of the deceased), so that a connection might be inferred with the manes. However, today at least, the ancestor spirits are of no importance in this place, and there is no indication that they may formerly have played a role here. On the contrary nobody will prepare a swidden on or near a graveyard.

According to the old rules, the *dim üa* should be performed shortly after clearing the swidden (for instance two days after the fire). Today, it is often performed together with the following sacrifice for the first seed (*liu klak*). The *dim* sacrifice itself must be repeated later (without the following *plang-kwak*, see below), in case the swidden rules were violated. It therefore seems to be a kind of cleansing sacrifice. A pig is brought along from home, put down at the *tur-tut* place, and with *tamma* (invocation, without calling the spirit by name) *mi tut ai-dam* (husked rice with wild ginger) is thrown at it, and the pig is stabbed to death with a sharpened bamboo (*cau*). A little altar or a bamboo sign are not needed. On the swidden edge, parts of the liver, heart and kidneys of the animal are cooked in a bamboo tube, called *pak-kia-pak-chak-röng* (pig-kidney-pig-heart-tube). At the *tur-tut* the cooked parts are poured with renewed invocation on a banana leaf, on which also are put the rest of *mi tut ai-dam* (husked rice with wild ginger) and the bamboo dart (*cau*) with which the pig was killed. The formula of the invocation (*tamma*) corresponds to that used at the time when the paddy is flowering (see J4c, only that *pak-laa* [pig] is used instead of *waa-ma* [hen]). After the *röng-thut* (pouring out the tube) the tube is split and an oracle thrown (*plang-kwak*). With the two parts the spirit is asked whether one will reap 300 *hari* in this field (1 *hari* ~ 10 kg). If the answer turns out negative, the bamboo pieces are thrown once more and the spirit is asked whether the swidden will produce 200 *hari*, and so on. The answer is positive when the one half falls with the outside up and the other with the inside up (Khamlai Ngarua, 23.04.1957). [In the case observed the answer turned out positive on the very first throw, which elicited an unexpected answer from Khamlai, who performed the sacrifice: "Dear spirit, this is very nice of you, but I fear you overestimate yourself."]

Before one can start with sowing (*liu-klak*), a chicken must still be sacrificed at the *tur-tut*. Formerly anyone who then sowed at the *tur-tut* himself and planted *ru-ke* ("white" taro, *Colocasia esculentum*) and *tam* (ginger) there observed a *khang* afterwards. Nowadays, only few people still observe this *khang* and the *tur-tut* place is determined only with the *üa dim.* Taro and ginger no longer are especially planted as *klaa liu* ("throw seed"), but only when they are generally planted in the swidden. The next sacrifice follows only when the first *ngui* (cucurbitaceae, see J3d) are ripe. It is called *khan-pon-thet* (see J4c) and only after this sacrifice is it permitted to gather fruits from the swidden and take them home (MK, 15.06.1957).

Khamlai and his nephew Dingte offering a piglet for the *üa dim*

The *üa dim* must be repeated if the seed grains from one man's field have been washed out into that of another, or if a cow or goat has invaded the field while the fruit is growing on the swidden. However, if the owner of the swidden agrees, the owner of the cattle or the goat needs to hand over to him only a chicken for the sacrifice, instead of the pig. Likewise an *üa-dim-pak* has to be given to the owner of the swidden, this time especially for the cotton yield, by someone who himself cooks meat or vegetables in a bamboo tube (*kan-röng* = side-dish quiver) at the brook below the field. If somebody prepares his meal in a pot, this does not have any consequences, the use of a *kan-röng*, however, ruins the cotton harvest (probably because only sacrificial offerings are cooked in such tubes below the field, cf. for instance the *pak-kia-pak-chak-röng* of the *üa dim*). If it is has been done by mistake, the wrongdoer will probably report it himself, if, however, it is done deliberately or nobody reports it, the swidden owner will offer a candlestick in a mosque on the grave of a Muslim saint (Bangla: *dorga*); by this, the wrongdoer will die. This is a method also used against theft (from the swidden), which, in the eyes of the Mru, will only be committed by Bengali people (MK, 11.09.1956).

J3 The seed

J3a) Species of rice

The dry rice species used by the Mru are distinguished by them as follows:

A *rai-rik-caa*, minute grains, early maturing, red (local Bangla: *kelong-dhan*);

B *deng-pli*, little grains, average maturing time, white; *rai-cangma*, id., red, very coarse grain (local Bangla: *sigon-dhan*);

C *caa-maa*, white variant: *pe-tui*, red variant: *taru-rik*, big grains, maturing last (local Bangla: *môta-dhan*).

There are three different subspecies of the big white rice (*pe-tui*): *ba-lum* with big "round" grains (*lum* = round), *mu rau* with blue-grey shoot place (*mu* = belly, *rau* = blue-grey) and *bön-cön* with short panicles. There is also a form of the big red rice with short panicles (*taru-bön-cön*) which shares with the white kind the advantage against the kind with long panicles (*taru-reng-chang*) that it can be trodden out very well, while the grains of the kind with long panicles are armoured with harder awns and sit more tightly so that treading out is more laborious.

All C types (*caa-maa*) mentioned require a very good soil to produce a sufficient yield; they are also less resistant to vermin. Where the soil quality is always too inferior, the cultivation of *caa-maa* must be foregone in favour of the smaller, more robust species, in spite of the preference for the better species: which are oilier and more palatable anyway, and this is also the reason why they are called *sugundi dhan* (pleasant smell paddy); people like to consume it in larger portions, which is a further disadvantage in times of scarcity. The yield of the red and white species is the same (the expected yield is 30–40 times the amount of seed). The white rice is soft and palatable, however, while the red is coarser, harder and less palatable, but more robust in the field. Which types and how much is left to the discretion of the peasant, but all species are sown, harvested and trodden out separately. For the storage of the paddy used for consumption, less attention is given to a sharp separation. This is all the more valid as today almost only B types are sown, which mature approximately at the same time, so that the paddy field can be harvested at the same time, too.

Apart from the species of dry rice mentioned, another spot of the swidden (about 10 % of the area cultivated with normal mountain rice) is always planted with sticky rice (*cha-tui caa*, local Bangla: *bini-dhan*); which is used not only for food, but also for religious purposes. The cultivation of sticky rice is regarded as a must, although the yield is expected to exceed the seed by only 10–15 times. It is never sown together with ordinary paddy, but on a plot of its own, usually close to the field house (on the edges of the swidden one would have to reckon with the birds pecking out too much). It is not considered necessary to make a selection of special seed grain (as in the case of normal paddy), so a part of the previous year's harvest is sown soon after the seed of the normal mountain rice, or new seed must be bought.

During the last few years, some people have also begun to sow Bengali water rice on the mountains on a small portion of their swidden. When sown in the dry fields this *kwar-tui-caa* (Bengali water rice) produces a considerably smaller yield than the mountain rice (approximately only a third; the expected yield is at best 12–15 times the amount sown), nevertheless it also has an advantage of maturing within three months (only two months are needed in the plains) and it can therefore already be harvested in the last few days of July or in the first days of August; it helps to bridge the hungry time up to the new harvest of the mountain rice in September.

J3b) Choice and storage

The choice of the seed already starts with the harvest. A swidden spot with particularly good yield and full and even mature paddy is harvested separately. This can happen before or after the general cut, it is important only that it has not rained for some time so that the paddy is not damp. It is trodden out then as soon as possible and dried on the platform of the field house. On a dry day full of sunshine, three to four hours will suffice for this (the other paddy must remain lying longer since it has ripened less evenly). Empty grains are eliminated by careful winnowing. At home the seed is filled into special baskets (*caa-liu-yöng*, see D2g) lined with *hap-ru* leaves (not identified): the leaves dry only in the basket (without damage to the rice) and protect the seed against pests and other outside influences. The full baskets are covered with a piece of mat, tied up with bamboo thongs and not opened again before the new sowing time. At any sacrifice which takes place in the house they are sprinkled with blood and fed with spirit titbits (*kom-pöt*).

Valley showing swidden fields everywhere

Before being sown out again the paddy seed is poured from the baskets on to the platform of the house on mats, exposed to the sun and fanned out, in order to remove remaining or newly formed empty kernels. As a rule approximately one third more seed than one intends to sow next year is always preserved. What is left of the paddy seed can be eaten directly. Seed grains are never sold, first for the practical reason that they represent the best paddy (and the paddy price only varies according to the weight), and also because of the idea that a sale would influence the yield unfavourably, i.e., reduce it (Kangku Catumma, 26.06.1957).

Anyone who does not have any seed, whether he has lost it by fire or because of his mismanagement, must therefore buy some consumption rice as seed paddy and has to be content with this quality. Together with the paddy, cotton, cucurbitaceae, beans and other fruits are sown. An exact mixing ratio is used only for the cotton (1:2 with good cotton seed), otherwise the addition is determined by the peasant's need and can be varied depending on swidden places. (For the cultivation of the other swidden fruits see J3d.)

J3c) *chot caa* (sowing paddy)

Sowing can start directly after the *dim üa* and the *klaa-liu* sacrifice, but as a rule one waits for the first rain that makes the hard soil a little sodden. Without the ground being softened – something accomplished by the first rains – it is extremely difficult to get the seeds into the earth. The ground is still too hard, the seed holes cannot be dug deep enough and crumbly surface earth slips directly into the hole before one can throw in the seed, the seed grains become easy spoils for the birds or they are easily washed out again after the rains have set in. One must try it nevertheless, in case the rains come too late and one must wait too long. At this point, in fact, the farmer becomes totally dependent upon those powers which he cannot control, even with the most intimate knowledge of the soil. In the hope of putting these powers or spirits into a favourable mood, the *dim üa* has been performed.

Sowing the seed in a working group

All members of the family who are able to work help to sow. On one and the same plot of ground several varieties of rice can be planted side by side; and on this selfsame plot a whole series of other fruits or vegetables is also to be found – no ethnic group practises monoculture on swidden farms. The question of what, and how much, is to be mixed in with the paddy is largely left up to the individual's own judgement. Only in the case of cotton – which, however, is not planted over the entire field, but only in the humid lower end – is a given proportion used for the mixture: 10 lb cotton seeds to 20 lb paddy kernels (see J3d2). Seeds of different types of squash and cucumber – the vegetables most commonly eaten by the Mru – as well as beans, sweet cane, and indigo, can also be mixed in with the paddy. The seed mixture is prepared at home and taken to the field in large baskets (*man,* see D2d). Crops such as Indian corn, Indian lentils, and rozelle are also planted; their seeds, however, are never mixed in with the paddy seeds, because their plants would overpower the paddy seedlings. Creepers – certain species of *Luffa* and *Momordica* – are likewise planted separately, in places particularly suited to them (see J3d).

On the swidden everybody puts on a little basket (*emca,* see D2b), filled with some seed, which is carried on the back at hip height, so that portions of seed can be taken from it with the left hand. One takes a special hewing knife, which may be called a dibbling stick (*charai kong*), in one's right hand – usually an old normal hewing knife which, however, is no longer sharpened on the side but like a chisel at its front tip –, stabs it into the earth and draws it slightly toward oneself so that a little hole is wedged open into which a couple of grains are dropped from the left hand. The blade is then pulled out of the opening again, a new hole stabbed a little distant from it and seed thrown in again. The planter may haphazardly cover the holes by pushing earth over them with his feet but most of the time, as the person proceeds up the hill, enough loose soil falls back into the hole of its own accord. One starts at the lower end of the field and advances slowly uphill while sowing, until one arrives at the upper end of the field and, if the *man* stands there, fills up one's *emca* with seed again, goes down once more and comes uphill again sowing the next column. If ten people work together in this way, they can plant 20 lb of seeds in less than two hours, thereby cultivating an area of about half an acre.

Two boys sowing side by side

When several persons work together and at approximately the same speed, the group moves uphill together; however, it is not the aim to arrive first but to sow the field as evenly as possible. Even if not everyone works at the same speed there is no danger of some places remaining without seed or being sown twice: the holes are still visible (until the next rain). This, however, does not mean that every spot of the swidden would be equally sown, since here we are not dealing with a level place in a tilled field, but with a swidden full of rough bumpy places with tree stumps, which must be passed by, and stony steep places, in which only a few seed holes can be stabbed and on which the paddy would not grow well – in contrast to more level places, on which one must bow deeper in order to bring the seed into the earth, but on which one also can hope for a better yield and in which correspondingly more seed is sown by increasing the number of seed holes, while the number of grains thrown into them remains approximately constant. On soft, gently rising places, the number of seed holes can reach eighteen and more per square metre while, on hard steep places or such with many obstacles, the number may be at best ten holes per m^2 (but it can also decline to five and fewer); and on average one can expect fourteen holes per m^2.[453]

Instead of asking the sowers how many grains approximately they would throw in per hole, I slowed down their work in order to count the grains which they took into their left hand, and after that counted the holes into which they were distributed. Thereby I found out that about 1100 grains were distributed in 90 holes, which amounted to about 12 grains per hole. Twelve years later this very number was mentioned to me on mere inquiry and thereby confirmed by a Bawm informant (Pardo Sailuk), so that I gained the impression that it is valid for all ethnic groups of the CHT[454], though later I did not have any opportunity to find out to what extent I could have asked the sowing Mru directly. – However, for my investigations I counted still more grains; the results, for instance, with regard to the land area required per person, will follow later (see J5a).

J3d) Other plants sown, planted and grown in a swidden

Besides the three main sorts of rice already mentioned (normal hill-rice, [ritually necessary] sticky rice and [optionally] Bengali wet rice), the Mru may (but need not) also sow

1) millet (*yak-ma, Panicum*). Millet is not sown in holes, but scattered on the earth (usually near the field path). It ripens after the paddy, the panicles are broken off by hand, trodden out and cleaned like rice, but the kernels need no drying. It is cooked by itself and sweetened by adding red sugar (*chakka rik*) or by mixing it with sticky rice. Though formerly much used, today it is considered a less important food item.

2) maize (*pren-bong, Zea mays*). Maize can be sown from late April (until the beginning of June), and can be harvested (according to the species) after either three or four months, i.e., if sown in late April, by the end of July (*pren-lüa*) or by the end of August (*pren-long*). There are five sorts of each: white (*pren-ko*), yellow (*pren-chüa*), red (*pren-rik*), black (*pren-num*), speckled (*pren-apru*), and a very small sort which takes four months before it can be harvested: *palang rung pren* (bottle-stopper-maize) (MK, 19.08.1956). Maize is often dibbled in together with paddy, although only in small quantities, as it will hamper the development of the paddy plants. Therefore it is considered better to sow them separately, and anyone who wants a bigger maize harvest and an unhampered growth of paddy will dibble the maize kernels into a separate part of the field. When the cobs are ripe, the kernels may be eaten directly in the field, either fresh or after roasting them, still protected by the leaves wrapping the cobs, or after bringing them home, where they may be eaten immediately (as in the field) or cooked. They may also be preserved for later consumption by letting

---

[453] Bernot (1967, p. 249) for the Marma records a distance of the holes of 20–25 cm. Following Khisha & Khan with the Chakma the holes are "12 inches apart", i.e., about 30 cm.

[454] Bernot (loc. cit.) speaks of only 6–10 grains per hole, but he adds: "non compté".

them dry, first by exposure to the sun, then by binding two together by their protecting leaves and hanging the bundles up on a bar in the house. Since the kernels become hard after drying, they must be cooked before they can be eaten. The better cobs are preserved for the next sowing time.

3) cotton (*laa', Gossypium acuminatum*). Cotton is sown on sunny slopes in the lower parts of a field mixed with rice in the ratio of 16 seri (10 kg) of rice to 8 seri (5 kg) of cotton seed. Since the kernels of the latter are much bigger than rice kernels the ratio is 1:10 (calculated from a count of five handfuls of mixed seed, yielding a total of 139 kernels of cotton and 1,370 of rice). The harvest begins after the rice harvest, after which the field has to be weeded once more (see J4f).

4) pepper (*ling-kö'*, *Capsicum frutescens*). Chilli pepper is grown by every household in small gardens near the house or, where the houses stand close together, a little bit outside the village. At the beginning of March the soil is cleared and loosened up with a hatchet and on this plot firewood is burned and the seeds are sown into the ashes and sprinkled with water (no dung is used). Special baskets (*ling-kö cua*, see D4h) may also be filled with the earth mixed with ashes and the seeds sown into it. When they have started to grow, they are watered twice a day, more often in the beginning, less often when the plants grow bigger – until the rains set in. In June the plants are transplanted into the field and dug in. With the help of a *charai kong,* one man can plant an area that takes 10 kg of rice seed in two days, but the soil must be very good if the seedlings are to be planted in an already existing rice field, or otherwise the yield will be poor. Accordingly, it is better to plant in a special plot at a spacing of approximately 1 m apart, wider on good soil, closer on poorer soil, since the shrubs will not grow so big. When planted in an existing rice field, they develop fully only after the paddy harvest and the chilli harvest cannot begin before October; when planted on a separate plot, they develop earlier and the harvest can begin as soon as August. The pods are plucked approximately every ten days until December. On good soil the average yield per shrub is 1 kg, a plot for 10 kg of paddy seed (according to my calculation 2,500 m^2) will yield 2–3 maund per month, i.e., approximately 10 maund (~ 370 kg) in total. A good harvest may yield more. In 1956, Menkröi had more than 2,000 plants to be replanted in an area for 6–7 kg of paddy seed (i.e., 3,000 plants for an area of 10 kg). If he had planted it with rice only, the same plot would have yielded him approximately 200 kg of paddy, so by intercropping it with chilli, he could hope for an additional harvest of 200 kg of chillis.

The harvested chilli pods are spread out and dried in the sun on mats on the *car* (outer platform) of the house; if the pods have not fully ripened and if the sun does not shine enough they may partly rot, however. Therefore it is considered better to dry the kind that might rot for two or three days over the fire inside the house by placing them in *kham* (flat baskets) above the *krök* (see C3e). After drying they can be kept for the next year. New seeds are obtained by breaking the pods open and then removing the seeds by hand. The amount of dried chilli not used for daily consumption is sold in the market and helps to improve the family budget. (Buying chillis in the market I had to pay 3 Rs. per kilogram, that is at least double – up to nine times – the price for rice [see A4c], but I do not know how much the Mru could receive if they sold it.)

5) and 6) Pumpkins, cucumbers and gourds (*ngui*, *Cucurbitaceae*). Cucumbers and pumpkins are sown together with rice, though not haphazardly, but preferentially near tree stumps and felled trees that have remained unburned, so that the vines can develop their fruits in these otherwise unused spots. Notable are *kan-pen* (*Benincasa hispida* [winter melon]), *chang-pho* (*Cucurbita maxima*) – both serve rather regularly as a welcome side dish –, *ngui-tang* (*Cucumis sativus*), *ngui-üm* (*Cucumis sp.*), many varieties: long or round, white or red, *ngui-prin-ca* (a small kind of cucumber), *pang-lai* (*Luffa acutangula*), a climber, fruits edible when young, become spongy when older, *dum-ui* (a "squash") and *tui-yia* (*Lagenaria vulgaris*), a small gourd, fruits edible when young, but mostly grown for the hard outer shell which is used for water-bottles and for gourd-pipes (*plung*) and can also serve as large ladles when halved. In order to obtain seeds from cucumbers, pumpkins and gourds, the seeds are preserved (when the fruits are cooked or used otherwise) in special small baskets (*but thung*) on the house-wall on the *car*, where they can dry.

7) related: also climbing and cooked like cucumbers: 1) *cing-cer* (*Momordica charantia* [bitter melon], Bangla: *korola*) tastes slightly bitter. The seeds are planted near cut trees, for the vines to grow. The size of the fruit is similar to that of the *kan-cangma* (*Momordica dioica ~ cochinchinensis*, (Bangla *k■korol*), which grows in the jungle and can become bigger than an egg-plant. The fruit is outwardly covered with a kind of spines, the inside kernels are hard, but the fruit is rated as a good vegetable, which is why the Bengalis also cultivate it in their fields. 2) *klai-long-ma* (not identified, Bangla: *ber[e]la*), sown together with paddy and eaten as a vegetable.

8) aubergine (egg-plant) (*min-tau*, *Solanum melongena*) is sown broadcast. To obtain seeds, ripe fruits are cut open, the seeds are cleaned in water and afterwards dried. There is also a kind of wild egg-plant, called *paleu*, growing in the jungle, which is also eaten. Another vegetable growing in the jungle is *klom-bü*.

9) beans (*be*, *Phaseolus* spp), small bean (*be chong-raa*, *Dolichos lablab*; *chong-kro*, *Dolichos gladiatus*; Bangla: *shim*), vines climb and spread over the field. – Beans and seeds from egg-plants, *berla* and so on are preserved separately in special small gourds (*tui-yia-ca*).

Also sown together with paddy:

10) indigo (*cha-ram*, *Indigofera tinctoria*) used for colouring clothes (see F1c).

11) sweet cane (*prang-ke, Sorghum saccharatum*), chewed raw in September, together with the first paddy. The internodes are broken off, the hard skin is peeled off with the teeth, the soft stems are chewed and the rest is spat out. The umbels are preserved for sowing, tucked in under the roof to dry.

12) root crops ("potatoes", *mo,* and taro, *ru'*): There are many kinds of potato-like root crops, planted in the field: *pöng-mo* (can become long like an arm, *Dioscorea bulbifera*?), *mo-pu* (like a round cooking pot, *pu*), *mo co'* (the skin has a bluish colour), *chikua-mo* (white, long and thick), *mo-wai* (yam, batata), *mo-ding* (sweet potato), *mo-ching* (Bangla: *thanda alu*), *hap-ru-ca* (Bangla: *ar alu*).

More common is taro (*Colocasia esculenta ~ antiquorum*): *ru ko* (Bangla: *shada kocu*), *khok dia* (Bangla: *hati-teng kocu*, Bawm: *baa sai-kee*), *ru long* (long, not very thick), *ru chau* (light brown), *pak-ki ru* ("pig-shit-taro", somewhat black), *kor-pa* (Bangla word: *korpa kocu*). Starting from July *kor-pa* can be harvested repeatedly by cutting off those parts of the root which grow above the earth. These can be eaten raw (with pepper) or cooked as a vegetable (they taste best when accompanying meat). Equally growing partly above the earth; *man kocu* (Bangla word used, *Arum indicum ~ Colocasia indica ~ Alocasia macrorhizos,* "giant taro"), planted near the house, can become very large (20 cm and more in diameter, up to 1 m and more in length, can weigh more than 40 kg) and is sold to the Bengali merchants (40 kg for 4–5 Rs.). For home use the bark is removed, the inside part cut up, dried in the sun and then pounded to produce a kind of flour, which can be added to the sauce of side dishes. It is also used as a food for sick people but can also be cooked like other kinds of taro. Parts of the roots of all kinds of taro are used for replanting; propagation is never from seeds.

13) ginger (*tam, Zingiber officinale*). Like *mo* and *ru'* parts of the bulb are dug in all over the field, after two or three rains have wetted the soil.

14) Malabar spinach (*kwar-pau*; *Bassella rubra*). Sown like paddy, small vine, climbs up trees. The leaves are cooked with onions and chilli as a side dish, but they can also be eaten raw.

15) red sorrel (*kan-chur* [*chur* = sour], *Hibiscus sabdariffa*), several kinds, the most used being: *chur-wi-ma* [wima = blood-red], sown together with paddy, but only near the field path, because it spoils the paddy. The edible calyces may also be used to brew a sour "tea", which was formerly much used by the Maraa, who drank it at each meal, but not by the Mru. As well as the fruits, the leaves may also be cooked and eaten together with pepper or *ngaa-pi,* also used for *chur-ngar* (MK, 19.06.1957).

16) pigeon pea (*pleng*; *Cajanus indicus*; Bangla: *orohor*), to be sown separately, but also sown together with paddy, although it spoils it, since the plants can become rather big (up to 3 m). The pods are harvested from January to March and can be cooked fresh; the flat "peas" may also be dried and are much used by the Bengalis to cook their *dhal*, used daily as a side dish.

17) sown broad-cast for embellishment near the path crossing the field: *ramma* (*zinnia*?) (flowering in June/July) and *khet-pau* in two varieties: *khet taraama* (yellow) and *khet rikca* (red) (flowering after the harvest); *chong-prem* (*tagetes*) also in two varieties, yellow (*chüa*) and red (*rik*); *cha-pan*, sown broadcast for its good smell and worn in the hair-knot or the ear-hole: *chong-hir-ma, chong-ku, chong-klang* (not identified).

18) also sown broadcast: *kan-düng* (not identified); cooked as a side dish, the leaves taste very "hot".

19) tobacco (*chi*, *Nicotiana tabacum*). When it is to be grown on the hill, it is sown in June-July under the field house in a spot cleared for that purpose. It does not get watered or washed away by the rain, grows very slowly and reaches about 10 cm in October, when it is planted out. To prepare the place, the straw of the paddy harvest is spread on the earth to rot; when rotten, the seedlings are planted out (like pepper) with a *charai-kong* in October-November at a spacing of ca. ½ m; in March the leaves can be plucked.

When grown near a rivulet, tobacco is sown directly into the earth in October-November, (when it will not rain anymore). The small plants are then transplanted in January into small, fenced-in plots along the edge of the river – plots that have been specially prepared for that purpose. The leaves from these tobacco plants may also be picked from March on. The preparation of the small, fenced-in plots is extremely troublesome, because if there is any *Imperata* grass growing on the plots, the entire root system of the grass must be completely dug up. In return for this trouble, considerably higher yields are obtained from these plots than from the mountain fields. The tobacco from the special, riverside plots also has a better taste. When the plants are approximately 30 cm high, the lower leaves are taken off, though the opinion is shared that the upper leaves are the better ones, they become longer, especially after breaking off the top before they bloom. After the leaves have been harvested, they are skewered on a thin bamboo thong and hung up under the roof of the house to dry. In order to use a leaf, one removes the middle rib and then packs the rib, as well as the tips and edges of the leaf, into the better half of the leaf, which serves as a covering. The whole thing is finally held together by a small cotton string tied close to the thicker end – the end which is then placed into the mouth. Old and young alike enjoy such a cigar. More often than not, however, there is not enough tobacco to smoke even one cigar a day; a person must therefore occasionally make do with a pipe. Quite often a small piece of tobacco is used as a quid. Cigarettes are available on the market, but in the long run they are too expensive even if one buys the worst quality. Most people cannot even afford matches – during the rainy season they become useless anyway unless one keeps them constantly drying over the fireplace. A cigar or a pipe is therefore generally lit from a smouldering stick or ember from the fireplace. For getting new seeds, some of the umbels are kept until April, when they are broken off and preserved without further drying. Alternatively, the seeds can be separated at once and preserved in a small basket. The state has tried to tax those farmers who grow tobacco, first by the number of plants and later by maund harvested (1 md. [~ 37 kg] = 32/- Rs.), but the farmers reacted by hiding their harvest, bringing it to a friend in a neighbouring village.

The first seed is growing

J4 From sowing till harvest

J4a) Some general remarks on swidden work

After having sown the paddy, the swidden must be weeded three times (May to beginning of September). Between mid September and mid October the paddy is cut. If cotton is grown, a fourth weeding will follow. After the paddy has dried in the field hut, it is taken home. In November the cotton is harvested. In December houses are mended and rebuilt. Women collect firewood in the period from December to February. In January and February most of the taro is harvested (dug out), but not before the *üa-plai* (see J6) has been performed; *ngui*, especially *kan-pen*, can be fetched any time. Starting in the second half of January until the beginning of March, new swiddens have to be cut, to be burnt after drying in March until the middle of April.

In contrast to the Marma, the Mru do not take up residence in the field house. They prefer to return home in the evening. However, when the swidden is very far from the village and the work is pressing, the workers may pass the night in the field house, while the remainder of the family stays in the village. The Southern Chüngma, however, sometimes follow the Marma custom, and the whole family stays in the field house. The other Mru groups speak disparagingly of this custom, and blame the Marma for their laziness: "they must constantly cook, eat and sleep".

Valley showing swidden fields everywhere

Rotation of the swidden sites was formerly ten to fifteen years and nowadays is four to six years, the length of the fallow period depending on the last yield; if the yield is sufficient, the plot may be re-used earlier. Formerly the good places of a swidden could even be replanted in the next year (*khung waa*), but since such places are not to be found anymore, the practice has been abandoned. The dangers of quick re-use are well known: the quicker the rotation, the smaller the yield. This reduction in the length of the fallow period is the result of population growth, that is, it depends on the area available for making new swiddens. It is known that thirty years ago one could expect a harvest of 3,000 kg from an area sown with 40 kg of paddy, nowadays the yield is at best 50 times the amount sown. Equally well known is the importance of ashes: the more ashes the burning produces, the higher will be the yield. Excessive rainfall at the beginning of June will wash away the ashes, but it may also wash out the grains and may even cause land-slides, particularly on steep swidden sites. The earth remains best protected and will yield the highest amounts in plots covered with bamboo; a swidden for 60–70 kg made in such a place may yield enough crops for a family of five, but such places have become rare; in other places the yield will be lower and will depend on other factors that are not easy for the Mru to calculate.

In places re-cultivated every four to five years the bamboo may die out, if it is burned five or six times in subsequent years it will barely survive; what remains is shrub, especially "Assam-lota" (*Eupatorium odoratum*). The leaves of this vine are used for styptic purposes (they speed blood clotting) and, when preparing a swidden, the vines cannot be cleaned without great effort (see J2a).

Weeding, however, causes no problems, hence this invasive plant is not considered a danger, but in the long run it may give way to grasses of the *Imperata* family, which survive burning splendidly and grow back rapidly in their own ashes, so that the place is lost for swidden cultivation and can no longer be used for preparing a field in the years to come, since the *Imperata* roots are difficult to eradicate and the grass will soon overgrow and smother the growing paddy plants.In areas where the soil cover is too thin another danger exists: the development of karst

Karst development

J4b) Weeding and natural pests

Like cutting and sowing, weeding is done by starting from the bottom of the field and working upwards. Sowing begins (according to the weather and the setting in of the monsoon rains) in the second half of April and may extend until the beginning of May. While the last seeds are being brought in, those sown first may have already started to grow, and the fields must be weeded three times from the middle of May until the beginning of August.

Between May and the end of August, all swidden fields must be weeded three times. During the months of weeding it rains almost every day; nevertheless most people go to the fields unless it pours and storms excessively.

The paths are slippery, and in many places infested with leeches. Everything is wet; and when weeding, one must walk into the middle of wet vegetation and catch hold of it, without being sure that in spite of all precautions one will not catch hold of one of the leaf-green vipers, whose bite can be deadly. This period is called *mo tuk khin* (< *tu mo*, to weed), and the three successive weedings are called: *thak bot-ramma* (to cut off the roots of the weed; 1st weeding), *tu bot-wai* (to cut the weeds; 2nd weeding), *(tu) mo-rek* ([to cut] the climbing weeds; 3rd weeding).

Only by the middle of August are the young paddy plants strong enough to suppress the growing weeds by themselves. If the first weeding is slowed down by too heavy rains or if a worker from one of the families has to drop out – due to illness, for example – the resulting lag in weeding can, in fact, mean the loss of entire parts of the field, since the earlier vegetation sprouts again and quickly overruns the rice plants.

First weeding of the swidden (*thak bot ramma*)

This is especially true when *Imperata* grass threatens to invade the field. In the vicinity of existing patches, seeds of this grass may be blown in everywhere, and when they grow roots they are difficult to eliminate, since they develop a rhizome like couch grass. These rhizomes must be dug out as far as possible, which slows down the weeding of the paddy plants and can reduce their yield. Normal jungle vegetation is able to overgrow *Imperata* grass, but only if it is not in danger of being burned down, whereas every burning will enhance the chances of *Imperata* propagating, since the fire does not affect the rhizome of the grass, but only destroys the normal jungle vegetation. If the patch is burned every year, four years will be sufficient for it to be completely overgrown with the grass, which grows up to a height of 3 m. It can then be cut and used as thatch, but even hungry cows will not feed on it, since its sharp blades cut their mouths. But if the place is burned only once in five or six years, there is no real danger of *Imperata* taking over, and after 12–30 years free from burning even a place that is totally overgrown with this grass may revert to jungle growth.

Other natural pests threatening the paddy are *nget* and *yum*. *Nget* is probably caused by a paddy bug that leaves hollow grains; species with little hard grains are affected less than those with big soft grains. The Mru do not know of any means to fight this disease. *Yum* may be caused by a fungus; the plant turns brown and the grains remain empty. The disease is attributed to bad wind; a measure taken to prevent it is the *caa yau* at the time of *khan pon thet* (see J4c). Once the paddy has become affected, there is no remedy available. Some people plant bamboo tubes all over the swidden, to the top of which are tied either the bark of *ching-chür* (not identified), or young bamboo shoots (*kau-lung*), or leaves of *chön-ram* (not identified). The latter are also used at the *plon* (entrance of the hamlet) to ward off spirits. But most people do not believe in the efficacy of these measures.

When rats swarm over the entire countryside, stripping the fields bare, an inescapable and total crop failure may well result. This happens, however, only about once every fifty years, when the bamboo blooms and then dies.

Rat invasion 2007 (photo: Menpöng Ngarünau)

Between April (soon after the *rat üa*) and the beginning of August (shortly before the harvest) a field house (*bok*) will be erected. If the way between village and field is long, and if a family cannot keep up with the weeding in any other way, family members may occasionally sleep overnight in the field. The main reason for erecting a *bok*, however, is that it has to serve as an intermediate storage for the harvested paddy. If the village were near, so that the harvest could be brought directly home, such a hut could be dispensed with, but this would not enhance the reputation of the owner. (On the construction see C4a.)

Even without matches and during the rainy season, the Mru know how to light a fire in the field. All they need is half of a split bamboo cane, in which a small notch is cut, and a long, thin, but strong thong cut out of bamboo bark. The halved bamboo is held down by the feet with the opening face up and the thong, which passes through the notch, is pulled up vigorously with the hands, alternating right and left. The fine sawdust that is rubbed off in the process collects in the inside of the bamboo cane close to the notch; it becomes hot and finally catches fire – provided one keeps up the pulling, which becomes more and more vigorous, and stirs up the smoulder by blowing lightly just at the right time. Since during the rainy season there is almost always a small brook close to the lower end of a swidden, a person can also cook himself a meal while out in the field.

J4c) *khan-pon-thet* (the 2nd field sacrifice)

The full ceremony would be the *chia-dong-khang*, the rainy season *khang*. What follows here is a summary of a ceremony that can be held instead. It is held at the end of July or the beginning of August, when the paddy flowers. The sacrifice consists of two chickens, one near the brook (where the drinking water is fetched when working in the field), the other at the *tur-tut*. My data were collected on the 2nd of August 1956, when Dingte was in charge of the work. Shortly after 2 o'clock in the afternoon Dingte put the 2 chickens in a *pom* (see D4f) and started towards the field, accompanied by his younger brother Kangding and by me. On the way he plucked leaves of *ro* (lime fruit) and *prön* (lemon) and put them in his basket (*em*, see D2a), in which he also carried, wrapped in a banana leaf, *pok-pok* (puffed sticky rice [*cha-tui-caa*]) and *kom* (also made from sticky rice, but soaked in water and pressed into small cakes).

The first sacrifice (*klaa-wi* = shedding blood) was performed near the brook, where Dingte erected a *pak-plai-kimca* (it is always called that, even if no *pak* [= pig] is sacrificed), a small altar, consisting of four small bamboo sticks, adorned with tassels, and carrying a small platform in between. On this platform Dingte spread a piece of banana leaf, on which he gave some *pok-pok* and *kom* as food for the spirits, to which he added a hewing knife point full of earth. From the bamboo used for preparing the altar, Dingte also cut a *ri-chang-pön*, a bamboo tube for pouring out water. Then he took one of the chickens out of the *pom*, Kangding grasped it with both hands by the body and Dingte with the left hand by the head, while using the right hand first to pull out the feathers round the neck, then to cut the neck with his hewing knife. Then he took the *ri-chang-pön*, closed its mouth with his finger, and lifted his finger six times to pour out water over the cut throat of the chicken, while speaking the following *tamma*:

phiöi, pakat, pre-kat, chum-kat, tali-kat, tangaa-kat, tàruk-kat – lup-rak pa-kaa-la, laa-rak pakaa-la, taku ning waa-ma hai ang klaa wi khök co (phyöi, 1x, 2x, 3x, 4x, 5x, 6x – work-yield [that is paddy etc.] let us get, cotton-yield let us get, with a nine-year-old hen I have shed blood). However old the chicken may be, it is always said to be a nine-year-old hen, the blood is allowed to drip out on the altar, after which the chicken itself is put back into the *em* under the *ro* and *prön* leaves and carried home again to be cooked for human consumption. This offer is meant for the *o-reng-nam* (the river spirit, also called by the Marma name *khyong shang-ma*), but the spirit's name itself is not mentioned. The same sacrifice is repeated on the *tur-tut* place, where the same *tamma* is repeated word for word, except for one difference: it starts, instead of with *phyöi*, with *phyok-carok* and is (again without mentioning the name) addressed to *caa-kom-laa-kom-nam-ma* (paddy-much-cotton-much-spirit). The first word of the *tamma* serves to mark it as addressed (by *phyok carok*) to a good spirit or (by *phyöi*) to a potentially malevolent spirit. Another small change should be noted: instead of the neutral banana leaf on the altar by the side of the rivulet, on the *tur-tut* a *ru-ke* ("white" taro) leaf is used from a plant growing there (after having been planted).

Moreover, besides the *pak-plai-kimca* something else is placed: a *caa-yau-röng* (paddy-swim-tube) with *thau-rin-bong*, i.e., a bamboo tube "decorated" by shaving off straight strips of its green skin. In order to facilitate fixing, its lower end has been cut into a spur or thorn. This piece of bamboo is also called *khan-pon-thet-röng* (offer-?-sprinkle-tube), and in it is placed a *chit-pot*, a bamboo stick with a short tassel to sprinkle the liquid contained in it, the *khan-pon-thet-tui*, consisting of water and some of the blood of the sacrificed chicken, to which are added some *prön* and *ro* leaves. Anyone who wishes can also add some leaves of *rui-chingma*, a climber, the leaves of which are also used as *chi-khang* (a kind of "medicine" used for larger cattle sacrifices, see R3i). Moreover, a *prön* twig is planted at the *tur-tut* place. It is said to protect the paddy against *yum* (being spoiled by illnesses, vermin and too much rain, see J4b). If it has any real effect it may be to drive away some kinds of vermin that are repelled by the smell.

Dingte sacrificing a chicken at the *tur-tut* place (during the *khan-pon-thet*)

At the end of the ceremony the *tom* is thrown away and the second chicken is put with the first one in the *em*, to be taken home. The *caa-yau-röng,* together with a last *chit-wai* (of which nine have been prepared) is also taken along, but only up to the upper end of the swidden. The *chit-pot* is taken out of the *röng* and used to sprinkle the *khan-pon-thet-tui* right and left over the growing paddy. At the upper end of the swidden the *caa-yau-röng*, the *chit-pot*, and the *chit-wai* are planted separately into the earth on the upper side of the path, outside the swidden. Then a small piece of bamboo is cut and split into four pliable parts. The two ends of each piece are stuck into the earth so as to form small arches (*kung-kaa* = to get an end), symbolically barring the way into the swidden. They serve as signs that the swidden is *khang* (forbidden) to outsiders and to animals: wild pigs in particular are expected to respect it. Only the owners are allowed to enter, when they have to weed the swidden for the third time (*mo-rek*).

In the evening of the day of this second swidden sacrifice (which may replace the *chia-dong khang*, see P4b), the two chickens are cooked and eaten by the family.

J4d) *caa-müng-pok* (the first taking of paddy)

By the last days of July or the first days of August the first paddy, *kwar tui caa* (Bengali water rice), can be harvested, provided that some of it has been sown (the yield will be considerably lower than for the normal dry field paddy that is usually sown on a swidden).

By the middle of August, preferably on a Wednesday, when the first hill paddy is ripe – although the proper harvesting of paddy doesn't begin until a month later – the next festival, the *caa-müng-pok,* follows. Only after this festival can the first fruits of the swidden be eaten. (For an exception, see J4e.) Among other things, tender cobs of corn are gathered, which may be roasted in the hull, and the first cucumbers. The required sacrifice is one chicken and, when available, one pig as well, as big as possible. This sacrifice is not made in the field, however, but both animals are killed in the *kim-tom* (the main room of the house).

Among the Chüngma Mru only the chicken is killed in the *kim-tom;* the pig is killed on the *car* and a *cari-yöng* has to be performed for it, the *caro-ca* (see J6a) is not only decorated with *chit-wai*, but also with paddy ears (*caa püi*), which have been fetched (without any special ceremony) from the swidden on the preceding day, together with leaves of *ru-ke* ("white" taro), *tam* (ginger, if available) and *prang-ke* (sweet cane). Something of all of these is put in a *klai-puk* basket (see D2c) together with a sickle (*tang-cen*) and a hoe (*tim*) as special tools for working in the swidden (but no hewing knife, *charai*). In front of this *klai-puk* are placed: a *tui-yia* (a calabash filled with water) and a *hom-noi-pu* (a small round pot with a narrow neck) filled with *hom-noi* (rice soaked in water) as a substitute for rice-beer (*yu*), into which a siphon is stuck together with some paddy ears (no fixed number).

Among the Anok Mru I observed this feast in the house of Menkröi, where the ceremonies were conducted as usual by his elder brother Khamlai. On the evening prior to the sacrifice, several items were brought in from the field: a few ears of paddy, some taro leaves, a little ginger and some sweet cane (not sugar cane, but a kind of sorghum). Some of each of these four fruits was placed into a small basket *(klai-puk);* then a sickle and a hoe were added – the sickle being the harvesting tool, and the hoe, the weeding tool. A bottle gourd filled with water and a small pot with a drinking reed were placed in front of the basket. The latter has the same shape as the large, round beer pot but instead of beer, it is filled with a mixture of water and cooked rice, called *hom-noi.*

For the sacrifice, at first the feathers round the neck of the chicken were plucked out, the neck was cut with a hewing knife and the blood collected in a small black bowl (any bowl would do). The bowl was then placed on the *kri* above the *ta-ping* (hearth). The blood is used to anoint everything that is considered important and valuable: the drum (*tömma*) and the flat gongs (*ner*), the *khong-rau* (small stones, said to be the seat of *khong*, the spirit who cares for the welfare and well-being of the whole house and family, for property and the harvest), the spears (*re*) and the old large hewing knives (for ceremonial use only), as well as gold and silver jewellery. After some blood has been collected in the bowl, the neck is cut off and the remaining blood is let drip out over a *klai-puk* while a *tamma* is spoken: *phyok carok, pakat, pre-kat*, etc. (see above). At every *-kat*, Khamlai took some *hom-noi,* bit off a piece of ginger, sucked a mouthful of *hom-noi* from the small pot, and, invoking the spirits, spat six times on the chicken. The invocation was rather lengthy (with no fixed text), Khamlai asked for good health and a good harvest, before he ended with *taku ning waa-ma ...* (see above). The chicken was then laid aside, and the pig was brought in, with legs and muzzle tied, and killed with a spear (*re*). The blood was let drip out over the *klai-puk*, accompanied by the same *tamma* and spitting of *hom-noi* and *tam* as described above for the *klaa-wi* of the chicken.

Then chicken and pig were dismembered, but kept apart. Little pieces of every part of the meat (entrails, thighs, abdomen, ribs and head – the ears were cut off from the pig's head and roasted separately over the fire) were wrapped in banana leaves and cooked over the fire, and used afterwards with some freshly cooked rice for making *kom-pöt* (see below). For preparing this rice, newly cut paddy is dried and roasted over the fire (*thang yong*) and pounded in the house while still hot. Small portions of this *mi* (pounded rice) are then eaten uncooked by everyone in the house. But before it can be eaten, two small portions, to each of which one piece of *prang-ke* is added, are used to make another two *kom-pöt* (without meat and cooked rice); one of these is thrown down on to the earth from the entrance, the other up on to the roof from the *car* (open-air platform). Only after this special *kom-pöt* has been offered can humans partake of the *mi* and nibble on the *prang-ke* (sweet cane). If they tried to do it before, their mouths would "break".

All the *köm-pot* made at this festival should have been wrapped in leaves of the *ru ke*. However, since there were not enough *ru ke* leaves available, a large part was wrapped in banana leaves. The first *kom-pöt* made contained the tongue of the chicken (*waa kamca*). The uvula in the cleft of this tongue must be bent, if it were straight it would bring misfortune, and a new chicken would have to be sacrificed to avert this. This first *kom-pöt* is placed into the *klai-puk* with a repetition of the *tamma* (this time meant especially for the paddy spirit). The following *kom-pöt* are also put in the *klai-puk*, everyone is spat at, but without a longer *tamma*. Only the recipient is named: one *kom-pöt* is for the cotton, one for the paddy, one for the *ru-ke* leaves, one for the ginger, one for the sickle, one for the hoe, and others for any other object that may be in the basket. Then further *kom-pöt* are distributed in the house, that is the *kim-ma* and the *kim-tom*, in every place where things are lying, standing or stored, including bigger baskets and mats. The addressees are all good spirits, who care for the harvest and house. According to Menkröi, they are not called by their names, but I could not be sure whether this was really the case, since Khamlai murmured very much and I could understand only some *-ma*, which might have been the final syllables of names.

The next day but one, the contents of the *klai-puk* were emptied out (it might be on the *car*, but no place is prescribed) and the tools (sickle and hoe) were stored away. Among the Chüngma, the *cari-yöng* would have to be concluded with a *bong-kom* (see R1i); pieces of the cooked chicken and pig meat would be distributed among the inhabitants of the village (see P5c) (observations: 16.08.1956, additional information: MK, 18.08.1956).

It remains to be remarked that among the Anok (in distinction to the Longhu and the Khumi, see J7) nowadays no swidden work must be followed by a *khang* (when no work will be allowed, see P3c).

J4e) Harvesting

People who have finished their stored paddy before the new paddy is fully ripe and have no money to buy any and no means to borrow any, may try to eke out their sustenance by making *caa plak*. Half ripe paddy ears are cut and trodden out; the kernels that do not break off are thrown away. Those that are broken off cannot be husked at once as they would be smashed. Therefore they are first steamed in a big pot (10 kg with half a litre of water), then dried in the sun (or if it is not shining, over the fire of the hearth) until the kernels harden and can be husked. Since the kernels that do not break off when the ears are trodden out are lost, anyone who can afford to do so will wait until larger patches of paddy ears are ripe.

Once they are ripe, it is the time to cut them (*ngen caa*) as soon as possible, that is, if the weather is fine and no rain will fall. If rain is expected to fall, the cutting will be postponed, but it should not be delayed too much, since the paddy stalks might break down, which would make it more difficult to cut the ears, or the ears might fall out before being cut, which would reduce the harvest. In 1956 the rains had more or less stopped by the end of August, and on the 4th of September harvesting could be done without difficulty. Before the paddy is harvested, large woven mats are brought into the field house and spread out over the floor and all the lower parts of the side walls and the back wall; the paddy ears are piled up on these mats. For the harvesting itself, several families again join together in mutual-aid groups. In 1956, Menkröi's whole family worked together, and they were helped with the cutting by Menching and Klangwai, whom Menkröi and Dingte would help in return. Kyo Thwân Ong also lent a hand, and so the whole swidden could be cut in one day. All the cutting was done by the men; the women (Menkröi's wife Thanni, his eldest daughter Tharpau, and his adopted sister Kaiche) received and stored the harvest in the field hut.

The men started their work at the lower end of the swidden and proceeded upward by taking a bunch of ears in the left hand, cutting the stalks off (approximately 20 cm above the ground) with the sickle held in the right hand, and with the left hand, throwing the cut part of the stalks with the ears over their shoulder into the *te* (a big carrying basket with a wide opening, see D2f), which they carried on their back (held by a carrying strap which passed over their forehead). When the *te* was nearly full, they carried it to the field house (*bok*) to empty it out. If the ears had still been wet (from rain or dew) they could have been kept to dry in the field house for a few days. But since the weather had been dry, the harvested ears could be trodden out at once.

Before the *thua caa* (treading out the paddy) can be started, a pig (called *caa thua pak*) should be killed. The pig is brought along from home in a *pom* (see D4f). If none is available a chicken can replace it, or one might even dispense with any sacrifice, or do it or repeat it later when the harvesting extends over several days. The pig is killed with a sharpened piece of bamboo (*cau*) inside the field hut in front of the harvested paddy, where the paddy will later be trodden out. The *tamma* starts with *phyok-carok,* ends with *taku ning pak-laa hai ang klaa wi khök co* (cf. J4c), and is accompanied by repeatedly spitting out *hom-noi* (which has also been brought along) from a *teng* (a swallow-tailed bamboo tube), stuck in at the foot of the harvested paddy (and later thrown away). Some of the pig's blood is sprinkled on the harvested paddy. The sacrifice is dedicated to *caa-kom-nam-ma* (the paddy spirit) and is to increase the amount of paddy when the harvest is trodden out.

The treading out starts inside the house. Bundles of stalks are pulled down with the sickle on to the mat spread out in front of the heap in the front part of the hut, separated from the rear part by a bamboo bar fixed waist-high and serving as a railing, to which two, three, or four workers cling while using their feet to roll a bundle of stalks into a "ball", which is then "kneaded" thoroughly by stamping on it, all the time shoving it slowly backwards; then the ears are brought forward again and trodden out a second time, until the grains can be expected to have all separated from the ears. The remaining straw is then shoved back towards the door, but before it is thrown out on the *car* (the open-air platform) of the field house it is examined by the women, who pull the ball apart and shake it out. If the inside of the room becomes too crowded, the threshing can also be done on a mat spread out on the outside platform (in front of the door). Paddy need not be threshed in the same manner as wheat or rye, since the small stems of the panicle on which the grains hang dry up when the paddy is ripe and therefore break easily when bent. If the ears have not fully ripened, there may be some left that still contain grains. They are collected in a basket, kept for some days to dry and then trodden out a second time. Grains that are still green, however, are left on the stalks – as they should be, because later they would grow mould.

The paddy is trodden out on the *car* of the field house

The loss simply must be taken into account because if one were to wait until all of the grains had ripened fully, some of the stalks might well die and break off, causing the grains to fall out and the loss would be even greater. On the other hand, when the ears are fully ripe, the grains break off quite easily, and since paddy has no sharp awns, they do no harm to the callused feet of the Mru. To be sure, the work could also be done with shoes, but the constant stamping would be exhausting for the leg muscles.

The Longhu (Southern Chüngma Mru), like the Khumi and Bawm, use special square baskets (1 m x 1 m, 60 cm high, called *nun*, see D5e) for treading out the ears, but the Anok Mru prefer to do without these special baskets, collecting the grains on their tightly woven mats. If the harvested heap is small or when the heap has dwindled so much that one can stand on it, the treading out may even be started by trampling and jumping up and down on the harvested ears, provided that the mat lying on the floor will stand that treatment. Occasionally the treading mats get damaged; but if some of the grains escape through the holes, they can still be retrieved from the floor of the field house. If, however, the paddy is worked on the open platform and too many grains fall through to the ground, one tries to gather them up by hand as efficiently as possible. Then on the next trip to the field, a chicken is brought along to pick up what is left. If the weather is fine, the treading can also be done in the open air on the *car*. This has the advantage that the work can be done quicker. The women will turn a *te* upside down, clinging with their hands to its bottom and the men, after having cut the paddy in the field, join them doing the same, but holding on to a bamboo bar fixed to the wall in front of the house. Empty ears are shoved aside, examined and from time to time pushed down from the *car*.

After the treading out is finished, the grains are sifted by hand (*chù caa*). Small remaining stalks are collected in a basket, and shaken out in the open air a handful at a time. The grains fall back into the basket, the empty straw will be thrown away. All the grains are spread out on the mat inside the hut to dry (*reu caa*) for some days; the cleaned and dried grains are then scooped into baskets. A mat is spread out on the open platform, and two persons, holding shovel-shaped winnowing trays (*per-cing*, see D4a), place themselves on either side of the mat.

Winnowing the paddy on a mat in front of the *car* of the swidden house

As the trays are swung back and forth like fans, a third person shovels up rice out of the basket with both hands and – in rhythm with the fanning – throws it onto the mat. After they have been winnowed, they can be carried home (*ton caa*). For this transport the *te* will be used again. The baskets are not filled to the brim, since otherwise grain might spill out and they might become too heavy, but strong men may increase the load by putting a *kan-pen* (see J3e) on top of the paddy. At home, the paddy will undergo renewed drying processes before and while being stored in rice bins (*caa pam*), by being exposed to the sun on mats on the *car* of the dwelling house.

J4f) *caa-moi* (feeding the paddy)

Depending on how the paddy ripens, the rice harvest may take the whole of September and late paddy may not be harvested before the end of October. In between there remains ample time to carry the paddy home in stages. Before the last portion of the harvest is taken home another sacrifice takes place in the field hut, the first *caa-moi*. (*Moi* normally means "to provide food for work received", but in this case it implies a thanksgiving in the form of *kom-pöt* for the paddy spirits.) Anyone who wishes may sacrifice a chicken or a pig during the treading-out; but the sacrifice of a chicken is mandatory only after the last paddy has been cut, worked, dried, and made ready to be carried home from the field. The chicken is killed in the field house in front of the paddy. For the invocation (*tamma*) the beer substitute is again used, that is, the so-called *hom-noi* – it would be even better, though, to bring genuine rice-beer from home. After the invocation, two to three feathers from the chicken's wings are tied to a small bamboo stick, held in the fire and finally placed on the paddy heap and the same is done with the end of the intestine, cut in two. Then *kom-pöt* are prepared: portions of the chicken's meat together with *tam* (ginger, newly reaped from the field) and *hom* (cooked rice, brought along from home) are wrapped in bits of *ru ke* ("white" taro) or banana leaf and distributed over the paddy, the mats, and the baskets. A last parcel is thrown outside on to the field on behalf of the cotton, which will be harvested later.

After everything (including the baskets) has been brought home, the paddy harvest is brought to an end with a second *caa-moi*. The same ceremony as in the field hut is repeated at home, the chicken is killed near the *caa-pam* (rice bin). If possible a pig can be killed also, but the chicken is a must. A pig alone is not enough; not only are chickens more highly valued by the Mru (when they must be purchased, they are also relatively more expensive), but also mainly because this *caa-moi* is dedicated to the paddy spirits, who are said to exist in great numbers, some of which do not like pork. As was the case with the festival of the first fruits, chicken and pork are cooked separately but packed together into the *köm-pot* parcels which are then placed on the paddy and on all important possessions. As always, the greater portion of the meat is consumed by the people, but only after the ceremony is it permitted to eat any kind of side dish along with the new rice – until this *caa-moi* has been performed, it is not permitted for everything to be eaten together with rice cooked from new paddy. Forbidden are, for instance, *nga-ki* (Muntjac) – but bigger deer (*nga-hiu*) and wild pig (*pak-tia*) are allowed –, *nur* (eel), *rai* (bamboo shoots), *khua* (a kind of fish), *röp-khar* (fresh-water crab), and *tem-thu* (big round water snails) – *tem-cong* and *tem-büt* (long and small round water snails) are allowed, however. (Also newly married people are not allowed to eat these things.) The reason: all these foodstuffs are regarded as too "low" (even the *nga-ki!)*. However, any kind of vegetable is allowed. It is permitted to prepare rice-beer (*yu*) from the new harvest, but it is forbidden to distil it for making *arak*. Since spirits (*arak*) are always necessary for big celebrations, it is only from this point on that the larger celebrations (like marriages or cattle feasts) can be properly held again.

After the *caa-moi* all food taboos are lifted, but at the appearance of each new moon, or perhaps a little later, a sacrifice similar to the *caa-moi* (but without burning feathers and intestine) is expected. It is called *caa-chok* (to feed the paddy) and consists in sacrificing a big chicken (a small chicken will do only in case of extreme emergency). The *tam-ma* requires at least *hom-noi*, and the paddy should receive a *köm-pot* from the slaughtered chicken, so that the paddy supply will dwindle away as slowly as possible (MK, 15.11.1956). When the bin is empty (*pam khök*) a last sacrifice should be offered, if possible with a pig, if not a chicken will do. Among the Longhu the *caa-moi* is called *caa-hüng*. The *caa-chok* is called *wur caa* and the *pam-khök* is the *hua caa*. These will be described under J7.

As the trays are swung back and forth like fans, a third person shovels up rice out of the basket with both hands and – in rhythm with the fanning – throws it onto the mat. After they have been winnowed, they can be carried home (*ton caa*). For this transport the *tê* will be used again. The baskets are not filled to the brim, since otherwise grain might spill out and they might become too heavy, but strong men may increase the load by putting a *kan-pen* (see J3e) on top of the paddy. At home, the paddy will undergo renewed drying processes before and while being stored in rice bins (*caa pam*), by being exposed to the sun on mats on the *car* of the dwelling house.

J4g) Cotton harvest

If part of the swidden has also been sown with cotton, after harvesting the paddy the swidden will have to be weeded a fourth time (*laa krük*) in order to allow the cotton plants to develop fully, unencumbered by creepers and *wöi-la-ram* (see J1c). The cotton balls are plucked (*laa hau*) three times from the end of October until November, according to the ripening every 10–15 days. The left hand holds or bends the shrub, the right hand plucks the fibre and throws it into the big basket (*tê*, which is also used for the rice harvest) carried on the back. No special sacrifices are necessary, but during or after the time of *laa hau* the *üa plai* (see J6) will be held. The fibres (with the kernels) are brought to the field house for drying (they usually become wet from dew or fog). Fibres and kernels are separated at home by the women with the help of a cotton gin (see F1c), the better fibres are kept (and later spun and woven) and the best kernels are kept for the next sowing period. Surplus material is brought to Bengali traders in the market in large specially plaited bamboo baskets (*laa-par*, see D2k). Formerly it provided the main source of income (in 1957 I was told that one might get up to 35 Rs. for 1 maund ~ 37 kg), today the demand has considerably decreased.

Since no better seeds have been made available and the soils have become exhausted, yields have become minimal, so that it may be more rewarding to plant rice, sell it and buy cotton cloths from the market. But as long as Mru women still wear their own kind of skirts and prefer to sleep in their thick cotton blankets, they have to weave them themselves, and the Mru will have to continue to grow their own cotton.

Bangali boat transporting cotton down the Songu river

J5 Calculations

J5a) Mru data

In the previous measures (given in kg) I equated 1 *sher* with 1 kg, though 1 *sher* is calculated to weigh 932 gr, but since I compared seed quantities with yields, this could not change the ratio. In order to convert local measures into English measures and these into international standards, let me start with 1 gallon = 8 pt = 34.659 cubic inches; and since 1 inch = 2.54 cm, 1 cubic inch = 16.386 cm^3; 1 pt. = 567.957 cm^3, 1 gallon = 4543.66 cm^3 or roughly 4.54 l. In Bangladesh paddy is measured in *sheri* (measure of capacity, "dry measure") and *sher* (measure of weight), but since 10 *sher* are calculated to be equal to 16 *sheri* of paddy, *sher* may also be used as a measure of capacity. I took an empty cigarette box 1.8 x 4.6 x 7.3 = 41.8 cm^3, into which I could fill 3 *tola* of paddy (*camaa*). Since 80 *tola* = 1 *sher*, I could calculate that 1 *sher* would fill 1.648 dm^3. By using a 4 gallon (or 18.17 dm^3) kerosene tin comprising 4 gallon (or 18.17 dm^3), which could be filled with 11.9 (± 0.1) *sher*, I received 1 *sher* paddy = 1.515 dm^3. This makes it possible to calculate 1 gallon ~ 3 *sher*, or 1 *Hari* [pronounced "Ari" in local Bangla] (= 10 *sher*) ~ 3.34 gallon, and 1 gallon (~ 4.54 l) of *camaa* paddy would weigh approximately 2.8 kg.

The *camaa*-paddy just mentioned consisted of a mixture of *pe-tui* and *taru-rik* paddy (see 3a), and I counted 480 kernels per *tola* (the weight of an Indian silver rupee in 1900) = 11.65 gr, or 41.2 per gr. *Deng-pli* has 685 kernels per *tola* or 58.8 per gr, *cha-tui* (sticky rice) on the contrary has 325 kernels[455] per *tola* or 27.9 per gr., *rai-rik-caa* (the smallest sort) was not counted.

[455] Bengal knew an official grain weight (used to weigh "drugs" etc.) of 320 *dhan* (bini-dhan-kernels?) per *tola*. However, when used to weigh gold and

Since *deng-pli* is the kind mostly sown, let me use this figure of 685 kernels per *tola* in the following calculation. On medium soils, sowers make 40–60 holes per minute (in difficult places this number may come down to 30 and less) and throw in 12 kernels per hole (see J3c), i.e., they can sow 480–720 kernels (mean 600) or (working hard on reasonably good soil) 1 *tola* per minute or 1 *sher* (= 80 *tola*) in 80 minutes of continuous work. If our hard worker continued like that he would need 13.33 hours for 1 *Hari* (this figure may be regarded as the maximum); however, this maximum is impossible to reach, since he will not only need additional time to go back to the starting place several times, in order to start the next row, but he will also have to pause, for instance to eat some food and to regain strength, moreover he can only work during daytime, which lasts at best 13 hours in this region, and he will need additional time in the morning to walk to the swidden and back to his village in the evening. All in all I would conclude that one man would need approximately 20 hours to sow 1 *Hari* of *deng-pli* paddy, a figure confirmed by Menkröi, who told me that it would take two days for one (hard working) man to sow 1 *Hari*. Even if double the time were needed, the time needed for sowing would not pose a problem for survival.

A more important question is: How much territory does one person need to plant in order to receive a harvest sufficient for one year? The answer to this question will in large part depend on the ratio of seed to harvest. But we will first have to answer the question: what area is needed to sow 1 *Hari* of paddy? For the Mru this is no question, since the size is calculated in *Hari* only, they do not need to calculate in square meters or acres, etc. Bernot (1963:231) calculates for the Marma 1 *Hari* = 20 ares (= ½ acre). On the basis of my Mru measurements, however, I would assume that 1 acre (= 40.5 ares) = 1.3–1.4 *Hari* or 1 *Hari* = 29–32 ares. The higher figure is based on the calculation of the number of kernels sown per square metre (168; see 3c) [16,800 x 32 = 537,600] and the (roughly equal) number of kernels per *Hari* [668 x 800 = 532,000] (the number of kernels here is based on my counting during the sowing process, and is lower than that of 685 because we have to deal with seed kernels); the lower figure is based on an (admittedly) crude triangulation of a small swidden said to comprise rather exactly 2 *Hari*.

These figures would suggest that 1 *Hari* is roughly 0.75 acre. Returning to our question: "How much territory must one person plant in order to receive a harvest sufficient for one year", the answer will now depend on how much this person consumes per year. When rice provides 73 % of the caloric intake, people should eat 2,050 calories per day from rice, or 211 kg per year[456], but that's rice, not unhusked paddy. By husking, 16 kg paddy will be reduced to 10 kg rice; that is in order to receive 211 kg of rice, one

silver the standard was based on the weight of a silver rupee and in this case the *tola* was reckoned to equal 384 *dhan* (= kernels of the best "môta dhan"?).

[456] The International Rice Research Institute published in "Rice Today" (September 2002, p. 29) the following data about Burma (1999): milled rice consumption kg/capita/year: 211, total calories/capita/day: 2,803, rice calories/capita/day: 2,050, % calories from rice: 73. For the Mru, I assume the same figures for the moment, later I'll raise the percentage to 80 %.

should produce 338 kg of paddy. The Mru themselves hope for a harvest of 40 *Hari* (= 372.8 kg) per person, which seems reasonable in face of the fact that some part of the stored paddy will be lost during storage (eaten by mice, rotting, etc.) and some paddy may need to be sold (in order to buy medicine, for instance). More rice will be needed to make offerings, when some should be used to feed guests and for brewing beer (and distilling *arak*). Let us estimate these additional expenditures (including the losses by storage) at 10 % or 37.28 kg, and we are already below the IRRI figure. Let me add that I would have judged the Mru caloric intake from rice to be higher than 73 %, but if we start with 80 %, 40 *Hari* would already be too little. A person would need 231.2 kg of rice or 370 kg of paddy, that is to say, no loss by storage or additional expenses would be allowed!

Returning now to our original question, the answer will once more depend on yet another factor: the input-output ratio, that is, we must know how much a family of five members will have to sow in order to harvest 200 *Hari*. As already said, this will depend not only on the weather but also primarily on the quality of the soil, and it is well known that the more often a place is used, the lower the yield will be. At the end of the 1950s, the Mru reckoned on an average input-output ratio of 1:30, but they knew that the yield might be less than expected. They also knew that in earlier times their forefathers could reckon on a ratio of 1:50, and that if they made a swidden in a place where there had been a long fallow period of twelve years and more, the yield might even be as high as 70 times the input. Calculating with a ratio of 1:30, in order to harvest 200 *Hari* for a family of five, the total area of swidden for sowing should be $6^2/3$ *Hari*, that is five acres (= 202.5 ares or roughly 2 ha). In other words, in 1957 the Mru expected a yield of 932 kg/ha.

With an average fallow period of five years, the family will therefore need 30 acres in the long run (or: a single adult person should have six acres at his disposal). But not every acre in the vicinity of a village can be used for swiddening. Kangku-KP of Tapwüa-Kua belongs to the Galengya-Mouza. According to the Census of Pakistan 1961, this mouza comprised 10,240 acres and had a population of 1,095 persons. This number would need 6,570 acres. I had the impression that the productive capacity of the land had been reached or exceeded; that is to say that at best 65 % of the land could be used. But this type of calculation is not reliable. The swidden plots I measured were not even situated in the Galengya-Mouza, since Menkröi made them in the Horinjhiri-Mouza. The figures for Horinjhiri-Mouza given by the 1961 Census Report read: surface 6,400 acres, inhabited by 360 persons. These would need 2,160 acres, or roughly one third of the available area. However, to start from these figures would lead even further astray, since the Horinjhiri-Mouza comprises a lower and an upper part; between upper and lower Horinjhiri there is an area of very steep slopes, which are too steep to be reasonably used for swiddening (the land would slip when the rains set in), and the soils in the lower part are to a large extent unsuitable for making a swidden, since they have been overused. The yield might easily fall below fifteen times the amount sown, which would mean that the area to be swiddened would be greater than the area that could be expected to be weeded, and if it is overgrown with weeds the swidden has to be abandoned, since it cannot be harvested. All in all, I would assume that not even 30 % of Horinjhiri can be used for making a swidden. In the light of these changing conditions, all calculations become very uncertain and fraught with speculation.

J5b) Data from Banderban District

Taking the whole Banderban District of 1,665 square miles with 67,400 persons enumerated in 1951 and deducting the reserved forest area of 997 square miles with 3,120 persons, we arrive at a total of 668 square miles (= 427,520 acres) with a population of 64,280, i.e., 6.65 acres per person.:

Union	acres	persons
Banderban	51,228	8,637
Alikhyong	72,696	7,746
Taracha	132,096	12,643
Ruma	52,120	8,262
Pantola	55,680	7,166
Thanchi (ex reserve)	74,240	6,068
Lama	113,043	10,968
Alikodong (ex reserve)	110,997	9,877
Naikhyongchori	91,318	9,347
Tombru	21,491	5,217
total	661,878	85,931

The figures for 1961, however, show that is 7.7 acres (= 3.1 ha) per person. Which data are more reliable I cannot decide. The FAO/RAP-RDES statistics for Bangladesh give the area under paddy in Banderban District in 1993 as 51,490 ha (= 127,136 acres). The same source gives as "total production of paddy" for 1993 in the Banderban District 42,390 metric tons, and under "yield rate estimation" 3,052 kg/ha (which would yield 3.707 times more than the actual total production [the difference is not explained]). We may compare these data with those given for the Chittagong District 1993: area under paddy 968,920 ha, yield rate estimation 3,302 kg/ha, (which once more yields 3.707 times more than the "total production" of 863,160 metric tons[457]. Obviously the area recorded for the Banderban District is that under permanent paddy cultivation, and most probably does not contain data on swidden-cultivation. I mention these official data nevertheless to show that in 1993 wet rice cultivation in the Banderban District (especially in the Matamuri and Naikhyongchori region) had already overtaken dry rice cultivation on swiddens.

Returning to my data given above, in 1957 the Mru expected 932 kg/ha, while the production on the territory of 51,490 ha cultivated with wet rice in 1993 was said to have yielded 42,390 metric tons (= 823 kg/ha) – that is, the Mru expectation for swidden farming in 1957 would predict roughly 100 kg more than was produced on wet rice fields in 1993. In order to produce 372.8 kg/person/year for a total population of 85,000 persons in 1961 by swidden cultivation in the Banderban District, the area available should

[457] All of Bangladesh: 18,340,280, confirmed by MMIS, Ministry of Agriculture, Bangladesh, Statistics, table 5.03, which gives 18,341,000 metric tons. This total yield was harvested on an area of 35,151,439 ha under paddy, that is 729 kg/ha, while the FAO yield rate estimation give 2,703 kg/ha (or once more 3.707 times more than the recorded harvest!).

have been 34,000 ha. The area inhabited by these people comprised 661,878 acres = 268,060 ha, of which 51,500 ha had come under permanent cultivation in 1993. This leaves us with 216,560 ha, which would have had to suffice for six years, that is 36,093 ha/year, of which, however, only at best half of the area (let us say 18,000 ha – instead of the 34,000 ha needed) could be used for swiddening. The production on these 18,000 ha would have been 16,776 metric tons (sufficient for 45,000 people, living like well-to-do Mru), instead of the 31,688 metric tons sufficient for 85,000 people. If we add the 51,490 ha cultivated with wet rice to the area available for swidden cultivation, this would increase the latter by 8,582 ha/year, and the yearly production would rise to 24,774 metric tons (sufficient for 66,500 [instead of 85,000] people). That is, without wet rice cultivation the area could not have been self-sufficient, while, on the other hand, the wet rice production by itself would be more than sufficient (leaving a surplus of 10,700 metric tons) to ensure the yearly paddy supply for the whole population of the Banderban District.

Incidentally, the conclusion that the population density is too high to enable survival from paddy cultivation on swiddens could also have been reached by looking at the Mru figures alone. It was already stated that a family of five people will need to sow an area of roughly 2 ha/year. For one person this would mean 0.4 ha/year, and with a rotation of six years he would need 2.4 ha in the long run. Calculating that half of the territory can be swiddened, he would need an area of 4.8 ha – but there were only 3.1 ha/person available in Banderban District in 1961. In order to raise this figure to 4.8 ha/person the number of people living in the district would have to be reduced by 30,000 people to a total of 55,900. In 1901 the population of the district (at that time "circle") probably comprised at best 40,000 ($^{1}/3$ of the CHTs' total), and this would go well with a rotation period of 12 years with yields twice as high as those reached by 1957. These general figures for 1901, however, do not exclude the high probability that in more densely populated areas (as for instance in the Kornofuli valley) rotation periods had already been shortened to such an extent that scarcity set in, which in the following years indeed induced especially the valley-dwelling Chakma to introduce permanent wet rice cultivation.

J6 *üa plai* ("swidden dance")

J6a) Preliminary remarks

After the paddy and the cotton have been brought from the fields, and after a few ears of corn, some millet, indigo leaves, sweet cane, beans, squash, gourds, taro and yams have intermittently been harvested, a few tubers and late-appearing gourds are still available on into January. Anyone who wishes may also help himself to the flowers and fragrant leaves which were planted along the edges of the field paths. It is December, climatically one of the most pleasant months, and time for the harvest festival. The sizes of the celebrations vary, depending upon what can be afforded following the harvest; and the ceremonies considered necessary at this time likewise vary from region to region. At the very least, a chicken must be sacrificed; most households, however, sacrifice a pig – one still small enough to be carried easily to the field. The three plate-shaped gongs and the drum may also be taken to the field. Four people are needed to play the musical instruments; as we shall see, others are soon added. This means that it is not a single household that sets out for such a "field dance" *(üa-plai),* but rather an entire group. The group consists primarily of the men who assisted each other during the field work. The women normally remain at home, since sacrificing is a man's job. This, however, does not keep young boys and girls from tagging along. Among the Anok Mru this harvest festival is not a must, except when one family member falls ill during the harvest time and the household head has promised to hold one. The "dance" demands the preparation of several paraphernalia, but may be executed differently. I shall describe two types that I was able to observe in Tapwüa-Kua plus one *üa-plai* which I observed when on tour with the Rümma.

J6b) *dor chör*

Dor is the name of a *tui-lam* (a piece of bamboo containing water) and *chör* (to advance, shift forward) here indicates the slanting position in which the *dor* is fastened. If the *car* (open-air platform) of the field hut borders on the swidden, it is fixed in such a way that its upper end is tied to a support in the field while its lower end reaches the *car,* where it is tied to another support, so that a *ci-ö-ca* (a small round clay pot) can be placed below it as if the pot (for this occasion decorated with a bamboo thong wound around it) was meant to receive the water from the *dor*. The supports are two *caro-ca*, bamboo poles, the upper segments of which have been split up and deprived of their outer hard green skin, so that they can be spread apart to look like a white crown. The segment below is shaved up to the internode and adorned with four *chit-wai*, bamboo sticks carrying tassels at their upper end. To be more exact, the lower end of the *tui-lam* was clamped in another split bamboo, placed directly besides the lower *caro-ca*, the upper end hung up in a sling tied to the upper *caro-ca*. On the lower *caro-ca* hung a *keng-ko* (a "silver" string used by the women to fix their skirts round the waist), a *kwai-che* (a bracelet), and a *tom-preu-kok* (an "attached spotted [sword] sheath", made of bamboo). The *dor* or *tui-lam* itself had been decorated with *yan-kia-bong* in the middle (made by shaving off the green skin in wavy lines) and *thau-rin-bong* (made by shaving straight lines). Its upper end was plugged up with *ram-nget* leaves (not identified). The water it contained was, however, never let out into the *ci-ö-ca*, which had been filled with *hom-noi* (pound rice mixed with water), to be sucked up from it through a small bamboo tube, and used for the *cho* (the spitting out) at the occasion of killing the chicken, the pig, and before leaving the swidden.

But that's not all, three more places have to be prepared. The first two also on the *car*, a *pak-rung* (pig-yard) and a *waa-rung* (chicken-yard), both consisting of a *caro-ca*, approximately 2 m high, with a broken *ci-ö-ca* (as *hom-noi pu*, sham beer pot) at its foot, placed on a straw pad. The *pak-rung* pot had a bamboo thong round its neck. The *waa rung* bamboo was accompanied by a small stand with a funnel like a chicken basket (on one of the bamboos of the funnel a piece of ginger was skewered), and on the bamboo hung a small carrying basket (*em-ca*, see D2b), containing a sickle and some straw, meant to represent cotton (which, however, was no longer available on the swidden). At its foot, besides the *ci-ö-ca*, a piece of taro was placed. The sickle was used to behead a small chicken (under *tamma*).

Moreover, a *teng* (a swallow-tailed bamboo) was stuck in all places (though it was never used). At the *pak-rung* a small bow was added, at the *dor* a *chia* (an iron "dart"), with which the pig, brought along from home and now placed between the two *ci-ö-ca* below the *dor* and at the *pak-rung*, was symbolically shot (after the chicken had been killed), in order to be killed afterwards with a *chin-re* (an iron spear), which was tied to its own bamboo post near the *dor*. At the places of the *pak-rung* and the *dor* a small mat had been laid out and on it were placed *ram-nget* leaves containing *kam-bor kom-pöt* (termite spirit food), ginger and taro. The *kom-pöt* were not specially filled, but just dipped into mud, made in masses, and distributed several times in all important places. Moreover four *li-cong* ("wind trap"), looking like hour-glasses made from bamboo, with a *chaling-chalap-ma* (a "butterfly", a small thread-square) tied to their middle, had been prepared, two were fixed to the *dor* and one each to the *caro-ca* of the *pak-rung* and the *waa-rung*, and pieces of the pork were put on four *cher* (skewers), accompanied by a piece of ginger on one side and a piece of taro on the other, called *ru-büng tam-büng* (taro-closing ginger-closing), and distributed like the *li-cong*.

The *dor* above the platform of the swidden house

The fourth place (the *tui-lam ne-raa*, "water-tube incline place") is made in the swidden, but receives little attention while executing the ritual. Twice two *chit-cüng* (bamboos with tassels) are rammed in obliquely so that they cross each other, over them are laid two more *chit-cüng*, one sticking out to the right side, the other to the left, and against this stand two *tui-lam* (without any decoration, but also with plugged-up upper ends) are leaned. Maybe one *li-cong* and one *cher* should have been deposited here instead of being twice hung on, or deposited at the foot of, the *dor*. The execution of the ritual depends much on the *charaa* (master of ceremonies). In the case observed (on 12.12.1956), Karbari Kangku Catumma acted as *charaa* for his son Mowai on the latter's swidden and did the whole job within a few hours. According to Menkröi and Khamcöng (10.02.1957) the ritual is especially elaborate among the Dengwüa, who need a whole day for it and sing a special song when calling the spirits.

J6c) *cam-plai*

1) General remarks

Cam here stands for *bok* (the swidden house), though it is called by that name only on the day when the ritual is performed. The owner of the swidden was Yenpau Khongtör, whose son Menlong had fallen ill while working in the swidden, and Yenpau had invited Karbari Kröilö from the neighbouring village to act as the *charaa* (master of ceremonies) on the 30.12.1956. The following report is based on my own observations. Menkröi was absent that day, but he, Khamcöng and Yenpau gave me additional information on 10.02.1957.

Two pigs and one chicken are necessary, and are sacrificed on the open-air platform (*car*) of the field house. The dance is accompanied by a *tömma* (drum), three *ner* (flat gongs), a *mong* (big gong) and three *tu* (special gourd-pipes with only one pipe), but no *tu-charaa* (*tu* master) is necessary, that is, they are not blown and one man may be sufficient to carry them around. Concerning the paraphernalia, all those are needed which are required for the execution of a pig festival called *tam* (see P5b). *Li-cong* ("wind trap", see J6a) and *chaling-chlap* ("butterfly", thread-square) are fastened to a special bamboo, called *krou* (see below). Furthermore nine *kau-chung* are needed (bamboos with leaves kept, nine is the usual number, otherwise only five or fifteen are possible), and (when the spirit is called to eat) a bamboo carrying 15 (or 9) *khö*, special offering baskets, into which, however, nothing is placed. Of these *khö* 13 (or 7) are simple ones, two (called *khö-kröng*) are double ones and the latter are fixed above the simple ones.

2) The paraphernalia

At first the pigs are prepared for being carried to the swidden. The smaller pig is just placed into a *pom* (a basket plaited in open wickerwork and usually used for the transport of chickens, see D4f), while for the bigger one a special "cage" is made by placing some bamboo bands on the earth, interweaving them on their ends with bamboo thongs, putting the pig on it (with the snout and the legs tied), pulling the thongs up, so that the pig will be wrapped in, and intertwining them, finally wrapping three more thongs round the parcel and tying them up, so that a bamboo pole can be pushed through, on which the pig is then carried. The chicken, being just a chick, is carried along with the pig in the *pom*. Since Yenpau is a poor man and his swidden is far away (more than one hour to walk), he dispenses with the *mong* (buckled gong), but he sees to it that paddy and cooked rice are taken along too.

Only five *kau-chung* (full bamboos) are set at the border of the *car*. In the middle of them a *krou* is erected. The *krou* consists of a bamboo pole, ca. 20 cm above the *car,* split into four parts, each of which is, at a height of ca. 1 m, again divided into three parts, the middle one of which is broken out, the remaining 2 parts (that is, in total 8 parts) stand free up to a total height of ca. 3 m. Where the middle parts have been broken out, a square bamboo wickerwork (*krou phai*) is inserted, and kept in place by inserting two *chit-wai* (bamboo sticks with tassels) in the four corners. The tassels hang outside and their sticks are tied to the four split parts of the *krou* by three bamboo thongs, the first and the third one green, the second one white. Over the front left *chit-wai,* a *keng-ko* (a "silver" string used by the women to fix their skirts round the waist) and in this a *kwai-che* (a bracelet) are hung. On the *krou-phai* two *röp* (small crabs) are placed, along with a *tom-preu-kok* (see J5a), a piece of bamboo to represent the sheath of a *chai-dong* (an old ceremonial sword), with which formerly the *charaa* was to dance, and a *büi-lu* ("head of a bamboo rat"); the *büi* should be caught in advance, but since none was available (and its meaning and purpose were unknown anyhow), it was replaced by a piece of ginger. On the two *chit-wai* of the right side, two *cher* (bamboo thongs serving as skewers) are hung, one on the back side with *kam-bor kom-pöt* (termite spirit food) and one on the front side with *ru-büng tam-büng* (see J6b and below).

At the left side of the *krou* two more short bamboo poles are implanted. To the one, called *lim-yung*, a spear is tied, on to the other the *khö* (6 normal and 2 *khö-krong*, see above) are stuck – one above the other, with their opening downward. Then at the foot of both *krou* and *lim-yung* a *pam* ([paddy] bin) is erected. A piece of an old mat (ca. 20 x 80 cm) is rolled together and fixed with two thongs so as to form a tube. Into the *pam* first a layer of straw is put, then roughly 1 kg of paddy is filled in and on top of the paddy are put some cotton seeds; later *kam-bor kom-pöt* and *ru-büng tam-büng* will be added.

At the left side of the *krou* two more short bamboo poles are set in. To one, called *lim-yung*, a spear is tied, and on to the other the *khö* (six normal and two *khö-krong*, see above) are stuck – one above the other, with their openings facing downwards. Then at the foot of both *krou* and *lim-yung* a *pam* ([paddy] bin) is erected. A piece of an old mat (ca. 20 x 80 cm) is rolled together and fixed with two thongs so as to form a tube. Into the *pam* first a layer of straw is put, then roughly 1 kg of paddy is filled in and on top of the paddy some cotton seeds are placed; later *kam-bor kom-pöt* and *ru-büng tam-büng* will be added.

Two bamboo tubes (approximately 1 m in length), adorned with *yan-kia-bong* (made by shaving off the green skin in wavy lines alternating with straight lines) are prepared as *tui-lam* (water tubes), to be plugged up with bamboo leaves. Kröilö himself also prepares two shorter tubes (*pön-po*; *pön* = their name, *po* = tied together), which he sticks into the wall of the field house, decorated with *thau-rin-bong* (shaved in straight lines) at their end, tied together at their upper and lower end with black and red cotton threads, with *chong-klang* leaves at their lower end and the upper opening plugged up with *ram-nget* leaves (not identified).

3) The ritual places and the election of the "officials"

Four ritual places are used: 1) *pak-rung* (pig-yard) on a mat in the middle of the *car*, 2) *pam-nam* (bin-place) at the border of the *car* in front of the *krou* and the two *pam*, 3) *ria-ca khüm-ra* (place to select the helpers) within the field house, and 4) the *waa-rung* (chicken-yard) in the swidden at the *tur-tut* (see J2e), but nothing is erected or placed there.

At the *pak-rung* two little round pots (*ci-ö-ca*) are put up, one of which is called *ron-pu* ("promise-pot"). It was wrapped in banana leaves and contained some uncooked rice. With it the spirits had been promised the *cam-plai*. Now the wrapping was removed and the rim wrapped with a new bamboo thong. The rim of three more *ci-ö-ca* are wrapped with straw, some cooked rice (brought along from home) filled into the pots, on which two *ram-nget* leaves were put, and water poured in, first by Kröilö some drops out of the *tui-lam*, then the water is filled up from *tui-yia* (bottle-gourds), and a small drinking tube is added. The pots now can function as *hom-noi-pu* and are distributed in the three ritual places, where they are set on a ring of straw. Beside the pots are placed some paddy, banana leaves, three kinds of flowers (*chong-prem* [tagetes], *ca-pan*, and *khet-pau*), and twigs from three kinds of plant (cotton, *mrut-ma* and *cang-ma*). Moreover, two clods of earth were laid in the *pam-nam*.

Into the pot in the field-hut, as well as the drinking tube a drinking sign (*pleu*) is also placed. This *pleu* has a special form used only during the *cam-plai*: the upper part of the little bamboo stick is split in two, the two branches forming a V, before they are bent upright again. If it had been a full ceremony, a *yu-kong* (a beer pot) should have been used instead of the *hom-noi-pu*. At this place, the *ria-ca khüm-raa*, the *charaa* (master) first has to select two *ria-ca* (helpers) by putting some uncooked rice on the tip of a sickle, with which he touches the forehead of the elected; he then elects in the same way three *pin* ("invited guests") of the feast-giver's *pen* (son-in-law sibs, see L3b), one *krong-töng-yaa* ("earth invocator") and one *poi-coi-yaa* ("feather plugger"). These five persons have no special function to fulfil apart from beating the musical instruments (which could be done by anyone). They will be required at the *pang-cang* (see below), however, after returning home, and they will receive a *bong-kom*. One of the *riaca* has to kill the *moi-ma-pak* (one of the pigs, see below) and the other has to kill the chicken.

After the distribution of the *hom-noi-pu,* the *lük-chur-ngar* (*lük*: to smear [on the forehead]) is prepared: *chong* (shrimps) and *röp* (small crabs) are cooked in a *röng* (a bamboo tube used for cooking) and then pounded together with *kan-chur* (red sorrel, see J3d-15) and water in a *khan-pon-thet-röng* (see J4c). With the help of a *chit-pot* (a stick with a short tassel) the mixture is then dabbed on the forehead of everyone present. Before this it is forbidden to drink water in the swidden.

4) Preparation of *kom-pöt* and killing the animals

To prepare the *kam-bor kom-pöt,* a big clod of earth (if possible from a termite hill) is mixed with water, some ginger and cooked rice, and into this mixture two *ram-nget* leaves are dipped, folded and skewered on a long thong. After two thongs-full (*cher*) have been prepared, those of the first thong are distributed by the *charaa* on every specially prepared bamboo, every instrument hung up or placed somewhere, on the walls, the corners, the roof, the baskets, and the mats in the field-hut; the second *cher* is fastened to the *krou* (as described above).

Next the small pig in the *pom* (the wickerwork basket in which it has been brought) is killed with a *cau* (a sharpened bamboo stake). This pig is called *mür* (literally: the good smell that it gives off when cooked) and can be killed by anyone. While it is cut up and cooked, cucumbers are cut up, chillies are sliced and turmeric and ginger are pounded in a *röng*. Then the 7+2 *khö* are taken down from their pole, one of the *khö-krong* is placed in front of the *krou* and is regarded as a *pui-tang* (table), the other one is placed in the field house. The seven simple *khö* are placed between the *pak-rung* and the *pam-nam*, and banana leaves are spread at a short distance from it, at the side of the *pak-rung*. Into the *khö, ram-nget* leaves are laid, and on them (and the banana leaves) some cooked rice (brought along from home) is distributed, to which little crabs and shrimps from the two *röng* are added, with taro and chillies, ginger, turmeric and pieces of the *mür*-pork. One thigh of the pig is placed on the *krou-phai*. Now Kröilöi takes six bites from a piece of wild ginger root (*ai-dam*), takes a sip of water from one of the *tui-lam* and spits it over the *krou-phai*, calling the spirits to eat. His invocation is uttered with a loud falsetto *hü* (a tone so unfamiliar in daily Mru life that the children start to laugh). Menlong must eat some of the food spread on the banana leaves near the *pak-rung*, after which all *khö* are emptied out, not put back on the bamboo pole, but collected in a heap; the *tui-lam* which Kröilö had used for his *cho* (spitting) is leaned against the *krou*.

Excursion to the *waa-rung* in the swidden

Now the bigger pig, the *moi-ma*, is placed on the mat of the *pak-rung*, and five men start to dance around it, one beating the drum (*tömma*), three beating the flat gongs (*ner*), one carrying (not blowing) the three *tu*, and one (Menlong, elected as *ria-ca*) carrying a basket (containing a chick and a sickle). After three circuits, the five men leave the *car* and proceed to the *waa-rung* in the swidden. Menlong takes the chick out of the basket and cuts its throat with the sickle, then the chick is thrown back into the basket and the sickle is used for cutting what is available here: some straw, cotton, leaves (of weeds). Some taro (*ru'*) should also be dug out (but none was available). All the things are put into the basket, and after making three circuits of the *waa-rung* the five men return to the field house, where Kröilö touches their foreheads with *ai-dam* (wild ginger) before they once more make a circuit of the *pak-rung*. After three circuits to the right and three to the left, Yenpau "shoots" the pig with a *chalang-baa*, a toy bow and arrow.

The arrow is a strong bamboo dart, which, however, is used less for shooting than for piercing the pig (at first to be done by one of the *ria-ca*, who had been elected for this job). The dead pig is placed on its stomach, the dead chick is held on its right ear (by Menlong as the *ria-ca*, who also grasps the left ear, the other *riaca* grasping its tail), and the two men now carry the pig round the *pak-rung*, accompanied by renewed circuits by the remaining four men, three to the right and three to the left. Formerly the *charaa* would also have had to join this dance, brandishing a *chai-dong*.

Yenpau "shoots" the pig with bow and arrow

The chick is roasted; the tongue, the feet and the tail of the pig are cut off and put into one of the *khö*. The remainder of the pig is then cut up and cooked. The cooking water is fetched in long bamboo tubes, the internodes of which have been pierced with the point of the *re* (spear, used for this purpose only). New *kom-pöt* (*ru-büng tam-büng*) are prepared, one *cher* with pork (*pak ngaa cher*), the other with the meat of the chicken (*waa ngaa cher*), to be distributed by the *charaa* like the *kam-bor kompöt*, though in lesser numbers.

5) *pong-chak*

Finally, Kröilöi takes the pig's tongue, feet and tail out of the *khö*, and puts two *ram-nget* leaves into it, on which are placed cooked rice with one egg in the middle, then he puts back the pieces of pork to which are added two little *röng* filled with blood, binds two sickles together so that their blades form a circle, puts the *khö* on this "ring", takes one *tui-lam* and carries these things down into the swidden, accompanied by Menlong who carries a small dog and one *cher* with mixed chicken, pork, and termite *kom-pöt*, hung on a *tim* (field hoe). During the whole excursion the drum and the flat gongs are beaten. The reason for the excursion is the *mau-ngak hom chör* (the presentation [*chör*] of food [*hom*] to the "bad bamboo" [*mau ngak*], a bad field spirit). At a "bad" place (it might be any place) in the swidden, a hole is dug with the hoe, at the bottom of which a *nam-ria* (which has been fetched from a tree on the way) is laid, the *khö* is placed on it, the *kom-pöt* are distributed around it, the small dog is killed and left lying there, the hoe is placed in the cooked rice of the *khö* and left there, and a *choi* (vowing) with water from the *tui-lam* ends the *pong-chak* (name of the ceremony).

Before cutting the swidden (it is assumed), a monkey, a hornbill, a small deer, or whatever animal may have died (*pong*) in this spot, and its ghost must have attacked Menlong while working in the swidden and caused Menlong to fall ill. The sacrifice had been promised with a *choi* and now it had been fulfilled. Whatever had perished (*pong*), *mau-ngak* or any other bad spirit would now be placated (which is why every kind of *kom-pöt* had to be distributed) and, as long as the iron (the hoe) stays in the earth, it should not be able to attack any human anymore (*nam-ria*, besides designating the plant, can also be interpreted to mean "spirit" [*nam*], "to take another way, to pass by" [*ria*]).

After returning from the *pong-chak,* Menlong has to perform a *caa-kwak* (throwing the food away). He must squat down near the *pak-rung* and, putting the right arm across the left with two *ram-nget* leaves in each hand, he receives some cooked rice and meat (cooked in a *röng*) on each hand and Kröilö spits at him with ginger and water from the *pön-po* (see above). Menlong has to take up the food from his hands with the mouth, his father gives him water to drink, which, however, he will not swallow, but must spit out again with the food. This is repeated three times, then Menlong sits (like a woman) on his knees with the legs backward to the left, crosses the left arm over the right and receives food and water three times, which he has to spit out as before. During this time the flat gongs and the drum are beaten again. Finally Menlong receives a *bong-kom* of red and black cotton threads, which are tied to his right wrist by Kröilö, who afterwards binds (without any special ceremony) a similar *bong-kom* thread to the wrists of the other "officials" also (whom he elected at the *ria-ca khüm-raa*) and of Yenpau, Menlong's father.

Then everybody takes food, and after it they leave the swidden. Before leaving, everyone has to take a sip of water from the *ron-pu*, into which all remnants of water have been emptied out, and to spit it out again on the remainder of the food, which is to be kept lying beside the *pak-rung* (it is forbidden to throw it down from the *car)*, and to which the *chalang-baa* is added, with the dart pointing obliquely upwards. The uncooked thigh of the pig, the *keng-ko,* and the *kwai-che*, on the other hand, are removed from the *krou* and taken back home like the *re*, one of the *tui-lam* and one of the *cher*. The drum and the flat gongs are beaten the whole way home, but before the players can enter Yenpau's house and stop beating their instruments they must make a *cho* (spitting out) with water from the *hom-noi-pu* put up at the *dong-lu* (the head of the log staircase).

6) Final ceremonies at home

In the house all are welcomed with a glass of *arak* (distilled rice-beer), and two beer pots are put up, a *choi-ma-yu* and a *pin-yu* (the latter is also called *pang-cang yu*). From the *choi-ma-yu* (the "vow-beer"), all (including the young boys and children) can drink; to the brim of the second (the *pin-*

yu, the beer for the invited guests) two thongs are tied, which Klingtui, who functioned as *ria-ca*, has cut out of the *tui-lam* (the water bamboo tube brought home from the field, the rest of which is stuck under the roof). Yenpau is the first to drink from this *yu*, followed by the other helpers. While they drink, Yenpau distributes in the house the *kom-pöt*, brought home from the field, and gives (without further ceremonies) *bong-kom* to the females and youngest members of his household who stayed at home during the *cam-plai*. Then follows the *pang-cang*. Accompanied by renewed beating of the drum and the flat gongs, Kröilö takes some cooked rice out of a bowl, which has been put in front of him, covered by a banana leaf, on an *em* (see D2a) turned round, sips some water out of the *hom-noi-pu* (which now has been brought into the house and placed in the middle of it beside the two beer pots) and spits it at the drum, the three flat gongs, and right and left of the *hom-noi-pu*. Then the two *riaca* (Klingtui and Menlong) and the three *pin* take a sip from the *pin-yu* and spit it (together with some cooked rice) right and left besides the *yu*-pot, while Dingte (as one of the officials) pours water (out of a halved calabash) over their necks.

Klingtui is the only one of the guests who receives cooked rice and pork to eat, since he has also to function as *chaa waa-ma* ("accompany the *charaa* = *waa-ma*"): when the ceremonies are finished, he must accompany Kröilö home (at least down the stairs of Yenpau's house). But he is ashamed and eats only a little. Before this Kröilö has a last job to fulfil: Yenpau sucks some beer from the *pin-yu*, spits the beer into a small *röng* and puts this in the bowl, which is now placed into the *em* together with the nearly empty *arak* bottle, a sickle, cooked rice and pork (wrapped in a banana leaf) and a piece of wild ginger. Kröilö takes the *em* and carries this *krong-ma nam-ma cak* (earth spirit's food) in front of the door, displays the rice and pork, empties the beer from the *röng* into the bowl, bites off a little piece of the wild ginger, takes a sip of beer from the bowl and spits this six times on the food and the bottle, "bites" into the sickle, spits again, empties the remaining beer out of the bowl, and pours the remainder of the *arak* from the bottle into the bowl, shortens the *tamma* to *phyöi taruk-kat*, once more bites off a little bit of wild ginger, takes a sip of *ara*k, spits it out, "bites" into the sickle, spits again, then puts his foot on the sickle, and all lookers-on must leave to the right side. Finally the *em* is brought back into the house, with the bowl, the empty bottle and the sickle.

Next day some *tür-ram* leaves are thrown away with a *cho* (a spitting-out) of the remaining beer. This definitely ends the *cam-plai*. This use of *tür-ram* leaves and the ceremonies called *pang-cang* recall a bigger cattle sacrifice (see R1i and R3f). Among the Anok Mru no immediate connection exists, among the Khumi and Longhu, on the other hand, big cattle sacrifices must be followed by a swidden festival the next summer (*lau-reng*, see R5h).

J6d) *üa-plai* of the Rümma

The ceremony was held rather late, that is, after the next swidden cycle had already started, on 14.03.1957, in Rengtan Rümthu-KP. The karbari himself acted as *wia-ma* (= *charaa*) and as my main informant. The swidden was a double one (with one field house) and the ceremony was also a double one: a *plai-tük* (held in the field house) and a *rengma* in a place cut free near the brook separating the two swiddens. Apart from the fact that more paraphernalia are used, the *rengma* resembles the *dor chör* of the Anok (see J6b); the *plai-tük,* on the other hand, could be called a rudimentary *cam-plai* (see J6c). This combined ceremony, however, is not the biggest *üa-plai* of the Rümma. For this three pigs would be required (but I did not get a description of it, since the next day I visited a cattle sacrifice of the Dömrong, and when I came back to Rümthu-KP, the hamlet held a *khang*, so I could not enter it once more).

1) *rengma*

First to be prepared are seven small (1 m high) *caro-ca* plus six *lu-kut* ("bald head", the upper end is not split but only decorated with *thau-rin-bong* [straight shaved lines]) with four *chit-wai* each. Of these thirteen bamboos eight are planted (ca. 80 cm apart) in one row (alternately from right to left one *lu-kut* and one *caro-ca*). Two pairs of one *lu-kut* and one *caro-ca* are first tied together and then tied diagonally ca. 10 cm above the earth to the two outer pairs of the *caro-ca* and *lu-kut* row, with the *chit-wai* jutting out to the sides. The seventh *caro-ca* is about 2 m high (higher than the rest) and is held in upright position by three additional bamboo rods. It carries (in a sling fixed to it in a hole) the upper end of the *dor* (see J6a), which is plugged with straw, while the front end lies on the right diagonal *caro-ca lu-kut* pair. To the first *lu-kut* another bamboo tube, 1.50 m in length, called *waa-dör* (chicken basket), is tied. Its upper end is split like a *caro-ca*, but interwoven with thongs, into which some chicken feathers (brought along from home) have been plugged. Into the "basket" a *waa cher* (a chicken wing; also brought along) is placed, and a *keng-ching* (a string of green beads, instead of the *keng-ko* of the Anok), carrying a *khen-tang* (an iron bracelet; instead of the *kwai-che*), is hung. Furthermore at the *waa-dör*: a *tom-preu-kok* of bamboo (see J6a) and a *büi-pang* (thigh of a bamboo rat), but since no bamboo rat could be caught, it was replaced by some *büi-öt* (earth thrown out by a bamboo rat), wrapped in *ram-nget* leaves. More eye-catching than these smaller things is a big *li-cong* ("wind-trap") placed over the *dör*. (It will be described later.)

Set in at the foot of the first *lu-kut* is a spear (*re*) and also tied to this first *lu-kut* is a *pön-mum-pön-büa*, consisting of three thin bamboos of ca. 40 cm, the upper end of which is cut just above the internode, thereby producing three little cups, into which some ginger (*tam*) and some beer yeast (*chang-ki*, brought along) are placed. This bundle of *pön* is tied to the *lu-kut* by means of three lashings (alternating white, green, white) called *mia-kom chum-kom* (lashings-three). Plugged into these lashings is an *üng*, similar to a small *teng* (swallow-tailed bamboo), but with only one long "leg". An *üng* is normally used for siphoning off beer from a pot, but now it is not used at all. The reason given for its making was, like for that of the other paraphernalia, "*athung*" (custom).

Between the fifth and the sixth bamboo pole of the row a stone is sited (*hua ai*), and in front of each of the bamboo poles (except No. 1 and 4) two *teng* (swallow-tailed bamboo) were placed. In front of the first (a *lu-kut*), instead of the second *teng,* a *nom* (a small water-pot with a decorative lashing tied up and down over its belly) was placed below the front end of the *dor*, in front of the fourth (a *caro-ca*) a *ci-ö-ca* (a blackened small round pot) was put on two sickles tied together to form a ring. Each of the *teng* and the two pots are filled with some *chang-ki* (beer yeast) and water, and a small drinking tube and a *pleu* (drinking sign, the upper end of which was split up and bent in such a way as to form a rhomboid) are added (cf. the *pleu* of the Anok *cam-plai* [J6b3], which, however, had the upper ends standing up). Moreover, in front of the pots 6-cm-long striped bamboo pieces carrying a little piece of earth are placed. They are called *rengma cöm-töm* ("field-spirit sitting-place"). In front of the eight bamboo poles small (only 5 cm high) plaited "tables" (*römca*) are set up, a *ram-nget* leaf is laid on them and on it cooked rice is put, to which are added (from three *röng* bamboo cooking tubes, brought along from home): *chüa wai hom* (yellow-?-rice), *chur ngar* (small fish, shrimps and crabs), and *waa but* (boiled chicken).

The *dor* with the Rümma

Then four *li-cong* ("wind-traps") are made, each carrying in its middle two *chaling-chalap* ("butterflies", thread-squares, with orange-red and black threads (*pring-ing*) wound on to a 4 x 4 cm cross of two bamboo pieces. The threads with which the thread-squares are fastened to the *li-cong* have to pass two 3-cm-long bamboo pieces, called *ram-cheng* ("earring", probably representing women's ear-plugs). Two of the eight slender bamboo sticks, of which the *li-cong* is made, are left longer, jutting out left and right. On them is stuck a *kam-bor kom-pöt* (prepared here), enclosed by two pieces of ginger. More *kam-bor kom-pöt* (here called *kam-bor lep*) are made and distributed: one each right and left into the funnel of the *li-cong*, but also into or on to everything mentioned so far: *waa-dör*, *waa-cher*, *re*, *khen-tang*, *keng-ching*, *pön-mum-pön-büa*, *üng*, *büi-pang*, even the *mia-kom-chum-kom*, and also for the *chaling-chalap*, *ram-cheng*, *pring-ing*, *tam*, *ram-nget,* etc., finally one for each of the bamboo rods themselves. Then the *li-cong* are distributed: one (as already mentioned) on the *waa-dör*, one each to the fourth and the eighth (last) pole to the left (both *caro-ca*) and one to the *caro-ca* at the upper end of the *dor*.

Before the *kom-pöt* are distributed a chicken is killed at the *waa-dör* and its blood is distributed to any *lu-kut* and *caro-ca* and dripped into any *teng*. Five of its tail feathers are tied to a little stick, singed over the fire, and distributed on the *röm-ca* outside right and left, one to the *ci-ö-ca*, one to the *hua ai*, and one to the *caro-ca* carrying the back end of the *dor*. Later on a pig is killed over paddy, cut up, and pieces of its flesh are skewered on five *waa-ngaa-pak-ngaa-cher* (chicken-meat-pig-meat-skewer), 20 cm long, with a piece of ginger at each end. These *cher* are distributed in the same places as the singed feathers. While the *cher* are prepared, a second man distributes the *kom-pöt* as described above. Finally all members of the family who have come along are called together: first the youngest son takes a sip from the *nom* (the pot over which the *dor* ends) and spits out six times over the *röm-ca* (the miniature table) near to it, the others follow, making one *cho* (spitting-out) each. Then they all withdraw to the field house to take a meal.

Excursion into the swidden

2) *plai-tük*

After arrival in the field house, some *chang-ki* (beer yeast) and *tam* (ginger), brought along from home, are put into all pots (also brought along). All family members dip a piece of ginger into the *chang-ki* and dab (themselves) a drip of it on the middle of their forehead. Thus protected they can begin their work.

On the border of the open-air platform of the field house, called *hau-kom-raa* (*hau* = paddy straw, *kom* = to tie, *raa* = place), four *kau-chung* (bamboos with leaves kept) are erected. No *krou* and no *pam* (see J6b2) are needed, but an additional bamboo pole is tied to the first *kau-chung* and on to this some paddy straw, a "wing of a chicken" (*waa-lap*, actually two feathers), a little twig of a cotton plant, and a stalk of a ginger plant, plus a *chia* (an iron dart) dangling on a *pring-ing* (red and black) thread. In front of this bamboo two *teng* (swallow-tailed bamboo) and a *röm-ca* (miniature table) are put up, the *röm-ca* covered with a *ram-nget* leaf with cooked rice and *chur-ngar* (also brought along). – This place corresponds to the *pam-nam* of the Anok.

In the middle of the *car* is the *pak-chot-raa* (place where the pig is stabbed), also equipped with two *teng* (one of them with a drinking tube and a *pleu*) and a piece of an old mat, on which some paddy was heaped. The third ritual place is inside the field house and called *waa-bet-raa* (chicken-throttle-place), equipped with one *teng*, one *ci-ö-ca*, and a *röm-ca*. A fourth place is in the swidden and called *krong-töng-raa* (earth-?-place). Here two large *caro-ca* are set up, the smaller one of 1.80 m carries four *chit-wai* (bamboo sticks with tassels at their end), the larger one of 2.50 m with 4 x 4 *chit-wai,* to which are tied: *chong-prem-chüa* (yellow tagetes) and *khet-pau* (giving the image of drooping orange tassels). At the foot of these two bamboos: 1 *ci-ö-ca*, 1 *teng*, both with drinking tube and *pleu*, and a *pak-plai-kim-ca* (a miniature altar), on which are placed husked rice (*mi*), ginger (*tam*) and a hoe (*tim*). Beside this altar: a *röm-ca* (with *ram-nget*, cooked rice and *chur-ngar*), a *chia* with *pring-ing* threads, and an *em* (small carrying basket, see D2a), containing a *pön-chia* (two small bamboo pieces tied together, which will be used for the *pang-cang* at home, see below), a *tui-yia* (water-bottle, used to fill up the water in the *ci-ö-ca*), two *kop-kap* (see below), one *tu* (gourd-pipe with one pipe only, adorned with a *tür-ram* [*Didymospermum*] leaf).

The *kop-kap* is a bamboo "rattle", 60 cm in length, cut off 10 cm above and 5 cm below its two nodes. The upper end is decorated by a *thau-rin-bong* (straight shaving lines). Before the part below the lower node was cut off, the tube was shaved upwards to the node, leaving a tassel right and left, then the lower node has been cut out at the front and the back and the internode split up. The skin of the internode has also been shaved upward to form two tassels to the right and the left. Grasping these tassels, one can draw apart the internode, but it will close again by itself, making a snapping sound, once the pull on the tassels is slackened. The two *kop-kap* are "played" when dancing round the *krong-töng-raa;* one of them is left behind at the spot, the other one is taken home to the village. (The Anok too make a *kop-kap* for one type of their harvest festivals.)

Another structure is erected 1.50 m away from the *krong-töng-raa*: a *lam-kia-lam-ne*, a stand to which a *tui-lam* (an undecorated bamboo tube for carrying water) is leaned (*ne*). The stand is made of 2 x 2 bamboo poles: one *caro-ca* and one *lu-kut* are placed 50 cm apart and another pair is tied (20 cm above the earth) across them. On to these the *lam* is leaned, but not used in any ceremony; it is just "custom".

After everything has been prepared, the *wia-ma*, accompanied by five men playing three *ner*, one *tömma* and one *kop-kap*, goes to the *krong-töng-raa*, scratches his arm with the hoe (this will protect him against spirit attacks), takes a sip from the *hom-noi-pu* and spits it on the instruments. All players shout six loud *hü* and start walking round the two *caro-ca* six times, while the *wia-ma* squats in the middle. Then follows a second series of circuits, during which, however, the players move back to the field house, where the *wia-ma* kills a hen at the *waa-bet-raa*, carries it out to the *pak-chot-raa* and lets some of the hen's blood drop on the paddy on the piece of old mat. Meanwhile the players continue to beat their instruments, standing on the open-air platform of the field house, on which a pig is placed beside the *pak-chot-raa* and some water is poured over its bound hind and front legs and snout before it is stabbed to death with a sharpened bamboo, after which the music ends.

While the preparation of the hen has already started, the pig is cut up and its meat cooked. Some of the meat is cooked in a *röng* (cooking bamboo), from it *kom-pöt* (but no *cher*) are prepared and the *wia-ma*, accompanied by renewed beating of the instruments, eats some bits of the meat from the *röng* at the *krong-töng-raa* and distributes some *kom-pöt*. Then all assemble to eat the meat of the hen and the pig. After they have eaten, the *wia-ma* and the players go once more with their instruments to the *krong-töng-raa*, the *wia-ma* takes a sip from the *ci-ö-ca*, spits it on the instruments, and after the players have called *hü* six times, he speaks a *tamma* (invocation). At the same time another man (it can be done by anyone) makes a *cho* (spitting out) and says a *tamma* at the *hau-kom-raa* (the border of the open-air platform of the field house).

During his last visit to the *krong-töng-raa,* the *wia-ma* had taken along the *em* (the small carrying basket), into which he has now put a small chicken (brought along from home). The small chicken, called *waa-rung*, is now carried around by the *wia-ma*, but not killed (*bet*) before being taken home again. To its leg the *chia* (iron dart) with the *pring-ing* thread (from the *krong-töng-raa*) is tied together with a *kom-pöt*, and into the *em* is also placed another *ci-ö-ca,* serving as *hom-noi-pu*, hitherto left unused in the field house. From this pot the *wia-ma* now takes a sip to make another *cho* at all instruments, then each player makes a *cho* spitting besides the pot, accompanied by a *phyok-tamma* (addressed to friendly spirits). For these *cho,* water is filled from the *tui-yia* (calabash) into the *pön*, with which the pot is refilled after each *cho*.

Then the instruments are played again and all players set out towards the *tur-tut*, while the *wia-ma* takes along the smaller of the two *caro-ca* from the *krong-töng-raa*. At the *tur-tut* some *kom-pöt* are thrown away together with the *pön* (tube for pouring water), into which, shortly before the visit of the *tur-tut*, some of the blood-stained paddy kernels from the *pak-chot-raa* had been filled. After taro and ginger have been dug out and taken along, the group returns to the field house, where in the meantime a chicken (*waa lot*), some husked rice (*mi*) wrapped in a banana leaf, and a pumpkin (*kan-pen*) have been assembled; after everyone has touched them, the parcel is taken along into the swidden, the *kan-pen* is cut open with a hewing knife, the husked rice is spread out, and the chicken is placed on it (*lot waa*, setting free a chicken) with a *phyok-tamma* (invocation of good spirits), all players call out *hü*, after which the *phyok-tamma* is repeated, the group moves uphill, all players say a third *phyok-tamma*, while the *wia-ma* touches their forehead with piece of ginger, which he has brought along from the *pak-plai-kim-ca* of the *krong-töng-raa*. All return to this place, where the *phyok-tamma* and the touching of the forehead with a piece of ginger are repeated, all pots are emptied out, the *wia-ma* puts the smaller *caro-ca* back into its place, now taking along the other, bigger one with its orange flowers.

In the meantime everything in the field house has been packed up, and the group now leaves the swidden to return home, while the players continue to beat their instruments (with the exception of a short halt during which everyone takes a bath in the brook below the village) until, after a last *cho* before entering the village, they reach home, where the *wia-ma*, before he enters his house, implants the *caro-ca* (with the orange flowers) beside his stair. The group set out at 8 a.m. and returned 3.30 p.m.

In the house the following should be put up: in the middle of the *kim-tom* two beer pots (one *wia-ma-yu* and one *pin-yu*). The *wia-ma yu* has a *nam* (a carrying strap), with a piece of ginger at its end, wrapped round its rim. Into it are planted two signs: a *pleu* (with the rhomboid top also used in the field) and a *paing-lak* (lying flat with a little thorn in its middle reaching into the beer). After the *pang-cang* (see below) the *paing-lak* will be thrown away and the *pleu* has its rhomboid part broken off and instead of it a horizontal piece of bamboo is inserted, which serves as a normal drinking measure, like that of the Anok Mru (see K2b). Since the *pin-yu* would be of no importance here, it has been replaced by a *hom-noi-pu*, while a third pot (called *waa-bet-yu*) was put up near the *caa-pam* (the rice bin) at the entrance of the *kim-ma*. Here the *waa-rung*, which had been brought to the swidden, was killed, and carried around dropping some of its blood on everything of ritual importance in the house.

In the evening follows the *pang-cang*, the guests drink from the *wia-ma yu*, and water is filled up by means of the *pön-chia* (which had been prepared in the swidden and placed in the *em* at the *krong-röng-raa*). The *pön-chia* itself is filled four times with water from a *pön-ma* (a big *pön* [tube for pouring water]). It is necessary to fill it four times, not because the *pön-chia* is so small, but because otherwise the paddy would dwindle too soon. And eight *pang* ("invitations") are necessary, the first three by the *poi-yaa* (the giver of the feast), then four by a *ria-ca* (who is to be determined on the spot; he has nothing to do before), and finally one by the *wia-ma* (who in this case is the same man as the *poi-yaa*), and before each *pang* a *chia-lu* ("washing the head") has to be done, that is, someone has to pour water over the neck of the man who does the *pang*. (If anything were done wrongly, the man would die.)

During the whole time *ner* (a bigger and a smaller flat gong), *tömma* (drum) and *kop-kap* (see above) are played; at the end follows the *krong-ding-döng* (a reversal of the *ner-tep*, the beating of the flat gongs). Instead of beating the bigger one four times and the smaller one twice, now the bigger one is beaten twice, the smaller one four times. Then the *wia-ma* makes three *cho*, and the rhomboid pinnacle of the *pleu* is broken off (*chuk pleu*). For the *pang* a special side dish (*pang-kan*) has been prepared: without chilli, only with ginger, salt and rice flour. The guests should receive one portion of *pang-kan* and one portion of normal *kan* (side dish). To be correct, the food should be given after each *pang*, in this case, however, the whole procedure was shortened to serving out the food only at the end of the ceremony. The *wia-ma* (unless he is the *poi-yaa* himself) is rewarded for his services with a turban, a hewing knife, and a bottle of *arak* (distilled beer), called *pleu-chuk arak* (see *chuk pleu*, above), which was drunk by all who had participated in the *üa-plai*.

J7 Longhu and Khumi (short notes)

J7a) Longhu

These short notes in part repeat what has already been mentioned. For the *üa-dim* (see J2e) a *dim-pak* (appeasement?-pig) is killed on the intended *tur-tut* place and is meant as a protection against, or is killed for, animals (snakes, bees, etc.) which died when burning the swidden. The meat can be taken home and eaten there. The *dim-pak* can be replaced by a chicken, but when sowing and planting begins the sacrifice of a pig (called *dup-pak*, planting-pig, see J2e) is obligatory. It is to be offered at the *tur-tut*; part of the entrails is dug in. If part of the meat is cooked in the field, only salt may be used; the remainder of the animal can be taken home, to be cooked there (but again only with salt). If a *liu-klak-khang* is observed, it means that for six to eight days only rice with some salt can be eaten and only water can be drunk.

If a head of cattle was sacrificed last winter, an *üa-reng* may follow, but this may also be offered in case of illness (see R6c).

Before beginning to harvest the paddy, a chicken is killed in the field house and a dog is clubbed to death. A few ears of all kinds of paddy are collected, spread on the open-air platform, and on it a pig (*caa-tlai-pak,* paddy-collect-pig) is stabbed to death, In the field house a *ca-pam* (paddy-bin) is made and the *khong-rau* (the ritual stones supposed to ensure luck) are brought along from home. On these *kom-pöt* from chicken and pig are put (the dog's meat is only for human consumption) and chicken and pork are cooked with salt, pepper, and ginger. Thereafter a *caa-ngen-khang* (paddy-cutting-ban) is to be observed, i.e., work in the swidden is forbidden for one day. The *caa-tlai* (for which no Anok parallel seems to exist) can also be postponed for a little while until after the harvest has begun.

The *caa-hüng* (corresponding to the *caa-moi* of the Anok Mru) is held twice, once after the paddy has been harvested and collected in the field house and once after the paddy has been brought home. For the first *caa-hüng* a pig and a chicken are killed, accompanied by beating of the drum, at the paddy-bin of the field house (*bok ca-pam*). The second *caa-hüng* is a big feast: in the village house a chicken and a pig are killed at the *ca-pam* and a dog is clubbed to death on the *car*. (Its meat is used for human consumption only.) *Pen* and *tutma* (see L3b) can be invited. The *pen* bring along a *re* (spear) and a chicken, and are expected to offer from 1 to 10 Rs., which they will pin to the *pam*; for this "*pam-chot*" they in return will receive 5 Rs. and pork or (even better, if brought along by the *tutma*) dried jungle fowl. (The ban on receiving and eating chicken from the *tutma* is valid for domesticated fowl only.) The *tutma* will bring a turban, which will be wrapped round the bin (*khen pam*) and in return he will receive, besides chicken meat, a 6-Rs. spear for a 5-Rs. turban. Also at this festival at home, only the drum (but no flat gongs) will be beaten, a beer pot will be set up at the *pam* (the paddy bin), *arak* (distilled beer) will be served out and songs will be sung. If available, another pig may be killed the next day (without ceremony) for food only.

To take out paddy from the bin (*wur caa*) is allowed only once a month, and it will be accompanied by a *cari-yöng* (with sacrifice of a pig, beating of drum and flat gongs) and a concluding *bong-kom*. If a pig cannot be afforded, a chicken will do, but no *cari-yöng* will be held and no *bong-kom* will be distributed. When the last paddy is taken out of the bin (*hua caa*, among the Anok called *pam khök*, see J4f), once more a pig should be sacrificed, but in this case too, if a pig cannot be afforded, a chicken will do (Phungkri Patlaica in Uikük-HP, 14.01.1957).

J7b) Khumi

The Khumi do not make "*üa dim*", though they kill a *dim-pak* on the occasion of the *khang* held in spring (see P3e). On the occasion of the first sowing, however, a pig can (but need not) be killed in the field (K: *paheong athüm-na*). Of all things that are to be sown or planted, at least one kernel or one root will be assembled at the *tur-tut* (K: *pai-hong-küüng*). The pig will be killed, dissected, cooked and eaten in the field. Unlike with the Mru (Longhu included), no parts of it can be taken home and what cannot be eaten will be buried at the *tur-tut*. Back home a chicken may be sacrificed, or if a pig has not been sacrificed in the field, a chicken must be killed.

After everything has been sown or planted, another chicken is killed at home. Some of the paddy seed must be preserved, and *hom-noi* is prepared with it. This will be needed for an invocation on the occasion of killing the chicken, which is sacrificed for a good growth of the seed and as a protection against its being eaten by jungle fowl. A *lau-reng* may follow (see R5h).

On the first day when the first ripe paddy is cut, a pig is killed in the field house; the next day a *caa-ngen-khang* (K: *co-cüng-tü na*) is observed.

After the paddy harvest has been brought to the village, one pig should be sacrificed. Some of its blood will be sprinkled on the paddy bin, beer has to be prepared, and everyone from the village can drink from it. A dog will be clubbed to death; it is not sacrificed, its meat just serves for human consumption. Anyone who cannot afford the pig need not do anything.

Taking paddy out of the bin is allowed only once a month. On this occasion a chicken will be sacrificed to "feed the paddy" (*caa-chok*) (Karbari Müilün Tamchaa, 25.03.1957).

K Food and drink

K1 Food

K1a) The main dish (*hom*: cooked rice)

The concern for the paddy is well-founded; for all too often a household begins to run low on food supplies before the next harvest. Buying food at such times is quite expensive. After the harvest the prices go down and creditors press for the payment of debts. One must therefore sell one's own rice – and sell it cheaply. This, of course, only increases the prospects of again having an empty storage bin prior to the next harvest. Rice is the main staple food; the Mru eat it morning, noon, and night. Everything else, such as meat and vegetables, is regarded as a side dish (*kan*), which one values but which one can also do without. The amount of paddy needed for each person is correspondingly high. If a family has 800 lb of paddy per family member in storage, its members can look confidently towards the coming year; for not only will each person have enough to eat, but the family will also be able to entertain guests, brew beer and distil liquor, and keep a few chickens. The family may, in addition, be able to sell some paddy, so that it will have money for occasional shopping. With such a reserve, one can even manage an unexpected illness which necessitates large expenses in connection with a sacrifice, without having to dip into the reserved seed grains. If the only concern is one's own consumption, one can also make do with 550 lb of paddy per person. In such a case, however, there is no room to maneuver – no reserve for pleasant or unpleasant events. For the Mru, paddy is a staple food, not a cash crop. The latter purpose is rather served by cotton. Should then the cotton harvest fail, most families have little other than paddy which could be converted into money. Only a small part of the paddy which goes from the hands of the Mru into the hands of the Bangali merchants actually enables the Mru to buy something; the larger part serves to cover debts with their exorbitant interest rates – debts which were incurred prior to the previous harvest during some emergency situation.

In Bangladesh official statistics reckon with 200 kg (440 lb) of paddy per head per year. Before paddy can be used as food, it must be hulled; this removes the husk and at least part of the silver skin. After it has been hulled, 1 lb of paddy gives about 0.6 lb of rice – the amount varies somewhat depending on the type. 440 lb of paddy per person for one year means, therefore, a daily ration of about 12 oz of rice. The nutritional value of this amount corresponds to about 1200 kcal.; with 250 kg (550 lb) per year, the Mru may have 1,470 kcal. per day. If we add a maximum of 250 gr of vegetables (= 50 kcal.) and an average of 30 gr of meat (= 80 kcal.) – there is normally nothing else available – we have a total of 1600 kcal. per person per day. Since we are dealing here with average figures, the rations for adults could be about one-third higher. (In chapter J5a I calculated with 370 kg paddy per adult person and year, a figure given me by the Mru themselves.)

According to the Mru, women eat more than men. As we have seen in relation to the farm work, men take over the most demanding short-term tasks; women, however, must carry water daily (usually two times), and daily they must pound the paddy. Additionally, almost every day during the cold season women must carry a load of firewood. Men are not obliged to do such hard work day in and day out, which may well explain why women – as is claimed – do indeed have a higher consumption of calories. The version offered by Mru men, however, is somewhat different. They say that when women are hungry, they put more rice into the cooking pot; but when men are hungry, they go visiting. If this is true, men may well eat less at home, but they do not eat less in general – in the household budget, though, the extra amount consumed by the men is charged to the "visitors' account", not the men's. The men's point of view, then, reflects once again the unequal division of labor; for while the women are working – the women must do the daily cooking – the men may take a stroll. (All men do know how to cook, but except for just helping out, they cook only for celebrations and festivities.)

Women pounding and hulling paddy (photo: C.-D. Brauns)

Mru women get up earlier than the men in order to hull the daily ration of rice (the men, however, cannot sleep any longer). During the cold season, the women pound the paddy in front of the house; during the rainy season, inside the house *(kim-tom).* The only thing the men contribute to this activity is the manufacture of the heavy, wooden mortars and pestles and the trays used for winnowing. The paddy is put into the mortar and pounded with the round end of the pestle, in order to remove the husks; then it is winnowed over an old basket in which husks and bran are collected. After this initial process, the paddy is skinned by a second round of pounding; this time the pounding is done with the flatter end of the pestle. A second winnowing finally removes the bran.

At this point, the women rekindle the fire which has been kept alive overnight in the fireplace by means of a thick log. The fireplace (see C3e) is situated on one of the side walls of the *kim-tom;* it consists of a mud square framed with bamboo boards, in the middle of which three conical-shaped stones have been placed. The stones serve as a pot stand; occasionally, however, an iron tripod is used in their stead. The everyday metal pots (see D6a), which the Mru purchase at the market, have a round belly and a neck; with such shape they rest firmly on the stones and can be taken from the fire with bamboo tongs. After the rice has been washed in a special basket (*mi-chau-tum*, see D2i), it is placed with a little more than double its amount of water into one of these pots; it is then brought to a boil, while one stirs occasionally. When the water has been absorbed, the pot is removed from the fire, covered, and placed next to the embers. It will then be turned from time to time. The secret to successful rice-cooking lies precisely in this art of turning the pot: the pot must be rotated just at the right time, so that the rice swells up evenly and becomes dry, not pasty. When the rice is done, it is dumped out onto a clean banana leaf which serves as a platter for the entire family. The women eat kneeling, the men squatting. Water is drunk directly from the gourd bottles; as a condiment, there is salt and, perhaps, a green chilli pod. At lunchtime one eats the leftover rice which was cooked in the morning; this time it is eaten cold. The evening meal may be a little more varied, but here again rice is the main dish.

Mother with child, cooking in the house (photo: C.-D. Brauns)

K1b) Side dishes (*kan*: vegetables, fruits, meat and fish)

Between August and January, that is, about half of the year, fresh vegetables can be brought home from the swidden; a few types of cucumbers and squash can also be preserved at home over an extended period of time (an overview was given in chapter J3d). The rest of the time, however, one must make do with the leaves of wild plants. At the beginning of the rainy season, there are young bamboo sprouts. A wild-growing root-plant is *makong*. It is dug up, the skin peeled off, beaten with a chopper or a piece of firewood until its fibers are laid bare, rinsed in water and the dirty scum poured away, then dried and before using it minced with the hewing knife and finally cooked together with rice. Also banana trees (up to 10 cm diameter) are edible. After peeling off the outside, so that only $^1/3$ of the inner stalk remains, this is cut into pieces with the chopper and the inside pulp stirred up with a small stick in order to remove the hair-like strings. This pulp is preferably cooked with oil, in order to make it tastier. Most vegetables, roots and leaves, however, are cooked with water and salt only. More extravagant ways of food preparation are known; but as soon as oil and spices are called for – things which must be bought –, they become financially unbearable for everyday cooking.

Eating a festival dish (with meat) in the house

The availability of fruits is even more seasonally determined than that of vegetables. Most frequently available are bananas – every village has a small banana grove. Papaya are found primarily in the valley villages. Mango and litchi are rare, but in many villages there is a jackfruit tree. Pomeloes are part of an old cultural heritage; in contemporary villages, however, it is rare to find more than one or two such trees. In contrast to the land, these trees are the private possessions of the persons who planted or inherited them. Rarely, however, can one harvest much from them, since the fruits are usually looted prematurely by children. If an owner protests such looting, he finds little sympathy, for the fruits are considered to be children's food, rather than food for adults. Mandarin oranges have only recently become common, as whole plantations of the fruit have been planted. They are sold to Bangali peddlers who take them to the urban agglomerations of the plains, where affluent Bangalis buy the fruits by the piece for a price many times higher than they brought in the mountains. Because of this middlemen's profit, the Mru would very much like to carry their fruits down to the plains themselves and sell them themselves; yet they fear – and rightly so – that they would immediately be surrounded and robbed of their goods or money. As mentioned, it has long been customary to consider these fruit trees the property of the person who plants them. He who plants a grove of mandarin oranges, however, incites the anger of the other villagers who do not have the money for such a project; for the villagers see that such an undertaking permanently reduces the land area which traditionally was at the disposal of everyone and that the yields of the field now benefit one individual family only.

Whereas mandarins were brought to the Hill Tracts by Lushai missionaries, limes have been growing wild in the jungle for a long time. Since limes, however, are available for only a short time each year, they cannot serve as a standard ingredient of indigenous cookery. Tea, which must be bought at the market and which on the plains is served with a lot of milk and sugar, is usually spiced with salt only – sugar would be too expensive. Only a few persons know how to use limes to prepare a tasty condiment: a kind of pickled cucumber chutney. Very common as a condiment are green chillies – but they too are available only part of the year –, and a popular fish paste called *ngaa-pi.*

Ngaa-pi must be bought, but it is relatively cheap; and although the paste is prepared by Bangali, who also call it *ngaa-pi,* the name is of Burmese origin and would literally translate "rotten fish". Fish caught in the ocean are spread out on seaweed and mud and left in the sun to dry. As the fish decompose, they are pounded from time to time until they have formed, together with the seaweed and mud, one pasty mass. This mass is then formed into large, light grey or dark grey balls, depending on the quality; and after it has been transported in this form into the interior of the country, the fish paste is sold by the piece in the markets. One needs only a little of this paste to conjure up a fish dish in the mouth, but the condiment smells strongly offish and does not suit the taste of every Mru. Dried fish is much more popular, although it consists of little more than skin and bones even after it has been cooked; in addition to that, it also must be purchased.

It is cheaper for a person to go fishing himself. Traps are very common and are sometimes used in conjunction with the diking and barricading of a waterway (see H8). The earlier practice of fishing with plant poison is now forbidden; today men try their luck with bought hooks or casting nets. An old method is still preserved by the women, though: they wade – often in groups – through shallow brooks, lifting up the stones, and catch in baskets the tiny fish and shrimps that try to swim away. Since most of the Mru do not live close to the large waterways, this traditional way of catching fish is perhaps the one most often utilized. The size of the catch, however, is quantitatively small; so although the Mru like to eat fish, it does not make up a meaningful part of their diet.

As mentioned, meat is eaten almost exclusively on festival days, unless one is forced by an emergency to slaughter an animal, or unless hunting or trapping provides a catch. Faced with such scarcity, the Mru are not choosy. Even insects, such as tasty beetles and maggots, are not scorned. Meat is always roasted or boiled. Cattle and deer are skinned, and goats likewise – although goats are very seldom kept up by the Mru themselves because of the damage they might cause. Pigs are singed; dogs as well, in the few instances when a dog is needed as a sacrifice. Chickens are plucked and singed (for more details see chapter G). After an animal has been carved up and gutted, the intestines are cleaned and cut up; everything else – except the skull, the hooves, and similar inedible parts – is also chopped or cut up into small cubes. (Even with larger fish one proceeds in this manner, paying no attention whatsoever to the bones – an indication that larger fish are only seldom available.) Together with some spices (at least some chilli pods) which have been freshly crushed in a bamboo quiver, the meat is then cooked in a pot.

Women fishing with their baskets (photo: C.-D. Brauns)

Men cooking a festival dish under open air

With a calabash ladle, some of this "curry" is spooned up for every person and placed next to his rice. Each one may then mix the proportion he desires, but always with enough rice to form a small ball. This ball is shaped with the first three fingers of the right hand – never with the left – and then popped into the mouth. No one is entitled to receive "the best parts". (We may recall that the spirits, too, receive little bits of each part of the animal. Their food, however, should never contain any spices; it is therefore cooked separately.) Whatever is chewable is gladly eaten by all who have teeth; whatever is impossible to eat is left for the dogs, which are generally already waiting behind the eaters. If the dogs venture forward on their own initiative, however, they will receive a blow on their nose until they learn to control their hunger.

Men eating a festival dish in the open

Meat which is not immediately cooked may also be dried to preserve it. For this it is cut into pieces (the size of a match-box), skewered on a bamboo thong and hung up below the *krök* (see C3e) over the fireplace. For the fire wood should be used (bamboo would produce too much soot). In order to dry small quantities of meat the time needed for cooking rice will be sufficient, for larger quantities the fire will have to burn a whole evening. The dried meat is then preserved in a *waa-pom* (see D4f) or a pot, but not covered or wrapped in banana leaves; the *pom* is hung up near the hearth and may be preserved up to 3 or 4 months, when it is not attacked by vermin.

K2 Drink

K2a) Water and tea

The daily drink is water which is fetched by the women in bottle gourds (*tui-yia* = water-bottles) from a nearby source (a spring or just a brooklet) and carried up to the hamlet in special baskets, called *yo* (or, especially in songs, *par*, see D2e). At the end of the hot season it may happen that the source dries up and the women must walk a long way, before they reach another one. To fetch water as long as the supply is ample, is no problem (the bottle is just plunged into the water, or even just hold under a little waterfall); to draw water from a source which is drying up, however, requires more work: a halved bottle (*yom*) must be taken along, in order to scoop out the water and to carefully empty it into the *yia*. There is no possibility anymore to take one's daily bath and children no longer can play in the water. But the men, at least after they have burned and cleaned a swidden in April and afterwards look like black goblins (see J2c), should wash themselves to get the soot off their skin; for the remaining days the water should be kept as clean as possible. The drinking water is never boiled before it is drunk (mostly just by putting a *tui-yia* to one's mouth), even at a time when definitely it would be better to boil it before drinking, especially when the rains have set in and "fresh" water becomes available again. This is the time when nearly everybody has to suffer from dysentery and amoebiasis; the reason being unknown to the Mru: the rains wash out all germs from the soil, and the defecation place is usually near to or on the way to the water-place.

Nobody understood that I preferred a cup of tea instead of taking just a sip of "fresh" water, and unless I brought my own tea-leaves along they were rarely available. Tea was available in tea-stalls in the market (cooked the English, i.e., the Indian way, with a lot of milk and sugar). In case a Mru household should have the means to buy tea leaves, their tea was normally a brew of last quality, without milk and made tasty by adding salt. (Sugar was too expensive and mostly not available in the normal market, but could be bought only in the "black" market. During the rainy season a Bangali peddler offered me some sugar, happily I bought it and found out afterwards that it tasted terribly, since quite apparently it had come into contact with kerosene.)

Thus, despite my reluctance to drink "fresh" water, I had to drink it sometimes, with the result that after one year also I got infected with amoebiasis which during the last 6 months drained my body in such a way that I was rapidly loosing weight and finally had to leave the country exhausted. The Mru, however, seem to have got accustomed to that illness: they all get their attacks of dysentery after the rains have set in, but they take it as a passing event and (at least in the 1950s and 60s) nobody could understand why they should boil the drinking water in order to let it become cool again before they could drink it.

K2b) *yu* (rice-beer) and *arak*

Water is the daily drink, but not the only thing which the Mru drink. *Yu* is drunk mostly by men, but woman are not forbidden to drink it too. *Yu* is must for festivities and whenever possible it should also be offered to guests. But its availability depends on that of paddy; when paddy becomes rare (or when the harvest had been bad) people prefer to eat rice instead of preparing *yu* from the husked paddy. Thus it can happen that nobody in the hamlet prepares *yu* or even *arak*, which is distilled from it. In case somebody falls ill (and a sacrifice is required) or somebody has to host an unexpected visitor who has to be honoured, neither *yu* nor *arak* may be available, and even sending someone around to look for a bottle of *arak* in the neighbouring hamlets may not be successful. That is, a Mru cannot become a habitual drinker, which does not mean that he might not do so if a regular supply would be available, or that some men are not rather heavy drinkers who sometimes get drunk. Still, I never saw a man behaving roughly towards a woman even when he was drunk. (The situation is quite different with the Marma, where often the husband may beat his wife in a quarrel, especially when both are drunk.)

From *yu* one cannot really get drunk. It is just too weak, only the first sips taken from a *yu-khong* (beer pot) may be stronger, but whatever one may drink from it, the custom of pouring in water to refill the *khong* lets the beer become weaker and weaker, until finally the rest is just stale water which no one wants to drink anymore, so that the remainder is poured out into the pigs' trough. However, before *yu* can be drunk it must first be prepared. The amount of rice necessary varies with the size of the pot. For small pots one needs 2–4 kg, for big pots 10 kg and for (rather rare) very large pots 30–40 kg. The lowest part of the *khong* is filled with (unhusked) paddy, then follows a large layer of cooked rice to which some "wort" (or yeast) of a former preparation has been added.

In case none is available one must prepare new wort. For it, one uses: *thöng* (*Ervatamia coronaria*), *raing-yö*, *thöng-klang*, *krong-dak-mo-khetca* (a root), *büi-bai* (a plant called "bamboo-rat nest"), and some people add leaves of sweet-cane, chilli, and even oil. Some rice is first cooked, emptied out on a *per-cing* (see D4a) and put on the *car* to become cool and dry. (During the cold season it may also be used when still warm.) Then the herbs, roots, etc. are added and the mixture pounded in the mortar to a flour-pulp called *tua.* To this *tua* a little water is added and with it round flat dough-cakes are formed, which are placed on a layer of *chit-mu* (bamboo shavings, also used for scrubbing pots). Over them a second layer of *chit-mu* is spread and the whole covered with a cloth, to keep it warm and to protect it against insects. It does not ferment. During the warm season it is kept like this for 4 to 6 days (during the cold season up to 14 days), then it is unwrapped and dried again for 2 or 3 days (in the sun or over the fireplace); it becomes hard and can be preserved like this.

Dingte sipping *yu* from a beer pot standing on the ground

For preparing *yu* a few of the dried *tua*-cakes are taken and again pounded to flour in the mortar. This flour is mixed (without adding water) with freshly cooked rice at a rate of 2 *sher* rice to 2–3 *tola* of flour (Bangla weight: 80 *tola*

= 1 *sher*); i.e., approx. 15 gr (½ oz) flour for 1 kg rice. This mixture (called *chang-ki*) is put in the beer pot and a little water added. The pot is covered with banana leaves which are tied with a special strip (*khong-ok-nam* or -*leng*), made from the inner bark of the *sterculia*-tree, round the neck of the *khong*. This strip has to remain always there (see below). When the preparation of *arak* is intended, there need not be a lower layer of paddy, though this is necessary when brewing beer, as otherwise the fermented rice will block up the lower end of the tube by which the beer must be sucked. In the warm season the fermentation process will take only three days, when the weather is cold it may take double the time. After fermentation the product is called *cham-pe* or when *bini* (sweet sticky) rice has been used *cham-por*. This can be eaten like this, and the white milky fermentation water, called *achi*, which has a sweet taste, can be drunk, though this is rarely done as one normally will soon go on to prepare *yu* or *arak*. In both cases the pot is filled up with water, for *yu* the banana leaves are removed and a drinking tube is inserted, for *arak* the pot, after the water has been filled up, is left standing covered another two days, then it is distilled. For distilling also *yu* (from which something already has been drunk) can be used, but the quality will be correspondingly minor.

The distilling apparatus looks like this (drawing by C.-D. Brauns):

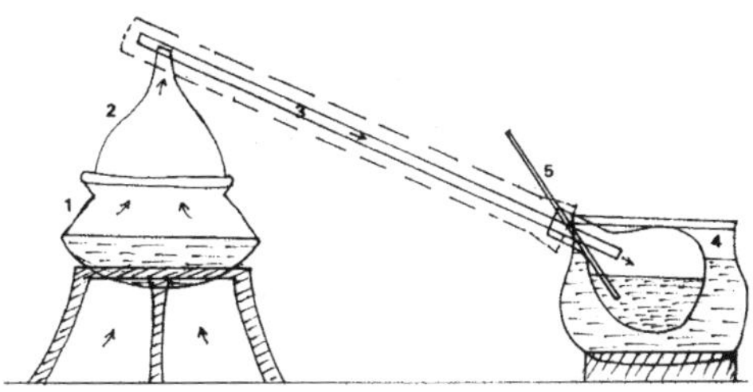

1) *yu-khong* on tripod
2) bottle gourd without bottom
3) cooling tube wrapped in cloth
4) receptable in water-bath
5) testing tube

The halved bottle gourd is colled to the *yu-khong* by means of *cham-pe*. The cloth by which the cooling tube is wrapped from time to time is soaked in cold water. For distilling 1 bottle of arak one needs roughly 1 kg rice; from a small *yu-khong* of *cham-pe* one expects to receive 3–4 bottles of *arak*. The main difficulty is the regulation of the heat of the fire under the *yu-khong*: if the fire is too weak the spirit will not be distilled, if it is too strong, the *cham-pe* may get stuck and charred at the bottom of the pot (and the *arak* will taste burned). Skilled women may even take 3–4 bottles of *arak* and distil them a second time to produce alcohol of a high percentage. (During World War II, when paddy was still amply available, high-percentage alcohol was even used to replace kerosene.) Many Mru women, however, are rather unskilled in distilling and they are not learning how to do it better, since they rarely drink their *arak*, they leave this part of the business to the men, who, in their turn, never distil. If one presents a bottle of *arak* to a guest or as a contribution to a festival, custom requires that it is full.

With the Dömrong, however, girls may distil and offer a water-glass full of *arak* to a young man of their choice, who has to gulp it down in one go. After he has drunk it, he has to pay 1 Re. to the girl, who will refill it, but this time the young man may offer it back to the girl, and when she accepts she may taste what kind of horrible brew she prepared. Therefore the

Dömrong *arak* is a little bit better than that of the Anok; the best *arak*, however, is prepared by the Marma women who are not afraid of tasting it themselves.

At big festivals the preparation of several bottles of *arak* is a must, one cannot hold a cattle feast without it and numerous pots of beer; but *yu* is also necessary for any kind of smaller festival, be it a *khang* (see P3), a *khan-pon-thet* (see J4c), a *caa-müng-pok* (see J4d), an *üa-plai* (see J6), a *cari-yöng* (see P5b/c), or any festival for which a *bong-kom-yu* (the beer used for the ceremony of binding a cotton-thread round the wrist of the family members) is necessary. One is also expected to present a bottle of *arak* when receiving the visit of a guest to be honoured (as for instance one's *tutma* [see L3b]). For a marriage feast the Rümma Mru calculated a minimum of 16 bottles of *arak*, the Anok told me that for a big marriage feast at least 80–100 bottles would be needed – but such big festivals (see M4) have become very rare. At Bulum's *lou-machi* festival 2 bottles of *arak* and 2 pots of *yu* were all what his father Klingtui Khongtör could offer (see M5b). When no cattle feast or marriage is intended, a household will reckon approximately 1 Hari (10 kg) of paddy per person and year for preparing *yu* and *arak*. Compared with the total amount of paddy (370 kg) needed per adult person and year (see J5a) this is less than 3 % of the harvest which will be calculated for brewing and which will be reduced even more in case the harvest was poor.

A note may be added concerning the *khong-ok-nam*. This strip is tied round the neck of the *yu-khong*, when it is used the first time. To present a *khong* without it, is treated like not honouring the one to whom it is presented. He may ask to be reconciled by a payment of 1–3 Rs. or a bottle of *arak*. The *nam* is especially guarded if the pot contains the *bong-kom-yu*, since this *yu* is regarded as *lüm-laa-yu* (*yu* for the soul). Loosening or damaging it is strictly prohibited; when a drunken man does this he must pay a fine in the height demanded.

No *yu*-pot can be presented for drinking without a drinking tube and a sign (*pleu*), of which there are two kinds: *chong-khok-pleu* and *cam-pau-pleu*. A third type is called *paing-lak*: a small bamboo laid over the pot with a tooth pointing downward. It is used only for death ceremonies. The two other kinds are put vertically into the pot, the *chong-khok* is split at its top in two parts which are bent in such a way that they form two opposite triangles which together look like a rhombus. It is used at cattle feasts and harvest festivals. After the *charaa* and the *riaca* have drunk from the pot, the *chong-khok* is broken off and thrown away (*chuk pleu*), the remaining sign is split up and a "tooth" spread sidewards below the surface of the *yu*. Those who drink have to take at least three sips until the "tooth" lies on the surface of the *yu*, then fresh water is filled up and the next person may drink. The *cam-pau-pleu* looks like shown in the small photograph. It has four wings, is coloured red (by taking the colour in the mouth and spitting it over the *pleu*), and seen from above it looks like a star. It is used by those Anok families who intend to give a *cam-plai* until the *pang-cang* (see J6c) has been performed.

When the feast is over, the pot with the remaining beer can be placed on the ground before the house, and everyone can then drink from it.

L Kinship

L1 The family

L1a) Biological and social unit

The basic unit in Mru society is the "small family", consisting of a husband, wife and children. A family can consist of a married couple without children, and even single persons, such as unmarried adult men, can form an economic unit of their own, but are not given full social recognition as long as there are no children. I have not encountered an example of a childless, independent old woman. Children are orphaned rather frequently, but never remain independent, being taken in by their closest family kin, according to rules still to be discussed. Cases where the mother has died and the father cares for the children, or where, after the death of the father, the mother alone assumes the task of heading the household are treated like an intact small family.

However, I'd better admit from the start that this concept of a "small family" is a construct of the social anthropologist, and that the Mru themselves have no term for it. They just speak of the inhabitants of a house (*kim*), however many people live in it. All (even the animals, like dog and cat) are included in the term *kim-reng-cak* (those living [at present] permanently in the house), but the proper meaning is revealed in the expression *wöi-kim-khök* (to be a house completed), signifying that a child has been born (see L1b).

In many cases the small family can be extended by the parents of the man or the children of the son. Although these cases are biologically the same, that is to say the family is made up of three generations, and also in practical life the transitions cannot be drawn easily, for social life different consequences arise from the two possibilities; when the father retires from the work process, leadership of the household is transferred to the son, but there is no tendency toward patriarchal sibs to be recognised here. It is characteristic of the defining role of the small family that, in such cases, from the moment the son has children of his own he belongs to the "elders" and has the same right as his father to speak and to be heard during village meetings, so that the community so to speak now acknowledges two families, or two family fathers, although it is actually only one sib. The same legal position of course is also accorded to all further sons and their male descendants once they have children. Unmarried girls with children remain dependent, however – unlike women whose husbands have died and who go on to head the household. These in fact are the only cases in which a woman takes the man's role.

Families vary considerably in size: in addition to childless marriages (which may be regarded as families, but can never truly fulfil the Mru criterion of a family) there are others with six, seven and more children. One clue about the average number of children is given by the sib lists of Tapwüa-Kua, in which, to the extent that they were remembered, the names of relatives in the male line were collected. It is, however, to be remarked that stillborn children or those that die shortly after birth are not counted at all and all those who die young soon fall into oblivion. Memory does not extend beyond the great-grandfather of the informant and includes only him – or it may even go back only as far as the grandfather, whereby the name of his wife (as also the names of his brothers' wives – and this is remarkable) is better known than the names, yes, even the existence, of his sisters. Little is known either about relatives with no known descendants. This forgetting of "unimportant" persons in the older generations may be the reason that the average number of children increases in the younger generations, although a general increase in births cannot be excluded as a second explanation.

The recollection of the individual informants, by the way, is definitely different, but comprises a not inconsiderable group of people, going far beyond normal European conditions, since not only the names of the agnatic relatives are known but to a large extent also the cognates and their families. That way I got, e.g., the names of the ancestors of Elai Atwang in cooperation with Menkröi Ngarua', whose mother was Elai's sister; however, many of the names of the Atwang grandchildren were mentioned to me by Chekkre, the wife of Elai's son, a woman who was originally foreign to the sib. Similarly, I got the names of the descendants of Hompo Khongtör not only from their relative Klingtui but also from Menkröi, who (although only very remotely related by marriage) knew them from having lived a long time in their residential area. Personal contact is an important factor, which is why the numerous relatives and descendants of Yenpau Khongtör that live far away in the south are absent from the list.

In evaluating the sib lists it must be taken into account that many of the families in the fourth and fifth generations are still young, their number of children cannot yet be regarded as definite and therefore cannot be used straight away for a comparison.

The relationship children: families according to the sib lists

G	A	K1	K2	Ng	Ng	C	T	M
1	5							
2	1	2:1	5:1	2:1	1:1	2:1	27:9	3.
3	35:6	8:2	13:	7:1	2:1	4:1	69:15	4.
4	53:14	21:	12:	12:	2:1	7:1	107:2	4.
S	108:2	31:	30:	21:	5:3	13:	208:5	4.

G = generation, A = Atwang, K1/2 = Khongtör, Ng1/2 = Ngarua', C = Catumma, T = total, M = medium size (1 = family 1, 2 = family 2); S = sum

From the table an average of exactly four children per family would result. The probability for this relationship can be enlarged above the 52 families examined to all those who gave their daughters in the marriage into these families. Since in the above "generations" persons of most dissimilar age had to be combined, the table cannot give an exact temporal clue to the growing number of families. If the first parents (omitting the first Atwang) were regarded as the first generation in the list, an increase of the complete family number from 6 :12 :18 : 30 would result for the first four generations, i.e., every generation is as large as the two previous ones taken together. Of course this numerical ratio is not a rule, but other considerations also lead to similar rates of increase, for instance when the numbers of the sons and the families founded are compared. When we omit the first to the third generation – since here, as said, the prematurely deceased or such men that did not establish a family for other reasons may be forgotten – and regard only the fourth and fifth generation (excluding those who have not yet reached marriageable age) we arrive at a ratio of 77 male descendants to 66 families founded by them. With an average number of four children per couple, a population increase of 71.4 % per generation would result, and the systematised family increase rate would be 6 : 10.3 : 17.6 : 30.2. To what extent this rate of increase also applies to other areas and sibs could not be asserted, but in my opinion significant differences need not be expected.

L1b) The household

As already mentioned, the language of the Mru does not have a real equivalent for our term "family", the usual term "*kim-reng cak*" (those living permanently in the house) could as well be translated by "household". "To have founded a family" is called "*kaa kim khök*" (get house [perfect]), i.e., to have got a household, and in the form "*wöi kim khök*" (there-is house [perfect]), i.e., to have become a household, is used (euphemistically) for the fact that a child has been born; the expression emphasising the role of children in the full recognition of a family. But this "coming-to-be-a-family" also quite realistically expresses the fact that the young family is entitled to a house of its own or, if it still lives in the house of the father of the husband, to a *kimma* of its own, because a *kimma* is the "real house". The erection of a *kim-tom* of one's own actually has to do nothing at all with the constitution of the family, the *kaa-kim*, but is a pure question of the room and compatibility. Everybody who has a *kimma* of his own is his own lord, and – irrespective of the actual command conditions in the family connected by the father-son relation and the common *kim-tom* – in the village community every *kimma* owner will have equal rights. The new *kimma* is best built before the young mother gives birth to her first child. A new *kim-tom* is not necessary for this. By setting up a new *kimma*, the splitting of the family is performed. Two married couples will never use a common *kimma*. But it can happen that a solitary father does without a *kimma* of his own and sleeps with the younger children and grandchildren in the *kim-tom* and thereby dissolves his own household: then he becomes a member of his son's family and has practically nothing more to say. The extent to which he is still respected depends only on his abilities.

But married brothers, also, whether their parents are still alive or not, can have a common *kim-tom* (never a common *kimma,* however) and run a common economy. Then every brother (whether elder or younger) actually has the same right and mutual good agreement will always be a prerequisite for such a community. Anyone who has a mind of his own and wants to run a household at his own discretion can become independent at any time. No change to his previous legal situation, however, expects him to be the owner of a complete house of his own, either within the village community or within his sib.

An unmarried man can also make himself independent, as soon as he is capable of being self-supporting and building his own house. But he will not be said to have "got a house", he remains an outsider. A young man will prefer to stay in the parental home, even if he does not get on well with his father. The situation is different when the parents have died. A good example is offered by Khamlai Ngarua', the elder brother of Menkröi. When their father died, Khamlai as the eldest son, at an early age, already had to help his mother to support the four younger brothers and sisters, and when the mother died, too, he was just old enough to head the household alone and to look after his brothers and sisters who had not yet grown up. But finally the brothers founded households of their own and the sisters were married. Khamlai did not then find a wife, however, at first because of his poverty, then because of his advanced age and perhaps also because of certain strange habits which he had developed. When I became acquainted with him, he was 52 years old and still looking for a wife, now more and more for a widow who if possible already had children, since he did not think he was still able to father children. He did not feel quite at ease alone in his house and awaited full of bitter expectations the time when he could no longer work enough to support himself.

Menkröi had offered to move into his house to collaborate with him, so as to have somebody who would feed him later, but Khamlai did not want that: he would then no longer have been his own master, but as an unmarried man only an appendage of Menkröi's family, and not even a respected one, because the fact that he was eight years older than Menkröi would not count, but only his position "without a house", virtually that of an underage person, which was why Dingte, Menkröi's son, could permit himself to speak of his uncle in a by no means respectful tone.

L1c) The household community

Economic independence guarantees a man restricted recognition as an autonomous member of the community in so far as he can claim, e.g., a certain area for cultivation. Here it is the economic units which count and hence a single man is less of a problem than those families who have but one *kim-tom*, because these of course appear to a greater extent as working and economic units rather than related families in separate houses. When I tried to record the harvest yields, I always received the total in houses with several families, however in one case, where a separation into two of the households had just taken place that year (because the inhabitants had become too many), one declared his yield separately, while the other told me the previous totals.

There is no regulation as to whether the harvests of connected families will be kept in one or several common paddy bins or separated according to small families; the decision on this is simply a question of good agreement, but since it is of benefit to the household economy to stay together and work as a team which does not carefully bill the working days in the fields, it is also reasonable to consume the products in common. Times of economic crisis, however, can promote egotistical tensions in the community and lead to their disintegration, even though the larger collective has better possibilities for reward (different yields in the different swiddens). The number of the economic units of a community, though generally based on small families, will be therefore mainly determined by the number of independent houses. The possible economic common interest groups of young girls (see J2a) do not change this.

The religious community of the family corresponds to the rules of the economic unity, although it tends to include other relatives as well, outside of the house. As concerns the contributions for common hamlet sacrifices, the houses count as units; house sacrifices are performed in the *kim-tom* and all house bans for foreigners isolate all persons living in the house. A certain small family may be obliged to perform a certain sacrifice within the house; the participants, however, are all those living in the house and all receive the *bong-kom* during the final ceremony (see for instance P2e and R1i), and even near relations living in separate houses can be included in it. The ceremonies will always be conducted by the oldest person, and in the example mentioned of Khamlai and Menkröi, it was always Khamlai who performed the sacrifice. If the course of the ceremonies requires some special knowledge, even a very remote relative of one's sib or a member of the brother-in-law sibs (p. L2b) can also function as a leader: in this case, on the one hand the narrow frame of the family is expanded and the religious community put on a broader basis, while on the other hand an individual (as a cause, giver or leader of the ceremonies) plays the leading role.

L1d) The family among the Khumi

The same basic principles of the life of the small family also apply to the Khumi, though the form of their house, which offers less room than the Mru house, supports the founding of separate households and thus separate economic units. If therefore there are fewer possibilities for a common management of brothers, it is still the rule for a son (mostly the youngest) to live together with his parents. They will find their place in the side house, leaving the main house to their son and his family, thus acknowledging him in his role as the head of the family. In these situations, their fundamentally anti-patriarchal attitude is expressed, just as in the case of the Mru.

L2 The sib

L2a) The individual sib

Beyond the small family, the individual is bound by the relations he has to the relatives of his parents and their children. What could already be recognised in the expansion of the small family now emerges clearly: the biological unity is overlaid by a social formant, an artificial order: the relatives are subdivided into social units. Defining for these is the relationship in the male line, all agnates of an individual combine for it to an exogamous unity: the sib. These sibs are predominantly but not strictly related by blood, a regulation about extramarital and orphaned children (M8b) in certain circumstances includes even cognates and others, definitely not related by blood, in the small families and thus in the sib.

The size of the sib is determined on the one hand by the regeneration of the small families forming it, on the other hand by tradition. What determines the kinship group is not a communal spirit held alive by the memory of a common ancestor but a traditional sib name. Members of a sib living far away who have never seen and never known each other will have to recognise themselves immediately due to their common sib name and address themselves as brother, father or son, etc., depending on the age relationship, and take on the corresponding obligations. It should be pointed out, however, that we must expect sib dividing, i.e., breaking-up of the traditional exogamous association. However, these are arbitrary interventions into the traditional order, not breaks at predefined junctions, because the relationship could not be proved any more. To be precise, therefore, the sibs of the Mru and Khumi consist for every ego of a smaller (genealogically provable) connected (predominantly) agnatic group and a larger group no longer connected by being genealogically provable but by identity of the traditional sib name. However, since this distinction does not play any role either in formal or in actual behaviour, the simplest and also most feasible way to describe the whole complex is to follow Murdock's terminology[458] and call it a sib.

L2b) Size and distribution

I cannot make any statements about the approximate number of the sibs of the Mru. My efforts to obtain the names of all sibs were always answered by a pitying smile, although an official inquiry could easily enumerate them. I noted about 100 sibs, and the total number for the Chittagong Hill Tracts may be higher by half as much again. The residential areas of the sibs cannot be stated exactly either: some live rather close together, others are scattered, and others inhabit a narrow area of distribution but with far-removed splinter groups. It is therefore almost impossible to get a reasonably sure idea of the size of the individual sibs. It is, however, certain that in the course of their history the sibs can experience times of strength in which they are numerous and powerful and also times of weakness or dying out. An example for this are the Ur, which was once a great sib with special prerogatives (L7c), while now the whole sib consists of only six houses scattered over the whole country. In this particular unfortunate position it is also the only sib that I could completely cover with regard to its distribution, but without knowing the history of its distribution.

But the examination of the distribution of the sibs offers the only clue for a tentative reconstruction of the migrations of the Mru in more recent times and moreover provides much interesting information on the relations between the Mru and the Khumi over the last centuries. However, no conclusions can be drawn on the magnitude of these movements from the present distributions and the information obtained from people on where their ancestors dwelt, since it is possible on the one hand for the descendants of a few men or even of one family to populate a whole area in a few generations (as for instance the Atwang around Empu), but on the other hand that only a few remnants are still to be found of a once great multitude (as with the Ur in the Chüngma area). The results of the surveys I made on the tours on sib distribution and migrations were recorded, but the immigration data can be no more than approximations. What conclusions can be drawn from these distributions will be discussed when the groups are examined (L8).

List of the noted Mru sibs

Abbreviations used; A: Anok, C: Chüngma (SC: Southern Chüngma [Longhu], NC: Northern Chüngma), DP: Dopreng, DR: Dömrong, Re: Rengmitca, Rü: Rümma, T: Tamcha

H: Headman, R: Ruaca; K: Karbari, P: Para, HP: Headman-Para, KP: Karbari-Para

[458] Murdock, G. P. 1949, Social Structure

Name	Group	Mouza	Para	Informant., date
Akhömca	DP	Luleng	Rümputma	K Kangpom Rümputma, 22.12.1956
Akhuaca	Re	Twainfa	Haiwön-HP	H Haiwön Macar, 21.03.1957
Akungca	A	Galengya	Thanglong-KP	Baiwai Atwang, 15.07.1956; note: see Kung
Alau	SC	Kwaimkhyong	Uikük-HP	Renghen Kwai, H. Uikük Nöngnau, 14.01.1957
Amangca	DR	Kamichora	Mangpung-KP	K Mangpung Pahönca, 16.03.1957
Amtham	DR (A)	Songkö	Nongneng-KP	K Nongneng Krimchang, 11.03.1957
		Kwainjhiri	Kukcöng-HP	R Tharngak Röngcha, 15.03.1957
		Dokhinhangor		MK, 11.03.1957: The Amtham in Dokhinhangor became Anok
Atwang	A	Horinjhiri	Atwang_wüa	
			Thonplem-KP	K Thonplem Atwang, 8.05.1956
			Rengnok-KP	
		Galengya	Thanglong-KP	(see Akungca)
			Chinglök-KP	
		Doluchori	Krongchüng-KP	
		Soroi	Mang'ing-KP	
		Luleng	Rümputma	/see Akhömca)
		Thhanchi	Yungreng	Angka Atwang, 27.03.1957
Boica	DP	Dordori	Chingkhung	K Rengdon Changpröca, 2.12.1956
Büiting	C	Kwaimkhyong	Tuktöng	Khuchi Patlaica, 15.01.1957
		Pantola	Chantai-KP	K son Rontui Ngaplo, 31.08.1956
Cakbümca	Re	Twainfa		(see Akhuaca)
Cala	A	Galengya	Kröilö-KP	K Kröilö Ngaplo. 22.06.1956
		Lama		

	(Rü)	Songkö	Chamklaa-HP	H Chamklaa Tou, 10.03.1957; note: From the Anok area, merged into Rümma
Catumma	A	Galengya	Tapwüa Kröilö-KP	(see Cala)
		Dokhinhangor	(Bali-P)	Mills ("Shituma")
		Lemupalong	Moloi-P	K Kongyeng Catumma, 7.12.1956
Catung	A	Lemupalong	Longron-P	MK, 1.12.1956
Chaicanau				
Chaitom	A	Galengya	Waling-KP	K Rengröng Klenchangnau, 19. 11.1956
		Lemupalong		
Changkan	A	Horinjhiri	Langrui-HP	K Langdi Tamtuca, 7.12.1956
		Luleng	Michi-KP	K Michi Changkan, 7.12.1956
		Lemupalong	Moloi-KP	(see Catumma)
			Rekca	Mangrui Rekca, 26.12.1956
		Takerpanchori	Chamklaa-KP	
Changpa	A	Pantola	Yöngtu-HP	MK, 28.04.1956
Changpröca	DP	Dordori	Chingkhung	K Rengdon hangpröca, 2.12.1956
		Naitkhyong	Khampuk-KP	K Khampuk, 3.12.1956
		Prumma	Lukha-KP	K Lukha Changpröca, 3.12.1956
			Taa-uk	K Menrüm Patönca, 3.12.1956
			Langrui-KP	25.12.1956
		Lama	Rünghim-KP	K Rünghim Pakröngca, 22.12.1956
		Luleng	Rümputma	(see Akhömca)
		Songkö	Chamklaa-HP	H Chamklaa Tou, 10.03.1957
			Nongneng-KP	(see Amtham)
Changpröng	A	Takerpanchori	Rengröng-HP	H Rengröng Bing-Tang, 3.05.1956
Changruica	DP	Luleng	Rongdim	K Khaipun Rongdim, 22.12.1956
			Rümputma	(see Akhömca)
Chatu	DR	Songkö	Chaman-KP	K Chaman Kung,

				13.03.1957
		Twainfa	Haiwön-HP	(see Akhuaca)
Chaula	A			note: see Ngarua
Chimlung	A	Dokhinhangor	Klangwai-HP	
			(Bali)	Mills ("Shimlung")
		Lemupalong	Rekca	(see Changkan)
		Bommu	Menching-KP	1.12.1956
		Coukhyong		note: see Ngarua
Chüngpongca	Re	Twainfa		(see Akhuaca)
Chüngprim	C	Galengya	Waling-KP	(see Chaitom)
		Pantola		
		Tindu	Müilün-KP	K Müilün Pahönca-Tamchaa (Khumi), 26.03.1957
		Thanchi	Laichia-KP	K Laichia Rongdim, 27.03.1957
Cöngnau	A	Dokhinhangor	(Bali)	Mills ("Chingnao"), note: see Ngarua
Deng	A	Rainkhyong		note: to the Dengwüa belong also the Khongtör und the Premchang
	NC	Alekhyong	Chapai-KP	MK, 18.03.1957, note see L3
Honca	A	Galengya	Kröilö-KP	(see Cala)
			Rengröng-KP	K Rengröng Klenchangnau, 19. 11.1956
Hualing	DR	Kwainjhiri	Kukcong-HP	Rungdön Hualing, 12.03.1957
Iaca	DR	Kamichora	Boli-HP	H Boli Pürca, 16.03.1957
Kanbök	SC	Kwaimkhyong	Uikük-HP	(see Alau); note: also Khumi, see Kanbeo'
Khatpö	A	Galengya	Tapwüa	
		Alekhyong	Thümru-KP	K Thümru Khatpö-Ngarua; note: see Ngarua
		Thanchi	Laichia-KP	(see Chüngprim)
Khongtör	A	Rainkhyong		(see sib list of Tapwüa)
		Shualok		"
		Dokhinhangor		"
		Pantola		"
		Daprutkhyong		"
		Galengya	Tapwüa	

		Horinjhiri	Tapwüa	
		Lemupalong	Longron-P	(see Catung)
		Luleng	Atüng-KP	K Atüng, 6.12.1956; note: see Deng
Khongwoica	Re	Twainfa	Haiwön-HP	(see Akhuaca)
Kilöngca	SC	Thanchi	Laichia-KP	(see Chüngprim)
		Twainfa	Haiwön-HP	(see Akhuaca); note: here classed by H as DR
Kitca	DR	Kwainjhiri	Kukcong-HP	(see Amtham)
		Kamichora	Mangpung-KP	(see Amangca)
Klenchangnau	A	Galengya	Waling-KP	(see Chaitom)
			Rengröng-KP	(see Honca)
			Chikcharon-KP	K Kangku Catumma, 30.03.1957
Kou	A	Alekhyong	Rengyong-KP	
Kria	A	Lemupalong	Moloi-P	(see Catumma)
			Rekca	(see Changkan)
Krimca		Pantola	Lukün-KP	31.08.1956
Krimchang		Ruma (?)		
	Rü	Songkö	Nongneng-KP	
			Chamklaa-HP	(see Cala)
Kuichengca	DR	Kwainjhiri	Kukcong-HP	(see Amtham)
			Rengtan-KP	K Rengtan Rümthu, 13.03.1957
		Twain	Chaman-KP	K Chaman Thangbüca, 18.03. 1957; note: K maintains that the Kuich. are not DR, but belong to the Pongmi.
Kung	Rü	Songkö	Chaman-KP	(see Chatu)
		Kämichora	Boli-HP	(see Iaca); note: the Kung probably are identical with the Akungca, for more see L3b
Kwai	SC	Kwaimkhyong	Uikük-HP	(see Alau)
			Tuktöng	(see Büiting)
		Thanchi	Laichia-KP	(see Chüngprim);

				note: also sib of the Khumi (Khoi)
Macar	Re	Twainfa	Haiwön-HP	(see Akhuaca)
Namlang	DP	Prumma	Rümpöng-KP	Rengnap, 25.12.1956
			Langrui-KP	"
Nang	A	Lemupalong	Thonru-KP	K Thonru Nang, 30.11.1956
Nöngnau	SC	Kwaimkhyong	Uikük-HP	(see Alau)
		Thanchi	Laichia-KP	(see Chüngprim)
Ngaica	A	Lemupalong	Longron-P	(see Catung)
Ngaplo	A	Galengya	Kröilö-KP	(see Cala)
		Pantola	Chantai-KP	(see Büiting)
	NC	Alekhyong	Chapai-KP	(see Deng); note: see L3
Ngaringca	A	Pantola	Yöngtu-KP	K Yöngtu Ngaringca, 29.06.1956
			Menu-KP	(" ")
		Dokhinhangor	(Bali)	Mills ("Nirincha")
		Takerpanchori	Rengröng-HP	(see Changprüng)
Ngarua	A	(see Khatpö)		note: Khatpö, Chimlung, Chaula, Cöngnau and Dopö belong to the Ngarua. see L2f
Ngarünau	A	Lemupalong	Rüm	Menyong Rüm, 30.111956
		Bommu	Ingrau-KP	K Ingrau Phen, 1.12.1956
Pachuica	DP	Luleng	Khaipung-KP	(see Changruica)
		Prumma	Rümpöng-KP	(see Namlang)
			Langrui-KP	(see Changpröca)
			Ta'uk	(" ")
	(T)	Coukhyong		MK, 19.03.1957
Pachui-Ngongklam	T (DR)	Kamichora	Boli-HP	(see Iaca)
Pacongca	DR	Twainfa	Haiwön-HP	(see Akhuaca)
		Naitkhyong		note: according to K Menyeng the Pacong belong to the Maranca (former Marma?)
Pahönca	T (DR)	Kamichora	Mangpung-KP	(see Amangca); note: K Mangpung Pahönca classified himself as DR

	(T)	Coukhyong		MK, 19.03.1957
	T	Tindu	Müilün-KP	(see Chüngprim); identified as Khumi
Pahüaca	T (DR)	Twain	Chaman-KP	(see Kuichengca); note: part of (?) the Pahönca
Pakönca				K Menyeng, 19.03.1957; note: according to K belonging to the Maranca (see Pacongca)
Pakröngca	DP	Lama	Rünghim-KP	K Rünghim Pakröngca, 22.12.1956
		Prumma	Lukha-KP	(see Changpröca)
Palau	A	Lemupalong	Longron-P	(see Catung)
			Rüm	(see Ngarünau)
		Luleng	Lungwai-P	K Lengklang, 6.12.1956
Pataca	A	Horinjhiri	Mangking-P	
Patlaica	SC	Kwaimkhyong	Tuktöng	(see Büiting)
		Thanchi	Laichia-KP	(see Chüngprim)
Patoca	DR	Kwainjhiri	Kukcong-HP	(see Amtham)
Patönca	DP	Prumma	Ta'uk	(see Changpröca)
			Rümpöng-KP	(see Namlang)
		Naitkhyong	Khampuk-KP	(see Changpröca)
		Twain	Chaman-KP	(see Kuichengca)
Plongkuma-Patönca		Kwainjhiri	Kukcong-HP	(see Amtham)
Phen	A	Luleng	Mik'au-KP	K Mik'au Rüm, 1.12.1956
		Bommu	Ingrau-KP	(see Ngarünau)
			Menching-KP	(see Chimlung)
	(Rü)	Songkö	Chamklaa-HP	(inf. and note see Cala)
Ponca	DR	Kamichora	Mangpung-KP	(see Amangca)
Pongmi	DR	Kamichora	Mangpung-KP	(" ")
		Kwainjhiri	Kukcöng-HP	(see Amtham)
Premchang	A	Rainkhyong		note: see Deng
Prencü	A	Takerpanchori	Chamklaa-KP	Kröilö Prencü
		Galengya	Thanglong-KP	(see Akungca)

			Inglai-KP	MK, 14.04.1957
		Lemupalong	Moloi-P	(see Catumma)
Prütchangca	A	Luleng	Atüng-KP	(see Khongtör)
Pürca	DR	Kamichora	Boli-HP	(see Iaca)
		Kwainjhiri	Rengtan-KP	(see Kuichengca); note: according to H Boli Pürca the P are the same as the Tom
Raha		DR area and Bommu region		K Mangpung Pahönca's mother and MK, 15.03.1957; note: see L2g
Rekca	A	Horinjhiri	Mangking-P	
		Lemupalong	Rekca	(see Changkan)
Rengkrum	Rü	Songkö	Chaman-KP	(see Chatu)
		Alikodong	Cöngpat-KP	(see Miyauca)
Rengnge	Rü	Alikodong	Cöngpat-KP	(" ")
Rengphörca	DP	Dordori	Chingkhung	(see Boica)
		Lama	Rünghim-KP	(see Changpröca)
Rengwia	Rü			K Chaman Kung, 13.03.1957; note: accordimg to K same as Kung, but see L3
Rongdim	DP	Luleng	Rongdim	(see Changruica)
		Prumma	Rümpöng-KP	(see Namlang)
			Langrui-KP	(see Changpröca)
		Thanchi	Laichia~KP	(see Chüngprim)
Röngcha	DR	Kwainjhiri	Kukcong-HP	(see Amtham)
		Alikodong	Cöngpat-KP	(see Miyauca)
		Naitkhyong (?)		note: noted as Aröngca.
Rüm	A	Lemupalong	Rüm	(see Ngarünau)
		Horinjhiri	Mangking-P	
			Thonplem-KP	(see Atwang)
		Luleng	Mik'au-KP	(see Phen)
	(Rü)	Songkö	Chamklaa-HP	(inf. and note see Cala); note: the Rümputma also belong to the Rümwüa, compared with them the Rüm in the narrower sense are also called Rümchengca. They

				have nothing to do with the Rümthu and the Rümma.
Rümput ma	DP	Luleng	Rümputma	(see Akhömca)
		Alikodong	Mangpre-KP	K Cöngpat Miyauca, 20.03.1957; note: see Rüm
Rümthu	NC	Galengya	Reng'e-HP	H Reng'e Rümthu
			Thanglong-KP	(see Akungca)
		Luleng	Rongdim	(see Changruica) (see Kuichengca); note: according to K Rengtan Rümthu the local R came from the Pantola region and merged with the Rümma
Tamchaa				note: the Pahönca belong a. o. to the Tamchaa, see L3c
Tamtuca	A	Horinjhiri	Langrui-HP	K Langdi Tamtuca, 7.12.1956; note: the Tamtuca form a part of the Prencü
Tang	A	Takerpanchori	Rengröng-HP	(see Changprüng)
			Dailo-KP	MK, 11.07.1956
		Pantola	Yongtu-KP	(see Ngaringca) 29.06.1956
		Galengyä	Waling-KP	(see Chaitom)
		Lemupalong	Longron-P	(see Catung)
Tapurca	DP	Dordori	Chingkhung	(see Boica)
		Prumma	Ta'uk	(see Changpröca)
			Langrui-KP	
Thangbü ca	DR	Twain	Chaman-KP	(see Kuichengca)
Thengma ng	Rü	Songkö	Chaman-KP	(see Chatu)
		Alikodong	Mangpre-KP	K Mangpre Thengmang (see Rümputma)
Tingca	DR	Kamichora	Mangpung-KP	(see Amangca)
			Boli-HP	(see Iaca)
Tom	A	Luleng	Atüng-KP	(see Khongtör)
			Lungwai	(see Palau)
			Rongdim	note: see Pürca
Ton	Rü	Songkö	Chamklaa-HP	(see Cala)

			Nongneng-KP	(see Amtham)
Tou	Rü	Songkö	Chamklaa-HP	(see Cala)
			Nongneng-KP	(see Amtham)
			Chaman-KP	(see Chatu)
		Twain	Chaman-KP	(see Kuichengca)
		Kamichora	Boli-HP	(see Iaca)
Uiche	DP	Dordori (?)		MK, 14.04.1957
		Galengya	Inglai-KP	("), K Inglai Uiche
Ur	NC	Takerpanchori	Niapiu-KP	MK, 18.03.1957, Chingtön Ur
		Ruma		(" ")
		Bommu	Ingrau-KP	(see Nagrünau)
		Lama		MK, 18.03.1957
		Monggu		K Chaman Thangbüca, 18.03.1957
		Twain	Chaman-KP	" " (see Kuichengca); note: 1 house each
Wiaca	DP	Dordori	Chingkhung	(see Boica)
		Naitkhyong	Khampuk-KP	(see Changpröca)
		Prumma	Lukha-KP	(" ")
			Langrui-KP	(" ")
			Rümpöng-KP	(see Namlang)
Yaca		Luleng		Menyong Rüm, 1.12.1956
Yömre	A	Lemupalong	Thonru-KP	(see Nang)
		Galengya	Waling-KP	(see Chaitom)
		Horinjhiri	Thonplem-KP	(see Atwang)

List of noted Khumi sibs
(Abbreviations used: R: Areng, W: Awa; H: Headman, K: Karbari, P: Para)

Amchang	R	Saimfra	Inghoi-HP	H Inghoi Amchang, 22.03.1957
		Tindu	Kangnong-KP	K Kangnong Kanbeo', 23.03.1957
			Renglün-HP	H Renglün Amchang, 25.03.1957
			Yeopo-KP	K Yeopo Deomkhang Sohn, 24.03.1957
			Müilün-KP	K Müilün Tamcha, 25.03.1957

Amcheng	R	Tindu	Renglün-HP Yeopo-KP	(see Amchang) (" ")
Balang	R	Remakri	Krabu-HP	H Renglün Amchang, 25.03.1957
Deomkhang	R	Tindu	Renglün-HP Yeopo-KP	(see Amchang) (" ")
Camthang	R	Betchora	Lenten-KP	K Leuten Camthang, 17.02.1957
Ceorkubeo'	R	Betchora	Lenten-KP	(see Camthang)
Kanbeo'	R	Tindu	Kangnong-KP	(see Amchang); note: Mru sib, see Kanbök
Kiaco	W ?			H Renglün Amchang, 26.06.1957
Langwuico	R	Saimfra	Inghoi-KP	(see Amchang)
		Alekhyong		K Lenten Camthang, 17.02.1957
		Ruma		" " "
		Mibokkya	Ongpha-HP	H Ongpha Langwuico
Nemlau[1]	W	Ruma		H Renglün Amchang, 26.06.1957
Tamchaa	R	Tindu	Müilün-KP	(see Amchang); note: Mru sib, see Pahönca

L2c) Sib names and their significance

Every sib has a name which is characteristic of it. These sib names would almost correspond to our European surnames if there were not occasional parts of sibs that still have their own separate name as well as their sib name. The role of these subdivisions will be examined below (L5a). The sib names are also similar to our surnames in that their real meaning is mostly unknown to their bearers, so that it was mostly impossible to obtain a translation. Hutchinson (p. 42) mentioned five sib names with translations: each supposedly refers to a plant species. He regards these names as [the] names of five "septs", into which all Mru were divided. In their traditional enumeration the Mru actually do divide themselves into five groups (L8), but this agreement is purely accidental. The names given by Hutchinson have nothing to do with the names of these groups, but are obviously sib names, even in their quite corrupted spelling, so that only two names can be identified for certain. Namely: "*Dengua*" stands for Dengwüa and "*Premsang*" for Premchang. "*Kongloi*" is probably Khongtör, especially since *-tör* corresponds to the given meaning "wild banana"; "*Gnaroo Gnar*" can hardly be anything else than Ngarua', although the given meaning "mango" cannot be correct, because the mango is *ui phum* in Mru, while *rua'* could mean "goat" and *nga-* is a prefix which is found with wild animal names, although a wild animal, called *ngarua*, is not known (the wild goat or serow [*Nemorhoedus bubalina*] is called *ngayaa*). Finally, Hutchinson's "Naizar" meant nothing to my informants: it is probably not a sib name at all but a relationship term: *nai-caa,* cousin's child. The given meaning "jack-tree" does not help either because the Mru expression for this is *latma*, and there is no such sib. Hence, the given translation is correct only in one case: the banana is called *deng*. But this does not necessarily mean that the sib name really means “banana”, because the same pronunciation does not guarantee the same meaning. However, my informants also thought the name probably meant “banana”, particularly as the Khongtör sib (with *tör* for “wild banana”) formed part of the Dengwüa.

Let me add the few sib names which could be translated easily and appropriately: Tom = bear, Ur = mouse, Kwai = bee, Krimca = hunter, Rümca = forest people, Wiaca = charcoal burner?; still conceivable: Büiting = stout old bamboo rat, Kuichengca = gluttonous curs, Bürca = spawn, Kouwüa = stirring stick people, Nang-wüa = horn people; less reasonable seem Honca = bender, Rekca = screwer, Rengphörca = those who always fly up, Tingca = those who arrive, Miyauca = rice washers, Chüngprim = old hilly country, Prencü = earthquake, Rengnge = remain crooked, Pongmi = decaying. The majority of the sib names, as stated above, cannot be translated; it is clear, however, that animal or even plant names are found only in an insignificant minority. The previous list gives a complete summary of the sib names noted.

There is therefore no basis for a hasty inference of a totemistic derivation of the sib names (as Hutchinson does [p. 49]). Nevertheless there are single dish taboos for certain sibs that were still followed by most people as traditional rules. So the Khongtör, Prencü, Tang and Rümthu do not eat a fish called *nur* in Mru (a kind of eel), because in the past they have drunk its milk (MK, 14.09.1956, K Rengtan Rümthu, 13.03.1957), furthermore the Tang and Tamtuca-Prencü do not eat bear meat and are not allowed to see a dead bear brought to the village before it has been covered with leaves. It is said that in the past the bear was their *tainau* (“brother”). The Ngaringca cannot eat meat of cattle killed by a tiger; since the tiger was their *tainau* in the past. The Honca divide themselves into two halves, the Carca-Mia and the Rangpangma-Mia. A meaning for *carca* could not be obtained, *rangpangma*, however, is a hornbill: the Honca of the hornbill branch do not eat the hornbill, because they have drunk its milk earlier (MK, 14.09.1956). The legend of the hornbill is found also among the Marma, among whom the Anok-Kolaksa do not eat the hornbill because in the past it raised their primary parents. Allegedly there was once a document in brass, on which the agreement not to eat the bird was written down (Marma in Nongneng-KP, 12.03.1957). Animals as foster parents were also possessed by the Khoi (“bees”), an earlier Mru sib (Kwai), which was assimilated by the Khumi: after their mother had died, the children (the progenitors of the Khoi) were raised with bees’ honey (Karbari Lenten Camthang, 17.02.1957).

L2d) Help between sibs

Even if there is no formal distinction between relations with near and distant members of one’s sib, where cohesion and assistance are concerned, the closeness of the blood-like relationship, intimacy of acquaintance and personal goodwill play the leading roles, which seems absolutely understandable. Even those who do not belong to the sib, but who are close kin, like the families of the mother, wife, sister or daughter, are considerably closer in emotional terms than any unknown person classified as father, brother or son of one’s sib. Someone in distress is therefore unlikely to seek help from strangers who are close to him formally; close acquaintance and real blood connections will be more decisive, and one is more likely to find understanding within his hamlet than from near relations who live far away, because the obligation to help a relative is not a written law, even if it is also enshrined in good custom. But often egoism proves to be stronger here: "If you are rich, everybody is closely related to you and your good friend; if you are poor, however, you stand alone in the world and nobody knows you." This proverbial wisdom, so often learned already, was told to me by Khamlai Ngarua', too.

Nevertheless, the importance of help from the sib should not be underestimated. If a whole hamlet burns down and not only the paddy needed until the next harvest but also the seed is destroyed, this help is the only rescue for the persons affected and, if the same hamlet is rebuilt next year, this cannot happen without sib help. Admittedly this alone will not be sufficient, and behind the reconstruction stands a perhaps endless indebtedness, in which the relatives also have their claim. It would be wonderful if this debt were only a moral obligation. But after all, the relatives will not demand usurious interest, and it will be possible to delay repayment of the debt for some time. Better rates are paid to relatives who work for their debts or for one's use. In the end, anyway, the sib base, the bonds of relationship on which fairness and kindness still prosper, is the basis on which a person knows that he can expect some return for help and favour, or that it may even be delivered without obligation.

L3 The marriage relations

L3a) Exogamy

Every sib is exogamous, which is to say that marriages within the sib – also between persons whose genealogical relationship is not provable – are against the rules. Relations to maternal relatives are not touched by this rule. Especially forbidden are connections between sibs that stand to each other in a "brother" relationship (L3c). On the other hand, there is no restriction by local exogamy, and marriages within the hamlet are freely permitted; if it is nevertheless almost the rule that the partner is taken from another village, this is a result of the actual possibilities, not of a custom. Not only is marriage banned between the members of one's own sib and its brother sibs, but also, in half of all other sibs, a young man will not find a woman for himself according to valid custom. Therefore the chances are not very great of finding a girl of roughly the same age in one's own hamlet, which generally has only ten houses, and, moreover, since affection and aversion play an undeniable role, it is only understandable that a young man looks for a marriage partner in the hamlets of his neighbourhood.

While, on the one hand, a man cannot take a woman from one half of all other sibs, on the other hand it is just the male sib members of this half whom his female sib relatives can marry. This is a characteristic of the relationship rules of the Mru and Khumi: every sib can stand to every other only in a one-sided marriage relation. Thus an individual of a sib must distinguish firstly between sibs (*cen*) that are in a brother relationship (*tai-nau*) to his own, and sibs that are in the marriage relationship (*can*) to it and, secondly, he must distinguish the *can* (brother-in-law sibs) according to whether they provide girls for his own sib (*tutma)* or whether they marry girls from it (*pen*). In other words: a man of a given sib will be able to marry a girl only if her sib is in the *tutma* relationship to his, and a girl will be able to marry only men whose sibs are in the *pen* relationship to hers.

L3b) The brother-in-law sibs (*can*)

For a man, all sibs to which the brothers of the grandmother, mother, wife, daughter-in-law, etc. belong are in the *tutma* relationship, but all those sibs to which the husbands of the sisters of his father, his sisters, his daughters, etc. belong are in the *pen* relationship to him. Since he can marry any girl who is in the *tutma* relationship to him, he also can take his wife from the same sib, even from the same family, as his father; a marriage with a daughter of the mother's brother, i.e., with a maternal cross cousin, is considered to be especially good, though it is hardly ever concluded. (But a marriage with the cross cousin on the father's side [father's sister's daughter] is excluded, even though she belongs to another sib.) Thus it is definitely possible theoretically for all *tutma* relatives on the one hand to belong to one and the same *tutma* sib and all *pen* relatives on the other hand to belong to one and the same *pen* sib; in practice, however, this is never the case.

When, however, a man is asked for his *tutma*, he always will mention only one name and this is the name of the sib of his mother's father and brother; questioned about his *pen*, he may even say that he has no *pen*, because he will think he has been asked for the name of the sib of his son-in-law. Only new questions can bring to light which other sibs still stand in these relations to his. In the narrowest sense of the word, *tutma* and *pen* describe only the persons of the maternal uncle and the son-in-law, although terms characteristic for them are available. Thus one should distinguish between a person's real *tutma* (the sib of his mother's brother) and *pen* (the sibs of his sons-in-law) and such sibs which stand in *tutma* or *pen* relationship to the whole sib. This use by the Mru themselves is an interesting example of their strongly individually oriented mind, since these terms would already receive their more comprehensive significance if one started from the collective of the small family, but this appears only secondarily. In order to escape from possible confusion caused by this ambiguity, in further examinations, I prefer to use the Mru concepts *tutma* and *pen* for the relations of an individual to his brother-in-law sibs (which at the same time corresponds to their common application), while I shall use the concepts father-in-law sib and son-in-law sib for the relation of the sib of an individual to its brother-in-law sibs.

L3c) The brother sibs

There are also formal and personal relations as regards the brother sibs. According to the definition just given I shall therefore distinguish here between brother sibs (in the broader sense) and *tai-nau* (the Mru expression both for brothers and for brother sibs). Details will still have to be given about the formation of *tainau* sibs (L6). The marriage ban between brother sibs is religiously fixed and is considered to be stricter than the rule not to marry within one's own sib, which is based on morality alone (MK, 6.08.1956). There is a belief that those who marry a member of a brother sib must die soon and that the children from this illegal marriage cannot prosper. Nobody who is entitled to influence the choice will give his consent to such a marriage. That such marriages can happen nevertheless is shown by the example of Khamcöng Khongtör and his wife Krapau Premchang.

Khongtör and Premchang are not only brother sibs but moreover, both are also part of the great sib of the Dengwüa (L5b).

The punishment for this fateful marriage seemed permanently to haunt Khamcöng, his wife and their children. Not only did the family almost never have enough to eat (for which, however, Khamcöng's aversion to work could also be blamed) but also that in the summer of 1956 his wife fell severely ill with a rheumatic paralysis and was possessed by spirits, so that Khamcöng, without rice or money, had to give three costly sacrifices on credit, among them a cattle feast, while he, kept back at home in order to care for his ailing wife and to look after the children, let almost his entire harvest for the next year spoil, and finally, because of her illness, Krapau could no longer nurse their most recent child, so that it died of hunger in her arms, while an older child had bad eyes; all this strengthened the other villagers in their belief that the justified dark consequences of the violation of the sib rules were manifested in these misfortunes.

The frequency of brother sibs and brother-in-law sibs within the sib relations of the Mru can be gathered from table at the end of L3d. According to it there were, among the 208 cases noted, 167 brother-in-law and 41 brother relations, i.e., about 20 % of all relations were brother sibs.

L3d) The ring formation

When we pursue the marriage relations beyond a sib it is clear that the *tutma* sib of a given sib in turn must have a *tutma* sib of its own, too. In which relationship this stands to the first sib, is not at all determined by the double *tutma* relationship; all kinds of relationship can be possible. Two sibs that stand in the brother relationship to each other can each also have their own relationship with a third one. Since a sib uses new marriages to restore its stock of women depleted by marrying out, it would undoubtedly be the simplest solution for three sibs to form a ring, so that sib A married B, B married C, and C married back into A. But since, as had been said, *pen* and *tutma* usually change from generation to generation and there are no regulations concerning which father-in-law sib one has to take one's wife from, such ring formation would be a coincidence. In a scheme of the theoretical possibilities, however, such rings (if the brother sibs are not taken into consideration) should form 25 %. An examination of sib relations noted for some hamlets shows that this relationship also applies roughly to the possible marriage relations between the Mru sibs (see the following table). Seventy groups of three sibs (excluding the brother sibs) show a ratio of about 2 to 5. The groups of three sibs with a closed marriage ring form 28 % of these possibilities; this is 3 % more than the frequency to be expected theoretically. This difference would hardly be sufficient to establish a definite tendency toward the formation of sibs in a closed marriage ring.

In the 220 marriage relations noted by me, the details from different hamlets contradict each other in four cases of relations between two sibs. This may be due either to a fault of my listing or to a mistake by the informants, but it is also definitely possible that other relations between the sibs have developed in different areas (see L5).

Examples of the marriage relationships between some sibs of different hamlets

(Sibs not belonging to the hamlet or names repeated for clarification of the relations are given in brackets, arrows point to the *tutma* sib; + and × = *tainau* sib; O = not asked or unknown;

Order of double signs: left-right or above-below).

1) Tapwüa-Kua, Thonplem-KP, Thonru-KP

Khongtör + Atwang ← Rüm O (Prencü) → (Ngarua) ← (Catumma) → (Atwang)

+ ↘↗ ↑ ↘↗ + ↘↗ ↑ ↘× ↑ ↖× ↓ ↘× ↑

Catumma →[459] Ngarua ← Yömre + Nang O (Khongtör) → (Changpa) + (Rekca)

2a) Moloi-KP, Rekca-Kua; 2b) Lungwai-KP, Atüng-KP

Catumma → Changkan ← Rekca O Palau → Tom ← Prütchangca

↓ ↖↗ ↓ ↖↗ ↑ ↘O ↓ ×↗ + ×O ↓

Prencü ← Kria → Chimlung O (Atwang) ← (Ngarua) ← Khongtör

3) Rümca-Kua, Ingrau-KP, Michi-KP, Mangking-P

Rüm → Phen → Ur O Rekca ← Rüm → (Kria)

↓ ↘↗ + ↘↙ ↑ O + ↘↗ ↑ O

Palau ← Ngarünau O (Atwang) ← Changkan O Pataca

4) Rengröng-HP, Yöngtu-KP

(Atwang) ←/→[460] Ngaringca → Tang ← (Ngarua) → (Premchang) + (Khongtör)

↓ ↘↗ + ↖↙ ↑ O× + ×O + O× +

(Nang) → (Prencü) → Changprüng O (Chaicanau) → (Rekca) + (Ngaringca)

5) Kröilö-KP, Chantai-KP

Ngaplo ← Honca + (Ngarua) ↗ Büiting → Ngaplo → (Khongtör)

↑ ↘↙ O ↘↙ ↑ (Catumma) ↑ ↘↗ ↑ ↘↙ O

Catumma ← Cala ← (Khongtör) ↘ Ngaringca → (Ngarua) + (Krimca)

6) Thanglong-KP, Waling-KP

Yömre + Klenchangnau → Atwang → Akungca + (Ngarua) → Chüngprim

↓ O× ↑ O↙ ↓ ↘↙ ↑ O↗ + ↘↗ O

---

459 1963 turned by a marriage in Tapwüa-Kua (see L4b)

460 see L5a

(Khongtör) → Chaitom Prencü + Rümthu → (Klenchangnau) → Tang

7) Chingkhung-Kua, Lukha-KP, Runghim-KP, Ta'uk-Kua.
Changpröca ← Tapurca + Boica ← (Changpröca) ← Pachuica + (Patönca)
↓ ↖↗ ↑ ↘↙ ↑ ↘↙ ↓ ↘↖[461]↙ ↓ ↖O ↑
Rengphörca → Pätönca ←Wiaca → (Rengphörca) ← Pakröngca → (Wiaca)

8) Langrui-KP, Rümputma-Kua
Tapurca ← Namlang + Changpröca ← Rümputma → Atwang + Changpröca
↓ ↘ + ↖× ↓ ↘O ↓ ↖↗ ↓ ↗
Pachuica + Rongdim → Wiaca O Akhömca ← Changruica

9) Rongdim-Kua, Rengtan-KP
Rümthu + Pachuica → Rongdim ← Rümthu
+ ↖O ↓ ↘↙ ↑ O↗ ↓
(Rüm) O Changruica ← Tom/Pürca → Kuichengca

10a) Chaman-KP, 10b) Uikük-HP, 10c) Tuktöng-Kua
Thengmang + Rengkrum Nöngnau ← Kwai Kwai → Patlaica
↙ ↑ ↘↗↓ ↘ ↓ ↘↙ ↑ ↑ ↖↗ +
Tou → Kung → Chatu ← Tou Kanbök + Alau Büiting + Lamia

11) Renglün-HP (Khumi)
(Langwuico) → Amchang ← Amcheng
O↙ ↑ ↘↗ ↑
(Kanbeo') O Deomkhang ← (Balang)

L4 The relation between sibs

L4a) The unbroken sib relationship

In social communication the *tutma* are accorded an essentially more honoured position than the *pen*. It seems as if the debt to have taken a woman from a sib cannot be cancelled by the conclusion of a marriage contract, but continues to exist, morally at least. The "hair money" [see Q1e] in case of a wife's death can also be interpreted as an expression of this relationship. Seen generally, the Mru and Khumi are not bound to any particular rules in their behaviour towards their relatives: everybody can meet everybody and speak with him casually. But there is quite privately something like shyness towards the parents-in-law, particularly at the time of the marriage: the bridegroom is usually too shy to take part in the ceremonies in the house of the bride's parents. Respect for the father-in-law

[461] see L5a

sibs, that is to say not only the *tutma* in particular, is expressed in its own way by the rule that chicken meat may not be accepted or eaten from the people of one's parents-in-law sib. In view of the extremely high esteem chicken meat enjoys among the mountain people, it seems reasonable to assume that this rule also reflects this shyness towards the parents-in-law: When I asked Kyaw Thwân Ong (a man of the Marma, whose sib relations differ very much from those of the Mru) whether he might eat chicken meat from his parents-in-law, he told me that he did not know (his wife was an orphan) but he thought he would hardly do it, as he would be ashamed if he did: what seems a natural reaction is in this case a rigid rule which must be observed absolutely among the Mru and Khumi. (It is valid only for house chickens and not for jungle chickens.) If it is broken, e.g., by inattentiveness in the hurly-burly of a feast, it must be set right by a special sacrifice (L4b); how much this rule can become a mere formalism is proved by the sophistry of my Mru companions on tours: they ate chicken meat with me without fearing any consequences, although the chicken was given by persons who stood in the father-in-law sib relationship to them, since the chicken had been presented to me and was now regarded as mine.

As it is forbidden for the *pen* to accept chicken meat, they in turn will try to entertain their *tutma* with chicken meat when visited by them, and to bring chickens to them as a special present during their feasts; the feast-giver, for his part, will entertain his *tutma* with them, because it is particularly at feasts that the sib relations are honoured by ceremonial exchange. The special role of the *tutma* among the Longhu and Khumi at their great feasts of merit will be discussed under R6; the exchange ceremonies will be described with the marriage customs (see M4); on the chicken distribution when the hamlet is closed cf. P3. A special privilege which can accrue to the *tutma* at feasts, especially with the Khumi but also with the Mru, is the *kom-bong*, "wrapping of the wrist" of the members of the family (R5b).

When several rice-beer pots are set up for any festive occasion, then the greatest one is the *tutma-yu* and it is a special honour to be the first to drink from this pot. Admittedly the *tutma* will not come empty-handed either, but even then (say on a simple visit on a journey), one should always be welcome with one's son-in-law people. It is regarded as a special honour if the son-in-law offers jungle fowl or bamboo rats, because anybody can have pigs and chickens at home at any time, but it requires more effort to be able to offer this game. Thus to the respect for the *tutma* on one side, goodwill towards the *pen* will correspond on the other side. In the real *tutma-pen* relationship between the families, a particular kind of father-son relation is enacted, which of course is the less, the more we have to do with purely formal relations between son-in-law and father-in-law sibs. Without private relations, two men of the same age, notwithstanding their sib relationship, will treat each other as "equals" (they use the same forms of address, anyway), whereas an unmarried man on a journey is usually better entertained by his father-in-law sib people, since, as a possible son-in-law, he can expect to be shown the goodwill due to him. Theoretically this relationship would be reversed for young girls, but this is of no practical importance, since women do not make any journeys if they can be avoided, or go only in groups, as when young girls visit a feast.

L4b) The reversal

The sib relationship makes the marriage relationship unequivocally certain; in principle a man can only marry a woman whose sib stands in a *tutma* relationship (or till now in no other provable) relationship to his sib. Anyone who nevertheless wants to marry a girl who stands in a *pen* relationship to him will neither get the approval of his own parents nor that of the girl's. If, however, the relation corresponds only to the tradition, there can be hope that the parents' mind can be changed. But when a current marriage is known and the young couple nevertheless does not change their mind, there actually remains only the secret flight (over the border to Arakan) or the common farewell to life (see M3b). To prevent the loss of his son the father of the bridegroom will have to give way if necessary and try to settle the matter with the bride's parents. No prospects of consent exist when the couple is closely related (so, if the girl for instance is the young man's father's sister's daughter). Otherwise there are two possibilities of this regulation:

1) When the bride's parents do not agree to the marriage and wish the retention of the previous sib relations, the father of the bridegroom must pay to them 70/- Rs. beyond the marriage price, called *wa-ngaa cak-cök* (chicken-meat meal-compensation). For the marriage the parents and relatives of the bride must be catered for with chickens. But they stick to the previous sib relationship, in accordance with which they are *pen* for the family of the bridegroom and do not take chicken meat from their *tutma.* The *wa-ngaa cak-cök* therefore forms the compensation money for the chickens not received. By this compensation the previous relationship lasts between the two sibs and will continue to exist for the children and descendants of the couple, too.

2) When the parents of the girl agree, the father of the bridegroom gives them (apart from the usual marriage money) Rs. 30/- and a hewing knife. This payment is called *wa-lu-pok* (chicken-head-chop-off). According to this agreement the bride's parents can eat chicken meat, since the payment cancels the old sib relationship. From now onward the bride's parents do not stand any more in a *pen,* but in a real *tutma* relation to the sib of the bridegroom, which henceforth stands in a *pen* relation to people of the bride. Henceforth, for the complete sib of the bridegroom, the marriage relationships to the family of the bride are reversed, but all this does not apply immediately to the bride's complete sib.

Here the relationship depends on the consent of the other families of the bride's sib. The bride's father has to distribute among all the families of his sib a part (not determined by rules but by demands left to the discretion of the families) of the money received (both the *wa-ngaa cak-cök* and the *wa-lu-pok*). In the first case (1) nothing changes in the relationship of the two sibs and everybody accepts the money willingly. In the second case (2), however, some people of the bride's sib may not consent to the reversal. They then do not accept any part of the *wa-lu-pok*, do not eat chicken meat from the bridegroom's sib and therefore keep their previous *tutma* relationship. The relationship is reversed only for those families of the bride's sib who get money and eat chicken meat, so that the bride's sib splits into a *pen* and a *tutma* part for the sib of the bridegroom.

A somewhat peculiar case of a partial reversal of the sib relationship is the relation of Khamlai Ngarua' to the Rüm. The Rüm are a father-in-law sib of the Ngarua'; Khamlai, however, claims that the Rüm are his *pen.* On the occasion of the wedding of a man of the Rüm he had eaten of the chicken meat. According to the rule, he ought to have sacrificed a *met-pak* (see P5g) to restore the old order; but, having neither the desire nor the means to do this, he rather accepted the abuse of his relatives and claimed from this time on that the Rüm were his *pen.* All other Rüm and Ngarua', Khamlai's brother included, however, continue to regard themselves as being in the old relationship to each other (Dingte Ngarua' in Rümputma-Kua, 26.12.1956). In Rümputma-Kua, however, Khamlai (who together with Khamcöng had accompanied me to carry my luggage) also found what he had been looking for all the while: a widow who consented to marry him. Unfortunately she was a Rüm, and when I came to know about this, I offered to give Khamlai the money to buy a little pig which he could sacrifice as *met-pak*, but he refused: the Rüm were his *pen*, and he was a honourable man and as such had to honour Mru custom, and no money would induce him to marry a woman who stood in a *pen* relationship to him!

L4c) The stability of the sib relationship

I cannot make any statements on the frequency of reversals and violations of the sib relationship. It is certain, however, that they are not an everyday event. Anyway, the young men, in view of the complications to be expected, will do their best not to think of a *pen* relative as a possible wife. The parents' refusal forms the next obstacle and even if they overcome their scruples, the "no" of the sib members who keep the tradition still ranks higher than a private love affair. An additional safeguarding of the sib relationship comes from the outside: the Bohmong permits himself to ask 200–300 rupees from the "guilty persons" in cases of *pen* marriages which come to his ears (MK, 11.07.1956). The only case of a full reversal of the sib relationship known to me happened by the marriage of Thonlök Ngarua' with Pronglai Atwang (cf. the sib lists of Tapwüa-Kua).

In the past, Atwang was a son-in-law sib of Ngarua', today Atwang is Ngarua's *tutma* (MK, 19.11.1956). The rigid maintenance of the traditional sib relationship, however, is in the process of dissolution where opium consumption has made greater inroads. Anything is sold for opium, even the old rules. This is how, for instance, the sib relations around Waling-KP (Galengya-Mouza) quite recently were almost completely dissolved. There the Rümthu and Honca were earlier *pen* for Klenchangnau, today the Rümthu are partly *tutma* for the Klenchangnau, partly *pen* and reversed; another part of the Klenchangnau has married Honca girls, without a full reversal. The Yömre are an old brother sib of the Klenchangnau; but even this relationship was disregarded and a part of the Yömre is *tutma* now (K Rengröng Klenchangnau, 19.11.1956).

There seems to have been a complete dissolution of the sib relations among the small northern Khumi group in the Betchora-Mouza, which got detached spatially from other Khumi. So the wife of Karbari Lenten Camthang belongs to the Ceongkubeo' sib, but the husbands of his father's sister and his daughter are also Ceongkubeo' people (K Lenten Camthang, 17.02.1957). In this area the lack of marriage partners (smallness of the group with only three sibs) may have promoted the dissolution of the old conditions.

L5 The sub-sibs

L5a) The dividing of sibs

There is no reason to assume that a complete or partial reversal of the sib relationship is a phenomenon confined to more recent times. The general validity of the reversal rules speaks for a certain age. Therefore a reversal may probably also be an explanation in such cases in which two parts of a sib stand in two different relations to another sib while no reason could be given for it. Pakröngca, for instance, is partly *pen*, partly *tutma* for Changpröca, and the same applies to Ngaringca in relation to Klenchangnau. While in these cases the two parts do not bear any particular name, in the relationship of Atwang-Ngaringca the two Atwang groups are named differently: the Kitu-Twang are *pen* for Ngaringca, the Kuletaing-Twang, however, are *tutma.* Kitu and Kuletaing are *tainau* for each other. But this difference in the marriage relationships of the Atwang could also be explained by a partial reversal, since normally two *tainau* sub-sibs stand in the same relation to other sibs. On my tours I could observe how my porters in places distant to their native place determined their relations to a strange sib straight away by analogy, whenever the relations to a *tainau* sub-sib of the other or their own sib were known to them. The assumption of a partial reversal of the formerly uniform sib relationships as a reason for the Atwang-Ngaringca relationship cannot be proved by this; but the reverse assumption, namely that the subdivision of a sib can be explained by differences in the sib relationships, must remain even more questionable. The thought of a possible development from a partial reversal of the marriage relationships via the separation into unnamed sib parts to a separation into named sub-sibs was not accepted by my informant (MK, 11.07.1956). The fact that for instance the division of the Honca with respect to their *tutma-pen* relations with the Klenchangnau (see L4c) has nothing to do with their partition into the *tainau* sub-sibs Carca-mia and Rangpangma-mia also speaks against such an assumption. The division into named sub-sibs is rather a traditional fact for which my Mru informants could offer no explanation.

L5b) The derivation of the names

I have called sub-sibs (in the case of the Honca *mia* “branches”) such parts of a sib which have a name of their own. What was generally said about the sib names (L2c) also applies to the names of the sub-sibs: they are inexplicable in most cases. Judging from the few explicable names, the partitions seem to date from different times. That of the Honca is mythologically founded because the Rangpangma-mia were brought up by a hornbill (L3c). The Atwang sub-sib of the Kuletaing does not have any real Mru name, but on account of the sound rule must be derived from the Marma; perhaps Kuletaing is the Mru corruption of Marma *ku’lâ tâm*, so that Kuletaing-Twang could be translated as “Bengali custom Atwang”. Today, however, there is no reason for such a name. In the case of the Prencü sub-sibs Tamtuca and Plamphai-kila we are dealing with an identification according to special occupations: *tu* is an instrument of importance for death ceremonies and feasts of merit (see Q1d–e and R3f–g). Its use among the Anok and its production is said to have been introduced by the Tamtuca (MK. and Khamcöng Khongtör, 7.09.1956); *plam-phai* describes the hardening and smelting of iron: in the past, the Plamphai-kila are said to have produced hewing knives from bought iron bars and scrap iron; today this knowledge no longer exists (MK, 15.07.1956). None of the names of sub-sibs I came to know seems to be derived from a personal name, and hence, even when the sub-sibs may be *tai-nau* for each other, we cannot infer that they could be derived from two brothers, two wives or some such people. A derivation from brothers, however, is not unknown, but then they are not sub-sibs but independent sibs. The four sibs of the real Rümma are one example: Rengwia, Thengmang, Rengkrum and Rengnge are derived from four brothers (whose names they carry) and therefore they do not intermarry (Chaman Kung, 13.03.1957, Chöngpat Miyauca, 20.03.1957). This brother relationship by no means always applies to sub-sibs; within the Ngarua' for instance, Khatpö men can marry Chimlung girls.

Differences also appear in the grammatical formation of the sub-sib names. In the case of the Atwang, for instance, the differential words are put in front (Kitu- and Kuletaing-Twang), in the case of the Rüm however (Rüm-chengca, Rüm-putma) they follow the main word, while with the Ngarua' the names of the sub-sibs (Khatpö, Chimlung, etc.) are completely independent and are actually not used at all together with the name under which they could be subsumed. Finally there are cases where only one part of the sib has a name of its own, for instance the sub-sib of the Bing-Tang inside the Tang (H Rengröng Bing-Tang, 3.05.1956).

L5c) Independence and emergence

The Mru themselves have no terminological problems with the sub-sibs: they describe all forms indiscriminately as *cen*. The *mia* "branch", as used in the case of the Honca, does not seem to have any general validity, occasionally *up* is used, but this can just as well describe an independent sib. If, however, a man is asked for the name of his *cen*, he always gives only one name at first. So the Atwang always describe themselves as Atwang (never mentioning their sub-sibs, Kitu or Kuletaing); people of the Dengwüa, however, almost always call themselves Khongtör or Premchang, etc., and hardly ever speak about Deng. The Khatpö Ngarua' always described themselves as Ngarua', but when they spoke about the Chimlung Ngarua' they always called them Chimlung. On the other hand, the Rüm-chengca were almost always called just Rüm, while the Rüm-putma (who are just as much Rüm), are called Rüm-putma. I asked a man of the Tamtuca-Prencü for his sib name and he answered "Prencü", but another man protested and objected: "This is not correct at all, he is a Tamtuca" (Moloi-Para 7.12.1956). Generally, perhaps one can say that the *tainau* sub-sibs, which are hardly ever mentioned and in whose names the specifying word precedes, are on one side of the distinctions by name, while the always mentioned *can* sibs (between which marriage relations are possible) are on the other side, with independent names within a "greater sib". Whether the existing intermediate forms are signs of a development must remain more than questionable. Judging from name formation and meaning, it seems highly probable that the emergence of sub-sibs was not a uniform process. At present it is not possible to observe anywhere either a process of splitting into sub-sibs (to be distinguished from the dividing of a sib discussed above according to marriage relations), or the formation of "greater sibs" (by the process of sub-sibs becoming independent or independent sibs forming confederations). In my view, the explanation for this lies in the fact that nowadays there is simply no motive for such sib dynamics. The sib has neither a common territory nor common interests and it is neither dependent on a leadership of its own, nor in charge of a common characteristic of its own. Today the only common elements are the rules of exogamy (differences arising in this respect can lead to dividing, but apparently not to splitting by name). In the past however, tighter forms of organisation existed (L7b) and migrations connected with wars may have imposed other requirements on the sibs. Today's "irregularities" probably date from those times. What particular events these were is forgotten today, however.

L6 Sib alliances

L6a) *cia mai lek choi*

Two sibs between which there are no marriage relations can be described as being in the brother relationship to each other. Actually, however, one should distinguish between sibs among which no provable connections exist but which could be tied by special ceremonies, and those between which marriage is strictly forbidden (see L3c).

The tying of a brother relationship is not a matter of a mere agreement, but must be produced ceremonially. Such an alliance between sibs is only possible, however, if, as far back as can be remembered, no marriage relations have hitherto existed between the sibs, i.e., brother-in-law sibs cannot enter into any pact. Most pacts date from the time of the warlike migrations; they commit both sibs to mutual support. They are called *krang-chaa-phak* (kind-transferring change) and are still remembered today. The obligation of support was concluded at the entrance of the hamlet (*plon*), by one representative each of the sibs to be connected, in a ceremony called *cia-mai-lek-choi* (cattle-tail-shaking oath) (MK, 6.08.1956). The same ceremony was also known to the Khumi (H Inghoi Amchang, 22.03.1957). A head of cattle was tied to a post and its throat cut with the hewing knife. The blood flowing out was collected and mixed with rice-beer and *arak* and the blood of a chicken. The beer for this celebration was prepared in a big round pot (*puma*), from which the two representatives of the sibs closing the pact had to drink first. The ceremony had to be carried out twice; first at the hamlet gate of one sib, then at the hamlet gate of the other. One half of the sacrificial beast went to the counter-sib, the other half was consumed by all who participated in the pact. The ceremony got its name from the fact that, for the oath, the cattle tail was dipped into the blood and swung to and fro. After the opening *tamma* (invocation of the spirits), amongst others, these words were spoken: "If you do not observe this pact, you will become like this cattle tail; if a young man of your sib marries into mine, he shall die; if I die without brethren and descendants you can use my property and the other way round I will use yours." The pact becomes effective with the first oath, but often there does not seem to be a great hurry with the reciprocation. So the Ngarua' have concluded such a pact with the Honca and Klenchangnau, but neither party has reciprocated the ceremony and to this day they are under an obligation to do so (MK, 11.08.1956).

L6b) *hom-bang choi*

Tainau pacts, too, can still be entered into nowadays. Although they are concluded by two single persons, the obligations of the brother relationship apply to all members of the two sibs. Therefore in this case, too, it is a prerequisite that there have been no provable marriage relations between the two sibs and that the sib members agree. The ceremony is called *hom-bang-choi* ("boiled-rice to-give-ceremonially-to-eat oath") and has to be carried out twice, namely in the house of each partner. In an arbitrary way and without special ceremonies, a pig and a chicken are killed in the house, and the meat is prepared as curry. In a third bowl, cooked rice is presented to the partner. The host mixes the food and puts some of it into the mouth of his guest. On this occasion a *tamma* (invocation) is spoken: if the partner does not fulfil his obligations, he will die. The invited partner must then eat the rest of the meal. Among these obligations, as with all *tainau* relationships, are to give help in any emergency and not to establish any marriage relations. Also, as a result of the *hom-bang-choi,* if somebody dies without relatives, the possessions of this person can pass over to the *tainau* sib. Especially for the two men concluding the pact by the *hom-bang-choi*, the intention is to found a friendship that goes beyond ordinary fraternal care if possible. This thought finds its ceremonial expression in the rule that on the marriage of a daughter of one partner the other one is entitled to receive Rs. 30 from the marriage price, called *mua-wui*. The obligation to pay *mua-wui* is mutual, but applies to one daughter only, with the result that only a temporary ceremonial financial exchange takes place (MK, 6.08.1956). Such a *tainau* pact was concluded in 1950 by Menkröi Ngarua' and Klangwai Changprüng. (For the role of the *tainau* at greater feasts [*tainau-pang*] see M4c.)

More weight can be given to the *hom-bang-choi* if it is preceded by a *pak-tan* (see P5d). The *tainau* receives the head and the right foreleg of the sacrificial pig in one piece. The *tainau* must be entertained with rice-beer and be given 5 to 30 Rs. if possible. The ceremony must also be repeated by the partner. If he does not do so and does not give everything back equally, he or his son will die and the debt will last for decades. In this case, the real oath is sworn while the partner is given arak to drink and is called *cang-pe-choi* (make-give-oath). Some blood from a small chicken, which may not be eaten, is dripped into the *arak.* This kind of oath can also be used with the simple *hom-bang-choi.* By the ceremony of the *cang-pe-choi* alone, a kind of fraternal alliance is concluded between two persons without any obligations arising from it for the sib members. It can be used both for the mediation of disputes and for the prevention of unwanted sexual relations (see M1c) (MK, 20.08.1956).

Among the Khumi, too, fraternal alliances are still concluded today. A head of cattle, a goat or a pig are killed in an arbitrary way and divided into two parts in the house. Every partner receives one part. For the oath proper, rice spirit, into which a little bit of the blood of the sacrificial animal has been mixed, is drunk (H Inghoi Amchang, 22.03.1957). Goats can be used for the *tainau* ceremonies of the Mru, too (Menrüm Patönca, 23.12.1956).

L6c) Pacts with foreigners

In earlier times, pacts similar to the *cia-mai-lek- and hom-bang-choi* were also undoubtedly used to conclude peace; not only within one's own tribe but also with foreign parties. So the Khumi entered into friendship pacts with Englishmen when the latter first penetrated into their territory. Lewin reports on this in minute detail (1870:228/29): "On one occasion I had to swear an oath of friendship with certain Chiefs among the Kumis, and sacrifice was then offered up as follows: – A goat was tied by the neck, the cord being held by me; another rope was fastened by the animal's hind legs, and held by the five Chiefs with whom I was concerned. The ropes were kept taut, so that the animal was thrown into an extended position. The head Chief bearing a fighting dâo, stood over the goat; and taking a mouthful of liquor from a cup which was handed to him, he blew it first over me, then over the Chiefs, and then a third mouthful over the goat. He then raised his dâo over his head, and addressed a loud invocation to the 'Nat', or spirit of the river, at the same time plucking some hairs from the goat, and scattering them to the wind. Then with one stroke of the dâo he severed the animal's head from its body. The warm blood from his weapon was afterwards smeared upon the feet and the foreheads of all who took part in the ceremony, with a muttered formula, indicating that anyone who was false or acted contrary to the object for the attainment of which the sacrifice was offered, could be slain without fault by his coadjutors. A grand feast on the goat's flesh concluded the ceremony."

This information is interesting in that the "spirit of the river" (probably *o-reng-nam* [see P2f]) is called and that the oath breaker may be killed by the partner. Neither is the case among the Mru: the punishment is left to the spirits which, however, are not mentioned in the *tamma* (invocation). The "Nat" mentioned above is not a name but the Burmese expression for spirit in general. Smearing the forehead and foot with blood was also reported by Tickell (1851:107), who concluded a peace pact with a headman of the Anu-Khumi: "The compact was ratified by the sacrifice of a chicken, which he and I held between us, whilst its head was cut off. He then dabbed my foot with the blood, and I marked his forehead in the like manner."

Although neither of the two authors speaks of a "brother pact", I think that it was such. I base this assumption primarily on an experience of my own, which is mentioned here because, although it did not lead to a pact being concluded, it can yield some information about the reasons for such pacts with foreigners. From the 23rd to the 25th of December 1956, I visited a buffalo feast in Rümpöng-KP (Dopreng Mru). The karbari of Ta'uk-Para, Menrüm Patönca, whom I already knew from a former tour, had accompanied me and during the whole feast did his best to care for my welfare. On the evening of the first day, Menrüm came to me, after he had been entertained with much *arak* as a brother-in-law of the feast-giver. He asked me repeatedly to excuse his condition, saying that he had impressed upon everyone that I should be honoured, but now he must make an important announcement to me: he wanted to sacrifice a goat and conclude a *tainau* pact with me. I was for an immediate performance of the ceremony, Menrüm, however, insisted on postponing it as he wanted to conclude the pact only on my next visit to his village when I could bring the tape recorder with me. Though I was younger than he, on account of my merits he wanted me to become his "elder brother". As I found out only later, in addition to my reputation of having supported needy persons, a pact with me was favoured above all by the fact that I was the best friend of Menkröi Ngarua' and his family, with whom Menrüm was also connected by a *tainau* pact (dating from help given by Menkröi, which had secured him a good name in Ta'uk), as well as by Menrüm's belief, which no protests from my side could shake, that I belonged to the same "sib" as King George: through my alliance I would be bound to urge the king to do something to improve the unpleasant situation in the Mru country. Under these circumstances, acceptance of the pact would have amounted to deception on my side, while a rejection would have given grave offence. Since I never again had the opportunity to visit Taa'uk later, I was saved from this embarrassing situation, but unfortunately at the expense of ethnographic data.

Pacts can (or could) also be concluded privately without involving the whole sib. The public, however, on coming to know of such a pact, will surmise that those who conclude such a pact have sinister motives and are making a kind of private conspiracy. The oath is called *long-haa-charai-cor-choi* (iron-chopper-[water]-running-down-oath). A lot of objects will be piled up (among them a hewing knife, a spear, the tooth of a tiger, a gun, the leaf of a ficus tree, etc.). Over these things water will be poured, the water running down will be caught in a bowl and drunk by the two men, who will speak the oath, containing the curse that the one breaking the oath shall die by being slain with the hewing knife, pierced by the spear, devoured by the tiger, shot by the gun, or become mad and wag his head like the leaf (the latter may be a remembrance of head-hunting days: the heads were hung up in a ficus tree, see also the end of the *khommi-kua-tuk-meng;* S2a). All objects have to be touched by those concluding the pact. No blood-sacrifice is performed. This oath is intended to secure the help of a partner necessary for a difficult job that one man cannot do alone.

A case still known was reported for the father of the headman of Tankaboti-Mouza. In his time there was still a little police outpost in this mouza, and the father of the headman, who was a big, strong fellow, concluded a pact with three other men to use this post for their own advantage by blackmailing other people with the police and to divide the profit arising from their activities among themselves. Thus the man terrorised the people, especially when he was drunk. But one day when drunk he started a quarrel with Bengali raftsmen, who killed him and threw him into the river. One of his sons was bitten by a tiger and died subsequently. Though it could not be proven that the headman's father had violated or not fulfilled the oath, people were convinced that these events were the result of the bad oath (MK, 22.08.1956).

L7 Rank formation between sibs

L7a) Respected sibs

In the previous chapters the sibs were regarded as equal in their mutual relations: these equal rights correspond to the present condition. The respect shown to the *tutma* does not give these sibs any general pre-eminence since they are *pen* with regard to other sibs and have to pay respect for their part. There also is no pre-eminence between the sub-sibs of a sib, the sibs of a great sib and the *tainau* sibs. Within the sibs all families are treated as equal; a rich man may well enjoy a good reputation, certain families (for instance those who have become addicted to opium) or even sibs in certain areas may well have a bad reputation: but these are questions of private esteem depending on time and situation. A headman, a karbari or *ruaca* may be respected, but only as individuals and then only if people are satisfied with them; no higher prestige therefore befits their sibs, no member of a headman's sib can expect preferential treatment due to this relationship, his social status is the same as that one of every other Mru or Khumi.

These equal rights, however, did not always exist, but are the indirect result of the policies of the English administration. Before the time of headmen (English word), karbari (Bangla word) and ruaca (Burmese word) appointed by the government, there were prerogatives of single sibs and within these sibs of single families. Unfortunately, the memory of this condition has already largely faded; although the descendants of the former reigning families still know with a certain pride about their distinguished descent, the information which they could give me about the old conditions remains very fragmentary.

L7b) The *ruaca*

Before the introduction of the administration system of today some – but not all – of the sibs had a ruler of their own. The Anok called them *ruaca*, the Rümma *chai-ria*, and the Khumi *angreng*. Today *angreng* describes a rich man, particularly a big feast-giver. The triple meaning of this word (ruler, rich man, feast-giver), becomes understandable when we realise that the ruler had the right to collect taxes, therefore permanently collected wealth, which he, according to the old custom, had to return to the community in great feasts. How closely connected these concepts of ruler – rich man – feast-giver still are today is shown by the fact that Karbari Reinnoi, who had acquired so much wealth that in 1957 he could give a *reng-rang*, a feast of merit which had not been given for decades among the Khumi of the Hill Tracts, became (unofficially) known in the whole region as "Raja of the hills" (Marma *tong-mâng*). In Arakan, too, the *Moi-A(ng)reng* (hill-rulers) of the Khumi were called *Tong-mâng*: "Each village has a chief or '*Toungya-min*', a title given by the Arakanese, in which language it means the chief of a hill ('toung' a hill, and 'min', a chief), a meaning which in their own language is conveyed by the word 'moiarain'" (Hughes 1881:11). The word *ruaca* is a loanword from Arakanese (*rwa-câ* = village eater), while *chai-ria* is a genuine Mru word.

Since the rulers of the Mru sibs had no functions of a "village eater" (see below), the designation *ruaca* may originally have been a title conferred by the Arakanese sovereign to the Mru chief who submitted to him. The Arakanese sovereign (formerly the king of Arakan, later also the Bohmong of Banderban) received tribute from the Mru ruaca and in return confirmed him in his dignity with a badge, called *kri-prak-chap* (brass-engraving-certificate). The *ruaca* themselves, as mentioned, already had the right to collect taxes. Tax collection today is the task of the headmen, some of whom are descendants of the earlier sib rulers. While the headman has authority over the population of a certain territory (a mouza), the place of residence was unimportant for the taxpayer of the *ruaca*: sib membership alone was decisive, and even in his own village the *ruaca* (contrary to the literal meaning of his title) had no sway over people who were not under his authority because of their sib membership (MK, 21.08.1956).

I was unable to obtain any details about how the tax was levied and collected or on other rights of the *ruaca* and his family. It is said that the ruler had himself carried by his subjects on their shoulders (MK, 15.07.1956). Whether there were rules for the succession is also unknown; the youngest brother (Rengnge) was the first *chai-ria* of the four brother sibs of the Rümma, the sib named after him was a "*ruaca-cen*" (K Chaman Kung 13.03.1957). The succession of the youngest son, however, was not necessarily the rule; in the case of the Areng-Khumi, among whom the *ruaca* belonged to the Amchang, the succession went to the most capable, as in the Arakanese and Burmese royal houses (H Renglün Amchang 26.03.1957).

L7c) Dependent sibs

The subjects of the *ruaca* were primarily the members of his sib, but also people of the dependent sibs. Within the great sibs only one sib was entitled to the *ruaca* privilege. Among the Anok, the Atwang, the Changpa and the Ngarua' were the *ruaca-cen* (ruling sibs). Within the Ngarua' the Khatpö provided the *ruaca*, the Ngarua' sibs of the Chimlung, Dopö, Chaula, etc. paid tribute. However, the Honca and Klenchangnau[462] were also tributaries (MK, 21.08.1956). These two sibs also concluded a *tainau* pact with the Ngarua', as mentioned above. This was probably a more or less voluntary submission: Honca and Klenchangnau were (and are) numerically weak units in comparison with the great sib of the Ngarua'. In the unsafe times before the English administration their pact with the Ngarua' perhaps secured for them the support of this powerful sib, the price for this protection being their recognition of the Khatpö ruler and the associated tribute payments.

But soon after this deal with the Ngarua' the Honca seem to have changed their mind; they did not answer the *cia-mai-lek-choi* and submitted themselves to the *ruaca* of the Ur, who also got tribute from their neighbours, the Rümthu (MK, 18.03.1957). Today, the fact that the Ur were once so powerful that they had a *ruaca* of their own and dependent sibs, while today there are only six houses left in the whole Mru area, may illustrate the length of time which has passed since those events. Therefore more detailed information about these times cannot be expected.

L7d) Disdained sibs

Nowadays there is no discrimination against the formerly dependent sibs; they are in no way regarded as inferior. Only such people who no longer respect custom and morals and have become impoverished through their own fault are looked down upon. There is one single exception to this rule, which was mentioned by chance by Menkröi Ngarua' and the mother of Karbari Mangpung Pahönca (16.03.1957) and on which they gave me information only very guardedly. Nowadays this case is largely superstition, although I suspect that social reasons originally played a role but could not obtain any details of it.

There are some few houses of the Rahaa sib in the Dömrong area and south of the Bommu. In both areas there exists the opinion that a part of the Rahaa is a "bad sib". The Rahaa are bipartite, divided into the Kim-Rahaa (house R.) and the Pökcheng-Rahaa (fisherman R.). Which part of the Rahaa is the bad one, is forbidden to tell, as otherwise people would get very angry; but anyone who marries a Rahaa girl becomes impoverished and no well-off man gives his daughter in marriage to a Rahaa. Therefore the Rahaa can marry only into poor families with low social standing and have few prospects of improving their material situation and their reputation.

---

[462] Perhaps the name of this sib dates only from this time; *klen-chang-nau* means to receive-help-wishing, i.e., the needy.

At the risk of incurring the anger of the Rahaa, I am inclined to assume that the fishermen Rahaa are the "bad" people. Nowadays their name has nothing to do with their occupation (because they eke out a sparse living on the cultivation of their fields like every other Mru); although the mountain population equates the fishermen (*sailya*) of the Bengali country with the Dom, the Hindu outcasts, I do not dare to decide whether the Mru have been influenced by the Hindus in their contempt for the Rahaa or whether the "fishermen" were formerly Dom who were accepted as tribesmen by the Mru.

L8 The group

L8a) Structure

According to the traditional count, the Mru are divided into five groups: Dömrong, Dopreng, Chüngma, Anok and Tamchaa. Three of these names can be translated: Dömrong Mru = plains people, Chüngma Mru = mountain people, Anok Mru = western people. Dopreng Mru can perhaps be interpreted as slope residents (*do* = downward, *preng* = set out). Since these names are no longer justified by where the groups reside today, they must refer to the areas in which the groups in question formerly lived and therefore still date from the time when the Mru still resided in Arakan. Probably at that time certain sibs were part of every group in its residential area: in theory this is still the case today, in reality, however, members of the same sib are occasionally found in different groups as well. Therefore the group cannot be understood as an association of sibs. Neither is a territorial delimitation possible, because members of different groups can dwell together in one hamlet. The groups have no political significance. The groups are distinguished from each other by single traits of their culture, such the patterns of the covers (*wanma*), details of house construction and primarily the performance of certain ceremonies (for the harvest festival and the feasts of merit, even though differences can also be found within the groups themselves), and finally also by variations in dialect (ia > a in Dopreng, prefix ta- in Dömrong, etc.). If one or a few families of a group settle in another group's hamlet, assimilation cannot fail to take place: the children will learn the new dialect in dealing with their playmates, their fathers will follow the local forms in their ceremonies (since all helpers know and execute only the rules of their own group), only the women show a certain trend towards inertia in their skills: if a cover with an Anok pattern is offered to a guest in a Dopreng house, he can be sure that the lady of the house originates from an Anok sib.

Where two or three groups border on each other, mixed forms will result and on the whole consciousness of group membership can fade to such an extent that the people do not regard themselves as belonging to a certain group and co-ordinate their rules with each other at their discretion (for instance in the Dömrong-Dopreng-mixing zone to the east of Alikodong).

*

The Khumi are divided into two groups: Areng and Awa. Almost all Khumi sibs of the CHTs are part of the Areng group; I did not meet any members of the Awa group. As in the case of the Mru, the differences between the Khumi groups are expressed primarily in special rules for the ceremonies and in dialectical peculiarities. But while the Mru dialects differ little from each other and therefore seldom cause difficulties of comprehension, the dialects of the Areng and Awa are clearly divided into two quite separate idioms with partly independent laws of sound development (cf. my essay on Khami/Khumi vocabularies). I have no material on the inter-relations between Areng and Awa sibs; none of the sibs I have noted in the CHTs was mentioned in the Census of India 1931 (XI, 1:249).

L8b) Group and sib

There are no barriers to marriage between groups (MK, 23.06.1956). Intermarriages cause no problems with regard to group membership: since all children belong to the sib of the father and the wife generally moves in with the husband, the children also belong to the father's group and will follow its rules as adults. No real problems arise even when a family of group A settles in the area of a foreign group B. The acculturation lasts at most two generations, but since nothing changes in the sib membership of the male descendants, in theory they remain members of group A, even though *de facto* they belong to group B.

But also in the general opinion, after some generations cultural membership is given priority over sib membership. This is shown by the fact that, even when not separated by long distances, parts of a sib can be regarded as belonging to different groups. So the Deng living in the Alekhyong-Mouza (Chapai-KP) are classed as Chüngma, while the remaining Dengwüa are regarded as Anok (and Anok sibs also live in Chapai-KP). The Ngaplo living in the Alekhyong-Mouza are also counted as Chüngma, while those of the Galengya-Mouza (Kröilö-KP) are Anok, even though there are also Chüngma (Rümthu) to be found in their immediate neighbourhood (MK, 18.03.1957). The Kilöngca living in Laichia-KP (Thanchi-Mouza) are part of the Chüngma there (Longhu), but Headman Haiwön Macar described the Kilöngca of his village (of Twainfa-Mouza) as Dömrong (21.03.1957).

The best examples of the change of group membership are provided by the sibs of the Rümma south of the Mranca hill range. The real Rümma are the four brother sibs of the Rengwia, Thengmang, Rengkrum and Rengnge (K Chaman Kung 13.03.1957). The karbari equated his own sib, the Kung, with the Rengwia. To Karbari Chöngpat Miyauca (19.03.1957), however, the Kung are not at all to be equated with the Rengwia, but rather, according to Menkröi Ngarua', with the Akungca in the Galengya-Mouza (an Anok sib, which, judging from its name, was perhaps a part of the Khami in the past). According to Karbari Nongneng Krimchang (11.03.1957) the Tou, Ton and Kung are part of the real Rümma, but Headman Chamklaa Tou described the Tou and Ton as Dömrong (10.03.1957). The Krimchang in the village of the Karbari Nongneng follow the Rümma rules. Approximately 100 years ago their ancestors still lived in the area of the Pantola-Mouza, today a part of them can be found in the Ruma-Mouza. The Amthang of the same village are actually Dömrong and some of them live in the Dokhinhangor-Mouza and have been absorbed by the Anok. The Changpröca who also reside in Nongneng-KP are still regarded as Dopreng, however.

Rümthu, Pürca and Kuichengca live in the hamlet of Rümma-Karbari Rengtan Rümthu: all follow the Rümma rule, as do the Pürca and Kuichengca, who are nevertheless considered to be Dömrong, probably since the neighbouring hamlet (Kukcong-HP) is almost completely peopled by Dömrong. The Rümthu are actually part of the Chüngma and migrated from the Pantola area into their present territory (K Rengtan Rümthu 13.03.1957). According to Headman Boli Pürca, the Pürca of the Dömrong are identical with the Tom of the Anok (16.03.1957). Those in the hamlet of Karbari Rengtan Rümthu immigrated two generations ago from the Kornofuli (and were therefore part of the Kaingsa group of the Anok). Karbari Chaman Thangbüca claimed that the Kuichengca (18.03.1957) were not Dömrong, but part of the Pongmi, but the Ruaca Tharngak listed the Kuichengca and Pongmi of Kukcong-HP separately and described them as Dömrong (15.03.1957). The Pongmi are also part of the Dömrong, according to Karbari Mangpung Pahönca (16.03.1957). Incidentally, the feasts of merit in the hamlet of this Karbari (Dömrong) follow the so-called Chüngma-Ri, but there does not seem to be any connection with the Chüngma of the Songu valley.

L8c) Group disintegration and formation

The Rümma group is absent from the traditional count, the Tamchaa group being mentioned instead, but the latter hardly has meaning today. Their main residential area is said to be to the north of the Matamuri, to the east of Lama; I myself have not visited that area, but I met with members of Tamchaa sibs in other hamlets. Karbari Chaman Thangbüca (18.03.1957) described the Pahönca (Pahüaca) living in his hamlet as Dömrong, and Karbari Mangpung Pahönca also said that his sib was Dömrong, without even mentioning the Tamchaa; the Khumi-Karbari Müilün (22.03.1957) gave Tamchaa as his sib name and when questioned specified Pahönca. Headman Boli Pürca (16.03.1957) listed the Tamchaa as one of the Dömrong sibs of his hamlet, but also explained that the real sib name of these Tamchaa was Pachui-Ngongklam. The Pachuica, however, were said to be Dopreng in all Dopreng hamlets. All these details give the impression that the Tamchaa are no longer to be regarded as a separate group, so that the name is no different from that of a great sib (like the Deng), while the former Tamchaa sibs have been divided up between the larger neighbouring groups, the Dömrong and Dopreng.

On the other hand, in this bordering area between Dömrong and Dopreng, the Rümma (forest people) have developed into a new group by assimilation of Anok and Chüngma who immigrated to their area: the 80-year-old Headman Chamklaa Tou still counted his own sib as one of the Dömrong, Karbari Nongneng Krimchang, however, described the Tou as an original Rümma sib.

In the past, the Rümma were probably a great sib of the Dömrong. According to their own tradition, however, they migrated from Boga Lake, which they call Rümma-Kan. This story is the only detailed legend of a sib origin of the Mru, however, and cannot in fact claim any originality since variants of it are also known to the neighbouring ethnic groups. There is no historical probability for a stay of the Dömrong at Boga Lake and the Anok also tell the same story, not of the Rümma, but the Dömrong sib of the Tarca.

While, in the case of the Rümma, the splitting-off of a great sib with assimilation of sibs from other groups led to the formation of a new group, in the case of the Tamchaa, the group was dissolved by being divided up among neighbouring groups.

L9 The ethnos

L9a) Delimitation

Neither Mru nor Khumi have any form of political organisation, on which a single person or group could base a claim to power over all sibs. The *ruaca* were subordinate to the Arakanese king, but even in the past wars might have never taken place between the Mru or the Khumi and their neighbours, for instance the Marma, since there could be no common supreme command. The Arakanese foreign rule therefore neither replaced an indigenous tribal ruler nor did it acquire stable rights over the complete tribal population and was not strong enough to cause a "national" unification. A reorganisation of the previous "administration" was achieved only by English colonial policy, which confirmed the power of the Rajas, supported them and delimited their territory. These boundaries did not take into consideration the distribution of populations but followed geographical lines: the boundary between the provinces of Bengal (Chittagong district) and Arakan (Akyab district) bisects the area inhabited by both the Mru and the Khumi. This involuntary state membership of the individual sections of the population is alien to the tribal spirit of the Mru and Khumi, however: a member of one's own sib continues to have the same rights and duties as any other member from the same area (Krabu-HP, 2.03.1956) and every tribal foreigner remains a foreigner, even though he may live in the same village (Marma in Ta'uk Para 3.12.1956). The most pre-eminent common bond is the language: everyone who speaks Mru or Khumi as his mother tongue is and will remain a Mruca' or a Khumi, no matter where he lives.

However, consciousness of being a community is not based on language alone; many other elements of the culture are regarded as typical of one's own ethnos. Through them a Mru or Khumi knows himself to be more clearly distinguished from his neighbours than a member of a European "people". Although ethnic groups also share a proportionally great number of identical or similar elements, each ethnos has its own peculiarities of material and intellectual cultural heritage to which it clings and sees no reason to exchange them for elements from another ethnos that serve the same end and purpose. Changes of cultural heritage are due less to borrowings from ethnos to ethnos than to the simultaneous acquisition of foreign goods – first Bengali and finally European. Even though these may also replace typical elements of the ethnic culture, they do not disturb the consciousness of separate identity. In order to characterise this consciousness of separateness in greater detail, it must be stated clearly that the barrier against penetration of elements from cultures that are on an approximately equal level, like those of the hill people of the CHTs, cannot be explained only as the result of indifference and the "continuation" of conventional forms or the failure to become aware of a cultural gradient, as a well-adjusted equilibrium which prevents the sliding up or down of cultural possessions, or even more strangely as the result of an inability to act differently. No, in many cases, we have to do with an active desire to avoid change, and the process of maintaining (or surrendering/abandoning) a distinctive cultural feature is definitely an active process, a process of mounting a defence (or adopting one), during which something may have to be sacrificed.

L9b) Typical cultural elements

As an example of reactions due to consciousness of separateness, I shall mention female clothing. The principal garment, the skirt, is produced by the women themselves and is typical of every ethnos in its form and patterns. The Mru women are very shy and restrained outside their hamlet, and most journeys to the market are made by the men. Where the market can be reached easily, however, the women also go there occasionally and then enter an environment where the "decently" dressed wives of the "more highly civilised" neighbours look at them askance, while the Bengali men, who are not used to such sights, do not always conceal their feelings toward these women, who to them are unattainable "barbarians". The result is embarrassment and sometimes they try to hide under a cloth cape, but they have no wish to change their dress, even though it is also inconvenient for them in their daily life (women normally sit on their knees, have to keep their skirt close when passing a seated man and must take care not to be followed by men when they are working in the field). Still: what is a Mru woman to wear, if not her skirt? A mid-calf length skirt and blouse? – She is not a Kuki. A *lungi* and chest cloth or little jacket? – She is not a Marma. A sari? – She is not a Hindu woman. The tribal work jacket? – This is worn only in the field. None of these is a possibility, because they all go directly against the consciousness of separateness. The best option is a novelty: a vest as worn in Europe by men doing gymnastics, which can be bought in the bazaar without knowledge of its original purpose.

In analogous situations the men can also be embarrassed: as was the case once when it was claimed that a film had been shown to the entire world in Rangamati, in which a Mru could be seen, clad only in a loincloth. Among the Twipra I twice met with decided disapproval when I wanted to take a photograph of them wearing the loincloth: they had no objection against my taking a photograph – but only if they were "decently" clad (Hamajon-KP, 23.02.1956, Noliram-RP, 17.01.1957). And for the men a Lungi makes few difficulties for the consciousness of separateness: The Bengalis, the Marma, the Khumi and some of the Chakma, Twipra and Mru wear it; so why not get rid of feelings of inferiority by wearing an "international" Lungi? And yet another example from the area of clothing: Dingte's mother made a new working jacket for him, which turned out five centimetres too long. Only a small flaw? No, strong protest: "I am not a Twipra, after all!"

But it is by no means always only "formalities" that lead to the rejection of an element typical of a parallel culture. A case in point is an episode that occurred during the building of my house in Tapwüa-Kua: my Marma boy claimed that I had been gloomy since the day before yesterday, I would not like the house, and since he was to blame also (he hadn't kept a close enough eye on the work), he wanted to have it broken down again and redone, especially the longitudinal walls, which would have to be put outside (in front of) the posts, while the transverse walls should be inwards. But to Kyo Thwân Ong's disappointment I disagreed on every point and with good reason (because of the wood insects) and favoured the present placement of the walls. Since he could not persuade me to change "my" view, the good boy thought that finally he must have the house changed; he could not stay in such a house, he would die in it; because it had the same form as a dead man's hut among the Marma. Perhaps I could live in it, and perhaps the Mru could, but he could not (30.04.1956). But since he did not get another house, in the end he had to dare to defy destiny under my protection, and left it still alive and in good health. Clearly the principles of one's own culture remain effective even in foreign surroundings.

Generally, however, recognition is refused to the "spirits" of another ethnos for as long as possible, as the following example will show: At the water fetching place of Tapwüa-Kua there is a big stone which is regarded by the Mru as a *nat-hua* (see P2f). Kyo Thwân Ong did not believe these things and when washing clothes always squatted on this stone. However, for some time he had had a *tinea* spot on his leg, which defied all treatment and was even spreading. Finally he began to think that this could not have a natural explanation and Chingkrat's arguments convinced him that it had happened because he always sat down on the *nat-hua*, defying all warnings. So Kyo Thwân Ong wanted to try to propitiate the angered spirit in the Marma fashion with a little rice. However, Menching, who had joined the two, thought he had better not do this: every spirit needs his particular sacrifice and if Kyo Thwân Ong did not do it properly, the illness might get even worse. In this case a *mri-naa* sacrifice (see P2e) was necessary. After long hesitation, Kyo Thwân Ong accepted this (30.09.1956) – and his *tinea* infection was cured.

In these two examples, a basic position of the consciousness of separateness is expressed: although in the end one's own body of thought is denied under the compulsion of "higher powers", the behaviour of "foreigners" is non-committal for oneself, the elements of the foreign culture obey their own laws, which one can know but need not appreciate. The following example shows perhaps most clearly this independence of mind when keeping what a member of the ethnos regards as obligatory for himself:

The Mru have their own way of singing, and all their songs are sung in this form and no other; but this does not mean that they could not sing otherwise and that they would not understand other melodies. Early in my stay in Tapwüa-Kua, people who wanted to have the "song box" demonstrated came again and again and the people of the hamlet also assembled again and again. In order to discover the attitude of the Mru to the music of their neighbours, I also let them hear records of other music from the CHT, although they showed real enthusiasm only for their own ways. One day I heard a girl of approximately ten years singing long passages from the Marma songs for a time: the melodies were also remembered by the rest of the audience, although it did not occur to them to sing a Mru song to this melody: "After all, I am not a Marma!" (cf. also the attitude toward one's own morality).

L9c) Tendencies towards change of ethnic membership

Besides the examples of stubbornly insisting on one's own cultural elements there are also examples of conscious adoption of a foreign cultural possession. Apart from the international waves of temporary fashions, which manifest themselves here too, without affecting the culture as a whole, there are also tendencies to exchange the inventory of one's own culture for another. Here the Southern Chüngma Mru are primarily to be mentioned, who to a large extent have adopted the culture of their Khumi neighbours. They are proud of having made this change and also express this in their own name by not describing themselves as Mru but by using the Khumi word "Longhu" for Mru. In answer to my question on how they could manage to turn away from their old culture, although they still spoke Mru, I was told (in Bangla) "mon to Khumi": "But in our mind we are Khumi" (Uikük-HP, 16.01.1957). Conscious self-perception is therefore decisive in this case too.

Admittedly this borrowing from another culture was made easy by the great similarity of the Mru and Khumi cultures, which has even led to the assertion that the two tribes had the same origin. "They are one family" writes Hughes (1881:11), and to be kept separate from the Chin. How far such data reflect the views of his informants must be left open; but the language and physical habitus obviously speak against such a thesis. However, the cultural relationship, derived from living side by side in the same environment for centuries, may also have contributed to a mutual understanding of the members of the two tribes and led to views (particularly when there is a mutual liking) that promote such ideas. Thus the Longhu clearly approved of their relationship with the Khumi; the Anok, however, who had a more highly-developed consciousness of identity, thought that the similarities were the result of the Khumi trying to imitate the Mru: sacrifices, feasts of merit, musical instruments, etc., which the Khumi wanted to take over from the Mru, though they did not understand the things properly and therefore converted them into a somewhat divergent form, which the Mru thought wrong.

Among the Khumi themselves, particularly during my stay in Tindu-Para (4./10.03.1956), I found a preference for the culture of the Marma (except for the isolated Khumi group in the Betchora area, who inclined rather strongly towards the Mru culture, see L9g). As an example they pointed out to "other" Khumi in Arakan who had already progressed much further in adopting the Arakanese way of living. Some Khumi men had adjusted completely to the Marma in clothes and hair dress and could speak Marma as fluently as their own language, and there was also a remarkable trend to master the Marma culture intellectually. An attempt was made to blur the tribal limits by describing the Marma not as such but simply as "Buddhists"; and it was said of an old Khumi, who could read and write, that he had read all religious books like a Marma priest. Even Karbari Roino, a rich old feast-giver (the "Raja of the hills") who, contrary to the general trend, still walked around in the old dress with a loincloth and proudly showed me his heirlooms, explained to my Marma boy one afternoon (8.03.1956) that he really would like to become a Marma: he had already tried it, and although it would not be so simple, he still intended to do his best in future.

L9d) Endogamy

The ethnos as an (almost) endogamous unity contrasts with the sib as an exogamous unit. But while the idea of a common descent for the sib makes exogamy obligatory, only acculturation opposes a marriage beyond the bounds of the ethnos; although it is possible, it is not usual to marry outside the tribe, because of fear of a foreign partner (regardless of sex), isolation, and the loss of so many features of one's culture to which one is accustomed from childhood. The frequency of interethnic marriages is therefore in inverse ratio to the magnitude of the cultural difference, i.e., for the Mru, the Khumi and Bawm are considered greatly preferable as marriage partners to the Marma or, even worse, the Bengalis. In rare cases this culture barrier can even become effective within the tribe, as in the only case known to me where a Mru man had gone to school and had received a position in the civil service. Girls he had approached refused to leave their native area and accompany a future husband to Rangamati, where they would have been completely isolated from their culture. (The man was Era Khatpö Ngarua, son of the Karbari Thümru.)

Concerning the possibility of getting to know a foreign partner, Mru girls most frequently get into contact with Bengalis who appear in the Mru villages every week or more frequently. Why should life as the wife of a Bengali not have something enticing about it for some Mru girl? But if the cultural barrier mentioned above is effective against anybody, then it is so against the Bengalis. Bengali culture and its representatives are not only considered as strange, but are often viewed almost with hostility, especially as concerns the moral aspect. The Bengalis view the Mru with contempt and the Mru are disgusted by the Bengalis. So not establishing marital relations with Bengalis has, so to speak, become common law: When a case occurred in Thanchi that a married couple addicted to opium wanted to sell their daughter (with her consent) as wife to a Bengali dokandar (shopkeeper), her kin intervened. The Bengali man did not marry the girl, but, because she had had intimate relations with him, she was not rehabilitated either: no decent Mru man would marry her any more. Meeting with a Bengali entails proscription for a Mru girl (MK, 24.09.1956).

Similarly, no Mru man concerned about his reputation will take a Bengali wife, although sexual intercourse is not excluded on a visit to the town. No case was known of such a marriage in the Anok area; on my Dömrong tour, however, a Mru who was employed as sweeper in the police station accompanied me as a porter from Lama. I thought at first that he was of mixed Mru-Bengali ethnicity, but in fact his ancestors had been pure Mru; he himself had once for a short time taken a Bengali woman as wife as he told us in Chamklaa-HP (9.03.1957), but he had soon sent her home again because of her "lack of ability". As said, the man lived outside his tribal community and apparently imagined nothing bad as he told the episode of his Bengali marriage. It did not speak for, but against the Bengali. The reaction of the audience, however, did not leave any doubt about its "esteem" of such a trial, mockery was still the most benevolent response.

I have not followed the marriage relations between Marma and Bengali. According to the observations of Dr. Kauffmann there can be no doubt that in the Marma villages, a favour is occasionally granted to the dealers in exchange for little presents and with the approval of the parents. Also, interethnic marriages do happen, but the fact that there are difficulties here, too, may be shown by the following case which I was told of by a Christian Marma who worked in the paper mills in Chandraghona: An executive of the mills, a Bengali Muslim, courted a Marma girl from a neighbouring hamlet and tried to win her favour with presents. Despite all his advances, the girl continued to reject him until her parents got into serious financial difficulties, with the result that she gave way to the pressure. But she accepted him only on condition that he undertook to care for her parents, herself and her possible children always and in case of a divorce to pay her compensation of ten thousand rupees. The employee consented to this and the contract was sealed in Rangamati and Chittagong. But now his team-mates and supervisors in the Mills took action, and explained to the man that it would be a disgrace for one of the executives to marry a "heathen" while there were plenty of pretty and rich girls of his own race and that he should either expel this woman or quit his job. He refused at first, but finally considerations of financial security were stronger than love and the Marma wife was dismissed and with the necessary tricks the compensation reduced to a few hundred rupees (Rajbila, 20.03.1956). Whether this story really happened as reported is unimportant for the basic position expressed by it.

L9e) Interethnic marriages

According to the patrilocal residence rule of the Mru and Khumi, there is little necessity for a man to do without the surroundings and contacts to which he has become accustomed if he marries a woman from outside. For a woman who moves in with a man of a foreign ethnos, however, this marriage means a far-reaching upheaval and, unless it is concluded in a bordering tribal area, complete separation from her own relatives. According to the general consensus, a greater percentage of Mru men than women take a partner from another ethnos. This may indicate that Mru girls are more closely rooted in their culture than the girls of their neighbours. If a Mru girl gets married outside her tribe, it is almost exclusively with a Khumi man; so for instance both grandmothers of the son of Karbari Nangyu Camthang, a Khumi, were Mru women (Lenten-KP, 17.02.1957).

I met a Khumi woman among the Mru in Thanchi (Uikük-HP, 16.01.1957); and in Müilün-KP (27.03.1957) a Khumi girl wanted to marry one of my Mru companions, an idea which, however, she had to give up since he was already married. As examples of marriages with "Kuki" girls I can mention the Karbari Kröilö Ngaplo whose personal *tutma* (father-in-law people) was Panghoi (Bawm) and Menkröi Ngarua', whose wife Thanvel was Lushai. Marriages with the Marma are extremely rare. In the Anok area, only four cases were known of a Mru man marrying a Marma woman and without exception these were people in prominent positions (missionary, employee, headman).

Contrary to the normal residence rule, in interethnic marriages uxorilocality also seems to happen. I met a Mru in the Khumi village of the Karbari Roino (6.03.1956) who had married a Khumi girl there. U Ba Myaing (1934:127) reports for the Ponnagyun Hills in Arakan of the Mru on the Yo-Khyong: "Some of the Mros are taking Kami (Ayaing) [i.e., Khumi] to wives and talking the language of their wives; while the Kami (Awa) have not done so." Unfortunately it is not indicated whether the Mru men also moved with their wives into the Khumi hamlets. It may also be possible that they had already joined them before their marriage. And finally there are examples in the Chittagong Hill Tracts of whole families taking up residence in a hamlet of a foreign ethnos, too. No objections will be raised when their children then want to marry into this ethnos.

L9f) Acceptance of foreigners

When a Mru man marries a woman from another ethnos, a foreigner is not admitted into the ethnos by the marriage, since this woman is still a foreigner because of her sib membership in the foreign ethnos; her children, however belong, as usual, to the sib of the father and therefore are born Mru. This relationship also remains the same, at least in theory, when the man moves into his wife's tribal area. The individual acceptance of foreigners therefore is (almost) equivalent to the question about admission into a certain sib. Since *de facto* nobody can change his blood relationship, also nobody can *de jure* change his sib and with it his tribal membership. But while women of whatever descent, by the ceremony of the *chuang-pe chak* (see M4d), are not incorporated, but "bound" at the wrist (by the *bong-kom*) to the sib of her husband, for strange men it is impossible to achieve incorporation by any ceremonial, even in case of uxorilocality. The only exception is formed by the so-called *u-toi-ca*, children who accompany the mother. If a Mru girl marries a tribal stranger, has children with him and, according to the rules of the other ethnos, can come back in company of her children to her parents after separation from her husband (by his death or by divorce) the children are also "bound" through a *bong-kom* to their grandfather's sib on the mother's side (MK, 24.09.1956). By this they are accepted as Mru, but still belong, according to Mru reckoning, to their father's sib and therefore remain half foreigners, so to speak. But if they continue to live among the Mru the feeling of the strangeness of their descendants will be lost as time passes. The process leads to the formation of a new Mru sib, however.

The majority of "foreign" sibs within the Mru and Khumi, however, probably cannot be explained by children who went with their mother; because foreigners who marry in or only settle in Mru hamlets can also be incorporated with time. Although this runs counter to all rules and therefore cannot be achieved overnight, the process already shown for the change of group membership also permits the assimilation of such foreign elements over the course of several generations. Hutchinson's statement (1909:43): "The Mros will admit no stranger into their tribe" therefore has to be revised.

L9g) Assimilation of strange sibs

The only reference to the foreign origin of a sib remains its name. In many cases a single person might have stood at the beginning of a development of which today only the result is ascertainable; but whole groups have also changed their ethnic membership. For some reason they moved to the area of the other ethnos – nowadays the reason may be a mere liking for the other group (or antipathy against one's own community), but also the need for more or better fields; in former warlike times, the slave economy and the invasion of foreign tribes also played a role –, settled down in their hamlets, adapted more and more to the local customs and after some generations the differences between the locals and the immigrants disappeared and were finally forgotten. Genealogical knowledge seldom reaches beyond the great-great-grandfather: in cases of isolated resettlement the feeling of not belonging is lost approximately by the third generation, in the fifth one perhaps still knows that grandfather's grandfather came from another area, in the seventh one at most still wonders why there is the same sib name in two ethnic groups. So the Khumi-Karbari Muilün Tamchaa, still knew that an ancestor in his male line more than one hundred years ago was a Mru; the Khumi-Karbari Kangnong Kanbeo', however, had no knowledge of a relationship with the Mru sib of the Kanbök (24.03.1957) (bök > beo-', according to the sound rules of the Khumi).

The larger the outside group, the slower the incorporation: the four sibs of the Rengmitca (three of them noted by me: Akhuaca, Khongwoica, Macar) must already have separated themselves more than 200 years ago from the Khami (Awa), but they gave up their own rules of performing sacrifices only two generations ago, and as well as using the Mru language they have preserved (fragments of) their own dialect at home to this day, but have no memory of their origin or any knowledge about its relationship with the Khami in Arakan. The first signs of the loss of their own culture as a result of isolation were shown by the Khumi of the Betchora-Mouza: for cattle feasts, for instance, they had to invite Mru chapels, since they no longer knew how to perform their own type of music. Communication with the Mru did not cause any difficulties, since the Betchora Khumi already learned Mru from childhood as well as their mother tongue (while the Mru made no effort); interethnic marriages were not rare here either (see above) (Karbari Lenten Camthang 17.02.1957). Some Khumi of this area seem to have been incorporated in the Mru a long time ago, because there is a sub-sib of the Dengwüa in the neighbouring Renikhyong-Mouza called O-püi-Khommi Deng ("down-river Khumi") (MK, 14.09.1956). I do not know why in this case Khumi could be subsumed under a Mru sib name: unless we are dealing with a nickname for a sib originally affiliated to the Mru or with the descendants of sons "who went with the mother", in the past there must have been a possibility of adoption that is forgotten today.

Another question is posed by the M(a)ranca: the hill range South of the Matamuri is called Maranca Hung, according to Karbari Menyeng, because the Mru sibs of the Maranca lived there (19.03.1957). In the past, the Maranca are said to have been actual Mran (i.e., Marma), who, however, later became Mru. To them belonged the Pakönca and the Pacongca. The fact that nobody could give me any other information on the Maranca, need not necessarily speak against the story, since Menyeng was a man of venerable age and therefore perhaps one of the few who still knew. On the other hand, the names of the two sibs mentioned are obviously of Mru and not of Marma origin, and the Pacongca were classed by my informants once as Dopreng, once as Dömrong. That, however, the Marma can also change their ethnic identity in order to assimilate themselves to a hill tribe was confirmed by Headman Calkhup for the Bawm (20.04.1957). Some Mru and Khumi also have become Bawm, and former Mru are said to make up half of the population in some hamlets. This information I got from Menkröi Ngarua', who himself had lived among the Bawm for a year; my attempts to obtain further details by my own investigations failed in the presence of their headman because of the refusal of the Christian Bawm to remember sib names abolished as heathen heritage. Particularly within the last few years, the introduction of the Christian lifestyle has advanced the process of assimilation with the Bawm; thus any member of a foreign tribe who confesses the new religion and settles down in a Bawm village is appreciated as a member of the community, his sib membership is annulled and no longer plays any role in the new community of religion and culture. – This at least was the picture that Headman Calkhup wanted to portray for me on my first visit. Only a check-up in 1964 brought the truth to light: Menkröi's statement was not correct (there were only two Bawm sibs that could be identified as Mru according to my name list (Buiting [<Büiting] and Demrong [<Dömrong]), and the picture drawn by Calkhup did not correspond to reality either: the sib names (altogether 30 in number) were still known and the headman only wanted to abolish them because the names reflected the old social stratification (by which his own sib was identified by its name as descendants of Khumi immigrants). And nobody became a member of the Bawm merely by immigration and acceptance of the Christian faith (see Spielmann 1968).

L10 Relationship terminology

L10a) Mru terms

To show the scheme and system of the relationship terms of the Mru, I start with the relationship terms familiar to us for the basic relations. I abbreviate them using English terminology as follows:

F = father, M = mother, B = brother, Z = sister, H = husband, W = wife, S = son, D = daughter. Elder and younger brothers and sisters are distinguished by e = elder and y = younger; m stands for male and f for female ego; when the sex of ego (the speaker) does not play a role or is obvious as in the case of H and W, ego is not marked. When the terms to be used vary according to sib relationship, (t), (p) and (a) designate *tutma, pen* and own (or *tainau*) sib. Forms of address are in brackets; when they do not differ from the form of reference I have just put an exclamation mark, when, however, the use of the personal name is usual, an (N). An asterisk (*) means that an unmarried female ego uses the same term as her brother for describing the relative in question, after marriage, however, the same term as her husband. In the following scheme, ordered by generation and relationship degree, the positions will be found determinable via the male primary relatives on the left side, those via the female primary relatives on the right. The numbers in the right margin refer to the reciprocal terms after the serial number in the left margin.

For the great-grandfather generation the terms of the grandfather generation recur, for that of the great-grandchildren that of the grand-children.

1)	FF, FFB	pu	(!)	MM, MMZ	Pi	(!)	(71
2)	FM, FMZ	pi	(!)	MF, MFB	Pu	(!)	(72
3)	FFZ	nako	(!)	MMB (p)	tarang	(!, N)	(77, 78
				MMB (a, t)	pu	(!)	
4)	FMB	pu	(!)	MFZ	Pi	(!)	(75, 76
5)	FFZH	tarang	(!)	MMBW	Pi	(!)	(79, 80
6)	FBWF	pu	(!)	FBWM	Pi	(!)	(81, 82
7)	FZHF	tarang	(!)	FZHM	nako	(!)	(81, 82
8)	HFF	tarang	(!)	WMM (t)	pi	(!)	(73
				WMM (p)	cu	(!)	
9)	HFM	nako	(!)	WMF (t)	pu	(!)	(73
				WWF (p)	tarang	(!)	
10)	HMF	nako	(!)	WFM	Pi	(!)	(74
11)	HMM	nako	(!)	WFF	Pu	(!)	(74
12)	F	pa	(apa)	M	U	(a'u)	(57
13)	FeB	pama	(ma)	MeZ	pama	(ma)	(62, 63
14)	FyB	pahi	(!)	MyZ	uram	(!, u)	(62, 63

15)	FeBW	pama	(ma)	MZH (a) e than F	pama	(ma)	(66, 67
				MZH (t)	Pu	(!)	
				MZH (p)	tarang	(!)	
16)	FyBW	uram	(!, u)	MZH (a) y than F	pahi	(!)	(66, 67
17)	FBWB	pu	(!)	MZHZ (a)	nako	(!)	
				MZHZ (t)	pi	(!)	
				MZHZ (p)	cu	(!)	
18)	FZ	nako	(!)	MB	pu	(!)	(62, 63
19)	FZH, FZHB	tarang	(!)	MBW, MBWZ	pi	(!)	(68, 69
20)	FZHZ	cu	(!)	MBWB (t)	pu	(!)	(68, 69
				MBWB (p)	tarang	(!)	
21)	HF, HFB	tarang	(!)	WM, WMZ	pi	(!)	(58, 64
22)	HFZ	nako	(!)	WMB (t)	pu	(t)	(64
				WMB (p)	tarang	(!)	
23)	HFZH	tarang	(!)				(70
24)	HM, HMZ	nako	(!)	WF, WFB	pu	(!)	(58, 65
25)	HMB (a)	s. 13,	(!)	WFZ	pi	(!)	(65
	HMB (t)	14	(!)				
	HMB (p)	pu	(!)				
		nako					
26)	HBWF	pu	(!)	WZHM (a)	s. 13,		(61
				WZHM (t)	14	(!)	
				WZHM (p)	pi	(!)	
					nako		
27)	HBWM	pi	(!)	WZHF (a)	s. 13,		(61
				WZHF (t)	14	(!)	
				WZHF (p)	pu	(!)	
					tarang		
28)	HZHF	tarang	(!)	WBWM (t)	pi	(!)	(60
				WBWM (p)	nako	(!)	
29)	HZHM	nako	(!)	WBWF (t)	pu	(!)	(60
				WBWF (p)	tarang	(!)	
30)	BWF, BWFB	pu	(!)	ZHM, ZHMZ	nako	(!)	(58, 59
31)	BWFZ	pi	(!)	ZHMB	nako	(!)	
32)	BWM, BWMZ	pi	(!)	ZHF, ZHFB	tarang	(!)	
33)	BWMB (t)	pu	(!)	ZHFZ	cu	(!)	(59, 58
	BWMB (p)	tarang	(!)				
34)	H	hua	(S's F)[463]	W	machi	(S's M)	(34
35)	HeB	taima	(tai)	WeZ	taima	(!)	(45
36)	HyB	nauma	(!)	WyZ	nauma	(!, N)	(44
37)	HZ	tuama	(N, !)	WB	nai	(!)	(44, 45
38)	HZH	nai	(!)	WBW	naicha	(!)	(38

[463] When there is no S, by (N).

					rma		
39)	HeBW HyBW	taima nauma	(!) (!)	WZH (a) WZH (t, p)	s. 40, 41 nai	 (!)	(39
40)	meB	taima	(N, tai)	feZ	taima	(N, !)	(41
41)	myB	nauma	(N)	fyZ	nauma	(N)	(40
42)	feB	rak	(tai, N)	meZ	tuama ta	(N)	(43
43)	fyB	rak	(N)	myZ	tuama	(N)	(42
44)	eBW	taima	(!)	meZH feZH	nai taima	(!) (!)	(36, 37
45)	yBW	nauma	(!)	myZH fyZH	nai nauma	(!) (!)	(35, 37
46)	mBWB	nai	(!)	fZHZ	(N)		(49
47)	fBWB	pu	(!)	mZHZ	cu	(!, N)	(47
48)	mBWZ	capenm a	(!)	fZHB	capen ma	(!)	(48
49)	fBWZ	–	(N)	mZHB	nai	(!)	(46
50)	FBS, MZS[464]	s. 40– 43		MZD[7], FBD	s. 40– 43		(50
51)	mFZS	nai	(!)	fMBD	–[465]	(N)	(54
52)	fFZS	capenm a	(!)	mMBD	capen ma	(!)	(52
53)	mFZD	cu	(!, N)	fMBS	pu	(!)	(53
54)	fFZD	–	(N)	mMBS	nai	(!)	(51
55)	SWF	nai	(!)	DHM (t, p)	naicha rma	(!)	(55, 56
56)	SWM	naichar ma	(!)	DHF	nai	(!)	(56, 55
57)	S	capa, ca	(N)[466]	D	cama	(N)	(12
58)	SW, SWZ	yauma	(!)	DH, DHB	mokca	(!)	(21, 24, 30, 32
59)	SWB	talung	(N, !)	DHZ	cu	(!, N)	(30, 32
60)	SWBW	talung ma	(!, N)	DHZH (p) DHZH (t)	mokca talung	(N, !) (N, !)	(28, 29
61)	SWZH (a) SWZH (t) SWZH (p)	taimaca talung mokca	(N) (N, !) (N, !)	DHBW	cu	(!, N)	(26, 27
62)	mBS fBS	taimaca *	(N)	mZD fZD	cu *	(N)	(13, 14, 18
63)	mBD fBD	taimaca ma *	(N)	mZS fZS	mokca *	(N, !)	(13, 14, 18
64)	mBSW	yauma	(!)	mZDH (t)	talung	(N)	(21, 22

464 Can be classified, if not (a), as (p) under 51–54 left and as (t) under 51–54 right.

465 u can be used.

466 (pa) and (u) preferred for small children

	fBSW	*		mZDH (p)	mokca	(N, !)	
				fZDH	*		
65)	mBDH	mokca	(!)	mZSW (a)	taimac	(N)	(24, 25
				mZSW (t,	a	(N)	
				p)	cu		
66)	HBS	taimaca	(N)	WZD (a)	taimac	(N)	(15, 16
				WZD (t)	ama	(!, N)	
				WZD (p)	yauma	(N)	
					cu		
67)	HBD	taimaca	(N)	WZS (a)	taimac	(N)	(15, 16
		ma		WZS (t)	a	(N, !)	
				WZS (p)	talung	(N, !)	
					mokca		
68)	HZS	mokca	(N, !)	WBB	yauma	(!, N)	(19, 20
69)	HZD	cu	(N)	WBS	talung	(N,!)	(19, 20
70)	HZDH (p)	mokca	(N, !}	WBSW	talung	(!, N)	(23
					ma		
71)	SS	cu	(N)	DD	cu	(N, !)	(1
72)	SD	cu	(N)	DS	mokca	(N, !)	(2
					, cu		
73)	SSW	yauma	(!, N)	DDH (t)	talung	(N, !)	(8, 9
				DDH (p)	mokca	(N, !)	
74)	SDH	mokca	(!, N)	DSW	cu	(N, !)	(10, 11
75)	mZSS	mokca	(N, !)	fBDD	*		(4
76)	mZSD	cu	(N)	fBDS	*		(4
77)	mZDS (a, p)	cu	(N)	fBSD	*		(3
	mZDS (t)	talung	(N, !)				
78)	mZDD (a, p)	cu	(N)	fBSS	*		(3
	mZDD (t)	yauma	(N, !)				
79)	HBSS	cu	(N)	WBSD	yauma	(!, N)	
80)	HBSD	cu	(N)	WBS S	talung	(N, !)	

L10b) The Mru system

The apparent complications which appear when we try to translate the kinship terminology of the Mru into our language largely disappear when we look at them in their own order given by the sib structure. The following scheme is based on the agnatic sib line: Brothers and sisters of a dissimilar sex are connected by bows (where different terms are available for brothers and sisters of the same sex but dissimilar relative age, they are written under each other), partners are connected by direct lines over the sib delimitation. The scheme is valid firstly for an ego of male sex, but secondly also for an unmarried ego of female sex (for the scheme of a married woman see below). Starting from the ego, the expression for any relative classified by the system can be read in any arbitrary order by following the lines connecting the individual relationship degrees; however, not always in any arbitrary direction, but only in the direction of the arrows, which means that for instance ZHMB or WFZ are not to be identified as a FB or MZ. The terms for one's own sib are also used for the members of the brother sibs.

All terms for distant relatives, who are unequivocally determined by the classification system, like for instance FMBSSD, SDHFFM, etc., but were not cited in the first list also can be gathered from this scheme. The descendants of the sisters of the relatives of the father-in-law sibs are not, however, classified from the start, since this is decided by the sib membership of the men only. When analysing the scheme its harmonious construction is striking: if we regard the cross formed by the grandfather grandson line of one's own sib and the brothers' and sisters' cousin row as middle axes, then the construction of the six terms on the right above corresponds to those on the left below, and that on the left above those on the right below. The only non-harmonious element of the system is the case *machi/capenma* by which the distinction is made between marriageable cousin on the one hand and the wife and her sisters on the other hand.

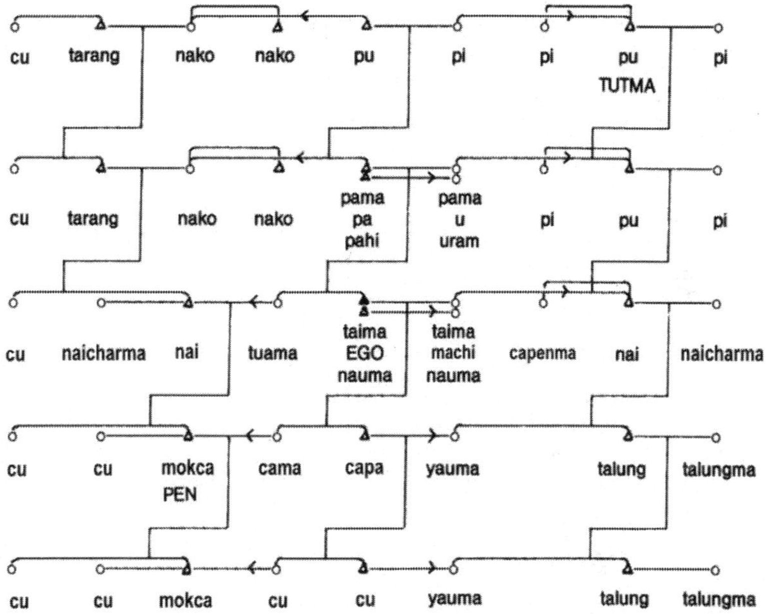

Compared with this closed system, that of a married woman seems somewhat more complicated, since regarding the younger generation she orients herself on the system of her husband, so that the *pen-tutma*-groups shift for her, the children of her brothers admittedly belong to her sib of origin, but are now terminologically to be classified as her *tutma*. The reorientation leads to "rejecttions" only regarding the female (classificatory) "cross-cousins" [entered in the system as (*)]: for a married woman's MBD and FZD there are no determinable terms, the relatives treated as equivalent to them are called as follows: (like the brother) eBW = *taima*, yBW = *nauma*, SWM = *nai-charma*; (like the husband) HZ = *tuama,* DHM = *nai-charma*. Like her brother and her husband the married woman calls SWF and (not entered into the scheme) DHF as *nai* and their wives as *nai-charma*.

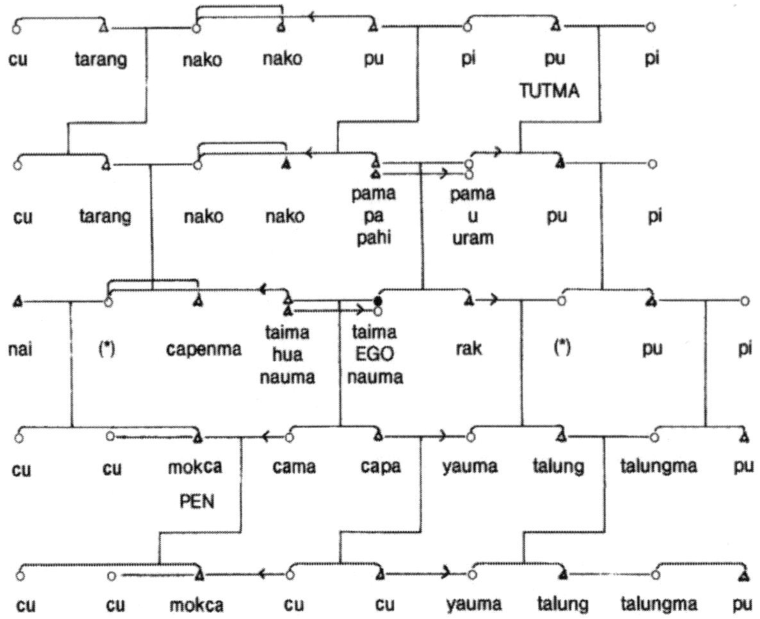

L10c) The kinship system and the sib

The structure of the patrisibs largely determines the application of the kinship terms. Thus all descendants of a man get a clear kinship name by their sib membership. The children of brothers describe themselves again as brothers; the husband treats his brothers' wives as equivalent to the sisters of a wife; there are no terminological differences between the parents-in-law, the brothers-in-law and the children-in-law of a man and his brothers. All relatives via sisters, however, can be further differentiated by the sib regulation; a woman's sisters' children for instance can be either *pen* or *tutma* for her; they would then only be terminologically equivalent to her own children if their fathers belong to the same sib as her husband. Moreover, the terms for all more distant female relatives are determined by the sib membership of the husbands. Therefore the terms for the *tutma* relatives still extend over the real sib limit; there are also correspondingly specific terms for the wives of the *pen* men provided that they were not taken from ego's sib. These are *nako, nai-charma* and *cu*, as wives of *tarang, nai* and *mokca.*

On the other hand, the membership of a woman in a sib hardly ever determines the term to be used for her partner: the only exception is MZH. But already MZS and MZD are again classified according to the sib membership of their father and therefore for ego can also be *capenma* (potential marriage partner). The special position of the mother's sister, however, is expressed not only by the fact that her husband is classified after her, but also by the fact that she forms the only exception in the marriage regulation by sib membership: although she is a *tutma* woman, ego cannot marry her (like also his own mother).

The second strong female position is the *nako* (FZ, HM, etc.). Provided that her brother does not belong to ego's sib, he (as *pen* or *tutma*) is not called *tarang* or *pu,* but by the the same term as his sister, namely *nako*. For the married female *nako* the reciprocal term is *talung* and *yauma*, for the male *nako* the reciprocal term is *cu* without consideration of the sib relationship. This, however, is not a breaking of the system, since the real reference axis lies between the husband and his *pu* (MB etc.) of whom the *pen* position of his *mokca* also determines for him that of the female persons (*cu*) connected with him. Hence married women, according to the terminology principle, belong only partly still to their birth sib. However, it would be a mistake to infer from this something like a "clan consciousness" (clan here understood in Murdock's sense): the same principle is also extended to the sisters of these women, who can be allied to completely different sibs.

L10d) Classificatory application

With few exceptions, the listed kinship terms are classifying, i.e., they serve as the term of more than one kinship degree. I, however, understand by "classificatory" the extension of the terms to persons and foreign partners who cannot be proved to be kin. Insofar as the terms can be used for all members of one's own sib and of all brother and brother-in-law sibs, the Mru system may be called classificatory, though in everyday usage a reduced system is used. So a man of approximately the same age as oneself of one's own or a brother sib is always called *tai-nau*, i.e., one does not distinguish between an elder and a younger brother, *taima* or *nauma*; besides *nai* the terms most frequently used are *pu, pi* and *mokca*, but otherwise relatives who are not related by blood are addressed mainly with their proper names, a use in which even children do not make an exception when addressing old people. Rules for a classificatory application of kinship terms for members of known sibs therefore cannot be given. This applies even more for communication with foreigners. If one meets an unknown person with whom one wants to open a conversation, one does not start it with remarks on the weather or something like that but asks for the sib of this person and tries to find out in which relationship it stands to one's own. If one has no success (which may happen in distant areas in view of the large number of sibs), a possible address is *tainau* (brother), however, as a rule *nai* (brother-in-law) is preferred (as also to tribal foreigners). Even when Bangla or Marma are used (both use as a general salutation the word for "brother", *bhai* or *ko*), one uses the corresponding expressions for brother-in-law (*shala* or *yokpha*). The equivalent for brother-in-law is also used when calling an unknown person. Children can be addressed by "*di ö*"; "*di*" expresses special affection and is also used in love songs. Persons who have any official position are addressed by their title (*Karbari, Hermaing* [= Headman]), but when they are related to the speaker also by their kinship name. As far as I could ascertain, these statements also apply to the Rengmitca and Khumi (using the kinship systems in question). I was unable to observe among the Mru the use of *pu* to express a special honour, as in the Bawm *ka pu* "my *pu*" ('uncle'), also used for younger persons (Headman Calkhup, 20.04.1957).

L10e) The kinship systems of the Rengmitca and Khumi

The kinship systems of the Rengmitca (Khami) and Khumi show a close structural relationship with the system of the Mru, although the terms themselves are only partly related. Deviations in the form of splitting and drawing together of the terms for single kinship degrees do happen. Unfortunately, I have failed to clarify an important question, namely, whether with the Khumi (as in the case of the Mru) the terms for mother's sisters' children depend on the sib membership of the father or not. In the following table these terms are treated as if a dependence of the sib membership of the father were not decisive. (This is the probable regulation since, contrary to the Mru, among the Khumi MZD cannot be married.)

Within my listings there are some differences, in a few cases, between the southern real Khumi area and the northern (isolated) group of the Khumi in the Betchora area; I mark the latter by *N:* and indicate in brackets if the term was taken over from the Marma. The Rengmitca system I got from Headman Haiwön Macar and his wife (21.03.1957), the Khumi system from Headman Inghoi Amchang and relatives (22.03.1957) as well as from the wife of Karbari Müilün Tamchaa (26.03.1957) and the terms of the northern Khumi primarily from Karbari Lenten Camthang (17.02.1957). I have incorporated other data found in the literature: The first, more detailed list is given by Webb (Census 1911), though the list contains a mistaken identification: the terms identified by Webb as "Mro" are from the Awa-Khami ("Rengmitca" of Arakan), those called "Kami" from the Areng-Khumi. The lists are not free from faults: indications like HB, HBW NaPe (*napi* = younger siblings) and HBS, HBD YaAt (*ya'ai* = elder siblings) are not fitting; but the possibility that the women also call the children of their siblings *napi* cannot be excluded, because of parallels in the Lakher system. U Ba Myaing (1934) copied Webb's list (without mentioning the source), corrected the identifications, but reduced the number of terms; the words themselves were corrupted by various misspellings (amongst others *u* and *b* occasionally have replaced *ee* und *a*). The data of U Ohn Pe On may be based on his own notations, however. His Awa Khami correspond with the Rengmitca system, his Ayaing Khami with that of the (Areng) Khumi. In the following list, the data of the two sources are marked with *W:* for Webb and U Ba Myaing and *U:* for U Ohn Pe. Wherever possible I have given a correction of the transcription in brackets. The terms collected by myself have been underlined.

Further terms can occasionally be found in other glossaries. For purposes of comparison data were cited from Phayre (*P:* for the Khumi, similar forms with Hughes and Campbell), Houghton (*H:* for the Khami, under Rengmitca), and Stilson (*S:* for the Khimi, under Rengmitca). With Shafer, moreover, (Saptawka's Khimi) the following forms are still found: *kang-sih* "forefathers", *kang-bäu-zäai* (*kang-böu-ya'ai*) "brothers", *ti-chaw* (*tico* < **tui-co*), "sister" and *ling-ngaw* "son's wife". For a better understanding, it should be noted that in Khumi, final -a changed to -ô.

In addition to the symbols noted under L10a, for the sake of brevity in the following list will be used: P = parents and C = children.

	Rengmitca-Khami		Khumi	
PF, WPF,	pa'ui	(apu!)	chi, *N*: ata	(!)
HPF	*W*: PaWe (pa'ui)		*W*: Aumsee (amchi)	
	U: pi-ee (? = PM)		*U*: na-si	
PM, WPM,	pi'i	(api!)	chi, *N*: ata	
HPM				
	W: PaE (pa'i)		*W*: Aumsee (amchi)	
	U: pi-e (pi'i)		*U*: na-si	
F	pa'ai	(apa!)	nga'ai, *N*: ya'ai	(ai-
	W: PaAt (pa'ai)		*W*: NgaAi (nga'ai)	o!)
	U: pa-è (pa'ai)		*U*: nga-an (nga'ai)	
	P: pa-ai		*P*: ampo	
FeB, MeZH	puma		mô	
(FB, MZH)	*W*: PaAtKho (pa'ai-		*W*: FB NgaAiHmo (nga'ai-	

	kho) *U*: pat-kho (pa'ai-kho)		hmô) *W*: MZH NgaAtKhi
FyB, MyZH	Ngakpi		acö
M	na'oi *W*: NaWe (na'oi) *U*: nè *P*: na-oi	(a'u!)	nei *W*: Neik (nei') *U*: nè *P*: amnu
MeZ, FeBW	puma *W*: NaWeTa (na'oi-ta)		mô, *N*: tô *W*: "MZ" Niketo (nei'-tô) *U*: "aunt" nein-hmoon (nei-hmô) *U*: "uncle" nay-to (nei-tô)
MyZ, FyBW	ata *W*: NaWeTa (na'oi-ta)		nei-cö , *N*: tô W: "FBW" ahNiTo (s. MeZ)
FZ, HM, FZHZ	napi *W*: NaE (na'i) *U:* na-i		(am)ni *W*: AhNee (ani)
FZH, HF	prang-pri *W*: PukPre (pak?-pri) *U*: ("great-uncle") pat-pri		prang *W*: prun (prang)
MB,WF	pa'ui *W*: PaWe (pa'ui)	(apu!)	pu (thus also HMB) *W*: Umpu (ampu) *U:* apoo (apu)
MBW,WM	pi'i *W*: PaE (pa'i)		pi *W*: Umpe (ampi)
meB, feZ	ya'i *W*: YaAt (ya'ai) *U*: ya-ah		yai *W*: Yike (yai') *U*: yaik (yai')
myB, fyZ	nau-pi *W*: NaPe (napi) *U*: na-pi		be *W*: Umbai (ambe) *U*: ah-kè (akei?)
mBW, fZH, WZ, WZH, HB, HBW	ya'i / nau-pi		yai / be *W*: Umbai (ambe)
fB	prak-pri *S*: pro-pri		(am)prô
mZ	ateca *S*: si-sa		tuicô
mFZS, mZH	khui-nai *W*: MaukPe (mok-pi) *H*: "B-in-law" mapi *U*: "cousin" nat-kaung (nai-kong?)	(khoi!)	naica *N*: yopha (< Marma) *W*: ZH Natsa (naica) *W*: FZC MoukSa (mou'ca = ZS) *U*: "cousin" naisa
mMBS,WB	khui-nai *W*: WB PaPre (papri = fB) *W*: MBC MaKon (ma-kung) *U*: nat-kaung (nai-kong?)	(khoi!)	naica, *N*: yopha (< Marma) *W*: Natsa (naica) *U*: nai-sa
mFZD, fMBS	lu-khin		lu-khün
fFZD, HZ	ateca *W*: HZ YaAt (ya'ai = eZ)		tuicô *W*: TokeSo (teu'cô)
mMBD,	niphungca		thaihu

mMBD, fFZS	niphungca	thaihu
fMBD, fBW	? *W*: BW KaMuk (kama') *U*: "Z-in-law" kama	mau *W*: BW AMoke (ameu') *U*: "Z-in-law" amauk (amau')
MBSW,WB W	naima U: "Z in-law" kama	mau *U*: "Z in-law" amauk (amau')
SWF, DHF	khui-nai *W*: SWP NatKaing (nai-keng?)	*N*: sama' (< Marma) *W*: SWP Natsa (naica)
H	tongpa *W*: NaPa (napa) *H*: na-pa	yawô *W*: YoOh (yawô)
W	tongpui *W*: NaPwe (napui) *H*: na-pwi, ni-pu	ayu *W*: Ahyu (ayu)
S, mBS, fZS	capa *W*: SaPa (capa), but fZC Nape (napi = fyZ)	capô *W*: SaPo (capô), jedoch fZC KaTu (katu = DC)
D, mBD, fZD	canu *W*: SaNu (canu), vgl. S	caniü' (thus also ZSW) *W*: SoNoke (coneu'), *cf.* S
fBS,WBS, SWB	linga (ling-nga?) *W*: WBC PaPre (papri = WB) *W*: fBC NaPe (napi = fyZ, myB)	leimô (leing-mô ?) *W*: WBC MokeSa (meu'ca = DH) *W*: fBC Umbai (ambe = fyZ, myB)
fBD, WBD, SW	linga (ling-nga?) *W*: SW NaSein (? Instead of naling?)	nga'oi, *N*: ngô'oi *W*: fSW LinNgo (leingngô) *W*: mSW SinNgo
mZS, HZS, DH	mokca *W*: DH MaukSa (mokca)	meo'ca *W*: DH, mZC MokeSa (meu'ca) *W*: HZC Umbai (ambe = fyZ)
mZD, HZD, DHZ	lu-khin *W*: mZC, HZC LoKhe (lukhin?)	lu-khün *W*: s. mZS
SC, DC	lu-khin *W*: TokPe (tu'pi), also WZC	katu *W*: KaTu, also WZC
Tainau	Tainau	yaibe, *N*: amnau
tutma	makung *W*: MBC MaKon (ma-kung).	pakiüng
pen	Pen	theo'

L10g) Comparative remarks

From the terms given by Webb, which do not fit in the system, no definite conclusions can be drawn. I am inclined to judge the majority of the terms as mistaken notations; there is no reason, for instance, why fZC should be equated once with the younger siblings, once with the grandchildren, while *katu* is less likely to describe WZC than fZC. Also the equal classification of all children not belonging to one's own line given by Webb for the Khumi (*mokca* for the man *ambe* for the woman) does not seem to be probable.

Some terms still have to be added from the literature for the Mru of Arakan. U Ba Myaing gives *MawMa*, fZH (probably mistaken for HZ) and *YawMa*, myZ (perhaps just a misprint for NawMa, i.e., *nauma*), U Ohn Pe, however, gives *YawLa*, Z-in-law, i.e., probably in the meaning of Khumi *mau*, for which a term is missing among the Anok Mru.

Among the Khumi, too, *mau* is in retreat; HZ normally is called by a term taken over from the husband's terminology: *tuicô*, Z, but also fFZD can be designated as *tuicô* – admittedly only according to data from Müilün-KP, a hamlet situated close to the Mru. On the other hand, DHM is still called *mau*, and this term is also used by the man (instead of *tuicô*), a use that might be based less on an adoption of the woman's terminology than on the reciprocity of the term. Possibly, however, the term *papri* for WB (extended to WBS), as reported for the Awa Khami, was taken over from the woman's terminology (*papri* = fB, Rengmitca *prakpri*).

The Khumi system largely lacks the extension of the terms over several generations characteristic for the Mru system. Only under an obvious influence of the neighbouring Mru *pu* can also be used as a term of address for fMBS (*lu-khün*). Contrary to the Mru system, the terms for the grandparents and grandson, *chi* (*N: ata* must be of younger date, since -a did not become -ô) and *katu*, are extended over the whole breadth of the generation. Overall, therefore, the Khumi system shows a form less influenced by the sib structure than the Mru system.

I shall desist here from a linguistic examination of the terms, but a reference must be made to the significance of the content of the terms as regards the brother-in-law sibs. Mru *tut* means "under, below" (*ching-tut* is the trunk down to the roots of a tree), *tutma* therefore are those at the base; *pen*, however, also means "give birth". Khumi *kung* analogously refers to being behind, *theo*' (< **thok*), however, to coming out. Thus one nearly could translate the terms by ancestors and descendants; correspondingly, the Mru *pu* and *pi* (at the same time "grandparents") in the *tutma* area and *cu* (at the same time "grandchildren") in the *pen* area are extended over three generations. However, these correspondences do not apply to the Khumi system. The terms receive a special meaning only when one takes into account that for Mru children *tutma* first of all are father's *tutma*, i.e., mother's parents, and therefore, on account of the reciprocity, *pen,* too, must refer primarily to the children of the daughter. So arises an apparent matrilineality, in which the "ancestors" are a woman's mother's parents, the "descendants", however, are her daughter's children. Viewed in a field of three generations, *tutma* can be seen as the parents of a woman as a marriage partner and *pen* the children of a woman as a sibling. Only in their application to the man's terminology system do the terms receive their contents as "givers of women" and "recipients of women".

M Marriage and married life

M1 Premarital relations

M1a) Behavioural pattern

Making love between the sexes starts with the beginning of pubescence, which transforms *ngiawüa* (children) into *klangwüa* and *malaawüa* (young boys and young girls). Similar playing before maturity is not taken seriously although it shows that the basic principles of sexual intercourse are known to the children, too. It is part of the natural behaviour of the domestic animals, which everyone can see daily. On the other hand, the childlike lack of inhibition in these matters fades noticeably with the onset of maturity and the young girls in particular begin to feel embarrassed in all possible situations. This embarrassment, as shown particularly by Anok girls, is natural only to a certain degree, however; observations among the other Mru groups and the Khumi prove that similar behaviour patterns are found there, too. This should not be taken to mean that their embarrassment is merely an arbitrary sham, a behaviour pattern into which the girls are enculturated, depending on the ethnic group, it rather seems so obligatory that an attempt to act contrary to it is not understood as a lack of embarrassment but as impertinence and is bound to fail because of the internal opposition of one's own people. The reactions shown in front of me admittedly are atypical to a large extent because I was not only a stranger but also differed completely in my appearance from the usual; but I found my conclusions largely confirmed by observations and remarks on the behaviour of the Mru among themselves.

Basically, the restraint shown by the girls is only part of the general behaviour: the weaker the self-confidence of the men, the stronger the tendency of the women to feel embarrassed. Confronted with authority the Anok Mru are kind, zealous and tolerant, and as a result, negatively viewed, often make a pathetic, even occasionally wretchedly servile impression. Among the Dopreng, however, in some cases, I was greeted with scant courtesy, the uninvited guest was made to know that they would prefer to see him gone, but nevertheless they deigned to comply with my wishes, in this case, too, without resistance to authority. My Anok companions did not like this type of behaviour at all, and Dingte for instance declared that the Dopreng were bad fellows. Undoubtedly the kindness of the Anok is considerably more pleasant for the visitor than the reluctance of the Dopreng, although it can also be seen as a greater independence of mind. I only seldom found frank pride, which honours the guest or openly says no without humbling oneself; it can also undoubtedly be found here and there among the Anok. The Khumi still proved to have the most backbone (although I also found some examples of the exact opposite among them). This gradation of male courage corresponds with the open-mindedness of the young girls: casual with the Khumi, unconventional with the Dopreng and timorously coy with the Anok. This difference was shown clearly in the photos I took of girls during the dances at feasts (I did not try to take photographs of them at other times, fearing that they would feel embarrassed): heads demonstratively turned away with the Anok, dark faces with the Dopreng, remarkably unembarrassed with the Khumi.

Among the Anok a young girl never addressed me on her own initiative and even in "my" hamlet, where I stayed over a year, only the girls under fourteen and the older married women came to me by themselves and unaccompanied, for instance when they needed medicine; Tharpau, Menkröi's eldest daughter, who now and then came into my house to fetch or to bring something, to give me some news, or even to show her friends around my house, one day, having become a *malaa*, suddenly needed her mother's protection to be able to have her fingernail treated. Among the Dopreng, however, it happened one evening that the daughter of the karbari of Rümputma-Kua eagerly asked me to drink alcohol with her and tried to persuade me to present her with a cape cloth which I had given to Dingte, who was accompanying me as a boy, after she had not succeeded in persuading Dingte himself. Dingte, at that time sixteen years old and in the Dopreng area for the first time, was astonished at the difference in the behaviour of the girls here and at home. He was not used to such "lewd talk" at home and he, an Anok *klangwüa*, seemed to be more embarrassed than the Dopreng *malaawüa*.

On the occasion of a buffalo feast which we visited in Rümpöng-KP (23./24.12.1956), on the evening of the 23rd, Dingte did not take part in the dance and on the following day he was still embarrassed, mainly because of one of the dancers, Ngenglong, whom he regarded as "extremely right (beautiful)". Since I knew Ngenglong from an earlier tour (she had given me a *kham-eng* in Taa'uk-Para), I asked Dingte whether I should call her for him, after which he ran away and it was quite a while before I caught him again. I warned him that if he didn't join the dance then (during the daytime) I would report it to Taa'uk Karbari Menrüm (who on the previous evening had welcomed Dingte as a "son" [of his *tainau* pact brother] and had put everything freely at his disposal), and tell him to arrange a meeting with Ngenglong for his "son": it then became extremely uncomfortable for Dingte and after a couple of pushes he finally joined the dance, and with persuasion and moral support from the local *klangwüa*, in the evening he even went to the *cam-leng*, the *cam-pua* of the Anok (see below).

M1b) First approaches

Young boys and girls are permitted to talk together and amuse themselves at their discretion. Admittedly it may happen that two young people meet on their way by chance and come to a quick agreement, but in general among the Mru conquests are not made overnight and the young generation spends a large part of the evening in flirting. This does not consist exclusively of mere preliminaries, but can comprise everything, from mere make-believe to serious advancement. This custom, which the Anok call *cam-pua*, is so much an integral part of their general culture that it has developed its own elements: of the musical instruments special gourd-pipes (*ri-naa-plung*), the bamboo zither (*ting-teng*) and the two-stringed fiddle (*taro*) resound only to the *cam-pua,* and nearly all newly created poems and songs are love songs.

After the work in the swidden fields has come to an end in the evening the *malaawüa* of the hamlet come together in one or two houses to spin; the *klangwüa* come to visit them and they talk and play instruments until late in the evening. Although the head of the household and his wife can protest against this disturbance of the peace, they would not win any popularity in the hamlet by doing so; a man with a marriageable daughter whose house is not visited by young people does not enjoy a good reputation. The *klangwüa* also pay visits to the girls of the neighbouring hamlets as well as those of their own hamlet. Except for visits at feasts and very occasional journeys, these are the only opportunities for the young boys to enlarge the circle of their acquaintance and to find marriageable girls. It is the task of the young boys of the visited hamlet to lead their visitors, who do not know their way round, into those houses in which young girls can be found. If the visitors are not looked after appropriately, they will behave in just such an unfriendly fashion on a return visit and the girls of their hamlet will also give the visitors the cold shoulder (MK, 27.09.1956).

Boys dancing with painted bourdons

These get-togethers also follow their own rules of decency: offensive talk and shameless behaviour are frowned upon, politeness and love songs are valued highly. Skylarking admittedly is also done with pleasure and everybody who does not want to be a spoilsport must tolerate jokes and teasing. The limits of the permissible differ depending on temperament, but there is no opportunity for intimate love play during the *cam-pua* and one can come to any agreement only by secret signs. A young man, for instance, can throw the girl coloured beads, which he usually carries wrapped up in a bit of paper on his body: if she consents, she puts the packet into her hair-knot. A little piece torn off from his loincloth can also be used as a "love letter" instead of a piece of paper, or a hairpin can be dropped, which the intended partner lifts and puts in (Dingte, 7.08.1956).

The colour (potassium permanganate) just mentioned is an almost indispensable component of the *cam-pua* among the Anok. It is even used to make paintings on the bourdons of the gourd-pipes. The young people use this colour to paint each other, first according to the rules of beauty, then of absurdity: and someone who gets a red spot painted in the wrong place, seeks to avenge himself as well as he can: everybody smears, throws and spits colour at someone else, so that real "colour battles" can develop. Due to the limited amount of material, this game never lasts long, but it is the only opportunity for the young boys and girls to touch each other. All other overtures of cheeky young men must expect disapproval. It would contradict the rules of morality to show openly to the public that one is a couple. Everything to do with more intimate sex life must be kept a closely guarded secret.

After the evening meetings, the young girls normally go back to sleep in their parental home. In their own hamlet, the young men also occasionally sleep together with their friends in a strange house. This custom, called *klong* among the Anok, seems to have been the general usage in Arakan because U Ba Myaing (1934:130) writes of the Mru and Khami (Areng and Awa) of the Ponnagyun Chin Hills (Akyab District): "The young men and women of marriageable age do not sleep in their parents' houses, but all the young men sleep together in one house and all the young women sleep together in another. The young men visit the young women, and marriage partners are selected as a result of acquaintanceships thus begun." The note by U Ohn Pe (1931:262) on the Mru of the same area seems somewhat dubious: "... there is a custom among the unmarried men and women to sleep together at night in a group in a house. They share blankets together." No such institutions resembling bachelors' and girls' houses are found with the Mru of the CHTs; it would be unthinkable for young people to sleep together publicly.

M1c) Restrictions

Even at feasts there are no more opportunities for dalliance than are generally offered during the ordinary *cam-pua*; because then there are so many people that it would be impossible for love play to remain unnoticed. During my stay I saw a large number of more or less drunken men but none of them made the slightest move to behave obscenely or to annoy a woman. I never saw a drunken girl. Among the Anok at least, no girl will drink without having been invited and then as little as possible; she afterwards entrenches herself again in the circle of her friends. Better opportunities are therefore offered by smaller, private feasts. But here, too, the chances are not that great, as in the case of a wedding celebrated in the smallest circle, two young people can get close to each other after midnight in the shadow of a post and the *klangwüa* can declare his love to the *malaa*, embrace her, kiss her body, pat her breasts and play with her skirt, until she, predominantly passive, is persuaded and excited enough that she consents, and the couple leaves the room separately and if possible unnoticed by those present. For anyone knows, both may have gone home separately, each to sleep alone? This is not talked about, since more intimate love play is taboo for the public, nobody must notice anything. It is unthinkable to publicise one's amorous successes. Menkröi had heard in Manipur that the Naga do something like that by sewing cowry shells to their loincloths: he was almost ashamed to ask me whether it could be true, as it would be too immoral (30.07.1956).

Nevertheless two lovers like to come together, but one may wonder how and where. In the two-roomed houses of the hamlet, one, the *kimma*, is reserved for the parents, and nobody can hide in the other, the *kim-tom*. The jungle does not offer any cosy places, the paths are not suitable either; the swidden houses remain the safest places, but since during the day somebody always knows what others are doing, this leaves only the night time. Ordinarily nightly walks are avoided as much as possible, because of the real and imagined dangers of the jungle at night, but neither the cold of the winter night nor the rain of the rainy season will prevent a genuine *klangwüa* from making an excursion. But what of the girls? Are the *malaawüa* any less robust? Why should a girl not come just as far by herself to meet a man? She is ashamed. The young man must give proof of his love and bravery first: till then the girl will remain at home. Perhaps he can meet her alone somewhere and try to conquer her heart more than the context of the *cam-pua* allows him; but girls do not ordinarily walk alone. In short, the whole thing is a terrible bottleneck and morals and opportunity considerably restrict the possibilities for premarital relations.

What is the *klangwüa* to do if his chosen one is embarrassed and will not come to meet him? He sets out, sneaks into her hamlet and climbs up into her house, careful that nobody hears him (any loud creak from the floor of bamboo-network can awaken the other sleepers in the house), careful too that he doesn't catch the wrong sleeping blanket (if there are three, four or more blankets hanging, this is not at all simple), and crawls to his sweetheart. If she protests he must flee and has to bear the mockery, if he is recognised, he is better advised to make sure of her consent beforehand and thus also his loving reception. And then it is up to the two to pay attention to themselves and the others; one loud movement – and the young boy must strive again to get away unrecognised.

Although the father does not have any right to intervene in the premarital life of his daughter with a young man, his treatment of the intruder and the daughter is left to his discretion. No well-founded objection can be made against someone who gives accommodation in his house to the two lovers but nobody will ask him for it either, since the couple prefers an unobserved place for more intimate play (MK, 11.09.1956). If the daughter leaves the house at night though being forbidden to do so, her father can rebuke her; if he is wise, however, he had better keep silent, because custom is custom. If he makes a row, the whole hamlet has fun, less because of the daughter (who has to be ashamed) than because of him.

Thus in Tapwüa-Kua (15.09.1956): Last night there was a man with a long hewing knife in Netkhai's house. One of his daughters will know who it was. Netkhai does not know it and scolds; his son Chingkrat heard the "thief", but sounded the alarm only after the intruder had left and taken to his kneels, leaving his hewing knife, worth at least 10 Rs., under the house – not particularly clever. He probably also shortened a bit of hanky-panky for another; because Karbari Kangku of the neighbouring house found a big stick under his house and in the evening Netkhai's daughter quarrelled with her brother since he found foreign jewellery with her. It was Netkhai's suggestion to look for it and since his daughters would say nothing, Netkhai wanted to set out with Kangku in search of the owners of the hewing knife and the jewellery. But even if they could find them and make their indignation and disapproval known in the right quarter, this would not do much good: a couple of days later their daughters would steal away from their house at night. The only possibility of stopping two lovers from having more intimate relations is to perform a *ria-pe-choi* (separation-giving oath). If both parents want to separate the couple, for instance because of the sib relationship and to avoid further complications (see L4b) before the love tie gets too strong, the boy and the girl are forced to take this oath (execution like *cang-pe-choi,* see L6c) and would die if they had further sexual intercourse.

M1d) Love spells

A boy who is a neat and strong *klangwüa*, knows how to talk well and to sing even better, will not have any difficulties in winning the heart of his chosen one; but even one who is not so brash and eloquent will find sympathy with some girl. However, if love remains unrequited and the most touching songs and sweetest looks leave the loved one cold, only sorrow and despair are left. Love spells are not highly thought of and a young man who is popular with the girls will turn up his nose at them. But a lad who knows no other way to help or comfort himself may approach a Bengali or Marma to obtain a magic potion. The Mru have no potions of their own. A potion called *bia* is bought from the Bengalis, and either made into a betel chip or, if she will not accept it, pinned to the *rai-cün* (banister, see C3d) at the side of the ladder to her house, so that the girl may come into contact with it.

A better potion comes from the Koladan area in Arakan, from where it is brought to the Chüngma by the Khumi. The Khumi call it *ploi amlau*, the Mru *cia-nöm oi-chaa* (goyal salt). As the name indicates, this salt is actually a means for the taming of the goyal: these animals cannot usually be led to another village and will return to their native hamlet, even if they are dragged there, as soon as they are released; they are therefore given the Arakan salt mixed with ordinary salt whereupon they tamely follow the one who feeds them. This drug analogously can be given to stubborn girls too. Its action is unfailing but the girl can die if the dose is too high. Risks for the marriage are also said to be connected with this drug: the children will not thrive and the couple will become impoverished.

A spell called *khek* can be got from the Marma. It only activates the girl's love, but does not help to get her in marriage. It is a sheet of paper on which the girl's name is written together with cabalistic signs. Spitting at it with ginger and uttering an invocation (*tamma*) in Marma makes the magic effective; the piece of paper is carried on one's own body (MK, 11.09.1956).

M1e) Initiation of love relations within the Marma

Continuing a conversation I had with Dingte Ngarua' about the love signs of the Mru (see above), my boy, Kyo Thwân Ong, who like me had made his own observations among the Mru, remarked that young people were much more restrained among the Marma. I represent his remarks here freely without comment or additions:

If the parents take part, young people mustn't, as in the case of the Mru, sit closely together and chat, but must mainly let looks speak and depend on stealthily thrown little letters, they can throw them to each other. They even keep their distance when the young girl is alone in the house during the boy's visit (which is seldom the case, however, because the grandmother is always sitting there somewhere) and mutual self-consciousness can be broken only if the girl pours him some rice spirit. Better opportunities of getting closer are offered by the night games or fishing for little crayfish, which also takes place at night. They roam around in gangs, engaging in horseplay occasionally falling in a heap and, since it is dark, a young man can also safely collapse with a girl.

Relations are fundamentally freer with the Palainsa Marma. Young people can squat together in the house as in the case of the Mru, even send away the old person in order to go to sleep in the girls' room. One can also approach a Palainsa girl on the way and, if she is willing, give her 2 Rs. (Kyo Thwân Ong, 7.08.1956).

In this connection it must be said that there is no "love for sale" among the Mru and Khumi. *Dons d'amour* may be given (for instances pieces of jewellery), but these are signs of mutual affection and not payments in any way. Konow, in the Linguist Survey (1904) gives a Mru version of the story of the lost son and translates a word written amath-amala with "harlot", but the given form must be read *a-mat malaa* and translated as "his girl(s)". Mru has no word for prostitutes and they are simply called *malaa ngak* ([morally] bad girls) if one wants to speak of them (the facilities are definitely known from visits to Chittagong or even from contact with Bangalis).

M2 Premarital pregnancy

M2a) Evaluation

I cannot make any statement about the frequency of premarital pregnancies with the Mru. To conclude from the very few data (particularly from Tapwüa-Kua), approximately a quarter of the women bring a child into the marriage, whether the child is already born or the mother is married in the pregnant condition. The percentage is very low in view of (theoretically) unrestricted sexual intercourse before marriage and the late age for marriage (see M3a), particularly since premarital children (*yang-caa*) do not cause discrimination. After the marriage of their mother they are regarded as the firstborn of their new father, belong to his sib and have absolutely the same rights and duties as children born during the marriage. There are, therefore, viewed legally, no half siblings and the children of a married woman are at the same time and primarily the children of her husband, whether they were fathered by him or not. The extramarital father is not entitled to his children and it is completely inconsequential for a man's reputation whether he is the genitor of his eldest children.

The non-existence of any discrimination against premarital children might to a large extent explain the fact that the possession of such a child not only does not decrease the marriage chances for a girl but that there are several cases in which men marry women impregnated by others even before the child is delivered (in Tapwüa-Kua for instance Netkhai Atwang's daughter Kungprüa and Cönglök Atwang's wife Thidon): the girl has proved her fertility and the man who marries her can expect her to give birth to further children for him, and this is important. Though marrying a pregnant woman is not to everyone's taste, as I think I can conclude from remarks on the occasion of Cönglök's marriage, there should actually be no reason for a girl to be ashamed of her premarital pregnancy, and no real reason to abort the fruit of a premarital sexual intercourse.

I say "real", because abortions do happen. This is part of the intimacies of which one does not speak, and therefore I did not hear anything more detailed, except that abortions are performed with certain roots that are eaten. This so-called *caa-klak chikhang* (make-drop-child medicine) is harmless up to the third or fourth month, but later associated with dangers for the mother, who might die or become barren for the following years (MK, 7.08.1956). As a reason for abortion "shame" comes into play. Although this feeling completely contradicts the men's opinions about pregnancy and premarital children, it fits in with the aforementioned taboo against making more intimate sex relations public. What should strictly remain secret is made known to everyone by the pregnancy: "people would grin". Considerably more weight is attached to the question about the genitor of the child, that is, about his sib membership. Premarital relations need not observe the sib rules, but a child of a man who belongs to one's own or a brother sib, makes public not only the sexual intercourse, but also an incest. An early abortion will conceal it. St. Andrew St. John writes about the Khami of Arakan (1872:239): "Till marriage the intercourse between the sexes is unrestrained, and it is considered rather a good thing to marry a girl in the family-way, even though by another man; if, however, a girl has a child before marriage, it is exposed in the forest." The latter may not be quite correct, since the author himself in another place gives a rule on the whereabouts of premarital children in case of divorce.

M2b) Compensation

A premarital pregnancy can therefore be viewed in two opposite ways: the value of the evidence of fertility (protected by the fully equal rights of the legitimate and premarital children) and the shame of public signs of premarital (particularly "incestuous") relations. Thus the question arises: is a wrong caused to the girl by the premarital pregnancy or not? The answer in the end will depend on the attitude of the girl, but the question concerns not only her but also particularly her father, who is legally responsible for her. And on this point (as already shown by the example of Netkhai [see above]) the Mru fathers are not unaffected by the double moral standards of the other men: if the neighbour's daughter gets a baby by his son, the father is proud of his offspring; if his own daughter should happen to be impregnated by the neighbour's son, the same father feels an offence to his honour. And in a way he really is dishonoured, since the full marriage ceremony with *paklaa kot* (se M4c) cannot be performed for a girl who is with child. The father of the girl therefore – if he had intended to marry her with a *paklaa-kot* – can demand a compensation payment, called *kan-khang-dong* (birth-Khang-payment) from the father of the young man who has led his daughter "astray". There is no claim if the girl was the driving force in the whole matter; this would be hard to prove, however; on the other hand the young man alone can never be the only "culprit", since although the *malaawüa* may be smaller in size than the *klangwüa* they seldom take second place in physical strength because of the work they do (for instance pounding rice), therefore there can be no talk of "rape". And such demands for compensation payment are actually made only when the parents are in dispute; a man who does not want to annoy the other leaves the matter in abeyance: finally these are matters for the young people, whose behaviour the parents can change little.

There are no rules about the amount of the *kan-khang-dong.* From the people of the son-in-law sibs one can ask up to 5 Rs., from those of the father-in-law sibs 7–15 Rs., if one is completely ill-willed even up to 100 Rs. (the question, however, is whether it will be paid); from people of the brother sibs or even one's own sib the amount will depend on one's embarrassment. The demand can be made as soon as the impregnation becomes obvious. In any case during the girl's pregnancy the father of the young man should sacrifice a pig for the people of her hamlet. The hamlet people can insist on it: without the sacrifice of such a *krong-dim-pak* (earth-Dim-pig), also called *yang-pak* (extramarital-intercourse-pig), the harvest in the fields would be ruined and the rice in the barns would dwindle. The father of the impregnator or the karbari of the hamlet will place the pig near the *ling-pu* (sacrificial post), throw *mi-tut aidam* (rice plus wild ginger) at it and stab it to death with a sharpened bamboo. Parts of the heart, liver and kidney are boiled over the fire in a small bamboo tube (*röng*) in the house of the girl and poured out on the *ling-pu* while speaking a *tamma* (invocation). In this *tamma* no spirit is invoked, the sacrifice (allegedly, and despite the implications mentioned above when not performed) is not meant for the spirits but for the people of the hamlet who hold *khang* on this day (MK, 7.08.1956).

For the Khami of Arakan the following details are found: "Rape of an unmarried woman is to be punished by a fine not exceeding Rs. 30." The fine is "to be accompanied by the cost of the animal (pig) slain to make the agreement binding." "Should a woman die in giving birth to a child before marriage, the reputed father must pay her value (that is, the full marriage price) to her nearest male relative" (St. Andrew St. John 1872:241/42). To this chapter probably also belongs the following note by U Ohn Pe (1931:263) on the Areng Khami of the Ponnagyun Chin Hills (Akyab District): "If a lad kidnaps a lass, the parents of the former have to give the latter's parents a pig and money, the most Rs. 20 and the least Rs. 15, as compensation for the loss of modesty."

M3 The marriage initiation

M3a) Restrictions and marriage age

The free choice of a marriage partner is restricted by rules and regulations of the relationship organisation (see L3a and L9e), depending on the memberships in the sib and in the ethnos. The sib membership determines the extension of exogamy (own sib and brother sibs on the one hand, son-in-law and father-in-law sibs on the other hand), the ethnos determines the endogamy (not completely closed). Thus, within the endogamy association, about 60 % of the possible partners are excluded by exogamy regulations.

Another limit to the choice of a partner is set by spatial restrictions. For the young people of a hamlet, the area for their choice of partner might be determined approximately by a radius of half a day's journey. Since it is desirable to get to know each other before marriage and to be together once in a while, most marriage relations are established between nearby hamlets. Only a few young people have the opportunity of making a longer journey in their country and the possibilities of getting to know partners in distant villages are correspondingly small. It is rare for the native hamlets of the partners to be more than a day's journey removed.

Finally, age is of decisive importance. The normal marriage age for girls is between 18 and 24, but cases also happen of girls not marrying until they are nearly 30. Men can remain unmarried even longer, but by then they are regarded as hopeless bachelors. If they decide nevertheless to get married one day, they must try to find a widow. The partners should have roughly the same age, marriages across the generations do not meet with approval. The men are usually a little older than their wives, but the reversed relationship also happens, mainly when the men get married while younger than 20. Age differences up to 10 years are accepted, however.

The variety of marriage ages can be represented by the example of the married couples of Tapwüa-Kua. The numbers are admittedly only approximate values; the older people are, the less easy it is to determine the exact relationship. Proportionally safe details (particularly for young couples) are given in bold, cases in which the woman is older than the man in italics.

husband	***16***	**17**	***19***	**20**	***20***	***21***	**25**	26	**28**	29	**30**	30	**31**	32	**53**
wife	***23***	**17**	***21***	**19**	***29***	***24***	**29**	23	**24**	25	**19**	22	**29**	22	**43**

M3b) The significance of love

The choice of the partner is mainly left to the young people themselves and therefore mutual affection plays a large role. It can be so strong that it defies all rational considerations and is ready to bear every consequence, but normally, rational reasons have a far-reaching effect as a corrective. The attitude of the parents with regard to the planned marriage is determined by the compulsion of the conditions, and the consent of the parents is extremely important (even if not inalienable). But also the young people usually submit to the external conditions given for them, and a good number of the marriages are more or less reasonable marriages. Nevertheless the significance of love should under no circumstances be underestimated: it causes young people to protest against the desires of the parents, even to wring a change of traditional rules out of them. The father of the young man will most likely be the first to give way since otherwise he runs the risk of losing his son. And so he must also try to obtain the consent of the girl's parents, if necessary against his own will. The hardest barrier against which young people run again and again is that of exogamy, which admittedly has only tradition on its side, but a breach of it can bring about disagreements within the sib and heavily damage its moral reputation. In the case of a *pen* marriage (see L4b) it is ultimately up to the parents how they settle the question. If, however, one side were to refuse its consent in principle, the lovers might elope if necessary (e.g., across the border to Arakan) or choose suicide. Two couples reached for this last solution for their desperate love in the Anok area alone in 1955: in one case by hanging, in the other by shooting themselves: Menkröi told me that he had heard that the loving couple sat down behind each other, fixed the gun in front of themselves, operated the trigger with the toe and so killed themselves with one shot (MK, 11.07.1956).

These examples should give pause for thought; both the strength of the love tie and that of the moral commandments are revealed when one hears that cases of suicide are not known for other reasons, sees how very attached to their children the Mru parents are, and considers that they had to expect this possible result of a denial of their consent.

M3c) Reasons for the choice

Within the limits determined by the social relationship and so on, the most important of all reasons to be taken into consideration for the marriage are primarily the economic conditions, especially those of the young man's parental home. Provided that there is prosperity, the young man has largely free choice: nobody will make many difficulties if he knows that his daughter will be accepted into a prosperous family (the conditions specified below presupposed). Though a rich father will want to make some demands with regard to his daughter-in-law, if the son insists on his request (in turn presupposed, that nothing else speaks against her) his parents also will agree on a daughter-in-law from poorer circles. It is different, however, when the young man is poor: he will already find it difficult to win the consent of the girl herself: the prospect of many years of poverty, hunger and drudgery puts love to a hard test, and even if the girl herself were prepared to take this on herself, her parents would try to prevent it in the interest of their daughter. The love songs often allude to this fact, for instance in a song by Menching (in free translation): "Oh girl do not think of a place where the grease of the chicken heaps up on the brass bowls; since what the heart desires, is that we are both together. Your mother selected a place for you, where the grease of the chicken heaps on the brass bowls, but consider that the heart does not get what it desires, when you hanker for such a place." But the poor man sings in vain, his beloved marries another one. For him only those remain who did not find a husband, are ugly, a little bit too old and themselves poor.

With this I have already reached another principle of choice: that of well-proportionedness. This cannot be understood purely aesthetically but includes both moral and physical qualities. Required are: a good shape of the body, not too thin and slight but sturdy and strong (fat young people cannot be found, permanent work seems to protect youth from obesity); good intellectual gifts, not foolish and idle, but skillful and hard-working (phoney and blethering people have no chances, everybody can keep a close eye on his neighbour to form a judgement); and beauty, a fair skin colour is a topmost trump. A round, harmonious face, long hair, well-formed legs (all qualities which are praised in songs) together do not offset a skin colour that is too dark; the ideal is the *u ko* (white, i.e., fair-skinned) girl, who is to be compared with *rau rök* (the rising morning, Aurora) and whose face is *tom la ko pung*, *tom ni ko pung* (like the picture of the white moon, like the picture of the white sun); and the beautiful young man is the *ko dok*, the white dear one. The purely aesthetic value of fair skin colour offsets many disadvantages even for the calculating parents, and a too dark skin colour reduces the chances of quite an efficient girl from a good house. (Menkröi was against the marriage offered him for his son by Karbari Inglai with the latter's daughter: the Karbari is a man looked up to and esteemed very much by him and the daughter an efficient girl, but nevertheless this would not be a good deal, the only reason being that her skin colour was too dark.) Dark-skinned girls must often wait for a long time for a suitor (both Cönglök's wife Thidon and Bulum's wife Kaiche until their 30th year of life).

Among the further reasons are social standing and reputation. The bare social rank of the father (headman, karbari) is respected and one likes to get married within the "better circles", but here the possibilities for the choice of a partner are too limited for the position of the father to be given much consideration. More important is the prosperity usually connected with it and the moral reputation of the family: and even if he were the son of a headman, a poor but honest father would refuse his daughter's hand, if he knew that the prospective son-in-law had become addicted to opium. And the chances of a pretty, efficient girl are also impaired if her father has not got a good reputation. One might also subsume these factors under that of the economic situation (laziness and opium consumption will not permit anybody to remain rich for long), but it will still be possible to distinguish between poverty due to a ruin for which one is responsible, and a crisis due to unfavourable external circumstances; reputation therefore remains of decisive importance.

While in all the above cases the pros and cons can be weighed up so that in the end everybody is married into the circle for which he is qualified, there still remain a few persons who are regarded as disqualified for a marriage. These are exclusively people with serious physical or emotional afflictions, especially madness and barrenness. Every girl who is not incurably ill will finally find a man, though, because of the sex ratio (men outnumber the women), this is not so for every man. Healthy unmarried old men are nearly always the victims of a former awkward economic situation.

M3d) The formal wooing

The first steps towards asking for a girl's hand can be undertaken by the father of the young man (or any other authorised person or near relation) or also by the girl's father with or without knowledge or approval of the children. These are done for mere information. They only acquire obligatory value when the preparations have led to a firm intention. But even this does not necessarily lead to "official" negotiations and any definite result is still left to the young people, who can decide whether they do not (yet) want to get married despite the encouragement of their parents or insist on a marriage despite the aversion of the parents. Provided that the parents act in agreement with their children, the following ceremonies are part of the proceedings among the Mru:

Lün-taa-machi (asking [for a] wife): the father (or a closely related sib member) of the young man goes to the father of the girl. He takes along a bottle of rice spirit (if he wants to do something extra also a chicken) and enquires about the marriage possibilities for the next year. If the prospects are positive, an "engagement", called *dong-pün-cia* (staircase-climbing cattle), follows between ten days and a month later. This time the father of the young man is accompanied by two or three elders of his hamlet (of arbitrary sib membership). Those to become engaged do not play any role. Formerly (as the name of the ceremony indicates) a head of cattle had to be taken along, today the following are necessary: 5 to 7 chickens, 1 *chin-re* (Chin spear), 1 *chia* (iron arrow), 1 *chai-dong* (sword), and three bottles of rice spirit. One of the chickens must be a mother hen, called *chia-waa* (iron-arrow chicken). All chickens are carried in chicken baskets (*pom*), *chai-dong* and *chia* also are in the *pom* of the *chia-waa*. The *re* is carried in the hand. On their arrival in the hamlet of the girl, the bridegroom's people kill the chickens (apart from the *chia-waa*) brought along by them from their hamlet and they present them to the bride's father, who in exchange has to slaughter a pig and entertain his guests with fish (if he has any). Apart from the chickens, the rice spirit and the weapons, the bride's father gets from 10 up to 30 Rs. as a deposit on the bride pledge (*mang-tang*). With the handing over of the presents and the payment, the girl is regarded as a *machi-kham* (wife promised), a fiancée, and can no longer be married elsewhere.

If this should happen nevertheless, the bride's father or the father of the new bridegroom must refund all expenses of the father of the previous fiancé and give back all received goods. If the girl fornicates with other young men during her engagement, the bridegroom's father can ask 30 Rs. and more from her lovers; in the case of grave consequences, he will receive the *kan-khang-dong* (see M2b) instead of the bride's father. If the fiancé marries another girl, the bride's father need not give anything back. If the girl dies during her engagement, her father must give the received money back; for the weapons the bridegroom's father receives the same value in goods as the girls dowry (*wan*) (MK, 24.08.1956).

The amount of the *wan* is also fixed in the engagement, however, and until the marriage ceremony further discussions follow between the future fathers-in-law. In these the size of the other expenditures (particularly that of the *pang*, see M4c) and the exact date of the celebration are discussed.

M4 The full ceremony

The full ceremony has become rare. In the following description of the *machi-pok-poi* (woman-fetching-feast) I use information which I elicited from Menkröi Ngarua' (23./25.08.1956), and my own observations which, following a special invitation by Karbari Rengnok Atwang (Horinjhiri-Mouza), the proud father of the bride, I was given the opportunity to make primarily in his house (19./21.04.1957).

M4a) *cam-pok machi*

On the day agreed upon (after many preparations) the bridegroom's father accompanied by people of his sib (once more the bridegroom himself is not required and stays at home) and their *pen* go to the hamlet and house of the bride's father in order to fetch the bride (*cam po machi*, go to fetch the wife). One starts in time so that one arrives in the afternoon, and takes along 15–20 chickens, 2 (or 3, if it is to be a great feast) *re* (spear), 1 *chai-dong* and 1 *chia* (sword and iron arrow). As in the case of the *dong-pün cia*, one of the chickens must be a mother hen, in the carrying bag (*pom*) of which also *chia* and *chai-dong* are carried (*chia-waa*, see M3d). Instead of the hen a cock can also be taken as *chia-waa* – it is not considered to be so good, however. While the other chickens are killed by the bearers for the sib people of the bride in the hamlet of the bride's father, the *chia-waa* is left alive: it is supposed to help to increase the bride's father's number of chickens; but not for a long time: if it should be taken by the goshawk or the wildcat, the accident would bring bad luck. All chickens must be faultless;

their tongue is examined (*mü wa-kamca*) and, if one is crooked or shows any other damage, the bride's father can demand a spear as a substitute.

On the arrival of the visitors the bride's father slaughters a pig, called *hom-choi-pak* (rice-pact-pig), and the guests are catered for with its meat. This pig is not sacrificed but stabbed to death on the earth with a sharpened bamboo or spear or, if it is too big and strong, shot with a gun. The bride's father also has to provide fish – big bride-fetching parties can eat 1–2 metric hundredweights during their stay. Pig and fish must be served at every meal and the visitors eat five meals (once in the evening of the first day, three times on the second day and once in the morning of the departure). The sib of the bride's father and their *tutma* should not eat any of the pig and fish; instead of these they get the chickens brought along by the *pen* (see L4a).

For the arrival of the guests the bride's father also has to keep sufficient rice-beer ready (up to 30–50 jugs), particularly a *chang-khi-tam lep-yu* (yeast-ginger chopped beer): pieces of ginger are soaked in the water, which is poured on the yeast. For a big feast a *puma* (big pot) is necessary as *chang-khi-tam lep-yu*, for smaller feasts a little jug will do also; it is very important in every case, however. If somebody damages this pot when drunk, he must pay a pig (*met-pak*) as a fine. This pig (a small one is sufficient) is killed on the side of the *puma* with a spear on *mi-tut aidam* (rice-plus-wild ginger) (*chot-met*), it can be eaten by everyone. At first all visitors have to drink from the *chang-khi-tam lep-yu*, then from all other jugs that are set up in the course of the feast; people of the sib of the bride and the other visitors from the hamlet and the surroundings may drink only after the inauguration by people of the bridegroom's sib. Additional special beer jugs must also be set up: one *malaawüa-yu-kham-lai-yu* (girls-beer-drink-call-beer), at which the young men call their *tutma*-girls (and vice versa) to a drink, one *war-ngeng-yu* for those who "cut through" the night, and finally also one *ni-wöi-yu*, a *puma* (big pot), for all people who still want to drink "next day".

The bridegroom's father himself need not bring any *yu* along (as reported by U Ohn Pe for the Mru and Awa Khami of Ponnagyun Township, Arakan, 1931:261, 259), but 3–4 bottles of rice spirit: if he does not bring them, he must buy them in the bride's hamlet.

After the meal in the evening of the first day, the bride's father has ceremonial presents (*pang*) distributed to the *pen* people. The amount of the *pang* must be negotiated beforehand, since the bride's father has to make corresponding acquisitions and the value of the presents must be refunded by counter-gifts in the house of the bridegroom. New pots are set up (in Rengnok's house they now numbered 17), the *pen* people squat around in a circle and every man receives a new turban (if required also several) of exactly the same length and quality of material, and every woman receives a cape cloth (*pahaa*). A man from the sib of the bride's father fetches the pieces one by one from the *kimma* (private room) of the house and puts them on the *pen* people, another one who can write a little bit notes the names of the recipients. The bridegroom's father examines the gifts, since he must pay for the equivalent.

In Rengnok's house the handing over of the presents dragged on for several hours until finally there was quite a row. The bridegroom's father, Karbari Michi Changkan, sat grinning amiably and drunkenly at his beer pot; the distributor, himself by no means sober, became more and more angry, stamped furiously on the floor, made threatening gestures with his fist and overwhelmed Michi with unfriendliness, but was held back by others. But then, he was suddenly beside Michi, pulled him up and started to beat him. Everybody rushed to separate the two fighters, there was much shouting and screaming; finally Michi sat grinning peacefully again at his beer pot, the furious distributor was dragged away and calmed down again. The reason for the row was that Michi today had demanded more cloths than originally agreed, Rengnok had to empty all his baskets, and when finally there was nothing more that could be given to Michi, he had declared that he was starting on his way home – he could pay, and if Rengnok could not do it, he should keep his daughter. The Changkan people had sided with Michi and when it came to the fight (the only one which I saw with the Mru, which shows how deeply felt was the offence to their honour), the Changkan-Headman had called out: Beat them, we mustn't tolerate anything from these poor, miserable Atwang. – In the olden days this could well have been a reason for a feud, here, however, the next day with clear heads everything was excused, the distributor (a member of Rengnok's family) paid an atonement of one rupee, invited Michi to a drink, praised his great qualities and gave instructions that the guests should be cared for in the best way. And this was done: the second day passed with eating, drinking and sleeping.

A feature of this caring for the guests, however, is that, according to custom, they have to suffer hardship: at noon their food is excessively seasoned, during the meal kilos of pepper pods are poured into the fire and then the doors are closed. The *pen* people must wait inside, their eyes watering and their necks scratching. In the afternoon preparations start for another hardship that represents the main fun of the whole feast: the *paklaa-kot* (pig-carrying).

kham yu in the bride's house

M4b) *paklaa-kot*

In the course of the second day the bride's father once more has two pigs slaughtered, one is cut to pieces, the other one, however, the *paklaa pur kung* (pig in front [and] behind), is prepared in a particular fashion; young men of the bride's sib do the work. They remove the entrails and clean it, then they again smear it with its blood and red colour and "spike" it all around with little bamboo tassels (*chit-wai*). The belly of the pig is filled with pumpkins (*kan-pen*) and ginger and large stones are added: the pig is meant to become as heavy as possible. The legs are tied and hung on a bamboo bar, which is kept as short as possible. The ends of the bar are smeared with pig fat, soot and crushed pepper; the length of the bar is adorned with a small construction of three vertical bamboo sticks with a red wooden top and tassel sticks (*chit-cüng*) plugged in. Finally the pig, thus prepared, is suspended with united strength from a corner of the topmost platform banisters (*yang pang*) where it stays until the next morning. Those who prepare the *paklaa pur kung* together receive 10 Rs. as *cin-tak-dong* (payment for the committal) from the bridegroom's father. Actually two such pigs should be prepared, so that one can be carried in front and one behind when the wife is led to her new home; whence the name: pig in front [and] behind.

On the same day a *rek-rüa plung* (special form of the gourd-pipe, see E1c) is prepared for each of the pigs. For every *plung* the bridegroom's father has to pay 1 Rs. These instruments with six pipes are played only by young men of the bride's sib during the *paklaa-kot*, from the moment when the bride leaves her parental home until the arrival in the house of the bridegroom. The melody played once had a corresponding text, which is never sung, however, so that nobody actually knows it any more. After the meal on the morning of the third day everything is prepared for the departure.

In the meantime the bags (*pom*) in which the guests had brought the chickens have been filled with parts of the pig that was cut into pieces, but by now these bags too have had grease, pot black and pepper rubbed in and the guests must carry them home. Two men of the bridegroom's sib or their *pen* must offer themselves to carry the prepared and adorned pig first. After it is taken down from the *yang pang*, it may not touch the ground in the bride's hamlet any more. None of the tassels should fall down either, but if it does happen, the bridegroom's father must pay one rupee for each. When those carrying the pig from the platform pass through the bride's house, they are first "fed": they have to drink rice spirits, and the man who hands them the glasses, first takes one mouthful himself, adds ginger to this and, if he is spiteful enough, tobacco juice, and spits the pungent mixture into the faces, if possible into the eyes, of the carriers. Then, as their only reward for their efforts (the bridegroom's father, however, also can give them a payment at his discretion), the bride's father binds a turban on each man's head. For this they must drink from the beer pot, while balancing the pig, because it may not be put down.

During this ceremony the bride is fetched. In Rengnok's house a shack had been built for her in the main room, which was now broken open by the *pen* against the resistance of the bride's people. Suddenly the bride's brother interrupted and hit his sister with a couple of firm slaps. His anger (*kunma rau*, literally "mind pain") is part of the custom, but in this case it really seemed too intense: the bride now really wept as never before, she yelled out, yelled like an animal and fled into the *kimma* of her parents. Only after a long time did she reappear sobbing and accompanied by the bridesmaids who carried her most necessary belongings; she still found the whole procedure "worth spitting". But the party left the house, went downstairs and headed towards the exit of the hamlet.

But this took a while: people of the village blocked the way and retreated only quite slowly. If the hamlet is a long one, this "battle" can last until evening. All the while further drinks are served, especially to the carriers of the pig, while all the time they have to carry the heavy pig on the ends of the bar, which are short and do not reach the shoulder, greased with grease, soot and pepper, and they mustn't even shift it quickly from one shoulder to the other, always taking care not to lose one of the tassels. Only after they have reached the hamlet exit can they hand their burden over to others. The group now goes quickly, accompanied by everyone from the bride's hamlet who is so inclined; especially invited are those who belong to the bridegroom's *tai-nau* sibs or to his *pen*. For their service the bridesmaids receive together Rs. 2–4 *chak-dong* (accompanying money).

paklaa-kot with (weeping) bride

The *paklaa-kot* in Rengnok-KP lasted for only a couple of minutes – to the great disappointment of everyone who had fled into the shade of trees and houses in order to observe the big "fun". After long negotiations about *ri* and *ri döi* (custom and not custom) it was decided not to make the departure too difficult for the guests; at the time it was just 12 o'clock noon and in the full glare of the late April sun, the hottest time, with temperatures above body heat even in the shade. Some people might well have been glad to have this torture for the guests, but their own cordon protested against this unreasonable demand, and they gave way.

M4c) *pang*

Also in the bridegroom's house a *chang-khi-tam-lep-yu* (see above) is kept ready. When the guests arrive in the house, the bride's father and the people of his sib are entertained with it first. Also to be provided are, besides a *malaawüa-yu-kham-lai-yu*, a *war-ngeng-yu* and a *ni-wöi-yu*, many further beer pots, primarily for the subsequent *pang* (handing-over of presents). After the first meal the bridegroom's father hands over the so-called *chum-mua* (see 4e) to the *tutma*, i.e., the mother's brother of the bride. For this *chum-mua pang* a *puma*, a big beer pot, is set up, beside it a *klai-puk* (basket) with its bottom turned upward, on which cooked chicken meat is placed. Then a sham negotiation starts: the bride's father demands 30 Rs. *chum-mua* from the bridegroom's father, the mother's brother himself wants to accept only 5 Rs., finally the bridegroom's father gives him 7–10 Rs. and a *re* (spear), lets him drink beer and gives him meat of the chicken to eat. Then follows the *yon-tui-cia-pang* (milk-cattle *pang*): the bride's mother also receives a jug of beer and a chicken, plus 10 Rs., called *yon-tui cia* (see 4e), and a new hewing knife.

Then follows the *tainau-pang*, which corresponds to the *pang* in the house of the bride's father; by *tainau* here the *tainau* of the bride's father are always understood and to be more precise the latter's personal brothers (if they have a household of their own), the brothers of his sib and the relatives of his brother-sibs with whom he is connected by an own *tainau* pact (see L6b), one each per household. They each receive (according to the value of the turbans and cape cloths distributed in the bride's parental home) a jug of beer and one *re* (spear). In the special case that the *tainau* sib of the bride's father is at the same time *pen* of the bridegroom's sib, according to the exchange rules (see below), the people are not allowed to receive any spears and have to settle for a jug of beer – a mere gesture (since afterwards everyone can drink from it). If there are particularly close relations with the *tainau*, the bridegroom's father can give them 30–40 Rs. in addition, for which he receives a corresponding gift in return (in gold, silver or other values) from the bride's father further to the bride's dowry (*wan*). These *tainau-pang* therefore exhibit not only the wealth of the fathers of bride and bridegroom but emphasise moreover the good relations to the respective *tainau* which feels obliged to help if difficulties should appear (return of the bride, non-payment of the *wan-chung* [see 4e]). If they were not invited to the *pang*, they, the other way round, would not feel under any obligation to assist.

This help is often already given beforehand, in that the *tainau* for instance help to bear the expenses of the *pang*. While turbans and cloths can be bought in the market, the spears, particularly as concerns the valuable pieces, are not available arbitrarily and the increase of these ceremonial weapons in personal possession happens mainly by their exchange. The larger the number of spears offered during the *pang* (it is up to the receivers to select and to act), the greater the reputation of the celebration giver. When the *tainau* provide him with spears, he must feel particularly obliged to them. Since they get only turbans and cloths in his house during the *pang*, he will also invite the *tainau* to the next *pang* at which he receives spears as *tutma* and then once more redistributes these spears to them.

It corresponds to the respect for the *tutma* (see L4a), that they get the precious spears, the goodwill toward the *pen*, that they are rewarded with turbans and cloths identical in value. Thus the system of the *tainau-pang* corresponds to a large-scale ceremonial present exchange in the long run: the father of the bridegroom gives spears to the father of the bride who, however, does not receive them himself but leaves them to his *tainau*; in return the father of the bridegroom receives the same value from the father of the bride in turbans and cloths, which are of benefit for the *tainau*; the respective *tainau* thus either receive a return for gifts already presented or are obliged to give something back on the next occasion be it by direct aid, or by invitation to other *pang*.

M4d) *anap*

To cater for his guests the bridegroom's father has to have some further 20–30 chickens killed, one of which is put next to a beer pot on the morning of the second day. Behind this jug, the *mang-tang yu* (bride pledge beer), the bride's father sits down; the chicken lies, as seen by him, on the right side of the jug. Beside it the bridegroom's father places the so-called *mua-pher*: 1 *re* (spear), 1 *charai* (hewing knife) and 1 *chai-dong* (sword) (this *chai-dong*, a *klaa-churi* or a *cia-ran* 10 Rs. worth, is the most important thing, without it the bride's father will not release his daughter) and finally the *mang-tang* (bride pledge, see below): an amount of 100 silver rupees. From the chicken meat the bridegroom's father gives the bride's father a little bit to eat, then he lets him drink from the beer. In return the bridegroom's father receives two turbans, one called *tangkaa-kom-cur* (rupee one-hundred tie up), the other *longhaa-kom-nam* (iron-wrap-strap). From the *hom-choi-pak*, the first pig which the bride's father slaughtered for his guests, some of the best parts of the meat have been put aside, cut small and cooked in two big bamboo internodes (*röng*); in one parts of the heart and kidney are contained, in the other sticky rice and belly grease. The containers are correspondingly called *pak-kia-pak-chak-röng* (pig-kidney-pig-heart bamboo-vessel) and *pak-chau-hom-plong-röng* (pig-grease-sticky-rice bamboo-vessel). The *röng* are locked and one bamboo strap each fastened at their upper and lower end so that they can be carried by this, and a turban of approximately 5 cubits wrapped around. In the bridegroom's house after handing over the *mua-pher* the *pak-kia-pak-chak-röng* is split with a hewing knife into two parts and the bride's father gives something of the contents to eat to the bridegroom's father and offers him a bottle of rice spirits, which he has brought with him or must buy here. The remaining contents of the *röng* are poured out on a banana leaf, the housefathers of the bridegroom's sib and their *pen* eat of them. The bride's father wraps the turban, called *pak-kia-pak-chak-chük-lapong* (pig-kidney-pig-heart-tying turban), around the head of the bridegroom's father, who presents him in exchange for it with a hewing knife, called *pak-kia-pak-chak-ai-charai* (...-split-hewing-knife), not necessarily the same one with which the *röng* was split.

The bride's father, however, does not receive this *charai* himself but has to pass it on to the man who has prepared and brought along the *röng*. The contents of the *pak-chau-hom-plong-röng* (see above) can already be distributed the same way shortly after the arrival in the house of the bridegroom or before the handing-over of the *mua-pher*: the bridegroom's father receives the turban, called *hom-plong-röng-chük-lapong* (sticky-rice-bamboo-container-tying-turban), and the bride's father a hewing knife, called *hom-plong-röng-chot-charai* (...-scrub-hewing-knife), which is to be handed over to the man who has prepared and brought along the *hom-plong*.

After the meal the fathers call their children for the *anap*. The bride has to sit down behind the *mang-tang-yu,* the bridegroom (who must appear here for the first time in front of his father-in-law) opposite to her, behind the *chang-khi-tam-lep-yu*. They have to do so less in order to drink, but because the bride's father now starts his wedding speech, as long as possible and with many good words. He first addresses his daughter and says among other things: "This man from now on is your husband, work for him, look after him, care for him when he is ill. He is taller than you, therefore you shall obey him. Also look after his parents when they become old. If you do not do it, look around for other men or run away, I shall be disgraced; therefore look after your husband and his parents …" The bride's father then addresses the son-in-law: "I have given you my daughter as a wife and for her received the bride pledge. If you do not keep her well or take other women, I will fetch my daughter back, but not give back the pledge and good, therefore you shall treat my daughter well, look after her and the children." Etc. After the speech he gives both to the daughter and to the son-in-law one rupee each. If the father of the bride has died, the one who receives *mua-pher* and *mang-tang* has to execute all ceremonies and to deliver the *anap* speech. The bridegroom's father does not give any speech.

The admission of the young wife to her new family happens by the *chuang-pe-chak* (enter giving life). The housefather organises a *cari-yöng* (see P5c) with the necessary *bong-kom* for which he also binds a thread to the wrist of the daughter-in-law. By this she is bound to the sib of her husband. The *chuang-pe-chak* is the only "religious" action which seals the marriage, though in the two houses sacrifices are already offered during the feast, on the one hand to ask the blessing of the spirits, but on the other hand also to protect against them, because the sacrifices are *nat-cang* with *phyöi-tamma* (see P2): a *pak-ngaa met* and a *wa-ngaa-met* (pork and chicken meat *met* respectively, see P5g). A little pig (*met-pak*) is stabbed to death with a pointed bamboo (*chot met*) by the head of the household on the open platform of the houses in question where the pigs or chickens are killed. In the *tamma* (invocation) the wish is expressed that the marriage may be happy, the couple engenders many children, does not get ill, is blessed with children again and lives for a long time. Of the pork the respective head of the household gives something to eat to the future father-in-law of his child paying a visit to him, he himself and his sib members eat the rest. When the relations of the two sibs are quite old only one pig can be sacrificed instead of the two pigs, one shares the expenses (3–4 Rs.), and kills it in the bridegroom's house. When the sib relations are newly established, the two pigs, however, are indispensable (MK, 29.08.1956).

With *anap* and *chuang-pe-chak* the real marriage ceremonies are completed. It is up to the young wife if she has the desire to return into the house of her parents some days after the marriage for a day; she then is led back again by her husband or (if he has not accompanied her) by the mother-in-law without any further ceremony. The bride good (*wan*) is also collected without ceremonies one to two months after the marriage. During the *cam po machi* itself only the most necessary clothes and blankets for the night are taken along.

M4e) The marriage money

Apart from the aforementioned material goods (spears, swords, hewing knife, turbans and cloths) the marriage ceremonial demands the handing over of amounts of money. Not counting the payments and smaller payments (for the bridesmaids, the pig carriers, etc.) these are: the *chum-mua*, the *mang-tang* with the *yon-tui cia* and the *wan-chung*.

The amount of the *chum-mua* (responsibility payment) is not fixed exactly; it is at least 5 Rs., but not more than 15 Rs. It has to be handed over to the (eldest) brother of the bride's mother before the bride pledge is handed over. It is not regarded as a part of the bride pledge and does not have to be paid for every girl, but only for the first of a man's daughters to be married. If this girl is separated and married a second time, the *chum-mua* for this second marriage does not go to the uncle but to the girl's father himself, who must repay it to the first husband.

The *yon-tui cia* (milk-water cattle: suckling money), as also the *dong-pün cia* (see M3a) probably consisted formerly of a head of cattle. Today a calf costs as much as 40 Rs., and for as long as anyone can remember the "cattle" are to be settled by a payment in the fixed amount of 10 Rs. The suckling money is received by the girl's mother, even if she is divorced by now. Only if the girl's own mother is no longer there (if she is dead, for instance) does the stepmother receive the money. If the girl has no stepmother either, her father (or whoever receives the *mang-tang*) can demand a compensation of 5 Rs. instead of the full *yon-tui cia.*

This *mang-tang* (literal translation uncertain, perhaps: going-away prostration), the bride pledge, is 110 Rs. in every case, regardless of any qualities of the girl. From this sum, the mother gets 10 Rs. *yon tui-cia*, the remaining one hundred rupees (*tangkaa-kom*) are received by the father (or guardian). They represent the real bride pledge and must be paid in silver rupees. Anyone who does not have sufficient silver can pay for this in paper money, but converted according to the actual exchange rate. In 1956/57 one had to pay 1.5 Pakistani paper rupees for 1 silver rupee, therefore for *tangkaa-kom* (one hundred rupees) 150 Rs. had to be paid in paper money. Such a conversion takes place in this case only, all other payments mentioned could be paid in the given amount with paper money.

Hence the information given by Hutchinson (1909:43) is wrong: "the marriage consists in paying the price fixed upon for the girl, which varies between fifty and two hundred rupees." Admittedly the amount submitted to the bride's father at the *mua-pher* may in principle exceed the 100 Rs., everything paid beyond the *tangkaa-kom*, however, is not part of a real bride price, but is *wan-chung* (possession price), the price for the dowry, which may contain the bare necessities only, but also goods up to 1,000 Rs., the value depending on the wealth of the parents. To the *wan* belong, besides blankets and other household objects like a *khong-töm* (see D2h), primarily silver and golden jewellery and other valuables like ceremonial weapons and gongs. No money has to be paid for jewellery that the young girl has bought with her own money, which she may have earned by joining with a friend to make a field on their own account, using the yield to distil and sell rice spirits (see J2a), because it continues to be her private property.

While the 110 Rs. of the *mang-tang* must be paid in the course of the marriage ceremony, the *wan-chung* can be paid later. Only with the express consent of the bride's parents may a part of the *mang-tang* be paid later. If one of the partners dies before full payment of the *mang-tang*, the following regulations are valid: if the young woman dies a normal death (i.e., by illness), the *mang-tang* debt remains; if, however, she should die an evil death (*char*, for instance in childbed, see Q2a) no further payment is to be made. When the man dies and the woman stays in his sib, her full price (if it has not yet been paid) has to be paid by the nearest sib relatives of the deceased; if she returns to her own sib, the rules mentioned for the divorce (see M8a) are valid (*ibid.* also the cases concerning the *wan-chung*).

It is not usual to work off the marriage monies in any form. The information of Lewin (1870:234) that "a young man has to serve three years in his father-in-law's home; or, if he be wealthy, he can dispense with this preliminary by paying 200 or 300 rupees down", the first part of which was repeated by Hutchinson (1909:43), must be wrong and may be due to a confusion with the *wan-chung*. Among the Mru of the CHTs, neither a "bride service", nor any other lump sum payment in the amount of 200–300 Rs. exists, and for Arakan, U Ba Thin (1931:255) also mentions the fixed sum of 100 Rs.

All payments are made by the head of the bridegroom's family, i.e., primarily his father. *Mang-tang* and *wan-chung* are received by the head of the family of the bride, again, in most cases, the father. If the father has died, the remaining family (mother and brothers of the bride provided that they still have not founded a household of their own) gets the sum for their common consumption. If both parents have died, but the children lead a household of their own, the money goes to the brothers of the girl, or to her sister(s) if there is no brother. Other relatives are entitled to the bride pledge only if the children do not lead a household of their own but live in the house of these relatives (e.g., the father's brother). The corresponding persons in the sib of the bridegroom are liable for the payment (MK, 14.07.1956).

M4f) The costs of the feast

Nowadays the full *machi-pok-poi* has become rare, allegedly out of one hundred weddings only one to two are feasts with *paklaa-kot*. The only reason for this decrease is that the costs are considered to be too high. And if today somebody nevertheless wants to do it for his reputation, he must, on top of the conventional expenditure, also so to speak pay for the rarity of the feast: nobody wants to miss this rare opportunity and everybody who comes to it must also be fed. The total expenditure for the celebration has been estimated at 3,000–4,000 Rs. (Rengthon Changkan 14.12.1956). This I would judge too high, but 1,000 Rs. might be reached (this would correspond to the value of the average total annual production of two families).

The amount in 1956/57 could be calculated as follows (for about 50–80 visitors):

Expenditure of the bride's father:		
25 Ari paddy (ca.100 kg rice)	Rs.	25
3 big pigs	"	80
20 kg fish	"	40
meal ingredients	"	5
30 jugs of beer	"	30

20 bottles of rice spirit	" 10
food together	Rs. 190
pang and other presents	" 100
wan new purchase	" 50
SS.	Rs. 340

Expenditure of the bridegroom's father:

40 Ari paddy	Rs. 40
50 chickens	" 125
drinks (as above)	" 40
ingredients	" 5
food together	Rs. 210
pang and other presents	" 100
cin-tak-dong etc.	" 20
mang-tang etc.	" 180
wan-chung	" 100
SS.	Rs. 610
Total expenditure approx.	Rs. 950

The expenses for *pang* and *wan* can be increased arbitrarily and consumption of food and drink also climbs with an increasing number of visitors, so that for this item alone the sum of approximately 1,000 Rs., could be arrived at with 300 visitors.

While services can be expected in return for presents and financial payments, either immediately or later, the expenditures for food and drinks are nevertheless actual expenses. A high number of visitors is an honour, however, and nobody can be told not to come. The only item that can be reduced without impairment of the obligations is the expenses of the bride's father for the *paklaa-kot*. If the bridegroom's father agrees, instead of the two (or at least one) *paklaa-pur-kung* (see 4b), only one pig is cut up and its meat put into carrying bags (*pom*). These are called *wai-pom-pron* (carry-[on the shoulder]-bag-distribute): approximately eight *pom* are made, each to be carried by a person of the bridegroom's sib (or their *pen*), there is no delay when leaving the bride's hamlet, *rek-rüa-plung* (see M4b) are not produced and blown; all other ceremonies, however, remain the same and the bridegroom's father has the same expenses. But if he should not have the necessary money (and nobody will lend him such large sums), his son would have to remain unmarried. If there is no way out, the whole course of the wedding is overturned: the marriage takes place first and then everything is settled as cheaply as possible.

M5 The small ceremony

M5a) *lou machi*

The *lou machi* (carrying off the wife) is the generally usual form of the marriage among the Mru in the CHTs nowadays. On the one hand it permits the expenses of the wedding ceremony to be kept low and on the other hand, it also allows a wife to be taken without the express consent of her parents. A previous arrangement by the fathers is not necessary, although it is advisable in this case, too, in order to avoid a quarrel later: of course the bridegroom can also kidnap his "fiancée". On the other hand the fathers, if they want, can also spare the costs for the *dong-pün cia*, the engagement, and, after the *lüntaa-machi* (asking for a wife, see M3d) has turned out positive, leave everything else to their children. The young man arranges an appointment with his lover and sets out on the night agreed on, mostly in the company of a friend, to "kidnap" the bride. The girl for her part has contacted some (or one) of her friends: they accompany her and together will receive for this 4 Rs. *chak-dong* (accompanying pay, a sum that is not fixed exactly, it can be as little as 1 Re. per person), like the bridesmaids; the bridegroom's companion receives nothing from the bridegroom's father. The father of the young man also knows and is already preparing everything at home for the reception: in the morning a *cari-yöng* is made, during which the daughter-in-law also receives a *bong-kom*: with which she is included in the family (*chuang-pe-chak*, see M4d).

In most cases the father of the bride is informed also and the kidnapping happens with his consent. If, however, his daughter is kidnapped secretly against his will, he can set out to fetch her back, after which the game is either repeated or comes to a sad end. Thus the father of the girl had better grin and bear it and communicate with the young man's father: he usually leaves his daughter in her new home but he also can take her home again until the real wedding ceremony (*anap*); this question arises particularly when the kidnapping happens during the rainy season and every house needs workers urgently: the young woman would then already be working in her new home for what she would consume the following year. In this case it undoubtedly makes a better impression if the bride is "kidnapped" only after the harvest, when the lofts of her new home are already filled and their contents provide a sure basis for the promise of the payment of the bride pledge. The ceremony itself usually still takes place at the beginning of the cold season as long as there are no more worries about the stocks.

Unlike the *cam-po-machi*, which is entirely a matter of the fathers' reputation (the bridegroom does not go along but waits at home by himself and the bride is led away whether she wants to go or not), with the *lou-machi* the young people perform everything unaided; the consent of the parents is a minor matter and only the agreement of the young people is important. But if the young girl has last-minute doubts and there is no promise by the fathers, no engagement to bind her, the wedding just cannot take place. The example of Bulum Khongtör (Tapwüa-Kua, 21.08.1956) shows that such changes of mind do happen. He had fixed a day with a girl in the Lemupalong-Mouza, on which he wanted to fetch her. As can happen in August, it had poured the whole night and therefore the "kidnapping" was moved from Thursday to Monday. In the meantime, however, the young lady had informed him that she wanted to stay at home a little longer and that Bulum would have to wait until after the harvest. The bearer of the message made a private comment on it: she had found another suitor. But Bulum gave little credence to this rumour and on Monday night set out in the company of Menching. Menching had previously proved to be a good friend by teaching Bulum some songs which had apparently also been successful. However, on Tuesday morning Bulum's father Klingtui waited in vain with the pig he wanted to sacrifice; the two young men came back only in the afternoon – without the bride. She was not willing anymore; somebody had described Bulum's poverty to her in too drastic terms and portrayed old Klingtui as an idiot, with the result that she now considered the other suitor to be better. Poor Bulum had had the same experience the year before: at that time it was Netkhai who ran and talked against Klingtui and his son. Bulum urgently needed a wife, since he represented the only full labour in his family of six (his youngest sister had just been born), which needed to reap twice as much to be able to settle its debts. But two months later Bulum finally found someone in Lemupalong. The girl was neither wealthy nor pretty and moreover ten years older than he. Bulum was not only a smart boy but also a very nice fellow, all the people of the hamlet liked him and tried to make the best of the small feast that Klingtui could give. And in fact it was the best of the three *chuang-pe-chak* ceremonies that I saw in Tapwüa-Kua.

M5b) Bulum's wedding

Toward 9:30 (4.10.1956) Bulum brought his girl in the company of two others; he led her into his (father's) house and came to me just afterward: he was pleased that I was still there (I wanted to go on a longer tour), they must have hurried, always fearing I might already have set out before they got back (after the failure of the last time this time hardly anyone had been informed). So I spent almost the whole day in Klingtui's house. For a long time nothing really important could be seen: the three girls (his new wife and her two companions) were in the *kimma* (the private room), Bulum did some housework and in between times slept a little bit (in the *kim-tom*, the main room). Some of the men in the hamlet, who had nothing to do, came in and chatted; Klingtui was in terrible embarrassment: he needed at least two bottles of rice spirit (*arak*), one for the housefathers of the hamlet, one for the companions, and two jugs of beer (*yu*), one for the guests and one for the *bong-kom*. This would be the least to which he could reduce the actually necessary drinks. However, he had only one pot of beer and no bottles of rice spirits at all, and also nobody in the hamlet itself could get hold of either the one or the other. Therefore two boys were sent to the hamlets in the neighbourhood and they returned empty handed at around 3 o'clock in the afternoon. This was in October shortly after the harvest, a good proof that nobody produces alcohol for mere consumption. Menkröi's wife Thanni had already sat down to work at noon and then thanks to her the rice spirits were ready in the evening and a second jug of beer also appeared, although of very poor quality. In the afternoon Klingtui slaughtered two small pigs, one for the *cari-yöng* in front of the staircase and one for the *tam* (see P5b) on his open platform. After the *bong-kom* followed the meal of the day in the evening.

Subsequently the companions of the bride, who herself had only shown up a short while before for the *bong-kom*, were then asked to drink from one of the beer pots, the *malaawüa-yu-kham-lai-yu* (girl-beer-drink-calling-beer). The young men of the hamlet sat in a circle around the jug and for every drink put two fingers (or also all four) of the right hand on the edge of the jug. First the two girls drank, but the pretty one, the fair-skinned Thithop, was really very embarrassed and had to start at least five times until she had drunk her "pint". I had to drink just after her and, in order to do it better, I tried to drink my portion in one go; only afterwards did I learn that everybody has to drink three times. After everybody had done his duty (the stuff tasted so bad that nobody wanted any more) the family fathers who had watched all activities gradually said goodbye, and after they had gone home only the young people were left behind. And so the way was free for a *cam-pua* with conversation, playing and singing. The songs are always invented *ad hoc,* they do not refer to the wedding but are addressed to the girls: "Why you want to go away again, leaving us alone, what sorrow you want to cause us as we like you so much." The young couple had already left by then.

Klingtui gives the *bong-kom* thread to Bulum; in the background Bulum's mother (Changlong) and his youngest sister Ngangli (see M7e)

M5c) *anap* and *cang-cho prüt*

After the necessary preparations on both sides, the bride's father sets out on a day agreed upon (several weeks after the *lou machi*) to "fix" his daughter (*nap camaa*). In the afternoon he arrives with the bridegroom's father with people of his sib and his *tutma* and brings with him in bags (*pom*) a pig cut into pieces (*wai-pom-pron*) and also *pak-kia-pak-chak-röng* and *pak-chau-hom-plong-röng* (see 4d). Among other things the bridegroom's father has to keep ready one *mang-tang-yu* (see 4d) and to have slaughtered some chickens. The next morning the handing over of the *mang-tang* and the wedding talk follow.

I witnessed the afternoon of such a day not in Tapwüa-Kua (I do not know how things transpired after Bulum's marriage in Klingtui's house) but in Thonru-KP (Lemupalong-Mouza, 29.11.1956). Here Kröilö Prencü from Chamklaa-KP (earlier Langthon-HP, Takerpanchori-Mouza) collected the marriage money (350 Rs.) in the house of a Yömre, into which his sister had been married a month before. He received from the Yömre spears to the value of approximately 120 Rs. Of these he kept a part for himself (in representation of his father) as *mua-phe*r (see 4d), he had the others given to his companions (*tutma* and *tainau*); this form of *pang* (see 4c) is called *mang-chok-mang-ploi*. The receivers sat at the jugs set up by the Yömre and let themselves be asked many times; either the spear was not good enough for them or the price was too high; everyone repeatedly declined the offered weapons, even if only to make it hard for the giver (the Yömre as *pen*) – this is custom. And nevertheless everybody must be persuaded to accept a spear, because (although anyone who does not come along has no claim to get anything) anyone who comes along and does not get a spear can afterwards complain and demand two. For the spears Kröilö received cloths to the same value (3–5 Rs. and more) later; he would distribute these again at the visit of the Yömre with his father among the latter's sib members, who for their part gave the bride's father the spears distributed today, so that the circle of the ceremonial exchange can be closed.

This visit of the bridegroom's father, on which he comes in the company of his *tainau* and *pen*, is called *cang-cho-prüt* (the literal translation "make-spit-drive-away", however, yields no sense because nothing is spat and nothing driven away). He takes with him: 1 *chin-re* (Chin spear in the value of 3 Rs.), 1 *chai-dong* (sword, *cia-ran* or *klaa-churi*) to the value of 10 Rs. and 1 *chia* (iron arrow), the last two together with a hen as *chia-waa* (see 4a) in a bag (*pom*). If he does not have these weapons (they have become rare), he can replace them by a payment of 13 Rs. The bride's father again has to set up a beer pot, to provide fish and to slaughter a pig (the *hom-choi-pak*, see 4a) for the guests. At this visit the *wan* (the dowry) of the girl is handed over. The guests returning home receive from the bride's father another pig (a small one will do, it is cut into pieces and put into *pom*). In the village of the bridegroom the *pen,* i.e., the sisters of the bridegroom's father, receive one leg each from it: the same rule also applies to the *paklaa pur kung* (see 4b) of the *machi-pok-poi* (MK, 25.08.1956).

The marriage feast takes the same course if the bride is pregnant or already has a child (in which case a *paklaa-kot* is impossible). Then, however, the bride is not usually "kidnapped" but fetched by the father (or the mother or brother, etc.) of the bridegroom during the day – this is also possible in other cases with the consent of the bride's parents. No ceremonies are necessary at this simple *po machi* (fetching the wife). If the in-law relation is traditional between the sibs, the representative of the bridegroom takes along a chicken and in any case he should get plenty to eat and drink in the bride's house. The real feast follows later according to agreement (MK, 29.08.1956).

M5d) The Mru in Arakan

On the marriage ceremonies of the Mru in the Akyab district we have only two reports. U Ba Thin writes for the Saingdin Chin Hills of Buthidaung Township (1931:255): "The bridegroom generally informs his father about the intended marriage, and the father, in company with his son and a few villagers, visits the house of the prospective bride with three fowls, a spear, and a dah [probably not a hewing knife, but a sword, *chai-dong*]. On their arrival, the spear and the dah are handed over to the bride's parents as presents and the fowls are given as food to the family of the bride. The bride's parents [= relatives], in return, slaughter a pig and give a feast to the bridegroom's party. The pork should not be eaten by the bride's party and the fowls should not be eaten by the bridegroom's party. The father consults his daughter and gets her consent to the proposed marriage. After obtaining the girl's consent, the father demands a dowry. This consists of, at least, Rs. 100, a dah, and a spear [these are *mang-tang* (without *yon-tui-cia*) und *mua-pher*]. The two latter are payable on the spot but the money can be promised later if it cannot be paid immediately. The bride-groom's father may not bargain over the dowry asked for [i.e., *tangkaa-kom* and *mua-pher* are obligatory in every case]. After the settlement of the dowry terms, the bridegroom's party should stay for three days in the house of the bride, drinking *khaung* [i.e., beer] and enjoying the feast, and, on the 4th day, the bridegroom's party, together with the bride, leave for the bridegroom's house."

The report connects (probably mistakenly) the preliminary negotiations (the *dong-pün-cia* of the CHTs) immediately with the handing over of the bride pledge and the fetching of the bride. But U Ba Myaing (see below) also speaks of a taking along of the bride after the first negotiations; if in fact no feast takes place in the bridegroom's house and the bride pledge is handed over in the bride's house, these customs would rather correspond to those of the Khumi (see M6).

According to an opening remark (not specified according to ethnic groups though) the bride is frequently kidnapped: "It is the custom for the spinsters of a village to collect together in a house to sleep and the lads to come at night to make love and win their hands. Most of the marriages are arranged by the young couples themselves, and the parents or relations of the girl are kept in the dark. The girls generally follow their lovers at night and the parents seldom interfere, though a girl's parents may later make a demand for the dowry from the father or relations of the bridegroom" (U Ba Thin 1931:254).

U Ba Myaing (1934:131/32), who also equates "parents" and "relatives", writes of the Mru of the Ponnagyun Chin Hills (Ponnagyun Township): "The young man's parents go to the girl's parents and after the girl's wish has been consulted, ask the girl's parents to give their daughter for their son. If they agree, the dowry money and the money for the marriage feast expenses are discussed. When an agreement has been reached, a date for the celebration of the marriage feast is fixed. The girl accompanies the young man's parents to their house." This section corresponds apparently to the *lüntaa-machi* of the CHTs, the immediate taking along of the bride is striking, however; it corresponds – as is also suggested by the order described in the followings – to that of the simple *po machi* of the CHTs, which can precede the small ceremonial instead of the *lou-machi* (see 5c).

"On the date fixed the bride's parents with their friends go to the bridegroom's house. On their arrival 'Khaung' [i.e., rice-beer] is given and fowls are killed and cooked. Two fowls are killed specially for the bride's father and a brother who lives with the father. If there is no brother, a brother by relation must take his place." [Most probably it is not the brother of the bride's father but that of the bride's mother, the *tutma*; the bride's father and uncle receive their chickens before the handing over of the *chum-mua* and the *mang-tang* (see 4d). On the other hand the "brethren" of the bride's father also might play a role, as on the *tainau-pang*; U Ba Myaing made his notes unaware of the role of the sib relations since he only describes the companions of the bride parents as "friends".] "After the fowls have been killed, the 'Sam-kar' (the tongue) is looked out." [I do not know how to identify "Sam-kar"; the action is called by the Mru of the CHTs *mü wa-kamca*: looking chicken tongue root, see 4a.] "If the tongue be strangled, it is a good sign. If the tongue be crooked, it is a bad sign and fresh fowls have to be killed until good signs have been obtained. The fowls then are cooked with all the ingredients except Ngapi." [For *ngaa-pi* see K1b.] [The reason for this ban may be the exchange rules: *tutma* (bride's father) gives fish to *pen* (bridegroom's father) and not the other way round.] "The bride's father and brother partake of the special curry. After dinner two more fowls are killed specially and when the good signs have been obtained the fowls are cooked. New 'Khaung' is given to the bride's father and brother who drink it with the fowl curry. It is at this stage that the bride's father demands from the son-in-law spear or dah which has to be given." [This probably is, as the chickens and the beer indicate, the handing over of the *mua-pher*.] "The next morning more fowls are killed and cooked. The bride's father and brother may now join in the feast with others according to the rules of commensality. Before the bride's father partakes of the food the dowry money and the money for the marriage feast expenses are asked by him. The dowry money ranges from Rs. 100/- to Rs. 200/- and the money for marriage feast from Rs. 20/- to Rs. 30/-." [Such amounts to cover the expenses are usual with the Khumi (see M6): the Mru have apparently taken over this custom.] "When all the money has been received justice is done to the feast. When the meal is over, the young bride carries some rice and fowl in a small basket [this coming along of the bride is not usual in the CHTs], the bridegroom carries the spear or dah for his father-in-law and they leave for the girl's father's house accompanied by the bridegroom's parents, bride's parents and many friends.

When the party arrives at the house of the bride's father a pig is killed on the open platform and its liver looked at [compare the liver inspection of the Longhu at the *mang-tang-pak*, M6a]. If the blood line goes straight it is a good sign. But if there be any obstruction in the blood line, it is a bad sign and fresh pigs have to be killed until the good sign has been obtained. The pork is then cooked, and the bridegroom and the bride have their meals with the pork curry. The bride gives the fowl curry and rice brought by her to her father. Many more pigs are killed and visitors fed with pork curry. 'Khaung' is freely drunk. After dinner, special 'khaung' and roasted jungle fowl are given to the bridegroom. Before he partakes of them he demands 'gaung baung' [turban] from the bride's father who gives it to him. The party drinks 'khaung' the whole night. The next morning more pigs are killed. Some pork is given to the daughter to take to the home of the bridegroom [compare the *wai-pom-pron*, M5c], the rest [is] cooked and the feast enjoyed. After breakfast, the young wife carries in her small basket some rice, pork curry and pork and leaves her father's house together with her husband. The marriage feast is over."

M6 Marriage among the Longhu, Khumi and Khami

M6a) Longhu

For the *lüntaa-machi* (asking [for a] wife, cf. 3d) one spear, one chicken and one bottle of rice spirit have to be taken along. Also in case of a rejection of the application the girl's father keeps everything brought, when however the parents of the girl agree, but the girl herself does not, then she brings these things (or substitutes for them) back into the house of the young man's father, and a marriage is not possible.

For the *dong-pün-cia* (staircase-climbing-cattle, cf. 3d) the bridegroom's father hands over one spear, one chicken, one jug of beer, and 10–30 Rs., to the bride's father (as a deposit for the *mang-tang*); the bride's father does not need to slaughter a pig for his guests. The rules for the *machi kham* (fiancée) correspond to those of the other Mru groups.

For the *köi-po-machi (*go-fetch-wife) the bridegroom can accompany his father, but, as in the case of the Mru, he can also stay at home if he is embarrassed. The bridegroom's father brings with him 7–10 spears, the same number of chickens and the marriage price. The tongues of all the chickens are examined (*mü-wa-kamca*), and for all faulty animals the bridegroom's father must buy a substitute. A *chia-waa* (see 3d and 4a) is required neither for the *dong-pün-cia* nor for the *köi-po-machi,* it can (but does not need to) be brought for the *hom-tang* ("boiled-rice-present"), the fetching of the dowry. For his guests (composition as in the case of the other Mru) the bride's father slaughters a pig. This pig has no particular name if it is small, but if it is thicker than five fists (approximately 40 cm), it is called *mang-tang-pak* (bride-pledge-pig). If the liver of such a *mang-tang-pak* has not grown cleanly, a second pig (a small one suffices) must be slaughtered, which then receives the name. On his way home the bridegroom's father receives another pig, it is called *i-müi-pak* (sleep-pig) and killed only in the bridegroom's house for the *bong-kom* (wrist-tying).

The amount of the *mang-tang* is always 150 Rs. (paper money) plus 10 Rs. *yon tui-cia* (milk-cattle) and therefore (see 4e) completely corresponds to the bride pledge of the Mru. For the dowry a corresponding *wan-chung* (possession price) has to be paid. If the bride pledge is not paid fully at the *köi-po-machi*, the bride may nevertheless be taken along in the meantime: the debts can be settled later. For this the young husband himself pays a visit to his father-in-law and brings with him a further chicken and a spear in addition to the money, for which he receives another *i-müi-pak* (sleep-pig) which he takes home for a new *bong-kom*. This process is repeated until the final settlement of the debt. The mother's brother of the bride receives 5 Rs. *chum-mua* (responsibility payment) without service in return, the bridegroom's father however also can pay 10–50 Rs. *chum-mua,* in which case he later receives a so-called *plau* for this. The *plau* is then fetched by the bridegroom's father (or the bridegroom himself) mostly in the house of the bride's uncle: as the bride's father had done, the uncle now slaughters a pig for catering and in addition gives a further *i-müi pak*, with which in turn a *bong-kom* is made at home. In addition, the bridegroom receives from the uncle-in-law, depending on the amount of the *chum-mua*, cloths (*wanma*), gongs (*mong*, *ner*), or also piglets for rearing (*pak-liu).*

After the handing-over of the *mang-tang* the bride's father and mother still receive 1 Re. *tor-ong* each. There is no *caran-tamu* (as in the case of the Khumi, see 6b). If a housefather from the bride's village escorts the bride, he receives from the bridegroom's father 1 Re. *kim-cong-dong* (house-supervising-pay), the companions of the bride (*toi-chaa-mi*) receive 1 Re. and 1 bottle of rice spirit as *chak-dong* (accompanying-pay). The bridegroom's companions (whether young or old) who are his *tutma* receive one spear each, those who are his *pen* and those of his own sib receive nothing. There is no *paklaa-kot*, no *rek-rüa plung* or anything similar. Since the marriage price has already been handed over in the bride's house, no particular ceremony follows in the bridegroom's house apart from the *chuang-pe-chak* (see 4d). The *köi-po-machi* of the Longhu thus corresponds rather to the Khumi than the Mru forms of the marriage. Ten chickens, ten spears and (apart from the marriage monies) some rupees – this is all the bridegroom's father has to spend. The Longhu ceremonial therefore requires much less effort than even the reduced ceremonial of the Mru (K. Laichia Rongdim and Angkaa Atwang, 27.03.1957).

M6b) The Areng Khumi (CHTs)

For inquiring with the bride's father the bridegroom's father brings with him a chicken and a spear, which the bride's father keeps, irrespective of whether he approves the marriage or not. If he agrees, a day is fixed on which the bride will be fetched. A question on which a special agreement must be reached is the amount of the marriage money: among the Khumi there are no generally binding rules for the amount of the bride pledge and the dowry, any sum can be demanded arbitrarily by the bride's father. It is, however, usual for the bride's father to also give a large dowry if he demands a high bride pledge: for 100 Rs. for instance, the bridegroom receives only the girl, a pig and the most necessary clothes, for 200 Rs. a full outfit, for 300 Rs. further valuables in addition, such as gongs and cloths, particularly *chamtu-dong*, 10 cubits long "Shendu-loincloths", worth 40 Rs. each. Before the bride pledge can be submitted the bride's father demands the so-called *caran-tamu* (sales tribute?) amounting to 10 % of the bride pledge. For this *caran-tamu* the bridegroom's father receives no service in return; the bride's father uses it to cover his expenses. The money, or at least a first instalment, is handed over in the bride's house when fetching the bride. The bridegroom himself must also be present (unlike among the Mru), otherwise he does not receive the bride. To be brought along are: 15–20 chickens, 20–30 spears and at least 50–60 Rs. On the arrival of the guests the bride's father slaughters a pig for the meals; another pig is given alive to be taken home by the bridegroom's people; this is not a kind of *paklaa pur kung* (see M4a), but rather an equivalent to the *i-müi pak* of the Longhu. The bride's father also has to slaughter chickens for his own *tutma* (Khumi: *pakiüng*) from his own and the bridegroom's hamlet. For this the *tutma* bring him rice spirits and beer (not so with the Longhu).

After the handing over of the bride pledge the mother's brother (*pakiüng*) of the bride receives from the bridegroom's father one spear, one chicken and 15–20 Rs. (*chum-mua*). The bridegroom receives for this in return (for the celebration or later at the visit of the uncle) a living pig (a boar) and a blanket. The mother of the bride receives no "milk-money" (see *yon-tui cia*), but the bridegroom can give 1 Re. each to the parents of the bride (cf. the *tor-ong* of the Longhu); all those from his hamlet who accompany him together receive 1–2 Rs. The amount of the *chak-dong* is the same as in the case of the Longhu.

In the bridegroom's house no particular feast takes place, also no *bong-kom* (unlike the Longhu). If the bride pledge is not paid fully, as is usually the case, further visits follow after the marriage in the house of the young wife's father; on these occasions the dowry is also fetched. As well as money, spears and chicken are taken along and the father-in-law once more feeds them with pork, beer and rice spirits. After the fulfilment of all duties the visits can also be continued on the invitation of the father-in-law, i.e., the father of the young wife: people of his daughter's husband or he alone or accompanied by his father bring along chickens and are catered for with rice and pork. As long as a man has not yet fully paid his marriage money, he may not add any turmeric to the rice paste painted on the forehead of the participants during cattle feasts. If he does not want to dishonour himself, the feast-giver therefore will have to pay his debts beforehand (K Müilün Tamchaa, 26.03.1957 with additions by K Laichia Rongdim, 27.03.1957).

Unlike among the Mru, if a girl becomes pregnant before the marriage, she must marry the father of her child, if necessary also against the sib rules (Nangyu Camthang, 17.02.1957). Kidnappings happen, but more rarely than among the Mru, because of the lower costs of the standard form of the wedding celebration. When the girl is brought into the house, the father of the young man kills a pig on the open platform of the house: some of its blood is dripped on the threads of the following *bong-kom.* When later the girl's father appears, another pig (*met-pak* for the calming of the evil spirits) is killed. The bride's father is entertained with chicken and rice spirit and gets a Chin spear. The bride pledge demanded by him must be paid in the amount he wants; it is handed over on the occasion of a visit in the house of the bride's father. The course of this feast completely corresponds to that of the normal wedding (K Müilün Tamchaa, 26.03.1957).

M6c) The Areng-Khami (Akyab district)

Concerning the Areng of the Ponnagyun Chin Hills, U Ohn Pe (1931:263) writes: "If there is a breach of promise of marriage by the lad or lass, he or she, who is at fault, has to give the other up to Rs. 15 as damages." According to U Ba Myaing (1934:131) the marriage is concluded as follows: "If a young man loves a young girl he informs his parents who would then go to the girl's parents taking with them two or three fowls and a spear. On arrival, they would kill the fowls, cook them and feed the girl's parents and give them the spear. The girl's parents in turn feed the young man's parents with vegetable and fish, if available. The girl's wish is consulted and the dowry discussed. Rs. 500/- is the largest amount given as dowry, and Rs. 120/- the least. The money for expenditure in the marriage feast is then asked by the girl's parents. Rs. 40/- is the most and Rs. 20/- the least [compare the *caran-tamu* of the CHTs]. The money for the expenses of the marriage feast must be given before the marriage ceremony is performed. The dowry money may be given gradually in two or three years' time according to the wealth of the young man and his parents. A date is fixed for the performance of the marriage feast. The young girl accompanies the future parents-in-law." (Such an early taking along of the bride is also reported by U Ba Myaing for the Mru of this area, cf. 5d.)

"On the date fixed, the young bridegroom and the bride, together with the bridegroom's parents, elders and friends go to the bride's parents. The bridegroom carries a fowl and a spear or 'gaung baung' [turban, Burm. *gong-bong*; since according to the rules of ceremonial exchange the bridegroom has to receive a turban, something must be wrong here, or we have to assume that the bridegroom passes this turban on to one of his *pen*, for instance during a *pang*, see 4c]. As soon as the party arrives at the house, the bride's parents kill a pig. The bridegroom party kills the fowl and the bridegroom gives personally the spear to his father-in-law who, in turn, gives him a 'goung boung' ... The bridegroom after spending a day at the house usually returns home with his wife. It is only after this feast that they become man and wife."

On the Areng Khami of the Chin Hills of Buthidaung Township, U Ba Thin (1931:255) writes: "The boy informs his father about his marriage with the girl he has selected and his father, accompanied by two village elders, should approach the girl and get her consent first. The father of the girl is then consulted about the proposed match and he generally does not interfere, if the girl consents. He then demands a dowry, which at least consists of Rs. 100, 30 spears, 15 fowls, and 6 bottles of country spirit, and the opposite party, without bargaining, must promise to pay the dowry asked for. Both the parties in consultation fix a date for the marriage. On the day fixed, the bridegroom, together with his parents and some villagers, go to the house of the bride, taking at least 15 spears, 15 fowls, and 6 bottles of spirit. The remaining dowry can be promised to be paid later if the bridegroom is unable to afford to pay immediately. On their arrival at the house of the bride, a pig is slaughtered by the bride's party to give them a feast and the fowls brought by them are also killed to furnish a banquet for the bride's family and relations. Like the other two races [i.e., Mru and Awa] the Ahraing Khami are also prohibited from eating what has been offered to the other party. The bridegroom's party stays for two days drinking khaung [i.e., rice-beer] and feasting, and, on the third day, the bride is taken away to the house of the bridegroom. The dowry, demanded by the father on the occasion of his daughter's marriage, is solely enjoyed by the girl's parents or her relations. On full payment of the dowry, the bridegroom gets presents of spears, gongs, silk turbans, and 'loongyis', in proportion to the dowry paid, from the father-in-law. Generally it takes years for the bridegroom to pay back the balance of the marriage dowry, and in this case the bride's parents and relations have to wait until the others are in a position to pay. The liability for payment of the dowry to the girl's parents or relations descends from father to son, according to the custom of these tribes."

M6d) The Awa-Khami (Akyab district)

On the wedding of the Awa of the Ponnagyun Chin Hills, U Ohn Pe (1931:259) writes: "The marriage may be contracted by the couple themselves with or without the knowledge and consent of the parents. If a lad kidnaps a girl to his parent's house, the parents take the couple with the village elders to the house of the girl's parents carrying three fowls and some *khaung* [rice-beer]. On arrival, the young man's parents ask the girl's parents what dowry they are prepared to accept. The dowry is then fixed and given. Sometimes the dowry is as high as Rs. 100, the lowest being Rs. 30. After handing over the dowry to the girl's parents, the fowls are given to them and the *khaung* to the visitors to be enjoyed there. In return, the bride's parents give a pig to the bridegroom's parents to be eaten then and there. Then the couple with the lad's parents return to their house. The girl goes to her parent's house on visits, occasionally, but she never returns there permanently.

If a lad agrees to marry on his own choice or on the advice of his parents, they go to the house of the girl's parents, where negotiations proceed. When they agree they fix a dowry and appoint a date for the marriage. On the appointed day, the same procedure, as in the case of marriage by kidnapping, is carried out. The couple are then taken back to the house of the young man's parents."

More details can be found in the report of U Ba Myaing (1934:132/33) from the same area: "If the young man loves a girl and wants to marry her, he informs his parents who would then go to the girl's parents taking with them 2 or 3 fowls and three gourds filled with 'khaung'. When they arrive there the fowls and 'khaung' are given to the girl's parents who, in turn, give them 'khaung' and pork curry or duck curry. No fowl curry must be given. At night the young man's father and the girl's father converse together. When they agree, the amount to be given as dowry is discussed. The dowry money ranges from Rs. 50/- to Rs. 100/-. The money may be given gradually little by little. If an agreement has been arrived at, the young man's parents return home the next morning after fixing a date for the marriage feast. On the date fixed, the bridegroom takes a spear and many fowls and dahs [*dah* here can mean either a hewing knife or a sword], if available, and accompanied by his parents and friends proceed to the bride's house. On arrival there the spear and fowls are given to the bride's father who, in turn, gives him 'khaung' and pork. The people who accompany the bridegroom are fed with pork or fowl curry according to the rules of commensality. The feasting and drinking go on the whole of next day. The next morning the bride accompanied by friends, parents and elders, follows the bridegroom to his house. On arrival there, fowls and pigs are killed and the people fed. 'Khaung' is freely given and drunk. The feasting goes on for another day. The next day the bride's parents and friends return home together with the bride. At this stage the bridegroom may follow them also or he may go the next day. When the bridegroom goes to the bride he must take with him a live fowl which is given to the brother-in-law on arrival at the house. The bridegroom sleeps in the visitors' room. The next day he takes with him the young bride and departs for his home. They then become man and wife."

The report may be short, but it shows that the ceremonies and the course of the wedding with the Awa are fundamentally different from those of the Areng (unlike the assertion of U Ohn Pe, 1931:263). While the Areng in the CHTs and in Akyab are celebrating only one great event in the house of the bride's father, the Awa like the Mru organise a double feast, first in the house of the bride's father, then in the house of the bridegroom's father. Pigs and chickens are slaughtered in both houses: this indicates that numerous people of the in-law sibs are invited. And these are big feasts lasting for days as the following report shows, in which also further traits reminiscent of the great feasts of the Mru can be found. The repeated return of the bride into her parental home for a day seems to be a peculiarity of the Awa.

From the Saingdin Chin Hills (Buthidaung Township), U Ba Thin reports (1931:254/55): "The man informs his prospective father-in-law, either personally or through a go-between, about marriage with the daughter and fixes a date for betrothal. The father of the girl may then consult his daughter and get her consent. If she raises no objection, the *khaung* pot and pigs are kept ready in the girl's house for the occasion. On the appointed day, the bridegroom's party, which generally consists of the bridegroom, his father, and a village elder, approaches the bride's father with a dah, 3 gourds of *khaung* and three fowls, one of which should be a crowing cock. [Apparently a kind of *chia-waa*, because:] On reaching the house of the bride, the crowing cock is set free to be kept as a pet in the house and, with the remaining 2 fowls and the *khaung*, a feast offering is given to the relations of the bride. The dah is handed over to the father of the bride. The bride's party then in turn offer pork and *khaung* to the bridegroom's party. The fowl should not be eaten by the bridegroom's party, and the bride's party should refrain from eating pork. If the bride's or bridegroom's party, through mistake, partake of the pork or the fowl, respectively, they are liable to pay a penalty of from Rs. 5 to Rs. 10. During the feast, the marriage dowry, which at a minimum consists of Rs. 30, 15 dahs, an axe and a spear, is asked for by the girl's father and the bridegroom's father has no option to bargain but to accept the dowry asked for. The bridegroom's party should stay for three days in the house of the bride and fix a date for the marriage ...

On the day appointed, the bridegroom's party, with 30 fowls, 20 gourds of *khaung*, and the dowry should visit the house of the bride. The fowls and the *khaung* are intended for a feast for the bride's party. The bride's party should slaughter a pig to feed the opposite party, and neither should eat what has been offered to the opposite party. The dowry, dahs, the axe and the spear [corresponding to the *mua-pher* of the Mru], together with at least Rs. 10 in cash, should be brought and handed over to the bride's father. The payment of the balance of the money may be promised later. The bridegroom's party should stay for 3 days drinking *khaung* and feasting. On the fourth day, the bride is brought to the house of the bridegroom together with two pigs [a kind of *paklaa-pur-kung*?], given by the father-in-law to the bridegroom as a present.

The new couple should stay for 3 days in the house of the bridegroom and on the following day they go to the house of the bride's father, where they stay for a day and then return to the house of the bridegroom."

M6e) Summary overview

For a simpler representation of the events connected with the wedding of the Mru, Areng and Awa, in the following table only symbols are representing events and localities. They mean:

A: Ceremony for receiving the bride in the house of the bridegroom (*bong-kom*)

B: House of the bride's father

D: collecting the dowry of the bridegroom

F: Feast

G: House of the (father of the) bridegroom

H: Bringing the bride home, i.e., fetching or kidnapping

M: Handing over of the marriage money, particularly the bride pledge

P: Preliminary meeting and engagement

R: Return of the bride into the house of her parents

Events which can be separated from each other by several weeks are separated by semicolon; actions taking place in an immediate connection are connected by hyphens. The details gathered from the literature are in lower case.

MRU big feast PB; FB, H, FG-M A; D
small feast (*lou*) H, A, PG; FG-M; FB D
small feast (*po*) PB, H, A; FG-M; FB D
Ponnagyun Hills pb, h; fg-m, r, fb, h
Saingdin Hills pb, fb-m, h
Longhu PB; FB-M, H, A; M D

ARENG, CHTs regular PB; FB-M, H; M D
CHTs kidnapping H, A, PG; FB-M; M D

Saingdin Hills pb; fb-m, h; (m)
Ponnagyun Hills pb, m, h; r, fb, h; (m)
AWA Ponnagyun Hills regular pb; fb, h, fg, r, h; m
Ponnagyun Hills kidnapping h, pb, m d
Saingdin Hills pfb; fb-m, h, r, h; (p)

M7 Married life

M7a) Authority

The marriage starts with living together and common management by the partners. It is therefore not necessarily bound to the fulfilment of the financial obligations arising from the marriage contract between the families of the parents; but it is based on the approval of these families and their mutual confidence that they can meet the liabilities. A one-sided breach of trust leads to the immediate dissolution of the marriage; for instance on denial of the bride pledge, the father of the woman is able to fetch her back. The marriage is based on the will of the partners to live with each other; if their mutual expectations are disappointed so that the result is a dissolution of the community of living and livelihood, the parents-in-law also withdraw their obligations. Therefore neither has the husband the right of disposal over his wife, nor can the wife disregard their common interests at her own discretion. The wife is by no means delivered without rights to the capriciousness of the husband, her opinions and wishes rather stand with equal right beside those of the husband, who cannot let her work for him and unconditionally ask more from his wife than her duty. Her duty is to carry out all work that falls to her in the context of job sharing between the sexes necessary to cover the needs of the family. If she deliberately does not comply with these tasks (i.e., not prevented by illness etc.), she damages herself thereby no less than her partner and to be more precise not only by the obvious defect in material regard but also in moral regard, as concerns her reputation in the community. The same also applies to the husband: not fulfilling his obligations to secure their livelihood damages his social reputation and affects helpfulness; anyone who accepts poverty out of laziness loses the respect of his neighbours on top of everything else. The general conditions of life are so meagre that in the long run it is simply impossible for a man or a woman to take over all the partner's tasks for their common livelihood. Husband and wife therefore depend on each other as partners and there is no room for domination and letting the other do all the work; a marriage without the will to co-operate is not viable. Everybody knows his daily and seasonal work and needs nobody to instruct him. And since the wife makes just as important a contribution to the essential work as the husband, he generally has no possibilities to become authoritarian. Whether somebody, either husband or wife, can give orders in the house depends more on the character of the partners than on a social claim. Domination is not necessary and does not exist in most cases either. And someone who is not forced does not need to defend himself: marital rows are rare with the Mru.

True, peace and harmony do not always reign, occasionally someone is annoyed and may express this with sharp tongue or ranting. Bickering women and raging men, however, or really violent arguments such as are usual among the Marma, are such rare events among the Mru that when they do occur they wake the interest of the whole hamlet and do not show their originator in a good light.

M7b) Family formation

After the marriage the partners at first live in the house of the family of the young man, uxorilocal residence is not allowed (MK, 6.09.1956). The young wife collaborates in the economic unit of her husband's family. She does not have to submit to her mother-in-law, but merely to adapts herself to the household and takes part in the work. As soon as time and means permit, the young couple gets a room of its own (*kimma*) which belongs to them alone. If the parents of the young man are already so old that no more intimate relations are to be expected, the young couple can in the meantime take over the *kimma* of the parents; single old fathers even usually sleep in the *kim-tom*. As long as only one son is married, the common household is usually maintained and, if the son is the only one, as a rule he stays on in the house, and later takes over the maintenance of his parents. Several married sons can also remain in the house; each couple receives its own *kimma*, while the *kim-tom* is used by all. Differences between parents and children do not need to lead to a *kim-tom* separation, since the residents have their own *kimma* for themselves. For several sons with family, however, the parental house soon will become too narrow and the elder sons will make themselves independent. A completely new house is built, either in the same village or far away, where the need for soil can be better satisfied.

Theoretically it is up to the young couple to build a house of their own immediately after the marriage in a place they prefer – even close to the wife's parents, a concrete example, however, was not known to me and might hardly happen unless in case of a flight marriage (see M3b). For gradually becoming independent there are, as said, no firm rules, a lot of different solutions seem suitable according to questions of living space (extension possibilities and size of the small families), mutual agreement (as also the age and surviving of the parents) and finally the soil needed for making one's fields.

Whatever the composition of those sharing a *kim-tom*, they have no common head of the household, and whoever may have a say, those who submit to him do it voluntarily. The basic unit in every case remains the *kimma*, the married couple in mutual responsibility. But these smallest units seldom exist for themselves alone: attached to them are the dependent members who need to be supported by the family, namely on the one hand the children until marriage, on the other hand those who have left the real work process, i.e., parents in need of care and attention and (in rarer cases) chronically sick persons. Every bigger dwelling and economic unity of several fully viable adults, however, is an interim solution; the individual small families are only conditionally responsible to each other and always have the right to become independent.

M7c) Bigamous marriage

Polygamy contradicts the principle of the right to self-determination in one's activities, co-ordinated by job sharing between the sexes. Since in practical life the wife is not under the authority of her husband, but is co-ordinated, a discordant relationship arises for two wives. A voluntary coupling of parallel activities between relatives is certainly possible, but to replace it by a compulsory division of independent tasks that tends to replace co-ordination by subordination works badly. No woman will let herself be married as a subordinate concubine for the performance of menial tasks and, if a man should decide to marry two women nevertheless, he had better do this in two houses, or even better in two different hamlets. Cases in which a man tried to marry more than two wives at the same time were not known, and the number of bigamous marriages is also extremely low; they were estimated at a few thousands. An aggravating factor is the fact that such marriages are not permitted or at least not custom in many sibs, for instance among the Ngarua', Catumma and Changpa; but they are allowed among the Atwang. On the sib membership of the two women there are no particular regulations apart from the usual rules; marriage is forbidden with two sisters, however. The rejection of bigamy among the Ngarua' is founded on the belief that one of the two wives would either die or go straight back to her parents (MK, 22.08.1956). The death threat might express the disapproval of bigamy by the spirits, the second part of the reason, however, shows the possible character of a second marriage as a veiled form of divorce. Since the second wife can be married without the consent of the first one, the man spares himself the divorce money that he would have to pay if he sent his first wife away (see 8a): the ban on bigamy therefore strengthens her position.

M7d) The right of the man

The fact that (in some sibs) the second wife can be married without the consent of the first one, fits ill with the right of the woman to manage domestic matters on an equal footing claimed till now. In a similar way the rules on the whereabouts of women, children and goods at divorce and death (see M9) convey the impression that the woman has no rights and that everything is settled only by the men according to their sib interests. Furthermore the regulations on adultery (see 7e) deal exclusively with the behaviour of the men, and whether the woman is at fault or not is unimportant. Nowhere has the woman any responsibility or claim, she does not seem to be a legal entity, only an object for whose negotiation the men have made rules. Moreover, I have explained that the old form of the wedding ceremonial was almost exclusively a matter for the fathers, although the condition of the bride had repercussions (no *paklaa-kot* with a premarital child), but this primarily damaged the reputation of the father. In the modern form of the wedding the young people alone are the central figures and the consent of the girl is the first prerequisite: she therefore determines her partner herself and is not married off to somebody any more as is so often lamented in love songs. Thus the traditional rules seem to indicate a social status of the woman which no longer corresponds to today's conditions.

For this reversal the changed political and economic conditions undoubtedly can be held responsible: the men have lost their warlike role and must increasingly collaborate in the fields to ensure survival. But this is only a simplifying theory (we do not know how warlike the Mru once were and whether working in the field ever fell mainly to the women), so it is better to start from the known facts. In the context of job sharing between the sexes, the preservation of rights falls to the men; it is up to the women how far they submit to this prerogative of the men or follow their own will.

In the traditional love songs quoted above, the girl is not asked to oppose her father but to follow her affection and not to marry the rich man her mother (!) has chosen for her. The prerogative of the men undoubtedly was confined not only to preservation of the rights but also covered jurisdiction and defence, but how far they found the support of their wives in this is another question.

Of my simplifying argument only so much is therefore left: Both sexes have essentially the same tasks in today's society; caring for the family by cultivation of the fields, the burden of work and responsibility fall equally to husband and wife according to their abilities. The progressive erosion of the soil lowers the yields and increases the need for soil; all forces must be used for the supply to be guaranteed. It is the workers rather than the spirits who determine the harvest yield; sacrifices lose meaning, the means for them dwindle, and the religious behaviour of the head of the household is no longer decisive for the welfare of his family. Here is only one example: In the past, the young husband after the marriage had to keep a *dong-thong-khang* (marriage *khang*) for the protection of possession and harvest yield: he was subject to a dish taboo (according to that of the *caa-moi*, see J4f) for seven years, muntjac meat and some species of fish and rice spirit were forbidden to him. Today, this *khang* is sometimes still adhered to for two or three years, but often also totally disregarded (MK, 10.09. and 15.11.1956). More important than the devaluation of the religious responsibility of the men is the diminished role of the sibs. The *tutma* still receive weapons during marriages, but they are only put in the corner and passed on at the next feast. The time of clashes with weapons, sib pacts, sib leaders, protections, war and indebted slaves is past. No man needs to use arms to protect his family, to take land, to win workers and wealth. Someone who was rich and strong at that time did not need to weed, he was the lord in the house over woman, children and slaves and a small piece of good land fed all of them. The sib protected him and he defended it. And the men of the sibs had to negotiate with each other when they wanted to get married or divorced. Only they could defend their right and therefore it was also valid only for them. Nowadays all these values of the men have disappeared. The men still negotiate but in the "eventuality of a war" it is not the strength of the sib that is decisive, but the money for the magistrate. But since the court costs exceed the traditional value of the objects of dispute by far, an agreement is reached, peacefully whenever possible, and somebody gives way. And when money is the concern, the wife does not have any less say than the husband. He still conducts the negotiations and refers to the tribal rules; but his whole right is based only on tradition, the true balance of power has moved.

Nevertheless, the women still adhere to the tradition even more than the men. The old rules do not give them any rights, but they do give protection: the agreements are at the same time obligations for their husbands and their parents. It would therefore be wrong to speak about initial equal rights of the women with the Mru. Their legal position with regard to their children for instance was different and is basically different from that of their husbands. But: the woman is (and was) in no way defencelessly surrendered to the capriciousness of the man. As long as it concerns the decisions of daily life, both partners have the possibility to push through their will; however, when it concerns the legal relationship of the marriage partners (especially conclusion of marriage and divorce), they are still subject to sib laws, which comprehend the families from the standpoint of the patrisibs, and assign the married woman, who doesn't belong to the sib of her husband and her children, for the maintenance of her right to the authorised agents of her own sib, i.e., primarily her father or her brother.

M7e) Adultery

All extramarital sexual intercourse is called *yang* by the Mru. While premarital *yang* is approved silently, it is strictly forbidden for a man to associate with married women outside marriage. Relations of a married man to an unmarried girl are not approved, but not punished either. The *cam-pua* (see M1b), for instance, is so well-loved by some young married men that they continue to walk around with the young men of the hamlet, pay court to other girls and occasionally also have intercourse with them instead of staying with their wives in the evening. In Tapwüa-Kua, Menching Atwang led such a life: he was a good companion and the best singer in the immediate surroundings, and since he was hardly more than 20 years old and certainly did not yet think of himself as an "elder", he preferred, instead of staying at home, to mix with the young people, among whom he enjoyed general popularity. Older people of the village did not find this laudable at all, but since he also complied with his domestic duties, there was little to object to. He may have been the father of a meanwhile deceased premarital child of Netkhai's oldest daughter, but perhaps this was only malicious gossip. His wife was some years older than Menching and had already brought a child into the marriage. She had to let her husband do what he wanted, because women have no legal leverage in such cases: Unfaithfulness of the man is not grounds for divorce. But this is not a sign of discrimination against the women: she can just as well have intercourse with another man. If it becomes known, her husband can make a row; but so can she the other way round. Not she, but the adulterer is punished. In no case does blame falls on the woman; this is not so for the man, whether married or not.

The fact is not particularly praiseworthy for anyone involved; a Mru man does not get any prestige by seducing a married woman. Especially the spirits of the hamlet feel insulted, and when adultery is uncovered there is much talk in the whole area. Such cases are not frequent; seen generally and officially nobody can complain about a lack of marital faithfulness in the Mru. Of course if the cage of respectability breaks, everyone has already known for a long time that something was not right. As in the case of Klingtui-Netkhai, which happened to everyone's indignation, during my visit. Klingtui's wife Changlong gave birth to her last daughter in May 1956 at the age of approximately 47 years. People would not believe, however, that old Klingtui, ten years older than his wife, was still capable of engendering this child, but he could not be disturbed by the talk and was very proud of his youngest. Why, then, did the spirits became angry and let the small Ngangli die in February 1957? Now Klingtui paid attention to the talk, which singled out as the father his hardly younger but stronger neighbour Netkhai, who had been seen together with Changlong here and there. Klingtui accompanied me as a carrier to Rengnok-KP, for the great wedding feast of Rengnok's daughter on the 19.04.1957, and it had to be assumed that we would return only on the day after next. Netkhai also assumed this. Klingtui, however, had already returned by the evening and crept into his own house in the darkness without being seen. And indeed after some time Netkhai appeared and, after he thought to have made sure that nobody saw him, went to the *kimma* of Changlong, passing Klingtui, who barred his way back. Noise and excitement followed in the hamlet and discussions between Klingtui and Changlong. Changlong maintained that she had been accosted in the field by Netkhai for two to three years, who several times wanted to give her tobacco and betel; however, she had not responded to these "annoyances". But this "who did what to whom" was not important; Netkhai was the guilty person and he had to pay 30–40 Rs. anyway. There is no rule for the exact amount of the penalty. If Netkhai refused to pay and was ready to risk a court case, Klingtui would be the fool, because he was as poor as a church mouse. On the other hand Netkhai would lose the last vestiges of his reputation in the hamlet if he did not pay; therefore Klingtui got from him almost 20 Rs. and another 3 Rs. later "because Netkhai was ashamed". On the 20th and 21st of April there must have been some lively discussions in the houses of Netkhai (particularly from his wife's side) and of Klingtui; I myself came back to the hamlet only on the afternoon of the 21st and only heard the end of the tirades. The outraged spirits of the two houses and the hamlet had to be calmed down: in the evening a *pak-tan* was performed in Netkhai's house. Thus normal relations were re-instituted and further violations prevented. A pig was killed at the staircase (*cari-yöng*), another one replaced by a dog (*kui-tan*): the bad spirits were driven away in the house by a *chit-ci-kuak* (see P5f).

In the literature I find only one piece of information on regulations in the case of adultery, it dates from the "Khami codex" of Davis (1867, quoted by Hughes 188l:26 and St. Andrew St. John 1872:241): "If rape be committed on a married woman, the husband is entitled to demand a sum not exceeding Rs. 60", the fine is "to be accompanied by the cost of the animal (pig) slain to make the agreement binding." "If a man commit adultery the wife has no redress."

M8 Divorce

M8a) The question of guilt and the bride pledge

The divorce (*taprai* = to separate) of a marriage happens with the Mru without any ceremonies. The wife returns to her parental home and can get married again. For the whereabouts of the bride pledge the deciding factor is who is to blame for the divorce. For clarification of the question of guilt the behaviour of the partners during the marriage is inconsequential, the only important thing is who makes the break. If the wife leaves her husband (*lan*), she is guilty, if the man sends his wife away (*khua*), he is guilty. If the separation happens by mutual consent, the man receives back the bride pledge (*mang-tang* including *yon-tui-cia* and *chum-mua*). In case of *lan* (departure) the relatives of the wife (i.e., primarily her father) must give the bride pledge back and in addition pay Rs. 30 *aprek* (separation). For the daughter of a karbari or headman the *aprek* is increased to 70 Rs. In case of *khua* the man does not have any claim to repayment of the bride pledge, instead he has to pay an *aprek* plus 10 Rs. *mang-lan-cia* (go-back-cattle), that is together 40 or 80 Rs., to the father of the divorced woman. In no case must the *wan* (the dowry) be given back, unless the father of the woman refunds the *wan-chung* and thus buys the dowry back.

If the guilty divorced woman wants to marry again, the *chum-mua* is not to be refunded to her earlier husband by her mother's brother but also by her father (or his inheriting holder); in the case of the re-marriage the girl's father receives the new *chum-mua* (see M4e). The repayment demands on the members of the woman's sib (if they are not solvent and the woman gets married again) can be transferred to the new husband, who then does not have to pay the bride pledge to the woman's father but to the divorced husband, which can give rise to many disputes (MK, 14.07.1956).

As in the case of the Mru, the question of guilt decides on the whereabouts of the bride pledge among the Khumi, too. Unlike the Mru, dowry and bride pledge are, however, coupled among the Khumi (see M6a): if the man is guilty he also loses all dowry with his wife. He receives no money back, but must pay Rs. 60 in addition to his wife's father. If the wife is guilty, the man keeps the complete dowry, the bride pledge is not refunded and moreover the wife's father, depending on his wealth, has to pay a penalty of 15–30 Rs. to her husband (K Müilün Tamchaa, 26.03.1957).

This regulation of the question of guilt with Mru and the Khumi has some repercussions on married life: if for instance the woman does not comply with her duties and the man sends her away, he is nevertheless the guilty person. Therefore if he does not want to punish himself for being stupid enough to have married a slovenly wife, he must try to bring her to reason until she either changes her mind or goes voluntarily, which, however, she can do only if her father is willing to pay. Conversely, it is much easier for a wife who has married an idler to leave her husband and go home, provided only that her family is ready to settle the affair. This case shows how urgently the girl needs the consent of her parents to the marriage: if she marries against the will of her father, her chances are reduced that he will help her in the event of a misalliance.

But the simple regulation of the question of guilt also can become a tragicomedy as in a case brought before the court: the mother of a young wife frequently fetched her daughter home on any trivial occasion and the husband constantly had to run to bring her back. Finally the thing became too stupid for him, and as his wife did not return to him once more, he sent to her parents, to say that they should give back the bride pledge. His wife came back immediately, telling him that she always wanted to stay with him. He, however, was furious and chased her away again. Thereupon her parents refused to pay anything at all, since it now was he who had sent his wife away. I do not know how the case ended.

Exceptions to the *lan-khua* regulation are made only in special cases: be it that one of the partners proves barren or must be avoided (leprosy), be it that he becomes a criminal or falls into an intellectual derangement or that the spirits annoy him/her in some other way (see P2j).

M8b) The whereabouts of the children

Regardless of the question of guilt, the rule is that if the wife returns to her sib relatives (on divorce or also on the death of the husband) the children belong to the father's sib and consequently have to remain there. If the father himself is not capable of providing their maintenance, this task falls (according to the rules valid for the inheritance) on those most closely related to him (see M9b). Little children who still need their mother, particularly babies, are an exception. The wife can (but she need not) take the youngest children with her and for every child the wife's father has to pay 30 Rs. to the divorced husband or the heir who has to look after the children. There is, however, no obligation of the father or his heir to leave the youngest children to their mother, so that the wife's father must offer more if his daughter wants to keep her children in the meantime. The situation is a bit different when the husband is at fault in the divorce: although the bigger children also stay with their father, the small children for the time being can be shared between father and mother, so that, if for instance there are two children, one child stays with the father and one with the mother, or if the mother takes both children her father has to pay only one deposit of Rs. 30. However, in no case can the relatives of the wife purchase a permanent entitlement to the children she brought with her by the payment of Rs. 30 (or more). As soon as the children have reached the age of three or four years or as soon as they can do without their mother, the father's side has the right to demand the return of the children. (This is also valid for the child the mother could take away freely in the case of a husband divorced by his own fault.) A prerequisite for the handing over of the children to their father (or his heir) is the repayment of the 30 Rs. to their mother's relatives as well as the refund of a sum not fixed by rules for the previous food and clothes of the children. If the 30 Rs. and the sum demanded are not paid, the child stays with its mother's family as *u-toi-caa* (mother-accompanying-child), but continues to keep the sib membership of its father (MK, 14.07.1956).

In the role of such *u-toi-caa* are primarily the *yang-caa* (premarital children), i.e., those children a woman already had before her marriage. Although such children are treated as equivalent to the legitimate children after the marriage of their mother, they are free to decide about their sib membership, and here the blood relationship usually dominates over the formal reckoning. The *yang-caa* keeps the sib membership of his father (related by marriage) until the remarriage of his mother. If the mother marries again, the old rule becomes effective again, namely that children of a wife are treated as equivalent to the children of her husband after their marriage, i.e., they receive the sib membership of their mother's new husband. If, however, the child prefers to stay with the father on the remarriage of the mother, it receives his sib membership. This possible sib change ends only when the *u-toi-caa,* the child accompanying the mother, has grown up and has founded a household of its own (MK, 31.10.1964).

Also, among the Khumi the children belong (independently of the question of guilt) to the sib of their father and therefore stay in his house. Small children up to three years in the meantime stay with their mother, who receives for this 10 Rs. per annum; they move into the house of their father later. When the daughters of a divorced wife are married, she receives 10 Rs. from every son-in-law. This sum does not have a name of its own, but corresponds to the *yon-tui-cia* (milk cattle) of the Mru, which, however, normally is not paid among the Khumi (see M6b) (K Müilün Tamchaa, 26.03.1957).

M8c) The divorce in Arakan

In Davis' "Khami codex" (St. Andrew St. John 1872:242, Hughes 1881:27/28) the following details are given: "If a husband wish to divorce his wife, he may do so and take all the children, but in doing so he will forfeit claim to dowry. If a woman have children by a former husband, she is entitled to them on divorce. – If a wife abuse or ill-treat her husband he may chastise her, but if on that account he divorce her, he forfeits claim to dowry. If the husband divorce the wife for (proved) adultery, he is entitled to receive the dowry (paid by him), and (he) may also demand from the adulterer a sum equal to the dowry in addition to a fine and costs [see 7e]. If a husband chastise or ill-treat his wife and she absconds in consequence, he is nevertheless entitled to receive back the dowry. A divorced woman must be supported by the male relative who receives her dowry, or his heir, until remarried."

From the Ponnagyun Chin Hills U Ba Myaing (1934:133) reports: "Amongst the Kami (Ayaing) if a man marries another man's wife, he has to pay compensation which is at most Rs. 100/-, to the husband. In addition he has to kill a pig and besmear the shoulder of the first husband with its blood. In case of Mro also money has to be given as compensation according to the wealth of the men. But no killing of the pig is necessary. Kami (Awa) however do not seem to have any prevailing rule. The Court is ordinarily sought for justice." It seems doubtful to me whether these details really deal (as indicated in the chapter heading) with "divorce". They apparently deal with adultery, and for the Areng Khami where, as the Khami codex says, adultery is reason for divorce, on the case (which is specified ibid.) that the adulterer marries the woman. That rules in case of divorce also exist among the Awa Khamis of the Ponnagyun Hills is proved by the data of U Ohn Pe (1931:259): "If the couple separate owing to the fault of the wife, she has to return half of the dowry to her parents-in-law. If the couple separate owing to the fault of the husband, the wife leaves without paying any part of the dowry."

More detailed data can be found only with U Ba Thin (1931:255). He writes on the Mru, Areng and Awa Khami of the Saingdin Chin Hills (Buthidaung Township): "On separation, either by death or mutual consent, the women of these tribes, according to their customs, cannot make claims on the estate or the children born of a marriage. All the property and the children, except the suckling baby, have to be left behind with the husband or the husband's close relations. The suckling baby is generally taken away by the woman at the time of her departure from her husband's house, but, as soon as the baby attains the age of about three or four years, it has to be returned to the father or his relations on payment of the feeding expenses. The Awa Khami and Mro generally pay Rs. 10 for this, but in the case of the Ahraing Khami, the amount varies from Rs. 60 to Rs. 70. On separation by mutual consent or otherwise, it is customary for the married couple of the Awa Khami and Mro to exchange small branches of trees as tokens of their separation, but no such practice exists among the Ahraing Khami. They divorce each other in the presence of village elders. After the separation the married couple may re-marry if they like."

"If the separation is caused by the fault of the woman, she, or her relations, have to return the dowry paid in the case of the Ahraing Khami. But, according to the customs of the Mro and the Awa Khami, the women have to pay a penalty of Rs. 60 and Rs. 30, respectively [but in a parallel place we read: "In the case of the Mro and the Awa Khami, the women ... have to pay a compensation of Rs. 30 to the husband"], apart from repayment of the dowry paid. – If the separation is due to the fault of the man, the dowry paid is forfeited by the Ahraing Khami, but the Mro and Awa Khami, in addition to the forfeiture of the dowry, have to pay a penalty of Rs. 30 to the woman." These rules apply to Mru and Awa independently of whether the payment of the bride pledge to the wife's parents is completed or not. For the Areng Khami, however, U Ba Thin gives further rules in case of "separation of husband and wife before full settlement of the dowry". In case the wife is faulty "it is customary among the Ahraing Khami to forfeit the dowry promised to the father of the girl, but she will be entitled to Rs. 50 from the husband for the children, if any, left with the husband. If it is only one child, she is entitled to Rs. 30 or Rs. 35. – If the separation is due to the fault of the husband, the Ahraing Khami women get a compensation of Rs. 15, and they are also entitled to Rs. 100 if they leave behind, with the husband, two or more children. If they leave behind only one child, they are entitled to Rs. 60 to Rs. 70."

I did not find out whether these special provisions are valid also with the Khumi of the CHTs. In comparison with the conditions among the Mru (CHTs and Akyab) and Awa Khami, the regulations of the Areng Khumi and Khami show a stronger position of the woman, expressed by the fact that if the man is guilty he loses not only the claim to the bride pledge but also to the dowry. The children born before the full payment of the bride pledge do not belong straight away to the sib of the father and as a result the bride pledge is necessary for retention of the continuity of the sib since otherwise the children would stay with their mother's sib.

M9 Inheritance

M9a) Death of a partner

Should a father or mother die, the inheritance goes to those who remain in the house, i.e., when the surviving parent with the children continues the household as before, there is no change of the old conditions. The widowed have the right to marry again. A woman can do this either in the sib of her deceased husband or in another sib, but she also can return to her own sib without getting married again. From the point of view of the sibs, there are therefore two possibilities if the husband dies: the wife remains in her husband's sib and with her children (whether widowed or married again to a man of her husband's sib), or she returns to her own sib (from where she can marry again or not) and leaves her children, who remain under the care of the father's nearest sib relative. In the first case the woman, whether she has children or not, also remains the administrator of the family property and can manage the household independently.

If she is remarried to a sib relative of her deceased husband, her father does not receive a new bride price but only 30 Rs. *kimma-thia* ([private house] roof-renewing). Of the deceased's brothers only a younger one can marry her, a marriage with an older brother of her husband is not allowed. On the other hand, after the death of his wife, a man can marry either an elder or a younger sister (however, as mentioned, bigamy with two sisters is forbidden). In case the woman returns to her sib, the rules mentioned under divorce (see M8b) apply to the bigger und smaller children. In case a returning wife is still marriageable, her father must give back 60 Rs. of the bride pledge.

The regulations in case the bride pledge has not yet been fully paid have already been mentioned when describing the marriage money (see M4e). If the wife takes her *wan* (the dowry) with her, its price, the *wan-chung*, in every case (whether re-marriage or not) must be fully repaid by her father or close relative. The money is received by the inheriting holder (*pöng-dong-thong-mi*), i.e., either the children of the deceased themselves or those of his nearest sib relative (MK, 12.07.1956).

On the regulations in case of death, U Ba Thin (1931:255) writes for the Saingdin Chin Hills (Buthidaung Township): "On the death of the woman, before full settlement of the dowry, the man has to pay the unpaid dowry to the parents or relations of the woman. This custom prevails among the three tribes (Mru, Areng und Awa Khami). If the man dies before the settlement of the dowry, the woman's parents do not get the unpaid balance of the dowry as it is considered that a fresh dowry can be demanded for their daughter on her next marriage. On the death of the father, the children remain with the relations of the father, but, in the case of the Ahraing Khami, the woman gets money from the relations of her deceased husband if any children are left behind with them." Corresponding payments of money were already mentioned in case of divorce, there also the rules about the care of the youngest children can be found (see 8c).

M9b) Inheritance of valuables

Valuables (ceremonial equipment, cloths, jewellery, cattle, money) and financial obligations (also debts) are inherited. Goods for consumption, including the house are not actually inherited, rather they simply continue to be used by the surviving family members (unless these goods, as personal clothes and work equipment, are burned with the dead body or placed on the grave) and the house, for instance, belongs to those who continue to live in it. What concerns the valuables, family property and personal property must be distinguished. The family property always stays within the sib provided that it is not converted into food and utensils (although it can be managed by sib foreigners, such as the wife as a mother). Personal property changes into family property with the death of the owner, but can be transferred (in the hand of a woman) from one sib to another. Thus, for instance, young girls are able to gain their own money (by making their own field and selling the produce) to buy with it their personal property (jewellery) (see J2a). This they take with them when they marry, and their husbands need not pay any *wan-chung* (dowry price) for it (MK, 9.04.1957). If the woman dies, these things become the possession of her husband or his successor, and his sons-in-law must pay for it, if their daughters get these valuables as a dowry.

The question of the division of the family property does not start only with the death of the previous administrator (on the occasion of which not the slightest change needs to occur on continuation of the household), but already when the children are moving out of the parental home. The daughters who move away when they marry receive a dowry for which the sons-in-law (or their fathers) must pay: so nothing is detracted from the possession. A married son, however, when he sets up a household of his own, must take a part of the previous family property with him. How much a son receives is left to his father's discretion. If the son departs in anger, the father can also "disinherit" him and give him nothing. There are no firm rules either for the distribution of the possessions in case of death (with dissolution of the household). Usually the one who stays in the house (ordinarily the youngest son) receives a little more because he has looked after his parents most; the father, however, can also bequeath more to his eldest son, depending on what he thinks of his sons. If one or several of the sons are already married and unmarried sons remain in the house, however, these receive everything, the married sons who moved away (and already got their share) receive nothing. Daughters receive the inheritance only as long as they are still unmarried. Unlike their brothers, married girls can never manage the inheritance, since they manage in a foreign sib, and if there are no male descendants in the family, the possession goes to the nearest sib relative of the deceased, be it back to his father (if he lived separated from him), be it the (eldest) brother or his son, be it the brother of the father or a descendant in his line or a sib relative removed by even further degrees. Such a transition to the next related family is in actuality also the case when unmarried minors inherit, since these children are then taken up by the most closely related. If the nearest relative is not capable of caring for them, the task falls on the next but one, who then takes the physical goods of the deceased, too.

Especially in cases of single families the inheritance is often enough minimal or really negative and it is not always the nearest relative who has pity for the orphans. Someone who shirks his duty to care for a relative, however, cannot also assert any claims later, for instance, to receive a bride pledge. If there is something to be inherited, it remains the property of the children: when they marry, the foster father has to hand it over to them, i.e., in the case of sons it becomes their family property, in the case of daughters it is taken over by the sons-in-law, who, however, have to pay the foster father the full price for it, with the result that the value of the possession remains in the sib of the deceased. The guardian does not receive a special compensation for his care of the foster children.

A special regulation is valid for the possessions of a married woman (personal jewellery and cloths): they remain property of the sib of the head of the household (i.e., as a rule her husband), but become the possession of her unmarried daughters. If she has no unmarried daughters, these things fall to the nearest relative (of her husband) with wife or unmarried daughters for their use. However, a nearer male relation (without wife and daughters) has the right to buy these woman things back when required against payment of their price (MK, 12./14.07.1956).

M9c) Inheritance with Khumi and Khami

I myself did not have any opportunity to observe deviations from the inheritance rules of the Mru among the Khumi. According to Karbari Müilün Tamchaa (26.03.1957) the inheritance rules of the Khumi correspond to those of the Mru: here, too, the younger sons usually receive a little more of the inheritance, but it is left up to the father to decide differently. Also, among the Khumi, the woman can stay in the family as an administrator of the possession. The statements are contradicted by Lewin's data (1870:230, 234/35), the correctness of which, however, I am inclined to call into question. On the Mru, Lewin writes: "On a man dying and leaving a young family, his eldest and nearest adult male relation takes the family and the deceased's wife to live with him. If a man has sons and daughters, and they marry, he will live with his youngest child, who also inherits all property on the death of the father." These data are in no way correct because the widow can only live with a younger brother of her husband (or a more remotely related sib member), an elder nearest relative will care for the family only if the woman returns to her parents. And the youngest one is by no means sole heir, any more than this applies to the Khumi for the eldest son: "Women among them have no rights of inheritance; the eldest son is recognized as his father's sole heir and representative."

Equally valid for the Mru, Awa and Areng Khami of the Ponnagyun Chin Hills are probably the following data of U Ba Myaing (1934:143): "Women have no rights of inheritance. When a man dies, all the properties left are shared between the sons, or if there be no son by the male persons who are related to the dead man. If there be grown up sons, the woman stays with them. If there be no sons, she goes back to her own parents or if they be dead, to the relatives of her father."

I found further data on inheritance only in the "Khami codex" of Davis (St. Andrew St. John 1872:241, Hughes 1881:26/27): "The debts of the father must be paid by the sons; a daughter never inherits, and cannot therefore be held responsible. If a man die without male issue, his estate is claimed by his nearest male relative; he, therefore, is responsible for the debts of the deceased, whether there be property or not. Should a man die leaving a son who is minor, the nearest male relative acts as guardian until marriage, when minority ceases, and the guardian is bound to give an account of his stewardship. A woman cannot inherit and is therefore irresponsible for debt. If a man die, leaving two or more sons, the property is divided as follows: two divide equally; if there be more than two, the eldest and youngest take two shares each and the others one share each." With the exception of this last rule, which exactly divides up the inheritance, therefore no contradictions to the rules of the Mru of the CHTs noted by me are found for the Khami of Arakan. But the statement that women do not have any right of inheritance might be correct for the Khumi and Khami also, only in so far as it is not possible for the women after concluding a marriage to claim a part of the inheritance from their sib or to transfer the value of an inheritance to another sib. If at a man's death his possessions fall to his wife or alone-inheriting daughters (unmarried and under care of a foster father), these women admittedly can, if they want, take with them all objects to the sibs of their parents or their future husbands, but then their families must pay the full price of these goods to the next (male) heir in the sib of the deceased, so that while women may actually be able to transfer the goods, they can never transfer the value of the heritage.

N From birth to maturity

N1 Entering the world

N1a) Birth

The expectant mother is not subject to any particular rules or restrictions. She takes part as usual in the life of the family and the village, and if possible she continues to work right up to the day of her delivery, though in the later months, she spends more and more time in the village or in the house. Arrangements are made with three or four experienced older women amongst the neighbours; they are informed as soon as the pains start, and will be present at the birth to act as midwives. One of them is the *pen-caa-kim-wöi-charaa* (*pen-caa* = "bear-child"; *kim-wöi* = "house be-there"; *charaa* = expert).

When enquiring about the birth, one would not use the term *pen khök*: that would be indelicate. One would use instead an indirect form of words, such as *wöi kim khök,* or even *wöi caa dia kha* (see also N5a). The father can be present at the birth, and in difficult cases he may even try to help. As a rule, however, he waits nearby, for example in the *kim-tom*. The mother gives birth in the *kim-ca*, which is the part of the *kimma* that is partitioned off. She uses the squatting position.

The newborn baby is bathed straight away and wrapped in a cloth. Practices such as shaking the baby to make it start breathing are not usual. The cord (*dai-rui*) is cut by the *pen-caa-kim-wöi-charaa*, using a bamboo knife (*chanaa*). Before the birth, the father makes the knife by cutting a piece from the *hungma-cüng,* using a hewing knife (chopper). The *hungma-cüng* is one of the central posts of the house (*klong-cüng*, see C3c); it may be made of any kind of bamboo, and it is built into the *kimma* especially so that the knife for the cord can be cut from it. This knife may not be cut from a *reng-cüng* (side post). After use it is thrown away, by the *charaa* or by anyone who lives in the house.

Mills writes: "At birth the navel cord is cut with a steel knife and hewn with a piece of bamboo. The baby is immediately placed on a plantain leaf" (1931:512). This does not correspond to the facts. Steel knives are not permitted. It is acceptable to lay the baby on a banana leaf rather than a blanket, or indeed to use both, but this is not a requirement.

The afterbirth, *bai* ("nest") is put into a small basket or *em* (see D2a). It is covered with *wia-chok* leaves (*Grewia microcos L.*), or indeed with any other kind of leaves, and hung from a tree far from the house. The *kwak-bai* (throwing away of the "nest") is the duty of the father, but if he is too embarrassed, his mother can do it instead. No other relative may perform this task (MK, 4.08.1956).

N1b) Birth *khang*

The young mother is not allowed to rest after the birth, because she must go and sit at the edge of the fire in the *kimma.* Two or three large logs are piled in the fireplace to make a big fire, which is kept burning constantly. The woman must spend five or six days sitting with her back to the fire; her back is often scorched so badly that scars are left afterwards. If she wants to sleep, she must put her head in her hands. From time to time she is washed with warm water; cold water is not allowed. It is believed that without this treatment – called *taheu* ("drying out") – the blood would rise to her head and she would go crazy.

For the first nine days after the birth, the woman may not leave the *kimma*; this is called *ta'ung* ("shutting oneself in"). After that comes the *döm-krong-lot-wat* ("climbing-down earth letting–out going-out"). Her leaving the house is not marked by any kind of ceremony. No house offerings may be made during the *ta'ung*, otherwise the rest of the family, including the child's father, are free of restrictions. The new mother must observe a *caa-dia-kan-khang* ("birth-food-taboo"), which differs according to her sib and the number of children she has. A Ngarua' mother, for example, is allowed only unsalted rice and warm water for a month after the birth of her first child; for fifteen days after the second; and for nine days after all subsequent children. Atwang mothers follow this diet for nine days after a first child and seven days after subsequent births. During a second period of the same length – for example a month for the Ngarua' or nine days for the Atwang – the new mother may eat salted rice once more. During the third similar period, she may have spices, fish and some leaf vegetables (*caa-theu*, "eating-vegetables"). After the fourth period, she may return to a normal diet. On the date of the *kan-khang-pöt* ("food-taboo-ending"), a *cari-yöng-tang* must be held, or a *cari-yöng* (see P2f) if a *cari-yöng-tang* cannot be afforded. Alternatively, the *cari-yöng* can be postponed for two to three months. It is intended for the child's benefit, to protect it from illness and to make it grow big and strong.

The day after the birth, the whole village celebrates the *bur-khang* (cf. *bur-vi* = menstrual blood). During this *khang*, it is forbidden to work in the fields. If the *khang* is not observed, the rice will not grow, and there will be a plague of mice. The village celebrates another *khang* day when the cord falls off; this is the *dai-klek-khang* ("cord falling-off *khang*"). Some member of the household throws the cord on to the roof of the house, or puts it into some piece of bamboo inside the house (*kwak-dai-rui*, "throwing-away umbilical-cord") (MK, 4.08.1956).

N1c) Bad omens and death

If the mother dreams before the birth of flowers, or of sleeping with another man, then either she or the baby will die. A child born in July (*lakauma-laa*) is unlikely to thrive. It is also a bad sign if a baby has two fontanelles: this means that it will have two marriage partners. Twins are not desired, because it is believed that if they both live, then either the father or the mother will die. Usually it is the second twin who dies. It is not killed, but it receives less care because of the perceived risks associated with it (MK, 16.08.1956). The death of a child does not carry implications of supernatural threat, and the body is buried in the way that is usual for children (see Q2b). Stillbirths require no particular ceremonies: the stillborn baby is simply buried. The mother, however, must still observe the *khang*.

It is seen as a very bad thing if the mother dies in childbirth (*kak-köng*, "receiving-dying"). Her death is regarded as a *char* ("bad death"). Until her body is buried or cremated, none of the members of her family may eat salt or side dishes; in other words, they must keep the *khang* that the newly-delivered mother would have observed. The various restrictions that apply in the case of a *char* (see Q2a) must also be observed. The mother's death is a *char* if it happens before the child reaches the age of three. After that point, it is a *rui-tha'*; I was not able to identify a definite meaning for this term, but the words could signify "[umbilical] cord-curse". If the full dowry has not been paid when the death happens, payment stops immediately, and no further claims are made.

If the mother dies during or after childbirth and the child survives but is not yet weaned, it can be given to another woman. However, it would then be lost to its birth family, because it would count as the child of the wet nurse, and as a member of her husband's tribe. The biological father cannot claim it back. ("Milk adoption" of adults is unknown.) Another woman would only take the child if she had none of her own, because if she did have children, they would fail to thrive and would die. It is therefore quite possible that nobody will be found to breast-feed the baby. Cows' milk is not commonly used: it is probably employed occasionally as an additional measure, but attempts to use it as the main way of feeding the child are unknown. Such an attempt would probably fail in any case, because even adult Mru cannot digest cows' milk.

The baby is also doomed to die in cases where the mother lives, but is prevented by illness from feeding it. This happened to the youngest child of Khamcöng Khongtör and his wife Krapau in Tapwüa-Kua (MK, 7.08.1956).

N1d) Naming the child

The morning after the birth, the child is named (*rük ming*). First a very small chick is killed; the method of killing it is not important. It is then plucked and cleaned, and cooked whole in a bamboo holder (*röng*). Its meat is not eaten; those who have touched the infant – the three or four older women who acted as midwives, including the *pen-caa-kim-wöi-charaa* (see 1a) – take part of its liver, heart and kidneys, together with some of the broth and some *wia-chok* leaves (*Grewia microcos L.*) on their right hands, then rub their hands together. The chick is called the *rut-chia-waa* ("hand-wash-chicken"). If the women did not perform this washing ceremony, they would not be able to work properly thereafter, and would meet with failure (MK, 4.08.1956). A member of the household then throws the rest of the *rut-chia-waa*, together with the bamboo holder, on to the roof (MK, 7.08.1956).

A name is then chosen by *kwak-plang* ("throwing-sign"). This ceremony requires either a turmeric tuber cut in two, or two cowrie shells (MK, 4.08.1956). The cowries (*preu*) are bought from the Bengalis, in Banderban for example, but otherwise cowries are not traded by the Mru. Outside the name-choosing ceremonies, their only use is with red beads (*keng-wi*) to make bracelets for small children. At one time, cowries were used as money. Their significance can be seen in the fact that they have a special form of numeral, which is used today for paisa: *pa-yun* means "one paisa". Tippera decorate their clothing with cowries; Bengalis tie them around cows' necks (MK, 6.08.1956).

Hutchinson (1909:43–44) writes as follows about the naming ceremony: "A Mro infant has to be named the day after birth. Selection is first made of three or four likely names, but the final selection rests on the throw of two cowries (small shells), or pieces of cut turmeric root. A name is mentioned and the cowries are thrown. If one cowrie falls with its face down and the other with the face up it is considered lucky and that name is selected, but if both the cowries fall with faces up or faces down it is considered unlucky, and the name is not selected. A fresh name is mentioned and the performance repeated. The same rule applies to the pieces of cut turmeric. If both cut ends appear uppermost it is unfortunate, but if one whole and one cut end appear the result is auspicious and the name is chosen."

Lewin (1870:233) mentions cowries, but not turmeric; the details he gives are repeated, partly verbatim, in the above quotation from Hutchinson. The *kwak-plang* is carried out by the father or the *charaa*. If the name is not received favourably by the oracle, another similar name, differing in one syllable, is usually suggested. If the family has a strong preference for a certain name, the oracle can be asked twice. The father, the mother and the *charaa* may all suggest names.

The parents have special priority when they wish to name the child after an ancestor (MK, 4.08.1956). This happens if the child has particular identifying marks, and especially if a deceased relative had been marked on some part of his or her body with dye (turmeric, or colour bought from the Bengalis), and the baby has a mark in the same place. In such cases, it is clear that the deceased person has been born again in the child (*wang-ru*, "come-grow"). There are no clear ideas about how the child receives the soul (*lüm-laa*) of this person, nor about circumstances surrounding conception in such cases. In the opinion of my informant, however, if an ancestor is reborn, it is probable that his or her spirit enters the mother at the time of conception. The ancestor may be a predecessor of either the father or the mother (MK, 7.08.1956).

N1e) Changing the name

An ancestor can be reborn in the absence of any outward signs. The parents may thus be unaware of the rebirth when they choose the baby's name, and may choose wrongly. If this happens, the child is sickly, cries a lot, and refuses to feed. A woman will then be consulted who knows how to determine the proper name with the aid of a cotton fluffing bow (*ba*). This woman is called the *püi-höng-charaa* ("ear [of rice]-swinging-expert"). She holds the bow by its cord, murmurs something that is her own secret, and utters names. When she says the correct name, the bow starts to swing (MK, 7.08.1956). In Tapwüa-Kua, the woman who possessed this skill was Changlong, the wife of Klingtui Khongtör.

Illnesses and dangers proceeding from other causes or spirits are not reasons for changing a name. Once a name is given, it cannot be changed later at will. However, many people are given a nickname during childhood, and this can take over from the original name. The nicknames often have a distinctly unpleasant meaning. They may be given by playmates, or by parents when they are cross with the child. A nickname is often a variation on the child's real name, or a parody of it, made by changing some part or parts. Ruirong, for example, becomes *kui-dong* ("dog-stairs"). The first part of boys' names is often replaced by *nia* (penis), and the first part of girls' names by *kai* (vagina). New nicknames can be given to people relatively late in life: an example is Niadöng ("penis-standing up"), who in view of his age was renamed Rengyua ("stay-friend").

Finally, a boy may grow to be a man before he starts to show characteristics that remind others strongly of an ancestor. In these circumstances, he is given the name of the ancestor somewhat belatedly, and without repetition of the naming ceremony. A young Atwang man from Thanglong-KP had been named Baiwai, but people called him Paring, after his maternal grandfather (MK, 15.07.1956).

N1f) Khumi und Marma birth customs

At the time of delivery, the Khumi fix a long rope to the inside of the house roof. The woman in labour holds tight to the end that hangs down and gives birth on her knees, in an upright position. After the umbilical cord has been cut with the bamboo knife, cord and afterbirth are put into a container (*bu-düm*?), which is closed with *nun* leaves (*nun*: Bangla word, unidentified) or banana leaves, and hung on a string from a tree (Bangla *ju*, unidentified). While climbing down from the tree, the person who suspended the container there holds his breath, pierces the bark of the tree seven times with his chopper from below, and asks for the child to be well and to flourish (KTO, 26.03.1960).

As with the Mru, the young mother must sit by the fire to dry out her body. She is allowed to go out again "as soon as her body is dry" and she feels able to work again. For the first ten days she must live on water and unsalted rice only. After that, she may use salt and have some vegetables. Spices follow only after a period of five months, and other foods after ten months, though she is not allowed to eat *ngaa-pi* ("rotten" fish) until three years have elapsed.

The child's name is determined by throwing cowrie shells. If the two shells come to rest on opposite sides, the name is regarded as accepted. The two cowries are then tied to the child's right arm with cotton threads (wife of Karbari Müilün, 27.03.1957). Lewin writes: "A child is named on the falling off of the navel string. In giving it a name, the mother binds seven threads round its wrist, saying, 'Be fortunate, be brave, be healthy!' The name given is generally one that has been borne by some progenitor" (1870:229).

On the day of the birth and the day when the cord falls off, the Khumi villagers do not work in the fields (Renglün, 26.06.1957). This custom parallels the Mru *bur-khang* und *dai-klek-khang*.

Marma women work less than Mru und Khumi women before the birth. For the birth itself, a woman who has some medical knowledge is summoned. In Bengali, she is called the *boidyo*. The mother gives birth lying down. After the delivery, relatives, friends and acquaintances are invited, and prayers are said to Buddha for the new baby. The mother then stays in the house for seven days, remaining awake at night and sitting by the fire. During this time she cooks for herself and eats separately from the rest of the household. Alternatively, the baby's father can cook for his wife if he wishes.

At the end of the period of seclusion, the used pots and dishes are thrown away. The young mother sleeps in the early morning, then gets up, takes the child on her hip, and goes to bathe. Friends and relatives are summoned, an offering is made to Buddha, and thanks are given to him for keeping the baby alive and well for these seven days. Prayers are then said for a long life for mother and child. From that day, the mother may eat chicken and various kinds of fish (Kyo Thwàn Ong, 26.03.1960).

N1g) Arakan birth and naming customs

U Ba Myaing (1934:134–35) describes as follows the ceremonies of the Mru of the Ponnagyun Chin Hills:

"The navel cord of the child is cut by a piece of bamboo pole at the head of the place where the mother sleeps. No other bamboo must be used. The midwife washes her hand with warm water in which a hen's egg or chicken has been boiled. The next day the midwife would suggest a name and taking in her hand two shells or two leaves (*taro rwak*) or two pieces of ginger, which are obtained by cutting a piece of ginger in the middle length-wise, throw them down on the floor. If one of the shells or leaves or gingers has its face up and the other shell or leaf or ginger has its face down, it is considered that the name suggested is a fortunate name and the child so called accordingly. If both the shells or leaves or gingers have the faces up or the faces down, it is thought the name is one which would bring bad fortune and ill-luck. Another name has then to be suggested and the process of throwing done until the lucky sign has been obtained."

The two shells mentioned in this extract are probably cowries. *Taro rwak* is Burmese; according to Judson (1953:472) it is "a species of linden-bloom, *Grewia*", and therefore probably the *wia-chok* leaves of the Mru of the Chittagong Hill Tracts.

A pig is now sacrificed; this ceremony is the equivalent of the *cari-yöng* of the Chittagong Hill Tracts. According to sources other than U Ba Myaing, this ceremony is carried out at other times too. I shall therefore leave U Ba Myaing's comments until I describe other pig sacrifices (see P5b/c). A special ceremony is described by U Ba Thin, of the Mru of the Buthidaung Chin Hills: "Soon after the birth of a child, four short bamboos, each measuring about a cubit, are fixed up on the bank of the stream, and a chicken is killed in honour of the 'nats' and the blood poured over the bamboos which are set close together. A prayer is then made for the welfare of the child. The chicken is then thrown away" (1931:252).

Writing about the Awa Khami of the Ponnagyun Chin Hills, U Ba Myaing notes: "There is no curious ceremony among these people at the time of child-birth. No killing of the pig or the tying of the string is necessary. The midwife who performs the work at child-birth smears her joints with a piece of ginger. The navel cord can be cut by any bamboo" (1934:135).

[The "tying of the string" refers to the *bon-gkom* ceremony, see P2e].

N2 Infancy

N2a) Mother and child

The baby (*ing-rik-ca,* "black-red-child") is tended mainly by its mother in the early months. Until its third year, the child is seen as so closely tied to its mother that if the parents divorce, it stays initially with the mother (see M8b), and the death of the mother in the first three years is seen as a "bad death" (see Q2a).

During these three years, the child may continue to take the breast; it is the breast milk that creates the legal bond between mother and child (see N1c). This does not mean that the child cannot be weaned before the third year, nor that it must be completely weaned during that year: children of five and six may still take the breast if they want to. In practice, however, they are often displaced at the breast before the third year by younger siblings.

Although the baby (*nau*) is breastfed until the mother becomes pregnant again, other foods are introduced soon after the first month in the form of rice that is first chewed by the mother then passed from her mouth to the child's. The chewed rice may also be stroked or pushed into the child's mouth by hand, as will happen later with unchewed rice. For practical reasons, the baby is fed the rice at family mealtimes; at other times, its hunger is satisfied at the breast. There is no particular schedule for breastfeeding, and one seldom hears a baby cry, unless it is ill. The mother carries it with her almost all the time in a sling. If it becomes restless, she will move it from her back to her breast and let it feed. The sling (*wan-li*, see F1d) is placed over the mother's right shoulder and supports the baby's back and bottom. The baby is tucked against the mother's left hip with its legs drawn up a little, almost in a squatting position. It is warmed by the mother's body and rocked by her movements. The baby is only rocked for its own sake or bounced in the sling at times when the mother is not working.

As soon as the child can sit in the sling and does not need to be fed so often, the older children or the father will take it with them. This is particularly likely to happen if the mother has to work hard, or do a lot of bending. At night, and during the day when nobody is available to carry the baby, the sling is used as a hammock. The cloth from which it is made is of a loose enough weave to allow urine to run away freely, so it does not have to be turned, changed or washed too often.

N2b) Care and training

As soon as the baby can sit up and crawl, it is allowed to move around the house by itself. It is free to hold things and put them in its mouth, though someone always keeps an eye on it. It wears no clothes, and the floor of the house can easily be rinsed, so there is no special need for toilet training. Dogs are sometimes encouraged to lick the small behind clean, or to lick up a small heap from the floor. Children eventually learn to go just outside the door on to the open platform, and later into the jungle, like the adults.

If children get too dirty on their expeditions, they are washed. When they are bigger, however, they are not washed if they resist. Kaichüa, for example, brought her son Menlong to me when he was just two, to ask me for something for his scabs. I asked her to wash the child first; the mother declared that he didn't want to be washed, and did not come back again. If a child has lice, older siblings or the mother remove them; however, since older children take bathing and hair care seriously, lice are not common.

Little trouble is taken to encourage the child's development during its first year. Things change when the youngest member of the family becomes a *rai-ring-caa*, and makes its first attempts at standing and walking: everyone is interested. The rest of the family encourage it and talk to it, which of course promotes language development too. "Baby talk", however, is not usual: it is left to the child's developing intellect to understand what the adults are saying. Only when a child is significantly late with walking and talking are deliberate efforts made to help it, and the advice of neighbours sought.

In general, children learn by observation and experience, and to a much lesser extent by instruction, admonishment or punishment. They are left to develop as their nature dictates; but this does not mean that they are left to themselves. They do not receive a great deal of attention, but they are always watched, and are always in the presence of the mother, father or older siblings.

N3 The ear-piercing feast

N3a) Chittagong Hill Tracts

It is customary for both girls and boys to have their earlobes pierced while they are still small. The piercing can be postponed up to the age of 15, but it can only be carried out in uneven years of the child's life, that is, in its first year, its third, its fifth, and so on. It is thought better if two children have their ears pierced at the same ceremony. If there is only one child, a dog must stand in as its partner; if there are two or more children, the dog is not needed. A partner shares the pain, so the child does not feel it so much.

Ears can be pierced only by a *param-rui-charaa* ("ear-pierce-master") or *param-rui-mi* ("ear-pierc-er"). Each village has one or two men who perform this task. They must be members of the older generation. When a *param-rui-mi* dies, any "elder" who is prepared to do so may take over the task. If someone who is not a *param-rui-charaa* were to conduct the ceremony, he would become blind, and his fields would yield no crops. The real *param-rui-charaa* on the other hand need fear no adverse effects on his other work.

The ear-piercing ceremony must be celebrated with a feast, which takes the same form as the *chot-cia* (see R1a). If the parents cannot afford this, however, it is enough if they sacrifice a boar, circumcised if possible. A sow is not acceptable. If a head of cattle is sacrificed for a *param-rui-poi* ("ear-piercing-feast") both its horns must be undamaged (MK, 25.06.1956).

The rule is that the feast must begin on a Monday evening. The ear piercing and the sacrifice of the cow take place on the Tuesday morning (Khamlai, 16.08.1956). On an occasion when I was present, the *param-rui-mi* came to the feast area about half an hour after sunrise. He had previously cut some small bamboo spikes and had stuck them into a piece of taro tuber (*ru*) like the spines of a hedgehog. He squatted on the ground, put down his "hedgehog", and prayed, facing the rising sun. A woman brought her son – about five years old – in her arms, put him down before the *param-rui-mi,* and held him firmly, with the help of some other men. The *param-rui-mi* took one of the little spikes and pierced first the right earlobe and then the left. Next he rubbed the child's earlobes with the cut edge of a piece of taro, then took some cooked rice into his mouth and spat that on to the earlobes too. He stuck one of the little wooden spikes into each hole, broke it off on each side, and spat on the child's ears once more.

Meanwhile, another woman had arrived with a child of about three. Unlike the first child, this one cried and struggled desperately, especially before the procedure began. This child's ears were pierced in the same way. Finally a dog was brought, and allowed a wooden spike to be pushed through its left ear, without making any fuss. The *param-rui-mi* neither spat on its ear nor broke off the ends of the sticks. He then wrapped up the rest of his ceremonial equipment in a banana leaf and took it away with him (Inglai-KP, 21.02.1956).

The rubbing with the taro is thought to relieve the pain. The wooden spike stays in the ear until the wound has healed. At the end of the feast, the child whose ears have been pierced gives the *param-rui-charaa* a folded banana leaf containing *mi-tut-aidam* ("rice-with-wild-ginger"), and says:

ang pe en mi-tut-aidam. The *charaa* answers: *rut nop* ("hand cold"); this means "May that which my hand has given be cold, so that the ear will not swell".

From the child's father the *param-rui-mi* receives a gift which differs according to the family relationship between them (see L3). If the *param-rui-mi* is a *tutma*, he receives a chicken; if he is a *pen*, he receives fish. If he is a *tainau* or belongs to the feast-giver's clan, he receives a chicken that he eats in company with the feast-giver. If he is a *can*, however, there is no shared meal.

Later, the girls have little rolls of *la-ia-ram* leaves (*Aglaomena hookerianum, Schott*) pushed into the holes in their ears, to make them gradually bigger. When the holes are big enough, the *param-cheng*, which is the jewellery that women wear in their ears (see F2b), can be inserted without any further ceremony. The ear-piercing does not bring any new designation of the child's stage in life (MK, 25./26.06.1956).

The Rümma Mru celebrate the feast rather differently (see R4d). The Khumi pierce their children's ears seven to eight days after birth; only in exceptional circumstances would it be delayed until the child's second year. It is not customary to hold particular ceremonies, but rich people may, like the Mru, give a *param-rui-poi* (Renglün, 26.06.1957).

N3b) Arakan

U Ba Myaing describes as follows the customs of the Khumi (Areng) of the Ponnagyun Chin Hills:

"On the third day after the child's birth, a pig is killed and cow-heads and other articles in the house are besmeared with blood. Pieces of thread are also besmeared with blood and tied round the right wrist of the members including the new born child of the house. The thread is tied on by an elder member of the house who, at the time of doing it, would bite a piece of ginger and say: "Phyauk! May all of us live together long. We have now killed a pig." Friends are invited and fed with pork curry. Before they have their meal, the child's ears are bored with the needle and a name is given to the child by a member – generally his father or grandfather – of the house. Some rice and curry are taken and a portion of it is placed inside the hall and another portion on the cow's head as offering to the nats." (U Ba Myaing, 1934:133–34)

It appears from U Ba Myaing's account (1934:135) that the Mru and the Khami (Awa) observe the same rules as the Mru of the Chittagong Hill Tracts. He writes:

"... a child's ears may be bored at any time within one year or on the 3rd, 5th and 7th year. Ear-boring is not allowed in the 2nd, 4th and 6th years."

U Ba Thin (1931:254) reports from the Buthidaung Chin Hills: "No custom prevails among the Ahraing Khami to hold an ear-boring ceremony for their children, as the ear-boring is done soon after birth, but the Mro and Awa Khamis perform an ear-boring ceremony for their children by slaughtering cattle in honour of the 'nats' ... The ear-boring ceremony is done soon after the dances. The children of the poor people in the village are allowed to do the ear-boring ceremony when a rich man celebrates the ear-boring ceremony for his children. Generally all the friends and relations are invited to enjoy the feast."

This author also indicates (page 251) that drums, gongs and cymbals and mouth organs are played at these feasts.

N4 Child-rearing

N4a) Parents and child

It is up to the child to adapt to its world and learn for itself the life skills that it will need. The parents adopt a passive role in child-rearing rather than an active one. The child learns from its siblings and contemporaries how to behave; its ultimate goal is to become – in its own way – like the adults. It is allowed to do everything that the adults and its older siblings do, just as soon as it is able. Children are not brought up in a special children's world, with its own taboos, privileges and ideals. No one requires children to behave differently from adults, or better than them; no one expects them to follow adults' wishes against their own desires. If the child misbehaves, that is excusable; it doesn't know any better yet. It is not bad, just naughty.

This does not mean that a child's behaviour never gets on its parents' nerves. When it does, they will try to quieten the child by frightening it with threats of a bogey-man (in Tapwüa-Kua, despite my many protests, I was often used as the bogeyman!). They also frighten children with tales of evil spirits, whose doings are so obvious that they can be used effectively to influence children's behaviour. Other punishments may be threatened, though they do not usually get beyond the threat stage. However, the child will register the parents' tension and displeasure, so this may have some effect, especially since it is only a matter of dealing with the child's demands. When more serious punishments are administered, it is not because the parents wish to impose discipline; the punishments represent lapses into undisciplined behaviour on the part of the parents, when provoked by the children's behaviour. These lapses are considered unseemly in an adult, as the following story shows:

Menkröi's youngest son, Menpaa, had an unpleasant habit of expressing his displeasure by bawling and whining. His mother Thanvel was forever saying *hang kakar* ("don't howl") but this was about as effective as saying *kui dom* to the dogs when one wants them out of the house (see G3a). One day, it became too much for his mother, and she smacked him. She was then so horrified at her own behaviour that she pressed a stick into Menpaa's hand – who by this time was howling louder than ever – and told him to beat his badly-behaved mother as hard as he could.

One could hardly find a better illustration than this of the belief in equal rights for children. Thanvel was also fairly sure that Menpaa's behaviour would settle down by itself: her eldest son, Dingte, had behaved in a similarly disagreeable fashion as a child, but had since grown into a nice young man.

Another story may serve to illustrate the Mru reaction to a children's world of make-believe. When I visited the karbari of Borpara (Pantola-Mouza) on 1.01.1957, he showed me a rare possession. It was a *baiskop* (bioscope), a peepshow box made in the USA. It had a number of picture discs, one about "Rural Quebec", one showing the Taj Mahal and Red Fort, and one about the Three Little Pigs (the children's story about the three little pigs and the big bad wolf). The karbari asked me what country this was, and what sort of people lived there, with pigs who wore clothes. He said that pigs don't look like that; I replied that in Europe, there is a different kind of pig, with pink skin. This, however, did not answer his question about why we put clothes on pigs. I said that we do not in fact dress pigs in clothes. The karbari pointed out that this was in direct contradiction to the pictures – so was it perhaps a different country after all? Or even if it was only a children's story, it was still not true! He said the Mru also had stories that were handed down through the generations, and it was possible that mistakes might have crept in. But to teach children something that one knows for a fact is not true – impossible!

N4b) Children: the social group

Parents put little pressure on their children, so there is little by way of resistance in return. However, it seems unlikely that the wish to equal the adults in abilities and achievements would be enough to suppress all antisocial tendencies, to the extent that friendliness and politeness would become the general rule in society. Even amongst the Mru, socialisation is not painless; however, the task of smoothing the rough edges belongs not to the parents, but to the peer group. It is no use saying "I'll tell my mother", because the principle of non-interference is even stronger where other people's children are concerned. Children from others families enjoy the privileges of guests, including, if necessary, the right to be protected against one's own children.

For better or worse, then, the child must take responsibility for its own behaviour. However, the peer group is not a closed unit; as children get older, they shape their behaviour increasingly according to that of the adults. The relationship of the older children to the younger ones to some extent parallels that of the adults to the older children. Jockeying for position is therefore rare. In cases of antisocial behaviour, for example using bad language that an adult might use to a dog at most, sanctions are likely to take the form of temporary rejection and exclusion. The general absence of specialised toys means that there is no strife or jealousy over particularly desirable items. (Articles with which I had finished constituted an unfortunate exception, and led to arguments.) Nevertheless, one can still have an argument over some bit of bamboo or other; and this is enough for children to learn it is not a good idea to take something from someone else. Everyone has the same opportunities to find things, and finding one's own bamboo stick leads to fewer social complications than taking someone else's.

Many children's games consist in copying the activities of adults (see N6a), so the childhood peer group is also an arena in which playmates can learn the principles of co-operation that they will need later. Usually children of roughly the same age play together; the younger ones often watch the older ones – and the adults too – from the edges. Since the villages are relatively small, there are rarely more than three children in each year group, and these do not spend a great deal of time together, since they also have the option of staying with their parents. If the parents have to work for long periods in the fields, they often leave younger children who are big enough to play independently with the others in the village. The children are therefore never away from the group for long, so they have no alternative but to get along with the others. They spend time in a group which has a fairly complex structure, and which imposes sanctions of short duration against antisocial behaviour. The individual child is thus exposed to many and varied life situations, which may be trivial in themselves, but which compel him or her to make use of the example of others, learning from experience how to manage such situations and to profit from them. The child learns from the reactions of its playmates to evaluate the success and failure of its different behaviours: it sees itself praised and blamed by its peers, and in the process it learns to distinguish right from wrong, honour from dishonour, according to the code that is practised by the adults.

N4c) Learning

To a large extent, Mru children bring themselves up. They learn skills and acquire knowledge through their own efforts. There is no systematic training: the child must take it upon himself to observe and to absorb what he needs to know.

When a child learns by observation, he sees the interplay of activities rather than the reason behind them. It is perhaps significant that the Mru equivalent of "why?" (*tong plong cö*) actually means "for what purpose, with what intent?" rather than "from what cause?". This implies that all questions about activities are to be answered in terms of their purpose, which in turn contributes to their "spiritualisation". If, for example, the purpose of a field is to yield an especially good harvest, it cannot be the field itself that pursues this goal, but a spiritual being that controls growth. "For what reason?" in the sense of "from what cause?" is not a question that is asked. This sets limits on children's knowledge, and behind these limits, the unseeable takes effect. I shall have reason to return to this later.

There is therefore no reason for children to doubt the justification for adults doing what they do. The task of the child is to imitate it. It is up to each child what areas of knowledge he wishes to acquire. When children play together, the interests of each are broadened by those of his or her playmates.

Even when children show an inclination to acquire certain skills, the parents do not express any expectations. They do not give the children bamboo strips or cotton thread so that they can learn to weave. Children can only observe the whole work process – and the adults do not help by slowing their work tempo – until one day they can put it into practice themselves, and can cut strips and weave them, or spin thread and weave cloth. If they make mistakes and ask for help, they are not told what their mistakes are; they are simply told and shown the proper way of doing it. If the child stops trying, nobody tries to persuade him to carry on. One day, he will start again by himself.

Children are therefore free to develop special interests as they wish alongside the everyday skills, and they can apply themselves to special areas of knowledge. The result is a multiplicity of "*charaa*", or experts, in different areas of adult life. There is no secrecy around the expertise of a *charaa.* Anyone can see and hear him putting his knowledge into practice; whether someone is interested and wants to learn is up to him. (In the case of the ethnographer, who also wants to learn, it is a case of "Look, listen, write down what you like. If you want comments or explanations, you will have to ask for them, because we don't know what it is that you want to know". As soon as people start to trust a visitor, they are perfectly ready to give information, but they do not volunteer it: one has to ask.)

N4d) Schooling

The style of learning described in the last section stands in clear contrast to a school system. The existing schools are far from home, and therefore operate as boarding schools (in Ruma, for example). The teachers are badly paid, so they are often badly trained, and the educational system is not one that is appropriate to the children's needs. The schools are therefore unattractive. Nevertheless, there are some Mru who will make efforts to secure a school education for their children. When I visited Pantola-Mouza on 1.02.1956 for a large buffalo feast given by Headman Yöngtu, he was talking about building a school in the village at his own expense. (This came to nothing because he lost his position prematurely on account of an illegal matrimonial situation.) Cönglök Atwang of Tapwüa-Kua had visited the Bawm mission school in Artha (Calkhup Headman Para), and Menkröi Ngarua' had sent his son there – but the son soon came back. He said that the teacher was a horrible man who beat the children. The habits of the old English school system are the greatest hindrance, even for children who want to learn. This throws a different light on a comment made by Hutchinson: "It is a boast of the tribe that they possess no education whatever and do not want any" (1909:43). This is also not entirely true, at least so far as literacy is concerned. I found people who knew the Roman alphabet, and even in areas like Tapwüa-Kua, where there is relatively little contact with the Marma, I found others who write with Burmese signs, although these are ill suited to rendering the Mru phonemes. I even met a man who had recorded Marma mantras in the old Chakma writing, which was no longer officially used (Menyom, 5.08.1957).

These people had skills that sprang from personal interest, and that did not estrange them from their communities. Knowledge that is acquired through formal schooling, and that leads to other things, is a different matter. Preceding generations had coped with their problems without the benefit of formal education; those who had acquired such an education had become estranged in the process from their culture. Karbari Thümru Chimlung of Alekhyong-Mouza was proud that his son Era (known in administrative circles as Hira Mrung) was the first Mru to become a road construction "engineer" (i.e., supervisor) in government service; but Era married a Chakma woman because he never found a Mru girl who was willing to become his wife: she would have had to move away with him, or stay at home alone. From the point of view of his work, Era was no longer a Mru. Karbari Leuten Camthang, of the Khumi of Betchora-Mouza, anticipated similar problems. He told me (17.02.1957) that now he had two grandsons, he would like to pay for a school education for one of them. Times were getting harder; it would do no harm for the boy to learn something, and it might turn out to be useful. The question was, once he became an "educated man", would he want to continue with his old way of life? Would he remain a real Khumi, and be useful to his own people?

N5 Growing up

N5a) Sexual maturity

The child passes without ceremony from *ing-rik-caa* ("black-red-child", i.e., newborn) through the stages of *rai-ring-caa* (toddler) and *nau-caa* (child that still needs supervision) to *ngiawüa*. At the age of 14 or 15, the stage of *tadök-löt* begins, and *taklang-ca* and *malaa-ca* (boys and girls) become *klang-wüa* and *malaa-wüa*. The girls start to menstruate, usually at 14, some not till they are 15. Breast development starts earlier, but is not complete until sexual maturity. The start of menstruation is initially referred to as *ui pau wöi khai* ("having the fruit flower"); it is not marked by any ceremonies. Later, menstruation is called *bur* (*bur-wi*: menstrual blood), but menstrual periods do not impose any restrictions on a woman, or bring her any particular consideration. Both sexes may begin to have sexual intercourse when they wish; there are no rules about it, but it is not spoken of in public. Intercourse is euphemistically called *taphua-tayua;* it would be vulgar to speak of *tabö'*. *Tayua* could be translated as "becoming friends" (*ta* is a reciprocal prefix). The stem *phua* is also used in the expression *wang phua khök* (come-*phua*-was), "a child has been born". For discussion of children born before marriage, see M2.

Once they reach sexual maturity, boys are called *plot-dong-khök* ("loosening the loin-cloth happened"). Once again, there are no special ceremonies to mark this stage, though changes can be seen in everyday behaviour. The occasional homoerotic games are replaced by a new tension in dealings with the opposite sex; this is noticeable mainly in the new behaviour pattern *ruk-rai* ("embarrassment") shown by girls; it may be manifested as reticence, silliness or flirtation, depending on personality. The *klang-wüa* distance themselves emphatically from the children, and try to impress by their appearance, though they do not assume foppish ways, as the Marma sometimes do. They try to show off, and especially at home they boss the women around, fluctuating between patronising magnanimity and physical control. Siblings are in a much better position than parents to observe each others' adventures, and get in each other's ways. The *nau-ca* still sleep with their parents in the *kimma*, but the *ngia-wüa*, separated into sexes, sleep with their older siblings in the *kim-tom*, and can observe the *malaa-wüa* and the *klang-wüa* during their nocturnal activities.

Children have their first clothes when they are 4 to 6 years old, girls usually somewhat earlier than boys. They are not yet aware of the purpose of clothes, and do not try to use them for modesty, but treat them as new toys. By the age of eight at the latest, children have become aware of sexual matters, by hearsay, observation and experience. They then follow adult patterns in the wearing of clothes. Boys of five and six will play quite openly with their genitals, but by the age of 8 or 9, they cover them with their left hand, as do the adults, when they take off their loincloth to bathe. Girls of ten, though wearing a *wan-klai,* still climb trees without any embarrassment. Clothing gains a certain charm only when children reach the age of *plot-dong-khök* ("after the loosening of the loincloth") that is, marriageable age. Exaggeratedly modest behaviour brings an awareness of clothing, and this behaviour can be used as a way of flirting.

N5b) Starting work

When they are about ten, children begin to realise that there are ways in which they can make themselves useful in the adult world, and that this gives them status. They can carry out daily household tasks at an early age. Girls of twelve can do the routine housework independently, and when necessary they can be left at home to take care of the younger children. Boys of the same age are probably able to do those things too, but they prefer to explore their environment, and the land, animals and plants with and on which they live. One might say that they play at being hunter-gatherers. Their fathers give them their first hewing knife (chopper); this will be a disused item that has become too small and too light for an adult through repeated sharpening. The boy will play by using the chopper to perform tasks that he sets for himself. No one expects him to make anything useful; his progress is driven by his own desire to prove that he is growing up, and can ask for better tools. The desire to be recognised for what one can do gains further impetus when a boy becomes aware that the girls are weighing his achievements: a bad worker will make a bad partner in life. Wealth and good fortune will not make up for it, because only a good worker will be able to keep those things (Karbari Chigohan, 5.12.1956).

As soon as work stops being a game, there is a social pressure toward work that comes from the family and from the community in general, because idlers are not respected. In the last analysis, it is a personal decision whether one wants to exchange a full stomach for more leisure; but nobody will willingly help an idler in times of need.

There are of course certain kinds of work that are not attractive, and that youngsters will leave to their parents for as long as possible. However, they do eventually assume those tasks and responsibilities on their own initiative, and generally before a stage is reached where they would be asked to do it, or where it would be expected as a matter of course. A boy called Chakching Khongtör provides a good example. He was a fifteen-year-old adoptive son who worked the field to support his lazy father, his sick mother, and his two small brothers aged 3 and 5. He worked so hard that he had no time for the usual pleasures of boys his age. The deciding factor that makes people behave in this way is the sense of responsibility, the security provided by the family, the give and take of co-operating, the understanding of the need to perform even arduous tasks if one wishes to live. That is how things were, how things are, and it is part of the culture, so shrinking from it would bring even greater hardship. "What to do, if I am not happy here? I have no means to reach the civilized parts of the world, though I wish to go there." This is the reply that U Ohn Pe received when he asked an old Mru whether he was happy and content (1931:263).

N5c) Self-awareness

The "antiauthoritarian" upbringing of Mru children favours rapid development of a sense of responsibility. Nobody takes it amiss if adolescents oppose their parents, and think that they know better than their elders. Dingte at 15 called his father's older brother, Khamlai, "a mean sort of fellow" for not going along with Dingte's request to carry out some improvements to my house without payment, after I had supported him generously. Chingkrat Atwang quarrelled just as bitterly with his father Netkhai, whose grown-up children had done all the work in the past, and who did not want to take on any hard work himself after his daughter's marriage, even though he was still perfectly fit and strong. His son said: "When I get married I shall get out of here, then you'll have to manage as best you can." Nevertheless Chingkrat, who was the only son, stayed on in the parental home after his marriage. Wilful behaviour only stands a chance of being accepted by adults if it is within the bounds set by social rules, and can therefore attract general approval. An individual whose personal wishes are against the rules is in a difficult position, and most people cannot manage the conflict with society. This is particularly true where choice of partner is concerned (see M3b).

Young people have to learn to play their part not only in their own family and their own village, but also when they come into contact with the world outside. Adolescent boys may parade their claims to equality in their own circle, but their self-confidence evaporates when they leave their usual surroundings and find themselves amongst strangers. They become timid (*chak-chöm*) and hopelessly shy.

It is best to meet strangers for the first time in the company of one's friends. Groups of young lads will set off together in the evenings to go visiting in neighbouring villages. Feasts are also a good opportunity to broaden one's horizons: they give young people from several villages a chance to meet, but unlike the evening outings to other villages, they provide an opportunity to encounter plenty of older people whom they do not know, as well as those they do. Visits to relatives who live some distance away are also helpful: meeting strangers on these occasions is unavoidable. These trips provide experience in getting along with people other than those whose ways one has known since childhood; this entails recognising and mastering difficulties that one's parents dealt with at home.

I had an experience in which a father of two children was afraid to have any dealings with me, and put me off until his own father came back in the evening (Lenten-KP, 16.02.1957). It is true, however, that there was an additional factor at work here. Up to that point, the younger man had only had to deal with people like himself; but if one wants to assume full adult responsibility, one will sometimes have to leave one's own culture, where one has known only the indulgent good will of one's parents, and come into contact with the world outside. There one may experience guile, ill will and scorn, and be obliged to put aside one's youthful pride and learn obsequiousness. If a young man has not had to cope with the demands of authority during his childhood and youth, he will not have learned how to submit to them, avoid them, or oppose them without losing his dignity. He will therefore stand before the demands of an outside authority not knowing how to react, and feel powerless rage, which turns all too quickly into a helpless fear and tries to deny its own existence.

N6 Play

N6a) Imitative play

Many of the games that Mru children play consist in imitating adults. All the things that adults do – such as working in the fields, holding feasts, eating meals, having sexual intercourse – can be mirrored in play. When particular objects are needed for the game, the children may use the real thing if it is available to them. If not, they will substitute something similar, or simply use "pretend" objects.

From 24.04.1956 to 17.05.1956, I stayed in the home of Menkröi Ngarua'. There I had plenty of opportunity to watch his three-year-old daughter Roupau at play. One day, she involved the other members of the household in a game of *caa-hom* ("eating-rice"). A broken plastic plate of mine and an empty tin served as crockery. Roupau set these things in front of her brothers and sisters, urged them to eat, and pretended to eat with them. Her mother said that if she wanted to play at being the mother, she should kneel like a grown woman. Much to everyone's amusement, Roupau said that in that case, she ought to have breasts as well! Nevertheless, she changed her position and knelt like a woman, and when she got up to fetch water to drink, she did a wonderful imitation of the "sailors' gait" that the women and older girls adopt. Finally, on her own initiative, she fetched a bamboo stick to eat like the *phalong*, that is, like me.

One of Roupau's favourite games was *kom-bong* (see R1i). She bound threads round her mother's and siblings' right wrists, reciting the proper words, and spitting on their arms and heads. At the end, someone had to put a thread round her wrist, too, and spit on her. On 8.05.1956, she played *rui-param* (see N3a) with a pointed bamboo stick, and was happy when it pricked. She was not quite sure when and how she was supposed to spit on her siblings' ears, so someone showed her.

On 7.08.1956, she was playing in front of the house with her brother Menpaa, who was two years older. They decided to play mothers and fathers. Roupau lay down on top of Menpaa, and he put his legs round her. Netkhai Atwang's wife Tumcong was looking on and laughing from the next house. Menkröi then came along. He was not amused, and tried to distract the children.

One of the favourite games of all the children was *plai* (dancing). All the village children up to the age of ten or so would play. Sometimes they would just dance; at other times, they would act out all the ceremonies of a cattle feast. This needed a certain amount of preparation, so some of the older siblings would take part. There were parents who did not like to see the children dancing around the real sacrificial post; they feared that it might put the spirits in the mood for a real feast, and the spirits might then make somebody ill, to make one happen. The children would therefore set up their own sacrificial area. On 8.07.1956, some older boys who could already use a chopper dug a hole and set two small logs in it. Two stones were rolled up to their bases, to represent the animals to be sacrificed. On 29.04.1956, a piece of banana log was produced; four bamboo sticks were stuck into it for legs, and horns were made from bamboo strips. After the children had circled the animal several times, they stabbed it with a sharp stick, and pretended to cut it up and eat it. A piece of meat (banana wood) was fastened to a wall of the house with a sharp stick. To represent the gourd-pipes (*plung*) used to accompany dancing, the children had thin bamboo sticks of different lengths, some of them more than three metres long; each boy took either one or two of these and held them in front of his stomach. They made the music with their mouths. Less often, they would play pretend percussion instruments, for which they would use all kinds of strange things, such as my empty cigarette packets. For cymbals they would generally use two pieces of coconut shell, which they would beat together and sing *ceng-ceng-cek*. On 8.07.1956, they played at holding a big feast with a *riyang* (see R3b). For this they even made a gong stand, on which they hung a metal lid, which they beat with a stick. The dancing followed the usual rules (see R2c). Sometimes the roles were mixed: on 29.04.1956, for example, Roupau danced with a gourd-pipe, but at the same time she went down on her knees, as girls do when they dance. Menpaa tied a cloth round himself and tried to dance like a girl, but he was not very good at it!

During the morning of 26.11.1956, some renovations were made to my house. The children were looking on. Kyo Thwân Ong gave them some rusks and told them to sweep up. Karbari Kangku Catumma saw what was going on, and told the children that now they had been paid, they would have to work the whole day. However, once they had had their fun stirring up the dust, the work held no great appeal. In the afternoon, they had a different idea, and decided to play *mran-pua* ("drunken Marma"). I also found Kyo Thwân Ong drunk in the house next door; it seems that even rowdy behaviour on the part of the children may actually be imitative play.

Another game in which the children all co-operated was building a *te* (wind shield, see 3.08.1956). The final article bore witness to their enthusiasm rather than their skill. It was not really useable, because to make a proper wind shield, one needs a sharp chopper and enough skill to gather the bamboos and cut them into strips. Anyone who has the chopper and the skills will be needed to do "something more useful".

However, in Menrong-KP on 3.04.1956, I found next to a full-size house a miniature one about 50 cm high, with a platform, a log staircase, and everything else that a house should have. The work was beautifully done. I was told that children had made it while playing, but I suspect that an experienced hand had also been involved.

N6b) Sports and other games

It is not always possible to make a clear distinction between sporting activities and the games described in the last section. Children may climb trees for different reasons. Sometimes they do it just for fun, especially the boys. At other times, they climb trees to gather fruit. Sometimes they start to play at being monkeys, and will suddenly start to screech like a gibbon. Poking around in a heap of dirt with a bamboo stick might turn into a shared game of pounding rice or sowing a field. In the rainy season, children may try to stick leeches on each other; this is imitative play, and it also develops dexterity.

Things that adults do as a matter of course have to be learned in childhood, for example, jumping down from high places. The log staircase to the house makes a good "springboard": first the child practises jumping from the lowest step, then from the next one. He can also progress to long jump, using a bamboo pole and jumping either from the log staircase or from the house platform if it is not too high.

The children play pole vaulting in groups. Each child takes a bamboo pole about one and a half metres long and grasps it in the middle with both hands. They jump while running, putting first one end of the pole to the ground, then the other. The whole group will run across the village square in this fashion, shouting "*huang-huang*" (4.08.1956). When they are tired, they may use the same bamboo poles for pounding rice.

There is a game that involves jumping from one hand and one leg; this takes real skill. The performer raises one leg in the air, and jumps from the other leg and one hand. He strikes his chest at the same time with the "free" hand. The two hands beat alternately against the ground and the performer's chest (Ruirong Ngarua', 19.07.1956).

There are no real competitive games or combat sports amongst children. Sometimes one boy will take hold of another and try to throw him on the ground; but that is just a kind of teasing, because one of them will soon run away, to get the other one to chase him. Sometimes they lift and carry each other. They will also stand back to back, put their arms round each other's bodies, and each tries to lift the other into the air. Alternatively, it can be done sitting; the aim then is to lift the other boy over your head. This only works if you can already do a somersault (19.07.1956).

N6c) Children's rhymes

The children often make up short phrases or word combinations on the spur of the moment, to accompany rhythmic games, or just as a kind of rhythmic chant. These chants are not always very polite; some girls of four and five, for example, invented one, and jumped up and down on a lid chanting: *malaawüa kai klipca* ("the girl's vagina is winking"). They kept chanting this for days, marching arm in arm across the village square.

Other verses are traditional, like one about pinching the dung beetle (*ki-töng cen*), which everyone knows and loves:

ki-töng cen-cen, röp-kar mak-mak,
ki-töng cen-cen, röp-kar mak-mak,
tam-tu-(ca) wang-po-la ki-nam-prun!

"Dung beetle pinch-pinch, land crab nip-nip, Tamtu brings stink of dung". –

While reciting this rhyme, the children sit in a circle and pile their hands in the middle, palms down, so that they can pinch the backs of each others' hands with their thumb and index finger. The pile of hands is raised and lowered on each repetition of the words *ki-töng-cen-cen.* The children say the last line very quickly, pull their hands back, and put them in front of their faces as though ashamed. Everybody then starts to laugh, and the game can begin again (MK, 31.07.1956).

Another rhyme is

Ting-tung khwök-khwök, pang-lai khwök-khwök,
or-raa en u en paa wang-po-la wa-cek-ca dui khök,
khwök carök-rak!

"Swing splish-splash, cucumber splish-splash,

Look up there! Your parents have brought little birds' eggs, splash your legs-in-the-air!"

Several children squat on the ground with their legs pulled in and sway, reciting the lines in chorus. If someone looks up in surprise on the second line, he will fall on his back with his legs in the air (Menching Atwang, 30.07.1956).

I heard two versions of the next rhyme; the second one does not make so much sense, but in Tapwüa-Kua it is the commoner of the two. The stress is on the first, third, and fifth syllables, and so on. The first version is:

O püi praa catim ca, o ce praa catim ca – rüm!

"Down the river he sways, up the river he sways – DOWN he goes!"

The second version is:

Li-ku praa catim ca, or ce praa catim ca – rüm!

"In the storm he sways, up there he sways – DOWN he goes!"

To play the game, one boy squats; a second one faces him, puts his feet on the shoulders of the first, and holds on by linking arms with him. The first boy holds on to the feet of the second, then stands up and sways him to the rhythm of the chant until the second cannot hold on any longer and falls off (MK and Menching, 31.07.1956).

There are other children's rhymes that are not part of a game, but which children simply recite quickly to themselves. Speed is of the essence, which means that the sequence of words is not too clear, and the adults are not always able to make sense of them either. On 31.07.1956, with Menkröi's help, I noted a rhyme that Langrung Catumma, who was about 9, was saying:

pi-pu ö daing-daing cong,
le-ma tui-kang chang ang dia,
ku-ning pak-la khem-a cho,
tüa pe töm-ma kök.

Some parts could be literally translated as follows:

"Aunt-uncle hey (drumming sound),
?~? water dried, won't you teach me,
nine-year old pig, singed, hisses,
? give the drum ?"

On 2.08.1956 I noted the following rhyme, recited by Menpaa Ngarua', aged five:

lip lu ö, ting krüa-nüa, cang lu ö, ting kriia-nüa,
alö katep khommi mong dür!

Menpaa's brother Dingte (15) corrected the last part; he said it should be *alaa katep khomi mong, khomi katep alaa long-um dür!*

Only the beginning is clear:

"Tortoise's head, hey! wobble to and fro, lizard's head, hey! wobble to and fro." The second line could mean: "That's how the Khumi beat their gong, they beat it in the house, BOING!"

I will end with a few *ta-o'-wan* (riddles) that parents ask their little ones, usually just before they go to sleep:

krong-reng hi-ta lu krüt, tong-cö? – caa ngen khök!
"The earth man's head is shorn completely bald.
What is it? – The earth when the rice has been cut."

Ra-khöng ma kap, Anok ma kap, tong cö? – klöng!
"It is fixed towards Arakan, it is fixed towards Anok.
What is it? – The cross-beam of the house."

cia-ma cia-laa tahau, tong-cö? – kim!
"The cow and the bull cover themselves.
What is it? – The house (which has a small *kimma* and a large *kim-tom*)."

kim bet mia tua, tong-cö? – ram-cheng!
"A bundle of wood strips on the outside of the house.
What is it? – Ear decorations."

klang-ca ria tan, tong-cö? – yüng!
"The young boy's hailstones stand in a row.
What are they? – His teeth."

kim klang klep nge, tong-cö? – dong-lu-kau!
"The long tail that hangs down from the house.
What is it? – The bamboo at the head of the staircase." (This refers to the decorative pole erected at cattle feasts).

klang-ca man bong chür, tong-cö? – na-kong!
"The pattern of the young boy's basket, turned around.
What is it? – His nose."
(MK and Kangku Catumma, 19.08.1956).

N6d) Sports

The older boys and the young men (*klangwüa*) take part in activities that are more like sports than children's games, though the younger children try to copy them. These activities include pole wrestling, high jump, walking on stilts, and football. Pole wrestling is the only popular combat sport. Young men often do it as a trial of strength when they meet at feasts and the girls are watching, though adults will also compete with the young men on occasion.

Two young men stand opposite each other at an angle, with their right shoulders pointing away from each other. They hold between them a strong bamboo pole, about two metres long, which passes under their right armpits. Their left hands cross on the pole. The aim is to push the other person around by forcing the pole upwards or downwards away from oneself, without losing ground.

It takes a lot of training to be a good pole wrestler, and even so, one's ribs start to hurt after a while. I was told of a keen pole wrestler who had one rib pushed in (28.09.1956). There is a variation in which one of the contestants puts the pole between his legs and holds it with his right hand in front of his body and his left hand behind. The other holds the pole with both hands in front of his chest and tries to push his opponent around. He usually fails, because the one who has the pole between his legs grips it between his thighs, and this gives him more power to throw his opponent down (30.04.1956).

High jump is an imported sport rather than a local tradition. Two bamboo poles are pushed vertically into the ground, and a thin bamboo lath is tied between them. The ties holding the horizontal lath can easily be pushed higher; if someone fails to clear it, it either breaks, or it bends and pulls out of the ties. A pole is sometimes used in high jump (19.07.1956).

Stilts are made from bamboo poles about the height of a man. Short pieces of bamboo ending in a node are set into the poles at an angle, about 30 cm from the bottom. Each pole is grasped between the big toe and the second toe; the short piece of bamboo acts as a support for the ball of the foot (28.07.1956).

The Mru do not seem to know the rules of "leg ball" as played by the Marma, much less those of international football. However, boys will set up two "goals" and play in pairs, each defending his own goal and attacking the other. This suggests that there may have been some external influence from football.

There are no balls. In July 1956, I saw boys using unripe grapefruit, much to the annoyance of the owner of the tree. The grapefruit were quickly destroyed, and a ball made from an old cloth wrapped in bamboo strips did not last long either (28.09.1956).

Gymnastic equipment is unknown in Tapwüa-Kua. At the urging of Kyo Thwân Ong, rings were made from tree roots bound with bamboo strips, and suspended by means of longer strips from a strong branch of the village acacia tree. The bamboo strips were later replaced by a rope. However, interest in the new piece of equipment flagged on the first day. Rope climbing is a well-known activity, but in practice lianas are used instead of rope (27.07.1956).

The Mru enthusiasm for climbing trees is often greater than can be explained in terms of its practical usefulness. They consider themselves the best climbers in the Chittagong Hill Tracts; there is no tree that they cannot climb. When climbing thick trees, they use their choppers to cut diagonal notches in a fishbone pattern as footholds. Alternatively, they may tie small pieces of bamboo crosswise on the trunk, using bamboo strips. These are also supported from below by vertical bamboos, and form a kind of ladder. A third method is to bind a short crosspiece to a long bamboo pole, hang it in a reachable fork in the tree, and climb it up hand over hand. Accidents are not uncommon. In the Tapwüa-Kua neighbourhood in the summer of 1956, a boy cut half his face open. Another – the son of Karbari Yöngtu – became mentally disabled as a result of a fall on his head (MK, 15.11.1956). On 31.12.1956 I was called to a man in Borpara who had fallen from a tree on to a bamboo stump, which had cut a piece of flesh from his thigh.

N7 Toys for young and old

N7a) Children's toys

I almost never saw toys made by adults especially for the children. The Mru do not engage in woodwork for art's sake, and the toys they make from bamboo fall largely into the category of puzzles or brain-teasers (see N7c), which are too advanced for the younger children (*ngiawüa*). The children play with things that they find for themselves and bring to life in their imagination. They find some kind of container in which to collect their treasures. Menpaa, for example, carefully gathered bits and pieces of mine – mainly scraps of paper – in an old tin.

The only common bamboo toy is the catapult with which boys between the ages of about eight and twelve like to play. The catapults are made by older boys and adults. They take one internode of a bamboo tube about as thick as a thumb and cut three holes in it, a small one at the back and a second one, about 4 cm long, at the front, about 15–20 cm away from the first. The third, which takes the form of a small notch, is cut into the underside, under the back part of the long hole. On the upper side, a bamboo strip is stretched like a bow from the hole at the back to the hole at the front; the strip is rather thinner at the front, and cut to a point, so that it can be placed in the notch on the underside of the tube. If it is pushed from below with a finger, the bamboo strip snaps forwards inside the tube against the front end of the long hole. While the bamboo strip is in the stretched position, one can put small stones, or sticks or seed balls into the front end of the tube; these are catapulted out when one "pulls the trigger". The catapults are small, and do not shoot with enough force to do serious damage. They are used for chasing chickens, dogs, pigs, and one's playmates.

In Yöngtu-HP, I was shown a spinning top. The headman himself had made it from a piece of wood. It had a short, thick section and a longer, narrower one; the result was more or less mushroom-shaped. A thread was wound around the stem and then pulled quickly away. The practice of keeping the top going by whipping it was not known, and in any case, its shape would have made this difficult or impossible. The adults demonstrated the top for me; the children tried a few times, mostly without success, and soon abandoned it (30.01.1956).

I never saw such a thing anywhere else. The headman had a good deal of contact with the outside world, so I suspect that the idea came from there. On the other hand, Lewin writes: "The boys play at 'konyon', or as they call the game, 'tsing khing'; and the peg-top is also a favourite plaything" (1870:234). The first of these games, "konyon", is called *khun-ñang,* or *entada* seeds", by the Marmnna. The Mru call it *ching-kling* ("shooting entada seeds"). For a description of the game, see Kauffmann and Löffler (1959). At the beginning of the 20th century it was still often played by children in the Chinbok area (Dokhinhangor-Mouza); today, the rules are forgotten (MK, 7.09.1956).

Children seldom involve animals in their games. Pigs are best left alone; they usually gallop off in any case, but sometimes they put up a fight. The children usually treat dogs in the same scornful fashion as adults treat them. They may pull their ears or tails for fun, or throw them around when they are puppies; but the dogs soon become more independent, and show their mistrust too clearly for children to seek close contact with them. Hens are even less interesting. Even their feathers are not of interest, unlike those of wild birds.

N7b) Toys for grown men: games of skill and small works of art

To show off their skill in bamboo work, the men sometimes make puzzles. There are two sorts of puzzle in which the task is to make a whole out of two similar parts that are linked together like parts of a chain, and cannot be separated. The first kind is made from a strong strip of bamboo wood and the second from a complete bamboo tube. The solution to one of these puzzles is shown in Figures a and b. To make the second, one takes a bamboo tube two internodes long and cuts out opposite sides so that the remaining parts form a chain. The art lies in breaking apart the centre of the node that joins the two internodes – which is notched in four places when the holes at the sides are lengthened – so that its top half forms an arch connecting the two remaining lower sides while its lower half connects the upper sides.

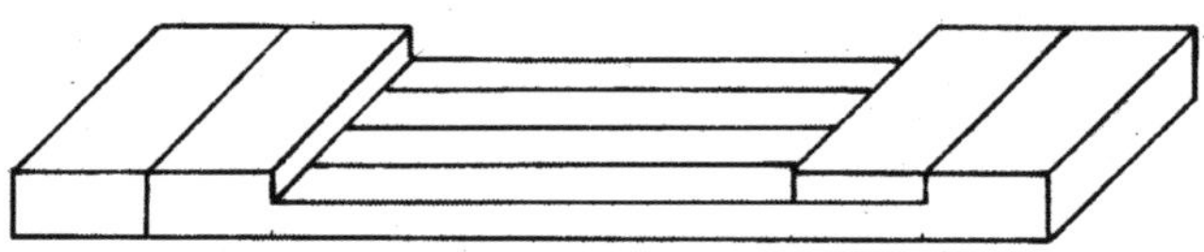

Fig. a

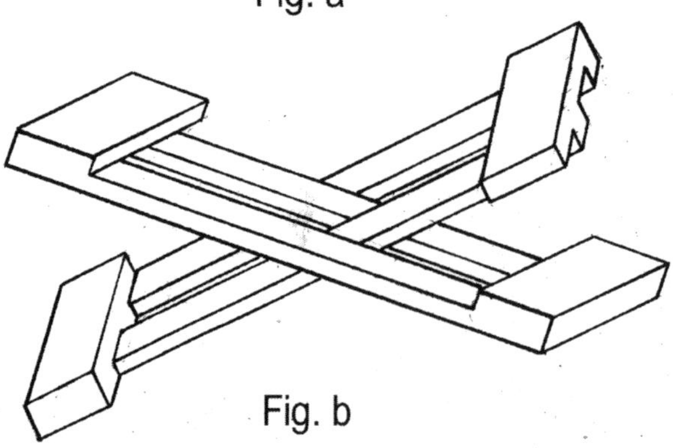

Fig. b

Another form of craftwork is to make a small ball from strips of wood. Khamcöng Atwang (30.09.1956) showed me an egg-shaped ball about 6 centimetres long, with a maximum circumference of 18 cm. The ball was woven from two strips about 0.8 cm wide, though it ought to be possible with just one, if one started in the middle and treated the two ends as two separate strips. Each strip is wound five times upwards and five times downwards, till the ends lie on top of each other. This forms a semicircle at the top and bottom, with a hole about one centimetre across at each end. The strips are wound over the previous rounds at the top, while below they are tucked inside and through them. The second strip passes over the first when it is wound downwards and under it when it is wound upwards, that is, the first downward round of the second strip passes over the first round of the first strip, and the first upward round passes under the second round of the first strip. The second time, it goes over the second and under the third, and so on. The ball was strong and stable, but I never saw children playing with it.

The Mru also weave a tube, which they call a "snake", from thin strips of wood. It has no practical use, but when it is stretched, it narrows, so if you put a finger into it, the more you pull, that tighter your finger is trapped. One needs two pairs of bamboo strips about 0.4 cm wide and 40 cm long. One strip is creased in the middle so as to form two arms with an angle of 45 ° between them. A second strip is placed crosswise through this angle and folded back around the two arms of the first strip so as to form two more arms; these cross over and run inside the first arms and parallel to them. The other two strips are prepared in the same way. The eight arms of the two parts are then woven together, over and under, to make a tube. The ends are folded over and tucked back into the weaving.

On 30.09.1956, Khamcöng Atwang brought two examples of his weaving to show me. One of these was a "snake", as just described; the other was a frog, which was also woven from narrow bamboo strips. The frog stood on its four legs and held its mouth open. Khamcöng also demonstrated his paper-folding skills: he folded a rectangular sheet of paper in two and two again, then once more from corner to corner, to make a triangle eight layers thick. In the two sides that had no open layers – the hypotenuse and one of the shorter sides – he used his chopper to cut a triangle parallel to the second shorter side from right and left almost to the opposite side. He then unfolded it and pulled it open; it made a kind of helmet (11.05.1956).

Another form of craftwork is to make a small ball from strips of wood. Khamcöng Atwang (30.09.1956) showed me an egg-shaped ball about 6 centimetres long, with a maximum circumference of 18 cm. The ball was woven from two strips about 0.8 cm wide, though it ought to be possible with just one, if one started in the middle and treated the two ends as two separate strips. Each strip is wound five times upwards and five times downwards, till the ends lie on top of each other. This forms a semicircle at the top and bottom, with a hole about one centimetre across at each end. The strips are wound over the previous rounds at the top, while below they are tucked inside and through them. The second strip passes over the first when it is wound downwards and under it when it is wound upwards, that is, the first downward round of the second strip passes over the first round of the first strip, and the first upward round passes under the second round of the first strip. The second time, it goes over the second and under the third, and so on. The ball was strong and stable, but I never saw children playing with it.

N7c) Brain-teasers

The first toy of this kind was given to me on 12.02.1956 in Longrau-KP. It is called a "bamboo rat" (*büi*). It consists of a bamboo tube 25 cm long and 2 cm in diameter, pierced by two central rectangular holes on opposite sides, and a notched hole at each end. A string made of twisted plant fibres is threaded through these holes with a loop passing through the two central openings.

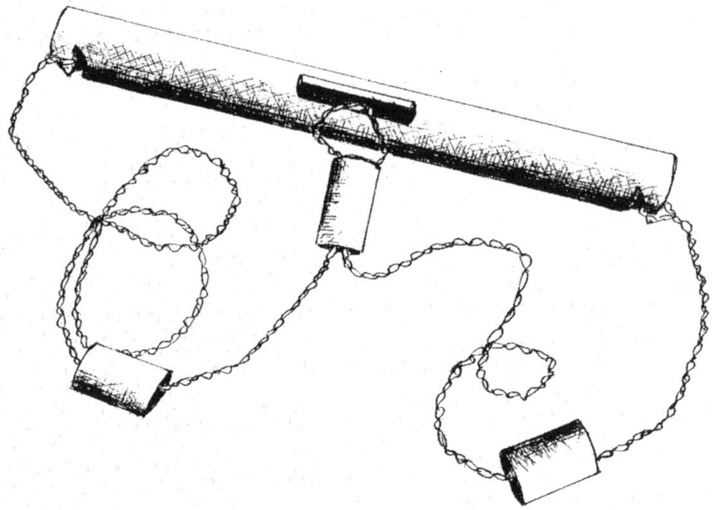

Büi

The top of this loop holds the ends, which are fastened in the notches at both ends and come up through the holes in the middle. This creates two side loops, on each of which is threaded a small bamboo tube 2.5 cm long and 2 cm in diameter. Both strings pass through a third small tube, 4 cm long and of the same diameter, in the middle. The task is to move the small tubes without releasing the thread so that two of them are on the same side, and the other loop is empty.

The solution is as follows:

1) put the middle tube and the side tube that you wish to move through the loop in the middle,

2) pull the head of the middle loop first back through the hole in the middle then through the middle tube, so that two loops now stick out from the middle tube,

2) push the side tube, which is now on what was the middle loop, through the pair of loops,

4) (reverse of 2) pull the pair of loops back through the middle tube and the central hole, and finally

5) (reverse of 1) pull the side tubes and the middle tube back through the (recreated) middle loop.

It seems simple once you know how, but people to whom I showed it in other villages (Yöngtu-HP, Menring-KP, Tapwüa-Kua) were not able to solve it without help. (The same applied in Germany!)

In Tapwüa-Kua, I was shown another brain-teaser (MK, 15.07.1956). It is called *pri-prang-lang* (tiger's leg). It consists of three bamboo strips tied together as shown in the diagram:

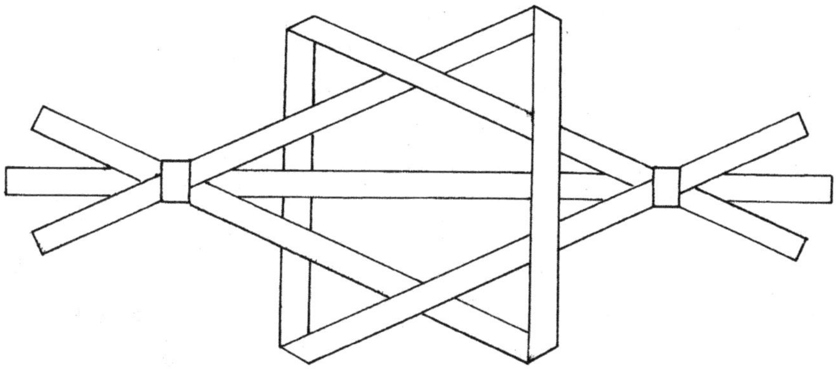

Pri-prang-lang (loop of thread not shown)

There is a loop of thread around the central axis; the task is to get it off.

The solution is as follows:

Pull the loop 1) from above over the left end, 2) between the central axis and the left side of the right-hand triangle, 3) from below over the right end and the right side of the left triangle, and 4) under the right side of the left triangle and out to the right.

N8 Children's drawings

N8a) General remarks

Mru children do not draw on paper because they do not have the materials. The only way they can draw is to scratch on bamboo and paint it with colours. Drawing in the sand would be another possibility, but I only saw this once, and I had probably prompted it myself, though without intending to do so.

In Tapwüa-Kua, during the afternoon of 6.09.1956, I happened to notice that Kangding Ngarua' and Langngi Atwang, both 8 or 9 years old, had scratched an oblong in the sand and drawn lines across it. In the spaces between the lines, and starting from the bottom left corner, they were drawing squiggles with bamboo sticks. They said the squiggles were *phalong-rui-mi* (Sahb writing). Next they decided to draw people, whom they called *philu-mang* (see P2b). They began each one with a long line for the body, put a circle on top for the head, and two short lines at the bottom for legs. To this basic shape they added two circles to the right and left of the head for eyes (*mik*); Kangding usually put his a little below the head. Under the eyes came a scratch of indeterminate shape; this was the *nakong* (nose). Below that, they put a small circle for the *nor* (mouth). Sometimes they added vertical strokes below the nose for *yüng* (teeth). They almost always added arms, too. Kangding added five fingers to the arms, counting them out as he worked. Langngi just made a cross to indicate the hand. If they drew hair at all, they put it inside the circle that depicted the head.

They had not been drawing people for long before they came upon the idea of giving the *philu* several heads. They did this by drawing a spiral: they counted the circular movements, each being an extra head. There were no ears or sex organs. Towards the end of the game, Kangding started to add *betma-krüa* (ribs).

On 7.09.1956 I asked for the two boys to come to me, and I explained through Menkröi what I wanted them to do. They had obviously forgotten their game of the previous evening, and when I suggested that they might draw on paper instead of in the sand, they were embarrassed at first. However, Kangding eventually had a try, then they both started to enjoy it, began to draw freely. By the end of the session, however, they were putting quantity before quality. From the beginning, they held the crayons in the same way as they had previously seen me hold a pen. Kanding's brother Menpaa, who was three years younger and whom I invited to have a try, did not manage to hold the crayon or to draw lines, and he soon gave up.

The drawings shown below are not always the work of the one boy or the other: they often added to each other's drawings. Nevertheless, one can pick out on the whole who did what, because Langngi was rather better at drawing lines than Kangding. The figures in the middle were generally drawn first, then the drawings were taken up again later and the figures at the sides were added. They are not, therefore, shown in the order in which they were drawn.

N8b) Notes on the drawings

The middle drawing on Sheet 1 was begun by Menpaa. He drew the rather misshapen circles inside the head, the lines to the sides of it, which were probably meant to be arms, and the horizontal legs. Kangding did some more work on the drawing, adding the circles under the head as eyes, the teeth – which he put on the chest (no mouth) – arms with five fingers each, and toes represented by three lines only. Kangding also did the drawing on the right: he counted out the toes and the fingers on one hand as he worked, but obviously forgot to do so for the second hand. The eyes are to the right and left of the head; the nose is a small circle under the head, with a line under it; the mouth and teeth are shown as a circle with lines in it, just over the arms.

The drawing on the right is Langngi's. It has two heads, which were drawn as a spiral, and eyes. Underneath the head are Langngi's "inventions": ears, shown as circles with a line in each (ear openings); the mouth quite small; and lower down, the ribs: he did not want to draw any teeth! He added a circle to the hands for the *rut-phaa* (palm) and a line across the arm for a *kuai-che* (bangle). The legs are bowed, and between them is a double circle for the *hang* (scrotum with two testicles) and standing out from it, the *nia* (penis). The hands and feet all seem to have six fingers or toes, except for the hand on the left of the picture, which has seven. However, for each extremity, the pencil was only put to the paper five times.

Both the drawings on Sheet 1 were done mainly by Langngi. The left one was his first attempt to make a drawing on paper. The body and head were done by Kangding. The numbers of fingers and toes vary, but this drawing has ears. The figure on the right is the same in essentials as the figure on the left of Sheet 1. There is no nose, but there is hair on the head. The palm of the hand is shown only on the right.

On Sheet 3, the middle drawing was done mainly by Kangding, even though the fingers and toes are just forks. Langngi added the ears; the armpits were added later by Kangding himself, after Langngi had "invented" them (see Sheet 4, not reproduced here). Ribs are typical features of Kangding's drawings; where Langngi copied them (Sheet 1, left), he put them in the wrong place.

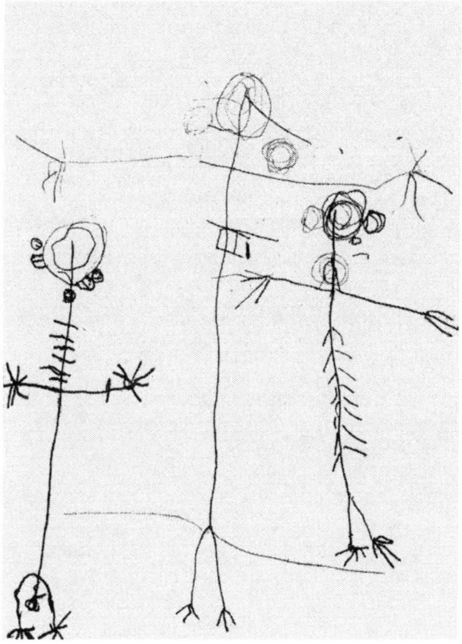

Sheet 1 (50 %)

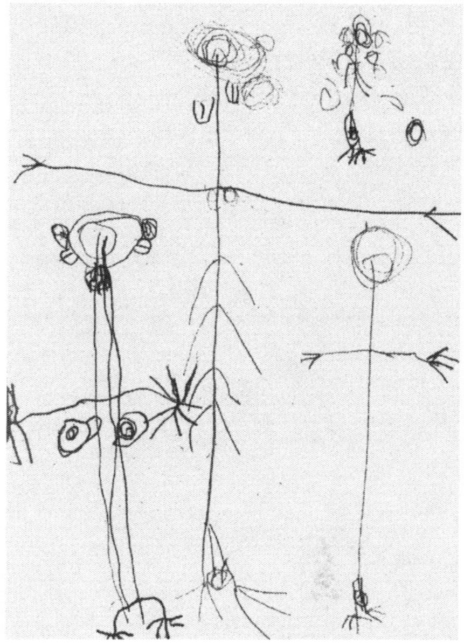

Sheet 3 (50 %)

Kangding also drew the two figures on the right. They are not early drawings, but later "standard models" in which detail was neglected. At the top on the right is a *philu-caa*, or *philu* child, surrounded by *phalong-rui-mi* (writing). At the bottom right is a *machi-wüa* (woman), recognisable as such only by the *kai* (female sex organs). The drawing on the left is Langngi's. It shows a *philu-ma,* a female spirit, which, unlike Kangding's woman, is identifiable by its breasts and a broadening of the body. It has two heads, a mouth with teeth, arms with innumerable fingers, and legs which seem to have five toes. The pupils are drawn in the eyes, and the breasts have nipples.

On the left of Sheet 4 is a four-headed *philu* drawn by Kangding. The figure has hair on its heads, this time drawn by Kangding himself; there is not so much of it on the right. Unlike Langngi, he has drawn the hair over the eyes, which are drawn separately below the head. Under this there are teeth (no mouth), below which two lines represent the jaws (*kamma*). The fingers and toes are again reduced in number. The penis is shown not as a separate line but as an extension of the body, as can be seen more clearly in the central figure on Sheet 3. The drawing on the right is also Kangding's, but he abandoned it after he made mistakes in drawing the legs, which are bent.

In the middle is a rather original drawing by Langngi, in "half profile": it has two eyes on one side, to the right over the ear. The mouth, teeth and jaws are below and also to the right. The left arm is bent round to the right. Here for the first time he drew *yak-dak* (armpits). In the middle of the head there is *rui-mi* (writing). The ribs were added partly by Kangding. There are five fingers and four toes. Below this figure (on the other side of the paper, which was turned over) Kangding drew a *cia* (bull) with horns, a humped back, and sex organs. Amongst the figures there is "writing".

When the children were drawing in the sand on the previous day, Langngi's efforts were significantly more basic than Kangding's. However, when they drew on paper, it was Langngi who made more innovations, and attempted to draw more difficult details, while Kangding produced a larger number of "standard models", to which he returned after making his one animal drawing.

P Medical knowledge, spirits, *khang,* believes, and pig sacrifices

P1 Medical knowledge

P1a) *rok-mi*

The Mru possess only limited medical knowledge. At first I thought people were telling me they knew nothing about medicine because they had heard I had a medicine box, and they came in the hope that I would give them a cure. I realised later that they also went to the Bengali market to buy medicines that they thought might help, although they rarely knew how to use them properly. An extreme case was a man who came to me because his kneecap would not heal. He had cut himself with a chopper, damaging the patella, and had pushed a pain-killing tablet into the cut, which of course prevented it from healing.

Up to the end of the nineteenth century, the Mru had *rok-mi.* The term literally means "healers", but "shamans" might be a better translation, because these men were also credited with the ability to rise up into the sky (*pün-kwang*). Their bodies remained lying on earth as though they were dead, but their *achak* (life spirit, soul) ascended, and could communicate in the sky with *Thurai*, the god of the Mru. A *rok-mi* would announce in advance on what day he would go to visit the people in the sky; perhaps it would be to attend one of their cattle feasts, to which all *rok-mi* and *ri-chi* (Marma: holy men) were invited. The *rok-mi* would also accompany dead persons to the sky and reunite them with their parents, if the latter had already died (see Q1n). No one seemed to know any more than this about the *pün-kwang,* or about the attendant preparations and ceremonies. I was equally unable to discover how one became a *rok-mi*, or whether the cattle feasts in the sky were linked to deaths on earth. All I could find out was that the *rok-mi* never committed any misdeeds on earth, and that they were able to identify and heal many illnesses. Their knowledge included the *ho-ra* of the Marma. They could also make medicines (MK, 16.08.1956). I could not discover when or why most of this knowledge was lost; it was said, however, that there were men amongst the Southern Mru who retained some of it.

P1b) Experiences in the 1950s

During my first stay among the Mru (1956–1957), I spoiled my chances of finding out about their medical practices by agreeing to share with them the contents of my medicine box. I did not make the offer myself; the people of Tapwüa-Kua simply came and asked me whether I could help, for instance when they had an attack of malaria. I had enough tablets, and I knew that I could buy more in Chittagong, so I did not hesitate to share my supply. I would give them two tablets free of charge, to show my gratitude for all that they so generously gave to me.

I also asked them what traditional remedy they would have used if I had not been there. The answer was: nothing, we have no medicine against the fever (*krau*). Later, however, I learned that there was in fact an indigenous remedy: the leaves of a certain plant – I was not told which one – would be boiled and the liquid drunk. The leaves could also be applied directly to the body. However, these procedures were not thought to be particularly effective.

People then came and asked me whether I had something for dysentery. I did. Again, they told me that they had no remedy of their own, except to take *bing* (opium). They asked me why the government sold them opium, which was a bad medicine, when something better was available. To be fair, the government did not simply sell opium; it was illegal to supply it without a prescription from a doctor. The doctor, however, would demand a large bribe in return for the prescription. An alternative was to buy opium from a dealer, who bought it from smugglers: most of the opium consumed in the Chittagong Hill Tracts was smuggled over the border from Arakan. The police did try to catch the smugglers; but it was common knowledge that when they did, they would make deals with them, usually wanting at least 50 % of the goods.

Of course, opium was not used only as a remedy for dysentery. It could be taken once or twice without ill effect, but anyone who took it more than that was likely to become an addict for life. Addicts had a bad reputation, though some were reliable people so long as they had the money for the drug. Getting the money was the real problem. Some would sell all their harvest and go hungry; without the drugs, they could not work in any case to produce the next harvest. They therefore sold whatever they could, from their labour to the sexual services of their own daughters. This meant that no decent woman would marry a drug addict, and people in general avoided areas where opium was used. Little wonder, then, that opium was declared a "bad" medicine, or that the people of Tapwüa-Kua were happy that I could provide them with a less dangerous cure for dysentery.

My reputation spread, and soon I woke up every morning to find patients from neighbouring hamlets waiting in front of my house. Menkröi normally had to be there to explain to me what their complaints were, and what they wanted. Sometimes I really could help, as in the case of a woman who came with a large suppurating wound on her upper arm. She had been caught by a local vaccinator, who had vaccinated her against smallpox. The Mru as a rule are afraid of these vaccinators, and flee from their houses to hide in the jungle until they have left. If the villagers are caught, they pay money to be let off the vaccination. When I saw this woman, I understood why they were so afraid. The vaccinators never used a sterilised needle – even the "doctor" in Ruma (see below) did not sterilise – and the arm often became inflamed.

I put some penicillin cream on the woman's arm and bandaged it. I told her to apply some more cream – of which I gave her a very little quantity, wrapped in a banana leaf – and to wash the bandage thoroughly the next day and put it back on. I asked her to come back in two days.

She did indeed come back. Her arm was almost healed, and she brought me a whole basket full of paddy, with a couple of eggs on top. I did not want to accept the gifts, because I did not take payment; but she insisted on leaving them with me.

The Mru were not generally afraid of vaccinations, and if there was a real smallpox epidemic, they might even ask the vaccinator to come. I could not discover why they believed in vaccination. Some men even came to ask me for an "inzikshen" to cure the aches and pains of old age. These men were some of the many people whom I could not help. Another was a man who had been to the doctor in Banderban with pains in the third finger of his right hand. The doctor had made an incision and severed "something". The pains had gone, but the patient could no longer move his finger. Now another finger had started to hurt, and he was afraid to go back to the doctor. He probably thought that the doctor would cut through another "something", and leave him unable to use his right hand. He was probably right. Unfortunately I could do nothing for him.

For myself, if ever I felt ill when visiting the plains, I never went to the doctor in Banderban. I preferred to walk the 40 miles back to my own medicine box, especially since Menkröi had told me that the sickness was to be expected: hill men always fell ill when visiting the plains.

I was also called upon to act as dentist. One man came to me complaining that he could not chew because it was too painful. When he opened his mouth, I saw that one of his teeth had come out, but had remained stuck between its neighbours. I used my tongs to lever it out. Another time, an old lady came who had just one tooth left. She probably hoped that I would give her dentures. I told her that the only thing I could do was pull out her last tooth. When I showed her my tongs, she fled.

I had to refuse to help when a man came with his leg covered in sores, and asked me to cut the leg off. Another man came, also to show me his leg; I assumed that he was a Marma, since he wore a slightly torn Marma lungi, and he called Kyo Thwân Ong to act as interpreter. The man smelled bad, and as soon as he lifted his lungi, Kyo Thwân started to scream, and ordered him to leave my house immediately. The poor man had leprosy; all I could do was tell him to visit the hospital in Chandraghôna, though it was said that the treatment there would probably not help. Lepers are treated as outcasts; they are not allowed to live in a village, but must build themselves a hut in the jungle.

Subsequently, on 17.08.1956, Menkröi told me of an old man living on the Mrungo who knew a lot about traditional remedies. This man had cured the wife of one of Menkröi's classificatory brothers by wrapping some leaves round her infected lower arm.

Another man came to see me because he could no longer walk without pain. He had had an accident in which he hit his kneecap with a hewing knife (chopper). He had limped down to the plains and asked in the local pharmacy for a remedy; they had given him MB tablets, which were sold as painkillers. He had put one of these tablets in the open wound and waited for it to heal, but it did not, so he had come to me. I tried to get the rest of the tablet out of the wound, cleaned it with antiseptic, and told him to go home and wait for his kneecap to heal. A few days later he came again, to tell me that my treatment had worked.

Another man walked right up the hill to see me, from a village near the edge of the plains. He wanted some medicine for his wife: she was at home, about to give birth to her first child, but the labour was not progressing. I told him that the only medicines I had to give were the ones I had brought for my own use, and since I was man, I had not expected to give birth. I said that I was very sorry, but I could not help his wife.

The young man would not believe me. I fetched Kyo Thwân Ong to explain the situation to him in Bangla, but he would not believe anything a Marma told him. I then called Menkröi, and asked him to explain; he too failed to persuade the young man to go home. His wife was at home with labour pains, but the baby would not come. She might be dead when he got home, and it would be a "bad death" (see Q2a); so please, would I give him some medicine to save them from the worst.

By now it was evening. I decided to give him one pain-killing tablet. He was overjoyed, and set off home at a run, while I started to wonder whether I had made a serious mistake. What if his wife did in fact die?

The next morning, the young man came back to tell me the happy news: less than an hour after taking the pill, his wife had given birth. Mother and baby were both well, and everyone was happy. He said he had known that I was a good doctor and could help his wife, and he thanked me very much.

Eventually my patients became a real burden on the local people, because the visitors would expect to be housed and fed in the hamlet. Furthermore, my supply of medicines started to run out. The villagers asked to talk with me about it. After a short discussion, we agreed that I would stop giving medicine to anyone who asked, and keep what I had for the local people.

From time to time I learned something new. One day my neighbour Yenpau came and asked whether I could help his son. The boy's spleen was swollen and extremely painful. I had no idea what to do, but I promised that I would ask the "doctor" in Ruma for advice next time I went down there. I had come to know this man during the early days of my stay; he was a Chakma in government service, and a very pleasant person. He had a little training in first aid, and was given a small supply of simple medicines that he was supposed to distribute free of charge. In fact he had to sell them, because it was impossible for him to live on the 20 rupees a month that the government paid him. An ordinary policeman, for example, earned a monthly salary of 60 rupees.

When I visited the "doctor", he gave me a heavy bottle containing a liquid, and said that the patient should take one spoonful after every meal. I passed this on to Yenpau, and asked him soon afterwards whether the medicine had helped his son. He told me that the boy could not keep it down: every time he took it, he vomited.

When I saw the doctor the next time, I told him what had happened, and asked him what he had given me. He replied that it was quinine tablets dissolved in water; the painful swelling of the spleen was caused by malaria, but quinine makes some people vomit. He suggested that I should give the boy some of my malaria tablets; these did indeed bring about an improvement.

On another occasion, I asked the Chakma "doctor" whether he had anything that might help with a serious ringworm (tinea) infection. My own medicines had not helped, nor had the usual Mru remedy. Infections of this kind are not uncommon amongst the Mru; their treatment is to rub the affected area with *plia-chi-khang* leaves (winged *Cassia*). In this case, the ringworm had spread all over the man's body, and he was having to face the fact that no woman would want to marry him. The doctor gave me a special powder which was to be mixed with kerosene and then applied to the affected parts; the patient must be very careful not to get the powder into his eyes. I gave the patient the powder and the instructions.

One day he came and told me that the medicine was working well, and the ringworm was fading. However, he had put the small box containing the powder on a rack above the place where he slept; the previous night, he had knocked the rack and the powder had fallen on to his face. His eyes were burning. I could do nothing except tell him to wash his eyes as often as possible. Eventually the burning stopped, and his sight was not damaged.

The treatment with *Cassia alata* leaves was almost the only local remedy that was described to me. It involved crushing the leaves with a knife and mixing the sap with minced garlic, soot and salt. The mixture was applied to the infected area.

Most people knew that *wöi-la-ram* (*Euptorium odoratum*) leaves had an astringent effect that could stop leech bites from bleeding. I used them several times myself, but I could not get hold of them everywhere. In the end I hit on a simpler way: I lit a cigarette and pressed the burning end against the leech. It would drop off immediately, without leaving a bleeding wound.

It took me some time to learn to cope with the leeches. The first time I made their acquaintance, I had climbed up to a house and was about to take my shoes off. The shoes were crawling with leeches, which I had never seen before. I was terrified, and flung my shoes across the room. My hosts were most alarmed, and enquired why I was behaving so oddly, especially since my feet were not bleeding. Eventually I realised why the leeches could not bite me: I was wearing the early version of knee-length nylon socks. These socks have not been available for many years now. They prevented the leeches from sucking, and they did not let the air through.

My woollen socks soon proved useless, partly because they offered no protection from leeches, but also because Kyo Thwân Ong washed them and tried to iron them. They shrank so much that they would have been about right for a two-year-old.

Scorpions can be a nuisance: they do not suck blood, but they sting. Fortunately they are rare. The Mru treat stings by trying to cut the wound open to wash it with potassium permanganate, which they buy at the market. They always have potassium permanganate to hand because young people use it freely to colour their cheeks or to paint a red spot on their forehead (see M1b). They also take *cang müt* (lizard gall) mixed with honey to lessen the pain. *Rua-lim-müt* (python gall) is regarded as a somewhat better alternative, but no other kind of snake gall will do. The Bengali use *ghee* (melted butter). However, when Menkröi's son Kangding was stung, neither *cang müt* nor *ghee* had much effect. These two medicines are also taken as a remedy for severe stomach pain (MK, 15.11.1956).

I asked what the Mru would do if someone was bitten by a snake. Again, the answer was that they had no cure. Not all snakes are dangerously poisonous, though the green viper is an exception. The doctors had no remedy either. At the Chittagong hospital, I was told: "If the victim gets here alive, then the snake was not a very poisonous one. If the snake was really poisonous, the victim will die on the way." Later I learned that this was not strictly true: the sufferer would not necessarily die, but there was a danger that the bitten limb would swell and start to rot. If that happened, the whole foot would have to be amputated.

I had brought from Calcutta two ampoules of a snake bite remedy. On 1.07.1956, Khaiyuk Khongtör, a boy from a neighbouring house, came running to me and told me that he had been bitten by a snake while weeding in a cleared area. He thought it was a green viper. I thought that the snake might not have been poisonous, so I told him to calm down while I tried to suck the venom from the wound. This was not very successful because the bite had gone into his knuckle. The boy then started to turn blue and wanted to vomit, though he was able to hold it back. It looked as though he might die, so I injected one of my ampoules. I was probably just in time to save his life.

On 1.10.1956 Menching brought his mother-in-law to me. She had been bitten by a snake while cutting bamboo shoots. This time I was able to suck blood out of the foot, using a small bamboo tube. I then injected half of my second ampoule, keeping the other half in case anyone else should appear with a snake bite. Happily, this did not happen. However, on 26.12.1956, while travelling in the southern area, I met a man who had been bitten some time previously. His hand was still badly swollen.

The Mru were affected by cholera from time to time. They call it *ri-naa*. This is a Marma (Arakanese) loan-word, which suggests that cholera was known to the Mru at the time when they still lived in Arakan. During the period when I was visiting the area, in the 1950s, cholera would spread from the plains every year between May and July.

Despite their long familiarity with cholera, the Mru had no effective remedies; they simply closed the village (see C1b). In fact this was not strictly enforced, because no one was prevented from leaving. Indeed, they would sometimes flee to neighbouring villages if there was a death from cholera.

The last severe epidemic was during the Second World War, when many men were employed as casual labour in the area above Thanchi. The epidemic greatly reduced the male population, but the overall effect of this was to restore the previous unequal sex ratio to normal. In the 1950s, if there was a case of cholera, it was no longer thought necessary to close the village, because outsiders would be too afraid to enter it. The people who lived in the village might stay there, but usually they would choose to go into the jungle, perhaps staying in their field houses, until the danger was over. This at least helped to keep the number of deaths down. People would also leave the village if someone had smallpox (*hoi).*

People who had died of cholera were usually buried. Death from cholera was considered a natural (i.e., not "evil") death, but it was not thought desirable to perform the whole *romma* ceremony (see Q1). No sacrifice was allowed following death from cholera. It was forbidden to eat fat, or to singe the skin of pigs, because the smell was said to increase the number of infections.

Sometimes people would try to protect their house against a spreading epidemic by preventing people other than family members from entering. To this end, they put out *kung-kaa,* which were thongs stuck into the ground at both ends and arranged in a semi-circle. The process of setting out the *kung-kaa* was called *tom-kim* ("binding the house"; Menkröi 30.11.1956). *Kung-kaa* were also used during a *khang,* to cut off the whole village from outsiders (see P3c). Céline Mouchet mentions a ceremony called *tom-kua* ("binding the village"). She gives no further details, except to say it was done because 17 people had died in the village during the previous two years (see P1d).

I did not expect that the Mru would have traditional remedies for venereal diseases; this expectation was fulfilled. I was told that some men had been infected during a visit to Chittagong, where they had had sexual relations with prostitutes. Not all of them seemed to know where the infection had come from; this was perfectly possible, since neither men nor women had any knowledge of modern diseases of this kind. So long as the disease was in the acute stage, sexual relations were avoided, but once the visible symptoms had disappeared, the sufferers were treated as if they were cured. This lack of knowledge might have contributed to the spread of venereal diseases; Menkröi reckoned that there was probably one case per village. Sufferers did not usually consult a doctor. In severe cases they would go to the hospital in Chandraghôna, but they were often afraid to do so because it was said that the headman of Renikhyong-Mouza went there, and had part of his penis cut off (MK, 7.08.1956).

I managed to gather information about a few more ailments. One of them was called *chaa-waa* ("hot-chicken") if the sufferer was a child, or nor *ngak* ("bad mouth") if it was an adult. The condition looks like a kind of pyorrhoea or thrush. The inside of the mouth is covered in white blisters; it looks as though the person has sour milk in his mouth. Eventually the skin peels off. If the disease reaches the throat, the sufferer may die. This is especially likely to happen if the disease occurs in babies under 6 months or pregnant women. Babies can be affected at any time of year, but the danger is greater during the cold season, when adults and older children also develop it. The illness usually lasts up to two weeks, though in stubborn cases it can last up to four. Menkröi told me that he had it every year for ten or twelve days. The Anok Mru had no remedy (MK, 7.02.1957); the Rümma told me they held ashes of ku-ram (Didymospermum leaves) and honey in their mouths (K Rengtan Rümthu, 14.03.1957).

I was told of a less common illness called *chak-tan,* which was said to be often fatal. I was not able to identify it; I learned only that it involved ulcers in the body, especially in the region of the back and the abdomen. If the sufferer recovered, he or she would pass pus in the urine or faeces, which would appear white in consequence. In Tapwüa-Kua, the *püi-höng-charaa* Changlong (see P2g) had pus coming out of her navel for five months before she recovered. In other cases, the disease had been known to last 6 to 9 years. No one knew the cause of the disease or how to cure it (MK, 15.04.1957).

One day a man came and asked me to help cure his broken collar bone. He said that he had fallen from a tree, but this seemed unlikely. Finally it emerged that he had tried to shoot some deer with his old gun, but had not held it firmly enough against his shoulder, and the recoil had broken the bone. From this I learned that the Mru still knew how to make guns of the old muzzle-loading type; but this was a big secret, not to be disclosed to any outsider. I promised not to tell anyone. Even so, the man would only tell me that when they wanted to make guns, they would go to the plains and buy gas pipes to make the barrels.

I feel able to break my promise now because of an incident that took place in October/November 1985. The Mru living near Alikodong[467] overpowered an encampment of Shanti Bahini; these were Chakma "freedom fighters", part of a guerrilla army fighting against the government. The guerrillas had brought a Mru girl into their camp and the girl had died. The Mru living in the surrounding villages were enraged, and they joined forces, took out their old guns, and attacked the camp one night. The Chakma fighters were equipped with modern arms, but they were taken by surprise and fled. The news of the attack spread, and the government praised

[467] Date and place according to Harry Belitz in "Delta" no 1, p. 42. Munich 1987.

the brave Mru and promised to arm them with modern guns in exchange for their old ones. For some years afterwards, young Mru men could be seen in Banderban, clad in their usual loin-cloth, but wearing a red ribbon round their upper arm and parading their new guns. However, these "Lal Bahini" ("Red Forces") never really served the government, and never saw a decent payment, so they finally "deserted" and returned home, taking their modern guns with them (Raja Ong Shwe Phru, end of 1990).[468]

During 1957, the last year of my stay, my reputation as a doctor waned. There were two main reasons for this. The first was that in the spring, I was called to a neighbouring village where a small boy had fallen into one of the big round cooking pots that are known to us by their southern Chinese name *wok*. A pork curry was cooking in the *wok* over an outside fire. I went and found the boy unconscious, his back charred, and the skin on the rest of his body, even on his stomach, so swollen with lymph that it was loose and could easily be torn off. I applied some powder, but had no great hope that it would help. The boy woke up for a moment and demanded some salt. This was given to him, and I was invited to drink some arak (distilled rice-beer). While we were drinking, we heard the boy's mother screaming; her son had died.

Even so, the villagers were grateful to me. If I had not come, they would have had to report the event to the police, because it would have been a case of "evil death" (see Q2a). However, because I had helped the boy and he had woken up for a moment, it must have been the bad spirits which had then caused his death. This did not have to be reported. When I returned, the people of Tapwüa-Kua had no doubt expected better news of the boy, but they also felt released from the obligation to close their village, which they would have had to do if I had come from a village in which a "bad death" had occurred. I, however, was upset about the incident, and I had to wait until I returned to Germany to ask a real doctor what he would have done. He told me that he himself could not have done anything in those circumstances; the only thing that could have helped the boy would have been specialist treatment in a clinic.

The second reason why my medical reputation began to lose its glow was that the first wave of Asian flu reached Tapwüa-Kua. I myself was spared, but several villagers were infected, and there was nothing in my medical kit that could help them. They therefore sacrificed a chicken, and for some days they wore blood-soaked cotton threads round their necks. This appeared to be a newly invented precaution against a previously unknown illness.

Finally my own health deteriorated – I suffered constantly from diarrhoea –and I did not have the money to stay any longer. I told Menkröi that I would have liked to stay, but that I had run out of money. He offered a solution: just in front of Tapwüa-Kua was the place where the borders of

---

468 [I add here a note from Céline Mouchet, which she sent me in 2007:] The official name is "VDP": Village Defence Party (they wear a dark red shirt). The Mru VDP (there are also Bengalis VDP in all the country) still exist. The brother of Aishom, Menrum, is a VDP commander, and I know that there is another Mru commander in Thanchi area. Menrum trains few Mru VDP in his village (actually his ex-village) and I saw on few occasions the training he was giving to them (mainly physical exercises, gym, etc.). Menrum told me that Mru started to be VDP in 1984. A commander is paid now 600 taka per month, the others receive 450/month. They receive their money from the army camps (but they are often late in paying them). Under Ershad, there were about 10,000 Mru VDP (Menrum said), and now they are about 600 …

four mouzas met. There they would build me a new house, and the people of the four mouzas would supply me with food, and money to buy new medicines. I replied that I doubted whether I was much use as a doctor. He then gave me another reason why I should stay. He said that since I had settled in Tapwüa-Kua, social relations amongst the villagers had improved, there were fewer quarrels, and they were not so afraid of the Bengalis, who had actually started to treat them with some respect.

In the end I decided to leave anyway, because my health was deteriorating. When I reached Germany, my weight was well under 50 kg, and I had to go into hospital for treatment for amoebic dysentery. The doctors told me that I had survived an attack of pleurisy that I had not even noticed. They hoped also that they would be able to eliminate the malaria from my system, but I did not have my first real malaria attack until the following summer. I treated it with the rest of the tablets I had used in the Chittagong Hill Tracts, and in the end it disappeared without further treatment.

P1c) Medical knowledge in the 1960s

I made a second short visit to Tapwüa-Kua in 1964. This was also my last visit, because the Chittagong Hill Tracts were by this time closed to foreigners, and the police cut short my stay on the grounds that it was illegal. In 1990, I was not even allowed to proceed beyond Banderban.

In 1964, I had intended first to accompany my students to the Bawm area, and to visit Tapwüa-Kua afterwards. However, when I reached Banderban, I met Menkröi, who told me he had dreamed that I would come, and had come down to the town to welcome me back. I therefore changed my plans, and went back with Menkröi to his hamlet.

I fell ill in the first few days, probably as a result of a leech bite, and one of my legs became swollen. Unfortunately, I had not taken my medical kit with me. It therefore fell to Menkröi, in whose house I was staying, to try to cure me. A chicken was killed, its feathers burned, and the ash was applied to my leg. I am not sure whether the treatment helped, but after a fortnight the swelling went down, and I was able to walk again. These events provided a good opportunity to ask about traditional remedies for other ailments. This time I amassed a little more information.

I was told about a new disease called *Kö-lak chak-tan;* the main symptom was chest pains. A remedy for this was made from *ram-klang* leaves (not identified), which were pounded in a bamboo vessel and boiled. The patient drank the resulting liquid. I also heard about another medicine called *ram-klaa chi-khang;* this would be made for small children who suffered from frequent fevers. It was prepared from a number of plants, namely, *rua-kok, chi-kung-kaa, mahaa-kaa, hua-pau, ngetca, nam-pramca,* and *ut-klui-mai.* I was not able to identify any of these. They were mixed and pounded in a small, hollow piece of bamboo sealed at the bottom by an internode. The resulting liquid was given to the child to drink.

The plant *bong-beu* (unidentified) was used as a remedy for snake-bite; its sap was applied to the limb that had been bitten. Its leaves and roots were collected and kept in the house as a precautionary measure: they were said to keep out snakes and bad spirits. I was not able to identify these plants because I would have needed the flowers in order to do so, and my short stay did not coincide with their flowering season.

I was able to identify two other plants that had medicinal uses. These were *waa-ko* (*Verbenacea: Clerodendron infortunatum*) and *chi-chaing-da'* (*Asclepiadacea: Calotropis gigantea*). *Waa-ko* flowers in late February and in March, and bears fruit in April and May. Its roots are crushed on a stone, mixed with water, and applied to the forehead as a remedy for headache and high fever. *Chi-chaing-da'* is used to combat rheumatic pain and swollen knees (*ceng-leng*).

Chi-chaing-da' appears to be a Marma term. I discovered that *Calotropis* is an old ayurvedic herb which is used as a cure for rheumatism and elephantiasis. The medicine is made mainly from the root bark, but the leaves and flowers are also used. It is also said to be effective in cases of heart disease, abdominal disorders and dysentery, leprosy, syphilis, and oedema, amongst other things. It is now used in homeopathic doses under one of its many Indian names (e.g., Madar) to combat the development of fatty tissue, and by models and film stars to keep their weight down.[469]

P1d) Current medical knowledge (notes by Céline Mouchet)

I did not manage to meet with a Mru who specialised in medical matters. I am therefore taking the liberty of quoting a passage from the MA thesis of Céline Mouchet (2001: pp. 34–46), who met two *charaa* (she writes *sra),* one among the Anok, the other among the Dopreng Mru. The latter was a very old man who had learned Marma, and had also made use of medical knowledge that he had acquired from the Marma people. It is not clear whether he was still using any of the old Mru knowledge, for instance about plants.

"Un sra ne se trouve pas dans chaque village et il est fréquent que les Mru doivent en faire venir un de l'extérieur. Je rencontrais un sra Anok dans le village de Menkom (situé dans Porda mauza – Thanchi Thana), habités par des Mru Chüngma. En fait, à peine arrivés dans ce hameau, nous dûmes en repartir très vite car le village allait se fermer aux étrangers dès le coucher du soleil et jusqu'au lendemain matin. Une cérémonie appelée *Tom-kwa*[470] devait servir à résoudre certains problèmes, car en deux ans 17 personnes du village étaient décédées, de mort plus ou moins étrange. Pour diriger cette cérémonie, un sra d'un village voisin fut appelé afin qu'il procède à différents rituels. Nous ne pouvions donc rester dans le hameau mais j'annonçais déjà au sra que je viendrais lui rendre visite afin de m'entretenir avec lui. J'appris plus tard que pour effectuer la cérémonie, le sra reçut des villageois entre 800 et 1000 takas[471], ce qui représente une somme importante pour les Mru. Toujours dans ce village, le même sra était également intervenu auprès d'une famille dont l'enfant était malade, et ceci afin de vaincre l'esprit Madala qui s'était emparé de lui. Le sra, cette fois-ci, n'organisa pas la cérémonie pour Madala mais il donna ses instructions à la famille. Pour prodiguer ses conseils, le sra était payé entre cinq et dix takas par la famille. Selon L. G. Löffler, un sra dans les années 50 et 60 était avant tout un spécialiste, et non un «professionnel». Je fus moi-même assez surprise que le sra reçoive de l'argent des Mru, au sein d'une communauté avant tout basée sur l'échange et où l'argent ne tient pas encore une place essentielle. Menkom, le Karbari du village, en me parlant des sra, me dit que

---

[469] Holisticonline.com: Arka; Medizin-News, 1.2.2006

[470] "enserrer le hameau" (L. G. Löffler)

[471] 1 franc = 7.50 taka environ

Menlong kua – le village du sra cité auparavant – possédait deux ou trois de ces spécialistes, que les sra avaient reçu l'enseignement de moines bouddhistes (il veut peut-être dire de Marma), qu'ils savent écrire marma[472] et qu'ils sont donc capables de lire les textes décrivant les actions à faire pour attirer la bienveillance des êtres invisibles.

Quelques jours plus tard, je pus retrouver dans son village le sra rencontré à Menkom para; ces retrouvailles se firent dans des circonstances très particulières que je décrirai plus loin en évoquant Kramma, la nouvelle religion mru. Ma discussion se restreindra uniquement à ce que le spécialiste me raconta concernant ses connaissances sur la religion traditionnelle. Le sra déclara avoir reçu l'enseignement de son oncle. Je demandais à Idui des détails sur le village fermé mais il resta très évasif. Il m'expliqua avoir écrit des formules magiques en marma sur des petits morceaux de papier qu'il avait ensuite enterrés à différents endroits autour du village. Nous avons demandé au sra de bien vouloir nous montrer un livre dont Karbari Menkom nous avait parlé, un livre qui lui aurait été donné par des Marma et qui contiendrait différentes formules magiques, des conseils, des actions à effectuer en cas de maladies, de problèmes avec les puissances, etc. Après avoir hésité, nous avoir montré plusieurs cahiers d'écoliers sans importance, il se décida finalement à sortir un petit «carnet» de quelques feuillets (certes qui n'avait rien d'un livre), assez vieux semblait-il et qui contenait quelques schémas. J'ai pu recopier tant bien que mal trois de ces schémas et inscriptions les plus lisibles. Après cette entretien, il nous sembla que les connaissances du sra n'étaient pas à la hauteur de sa réputation. Simple concurrence ou réelle connaissance? Toujours est-il que le sra Penköi, lorsque je lui montrais ces schémas, déclara que seule la première formule concernant Madala était correcte.

Le sra Penköi, un Mru Dopreng, est un vieux bonhomme âgé, selon ses dires, de 90 ans. Penköi est un être particulier, au regard perçant, au visage quelque fois menaçant; il dégage une tranquillité puissante et mystérieuse. Le sra habite avec sa femme, son fils remarié et son petit-fils issu du premier mariage. Penköi commença à être sra vers l'âge de 17 ans. Il se maria à 15 ans, mais environ deux ans plus tard, femme et fils décédèrent; Penköi fut très affecté par ces disparitions. A cette période, il pensa un moment partir en Birmanie pour y apprendre la langue. Puis il commença sérieusement son apprentissage de sra. L'enseignement fut reçu de deux sra, Longyen, son oncle paternel, et Chompat, son grand-père maternel. Le père de Longyen, Renkung, et le père de Renkung, étaient aussi sra. Penköi fit dans le passé une rencontre qui lui permit de devenir un sra puissant et capable de conduire des rites magiques: alors qu'il était jeune et pauvre, et qu'il se rendait au bazar, il croisa sur son chemin, un Richi (ou Toni Richi Pati, désignant une sorte d'ange, de saint). Le Richi est en fait un envoyé de Thurai, il «travaille» pour lui. Le Richi donna au sra des potions magiques pour qu'il acquière un pouvoir. Le Richi en partant déclara au sra que celui-ci ne le reverrait jamais; en effet, Penköi ne le rencontra plus. Penköi eut donc une révélation dont il retira une vertu permanente; il obtint ses pouvoirs aussi par tradition (une grand partie de sa famille était sra) et par initiation avec l'enseignement de ses «maîtres». Lorsque je demandais au sra si sa succession en tant que sra était bien assurée, il me répondit qu'il

[472] Le marma est une forme archaïque du birman.

enseignait son savoir à son fils. Je demandais s'il en savait autant que lui, mais il me dit que non, et que de plus, son fils était maintenant un homme qui pouvait décider par lui-même d'apprendre ou non. Le sra enseignait aussi avant à un autre garçon du village mais celui-ci voulait apprendre trop vite, ce qui avait le don d'énerver le sra qui choisit alors de ne plus rien lui instruire.

Penköi ne se contentait pas de diriger parfois encore des cérémonies religieuses, il était surtout réputé pour ses rites magiques, ses enchantements, sortilèges, guérisons ... La maison de Penköi contenait tout un bric-à-brac intéressant: petites fioles, objets mystérieux emballés dans des sacs plastiques (il y en avait une multitude accrochée aux murs), des cornes de chèvres sauvages, des cordons, des ossements, etc. Penköi nous montra une partie de ces petits trésors, comme une pièce d'un quart d'anna datant de 1897, ou encore de vieilles photos jaunies ... Penköi descendit d'une étagère une vieille malle poussiéreuse. Il l'ouvrit devant nous, celle-ci contenait un tas de petites choses: morceaux de papiers, fioles, ingrédients divers, et surtout des livres. Il ouvrit d'abord un emballage de papier contenant trois ouvrages anciens écrits en birman et contenant, selon le sra, des formules magiques, des règles à suivre, des potions curatives. Le sra savait lire et écrire birman et se servait des techniques et amulettes birmanes pour exercer ses pouvoirs. Ces livres lui auraient été donnés par son grand-père, eux-mêmes donnés par son père, etc. Penköi était même prêt à nous prêter ces livres pour que nous les photocopiions, mais nous refusâmes – connaissant le peu de soin des Bengalis à manipuler les ouvrages qu'ils devaient reproduire. Il était aussi fort possible que le sra, en nous faisant cette offre si rapidement, nous faisait subir une sorte de test en vue de savoir si nous étions pressés de lui emprunter quelque chose.

Le sra nous montra de vieilles inscriptions birmanes assemblées sur un ensemble de feuilles de palme formant un petit cahier long, étroit et rectangulaire. La couverture de ces petits livrets était en bois; un petit lien muni d'un morceau de bois fermait le tout. J'avais eu l'occasion de voir lors d'un séjour en Birmanie des livrets similaires qui contenaient en fait des horoscopes, des prévisions astrologiques. Penköi nous permit d'ouvrir ces livrets qui paraissaient très vieux, mais, ne connaissant rien au birman, il me fut impossible d'y comprendre quoi que ce soit. Plus tard, lors d'un autre entretien, Penköi me présenta un livre en birman vraisemblablement plus récent. Il me montra le dessin d'une amulette destinée à se protéger d'un sra ennemi. Ce dessin représentait une tortue avec en son milieu, à la place de la carapace, une grille d'inscriptions birmanes. Je demandais au sra la possibilité de recopier quelques dessins mais il refusa, pensant que cela pouvait être dangereux pour lui et pour nous aussi. Si jamais quelqu'un voyait ces dessins, cela pouvait être utilisé contre lui par un ennemi, par exemple. Le sra m'autorisa tout de même à regarder de nouveau brièvement le livre. Je n'insistais pas trop auprès de lui par respect et peur de l'offenser. Lors de mes derniers entretiens avec Penköi, je fis une nouvelle tentative pour consulter ses livres mais cette fois-ci, il refusa même de me les montrer. Il m'avoua craindre que, de mémoire, je recopie certaines formules; cela pouvait alors l'exposer à des risques divers. Lors du dernier entretien, en désespoir de cause, je demandais à noter juste le titre des livres qu'il possédait, sans qu'il les ouvre, mais il n'accepta même pas, par peur encore que cela soit utilisé contre lui ou le mette en danger si cela venait à tomber dans de mauvaises mains. N'aurais-je pas dû faire cette demande dés le premier jour alors qu'il était peut-être moins méfiant? Il y a également certaines précautions à prendre avec les livres du sra Penköi: en effet, le premier jour de notre rencontre, j'ai failli poser le livre sur mon pantalon, une chose à ne pas faire; il vaut mieux consulter l'ouvrage à même le sol pour éviter de le «souiller» au contact des vêtements.

Le sra n'était pas toujours rassuré par mes nombreuses questions concernant ses pouvoirs, les puissances, les ennemis, les sorts ... Un jour, après un entretien, il sembla un peu perturbé, hésitant, dubitatif, un peu inquiet sur nos intentions. Il me demanda aussi pourquoi je lui posais de telles questions. Jamais auparavant quelqu'un ne l'avait interrogé sur ses pouvoirs et le monde invisible. En fait, les connaissances de Penköi sur la magie devaient rester cachées à l'écart du grand jour et du public; Mauss nous dit «L'isolement, comme le secret, est un signe presque parfait de la nature intime du rite magique. Celui-ci est toujours le fait d'un individu ou d'individus agissant à titre privé; l'acte et l'acteur sont enveloppés de mystère.»[473] Aussi Penköi tenait à conserver ce mystère. Notre informateur et interprète, jeune Mru qui connaissait déjà bien le sra, nous avoua qu'il avait peur de poser toutes ces questions à Penköi, une personne qu'il respectait et craignait pour ses pouvoirs. Il redoutait que le sra se venge par quelques maléfices. Plus tard, les anciens du village me confièrent qu'il n'était pas très bon que j'interroge trop le sra sur la magie et les êtres invisibles, qu'ils connaissaient eux-mêmes toutes les puissances surnaturelles; ainsi ils me proposèrent de leur demander des renseignements à eux plutôt qu'au sra, qu'ils ne s'en sentiraient pas du tout dérangés et seraient même heureux de répondre à mes interrogations. Mais le sra accomplissait des rites magiques que les autres villageois n'effectuaient pas; l'exercice de la magie lui semblait réservé.

Je demandais à Penköi si, en tant que sra, il y avait des interdits pesant sur lui, des restrictions, comme par exemple sur la nourriture ou certaines actions. Il répondit: «le sra ne peut pas aller dans une maison où une femme accouche et il ne peut pas se rendre dans la maison d'un homme qui se meurt». Il me donna un autre exemple: «si une femme accouche, la maison tuera un cochon et un chien; si la naissance s'est passée sans problèmes, je pourrai manger de ce cochon et de ce chien. Mais si un incident se produit lors de la naissance, femme et enfant morts ou malades, alors je ne mangerai pas de ces animaux tués pour l'occasion.» Le sra ne peut pas boire un arak trouble et blanchâtre; il ne peut pas manger une variété de courge; il ne peut pas consommer ce qui m'a paru être une variété de fougère (celle-ci, mélangée à du curry est régulièrement mangée par les Mru). Je lui demandais quels risques il prenait à manger ces choses-là ou commettre ces actes, il me répondit qu'il risquait de perdre ses pouvoirs.

Pour Penköi, le rôle du sra en tant qu'homme médecin est très important. Il doit être qualifié et posséder de nombreuses connaissances pour arriver à soigner efficacement. Les plantes sont couramment utilisées par le sra. Il pense que Thurai lui envoie des plantes aux pouvoirs spéciaux. Pour faire des plantations de tek, le gouvernement bengali détruisit un de ses jardins situé dans l'ancien emplacement du village où, paraît-il, Penköi cultivait un très grand nombre de plantes médicinales. Maintenant, Penköi ne possède

---

[473] Mauss, M. 1950. *Sociologie et Anthropologie*. Quadrige, Presses Universitaires de France, 1999, op. cit. p. 15

plus qu'un petit jardinet près de sa maison. Il déclara que de nos jours il avait des difficultés à trouver certaines variétés de plantes; plusieurs avaient totalement disparu ou étaient introuvables dans la région. Penköi consultait un livre birman pour se servir de ces plantes. Le sra se servait parfois pour ses remèdes des graines trouvées dans la fiente de certains oiseaux, aussi des racines et des morceaux de bois particuliers. Penköi me montra sur sa terrasse une plante en pot aux longues feuilles vertes et effilés (impossible pour moi à nommer): il m'expliqua que lorsqu'un enfant avait la diarrhée, il fallait brûler un peu de cette plante, récupérer la cendre, la mélanger à du sel et de l'eau et administrer ensuite cette potion au malade. (On sait déjà que sel et eau sont efficaces pour réhydrater en cas de diarrhée.) Plus tard, je demandais au sra la permission de visiter son jardin de plantes mais il paraissait un peu réticent. Finalement, il demanda à son fils de m'y accompagner. Samong, le fils, me désigna d'abord une liane entourant un arbre, il en ouvrit l'écorce et me montra l'intérieur. A l'extrémité de la liane, l'intérieur formait comme une petite main. Je demandais alors ce que cette plante soignait et comment il fallait l'administrer. Le fils répondit qu'elle guérissait les problèmes d'allaitement et que pour cela, il fallait la porter en amulette autour du cou. Il fit une réflexion similaire au sujet d'une autre plante: quelques feuilles à porter en amulette. Moi qui m'attendais à découvrir une réelle connaissance des plantes médicinales, je fus plutôt déçue. Soit le fils ne voulait pas me révéler l'utilisation correcte de ces plantes (par peur de dévoiler les connaissances de son père), soit ces dernières n'étaient pas vraiment utilisées pour leurs éventuelles vertus thérapeutiques mais surtout pour leurs pouvoirs supposés «magiques». L. G. Löffler me dit que, lors de ses séjours dans les années 50, il avait abandonné tout espoir de trouver un Mru capable de connaître encore quelque chose sur les plantes médicinales. En dehors des plantes, le sra utilise toute une diversité d'ingrédients aux vertus magiques et supposées curatives. Il m'en montra plusieurs exemplaires, soigneusement enveloppés dans une multitude de plastiques ou conservés dans des fioles. Je vis ainsi des petites parties (indéfinissables) d'un cobra royal, des os de tigres, des dents de sangliers pour guérir les problèmes de toux et respiratoires, des pinces de scarabée pour soigner la chute des cheveux (des morceaux de pince sont alors mélangés à de l'huile) ... Le sra précisa que le cobra était utilisé pour rendre une femme de nouveau amoureuse d'un mari qu'elle n'aimait plus: une partie du cobra devait alors être mélangée à de l'eau, mélange que la femme devait boire durant la période de ses règles. Le sra me montra également dans un petit flacon du lait de tigresse solidifié, un médicament soi-disant très puissant dont on se servait notamment pour contrer Madala. Ce lait était alors déposé sur les lèvres de l'enfant et le sein de la mère. Le jour de cet entretien, un Bengali se présenta justement chez Penköi. Il venait consulter le sra au sujet de son fils paralysé depuis la naissance. Le sra lui prépara un «médicament»: il prit une grosse araignée morte et la broya avec un peu de gingembre et d'ail. Puis il dispersa cette mixture dans trois petites boules de coton. Il mit les petites boules dans un sac en plastique et les donna au Bengali. Une boule de coton devait être posée pour un instant sur chaque paupière de l'enfant paralysé, la troisième boule devait être avalée par ce dernier. Selon Penköi, le fils aurait vu une puissance malfaisante. Comme payement, le Bengali donna au sra des feuilles de bétel et des noix areca, un peu de riz et un billet de 20 taka. Pour terminer, le sra, tout en parlant au Bengali, entoura son œil gauche de sa main gauche, comme s'il regardait à travers un œilleton. Puis le Bengali, avant de partir, mâcha un peu de bétel et de noix areca que le sra lui avait proposés et la consultation se termina ainsi. Là encore, le spécialiste se fait payer pour ses services. Ses pouvoirs magiques réputés constituent pour lui un gagne-pain. Le sra observe des conditions pour pratiquer ses rites, il suit l'évolution de la lune pour administrer les «médicaments»; il ne les prescrit jamais un mercredi, connu comme un mauvais jour pour délivrer des «remèdes».

Nombreux sont ceux qui viennent consulter le sra pour des problèmes amoureux, notamment les Bengalis[474], femmes et hommes. Le sra m'expliqua que chaque jour de la semaine était associé à un animal ou un élément; si l'on naît tel jour de la semaine, on est associé à une certaine chose. Ces associations sont surtout utilisées pour vérifier que deux personnes prêtes à se marier peuvent former un bon couple (entente et fertilité). Le vendredi est associé au vent, le samedi au feu, le dimanche à l'aigle, le lundi au tigre, le mardi au lion, le mercredi à l'éléphant, le jeudi au rat. Par exemple, l'association du rat et du tigre est négative (car le tigre mange le rat), celle de l'aigle et du feu aussi, comme le vent et le tigre; en revanche, l'association aigle et vent est positive, ainsi que le feu et le tigre. Les Marma utilisent peut-être ces associations. Pour les problèmes amoureux, le sra produit également des enchantements, des sortilèges divers.

Je demandais au sra s'il communiquait directement avec les esprits; il répondit qu'il ne pouvait pas parler avec les puissances, mais qu'il avait la faculté de les voir. En outre, le sra ajouta qu'il arrivait à voir sur son ongle du pouce de la main droite les sacrifices à offrir quand un homme par exemple tombait malade. En le questionnant sur d'éventuelles visions sur l'avenir, il me dit qu'il pouvait avoir des visions sur le futur, mais sur une personne seulement et pas tout un groupe.

Je questionnais ensuite le sra sur les rêves. Selon lui, ils peuvent parfois instruire, révéler certaines choses et ont donc un certain pouvoir sur les hommes. Penköi rêva une fois qu'il donnait des médicaments à un homme qui s'était blessé à la jhum. Ce médicament, dans le rêve, parvenait à guérir cet homme. Plus tard, Penköi, qui se blessa dans son champ, mit en application sa vision dans la réalité; le médicament guérit Penköi. Mon ami Constantin confia au sra qu'il faisait régulièrement des mauvais rêves, le sra lui conseilla alors de tatouer ce motif dans son dos pour faire disparaître les cauchemars:

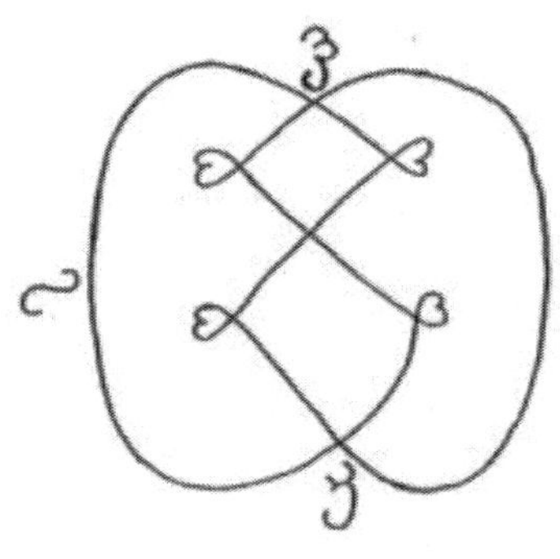

474 Les Bengalis sont appelés "Kula" par les Mru Dopreng et "Kwar" par les Mru Chûngma.

L'utilisation de la magie est double: le sra Penköi possède des pouvoirs bénéfiques mais aussi maléfiques. Certaines personnes viennent consulter le sra afin de s'attaquer à un ennemi: le tuer ou le rendre très malade. Le sra peut alors donner toutes sortes de plantes empoisonnées à mettre dans l'eau ou dans la nourriture de l'ennemi. Des poisons, selon Penköi, peuvent donner l'impression d'avoir un couteau dans la gorge (Penköi fit le geste). Les poisons donnés ne peuvent pas être contrés par les docteurs, c'est alors la mort assurée. (J'appris par la suite que le sra fut très embarrassé de m'avoir avoué cela). Le sra peut aussi séparer les couples, il peut empêcher quelqu'un de prendre un chemin, le faire changer de direction. Penköi sait user de la vengeance. Notre informateur Mru, qui connaissait bien le village, ses habitants et ses histoires, me parla d'une des femmes du hameau, Tumpöi, qui était devenue un homme. (Nous avons rencontré cette personne, il était évident de voir que c'était une femme à l'origine – elle portait maintenant des vêtements masculins et des cheveux courts). Cette femme était autrefois mariée et avait eu un enfant. Son oncle était le sra Penköi. Lors du mariage de cette femme et du don de roupies d'argent par la famille du mari, le sra ne reçut pas sa part (ce qui en effet est anormal, il aurait dû avoir quelques roupies). Pour se venger, le sra punit les deux familles en transformant Tumpöi en Pressun, c'est à dire la femme en homme. S'ensuivit un divorce entre les époux. Tout le village croit à cette version sur la transformation de Tumpöi par Penköi. L'imagination populaire fait la réputation du sra et de ses pouvoirs, construit l'image du «magicien»; à force d'entendre parler de lui, on finit par le voir agir et puis le consulter ... L'énormité des pouvoirs qu'on lui prête en fait une personne crainte et respectée.

Je montrais un jour à Penköi les tatouages birmans d'un Mru Anok que j'avais dessinés. Le sra me dit alors qu'il ne comprenait pas ces tatouages, dépourvus de sens selon lui, et que l'homme qui les portait s'exposait à de gros risques. Il ajouta qu'en regardant simplement ces tatouages sur le papier, il pouvait à distance blesser l'homme voire le tuer ou le frapper d'un grand malheur (j'eus le soulagement de voir que cet homme était toujours vivant en le rencontrant par hasard quelques jours plus tard ...).

Le sra, comme il peut punir, transformer, ensorceler ... n'échappe pas toujours aux pouvoirs similaires d'un ennemi. Penköi avait un pied extrêmement enflé, noir, semble-t-il presque gangrené; d'après lui, c'était le résultat de la vengeance d'un sra ennemi; il ajouta que ce dernier aurait aussi jeté un sort sur son petit-fils, un garçon d'environ 12 ans, malade et apparemment atteint de la poliomyélite.

Nous avons vu que le sra est un maître dans les rites religieux, collectifs, un intermédiaire dans les interactions avec l'invisible, mais il peut aussi pratiquer des rites magiques, plus secrets et mystérieux qui animeront l'imaginaire et les croyances populaires."

Penköi speaks here of competition amongst the *charaa.* This is perhaps surprising because there do not seem to be many of them: the Mru normally depend on the Bengali medicine sellers who have recently started to operate in all parts of the Chittagong Hill Tracts. I learned this from a brief report that Céline sent me in 2006:

"Now many pharmacies are flourishing in the bazaars (even the smallest), and needless to say that most of the Bengalis who run it have hardly any knowledge of what they sell. Worse, they understood that the tribal people developed an interest for their 'magic pills', supposed to cure so many diseases, and I heard endless stories of people being ripped off by so-called 'doctors', who will not hesitate to sell inappropriate medicines for much more than their real prices. Once, I met a Mru who fell on a bamboo stick, he described his pain and it was quite obvious (even though I am not a doctor) that he might have broken a rib. He was just coming back from the bazaar where a pharmacy man sold him for 300 taka (about 4 Euros, quite a lot for a Mru) some penicillin injection (advising him as well to come back soon to get more of it)! The man was going to administer it to himself with an absolute filthy needle. (It would have brought even more money for the pharmacy to cure – at best – an infection resulting from it!). I met quite a few Mru who believe that injections might be an efficient and powerful way of taking medicines. Ignorance is a great deal for all these medicine shops ..."

P2 Spirits

P2a) General

The Mru believe in spirits; the general term for them is *chüng-nam*. Most of them are not well disposed towards humans, and will harm them in various ways, causing illness or even death. However, there are some spirits which do help humans, for example those that make the harvest grow. The spirits have different names, but it is difficult for an ethnographer to know which is which, because the Mru themselves are generally not sure with which spirit they are dealing in any particular situation. The same spirit may be called by different names, and the same name may be used for different spirits. After the harvest, a thanksgiving offering is made to the paddy spirit, *caa-kom-namma* (see J4e). This may not be the same spirit as the one that made the plants grow, and to whom the *üa-dim-pak* was sacrificed after reaping, because the same sacrifice is often also intended to appease the potentially harmful spirits of the snakes, bees and other creatures that were killed during slash and burn operations (see J7a). The same kind of uncertainty surrounds the second field sacrifice (*khan-pon-thet,* see J4c), where a sacrifice of a chicken is made twice, once on the side of the stream below the field, and once in the field on the *tur-tut*. The first sacrifice is made to *o-reng-nam*, the spirit of the river (see P2c), and the second to *caa-kom-namma*, the paddy spirit. The only difference between the two ceremonies is in the opening words of the *tamma* (invocation): for the sacrifice by the river, the invocation starts with *phyöi*, which is the form of address used to a potentially harmful spirit; for the sacrifice in the field, it starts with *phyok caro*, which is the form of address for a benevolent spirit. The names of the spirits themselves are not mentioned during the ceremony. This may be wise when making an offering to potentially harmful spirits, because one can never be sure which spirit is responsible. It is thus left to the spirits themselves to decide whether or not they are the intended beneficiary, and whether they will accept the offering, feel pleased, and decide to help the offerer.

Even harmful spirits may decide on occasion to be helpful (see P2b); this is the reason why one can only speak of "potentially" harmful spirits. All spirits expect to be offered something by humans. If they receive no offering (usually a slaughtered animal), even normally benevolent spirits can become malevolent and potentially harmful. This is why animals are almost never slaughtered without being declared a sacrifice, even though it is not forbidden to kill a domestic animal without making an offering of it.

As a rule, humans cannot see the spirits. I did, however, meet a man who was surprised that people other than himself could not see them. Other people treated him as a madman, and said this was proved by the fact that he did not know his family name. He only knew that his name was Menru. Otherwise he seemed harmless and behaved normally, except that did not cultivate a field of his own, because he had no fixed dwelling. I met him while I was touring the Southern Mru area; he decided to accompany me to Tapwüa-Kua, where any family he chose would give him food and shelter until he decided to move on.

Before he left, he visited me once more, and lingered at my door. While he was standing there, his face lit up, and he said: "Look there!" I saw nothing unusual, and asked him why he was smiling so happily. He replied: "Don't you see the beautiful *chüng-nam malaa-wüa*? They are walking across the hamlet. How beautiful they are! Can't you see them?"

I told him that I saw nobody. He was disappointed that I, like everyone else, could not see the *chüng-nam*. After this incident, he left suddenly, and never came back. Perhaps he was trying to follow the beautiful and unexpected visitors.

P2b) Divining

When someone is ill, the Mru initially assume that the cause is a natural one. If somebody hurts himself, has stomach ache or headache, or an attack of malaria or dysentery, the expectation is that it will get better or heal up by itself, "as normal". Only when this does not happen do they start to think of supernatural causes. This means that many cases of sickness must ultimately be attributed to the activities of spirits. Medicines are rarely used (see P1b).

Usually people have some notion of which spirit might have caused the illness. The idea can be tested by promising a sacrifice to that spirit and waiting to see what happens. If the sick person gets better, the promise must be kept, or the spirit might take revenge. If the patient does not recover, one can try another spirit. However, it is more usual at this stage to seek out a person with special knowledge of healing, for instance somebody familiar with the Marma *ho-ra* practices.

One of these practices consists in taking a bronze bowl, spreading oil on it with one's finger, reciting the proper *montro* (mantra), and looking in it to see what appears. This will be not the spirit itself, but the animal it expects to be offered, which may be a chicken, a pig, or a cow. The worst scenario is when a human being appears in the bowl; this means that the sick person will die. A simpler kind of *ho-ra* can be performed by looking into half a bottle-gourd filled with water while saying the *montro*. Not everyone knows the correct *montro* for this method of divining, so someone who has special knowledge is consulted. The inhabitants of Tapwüa-Kua would go to an old man in Yöngtu-KP, Pantola-Mouza, who was said to be very knowledgeable.

Another method of divining is *kaing-ku-twak*. Someone cuts a piece of bamboo and puts two or three grains of paddy rice into its concave end, while reciting a *montro*. He then holds the stick in his outstretched hand and recites another *montro* over the lower arm. The spirit now takes possession of the person holding the stick: his whole body starts to tremble, and he is led by the stick, which draws him to the place where the sacrifice must be made. When he reaches that place, he will be unable to move, and water is poured over him while he recites another *montro* until he stops trembling. *Kaing-ku-twak* can also be used when one has lost something – money, for instance – in the jungle. In Tapwüa-Kua, Menching Atwang claimed to know *kaing-ku-twak*, and used it to find the proper place for a sacrifice when a son of Yenpau Khongtör fell ill. After the sacrifice, the child died, and many people stopped believing in Menching's ability to divine.

The Mru buy amulets (*la-phoi*) from the Marma to protect themselves against all kinds of illnesses. They do not know how to make the amulets themselves. They consist of a small wooden plate bearing Marma letters, and are tied with string around the upper arm.

Another way to find out which spirit has caused an illness is to consult a woman called a *püi-höng-charaa* (see N1a). She can use a bow to find out which spirit is responsible: when she says the correct name, the bow will move. If the sick person is going to die, the bow will become very heavy (MK, 16.08.1956).

P2c) *Cöng-cöi*

The spirit named *Cöng-cöi* did not seem to be female, although Kyo Thwân Ong, after hearing stories about it, identified it as a female spirit in which the Marma believed. The Mru cannot normally tell you whether a spirit is male or female, though *caa-kom-namma* may be presumed female because of the ending of the final word. There is little doubt that Cöng-cöi is a male spirit, and he is clearly also malevolent. I found out more about him one night (8.07.1956) when there was a sudden commotion in Menkröi's house. The pigs were squealing, and the people were shouting "*pri, pri*" ("tiger, tiger") and beating the floor. Finally Dingte came to ask whether I would lend him my torch, and went off to look. He did not find anything unusual, and the "tiger" seemed to have been frightened off. The next morning, however, it was discovered that a piglet was missing. It was found in the pigsty of Menkröi's neighbour Netkhai. Netkhai had also been woken up by the turmoil in Menkröi's house, but had stayed inside his four walls in order to protect his pigs if need be from the "tiger".

One thing soon became clear: there had been no tiger, and since no human being could have transported the piglet from one enclosed pigsty to another, it must have been a spirit. Menkröi's pigs must have seen the spirit and started squealing.

Menkröi mostly followed his Christian teachers and did not believe in spirits; but on this occasion, he was bound to accept that there had been one. I asked him what the spirit was called, and what would happen now. He said he would have to sacrifice a cow to Cöng-cöi, otherwise a member of his household would die, or at least go mad. However, he said he had no money to make the sacrifice, so nothing happened, until his adopted "sister", who was only a few years older than his son Dingte, started to dream about a young man coming to visit her in the night. Menkröi finally resolved to do something, and his brother Khamlai sacrificed a dog on the roof of Menkröi's house to protect her from the danger.

I was told more about Cöng-cöi by Menching Atwang. Two years ago, Cöng-cöi had put a child into the stomach of an old man in a neighbouring village. One could feel it with one's hands. To stop the child from growing any more, the man sacrificed a cow, and three days later the child had disappeared. According to another story, Cöng-cöi moved the cotton from one man's field into that of his neighbour. The latter used the extra harvest to pay for a cattle sacrifice, and thus managed to appease the spirit. From Kyo Thwân Ong I learned the following harmless story about Cöng-cöi, whom the Marma regard as female. A young lad met her in the jungle and went with her to her house. There were many young people there, both boys and girls. The spirit tried to persuade the young lad to take off his bangles, but he refused, and during the night, he thrashed spirit soundly until she released him and let him go back.

According to Menkröi (8.09.1956), Cöng-cöi is very big, and mostly lives in the jungle. From time to time people see his footprints, which are as long as a man's arm. He can also appear in people's houses, where he makes himself small, and plaits their hair while they are asleep. He is generally mischievous, but he sometimes helps humans by giving them unusually large harvests of paddy and cotton. In return, they must make him an offering of a buffalo feast. If Cöng-cöi did not receive this gift, the person who had shown ingratitude would die.

Menkröi told me a similar story about a man with a child in his stomach. Karbari Rengnok Kitu-Atwang in the Horinjhiri-Mouza was said to have made a pact with Cöng-cöi. Rengnok was the richest man in the whole area. He acquired his wealth during the last world war, when large quantities of money were dropped on to his field from an aeroplane.

Three years before Menkröi told me the story – so probably in 1953 – Rengnok thought there was a child in his stomach, because he could feel its feet. He was extremely anxious about how he was going to give birth to the child. Finally he took a tethering rope and swung it to and fro, while promising to sacrifice a buffalo if the child disappeared.

It did disappear. From then on, Rengnok sacrificed a buffalo every year, and Cöng-cöi granted him extraordinary harvests. Each year after the paddy has been harvested, Rengnok leaves it in the field house for a night before his men start to tread it out; during that night, Cöng-cöi turns the whole heap over and puts his footprint on it. Rengnok then has an unusually high yield.

Cöng-cöi can also drive people mad. Most people who lose their sanity will die sooner or later. This is the reason why Cöng-cöi is greatly feared.

P2d) Insanity

The Mru have beliefs about which spirits cause which illnesses. There are two spirits – Rukka and Cöng-cöi – that can cause insanity. Attacks of madness brought on by Rukka are easier to cure than attacks brought on by Cöng-cöi. This may be the reason why people choose initially to believe that the illness was caused by Rukka. If a sacrifice to Rukka is not effective, they will then believe that it was caused by Cöng-cöi, and will carry out a *tang-chak*. This progression from the one spirit to the other may be encouraged by the fact that mental illnesses often do not appear suddenly, but take time to develop.

The offering to Rukka is made at a *kan-cang-ching* (a tree bearing mistletoe). At the foot of the tree, a *mrinaa-töng* (see P2e) is erected, and a banana leaf containing a *bong-kom-röng* (see P2e) and some *pok-pok-kom* (puffed rice) is placed in its *khö-kröng*. A whole series of animals is sacrificed at or above the khö-kröng: one goat, three chickens, two pigeons, one duck and a dog. All these animals are killed with a chopper. A further chicken is killed by holding its head under water until it drowns. The water is brought in a *ci-ö-ca* (a small round pot) wrapped in cotton-threads. The pot is covered with a lid for the return trip. Some meat from all the animals is put into the banana leaf and cooked over a fire together with the *pok-pok-kom*. The meat is then divided amongst 30 simple *khö,* which are placed at the foot of the tree beside the *mrinaa-töng*. Water is sprinkled over the *khö* from thirty thin *pön* tied together, and a *tamma* (invocation) is made to the spirit. At the end of the ceremony the thirty *pön* are tilted six times to the left and six times to the right until all the water has been poured out. On returning to the hamlet, the people cook and eat all the remaining meat before entering their houses (Kangku Catumma, 19.09.1956). No further sacrifices are made to Rukka. If the sick person does not recover, offerings are then made to Cöng-cöi.

People are very fearful about performing a *tang-chak* for Cöng-cöi. The first step is to perform a kind of *cari-yöng tang tam-pük* (see P5b), with the difference that the *ner* (flat gongs) must be made from bamboo. The meat of the slaughtered pig may be eaten, though parts of the heart, kidneys and liver are kept and dried. The dog that is killed after the pig is not eaten. A special "medicine", *chi-khang*, is made for the sacrifice; some of the plants are identified in R3i. Those used here are *kom-ram*, *re*, *ching palu*, *waama ki-or*, *nam-pramca*, *la-ia-ram, prön-ram*, and *ro-ram*. The leaves are pounded with blood from the dog in a paddy mortar. Some of the mixture is then put into a *röng* and sprinkled with the aid of a *chit-pot* (a bamboo stick with a short tassel) over all the important items in the house. The rest of it is put into a bowl and smeared on the wrists, elbows, ankles, knees, hips, neck and head of the person who is in danger of insanity. The entrails of the dog are tied over the sufferer's back, crossed over his chest, and the ends thrown over his shoulders.

An expedition is then made to a *kan-cang-ching* (mistletoe tree). There must be an odd number (*tör-cöu*) of persons. If possible, there should be eleven men including the sick person: three to beat the bamboo ner; three to blow *tang-plung* tu; one to carry the *röng* which replaces a drum on this occasion, and beat it from time to time against the ground; one to sprinkle *tang-chak chi-khang* from the *röng* on to the sufferer's head throughout the walk; one to drag the dead dog on the end of a rope; and one to act as the *tang-chak-charaa* (master of ceremonies). The sick person takes along a new *em* (small carrying basket, see D2a), in which he takes everything necessary for the ceremony. He must wear a *körma* (shirt), and he is forbidden to turn round and look back while the party is on the way.

When they arrive at the *kan-cang-ching*, the area around it is fenced off by means of *kung-kaa* (small bamboo arches, see J4c). The sick person sits inside, takes off all his clothes, and puts on a new loin-cloth. His clothes stay inside the circle together with the *em* and thirty newly made *pönca* (a kind of *ri-chang-pön*), which are filled with water. Seven of them are tied together, while the rest are stuck into the ground at an angle. It does not matter if they fall over and the water runs out. Holding the bundle of seven *pönca*, the *charaa* makes a *choi* ("oath"), pouring out water seven times and addressing a *tamma* to the spirit Cöng-cöi, though his name is not mentioned. In the *tamma*, the *charaa* asks Cöng-cöi to free the sick person. He even tries to threaten him, by telling him that if he does not help, he will run out like the water from fallen *pönca*. The *charaa* then pours the water into a *ci-ö-ca* (a small round pot), takes handfuls of the sufferer's hair, and pours water over it from the pot. The bottom part of the pot is covered with white cotton strings which loop up and down. It is covered by a "helmet" made of a length of bamboo split upwards from its lower end. The "helmet" is also covered with cotton threads. The *charaa* then repeats the *tamma* with the seven *pönca*, the circle is opened, and the sick person leaves it. The circle is then closed again and the group leaves with the sufferer, who must not look back on the way home.

For the next nine days, the patient must keep *ta-ung*, that is, he must not leave the *kimma*, or even try to peep out. He may not wash himself because the dog's blood must stay on his body. He must simply sit down and stay where he is. All he gets to eat is rice, cooked without any salt or spices, and he may only drink water. These nine days are followed by a month of *caa-theu*, during which the sick person may eat salt again, *oi-lo-ram* (salty? leaves) and tree leaves. He may re-enter the *kim-tom*, but he may not come down from the house before a *kan khang phöt* (lifting of the ban on side dishes) has been held. This means holding a big feast for which two to three cows are sacrificed, or a buffalo if at all possible. The ban is lifted by means of a *cak kwak* (throwing away of food). Some dried meat (heart, kidneys and liver) from the pig killed for the first *cari-yöng* is used for this ceremony. It is cooked in a *röng* (bamboo tube) to produce a *pak-kia-pak-chak-röng* with which to perform the *cak-kwak*, in the same way as for the *pak-tan* festival (see P5d).

After the ceremony, the sick person may eat all foods once more. A *cia-chot* festival is then held, for which *chi-khang* ("medicine", see R3i) is prepared and the *tu* are blown. Two men from the sufferer's household –

though not the sick person himself – must keep *cia-chot-khang* (see R3h) (MK, 10.09.1956).

If all these offerings fail to restore the sufferer to sanity, hope is abandoned that he will ever recover and become normal again. So long as he is not dangerous to other people, he is allowed to roam freely, but if he becomes dangerous – for instance, if he tries to kill people with his chopper – he is put in restraints. His family continues to feed him, but not to care for him otherwise, and the expectation is that he will soon die. Unfortunately, such events are not infrequent, and the ceremonies are therefore fairly well known. However, not all families can afford to perform them, especially when it comes to sacrificing two or three cows. Mental illness can be a late consequence of amoebiasis, developing when the amoebae pierce the skin of the intestines, enter the spine, make their way up to the brain, and start to destroy it. Without modern medicine, there is no real remedy for Cöng-cöi's mischief.

P2e) *mri-naa*

Mri-naa is not a spirit; it is a form of sacrifice to *tui-ching-nam*, the spirit of a place where puddles of green water (*tui-ching*) are to be found. *Tui-ching-nam* is responsible for a number of disorders, for example sore eyes, eczema and toothache.

There are several kinds of *mri-naa* that may be carried out to placate this spirit. The ceremonies require an uncastrated young boar (*pak-hang*), a duck (*um-pai*), or a dog (*kui*). These animals are the same as those needed to placate *chüng-hi-nam*, the border spirit, which is responsible for coughs and, once again, sore eyes. The *püi-höng-charaa* (see P2b) can ascertain whether the offering should be made to *tui-ching-nam* or *chüng-hi-nam*. When she utters the correct name, her bow will move. She can find out in the same way which animal should be sacrificed.

If she finds that the sacrifice should be made to *tui-ching-nam*, the people making the offering will go to a *tui-ching*, if possible to one where there is a *kan-cang-ching*. There they erect a *mri-naa-töng*: a length of bamboo is set into the earth, and its top internode is split into four parts. Each part is bent down to the ground and fixed there with a *chit-wai* (bamboo stick with a tassel). A *khö* (see P5d) is set on the point where the four parts of the bamboo separate. In it are placed *pok-pok* (puffed rice) and *kom* (flat dough cakes), together with a small bamboo tube the length of a finger. The tube contains some water; it is plugged with ginned cotton, and wrapped in *bong-kom* threads. The threads are first wrapped round the four fingers of the right hand, then slipped over the bamboo tube and twisted together to take up the slack.

The person starting the ceremony pours water from a *ri-chang-pön* (a bamboo tube for pouring out water) and utters a *tamma* (invocation). The animal is then sacrificed, and some of its blood is dripped over the *khö* and the items it contains, especially on the *bong-kom-röng*. The tube is then taken back home and the *bong-kom* threads are distributed to all the family: each member has a thread wrapped around his or her right wrist, and a piece of the ginned cotton that was used to plug the *röng* is put into each person's hair-knot. The ceremony ends with a *plang-kwak*, in which the *röng* is split in two and thrown on the floor. If the two halves fall differently, i.e., one with its inner surface facing upwards and the other with its inner surface facing downwards, the omen is good, and the two parts are thrown out of the window. If the two halves fall with the same side up, however, the throwing is repeated until a good result is obtained. The two pieces must then be thrown down the stairs (MK, 24.08.1956). A sacrifice to *chüng-hi-nam* will be described later (see P2i).

A *mri-naa* may also be performed if a baby of two or three months is unsettled and cries at night. The problem may be due to the *tung-do-chüng-nam*, a spirit that normally lives under the house, but which may come out and frighten the baby. To make it stop, a chick is sacrificed under the house after nightfall. A *mri-naa-töng* is set up with *chit-wai* at its four corners, and on it is placed a simple *khö*, about eight centimetres in diameter, containing a *bong-kom-röng*. The white *bong-kom* threads are then distributed to all members of the family, including the baby.

P2f) *o-reng-nam* and *nat-hua*

O-reng-nam is the spirit of the running water in streams and rivers. R. H. Sneyd Hutchinson mentions it in his "Gazetteer of the Chittagong Hill Tracts District" (1909). On page 42 he writes: "'Oreng,' the spirit of water, is their most honoured deity. In the month of July the whole of a village community will go to the side of running water and sacrifice a couple of goats and from twenty to thirty fowls. A miniature altar of bamboo is erected in the water; on this are placed rice-flour cakes and parched rice, and the spirit Oreng is invoked to make the jums [swidden fields] yield a good harvest and keep away sickness or any other ill-hap from the village. All oaths to be of binding nature must be sworn by Oreng, and are by gun, dao and tiger."

My own findings are somewhat at odds with these remarks. According to my informants, *o-reng-nam* is a minor spirit, and would not be honoured by a whole village community making a major sacrifice of goats and chickens. The only sacrifice to this spirit that I witnessed was an individual offering of a chicken, to make the paddy grow (see J4c, *khan-pon-thet*). I did not hear of any binding oaths sworn in the name of *o-reng-nam*.

Hutchinson's statements on page 43 are more consonant with my own observations. He writes: "When a party of Mros are going on a journey to a distance from their village, each member of the party will pluck a piece of sunn grass, and, going to a stream, the senior of the party will enter the water and invoke the aid of 'Oreng,' after which each person will stick their piece of grass into the earth or sand at the edge of the stream and then set forth on their journey."

This *tang pram* (presentation of leaves) is also performed by individual travellers at confluences of streams or rivers. It is thought to ward off illness and to offer protection against a difficult journey. The traveller may present the ubiquitous "son" (*imperata*) grass (*ci-daa*), or *kan-kok* (fern) leaves.

So far, we have seen *o-reng-nam* taking a beneficent role. However, the *tamma* used at the beginning of sacrifices to him opens with *phyöi,* which is used to invoke malevolent spirits. The reason for this is that *o-reng-nam* can also cause stomach-ache, loose bowels and even fever: in other words, the symptoms of dysentery. To placate him, one may sacrifice a goat, a pig (it need not be castrated), a duck, a chicken or a pigeon.

Nat-hua ("spirit-stone") is a large boulder near the place where fresh water is drawn. This spirit stone is thought to cause the same sickness as *o-*

reng-nam. After the start of the rainy season, drinking fresh water may indeed cause dysentery, because the rains wash out germs from human faeces deposited outside the hamlet, often above the place where water is drawn. The *püi-höng-charaa* can determine whether the culprit is *o-reng-nam* or the *nat-hua*; however, this is rarely thought necessary, because the two spirits have more or less the same role, and the boulder may even be declared to be the home of *o-reng-nam*. The *nat-hua* can be also be invoked to send good paddy and cotton harvests.

The form of address used to invoke the *nat-hua* has an unusual feature. It begins with "*phu nat-hua*", that is, the spirit is addressed as "stone", which is what it actually is. The animal sacrificed is normally a pigeon, though the pigeon can be replaced by a small chicken. When people pass the boulder, they offer it a greeting and place a *kan-kok* leaf near it. If someone uses the boulder as a convenient place for washing dirty clothes, the boulder may become angry, and may avenge itself by giving the offender eczema on his legs.

Kyo Thwân Ong found this out the hard way. At first he would not listen to the warnings of the Mru: he thought that as a Marma, he could not be harmed by the Mru spirits, and therefore need not honour them. He regretted his behaviour when he developed a spreading *tinea* infection. He then decided to make a small offering of some grains of rice, but changed his mind on the advice of the Mru, who said he had better perform a *mri-naa*. He enquired whether a field could be established on the slope above the *nat-hua*, and learned that nobody would dare to do so because he would be sure to die. This finally convinced Kyo Thwân Ong to make a proper sacrifice, and he decided to offer a chicken. Menching agreed to help him to do this correctly, and Kyo Thwân Ong stopped washing his clothes on the *nat-hua* (30.09.1956).

The way in which goats are sacrificed to *o-reng-nam* or *nat-hua* shows possible Marma influence, because it requires the erection of a *tong-mang-rum* or "hill-lord's-enclosure". All three elements in this term are Marma words. The *tong-mang-rum* is a circular enclosure surrounded by fence, woven diagonally like the one around a Marma *phra-jedi,* with a "door" that is closed when one leaves. *Pok-pok* and *kom* (see P2c) are scattered inside the fence. Before the goat can be sacrificed, a small chicken must be killed, not in the *tong-mang-rum*, but always near the river, over a *chit-wai* or a simple banana leaf. The *klaa-wi* ("shedding-blood") is offered to the spirit of the particular stream or river.

The people making the sacrifice then return to the *tong-mang-rum* and cut through the goat's neck with a chopper. It is permissible to use more than one blow. At the same time, water is poured from a *ri-chang-pön*. The goat is sacrificed to *o-reng-nam* to secure his help in producing a good harvest; the *tong-mang-rum* is normally erected in the field above the water course.

When the *nat-hua* is to be invoked as well, a *tong-mang-kim* ("hill-lord's-house") is erected instead of a *tong-mang-rum*. The *tong-mang-kim* is also called *dö-ker-kim* (*dö*: to fall down, *ker*: to place crosswise). It is made as follows: Four pieces of bamboo about one metre long are set upright in a square in the river. Each is supported by two crossed bamboos standing outside the square. Two bamboo cross-beams are placed in the forks at the top, and smaller bamboo sticks are laid across them, forming a larger version of the *pak-plai-kimca* (see P5c). Four *chit-wai* (tassel sticks) are set into the upper end of each of the vertical bamboo poles, together with a *la-ia-ram* (a leaf of *Aglaonema hookerianum*). A banana leaf is placed on the platform and *pok-pok-kom* (puffed rice) is scattered on it. The goat is then sacrificed as described above. This offering may be made for a good harvest, or to protect against dysentery.

If a pigeon is to be sacrificed to the *nat-hua*, this must be done before the goat is killed, and a *pong-mö* must be made (*pong* = the arched roof of a Bengali-boat; *mö* = to roof over). The *pong* is eight centimetres high. It is woven diagonally from bamboo strips and fastened to two pieces of bamboo twenty centimetres long, which are tied together. The resulting "boat" is actually only the roof of it turned upside down. *Pok-pok-kom* is placed in it, and it is tied with a liana to the *dö-ker-kim* so that it floats on the water a little way below the *kim*. Three bamboo poles are set vertically to the right and the left of the *pong-mö*; their upper ends decorated with *chit-wai* and *la-ia-ram*. The pigeon is killed over the *pong-mö* and its blood is dripped on to the *pong* and the *chit-wai*. (For a description of another variant of the *pong-mö*, see P3c.)

If a pig or a duck is to be sacrificed, a *mri-naa-töng* (see P2e) must be erected. If it is only a chicken, a *nat-rang-töng* is needed (see also P2l), and the sacrifice, which is made near the river and is addressed to the *nat-hua*, is called *o-nat-rang*. The *nat-rang-töng* resembles a *mri-naa-töng*: one bamboo is set in the river at an angle, its top end is split on the upper surface, and a *khö* is fixed into the split. The upper end is also decorated with four *nat-rang-chit*: these are *chit-wai* with an unusually crudely shaved tassel on one side only of the bamboo stick. These *nat-rang-chit* are probably of Marma origin, and can also be used for the *tong-mang-rum*. A *la-ia* leaf may be added if desired. *Pok-pok, kom*, and a *bong-kom-röng* are put into the *khö*. While the chicken is being killed, water is poured as usual from a *ri-chang-pön*. At the end of the ceremony the white *bong-kom* threads are distributed to all members of the family.

If only a pigeon is to be sacrificed, a *ca-kom-kim* must be made. *Ca-kom* was said to be a Marma word. The structure looks like a tall, thin *pak-plai-kimca* (see P5c), and it has a small platform. The four vertical bamboo poles are about fifty centimetres high, and each one bears a *chit-wai* and a *la-ia-ram*. The same things are placed on the platform as when a chicken is sacrificed, and – as with any sacrifice to the *nat-hua* – the ceremony ends with the distribution of *bong-kom* threads to all members of the family (MK, 25.–26.08.1956).

P2g) *mo-tia-chot*

This is the name of a ceremony that was performed on 29.09.1956 by Khamcöng Khongtör, with the help of his neighbours Netnöng and Niadöng Atwang, and Klingtui Khongtör, whose wife Changlong had previously acted as *püi-höng-charaa* and had ascertained that a complicated sacrifice was needed to help Khamcöng's wife Krapau, who was always ailing and mostly unable to work on account of rheumatism. Less than three months previously, Khamcöng had held a *cia-chot-poi* (see R1h), but this had not brought the desired relief. No particular spirit had been named – at least I was not told of one – but it might have been *o-reng-nam*. The villagers believed that nothing could help Khamcöng and Krapau, and that at least one of them, or one of their children, was sure to die sooner or later because their marriage was considered incestuous (see L3c). They were neither siblings nor cousins, and they even belonged to two different sibs (Khongtör and Premchang), but these two sibs were thought to be *tai-nau* (classificatory siblings), especially since both were regarded as subsibs of the Dengwüa. Their parents had objected their marriage, but they had married nevertheless, been obliged to leave their hamlet, and had finally had been allowed to settle in Atwangwüa-Kua. Now, however, their inevitable fate had caught up with them. Krapau had not been able to suckle her youngest child and it had died in her arms shortly after it was born. Khamcöng had become very poor, and had had to borrow the money for the small cow for the *cia-chot-poi*. Nevertheless, he did not give up hope that Krapau would recover, and Changlong told him that he needed to sacrifice some more animals.

The following animals were to be sacrificed: two chickens (*waa a-cum*, a large one and a small one; two pigeons (*tur a-cum*); another small chicken; and a small goat (*rua mai-mak* = goat ?-?). They were to be sacrificed in that order. The offering was to be made at a large *kong-chok* tree (species unidentified) outside the hamlet. At the foot of the tree, a space was cleared between two prominent roots, and a small *tong-mang-rum* (see P2f) was erected. It was twenty-five centimetres high and was constructed in two parts, with an opening at the front. Four "trees" were planted inside this enclosure; in fact they were only sticks about thirty centimetres long that were stuck in the ground. The sticks were an *ui-pia-ching*, a *rai-kör-ching*, a *klang-kü-ching* (all unidentified), and a bamboo stick (*rai-tua*) to which *ku-ramca* (leaves of a palm, probably *Licuala peltata*) had been tied. The leaves symbolised the promised sacrifice. Once the sacrifice had been carried out, the leaves would be untied, symbolising the purpose of the ceremony. This was to release Krapau's blood and relieve her feeling that her circulation was cut off, which was paralysing her.

Four short bamboo tubes with one tail were set upright in the ground in a square. Inside this square was set a bamboo pole 1.20 metres long, called *waa-co-pre*. The pole was split into four sections throughout its length except for the uppermost internode, into which were inserted four *chit-wai* with crude tassels crowned by little balls of cotton. Similar *nat-rang-chit* adorned with cotton balls were stuck into the earth at the foot of each of the *waa-co-pre*.

In addition to this first "house", which was for the pigeons and the chickens, a *pak-plai-kimca* was erected. This second "house" was for the goat. It was roofed by a *pong-mö* (see P2f) consisting of three bamboo strips fifteen centimetres long and interwoven with bamboo strips. The middle strip was folded back at the ends, and the outer ones were fastened between the four verticals of the *pak-plai-kimca*. More *nat-rang-chit* with cotton balls were inserted at the lower ends of the verticals. A third "house" was made for the third chicken. This too was called a *pak-plai-kimca*, though it consisted only of four broader strips of bamboo lying on the ground. Its corners were marked by *nat-rang-chit* with cotton balls.

Finally, a *ri-chang-pön*, or tube for pouring out water, was placed at each house, the one at the goat's house being wrapped in white *bong-kom* threads. A *thi* (umbrella) was set between the first and second houses, near the tree. The *thi* was a length of a bamboo split open at its upper end, with a large amount of cotton spread over it.

Before the animals were sacrificed, banana leaves holding large quantities of *pok-pok-kom* and husked grains of *cha-tui-caa* (sticky rice, see J3a) were set down near the houses. Some feathers were plucked from the necks of the chickens and pigeons and were placed on the houses, then water was poured over their necks from the *ri-chang-pön*, and a *tamma* was recited, including *phyöi kat, phyöi pre-kat*, and so on, up to *taruk-kat*. Finally, the animals were handed over to Klingtui, who cut their throats with a chopper and let the blood pour over the houses, without saying any more *tamma*. The goat was then led into the *tong-mang-rum*, where water was poured over it and a *tamma* was said, and then it was led out again. Water was poured over it a second time, then it was taken to one side and Netnöng severed its head with a single blow. He carried the body back into the *tong-mang-rum* and let the blood drip on all three houses. The head was placed for a short time on the *pok-pok-kom* belonging to the goat's house, and then taken back out.

After the animals had been sacrificed, an additional bamboo stick was set upright in the *tong-mang-rum*, and the goat's tethering was hung on it. A notch was cut into the bark of a small tree that had been left standing in the enclosure and used to hang up the basket in which the pigeons had been carried. The ceremony ended when the contents of three *röng* (bamboo tubes) in which parts of all the animals had been cooked were poured out and a last *tamma* said. When the party arrived home, the *bong-kom* threads were distributed to the members of the family.

P2h) *po-pramca-laa*

This is a ceremony performed in honour of the spirit of the field used the previous year (*pramca*). This spirit does not seem to have a name of its own, but the offering has connections with *o-reng-nam*, so we may assume that he is the spirit intended. The ceremony is performed if someone is troubled by daily attacks of fever and cannot eat; it is presumed that the spirit has taken (*po*) his soul (*lüm-laa*), or, since humans have seven souls (see P2k), has taken at least one of them. It is hoped that the sacrifice will persuade the spirit to set the soul free.

The ceremony is performed by three to five men, and two chickens are sacrificed. The first chicken is killed in a *klak-wi* (blood shedding) ceremony, which is held on the stream below the old field. The second chicken is sacrificed in the old field. A small hole (*kwam*) is dug, and four *chit-wai* are set upright at its corners. Either five or seven pieces of *ting-ai-ram* (palm leaf) are placed in the hole with *pok-pok-kom* on them. The hole is then surrounded by *kung-kaa*. Depending on the lie of the land, two lengths of bamboo are placed along each of three sides, and three men sit on them. They are then fenced in again by *kung-kaa*. A few grains of rice are put into a *laa-rik-yia*, which is a small water-bottle for calling the soul. The man in the middle blows six times over the bottle and starts to call the soul, asking it repeatedly to return, until an insect – an ant for instance – appears in the hole, attracted by the *pok-pok*. This insect, or any other creature that appears, is caught with a cotton thread, wrapped up in it, and put into an *em* (see D2a). Another handful of *pok-pok* is then thrown into the hole, and the spirit is addressed in a *tamma*, which informs it that the soul has been captured and begs it not to take the soul back and make the person ill. The spirit is offered a chicken as a ransom. The chicken is not killed, but set free; it has been exchanged for the human soul, and is now regarded as the property of the spirit. The people who have taken part in the ceremony go home; they are forbidden to look back on the way.

When they arrive in the hamlet, the participants put down the *em* near the house, while a *cari-yöng* is performed (see P5c). For the final *bong-kom* ceremony, someone fetches the *em* containing the caught soul and the *laa-rik-yia*. The latter is blown again. The soul is placed in the hair of the sick person, and a turban is wrapped round his or her head to protect it. *Bong-kom* threads are distributed to all members of the family. Two *köm-pot* are prepared from raw liver, heart or kidneys take from the pig that was sacrificed in the *cari-yöng*. Both are thrown away outside the house, one to the west and one to the east. They are intended for the *mali-malaa-nam*, which are good spirits that protect humans. Nothing more was known about these spirits. The soul creature is not thrown away, but is placed with the *laa-rik-yia* in the *khong-töm* (see D2h) at a time when the sick person is asleep. For twenty-four hours after the men went to the field to recapture the soul, the house is *khang*: outsiders are not allowed in, though members of the household may leave (Kangku Catumma 19.09.1956).

P2i) Other spirits

The spirit that guards the kimma is called *kim-kwai-mik-nam*, or "the house spirit that stings the eyes", because it causes inflammation. Outsiders are forbidden to enter the kimma (see C3b), or even peep into it, though when acting as *riaca*, even people who are not family members may enter it freely when asked to fetch items such as spears or clothes (see R3i). If someone did glance into the kimma without permission, it would probably be unintentional, because he would be ashamed to do so deliberately. His eyes might nevertheless become sore. If this happened, he would only need to throw mi-tut-aidam (husked rice mixed with wild ginger) into the room; he would not have to make a special sacrifice.

The Longhu have no such prohibition. They have taken over the arrangement of the Khumi house, which has no special kimma, so the parents' sleeping place is always visible, though the view is somewhat obscured by stored baskets.

There is a more severe form of eye inflammation that is caused not by looking into someone else's kimma, but by a spirit called *chüng-hi-nam* (see P2e). This spirit must be placated by the sacrifice of a pig. A sacrifice of this kind, called *chüng-hi-dam*, was carried out on 23.08.1956 near Tapwüa-Kua for Menkröi's adopted sister Kaiche. Klingtui's wife Changlong had acted as *püi-höng-charaa* (see N1a), and had ascertained that *chüng-hi-nam* was responsible for Kaiche's eye disorder. Menkröi had promised to make the sacrifice if the symptoms disappeared, though he regarded himself as a Christian and had to ask Karbari Kangku Catumma to make the offering. Klingtui would lend a hand.

The place where the spirit was to be addressed was not in doubt because the borders of four mouzas – Horinjhiri, Takerpanchori, Galengya, and Pantola – met fairly close to Tapwüa. A piglet was wrapped and taken there, together with an *em* (see D2a) containing everything necessary for the sacrifice. Especially important were a little bowl and some *röu-kaa-pik,* or bark of the *röu-kaa* tree (unidentified), and millet, to which water from a *tui-yia* (water-bottle) was added. When the mixture had been prepared, some of it was sucked through a small bamboo tube and spit with a tamma over the piglet, which had been laid upon a layer of leaves stripped from a nearby tree. The tamma was recited by the karbari. It was rather long, because the headmen of all the four mouzas had to be included. It started with *phyöi kat, phyöi pre-kat, phyöi chum-kat* ..., and ended with mention of the usual *taku ning pak-la*, or nine-year-old pig that was the nominal offering. After this the real offering, a piglet less than a year old, was killed by stabbing it with a *cau* (a sharpened bamboo). Another *tamma* followed before the piglet was cut up on its layer of leaves. Its scrotum and penis were thrown away, the entrails removed, and small pieces of its liver, heart and kidneys were cut off and cooked in water in a (*pak-kia-pak-chak*)-*röng*, placed across a fork in a branch. The body of the piglet was then skewered on a sharpened branch and the hair was singed off its skin over a bigger fire. The skin was scraped clean and the piglet was held over the fire again until it was blackened. At this point its feet were cut off and the carcass was washed. While Menkröi's son Dingte performed all these operations, the karbari took the *röng*, which was now steaming, and poured its contents over the place where the piglet had been sacrificed, while spitting some of the *röu-kaa* water over it six times and more or less repeating his first *tamma*. The *röng* was cut in two on a banana leaf with some *mi-tut-aidam* (husked rice with wild ginger) and the two halves were thrown down on the leaf. They both fell with the inner surface upwards; this signifies non-acceptance, so the *plang-kwak* had to be repeated, with an extra *tamma*. This time it was accepted, which meant that the banana leaf and the halves of the *röng* could be thrown away and the rest of the *röu-kaa* water poured out. Meanwhile, Klingtui had chopped the meat of the piglet very small on half an old basket on the leaves where it had been killed. He had also cleaned and chopped a large cucumber, and was able to prepare a meat dish on the spot, using a pot and some chillis that had been brought from the hamlet. This dish was eaten with rice by all the participants when they returned home.

P2j) Spirits that take possession of men

The Mru believe in spirits that take control of people, though the Anok Mru do not seem to believe in their activities, even though one of them is called *ui-mukma*, which is a Mru name. The final syllable of the name suggests that this spirit is thought to be female; in any case, it only takes control of women. It is said that the Bawm commonly believed in it before they became Christians. The belief is still found among the Khumi, and perhaps also the Longhu. Menkröi (16.08.1956) remembered the case of a woman who, some two years previously, had been taken down three times by the police to Banderban. There she was examined by a doctor who could not find anything wrong with her, but the police had to intervene because the people in the woman's hamlet were saying that they would have to kill her. Menkröi did not know what happened to her after that. However, he was able to tell me something about the symptoms.

At night, the woman's body lay as though she were dead. Her soul left it and visited other houses, where it fed first on chicken droppings: it appeared there as a red, fiery glow. The soul then fed on the entrails of the people living in the house, causing them to experience severe pain. After the *ui-*

mukma returned home, her mouth was blood-stained. The Bawm believed that a woman who used the comb of an *ui-mukma* would become one herself.

If someone has severe stomach pains and suspects that they are caused by an *ui-mukma,* he will have to sacrifice a chicken or a pig to her at the dong-lu (the head of the stairs). The animal's head is cut off and falls to the ground; the beak or snout will point towards the place where the ui-mukma lives. If this is not enough to make a clear identification, all that can be done is to throw the head further in the same direction. The rest of the animal can be eaten.

Belief in were-tigers is common among the Naga. It is not known amongst the Mru.

P2k) *chit-ci*

The concept of *chit-ci* is most probably of Marma origin (cf. Marma < Burmese: "cacche" = apparition, spectre, ghost). A *chit-ci* can be kept by a person who knows the appropriate Marma *montro* (mantra), and it can be sent out to harm other human beings. *Chit-ci* can assume any form they wish and can appear in it, as human beings, monkeys, or any other creature. However, keeping *chit-ci* is not without danger to the keeper. If he does not give them enough victims and loses his power over them, they will start to kill his family members, and finally the keeper himself. One case was known in which a man lost first his brothers, then his wife, and then he himself died after a short attack of mild fever. If a Mru is attacked by someone who uses a *montro* to send him a *chit-ci*, no sacrifice will help him. The only effective remedy is to know who the sender is and to destroy him by means of an opposing *montro*.

There was an example in a neighbouring hamlet. A man called Eraa had a sister who married a man for whom she had left her former lover. After her marriage she suffered many troubles: the spirits threw stones at her, hit her with a chopper, urinated on her head and grabbed at her genitals under her skirt. Her new husband saw only one way out of his troubles: assuming that the *chit-ci* that attacked her were under the command of her former lover, and not knowing how to take control of *chit-ci* himself in order to launch a counter-attack, he simply sent his wife back. Menkröi did not know what happened to her after that (10.09.1956).

He also told me of another and more recent killing by *montro*. In Chapai-KP (Alekhyong-Mouza), five people were said to have been killed by *chit-ci*, and then a classificatory brother of Menkröi became seriously ill. Nobody knew who was the master of the *chit-ci*, or how or why he sent them.

Rengyok, the son of Karbari Kangku's wife's older brother, was supposed to command eight *chit-ci*. Rengyok himself denied the accusation, but nobody believed him; he was not allowed to settle in a hamlet, and had to wander around. Whenever he tried to stay somewhere overnight, the spirits would move about and make knocking sounds in the houses. In an attempt to drive him away, people would throw *kür-khaa-pleu* into the fire (*kür-khaa* is a very bitter fruit, eaten to ward off spirits). This would cause the man's body to become overheated, and he would run in and out of the house, unable to find rest, until he finally went away. If he was not driven away, his spirits would eat everything – the rice, the cattle, and finally the people (Dingte, 14.06.1957).

Rather like the *chit-ci*, but perhaps even more dangerous, are the *cheng-phak*. These too can appear in the shape of human beings, monkeys, cats or hawks. Anyone who is attacked by them will die on that day or the next. The only thing one can do – if indeed one is still able to do anything at all – is to perform a *chit-ci-kwak* ("throwing away of *chit-ci*", see P5f) using a counter-*montro* (MK, 19.09.1956).

A house can be infested with *chit-ci* by a malevolent outsider. It can be done after a cremation by taking some of the charred wood while saying the appropriate *montro,* putting it into a *röng*, and placing the *röng* below the stairs or the roof of the house. When this has been done, the house will vibrate from time to time, and the people who live in it will start to die. However, one cannot be certain that this is really the work of *chit-ci*, because many unknown and malevolent spirits hover around places where cremations are carried out. They may even include one of the seven souls of the human being who was cremated.

Some people believe that the souls (*lüm-laa*) of the dead can move back and forth with ease between the earth and the sky. Both men and women have seven souls (see Q1n); the smallest and youngest of them is kind, but another one – it is not clear which – is bad. The spirits that haunt burial places (see Q2b) are especially dangerous. These are the spirits of children that died below the age of two, very poor persons with no relatives to care for them, and people who died in an epidemic. The spirits of the Bawm, who occupied this part of Mru territory for a short time and left their grave-stones behind, are also mistrusted. They are suspected when small children fail to thrive, drinking milk but not eating or growing. The only remedy is *pong chak.* This does not involve sacrificing animals. The procedure is simply to take any kind of basket, a pot (*pu*), a water-bottle (*tui-yia*), and a bamboo tong (*kep*) and put them at the *dup-raa* (place of burial). It is in fact usual to put these things at a *dup-raa*, but they may have been forgotten or stolen (MK, 10.09.1956).

P2l) The role of women

All the spirit offerings described in this chapter so far were carried out by men. Women were involved only in their role as *püi-höng-charaa* (see P2b). This indeed is the normal state of affairs: women do not perform sacrifices, though they can be the victims of a malevolent spirit, and some spirits are regarded as female. These include *caa-kom-namma*; possibly *o-reng-nam*, which is less benevolent; and *ui-mukma*, which is certainly malevolent, and which only appears as a woman. When one asks a man why women do not offer sacrifices, the answer is simply that they do not know how to do it. So long as there is a man in the house, he will be the one to perform the sacrifice, even if he has to ask someone else for help or guidance. I asked whether a woman could do it if there was no man in the house. The answer was that so far as the bigger sacrifices were concerned, she would have to ask a male relative to do it for her. Regarding smaller sacrifices, a woman could carry out a *wi klak* ("blood-shedding"). She could also carry out an *o-nat-rang* (see P2f) for smaller children suffering from fever or diarrhoea; she could do this at the place where she drew water or on the swidden.

Nat-rang-nam is not the same as *o-reng-nam*: it is a small spirit that can be passed on to someone by means of a *montro*. This suggests that it may be of Marma origin. To make an offering to *nat-rang-nam*, one needs only a chicken, some *pok-pok*, a *nat-rang-töng* with a small *khö* containing four *nat-rang-chit*, a *bong-kom-röng* and cotton threads, and a *ri-chang-pön* (see P2f). When the person making the sacrifice returns home, the *bong-kom* threads are simply wrapped round the wrists of the family members without further ceremony. This can be done by a woman. However, if there is a man in the house, he will perform the proper ceremony, blowing on a *laa-rik-yia* and calling the soul.

One might expect that women would pay special attention to the fire spirit, *mai-nam*, since they use the hearth every day for cooking rice and preparing meals. However, *mai-nam* is not linked to the individual fireplace, and no offerings or special sacrifices are made to it. The only exception is that it receives a *köm-pot* at the bigger festivals.

If a marriage is childless, the wife is often, but not always, thought to be responsible. Since a degree of uncertainty exists, *tui-ching-nam* is blamed as the real culprit. The offering made to this spirit is called *mung-chak*: *mung* is a place believed to be under the protection of *tui-ching-nam* (see below). A *mung-chak* must be performed under a *kan-cang-ching* (see P2d); a *mri-naa-töng* is erected there (see P2e), and an animal designated by the *püi-höng-charaa* – a pig or a dog – is sacrificed. This, however, is only a supplementary offering; the main one requires a *cia-chot-poi* (see R1).

Tui-ching (green water) and *mung* (saltlick, where animals come to drink) are signs that the places where they are found may not be used for swidden. If someone disregards these signs, he will die.

P2m) *Thurai* and creation

Thurai (God) is not regarded as a spirit, nor even as the supreme commander of the spirits. *Thurai* is the creator of the world and everything in it.

Other stories also exist about the creation of the earth. One is that the earth was created by two brothers, *Krong-reng*, who made the plains, and *Waa-da-klaa-lama*, who made the hills. *Krong-reng* is known only by his name, but *Waa-da-klaa-lama* is believed to be a woodpecker who was not a skilled worker, and who made the hills to resemble his crest.

Another story is quoted if someone lays claim to a place that is already occupied. The original occupant will say to the claimant: "You did not make this place, it was made by *pak-charua-ma.*"

Pak-charua is a wild animal from the jungle. Its meat can be eaten, but is not highly valued because it has a bad smell. It is probably a hog-badger. *Pak-charua* is said to have made men from the earth, using material from the roots of a *mukma-ching*, a tree which has black sap. Places where this tree grows are highly valued; if one finds a *mukma-ching* growing on one's swidden, the place will be chosen as the *tur-tut* (see J2e). The spot where the tree grows can also be called *liu-klak-raa* ("seed-dropping-place"), and there is a story that when *pak-charua* made men, the first *liu-yang* (seed of fornication) appeared there. My informant, Kangku Catumma (19.08.1956), was not able to tell me how this happened.

There are songs in which *Thurai* is identified with fate, and prayers can be said to him; however, most people do not pray to *Thurai* because they do not believe that God will change their fate. There is a popular tale (see R8a) according to which *Cakraa-mang* (another name for God, taken from Marma) once sent rules to every tribe, telling them how they must live. To the Mru he sent a cow, bearing a palm leaf on which the rules were written. The cow swallowed the leaf, and told the Mru that they must weed three times a year and harvest once. The Mru felt that they had been treated worse than the people of the plains, and they went to God to complain. God told them about the cow and the leaf, and the cow had to confess. Enraged, God kicked the cow in her mouth, and told the Mru how they must kill her. However, he did not change their fate, and their rules remained as the cow had told them. This illustrates the fact that God could not or would not change anything that had been ordained by him, or stated by one of his messengers.

This story is told to children and foreigners, and I did not think that grown men seriously believed it. However, I felt less sure when I was told forty years later, in 1990, that a Mru prophet had taken up a detail of the story that I have not mentioned yet, namely, that God told the Mru they would find his original message for them in the cow's stomach after they had killed it. They did indeed find the message, but they could not read the writing. Then a new Mru prophet, named Menlê, came and read the script, and put it into a new form of writing that the Mru could learn. This prophet revised the rules according to which they should live, founding a new religion under which cattle sacrifices and all sacrifices to the old spirits were abolished.

It must be admitted that his new version of the Mru religion was described to me by a learned Bawm, and not by an adherent of the new religion. I do not know whether a member of the new religion would have told me the story in the same way. However, the religion spread rapidly, and with it a renewed belief in *Thurai*, who can now be addressed directly by believers. It seems possible that the old story contributed in some way to the success of the new religion.

P3 *khang*

P3a) *khang* against bad spirits

A *khang* is a special time during which something is forbidden. It may be kept by individuals or by groups. The *chüng-nam-khang* was described in Chapter C1b as the simplest reason for closing a hamlet to foreigners or, to be more precise, to everything and everyone coming from outside. The most important thing was to keep out unwanted spirits.

If it was thought that some spirits had entered and had brought cholera, or even if a number of people were suffering from fever or dysentery, the elders might decide that a *chüng-nam-khang* should be held. All the people of the hamlet would be informed, and a honuman monkey (langur) would be hunted, killed, and fixed to a bamboo pole at the entrance to the hamlet (for picture see C1b). If the hunters did not catch a langur, a dog of either sex could be used instead. One dog would suffice for the whole hamlet. It did not matter to whom it belonged; a man from one of the afflicted households would be sure to offer one.

All the men would then make a fortified gate for the hamlet. This gate was a sign that no one should enter. If somebody bypassed the gate, he would have to pay a fine, which would be the cost of repeating the *khang*.

All the villagers would also protect themselves with the aid of *chi-kung-kaa*, a medicine thought to ward off bad spirits. Leaves of the palm fern *Polybotrya appendiculata var. hamiltoniana* would be collected, chopped, and worn in a *laa-puk* (a small lump of cotton) somewhere on the body. The leaves were supposed to put to flight any spirits that might wish to attack the wearer (MK, 26.04.1957).

When all these preparations had been made, the hamlet would be closed in the evening and would remain closed the following day. If the monkey or the dog had been freshly killed, no further sacrifice would be needed. During the night, the outside walls of the affected houses would be pelted with earth and pebbles (*wet re),* and the villagers would shout "*pöt*" ("Go away!"), to drive out the harmful spirits that spread illness. The next day, no work would be allowed; preparing food does not count as work (MK, 26.02.1956).

P3b) *Khang* for the paddy

Two types of *khang, chatma-kang* and *chia-dong-khang*, were held regularly in all the hamlets in the Anok area. Twice a year the hamlets were closed for three days, once in March (*chatma*), in the hot season, and once in July (*chia-dong*), in the rainy season. On both occasions, two goats were sacrificed at the place where water was drawn for daily use. A *dö-ker-kim* (see below) was erected there; each household would contribute at least one *chit-wai*. The goats could be male or female, and castrated animals could be used. They were bought at the market; every household in the hamlet was expected to contribute one to two rupees. Each household would also sacrifice a small chicken. For the *chatma-khang*, each household would sacrifice a large chicken, or if possible, a pig. These animals would be killed in the house near the *ca-pam* (rice bin). For the *chia-dong-khang*, the head of the family was obliged to sacrifice up to three chickens for every swidden in order to sprinkle the paddy with blood (*yau caa*). One of the chickens would be sacrificed near the water course below the swidden. *Kom* were prepared for the ceremony. These are round, flat dough-cakes made from sticky rice flour *(cha-tui-caa*) or coconut or banana flour, with red sugar (*chakka-rik*). No dogs or honuman monkeys were required for these two khang.

The two goats are bought before the village is closed. Two pigeons are also needed; if nobody in the hamlet can provide them, they must be bought from outside. If some households are too poor to contribute anything, as happened in Tapwüa-Kua in 1956 and 1957, the assembly of the elders can decide not to hold a *khang*, because it is a condition of the *khang* that everyone in the hamlet must agree to it. If everyone does agree, the elders fix a date that they think propitious.

The evening preceding the *chia-dong-khang* is called *kom-tir-yak* ("*kom-soak*-evening"); the first day is called *tacöng-ni* ("begin-day"); the second day is *ung-cir-ni* ("stay-inside?-day"); and the third day is *ung-wai-ni* ("stay-inside?-next-day). This is the end of the *khang*: the following day is *talot-ni* ("go-out-again-day").

On the *kom-tir-yak*, husked *cha-tui-caa* is soaked in water. In the early morning of the *tacöng-ni*, the *kom* (see above) is prepared in every household. The two *khang* are described in detail in the next two sections, starting with the *khang* held in the rainy season.

P3c) The Anok *chia-dong-khang*

On the first day of this *khang*, every household goes to its own *üa* (swidden) to perform the *yau-caa* (sprinkling the paddy, see J4c). This ceremony is also called *po-caa-ram* ("taking paddy leaves"). The first chicken is sacrificed on the stream below the field (see J4c, *klaa-wi*), and the second at a *büi-chü* (termite hill) in the jungle. For this second sacrifice, a *pak-plai-kimca* is erected with four *chit-wai*, *pok-pok-kom* and a *pön* (bamboo tube for pouring water). The chicken is sacrificed to *caa-kom-laa-kom-namma*, with a *phyok-tamma* for a large cotton harvest. The third chicken is sacrificed on the *tur-tut* (see J2e) for the paddy. If a household cannot afford three chickens, or does not wish to do so, the first and third chickens will suffice. After the *caa-yau* (see below and J4c), leaves from the paddy, cotton, taro and ginger are taken home together with the chickens, which are eaten at noon.

After the meal, one man from each household goes to the *nat-tang-yaa*, a man who is elected to this office by the villagers. In Tapwüa-Kua it was Elai Atwang, the oldest man of the twin hamlet. They must take with them another small chicken and threads for the *bong-kom*. The *nat-tang-yaa* sets up a *yu-kong* (beer pot), which is called a *nat-tang-yu* ("spirit-dedicate-beer"). All the visitors drink from this pot. The whole group then goes to the spot where water is drawn for daily use. The last men to leave the hamlet close the entrance symbolically with *kung-kaa*; these are thin bamboo strips bent into a bow shape and fixed in the ground at the two ends. The men who go to the river to perform the ceremony (see below) are allowed to return, but nobody else may cross the *kung-kaa* to enter or leave the hamlet until the *talot-ni*. Anyone who does so could in theory be made to pay all the villagers' expenses for the *khang*. In fact nobody would be expected to be able to pay so much money, therefore a minimum payment would be one goat.

In the stream or river, a *dö-ker-kim* (also called *tong-mang-rum*, see P2f), is erected. *Chit-wai* are inserted into the open ends of its four rows of three vertical bamboo poles, and all the *bong-kom* threads are hung on one of them. A further *chit-wai* is then placed at the foot of each vertical bamboo. If desired, *chit-wai* may also be set at the points where the slanting bamboos are fixed. This makes either sixteen or twenty-four *chit-wai* in total. A *pong-mö* (see P2f) is tied to the *dö-ker-kim* so that it floats downstream. A small *ca-pam* (paddy bin) is placed a short distance above the river and filled with earth, into which an unspecified number of further *chit-wai* are set. Further uphill, a *tong-mang-rum* (see P2f and g) is erected. In its "yard", a thin stem of *klang-kü-ching* (unidentified) about 1.5 metres high is planted, and may take root. A number of *chit-wai* is then set in the ground.

The sacrifices may be performed by anyone who wishes to do so. First a *klaa-wi* ("to shed blood") using one chicken is carried out a short distance below the *pong-mö*, over a *chit-wai*. Two pigeons are then sacrificed over the *pong-mö*. Next, the remaining chickens are sacrificed over the *dö-ker-kim*; as each chicken is killed, a *tamma* is said by the person performing the sacrifice, not by the one who provided the chicken. A goat is then offered in the same place. It is not tied to anything, but simply held by its tethering rope. Before each sacrifice, *pok-pok-kom* is spread over the place where the animal will be killed. At the same time, *pok-pok-kom* is placed on the *ca-*

pam, together with an egg, but nothing else is offered there, and no *tamma* is said. Finally a good quantity of *pok-pok-kom* is scattered in the *tong-mang-rum*, and the second goat sacrificed at the *klang-kü-ching*.

After the sacrifices, the *dö-ker-kim* and the *tong-mang-rum* are closed off with *kung-kaa* and all participants return to the hamlet, taking with them the sacrificed animals and the *bong-kom* threads. The animals are taken first to the house of the *nat-tang-yaa*, who also distributes the *bong-kom* threads to all the members of all the households. Meanwhile, the chickens and goats are prepared, cooked and eaten. The meat is distributed according to sib rules (see L4a). Catumma and Ngarua are the *pen* of the Atwang, while the Khongtör are the *pen* of the Ngarua. All chicken meat will therefore be eaten by the members of the Atwang sib, while the members of the other sibs have to make do with the pigeons. The meat from the goat is distributed to all sibs.

The custom of distributing meat according to the sib rules is not followed everywhere. In Chinbok-Kua, for example, where Menkröi spent his youth, each household ate its own chicken. It was also the custom there to use a duck for the offering at the river. This duck would then be prepared, cooked and eaten on the spot by all who were present.

After the goats are sacrificed, it is forbidden to make noise in the hamlet. No drums, gongs or *plung* (see E1) may be played, nor guns fired, during the *khang*. The *khang* is meant to ensure a good harvest. The goat sacrificed at the riverside is an offering to *o-reng-nam* (the river spirit), while that sacrificed in the *tong-mang-rum* is for *tong-mang*, a Marma term meaning "lord of the hills". In Mru, this spirit is *chüng-rengma*, but Mru terms are not used in this context, which suggests that at least the goat sacrifices are borrowed from Marma customs.

The *khang* does not end with the sacrifices made on the *tacöng-ni*. The next day (*ung-cir-ni*), a private sacrifice, called *caa-ram nüm* ("paddy leaves soft") is made in every household. The leaves that were brought from the field on the previous day and the equipment used for the purpose are put into a *klai-puk* (see D2c). Some paddy and taro leaves are laid out on the floor and serve as a mat for a *yu-kong* (beer pot) which is prepared in advance. The mat has no further significance, and the leaves will be thrown away on the *talot-ni*. A rich man may place beside the pot a vertical bamboo pole, tied at the top to the *kri* (see C3e) or to the roof. To this pole, which has no special name, he ties his *re* (spears), his gun, and any other valuables that he owns. The *khong-rau* (see D2h) are put into the *klai-puk*. A chicken, called *caa-ram-nüm waa*, is sacrificed, and anyone who wishes to offer a pig and can afford it may do so. The whole ceremony resembles a *caa-müng-pok* (see J4d). *Köm-pot* are prepared and placed on all the items that were put in the *klai-puk*, or set beside the bamboo pole or tied to it.

The following day (*ung-wai-ni*) is a day of rest. No more sacrifices are made. In the evening, Chüngma Mru (but not the Anok) must drive away the evil spirits in a ceremony called *re-wet-khang* ("pebble-throwing-khang") (MK, 26.08.1956).

The *caa-yau* of the first day (*tacöng-ni*) may be performed in advance of the *chia-dong-khang*. If the *khang* is not held, it may take the form of the *khan-pon-thet* described in the chapter on swidden cultivation (see J4c). *Khan-pon-thet* and *caa-yau* are performed in the same way, but they have different purposes. In both cases the *röng* (bamboo tube) contains blood of the sacrificed chicken mixed with lemon and lime leaves. Before the mixture in the *röng* is sprinkled over the paddy with a *chit-pot* during the *caa-yau*, the following *tamma* is said: *caa-kom-laa-kom-namma, wang caa pon-pon, wang kham pon-pon, caa wang, caa wang huu* ("much paddy much cotton spirit, come and eat all, come and drink all, paddy come, paddy come – huu!"). The bad spirits are driven off at the same time: *yum nam, pöi nam, pöt!* ("spirits who spoil the harvest, move out!". *Caa-yau* thus has two aims, namely, to drive away spirits that cause the paddy stalks to go brown and the husks to remain empty (see J4b), and to summon the good spirits. The *khan-pon-thet* on the other hand is only meant for *caa-kom-namma*, so the *tamma* therefore begins with a *phyok carok*. After the paddy has been sprinkled, the *röng* and the *chit-pot* are set into the earth at the upper end of the field (see J4c).

P3d) The Anok *chatma-khang*

The *chatma-khang* may be held at any time between the end of January and April, before or after the swidden is cut. It too lasts for three days, which are called by the same names as the corresponding days of the *chia-dong-khang*. On the *tacöng-ni*, there are no individual sacrifices in the fields. Instead, a communal sacrifice is made in the hamlet, at the *ling-pu*, which is the place where the cattle are tethered at cattle feasts (see R1b). Any villager may lead the ceremony; in Tapwüa-Kua, Karbari Kangku acted as *ther-khak-yaa* (or *khak-ther-mi*) every year for both parts of the hamlet, though in 1957 he confined himself to supervising his grown-up son Mowai, who performed the sacrifices. At the *ling-pu* in the Horinjhiri part (Atwangwüa-Kua), men from both parts of the hamlet erected a *therca-kim*; this is a kind of miniature house without a roof.

On 19 January 1957, well before the usual *chatma* period had started, I arrived at the hamlet at 4 p.m. to find that the *khang* had begun. Since no sacrifice had yet been made, I was allowed to enter, but the men who had carried my luggage had to stay outside. The *therca-kim* had already been erected, and the sacrifice was just starting at the place where the daily water was drawn (*rüm kwam)*. I was too late to witness the sacrifice, so all I could do was take some photographs of the scene the next day.

The *tong-mang-rum* fence was no more than 20 cm high, and it surrounded only part of the village square. A *ca-pam* (rice bin) had been placed inside it. The bin was 30 cm in diameter and was woven in the same way as an open wicker basket. The spaces in the wickerwork were filled with pieces of banana leaf, though any kind of leaf would be acceptable. The walls were reinforced with short pieces of unsplit bamboo. Finally, the pam was filled with earth, into which were set:

1. a *laa-ui-töng*, which is an upright bamboo pole bearing six *laa-ui*. These are cotton "fruits", i.e., balls of cotton that hang from a short *röng* (bamboo tube) into which they are fixed by means of a long, flexible spike. The spike passes through the *röng* and is pushed into a hole in the bamboo pole. Because the spike is flexible, the *röng* hangs down like a bunch of fruit.

2. a bamboo pole with a hole at its upper end. A thong was threaded through the hole and tied to the *pam-khen* ("bin-girdle"), a ring 50 cm in diameter. It bore thirty-six *chit-wai*.

3. a *pai-leng*, also called *tang-khwön* (flag). This was a straight, slender bamboo to which were tied cotton threads with nine "knots" (places where the thread was not twisted).

4. a *thi* (umbrella), which was a bamboo with two holes in the top into which bamboo strips were inserted. This produced a similar result to splitting the top part of the bamboo lengthwise into four parts. The strips were not used to make a circle, but were bent over to make two semi-circles at right angles to each other. Loose wads of cotton were laid across these semi-circles.

The bamboo poles used to make items 1 to 3 were about 2 m high, while that used for item 4 was only 1.5 m. The top half internodes of items 1, 2, and 4 were decorated with stripes made by shaving the bark. Twelve longer and twelve shorter *chit-wai* were set inside the *tong-mang-rum*. Five chickens had been sacrificed there instead of the first goat.

The only goat to be sacrificed was killed last, at the *dö-ker-kim* (see P2f and P3b). The *dö-ker-kim* was built not in the water, but a short distance below the *rüm-kwam* (water hole), because water was already becoming scarce. This may have been the reason why the *chatma-khang* was held so early, well before the end of January. The stream had been dammed to make it easier to draw water every day; the dam was above the place where the sacrifice was held.

For the same reason, the *pong-mö* (see P2f) was not floating on water, but was built beside the *pai-leng*. Only one pigeon was sacrificed there. Two rows of three 70-cm bamboo poles, set close together, were erected so as to form an acute angle pointing upwards towards the source of the water. Notches were cut in the inner sides of the poles and lengths of thin bamboo rested in the notches along each of the two rows. Leaves were laid across these lengths of bamboo so as to form a small platform, on which *pok-pok-kom* was scattered. A *chit-wai* was stuck into the top end of each vertical, and another at its foot; this made twelve *chit-wai* in all. To make the *pong* (boat roof), 3+4+3 short bamboo strips were fixed so that they formed an arch across the tops of the vertical poles. Longer strips were then woven through them.

After the animals had been killed, *kung-kaa* were placed around the whole sacrificial area. The goat sacrificed to *o-reng-nam* was butchered in Elai Atwang's house, though all the villagers who had contributed to the cost later helped to eat it.

According to my earlier information, obtained in 1956, a *laa-ui-töng*, a *thi* and a *thang-khwön* should have been erected not only in the *tong-mang-rum*, but also at the *dö-ker-kim*, and a third set should have been built at the *therca*. In 1957, this did not happen. The only items erected were a *laa-ui-töng* 2.5 m long, with 12 *laa-ui*; a full bamboo 3 m long, with leaves; and a cut-down *ram-nget-ching* (unidentified) more than 2 m long. Two of its upper branches were left in place, and an *em* (see D2a) was hung on one of them.

The sacrifice at the *therca* started after nightfall, at 6 p.m. The following day I made my notes on the *therca-kim*, which adjoined the *ling-pu* on the north-eastern side, and looked more like a chute than a house. It consisted of 9 vertical bamboo poles arranged in a square, each side having 3 poles 30 cm apart. The house thus measured 60 cm x 60 cm. The top half internode of each pole was decorated with stripes made by scraping off strips of bark. The next internode was shaved upwards to form tassels. The row nearest to the *ling-pu* was 90 cm high, the next one 80 cm, and the last one 75 cm. The impression of a chute was strengthened by pairs of bamboo poles attached to the sloping sides of the house. The internodes of these poles were lightly shaved towards the upper nodes to form small tassels. Their higher ends were tied at a height of 25 cm to the tallest row of verticals, and at a height of 10 cm to the lowest row. The lower ends extended 50 cm beyond the shortest row and reached the ground. A real chute was formed by a "floor" made of bamboo poles split in half. This floor was fixed at a height of 10 cm at the higher end, while the lower end projected as before until it reached the ground. This meant that its slope was shallower than that of the pairs of bamboo poles. The presence of the central vertical poles meant that there was a gap in the middle of the floor; the gap was filled by three pieces of split bamboo, which rested on thin transverse bamboos tied to the verticals. The part of the floor which lay inside the house was framed by bamboo poles tied to the outside of the verticals. These poles were not decorated with tassels. The floor must slope, or bad luck would ensue.

The *khak-ther-mi* prepares a pot of beer which is placed in the *therca*, and from which all those who helped to build the *therca* will be the first to drink. Afterwards, the rest of the people in the hamlet may drink from it. At the ceremony that I witnessed in 1957, the *yu-kong* was replaced by a *hom-noi-pu*, which is a small pot containing some cooked rice soaked in water. After the sacrifice at the *rüm-kwa*, the *khak-ther-mi* began his duties by sacrificing a chicken. This chicken was his own, and would be eaten by himself and his family. It was believed that this would bring him a particularly good harvest. He then killed a pig, bought at common expense and eaten later by all the families. The entrails of the pig (*pak-ria*) were buried in the ground beside the *therca*; to prevent dogs or pigs from digging them up, a paddy mortar or a piece of thick tree trunk would be placed over it, and bamboo poles were set around the obstacle so that it could not be pushed over. On *talot-ni*, the *pak-ria* was dug up and thrown away. If the entrails disappear before the *talot-ni*, it is thought that the spirits have stolen them; this is a bad omen.

The next day, one of the pig's legs is cut off, roasted, and put into the *em* which hangs in the *ram-nget-ching*, together with the *hom-noi-pu*, complete with drinking tube; a sickle; some paddy; some cotton; a large taro root; and some *köm-pot*. On *talot-ni*, the leg of pork will be taken down, cut into small pieces, and put on skewers. This task may be performed by anyone who wishes to do it. Two skewers are given to each household in the hamlet. They are stuck into the wall that separates the *kimma* from the *kim-tom*, on the *kim-tom* side (see C3b).

The next day (*ung-cir-ni*), each household sacrifices at least a chicken beside the *ca-pam*. This chicken is the *caa-chok-waa*, or "paddy-feed-fowl". Those who can afford to do so sacrifice a pig. Karbari Kangku contented himself with a chicken. He had set up a beer pot in the *kim-tom*, tied a thong around the neck of the pot, and filled it up with water. At 7.30 a.m. he moved the *yu-kong* to the *ca-pam* in his *kimma*, in the left-hand corner of which he had assembled his spears, a gong, and other valuables. A small bowl stood to the right of the bin. He fetched the chicken and made a *cho* (ceremonial spitting) with beer. He then laid the chicken with its neck across a chopper and cut through the neck, letting some blood fall on the *yu-kong* and on his assembled valuables. He then let the blood drip into the bowl. After performing another *cho*, he carried the bowl, the chicken, and the *yu-kong* back into the *kim-tom*.

The first visitors then arrived and began to drink beer. This was the day on which everyone in the hamlet visits all the other houses, drinks beer and arak, and eats roast chicken and *kom* (round flat dough cakes made of sticky rice flour). People usually drink until they become sleepy, then have a short nap and start visiting again. The *long-tang-wüa* (older men) take the lead in the drinking, though the children also drink beer. Only the women stay sober: they have to do the housework, and take care of the inebriated men.

Portions of chicken meat are distributed to all the households that belong to a *tainau* sib (see L3c), or to the same sib as the head of the family. If there are two *tainau* sibs who are allowed to intermarry, two chickens must be sacrificed; alternatively, two households that belong to the same sib may divide their *tainau* sibs between them. In Tapwüa-Kua, the three Khongtör households had agreed that Klingtui would give to the Catumma, while Yenpau and Khamcöng would give to the Atwang (see L3d).

Some of the *kom* is placed on the cross-beam of the house for each of the Mru groups (officially these are Anok, Dopreng, Dömrong, Chüngma, and Tamchaa), and for all the other tribes. If any group did not receive this gift, its members might become angry. The other tribes are the Wakung and Balong. The Wakung are the Awa-Khumi; the Kung are a sib of the Awa. The Balong are not clearly identified; Headman Inghoi thought they were the Balang, an Areng-Khumi sib living on the upper Koladan River in Arakan.

I learned all this from Menkröi on 21.03.1957. I therefore asked the Rengmitca headman, Haiwön, in whose house I was spending the day, whether there were further sacrifices at which gifts were made to other Mru groups and other tribes. I was told that at every *cari-yöng*, they made *köm-pot* morsels of food that the Anok prepare for spirits only – for all the groups and tribes, including the Balöng (not Balong). Haiwön, like the others, was unable to identify the Balöng.

On the last day of the *khang* (*ung-wai-ni*), the men start to do small jobs at home and in the hamlet. Two Mru men came to the hamlet at noon; they were breaking the taboo (see C1b), but this did not seem to cause any problems. In the evening, two other men came and asked to be admitted. They were allowed in even though they were Bengalis, and the hamlet was not due to be opened until the next day (*talot-ni*) (MK, 27.08.1956, and my own observations, 19 to 21.01.1957).

P3e) The Rümma and Chüngma *khang*

The Rümma also begin the *chatma-khang* by sacrificing goats at the stream below the hamlet. The central sacrificial area is called *naik-kyok*, which is Marma for *nat-hua*. Upstream, the villagers prepare a *khyong-chang*; this term derives from the Marma *khyông-cang* which means "river-stage". The Anok would call it a *pak-plai-kimca* (see P5b). Downstream they build a *dö-ker-kim* (see P3b). Two goats are killed, one at the *khyong-chang* and one at the *tong-mang-rum*. At one time, a third goat would have been killed at the *dö-ker-kim*.

Like the Anok, the Rümma build a *therca* at the *ling-pu*. A pig and a dog are killed there, to the beating of a drum. The *khak-ther-mi* must observe a *kan-khang*; he may eat nothing but rice and pork with ginger and salt, but no pepper. Several chickens are sacrificed, either one by each household, or two or three for the whole hamlet.

For a *chia-dong-khang*, three more days of *khang* follow the *talot-ni*, on which day *rai* (bamboo shoots) are collected. The fifth day is *rai-dik-khang*: no one is allowed to leave the hamlet. It is believed that anyone who did so would be bitten by a tiger. The sixth day is *mung-plan-khang*, when the weeds that were cut in the swidden and put in a heap must be turned. The seventh day is *po-caa-ram-ni*, also called *caa-ram pok-khang*; paddy leaves may not be gathered on this day. On the next day, or at some time subsequently, another day's *khang* is observed. This is the *caa-nöm-khang*, when a pig is killed in every hamlet. It may be killed on the ground or in a house, and anyone may perform the deed. Each household receives some parts of the kidneys, heart, liver and brain, and cooks them in one or several *röng* (bamboo tubes), one *röng* for each swidden area. The tube is then taken to the swidden, where its contents are poured into a *ram-nget* leaf (unidentified). Some ears of paddy are dipped into the mixture, while a *tamma* (invocation) is uttered calling for unspoiled rice and a large harvest.

This is the Rümma form of *yau-caa* (see P3c), though there is another *khang*, the *chit-ci-yau-khang*, which is also given the name *caa-yau-khang*. This *khang* must follow the *chia-dong-khang*. One goat is killed in the hamlet, its hair is burned, and each household collects some of the ashes, together with some of the goat's blood and faeces. These are mixed on a plate (*long-pan*) with *prön* leaves (lemon), *ro* (limes), *waa-ma ki-or* (*Clausena heptaphylla*) and bark from the *ching-chür* (unidentified). The mixture is taken out to the swidden and scattered over the paddy. It is intended as food for the spirits, which might otherwise spoil the paddy. In this context, the spirits seem to be identified with the *chit-ci* (see P2k).

A dog can be killed instead of the goat, but it must have a brown coat: neither black nor white dogs are acceptable. No chickens are sacrificed (K Rengtan Rümthu, 14.03.1957).

Other groups also beat the drum as an accompaniment to the *chatma-khang*. The Northern Chüngma, for example the Rümthu in the Galengya-Mouza, also have dancing. The young people perform a dance at the *therca* while drums are beaten and plung (see E1a) are played (MK, 24.03.1957). Karbari Rengtan Rümthu told me that his sib moved from the Pantola area into the Rümma region (see L3d). It is possible, therefore, that the Rümma have inherited some of the customs of the Northern Chüngma.

The Southern Chüngma (the Longhu) have different *chatma-khang* customs. They too build a *ther(ca)-kim*, but it is destroyed after the three days of *khang*. To prepare for the *khang*, they pound sticky rice (*cha-tui-caa*) to make *kom* (see above). Each household in the hamlet contributes 4 annas, which is $^1/16$ of a rupee, or 2 annas if the family cannot afford 4. The money is used to buy a chicken and a pig. In the morning of the first day, the chicken is carried around the hamlet with 2 *pau* ("flowers", the equivalent of the Anok *chit-wai*) and a small amount of husked rice wrapped in a banana leaf. Everyone must touch the chicken before it is taken by any two people who wish to perform the task down to the place where the daily water is drawn. There the leaf is unwrapped and fixed to the ground by means of the two *pau*. There is a ceremonial spitting at the chicken, using water, after which its throat is cut with a chopper. It is then carried back and left to cool in the open in the middle of the hamlet; it may not be taken into a house. Anyone who wishes to eat some of the chicken may do so. However, before it is cooked, a pig, called *therca-pak*, must be killed, and the hamlet is then closed. The pig is sacrificed at the *ther-kim*, which is built not at the *ling-pu*, but at a separate place where a post made from *ram-hoi-ching* is erected. The tail and liver of the pig are tied to the post with string (*leng*), together with some *tam* (ginger). The other parts of the animal are cooked and distributed to all the households. When the *ther-kim* is dismantled, the post is left standing. It may fall down later, or it may take root and grow; its fate is of no importance.

During the *khang*, it is forbidden to eat *ngaa-pi* (see K1b) and chillies. My informants, Phungkri and Cöngreng of Uikük-HP (16.01.1957), said that this simple form of *khang* had been adopted from the Khumi. However, the information I obtained about the Khumi customs suggests that the picture is in fact rather more complicated (see P3f).

For the *chia-dong-khang* observed on *tacöng-ni*, the Khumi pound sticky rice and make *kom* (see P3d). I was told that the main activity of that day is *tung-kom* (pounding *kom*). No sacrifices are made at the place where water is drawn. Persons who have previously given a cattle feast, and any others who prefer to do so, hold an *üa-reng* (see R6c) instead of the *chia-dong-khang*. However, large *üa-reng* feasts must be held in advance of the *khang*, because only small feasts are allowed at that time. The closing of the hamlet on *tacöng-ni* is not strictly observed because young people who come to dance must be allowed to enter (Khushi Patlaica, 29.03.1957). It is only on *ung-cir-ni* that the *khang* will be strictly kept, because an *üa-reng* must also be followed by two days of *khang*. On *ung-cir-ni*, a *waa-bet-yu* must be placed near the *hungma-cüng* (the main house post) close to the *ca-pam* (paddy-bin). A hoe (*tim*) and a chopper (*charai*) are placed together and a chicken is sacrificed (*waa-bet*) over them. During the sacrifice, there is a ceremonial spitting (*cho*) of beer (*yu*) over the chicken. On *ung-wai-ni*, villagers go out to dig up bamboo shoots (*rai*), but no outsiders are allowed to enter the hamlet. On *talot-ni* there is a *tan-waa* (= Anok *ngen-waa*, "wringing a chicken's neck") in the swidden at the *tur-tut* (see J2e). The chicken is killed over some grass and other weeds that have just been pulled up. While it is being killed, there is a *hom-noi-cho* (spitting of water mixed with soaked rice) from a *teng* (a swallow-tailed bamboo tube). Two leaves of paddy and two of cotton (*caa-pre-tut* and *laa-pre-tut*) are taken from the *tur-tut*; the first pair of leaves is pushed with the *teng* into the *waa-pom* (chicken basket) and left in the swidden. The kidneys and the liver of the chicken are cooked in a *röng* and used to make köm-pot. The rest of the chicken, together with the *röng* and the second pair of leaves, is taken home. The leaves are fixed under the roof, above the door. The rest of the contents of the *röng* are used to make more *köm-pot*, which are distributed around the house (son of K Wöilaa, 28.03.1957).

P3f) The Khumi *khang*

Like the Mru version, the Khumi *chatma-khang* and *chia-dong-khang* both last three days. The Khumi have different names for the days of the khang (Khumi: *yana'),* but for the sake of simplicity I will continue to use the Mru terms.

The chatma-khang is normally held before the swidden is cut. On the first day, the hamlet is closed, and the women prepare kom. Strictly speaking, a goat and a chicken should be bought with money collected in the hamlet, and killed by the river. If there is not enough money for this, at least one chicken should be killed as a communal sacrifice. Everyone who can afford it may then sacrifice his own goat, plus a chicken for himself. Both the goat and the chicken are cooked and eaten by the river; cooked rice is brought from the hamlet for the meal. The women stay at home. Any food that the men cannot eat must be thrown away; it is forbidden to take it back.

The sacrifice is offered over a simple chit-wai (bamboo stick with a tassel) and some husked rice on a banana or la-ia (Aglaonema hookerianum) leaf. Even if a goat is sacrificed, neither a tong-mang-rum nor a ca-pam is needed. However, a very large therca-kim is built in the hamlet (see P3d), not at the ling-pu, but in a different place, and with no decoration. Everyone who can afford to do so must put a pot of beer there. A man of some standing in the hamlet, called the *nat-tang-yaa* (see P3c), kills a large sow, called the *therca pak-maa* (Khumi: *khu-kadam*). This sow is purchased by the community, and costs twenty rupees. No chickens are killed at the *therca*, but beer is drunk there from all the beer pots. The pig is killed not in the *therca*, but on a layer of banana leaves. Its entrails are not buried. Part of the pig is cooked and eaten by all the villagers near the *therca*. The remaining meat is distributed to all households. Some of it will be preserved, to be taken when the first swidden is cut and eaten there. If the swidden has already been cut, none of the meat is preserved. In addition to the pig, which is a communal sacrifice, each household may also kill a chicken at home. This is cooked and eaten with ginger and salt (K Müilün Tamchaa, 25.03.1957).

Headman Renglün gave me a slightly different account of the proceedings. He said that the Khumi, like the Mru, prepare the *kom* on the evening before the hamlet is closed. On the first day of the *khang*, every family that can afford to do so sacrifices a chicken by the river. This chicken is killed on a *pak-plai-kimca* with four *chit-wai* (Khumi: *capon*). Some husked paddy is scattered on the *pak-plai-kimca* and blood from the chicken is allowed to drip on it. The whole hamlet buys a goat, which is sacrificed at a *dö-ker-kim* on which *pok-pok-kom* is scattered. No more ceremonial items are built at the river, and there is no *ca-pam* or *tong-mang-rum*.

The hamlet is then closed, and every household that can afford it takes a pot of beer to the *ling-pu*. A pig bought at common expense is then sacrificed near the *ling-pu*, not at a separate place. The animal is killed on the ground with a spear. It is then prepared and eaten by all the villagers. No parts of it are buried or hung from a tree.

The village is then closed for two more days. The first day is dedicated to the people; no outsiders are allowed in, but the people who live in the hamlet may leave. The second day is dedicated to the domestic animals, that is, to the pigs and cattle. No one may enter or leave the hamlet. On the fourth day, all restrictions are lifted, without any further ceremonies (H. Renglün Amchang, 26.06. 1957).

On the first day of the *chia-dong-khang*, the hamlet is closed to outsiders, but the men from all the families go to the riverside and sacrifice a chicken.

The people of the hamlet do not make a common sacrifice. The chicken may be taken home and eaten there. The women start to prepare *kom*. The fire in the hearth is extinguished, and the ashes are thrown away, with a *tamma* calling on the spirits of poverty to leave and those of prosperity to enter. Water is then poured over the hearth. The men go to their swidden and sacrifice a second chicken. Rich people may also take a goat for this *klaa-wi* ("shedding-blood"). The sacrificed animal is then taken home, with some paddy leaves. In the hamlet, the chicken is prepared and the pen *men* take some of it to their *tutma* (see L3b), if they have any living in the hamlet. In return, the *tutma* offer fish or dried game to their *pen*. If they have neither of these things, they may offer them pork. Dried game is much more highly valued, because anyone can give pork, but game is rare. Bamboo rats are especially prized.

On *ung-cir-ni*, beer and arak are drunk, and no work is done. No more chickens are killed, though there may be meat left over from those killed on the previous day. The last of the meat, and especially the chicken cooked in a *röng* on *tacöng-ni*, is eaten on talot-ni. Anyone who is willing and able to give a *lau-reng* (see R5h) may hold it after the *chia-dong-khang* (K Müilün Tamchaa, 25.03.1957).

P3g) Other kinds of Mru *khang*

It used to be the custom that a man who had killed a chicken and performed the *klaa-liu* ("throwing the seed", see J2e) must observe a *kan-khang* (taboo on certain dishes) for seven years. This was called the *liu-klak-khang*. During these seven years, he could sow again every year without his *khang* being prolonged. While the *khang* lasted, he was not allowed to eat any game, meat from uncastrated pigs, or *khua* (a kind of fish). He was also forbidden to eat meat from domestic animals (poultry or cattle) that had been attacked by a tiger, wild cat or wild dog; and to eat anything in a house where a dead person was laid out. It was believed that if he broke the taboo, his harvest would be smaller.

There is another food restriction that is linked with the *caa-kom-kan-khang*, and which lasts from the beginning of the paddy harvest until the *caa-moi*. Certain food restrictions are in force at this time in any case (see J4f), but this *kan-khang*, which is kept to please *caa-kom-namma*, is observed more strictly. It applied to heads of families, but in Tapwüa-Kua, it was kept only by Yenpau Khongtör. The rule is that the head of the family should not eat pork, any game, fish, oil, onion, chillies, or almost any other kind of vegetable. He is not allowed to drink beer or arak. He is allowed ginger, *oi-lo-ram* (a kind of salty [?] leaf), *ching-ram* (? tree leaves), and *kan-deng* (? a kind of banana). Before starting his *khang* – that is, before starting to cut the paddy – the man rubs his hands with *kan-chur* (*Hibiscus sabdariffa*) leaves, to remove any unpleasant odour that might offend *caa-kom-namma*.

After the first paddy is cut, there is a *caa-wi-hok-khang* ("paddy-blood-fall out-ban"), so no work is allowed in the swidden. Strictly speaking, this *khang* should be kept by all those who are starting to harvest. There are no food restrictions. None of the people in the hamlet are allowed to work in the swidden when a wild pig or a muntjac is caught in someone's *chot* (spear-trap, see H4) for the first time that year. This *khang* is called *cak-kak-khang* ("game-receiving-ban"). (MK, 10.09.1956).

P4 The sun, the moon, and natural phenomena

P4a) The sun and the moon

Neither the sun nor the moon is regarded as a spirit or a god. People do not pray to them. This seems consistent with Hutchinson's statement: "They venerate the sun and moon, but do not make actual worship to either."

The sun and moon are not generally thought to belong to either sex (MK, 17.08.1956). On occasions when a gender is assigned, the sun tends to be male and the moon female; this is in line with the beliefs of the neighbouring Chin. U Ba Thin, however, when writing about the Buthidaung Chin Hills, says: "The Mro regard the sun as female and the moon as male. The Ahraing Khami and Awa Khami have no different genders for them" (1931:249–50). According to U Ohn Pe on the other hand, the Mru of the Ponnagyun Chin Hills think of the sun as male and the moon as female, though he also maintains that the Awa, the Mru and the Areng have no particular beliefs about the sun and the moon (1931:260).

The earliest source is Latter, who does not share this view, and who describes some definite beliefs about the sun and moon. He wrote of the Khumi on the Koladan: "They worship the earth as the author and giver of all they possess; the sun also, in its noon-day height, as the pledge of safety from their foes ..." (1846:63). He supports the second part of his statement by citing the beginning of a prayer: "Oh! spirit of the day-sun ..." Unfortunately he does not quote the full text; he does not say whether it was known to him. The prayer was probably addressed to *kani.* In 1872, Andrew St. John wrote: "The word ka-nie is used by the Hkamies to represent a spirit or dryad, also the sun and day; a spirit or ka-nie is supposed to reside in almost everything" (1872:237–38). Latter writes: "... in the Khumi dialect the term for 'sun' enters into competition for that of 'God'" (1846:78).

In his glossary, Latter gives the term "*k'ni-y'lo'ng*", which is probably the equivalent of *kani-along.* According to Neapan (1923), "*kani-kahlong"* means "sky and earth", and is used as an exclamation in situations where we might say, for example, "Good Lord!", or "My goodness!". The Reverend Stilson translates "God" in Khami as "*ki-nî-k'ô-mä"*, with "*k'ô"* appearing again in "*k'ô-s'i-nî"* = sky (1864:225). According to my notes, the term *kani* also exists in Khumi. I did not pursue this question with the Khumi of the Chittagong Hill Tracts, so I am not in a position to assess whether there is, or was, a conceptual link between the uses of *kani* in the two languages, or whether they are simply homonyms. It is possible that there is a tonal difference in pronunciation between the two words.

In his discussion of the Areng Khumi of the Ponnagyun Chin Hills, U Ohn Pe writes: "They think that the phases of the moon are due to its being buried in the sky and appearing again" (1931:262). The first crescent seems to be the only phase of the moon to which any importance is attached: it is considered propitious when the right tip points upwards, but unpropitious when the left tip does so, which in fact rarely happens. If at all possible, cattle feasts should be held in the first fifteen days of waxing moon; the period when the moon is waning should be avoided. Paradoxically, most of the feasts I attended were held after the full moon. The timing might have been planned otherwise, but delays probably arose because preparations such as buying the animals for sacrifice can take longer than expected.

A bride should be fetched only when the moon is waxing. The phases of the moon are not thought to have any importance for sexual relations (MK, 16.08.1956).

Regarding the months, the whole of *la-kauma-laa* (July) is not considered a time of prosperity. Families may be short of provisions, and waiting anxiously for the next harvest. A child born during this period is thought unlikely to become rich.

The Mru take the names of the days of the week from the Marma, but attach less importance to particular days than do the latter. Most Mru do not usually know what day of the week it is. Those who do take an interest sometimes have beliefs about which days are propitious for certain activities. Tuesday and Friday, for example, are good days for moving into a new house, but Saturday is unlucky. Tuesday and Wednesday are the best days for making a sacrifice on the swidden. Tuesday and Friday are the best days for a *cia-chot-poi* (cattle feast), while Saturdays should be avoided. Bodies should not be cremated on Saturdays, because the fire would grow too hot. A *param-rui-poi* (ear-piercing festival, see N3a) should start on a Monday evening, and the animals should be killed and the ears pierced on the Tuesday morning. The festival at which the first rice of the new harvest is eaten (*caa müng pok,* see J4d) should be held on a Wednesday; Thursday and Friday are possible, but not Saturday (Khamlai, 16.08.1956). The Mru share the Marma belief that Saturday is not a good day for doing something important. However, they do not share the belief that Thursday and Sunday are propitious, and they show no reluctance to start something, a journey for example, on a Wednesday.

The Mru do not have the notion that there is some kind of being on the moon; they believe that the moon's markings are made by the sap of a fig tree. I could not find a story to explain this, but U Ba Thin, writing about the people of the Buthidaung Chin Hills, offers the following: "According to the Mro and Ahraing Khami, the moon was brighter than the sun before, and the moon by its scorching heat had once killed a man, whose son in revenge defeated her [the moon] in a duel and then besmeared her with the juice of the banyan [ficus] tree. Thus the moon lost her power and heat. They regard markings on the face of the moon as juices of the banyan tree. The Awa Khami regard these markings as shadows cast by a banyan tree, which is supposed to grow around the moon" (1931:250).

Of the Khumi (Areng) of the Ponnagyun Chin Hills, U Ohn Pe writes: "... they think that in the early ages of the world, both the sun and the moon were very hot. So the people of those days put a banyan tree with its juice on the moon to keep off the heat from her and allow only the sun to shine" (1931:262). Writing about the Awa Khami, the same author says: "They take the markings of the moon as a banyan tree" (1931:258). He was not able to obtain information about the beliefs of the Mru living in that area (U Ohn Pe 1931:260).

P4b) Eclipses

The period between a lunar eclipse and the next new moon is regarded as unsuitable for holding a feast or for marrying. The Mru believe that the eclipse (*laama kaa mu)* is caused by an ogre, called by the Marma name *Philu,* who swallows the moon. When the moon comes out of his mouth again (partial eclipse), it is a good sign, but when the moon is digested and re-emerges through Philu's anus (total eclipse), bad times will follow. The same applies to solar eclipses. Philu can appear in a variety of forms, including that of a human being, but he can also swallow humans, which kills them (MK, 17.08.1956. For a description of Philu's role in founding the cattle feasts, see R8b).

Even as children, the Mru know Philu's name, and assimilate beliefs about him (see N8a). Both the name and the concept are clearly borrowed from the Marma. According to Judson, "*bhilu*" is "a kind of monster which eats human flesh, and possesses superhuman power; said to be red-eyed and to have prominent, long eye-teeth" (1953:727).

Another Mru explanation of eclipses is that *Angum-Angia* wishes to impose a tax on the sun and moon. The tax consists of one Ari (< Bangla *hari,* roughly 10 kg) of red eggs and one Ari of white eggs. The moon is unable to deliver the tax, and Angum-Angia puts it in jail. When there is an eclipse, people who have money throw between one and four anna coins into a water-hole and ask the spirit to release the moon (Kangku Catumma, 19.08.1956).

U Ba Thin records a similar story told by the Mru and Khumi of the Buthidaung Chin Hills: "... the sun and moon could not pay their taxes to the king of 'nat', who arrests and keeps them in a dungeon for some time and thus causes an eclipse." "According to the Awa Khami, the sun and moon had once borrowed rice from a 'nat' called Sa Aung and they could not repay the rice; for this debt Sa Aung, when he finds an opportunity, seizes the sun and moon and swallows them. This causes an eclipse. During an eclipse, the Awa Khami beat gongs and scatter rice on the ground by way of helping the sun and moon to repay their debt and they also do not pound rice or do any other work, save household affairs, for a couple of days" (1931:250).

Of the Awa Khami in the Ponnagyun Chin Hills we read: "They think that the sun or the moon, as the case may be, is swallowed by a tiger whenever an eclipse takes place" (U Ohn Pe 1931:258).

Of the Mru the same author recounts a completely different story: "Once upon a time, a woman of their tribe gave birth to a son without a father. As soon as the son was born, he dug out 7 rats from the ground and ate them. Then he asked his mother who his father was. Feeling ashamed of the non-existence of a husband, she falsely told her son that his father was devoured by a tiger. So the son went into the jungles and killed a tiger with a spear. He brought the tiger's head and slept a night keeping it under his head and beseeching it to show him his father at night. Though the day dawned, there was no sign of his father. So he asked his mother again as to why he could not see his father, though he had killed the tiger and had slept with its head under his. His mother then gave him a new tale of an elephant having killed his father. So the son did the same act as in the case of the tiger. When he did not see any signs of his father, he asked his mother again. Then his mother gave him a new story saying that his father died on account of the heat of the sun. So he told his mother and the people that he would go and wage a war against the sun and instructed them to watch and to follow him when they should see him fighting with the sun. With this belief the Mro raise cries as war songs whenever they see an eclipse of the sun. They believe that the moon, being the sister of the sun, the son of that woman went and waged war against the moon, when he found that he could not overthrow the sun" (U Ohn Pe 1931:260).

P4c) Stars

The stars do not play an important part in Mru beliefs. I was not able to discover whether the Mru have names for any of the constellations. I asked three informants, but they did not know, and did not even have a way of identifying Venus (22.08.1956). U Ohn Pe writes of both the Mru and the Khumi of the Ponnagyun Chin Hills: "They do not know the Great Bear or any other stars" (1931:260, 262). Of the Awa Khami he says: "They call the 7 stars towards the tail of the Great Bear 'stars of 7 days' people.' They cannot explain the meaning ..." (1931:258)

U Ba Thin had a little more success with his enquiries in the Buthidaung area: "All the three tribes have very little knowledge of the different constellations, except Orion's Belt and the Great Bear and the star Sirius.

Orion's Belt. – The Awa Khami call this constellation, Kala-papo, because it resembles a man carrying two baskets on his shoulder. The Mru call it Uraitikedawyaung (3 steps of a stair). It is so named because it resembles 3 steps of a stair. The Ahraing Khami call it Ayevolokinyauk (2 men carrying a hog). It is so named because it resembles 2 men carrying a hog. [Only the word given for the Areng-Khami can be identified as *eo'-lô kayong* = a male pig, carried (on a pole by two persons)]

Great Bear. – The Mro call this constellation, Tabrikhare (tiger [probably: *tapri-karek* = 'tiger-star']), because it resembles a tiger, 4 stars forming the 4 legs and three other stars forming a tail. The Ahraing Khami call it Kachidaung. It means a square formed by 4 stars with a tail of 3 stars. [Apparently not, since *kaachi* = 'star'.] The Awa Khami call it Naga because it turns its head according to the seasons.

Sirius. – The Awa Khami call this star Achimanye (burning charcoal [?, *achi* = 'star', mai = 'fire' also in a composite word for burning charcoal]). It is so named because it resembles a burning charcoal. The Mro have no name for it. The Ahraing Khami call it Kachitakike (tiger's eyes [probably *kaachi-takai* = 'star tiger']). It is so named because it sparkles like the eyes of the tiger." (1931:250)

P4d) Rainbows and thunder

The Mru term for the rainbow is *tum-pa-wia.* I could not discover the meaning of *tum-pa; wia* probably means "bow" or "bank". When only half of the rainbow is visible, it is said to be a goat (*rua*), going to a stream to drink, but when it gets to the stream, it is eaten by small crabs (*röp*). One must not point one's finger at the goat. *Mintau-chüa-malaa*, the yellow egg-plant girl, once went to the bank of the stream and saw something resembling a goat. She tried to point at it, and her finger broke in two (Kangku Catumma, 19.08.1956).

A double rainbow is said to be a man and a woman having sexual intercourse. When it is fully visible, it is said to be the *char-dong*, or the ladder (*dong*) of the evil death (*char,* see Q2a). During cattle feasts, *char* is kept at bay by a *chaling-chalap* ("butterfly"), which is a square made of threads (see R4a and R6c). The evil death descends from the sky by means of the ladder; when this happens, someone will die.

Thunder (*muk-tün*) is the noise of a feast in the sky. It makes the mushrooms grow (*thum-pau*) (MK, 17.08.1956).

The (Areng) Khumi of the Ponnagyun Chin Hills "call the rainbow 'Saung-yat', and think that it drinks water like a human being" (U Ohn Pe 1931:262). For the Khumi of the Buthidaung Chin Hills, the rainbow is "a fairy who had a lover on earth before she became a fairy and ... she appears in the sky as a rainbow to show herself to her lover. The Awa Khami explain that once upon a time a boy, who was ill-treated by his grandmother, left her house for heaven and when he reached his destination, he spread out his loin cloth just to show his power to the old lady. The Awa Khami regard the rainbow as the loin cloth of the grandson in the heavens. The Mro regard the rainbow as a bridge, by which the 'nats' from the heavens descend to earth" (U Ba Thin 1931:250).

On the subject of thunderstorms, we read (loc. cit.): "All three tribes explain that lightning is caused by the flashing of the dahs [choppers] of the 'nats' in the heavens during the course of dances." "It is explained by the Awa Khami that thunder is the beating of the drums of heaven. The Mro explain that it is the roaring of a cannon on the mountain in a competition between earthly and heavenly 'nats'. But some Ahraing Khami stated that it is the roaring of a heavenly gun by which the wicked on the earth are shot." In the Ponnagyun area, the Mru were the only people to offer U Ohn Pe an explanation, telling him "... that the thunderbolt is thrown by a powerful 'nat' to a less powerful 'nat' in charge of a certain tree, which the thunderbolt strikes. In the case of the death of a man from lightning, they think that the powerful 'nat' strikes him with the thunderbolt as he was mischievous" (1931:260).

P4e) Earthquakes

Minor earthquakes are fairly common in the Chittagong Hill Tracts. The Mru term for them, *li-höng* (?-shake), is also the name of a man who carries the world of humans (*lu-cha*) on his shoulders. *Lihöng* is visited by a scarab (*ki-töng*), who tells him that there are no longer any men on earth. If this were true, it would release *Lihöng* from his duty of carrying the earth. First, however, he must make sure, so he shakes the earth. The men on it call out: "*lu-cha wöi reng, lu-cha wöi reng*" ("The world of men still exists!"). Some of them also beat large gongs and small gongs and make the pigs squeal, so *Lihöng* can hear that there are still men on earth (Dingte, 19.07.1956, Kangku Catumma, 19.08.1956).

Two people told me this story on separate occasions, but I never heard anyone call out during the earthquakes that happened while I was there. Kyo Thwân Ong once shouted "*hnyâng-hnyâng*"; this is a Marma term, and probably means "calm, calm!" He said that this was a Marma custom, but he could not tell me what lay behind it (KTO, 12.07.1956).

"As regards an earthquake, they think that a man of supernatural strength from underneath the earth shakes it to find out whether or not there are still human beings on it. Believing this, they shout out, whenever an earthquake takes place, that they are still in existence. They think that this man of supernatural strength will turn the earth over, if they do not shout out", writes U Ohn Pe of the Khumi of the Ponnagyun Chin Hills (1931:262). Of the Mru he says: "They believe that the cause of the earthquakes is that a supposed Dragon shakes the earth to find out whether people are still in existence" (1931:260).

U Ba Thin provides a more detailed account of the beliefs of the Buthidaung (Saingdin) area: "According to the Mro, the earth is carried on the shoulders of a Naga residing underneath this earth and a beetle reports to the Naga that the human beings are no longer in existence and the Naga, in order to verify the report of the beetle, shakes the earth, thus causing an earthquake. Awa Khamis and some Ahraing Khamis stated that the earth is shouldered by 2 or 4 giants and that when they feel tired, they change shoulders and thus cause the earthquake. Some Ahraing Khamis said that there exists another world under the present world and that a post resting on the netherworld supports the upper world. The people living in the lower world, in order to find out if the people living in the upper world are living or dead, shake the post and cause the earthquake. It is customary among the three tribes to respond to an earthquake by saying 'We are alive'" (1931:250).

P4f) Dreams

Dreams are particularly important when one is choosing a place for some purpose. It might be for establishing a swidden, or for building a new house. When seeking for a new site for a swidden, it is lucky to dream of porpoises or lizards, because they are signs of a large cotton harvest. It is also good to dream of clear water or fish, because fish portend large amounts of money. If, however, one should dream of fire, or of killing a chicken, it is better not to proceed with one's choice, otherwise someone in the family will die.

The same applies when building a house. To dream of killing a chicken is a bad omen at any time. Dreaming of rain presages bad luck: if one is planning to build a house at the time, it means that one day the house will burn down. It is good, however, to dream of new white cloth: this stands for large amounts of money and paddy, and foretells prosperity. To dream of mice is bad; annoyance and trouble will follow. Dreaming of fire means hardship, and trouble with other people. Money means trouble and illness; faeces mean deep sorrows; rain means that someone will die, though not necessarily in one's own family. It is also bad to dream of elephants (someone will die), or of a hawk (a spirit will seize someone). A dream about a tree falling on one's house means that the house owner or his brother will die. A dream about a canine tooth falling out means that it will be the house owner. Dreaming about game or jungle fowl is less dangerous, but it does mean that the dreamer will experience shame.

When an expectant mother dreams of flowers or of living with another man, either she herself or her child will die during the birth. When an unmarried girl dreams of flowers or of living with a young man, she will have attacks of dizziness. To stop them, a dog must be sacrificed on the roof (*kim-luk-pün*, see P5a). However, if a young man dreams of sleeping with a girl, it is not a bad sign, and if he dreams of fetching a girl from another hamlet, it is good, because it means that he will catch a deer or a wild pig in a spear trap. If a cow or a goat appears in one's dreams, it means that a spirit is demanding a sacrifice of whatever kind of animal is seen. Dreaming of paddy has no particular meaning (MK, 16. and 17.08.1956).

P5 Pig sacrifices

Pigs, whether sows or castrated males, are nearly always killed as sacrifices. Many are killed while they are still small piglets. A few sows survive a little longer, but even they are killed before they reach full adulthood, when they may become dangerous (see G2).

There are many occasions when a pig must be sacrificed. Pigs, or at least a piglet, are particularly necessary for the death rituals (see Q) and at all larger cattle sacrifices (see R). Amongst the Khumi, dogs must also be killed when cattle are sacrificed. The Mru dislike dog meat, and rarely sacrifice dogs. However, in the next section I shall describe an occasion when some men who were performing a sacrifice were obliged to eat dog meat.

P5a) *kim-luk-pün*

Menkröi maintained that he was a Christian and did not perform sacrifices. One evening, however, he had to sacrifice a dog, because his adopted sister, who was much younger than himself and still unmarried, was dreaming repeatedly of a young man coming to visit her at night. The visitor was thought to be a bad spirit; it could be stopped from reaching her by sacrificing a dog on the roof of Menkröi's house. Menkröi initially disputed this, but his sister continued to be troubled by the "visitor", and in the end he agreed to sacrifice the dog.

First a halved bottle-gourd (*nom-praa*) was filled with water, into which were mixed pieces of the bark of a *ron-kha* tree (unidentified) and millet (*yakma*). A *char-dong* ("ladder for the evil death") was made from four double bamboo strips with a cross between them made of two pairs of red and black threads. The sacrifice can be promised in advance, in which case it must be carried out at the next new moon. The *char-dong* must be prepared when the promise is made, then kept in the house. There are no particular ceremonies attached to the making of the *char-dong*, except for pouring out a small amount of water.

Khamlai and the karbari kill a dog on the roof

When it is time to perform the *kim-luk-pün*, the household dog is caught, tied up, and clubbed to death in front of the house. A ladder is placed against the roof of the *kimma* and two men climb up, in this case Menkröi's elder brother Khamlai and Karbari Kangku Catumma. They took with them the dead dog, tied to a long strip of bamboo; the *nom-praa* filled with mixture described above; and the *char-dong*. The dog was laid on the roof ridge and the water mixture was spat at it six times. The spitting is called "*cho-yu*" ("spitting beer"), even though beer is not used. A *tamma* is said, the content of which I could not discover. Menkröi told me that he himself did not know it, but that it contains the message "we have promised and sacrificed all", plus a plea to the spirit to set the woman free. It probably also contains a promise of a future sacrifice; this depends on which spirit is thought to be responsible and thus needs to be appeased before it will agree to stop visiting the woman.

Finally the dog was thrown down; the *nom-praa* and the *char-dong* remained on the roof. The two men descended, and Khamlai singed the dog and cut it up. Menkröi's wife Thanni cooked it. The members of the household are usually obliged to eat it, though on this occasion, Menkröi and his son Dingte refused. Anyone else who wants to eat some of it may do so, though I was not allowed to taste it.

The next day the house is *khang*, i.e., it is closed to outsiders. The entrance is barred by three *kung-kaa* (bamboo strips bent into semicircles) set into the ground around the foot of the tree ladder. The members of the household are forbidden to eat. Menkröi, however, set up the *kung-kaa*, then instead of going back into the house, he went to buy more pigs for the sacrifices that were to follow.

P5b) *cari-yöng*, *tam* and *tang-pük*

On the sixth day after the *kim-luk-pün*, three further offerings are made: a *cari-yöng*, a *tam* and a *tang-pük*. In this case, the sixth day was 10 July 1956. Men from different houses worked together to prepare the equipment and carry out the sacrifices. Yenpau Khongtör and Karbari Kangku Catumma were amongst them.

The three ceremonies can be held together or separately; at least one pig is needed for each. Menkröi had decided to offer one large pig for the *cari-yöng*, and two small ones, one each for the *tam* and the *tang-pük*. When the three sacrifices are performed together, this is called *kham cia-chot* ("to promise a cattle sacrifice"); the *poi-yaa* (feast-giver) is then under an obligation to hold a cattle sacrifice during the same year or the following one. Menkröi, however, did not intend to do so. He told me that a cattle feast can be given without the preceding three pig sacrifices; also that the three pig sacrifices are regarded as almost equivalent to the cattle feast. For a cattle feast, more paddy is needed to feed the guests and to make beer (*yu*) and arak.

A *cari-yöng* can be held separately; a *tam* only in conjunction with a *cari-yöng*; and a *tangpük* only in conjunction with the first two. I did not manage to discover the literal meanings of the names of these sacrifices, though *tang* appears also in the term *reng-tang*; this is a big Longhu cattle sacrifice, and is the equivalent of the Khumi *ku-ram bung* (see R5h and R6b).

Amongst the Mru, the two kinds of *tang* sacrifice have a common feature, namely, that if there is dancing, a special kind of *plung* (see E1c) is needed. Twelve of these are required for a Longhu *reng-tang*. At a pig sacrifice, there must be four *plung*, but they need not be played. For a Khumi *tum-tang* ("playing the tang-pipes") six *tang-plung* are sufficient (see particularly Chapter R6j).

1) The *cari-yöng* started with the making of a *caroca*, which is a decorative bamboo pole. A piglet was caught with the aid of a net and kept in a *pom* (see D4f). The *caroca* was erected at the side of the log staircase away from the house. It was about 4 metres high and bore three sets of three *chit-wai* (tasselled bamboo sticks). The top internode was split. A *pak-plai-kimca* ("little pig-dance-house") was built; this consisted of four bamboo sticks 15 cm long, set in a square and split at the top to hold two side bars, which supported a small platform made from bamboo laths. A piece of banana leaf was placed on the platform and some rice kernels and six small pieces of curcuma and wild ginger root (*ai-dam*) were put on it. Next to the *pak-plai-kimca*, four *teng* – small swallow-tailed bamboo tubes – were set into the ground and filled with water mixed with cooked rice, curcuma and wild ginger. (Two *teng* are all that are required, but on this occasion four were used.) A *hom-noi-pu* (small round clay-pot) with a drinking tube was placed next to the *pak-plai-kimca*; it was filled with water in which cooked rice and yeast had been soaked.

The pig may be killed by anyone who wishes to perform the task. Shortly before the killing, the person who is going to do it sips some of the *hom-noi*, takes some rice kernels and pieces of wild ginger into his hand and throws them over the pig, while spitting *hom-noi* at it six times and uttering a *tamma* to invoke a benevolent spirit that promotes well-being. The spirit is not addressed by name.

The large pig was then killed with a *re* (spear); alternatively, a *cau* (sharpened bamboo) could be used. During the killing, a drum (*tömma*) and three flat gongs (*ner*) were beaten. This was done during all three sacrifices, and the same *tamma* was repeated. Some of the pig's blood was smeared from the tip of the *re* on to the lower part of the *caroca*; blood was also sprinkled on the platform of the *pak-plai-kimca* and into the *teng* that stood beside it. Before the meat from the pig was cooked to make pork curry, three *köm-pot* (morsels of food for the spirits) were prepared. These contained only a piece of turmeric and a small amount of raw liver. Two of them were thrown on to the *pak-plai-kimca* and one was kept in the house. The pork may not be eaten until the *köm-pot* have been offered.

2) Then followed the *tam*, during which one of the piglets was killed on the *car* (the platform of the house). In preparation for this sacrifice, the following items were arranged at the edge of the platform: 1) to the left a *kau chung*, which is a bamboo pole with the leaves left on; 2) a *krou*: this is a bamboo tube about 1.90 metres long. It is split down the middle into four parts from the top to a point about 50 cm above the *car*; and 3) to the right, a *lim-yung* (see below). The four parts into which the *krou* was split were all split again into three sections at a height of about 1.30 metres. The central strip of each set of three was then broken off, leaving four long forks (*ke-bor*) consisting of two strips each. The forks were about 60 cm long. One half of each fork was then bent over and attached to another fork, working from right to back, back to front, right back to left back, left back to right

back, and the remaining one to the right front. The *ke-bor* that were bent over to the sides were tied to the green outer part of the stem by means of five bamboo strips; three would also have been acceptable, but not four. The knots were twisted not to the right, as would be usual, but to the left. The middle strip was wrapped around with its green side facing outwards, while the rest had the white side outwards, as is usual. At the point where the pole was split to form the *ke-bor*, a small oblong platform (*krou-phai*) was woven between the four main strips; the ends of the transverse strips were bent downwards then fixed upwards at the back left-hand side.

Two small bamboo tubes (*röng-buk*) were laid on the platform. They were 4 cm long and ended in nodes. The sides were decorated with *thau-rin-bong* made by scraping off strips of the green skin lengthwise, and with *röng-praa* (halved tubes) of equal length, into which were put some rice kernels and small pieces of curcuma. A *plaa-tong* ("soul place"), which is a small pot covered by a banana leaf, was placed on the *krou phai*. In the *plaa-tong* there was another *röng-buk*, and a röng-pra containing rice kernels and curcuma. Later, a *köm-pot* (morsels of food for the spirits, wrapped in a piece of banana leaf to make a small cone) was added. In the evening, this pot was thrown away at the *plon* (entrance to the village) on the way to the swidden. Another *köm-pot* with rice kernels and curcuma pieces was pushed between the ascending and descending *ke-bor* at the right front; a *kwai-che* (arm bracelet) and a piece of *keng-ko* (chain of yellow glass beads), both belonging to Menkröi's smallest daughter Roupau, were also hung there. No one could tell me the reason why, but I was told that without them, the piglet could not be sacrificed. After the ceremonies, they could be thrown away or taken back.

The *lim-yung* is a bamboo tube about 1.30 metres high. Half way up, a piece of an old bamboo mat was pinned to it. On this mat were placed another pair of *röng* and a *köm-pot*. At the bottom lay a large banana leaf, on which more *köm-pot* were laid, together with the cotton threads for the final *kom-bong* (binding the wrist).

At the foot of the *krou* stood a *röng-tung*, an empty bamboo tube closed at both ends by a node. The upper end was decorated with *thau-rin-bong* (straight stripes made by shaving). Before the piglet was killed, someone banged it six times on the *car*. At the foot of the *kau-chung* was laid another piece of mat on to which some earth was poured. This was called the *taping* (hearth).

The piglet was killed at the foot of the *krou* and some of its blood was smeared with the spear on the bottom end of the *krou*. The piglet was placed on the earth and singed, then cut up on the *car*. Its skull was split with one stroke of the chopper and the brain taken out. A small fire was lit on the *taping* and a piece of the piglet's brain (*pak lu-klok*) was burned in it. Other pieces of brain were put into five *nga-om-röng*, which were made at this point. They were thin pieces of bamboo tube, about ¾ cm in diameter. A slender section about 20 cm long was left standing to form a spike, while the rest was almost severed just above a node. To this spike was tied another strip, with its lower end – which pointed to the node – bent upwards slightly, so that it could be slid downwards to close the main part of the tube. With the aid of this slide, a small piece of *pak lu-klok* was pushed into the tube and baked over the fire-place in the house. This meant that the lower end of the *röng* was somewhat scorched. Finally, the five *nga-om-*

röng were tucked over the door to the *kimma*. They were regarded as "medicine" that would prevent evil spirits from entering. Over the *ca-pam* (paddy bin) in the *kim-tom*, a chicken was sacrificed and a *tamma* (invocation) uttered; a *cho-caa* ("spitting at the paddy") was performed.

3) While some of the men were still busy preparing the *nga-om-röng* on the *car*, the third pig was sacrificed for the *tang-pük* in the *kim-tom* (main room of the house) next to two *per-lum* (round winnowing trays, see D4a). On one of these trays were pieces of banana leaf for making *köm-pot*. On the other were four old water bottles of different sizes, each having a short bamboo tube set in one side. There was also one *laa-rik-yia*, a very small, yellow water-bottle used for calling the souls. It contained some paddy kernels.

Shortly before the pig is killed, someone blows into the *tang-plung* once or twice, but so as to produce no sound. A chick must also be killed by wringing its neck over the *per-lum*.

During the sacrifice, a drum (*tömma*) and three flat gongs (*ner*) are beaten. A "butterfly" (*chaling-chalap*) must be tied to the largest water-bottle. The "butterfly" is a square made by wrapping red and black threads round two crossed bamboo strips. Without it, the pig cannot be killed, though no one was able to say why. The *chaling-chalap* is important at other ceremonies too: at a swidden festival, for example, a *chaling-chalap* and a *li-cong* ("wind fish-trap") are fixed to a *krou.* For more details of the swidden festival (*cam-plai)*, at which two pigs are sacrificed, please see J6c. At death ceremonies, a *chaling-chalap* is fixed into the roof over the head of the dead person (see Q1a).

Preparation of the *tang-plung*, the *laa-rik-yia* on the *per-lum*, and the *köm-pot*

After the three pigs had been sacrificed, pieces of their liver (*plong*), thighs (*thung*), shoulder (*pang*), and entrails (*ria*) were cooked in a bamboo tube *(röng)* over the fire, then chopped finely, mixed with some cooked rice (*hom*) and ginger (*tam*), and wrapped in pieces of banana leaf to make *köm-pot*. The same was done with the chicken meat, but this was wrapped separately. The *köm-pot* were then put in all the important places in the house, and especially on all the baskets and the instruments.

The *köm-pot* containing the meat from the pigs slaughtered for the *tam* and the *tang-pük* must not come into contact with the paddy; for this, only *köm-pot* made from the *cari-yöng* pig and the *cho-caa* chicken can be used.

When the *köm-pot* had been put in the proper places, Menkröi's children and their adopted aunt Kaiche assembled in the *kim-tom*. Khamlai (Menkröi's elder brother) blew over a *laa-rik-yia* to call their souls, took a sip of *hom-noi* (beer substitute) from a bowl, spat it over their heads, and made the *tamma* (invocations). Another man wrapped the *bong-kom* threads around their wrists and dabbed curcuma on their forehead. Menkröi and his wife Thanvel were left out because they were "Christians". The children and Kaiche sat in a semicircle and Khamlai swung the *per-lum*, from which the threads, the curcuma and the drink had been removed, over their heads. While this was going on, the pork was cooking outside on the ground, and Thanvel was cooking rice in the house. All who had taken part and helped – there were a good many men – were invited to drink *arak* (distilled beer) before eating the rice and the pork curry. In the evening followed the *plon-köi kwak-a* ("throwing it away at the plon"; *plon* = village entrance): the *röng-tung,* the *pla-tong* and the two *röng* used for cooking the meat of the pigs and chicken were thrown away at the *plon* without any accompanying ceremonies. Anyone who wished could take part in throwing the things away.

P5c) The simple *cari-yöng* and rules for meat distribution

Eight days after the big festival, Elai Atwang held a *cari-yöng*. A *caroca* was erected at the foot of his log staircase and a *pak-plai-kimca* next to it. To make the "little pig-dance house", four bamboo sticks 15 cm long were stuck into the earth in a square. Their tops were split, and two sticks were placed across them so as to form parallel bars on two sides of the square. Several lengths of split bamboo were laid across the bars and a piece of banana leaf spread over them. This made a small "altar", at the base of which were set four *teng* (narrow, swallow-tailed bamboo tubes about 10 cm long; the tails were cut from the lower internode). When the pig was killed, only two *teng* were there; I was told that children had added the other two.

Kernels of husked rice and six pieces of curcuma and wild ginger (*ai-dam*) were scattered on the banana leaf of the *pak-plai-kimca*. Water mixed with cooked rice, curcuma and *ai-dam* was poured into the *teng*; later, some blood from the pig, which was killed with a *cau* (sharpened bamboo) was added. Next to the *kimca* was placed a small, round clay pot (*hom-noi-pu*), with a narrow bamboo tube to use as a straw. The pot contained a beer substitute, consisting of cooked rice, yeast and water.

Any man present may kill the pig. Before doing so, he takes some *hom-noi* into his mouth and some *ai-dam* and rice kernels (*mi*) in his hand. He then spits the *hom-noi* and throws the *ai-dam* and rice six times over the pig, while uttering a *tamma* (invocation). The *tamma* is addressed to a benevolent spirit; no name is used, but the spirit could be *caa-kom-laa-kom namma* (see J4c), or any other spirit that brings health and riches. The person saying the *tamma* counts to 6 ("*pa-kat, pre-kat, chum-kat, ... taruk-kat*"), then tells the spirit that it is being offered a nine-year-old pig (*taku ning pak-la*). In reality, the offering may be just a piglet. Three *köm-pot* were made, containing only pieces of curcuma pieces and small portions of the pig's liver. Two of them were thrown over the *pak-plai-kimca* and one at Elai's house. If this had not been done, the pork could not have been eaten.

After the pig is killed it is cut up, and if the villagers are not invited to come and eat it, portions of meat are taken to all the households. The meat can be given either wrapped in a banana leaf, as *chur-kan*, or in a small bowl, as *kwai-kan*. The first way is considered better, but certain rules apply. Meat from the shin may not be given, because anyone accepting and eating it would become poor. If someone does receive this cut, he can either return it to the giver, who would be shamed, or, if he thinks the meat has been included by mistake, he can give it to his dog. If he thinks the meat was given deliberately and with ill intent, he can hang it over his fire-place, and the giver will become poor.

Pork from an adult pig or from a cow should be given to all families, or at least to those who have a *can* or *tai-nau* relationship (see L3a) to the giver. If only a piglet has been sacrificed, meat need not be given to anyone other than the *tai-nau* relatives. Meat from a chicken need only be given to *tai-nau*, and to the *tutma* if there are any living in the same hamlet. This is called *tö wa-ngaa kom pan* ("to make chicken-meat rice-powder gift"). When chicken is given as *chur-kan*, it must not contain any pieces from the neck or wings; if it does, the recipient is entitled to react as in the case of a pig's lower leg.

P5d) The *pak-tan*

A *cari-yöng* must be held, and an extra pig sacrificed, in the following circumstances: when someone sees a tiger, otherwise he will die; when the house is excessively infested by vermin; and after extramarital intercourse. A *pak-tan* can be promised if someone cuts himself and the wound does not stop bleeding.

The pig is killed in the house during the afternoon, between 4 and 5 o'clock. The killing is accompanied by a *pak-tan tömma long*, which is a special sequence beaten on a drum. A dog may be killed instead of a pig, in which case the sacrifice is called *kui-tan*.

A *pak-tan* can be held without a preceding *cari-yöng* to establish a *tai-nau*-relationship between two sibs. In this case, none of the usual *pak-tan* ceremonies need be performed. A pig is killed with a spear; no *tamma* is said, but the drum must be beaten in the *pak-tan tömma long* rhythm.

For a *pak-tan* proper, four small pots filled with *hom-noi*, or preferably *yu* (rice-beer), are placed in the house. These are 1) *a waa-bet kong*, 2) *a pak-tan kong*, 3) a *cak-kwak kong*, and 4) a *pang-cang kong*. For the first, a small chicken is killed *waa-bet*, that is, by wringing its neck. A *phyöi-tamma* (invocation of a malevolent spirit, see P2a) is said and *hom-noi* is spat at the chicken's head, while its tongue is pulled out and the root (*waa-kamca*) examined. If the root is straight, a family member who is ill will recover. If it is crooked, someone may fall ill. The chicken is not eaten, but thrown away.

At a *pak-tan kong*, the pig is laid down and *hom-noi* or *yu* is spat at it, then its neck is cut through with a chopper. The person for whom the offering is made must step over the spilt blood (*kan*). The *kan* is regarded as essential for anyone who has lost a large amount of blood or fallen from a tree. If the person cannot walk, he or she must be carried over the blood. During that night and the following day and night, the patient must stay in the house and keep *kan-khang*, that is, he or she may eat only cooked rice and drink water (*tui lök, hom lök*); no accompanying dishes are allowed.

For the *cak-kwak* ("throwing away food") a *khö* (open-weave bowl) is made and lined with a *yaa-dai-hap* (a special banana leaf). Some cooked rice (*hom*) and a piece of the pig's intestines (*pak-ria*) are placed on the leaf. Some of the other pieces of meat are cooked (*cak-kwak-kan*) in a clay pot. The *akhangmi* – the person who is keeping the *khang* – is then joined by any family members who wish to participate, forming a group of three, five or seven people. They squat or sit near the log staircase, cross their arms with the palms turned upwards, and receive some *hom* from the *khö* in their right hand some *cak-kwak-kan* in their left. These they throw down the ladder, saying "*phyöi*". This is repeated three times. Finally the rest of the *khö*, a *tui-yia* (water bottle), a *tim-chük* (grinding stone), a *re* (spear) and the *cak-kwak-kan-pu* (clay pot in which the special pork curry was cooked) are thrown down the staircase. The *tim-chük* and the *re* are picked up later and brought back into the house, but not the *pu* and *tui-yia*, because they will be broken.

The rest of the pig is cooked for eating, except for some pork that is placed on a banana leaf to the right of the *pang-cang yu*. Those who took part in the *cak-kwak* take a small piece, then suck a small amount of *hom-noi* or *yu* into their mouths and spit it to the right and left of the *kong*. This is called the *pang*. A full *pang* is performed at the end of all cattle feasts (see R1i), and at large weddings (see M4a).

For the ceremony that ends the *kan-khang*, some *nam-ria* leaves (unidentified) are crushed in a *rön*, and some of the juice is poured over the *akhangmi*'s hair bunch and collected in a *nom-praa* (halved water bottle). Both the *röng* and the *nom-praa* are then thrown away at the foot of a tree; any tree will do.

During the killing of the pig, the *cak-kwak* and the *pang-cang*, the drum must be beaten in the *pak-tan-long* rhythm. This was said to match a sequence of words, but they are never spoken or sung, and nobody seemed to know them.

P5e) *kwak-cung*

Literally this means "throw away the *cung*". *Cung* was said to be a short form of the Marma *cûm-krû*, the name of a spirit resembling the Mru *ui-mukma* (see P2j). This offering is made if someone has pains in their stomach, back and the ribs, and swollen testicles.

A *dö-ker-kim* (cf. P2f and P3c–f) is built at the bottom of the stairs. This consists of four upright poles 50 cm high and about 10 cm apart. Diagonal poles are tied to them so that seen from the inside, those that point to the right are tied to the outside of the vertical poles, while those pointing to the left are fixed outside the ones that point to the right. Both sets of poles are fastened with the same bamboo strip. The tops of both the vertical and the slanting bamboo poles are decorated with *thau-rin-bong*.

A small platform is created between the vertical posts. First, some bamboo poles are cut in half lengthways. Four of these are then placed so that they lie north to south, then four more are placed on top of them lying east to west. A further four are added lying north to south, two on the right and two on the left. They are all tied to the bottom layer with bamboo strips. At the foot of each vertical pole is set a bamboo stick 20 cm long; these sticks are decorated with shavings that hang down and are smooth rather than shaggy. *Teu-pau* – a decoration used only for the *kwak-cung* – are hung between the tops of the vertical poles. A bamboo tube for pouring water (*ri-chang-pön*) leans against the *dör-ker-kim*, and a *pöng-pai* (a loom sword) is stuck into it; this bears the *bong-kom* threads that are given out to the family members at the end of the ceremony. To the side of the *dö-ker-kim* is built a *pak-plai-kimca*, each of its vertical poles having a *chit-wai* at the top.

Before the ceremony starts outdoors, the person representing the household makes a *ron-nat* (a promise to the spirits) inside the house. On the occasion I witnessed, the representative was Klingtui Khongtör. The ceremonial equipment, which includes a piece of cloth, is assembled in a *klai-puk* basket or a *per* (winnowing plate). Water is poured from a *tui-yia* (water-bottle) while someone says "*pakat, pre-kat, ... taruk-kat*". This is the first *tamma*, though no animal is sacrificed yet.

The *klai-puk* is swung over the heads of the assembled family. It is then carried outside. A bowl filled with water is placed on the platform of the *dö-ker-kim*, and the cloth is spread over its *teu-pau*. Small balls of raw cotton are stuck on the *chit-wai* that do not have the shaggy shavings, then placed on the water in the bowl. *Pok-pok-kom* is scattered over the cloth and the *pak-plai-kimca,* while a second *tamma* is said. A chick is taken out of the *klai-puk*, water from the *ri-chang-pön* is poured over it, and it is killed by wringing its neck over the *pak-plai-kimca*. A piglet is then killed, unless the people do not eat meat from castrated animals, in which case a duck can be substituted. The piglet's legs and its snout are tied together, it is pierced through with a sharpened bamboo, and finally its throat is cut with a chopper, while someone pours water over it and says a third *tamma*. Its blood is allowed to drip on the *dö-ker-kim*, into the bowl, and on the *pak-plai-kimca*. The chick is also threaded on to the sharpened bamboo so that its head hangs down into the cut neck of the piglet. The two creatures are then swung around three times to the right ("*pakat, pre-kat, chum-kat*"), three times to the left "(*tali-kat, tanga-kat, taruk-kat*") and once more to the left, and the *tamma* is said. All the time the piglet is being carried, the drum is beaten in a particular rhythm, but not that of a *long*, or drum verse.

This ends the first part of the ceremony. The entrails of the piglet are taken out and it is singed, washed and cut up. The chick is thrown away, and the chickens come and peck at the *pok-pok-kom*.

Some of the meat from the pig is cooked in a *röng* (bamboo cooking tube) and used to make *köm-pot*. Klingtui spread some of these around the house, and threw four more over the roof towards the four points of the compass. Every throw was accompanied by a drum beat. Karbari Kangku Catumma, who helped Klingtui to perform all the ceremonies, stood at the *dö-ker-kim* and said a *tamma* every time, adding the question: *yung prang khök ü?* ("Was it really good?"). Klingtui had climbed on the roof; he was supposed to throw away one *köm-pot* each time and answer "*yung prang khök*". However, he had not understood properly, and he did not throw them until Kangku had finished the four *tamma*, at which point he threw them all.

When he came down he collected the bowl, the cloth and the raw cotton balls sprinkled with blood and brought them back indoors, before dismantling the *dö-ker-kim* and the *pak-plai-kimca* and throwing them away behind his house. All the members of the family then had to wear some cotton from the balls in their hair-knots.

P5f) The *chit-ci-kwak* and *mri-naa cung praing*

The *chit-ci-kwak* is also called *ban-nön-kwak.* It is performed if someone develops a fever on the way to the swidden or to another hamlet. This is thought to be a direct attack by *chit-ci* (see P2k) lurking in the jungle; it is not attributed to a person who bears the sufferer ill will and knows the *montro* for bringing on the fever.

To make the *chit-ci* go away, people perform a *mri-naa-töng* (see P2e) with a duck (*um-pai*) at the *plon* (entrance to the hamlet). A *khö-kröng* is prepared at home and put on the top of the *mri-naa-töng.* Rice is cooked, also at home, with leaves of some kind, to make *chit-ci-kan* (curry for the *chit-ci*). This is put in the *khö.* For the *ron-nat* (see P5e), strips are torn from an old cloth, tied to a bamboo stick (*chit*) held by the person making the offering, and burned. The smoke is thought to please the *chit-ci.*

A group of men goes to the *plon,* where they put the *khö* containing *chit-ci-kan, pok-pok-kom,* seven flowers of any kind, and (optional) a *bong-kom-röng* (see P2e) on the *mri-naa-töng*. They place four *chit* (tasselled bamboo sticks) upright at the four ends of the *mri-naa-töng,* which are bent downwards, and surround the whole structure with cotton threads. Loose cotton balls (*laa-puk*) are laid on the *khö* and placed on the tips of the *chit.* A pig is killed and some of its blood is dripped over the *mri-naa-töng*. Some of its intestines and one of its claws are also put into the *khö.*

When the men return to the house, each has to ask someone who stayed there: *yung prang khök ü?* ("Was it all really good?") and the other person must answer*: yung prang khök.* If a *bong-kom-röng* has been brought, the *bong-kom* distribution follows. Each family member must also wear some of the *laa-puk* in his or her hair-knot (14.09.1956).

Five days later, Yenpau Khongtör performed a kind of *cung kwak*. He did it for the same reason as his neighbour Klingtui, though Yenpau Khongtör erected a *mri-naa-töng* instead of a *dö-ker-kim*. This kind of sacrifice is called *mri-naa cung-praing* and must be performed after the sun goes down (after 7 o'clock).

In addition to the *mri-naa-töng*, a *pak-plai-kimca* is built. A *pöng-pai* (loom sword) bearing the *bong-kom* threads is leaned against it. A small *khö* was placed on the top of the *mri-naa-töng*; it contained a chicken egg, since a duck egg was not available. The egg was thrown away at the end of the ceremony. Some *pok-pok-kom* was placed on a banana leaf and was also put in the *khö*. To this was added some *chit-ci-kan*, made by toasting *wöi-la-ram* [*Eupatorium odoratum]* leaves over the fire. The *khö* was then covered by a cloth, which was called a "roof"; the cloth was taken back into the house after the ceremony.

In addition to the *khö*, five small *pri-kung-khö* ("tiger-footprint-baskets") are needed. They are short plaits made from five bamboo strips. One was pinned to each of the four bent ends of the *tong*, and one was placed beside its central bamboo pole. The duck was then sacrificed by cutting its throat with a chopper over the *pak-plai-kimca*. The drum was beaten three times during the *wi-klak* (shedding of blood).

Small pieces of the duck's intestines were wrapped in pieces of banana leaf with some *kom* (rice flour) and *pok-pok* (puffed rice) to make a kind of *köm-pot*. These were cooked over the fire and then put in the places (the *khö* and the *pak-plai-kimca*) where the *chit-ci-kan* had been put earlier. To make the proper *köm-pot*, part of the duck's heart, kidneys, and a small piece of the thigh were cooked in a pot with some oil, wrapped in pieces of banana leaf, and four of them were thrown out at the corners of the house. This could be done from either the inside or the outside of the house, but not from the roof. Next, some of the milky fermentation liquid from beer that was partly brewed was poured on each of the *pri-kung-khö*, while a *tamma* was said. The *mri-naa-töng* was then dismantled and thrown away, and the family members received the *bong-kom* threads. It is only after all this has been done that people who are not family members may enter the house again, and may stand on the sacrificial area: the way had previously been barred by *kung-kaa* (bamboo strips bent into a bow shape).

Only three or five other people may take part in the ceremonies; even numbers are not allowed. Since Yenpau performed most of the work alone, I was allowed to be present and to watch.

P5g) *met pak*

A *met-pak* may be needed in various situations (see, for example, R6d, i and j). In the present context, the term is probably best translated as "atonement pig". We shall meet this again in Q1h. When a dead body is carried out of the house to be burned or buried, the *hom-kho*, containing the *la-luk hom*, is also carried out. It can only be carried by a member of the deceased's family. If it is carried by another person, either because he knows no better or because no family member is available, the family of the person who carried the *hom-kho* can demand that a relative of the deceased must slaughter a large *met-pak* to protect the carrier's life from the wrath of the spirits. The pig must be laid on *mi-tut-aidam* (husked rice and wild ginger) and killed with a *cau-cem* (sharpened bamboo) by the carrier or one of his relatives, who must say a *tamma*: "The pig has been given and the carrier should be saved."

A situation of this kind arose when Khamcöng's child died and Menkröi's son Dingte wanted to help. Menkröi, however, would have been too embarrassed to ask a poor man like Khamcöng for a pig, so Dingte had to risk the spirits' wrath (MK, 18.07.1956).

P6 Menlè, Kramma, and Riyen

I think this is the right moment to return to Menlè, the prophet I mentioned at the beginning. He turned up in the years before 1985, and I came to hear of him when I became the chance to visit Banderban for one day; a longer stay was not allowed by the police, In 1998 the Hill Tracts had been reopened to foreigners, and Céline Mouchet was one of the first social anthropologists who could visit the Southern Hill Tracts, and she, so to say, had to pass through Malau-O-kua, the central place of the new religion which had been founded by Menlè. This therefore is the moment to pass my word on to her.

P6a "Menle, Messie Mru, un nouveau ordre moral

Au cours de cette partie nous retracerons, de son enfance jusqu'à sa disparition, le parcours de Menle, à travers d'une part les paroles de ceux qui l'ont connu et côtoyé, et d'autre part le corpus de textes qu'il laissa derrière lui.

Une écriture et de nouvelles règles de vie seront révélées par Thurai au jeune messie afin qu'il les livre à sa communauté, et nous verrons de quelle manière celui-ci transmettra le message divin, s'élevant lui-même au rang de personnage céleste, et établira les bases de la religion Kramma.

Menlè ["kramma kongmu"] (copy from a booklet)

Tous les témoignages que j'ai pu recueillir sur Menle et son parcours sont souvent confus ou imprécis au niveau des événements ou des dates, voire contradictoires, même lorsque que j'interrogeais la famille proche du messie, son père ou ses frères. A travers l'itinéraire de Menle que je vais retracer ici s'entrecroisent en effet l'histoire réelle mais aussi une dimension imaginaire véhiculée par des partisans Kramma ou des personnes qui croisèrent le jeune messie. Il n'est pas toujours possible de dire avec précision si certaines caractéristiques du messie constituent des éléments de la réalité sociale ou de composantes surnaturelles. Cependant l'intérêt n'est pas de définir si les représentations messianiques renvoient à des faits avérés ou imaginaires, mais plutôt d'essayer de comprendre les significations qui se cachent derrière l'affirmation de ces faits, et d'exposer les traits principaux qui distinguent le messie Menle des autres hommes. Christian Culas, dans son ouvrage sur le messianisme hmong, explique ainsi la difficulté de saisir les moyens par lesquels s'exprime l'état de messie, qu'il nomme "messianité" : «(...) l'idéal aurait été de pouvoir comparer certains éléments de l'identité des personnages avant et après leur accès à la fonction messianique. Malheureusement, les données sur la situation préinitiatique sont souvent fondues dans les représentations du messie ou de ses disciples et ne sont plus des « données initiales »mais déjà la construction a posteriori de données initiales dans un but de légitimation[475].»

Menle est né dans un petit village mru situé au nord du district de Bandarban, dans le mauza de Renikhyong. Ce hameau, appelé Mälau O, existe encore (bien qu'il ait été légèrement déplacé de son lieu d'origine). Les Bengalis connaissent mieux le village sous le nom de Pura para ; pura signifie "brûlé" en bengla (et para, village). Le hameau de Menle brûla en effet il y a plusieurs dizaine d'années.

Menle serait né entre 1962 et 1965. Je n'ai pu obtenir, des membres de sa famille, la date exacte de sa naissance; en effet, la plupart des Mru ne retiennent pas vraiment ce genre d'événement et ne peuvent dire avec précision ni leur âge, ni celui de leurs enfants (qu'ils ont souvent en grand nombre). Ils le calculent approximativement en comptant le nombre de jhum passés depuis leur naissance. S'ils ont cultivé par exemple sept fois depuis la naissance de l'enfant, ils en déduiront qu'il a environ sept ans. Cela reste à peu prés facile à estimer tant que l'enfant est jeune, mais dès qu'il atteint l'âge d'adulte, personne n'est plus sûr de rien ! Cependant, grâce aux écrits du messie et de certains de ses disciples (et malgré les imprécisions et contradictions), il est permis de dater à peu près l'année de naissance de Menle.

A plusieurs reprises, dans ses textes, Menle mentionne la date 05-05-19 comme celle de l'annonce officielle de la religion Kramma. Le 05-05 correspond au jour et au mois de l'annonce, c'est-à-dire le 5 mai; quand au chiffre 19, il équivaut à l'âge du messie lors de cette annonce. J'ai eu beaucoup de difficultés à obtenir des réponses concordantes quant aux dates des divers voyages de Menle, sa disparition, l'année officielle de l'annonce, etc., et en fait elles ont presque été tout le temps différentes. Le frère cadet de Menle, Menya évalue la disparition de Menle à 1985, ou 1986, alors que le messie avait 19 ans. Menya me dit plus tard, en décembre 2000, que son frère aurait maintenant 36 ans. De nombreux autres Mru que j'interrogeais avaient généralement tendance à surestimer les années d'existence de Kramma. Ainsi il était courant qu'ils se disent, en l'an 2000, Kramma depuis 25 ans. Autre contradiction: sur un des manuel de mathématique destiné aux enfants Kramma[476] figure la mention 41, or il n'a pas pu être écrit après 2003, l'année où j'ai recueilli le manuel. En admettant qu'il ait été rédigé début 2003, ou en 2002, cela signifierait que Menle serait né en 1962, ou 1961, une version peu probable (il aurait alors débuté sa scolarité à presque 20 ans!). La date me paraissant la plus fiable est celle figurant dans un des textes écrits par Menle. Racontant son voyage en Birmanie, il fait figurer, à côté de la mention «21-10-19» (donc un 21 octobre), l'année 1984, ce qui équivaudrait à dire qu'il est né fin 1964 ou en 1965. Menle a bien sûr pu se tromper ou vouloir déformer la réalité, mais les divers témoignages

[475] Culas, C. 2005. Le messianisme hmong aux XIXe et XXe siècles. Op. cit. p. 167.

[476] Des Mru de Mälau O ont écrit divers manuels scolaires destinés à l'éducation des enfants kramma, et écrits en alphabet kramma.

tendent à s'orienter vers ces dates. Les difficultés rencontrées pour obtenir des informations précises sur la vie et les actes de Menle, ayant pour conséquence le caractère parcellaire des sources, ne sont pas anodines et j'ai souvent pu vérifier les propos de Culas : «les histoires messianiques tolèrent mal le questionnement ethnologique et l'ambiance n'est pas à la plaisanterie»[477]. J'ai senti, chez plusieurs interlocuteurs, et notamment les plus proches de Menle (par exemple son jeune frère ou des "prêtres" kramma), une certaine réticence à parler de leur messie, une volonté de garder une part de mystère sur une personne considérée comme divine. Cette méfiance trouve peut-être son explication dans la crainte que je mette en doute leurs paroles, pointe leurs incohérences et leurs contradictions, et donc remette en question le caractère surnaturel de Menle, mais, surtout, «cacher ou faire croire que l'on cache des choses essentielles au non-initié est l'une des stratégies bien connue pour donner du sens à des actes ou des paroles qui une fois révélées paraissent souvent bien banales».

Menle, appartenant au groupe des Mru du nord, les Anok, est du clan (cen) Ngrünau. Il est l'aîné d'une famille de cinq enfants. Suivent après lui deux frères, Kir Wön et Menya, et deux soeurs, Wanlong et Tonrung. Ses parents, Mensing le père, et Tumte, la mère, avaient peu de terres à cultiver et étaient pauvres. Le père, en dehors des travaux agricoles de la jhum, devait vendre des paniers pour faire vivre la famille. (La pauvreté est un critère commun à de nombreux messies, qu'ils viennent d'Asie ou d'ailleurs, nous en reparlerons plus loin). Menle, jeune garçon, s'intéressait peu aux travaux agricoles, et rêvait d'aller à l'école et de s'instruire. Pourtant l'accès à l'éducation au Bangladesh dans les années 70 était extrêmement limité, d'autant plus pour ceux vivant en zone rurale, et appartenant à une minorité ethnique. Parmi ces dernières, seuls de jeunes Chakma, Marma ou Tippera, ainsi que quelques Bawm (surtout grâce aux écoles créées et financées par des missionnaires chrétiens) pouvaient espérer suivre à l'époque une scolarité régulière. En 1981, cependant, l'UNICEF, conscient qu'une grande partie de la population tribale restait totalement exclue du système éducatif national, établit non loin de Bandarban une école destinée exclusivement aux Mru[478], ethnie du district numériquement majoritaire derrière les Marma.

Menle, alors adolescent, avait, comme tous les Mru de la région, entendu parler de cet établissement, doté en plus d'un pensionnat, et situé à deux heures de marche environ de son village. Il se mit en tête d'y entrer, comme l'avaient déjà fait quelques enfants du village, mais son père s'y opposait, le trouvant trop âgé. Katang, un habitant du hameau (devenu une autorité religieuse importante de Kramma), me raconta que Menle jeûna pendant deux ou trois jours pour manifester son mécontentement. Katang, qui connaissait l'écriture et la langue birmane, lui proposa alors de les lui enseigner, ce qui parvint à adoucir un peu la colère de Menle. Mais celui-ci (toujours selon les dires de mon interlocuteur) apprit le birman en seulement trois mois (alors qu'il avait fallu 6 ans à Katang). Le futur messie est présenté ici comme possédant

---

[477] Culas, p. 92

[478] L'école de Swalok compte maintenant une vingtaine de Khumi, sur un total de 220 écoliers. Elle est en grande partie financée par le gouvernement (à 90 %) et l'UNICEF, à l'origine de sa création, s'occupe de la partie restante.

déjà les attributs d'une personne plus intelligente et avancée que les autres, une nature supposée extraordinaire, se situant en marge de la temporalité humaine. Ainsi Menle apprenait vite et bien, et était doté de facultés intellectuelles hors normes. Le jeune homme avait donc soif de connaissance, mais Katang n'était pas en mesure de répondre à ses attentes et il lui conseilla ainsi de se tourner vers un temple bouddhiste, pour apprendre avec des Marma, ou d'aller à l'école… Mensing, le père de Menle, refusait toujours mais dut céder après que son fils menaça de nouveau de ne plus s'alimenter.

Menle, âgé d'une quinzaine d'années, débuta sa scolarité à l'école de Swalok. Il passa directement de la classe I à la classe III et eut comme professeurs, un Marma, Ti En Mong, et deux Bengalis (l'un hindou, l'autre musulman). A Bandarban, je pus rencontrer Ti En Mong, à la retraite, et le questionner sur le jeune Mru qui fut son élève il y a une vingtaine d'années. Celui-ci me décrivit Menle comme

- « un garçon très calme et réservé, ne se mêlant pas beaucoup à ses camarades, mais intelligent et capable d'apprendre vite. Il cherchait à comprendre les croyances et les pratiques rituelles de sa communauté, et il se montrait sans cesse préoccupé par la recherche de la vérité, la vérité du monde. Un jour il me demanda si l'histoire de la vache mangeant l'écriture mru était véridique ; je lui ai répondu que "cette histoire est l'invention des hommes, et sacrifier une vache à cause de cette histoire n'a pas de sens et constitue un pêché. Je ne sais pas pourquoi les Mru n'ont pas d'écriture, mais les hommes qui ont inventé cette histoire sont des malins…Sacrifier des animaux n'est pas une bonne chose" et en plaisantant j'ai ajouté, "si, par exemple, mon fils venait à tomber malade, alors je te sacrifierai, toi"… ».

Cela fit semble-t-il beaucoup réfléchir Menle, et peu à peu, poursuivit le retraité Marma,

«il se mit à changer et à tenir des propos bizarres. Il raconta aux professeurs et aux élèves que s'il prenait une poignée de poussière ou de terre et la jetait par terre, il provoquerait une explosion et des gens pourraient être tués par celle-ci. Tout le monde, écoliers et instituteurs, commencèrent à avoir peur de lui, de son caractère mystérieux, et de ses prétentions à accomplir des actions extraordinaires. Ils étaient tous très impressionnés. Une autre fois, il annonça qu'il viendrait un jour habillé de blanc, sur un cheval, portant un turban blanc et un sabre. Pour tenter d'apaiser la situation, j'ai expliqué aux élèves effrayés que Menle ne se comportait pas bien, qu'il n'était pas bon de vouloir se prendre pour Dieu et de raconter toutes ces choses. Je leur ai dit que s'il venait un jour sur un cheval, alors cela confirmerait sa folie ».

Ces témoignages d'un proche de Menle, Katang, et de son ancien professeur, exprimant une réalité sans doute parfois sublimée, transformée ou revisitée, annoncent cependant déjà l'apparent décalage de Menle par rapport à la « normalité » (son intelligence, ses facultés intellectuelles), et sa rupture avec l'ordre terrestre (ses prétendus pouvoirs surnaturels). « Avant de tenter d'obtenir un début de reconnaissance sociale, certains messies s'ingénient à montrer, à prouver, voire parfois à exhiber, leur supposée proximité avec le Ciel par des actes plus ou moins extraordinaires »[479].

Sa rupture définitive avec la sphère sociale et son entrée dans la messianité se manifesteront sous la forme d'un appel surnaturel, une révélation, et notamment la révélation d'une écriture. Les témoignages relatant les circonstances de cette révélation présentent des variations et restent donc floues ; les disciples ne s'accordent en effet pas tous sur les détails de la révélation et la confusion, parfois voulue, des discours règne. Le père de Menle me raconta que son fils, alors qu'il était malade et presque inconscient, vit le script dans son délire et se sentant ensuite mieux, travailla à l'élaboration de ce nouvel alphabet. Un autre Mru me dit que Menle fit un rêve où il vit Thurai lui proposait trois alphabets ; le premier était en roman mais Menle le refusa, tout comme le second, en arabe. Il choisit le troisième qui était en kramma. Cette version des faits tend à vouloir mettre en évidence le fait que Menle désirait une écriture nouvelle et originale, qui ne ressemblerait à aucune autre, et qu'il cherchait de surcroît à marquer son indépendance vis-à-vis des grandes religions universalistes comme l'islam et le christianisme (nous reviendrons plus tard sur cette question). Quoi qu'il en soit, ces témoignages, malgré leurs variations, s'accordent tous sur le fait que Menle reçut une écriture mais aussi un message salvateur de Dieu même, Thurai.

Le père du messie prétend que Menle aurait commencé à travailler sur l'écriture mru en 1977, donc bien avant qu'il n'aille à l'école, puis en aurait parlé uniquement aux gens de Mälau O en 1982, avant d'en parler officiellement en 1984, annonçant en même temps à tous l'avènement de la religion kramma. D'après Ti En Mong, Menle ne connaissait ni le marma, ni aucune autre écriture en arrivant à Swalok – ce qui vient donc contredire les propos de Katang cités précédemment, et aussi ceux du père de Menle. Le professeur aurait encouragé Menle à apprendre le script roman et bengali à partir de la classe III, et à maîtriser la grammaire. Après 5 ou 6 mois de cours, Menle commença à créer une écriture pour les Mru, qu'il acheva au bout de deux ans (c'est-à-dire à la fin de sa scolarité). Il la présenta à son professeur afin d'obtenir son avis et ce dernier lui répondit que si les Mru pouvaient la lire et la comprendre, « alors pourquoi pas ? ».

La formation messianique de Menle peut suivre le schéma classique des rites de passage de forme diachronique tripartite[480] définie par Van Gennep[481] et développé par Turner[482]. Tout d'abord, en premier lieu s'opère une phase de séparation : au cours d'un rêve, ou d'une maladie avec fièvre délirante (marquant une perte de contact avec la réalité humaine et une mise à distance avec le monde terrestre), le jeune homme va se détacher du cadre social, religieux et symbolique habituel pour recevoir la révélation. Son changement de statut, son accession à la messianité, interviennent après que Dieu soit venu lui annoncer sa mission parmi les hommes (confirmant par la même son ascendance terrestre), et lui livrer les règles de vie et la nouvelle écriture. Notons qu'avant la

---

479 Culas, p. 104

480 Culas, p.215

481 Gennep, A. Van. 1909. Les rites de passage. Paris, Edition Nourry.

482 Turner, V. W. 1990. Le phénomène rituel. Paris. PUF.

révélation, nous l'avons vu, Menle appartenait déjà à une sphère surnaturelle et présentait des signes annonciateurs de la messianité. Le contact divin achève de concrétiser son statut de messie.

En second lieu, lors de la phase liminaire ou de marge, le messie se tient à l'écart pour se consacrer aux aspects scripturaux et rituels liés à sa nouvelle fonction, et ainsi élaborer la nouvelle religion et ses règles de vie. Menle, d'après ses proches, aurait gardé le secret sur sa révélation pendant plusieurs années et aurait travaillé à l'élaboration de textes, avant finalement de dévoiler l'écriture et les lois Kramma.

Enfin, au cours de la phase d'agrégation, le messie retrouve une proximité avec les hommes et commence alors à intervenir dans la vie sociale, religieuse et politique de la communauté afin de transmettre le message divin, instaurer les nouvelles lois et les diffuser, enseigner l'écriture. Menle parlera d'abord de la religion et du script aux gens de son village, avant de les annoncer officiellement à toute la communauté Mru, en effectuant notamment des pèlerinages, allant jusqu'en Inde et en Birmanie. Il finira par disparaître peu de temps après avoir annoncé Kramma."

P6b) My first notes on Menlè

When I stayaed the last time in Banderban (1990) and first heard about Menlè I asked a man who knew Bawm and Mru, to bring me a proof. I would pay him for that evidence. He brought me twelve pages written in a new script and gave me a translation, which I could controll afterwards. Thus I got the first knowledge on the script and the contents. The script had been invented by Menlè who also gave it a name: he called it "Riyen". The script was afterwards used not only by the Mru themselves, but also by American missionaries. The already baptized Mru, however, would not accept it; they used the Roman script introduced by pastor Dala and thaught that the new script must have been invented by the devil. Since those who accepted the new script were also baptized in a new creed, there now exist three groups of Mru: those following the old rules, those who had become Christians, and those who follow the new religion, invented by Menlè, that is the Kramma. At first, the three groups did not intermarry, but in the meantime this separation has been broken down.

To give an impression of the new Kramma belief, I reproduce here the contents of the twelve pages, which the new priests used also to form new rules for the new religion, when and after it became clear that Menlè had disappeared.

Turai öm ri, kramma ri
Thurai Öm Ri. Kramma Ri.
God believe rule. Kramma rule

en-ni äri bök-mi, en-ni long-tang-wüa, mä-tang-wüa,
you (followers) of the rules, you aged men and women,

ngeu-cä, ngeu-wüa-wüa, äri bök mi äyang bök mi,
small children and aged children, all (following) rule and custom

t'ra up-mi, s'khang up-mi, äri up mi, ärip up mi,

administrators of law, of medecine, of rules (of comportment), of reprimands

caa up-mi, sang-mi-wüa, en-ni ngeu-cä-wüa ngeu-wüa,
administrators of writing (= teachers), learned people; you small and (growing up) children,

long-tang-wüa, mä-tang-wüa, kim-reng-cä-wüa, bök-bök
aged men and women, middle.aged persons, all

en-ni pren kom-mä cak, kram-ma ri köi öm khä-böt-mi
you people from many countries, who must believe in the Kramma rule

dä p'dä, kramma lai-klö en-ni-mi sün-ta' dia.
concerning this, hear the voice of Kramma.

ang ätek tä-pöm yong dä, nö thik köm, nö na köm,
the way of my speach, it will be correct it will be true

thurai tut ang po pee äna-mi c'khai, en-ni-mi
what I take to give under God is true, you people

hang kö tä-üa, hang kö tä-nok lä büa,
do not challenge (fight with) each other, do not measure (? compete with) each other

hang kö tä-caing lä büa. cä-cee dä m'yong cü-mü
do not test (ecperiment with) each other ..Therefore, in this way, however,

kramma ri dä ang nö tek lä en-ni, ang nö ma
the Kramma rules which I now tell you and give you

lä en-ni sang. sün-ta büa dia. bök-bök
for remembering learn them, listen to (them). Not even
.
en-ni mi lök poo hang prum prang dia.
one of all you people ahould not be (? rude ?)
,
älök poo hang tä-ngok dia, älök poo hang tä-
not even one should make fun of someone, not even one

krau dia. en-ni nä-e kö wöi-mi ma-ten mi
should tease (?) someone. All you people who came to be here

mia ngai cö, en-ni sün-ta-mi ang dä ang tuk döi.
who are those listening to me, I do not know

nö en-ni lök poo hang kun-rau dia, en-ni-mi
and none of you should be angry
,.
lök poo hang sak-duk dia, en-ni-mi lök poo
none of you should be grieved, none of you

hang klöi p'ram dia, bök-bök äkrüp sang ängi sang
should misunderstand (?, turn his ear away?), learn all well and smoothtly

en-ni-mi sün-ta büa dia, ri-yang krong c'kha[i]
you people do listen, as the rules and customs are elevating (?)

cee, n'mi ri dä p'dä, kramma ri.
Regarding these rules, they are the Kramma rule

kramma ning len 19 ning lee plau khee kö,
Kramma, once when 19 years had passed?,

äwong[= äwang] kla mi, kramma kong mu' ri kä kö' kramma ri,
kramma religion rules which came down, called (in short) kramma rule
.
kramma ri thurai ri, kramma ri thurai c'tang ti'
Kramma rule (is) God's rule, Kramma rule God's message (is) this.

- kramma ri -

ngeu-wüa phua, äwön ming tang t[ong]-yong cö ap ä köm-mi
the child is born, how is the first proper name to be bestowed

ca köi, n'mi dä p'dä m'yong sün-ta dia. en-ni
to the child? As concerns this, listen this way,

tuk dö kä-ca-lö nö, m'yong tek ä köm.
in case you don't know, this way you will say it (should speak):

'oo, pa thurai{y}ö, cä-cee krong-mi di kramma
-O, God father! Being an earthly kramma (as one kramma adept on earth)

sak sai hön p'dä, en thurai kä-e ang 'or
with pure heart, so to you God, I hand (it) up.

ap en. m'mi ca dä ang ca na döi. mä-mi ca dä,
to you. This child is not my child, this child

mä-khöt ni, mä-khöt wor[=war] khin{g} köi, en thurai
for many days, many nights, is what you God

wong[=wang] yu tö ä mi, cä-cee dä ang ca p'mi dä, ang
have put (let come) down, therefore this my child, I

kham döi; en thurai wong[=wang] yu tö ä dö[i]-e dä wöi dö.
do not claim (?), there is nothing what you God did not put (let come) down
[a phrase recalling a similar one at the beginning of St. John's Gospel]

en thurai wong[=wang] yu tö ä kö tö cee dä n'mi ca dä
since this child which you God have put down (= created)

ang ca na döi, en thurai ca c'khai, süng-nam-mi
is not my child, it is your child. Do not enable the hill spirits to

hang pä-ya' ken pee kun, hang pä-ya ken pee ko'.
open up his "stomach" (feelings), let them not open up his "throat" (speach)

hang pä-ya lim-ma ä lä, hang pä-ya kwai-cai ä lä.
do not enable them to deceive/betray him, do not enable them to be malicious (?) to him

cöng köi 'or yuk khä-böt nö, kuk dam-m'ra, lup ngak
when grown up (= in life up big have become and), stealing and robbery, alll kinds of

la' ngak m'yö-mi, cang-sak hang pä-wöi lä, kun yung
bad work and so on, let there be no intention, a good mind,

ko' yung, sak sai plong sai, cöng dong sak klang,
a good speach, a pure heart, a pure liver (emotions), a long (?) life, a brave (?) heart,

ru yung phua yung-mi pä-wöi lä lö nö, en thurai-mi,
the well-born shall stay (as they want ?), and you Godly

äsai-mä hön, en wong[=wang] yu tö ä dit, äko hön-ya'
whom with purity you let come down please, from now on

en wong[=wang] yu pee ä äsai dit, cöng köi cang lä lup ngak
you please give him down purity, in life not to make bad work

la' ngak dö[i] köi, ni' p'ya köi, cöng p'ya köi lä
and bad deeds, to be able today, to be able for life,

en wong[=wang] pee ä, sak sai plong sai yung dit, wong[=wang] yu
you provide him please with a pure heart, pure emotions, give him

pee ä, kh'ti yung dit, en wong[=wang] yu pee ä, sek-ca yung dit,
down good correctness, you give him down please good humility

en wong[=wang] yu pee ä, sak cer plong cer, en wong[=wang] yu pee ä,
you give him down an enlightened heart, an enlightened mind, you give him down

sak kum sak mat, en wong[=wang] pee ä, kong mu' yung.
a subtle and mindful heart, you give him a good religion (= dharma),

en wong[=wang] yu pee ä, kh'ti yung sek-ca yung t'ra yung dit,
you give him down good correctness, good humility, good justice, please,

cöng kä-e p'ya nöi pren, p'ya phia pren lä
during life let him enlighten the country, let him (enlarge?) the country

en thurai kong mu' t'ra p'ya cia lä p'ya prö lä
you God enable him to preach and to tell religious justice,

p'ya tek lä p'ya lä lusa krong mru lä, en wong[=wang] yu
let him speak to (?) the people of the earth, you give him down

pee ä prengya yung pöing-ya yung dit, cä-cee dä
good wisdom, good ability, please, therefore

en thurai ca c'khai, sak sai hön dä, en wong[=wang] yu tö ä
he is you God's child, with pure heart, you give him down

dit, nöm-nöm-cä, en thurai köi, öm en kö[i], en thurai
please, truly, believing in you God, in your God's

ca' pra köi, en thurai köi 'or ap en cee dä
child's side (?), since (I) entrust (it) up to you God (it is)

äpa thurai saing-sa cä-kra cen cu'.
father God's (...?) heavenly tribe's grandchild (heaven's offsping) (?)

nö m'yong p'dä nau ap nö wöi köm, nau ap khee dä
then, this way (you) will entrust the baby, when you entrust the baby

n'yong dä nö tek äköm c'khai nö, en-ni-mi sak kö[i]
that way (you) will to have speak, and in your heart

mat ä büa, bök äri bök äyang-mi wüa, nö,
remember it, all rules all customs, then

ca ap khee 7 [rä-nit] ni tö ci khök khee, ap nau khök khee
7 days after entrusting the child, after entrusting the baby,

cä-yang ming poo, en-ni nö rük lä pee köm, ming dä ming
also a name you will have to select and give, and the name

yung nö rük äköm, ming ngak dä hang kö rük, ming yung dä rük
you bestow should be a good name, do not bestow a bad bame, bestow a good name.

n'mi ngeu-wüa dä, nö 'or yuk, yuk khök, 12 [lai pree] ning
that child, it grows up, has become big, after it has become 12 years

tä-ci khök nö, nö tang pee phung yok phung tang p'mi dä,
old, then dedicate again the one to be dedicated.

cee kramma cap, kramma kong-mu' kö[i] nö swang ä köm-mi.
since he is a kramma being, in the kramma religion he should be entered,

cä-cee, äni, ni hön-ya böt nö:

therefore, today, from this day onward

äpa thurai{y}ö, mä-mi mru kä-e ang ätang pee
God father !, to this man I give the "dedication"

lümla' sai cä-böt cee dä, en thurai rut sai
a pure soul henceforth, to your God's pure hand and

khok sai köi, ang ä 'or ap tö cä-böt en-mi
pure foot, I entrust and put him henceforth up to you
.
wong[=wang] yu pee ä sak sai plong sai dit, cöng kä-e
give him a pure heart and pure emotion, please, in life,

lup ngak mä-yö cang-sak hang pä-wöi, lup yung mä-yö
he shall have no intention to do all kinds of bad work, all kinds of good work

cang-sak pä-wöi, lup yung mä-yö cang sak pä-wöi nö,
he shall intend to do, he shall have the intention to do all kinds of good work, and

süng-nam-mi hang pä-ya lim-ma lä, nö, en thurai-mi,
do not enable the bad spirits to deceive him, and you Godly

wong[=wang] yu ken pee kun, wong[=wang] yu ken pee ko'.
open up his "stomach" (feelings of approval or refusal), open up his "throat" (abilities of good speach ?)

wong[=wang] yu ken pee plong, wong[=wang] yu ken pee sak.
open up his "liver" (feelings of love and hate), open up his "heart" (intelligence, intention)

en thurai-mi wong[=wang] yu' pee sak sai nö, äya'
you Godly provide him with a pure heart, and when he

tuk dö[i] tak khee nö, tong lup ngak cö, cang tä-
does not know yet which work is bad, makes

khä [?=khai]-mi dä, en lot dia. nö ko-mat hön-ya nö
(??) one to be considered (= judged), you release (him). And from now on

kh'ti yung sek-ca yung, en wong[=wang] yu pee ä, sak sai dia
good correctness and good humlity you will give him, purify the heart.

kongmu' t'ra hön dä pä-ya wöi äpa thurai,
with religious justice enable him to live as God father's

saing-sa cä-kra cencu'. p'dä phung tang won[=wan].
(??) heavenly offspring. This is what concerns the "dedication."

nö wöi yok. khöt-nö n'mi mru dä nö 'or yuk,

And there is more: afterwards this man grows up

yuk khök nö cang kim nö cang üa nö po m'si ngiu
when grown up, he makes a house (= marries), makes a swidden (= produces), takes a wife (...?)

nö m'si pok krang dä p'dä', ci-ni yak hä[=hai]
and regarding the way to take a wife, Saturday evening and

but-tu yak hön hang kö tong ä, en-ni cam po m'si
Wednesday evening do not marry. When you are definitely going to take a wife.

khä-böt nö tom[-ng] plong t'lang-ku-ni hä[=hai], kras'ti ni hä[=hai]
why [not!] on Sunday and Thursday?

t'lang-ku-ni hä[=hai], kras'ti-ni hä[=hai] en-ni tong khä[=khai] kä-ca-lö nö,
In case you marry on Sunday and Thursday,

t'lang-ku-ni kras'ti-ni ya sai äsak dö[i], nö,
Sunday and Thursday you cannot kill life, and

m'si pok poi dä, en-ni äcang tä-kla köm c'khai.
a marriage feast., you will have to perform.

cä-cee dä, n'mi ni, 2 [pre] ni mi kä dä, hang kö tong ä
Therefore these days, on these two days, do not (bring them=) marry.

büa. nö po m'si khä-böt, wong[=wang] po {~~lo~~} lä m'si khä-böt.
And when definitely taking a wife, in case you come to take a wife,

kä-ca-lö nö, wong[=wang] töm nö m'si pok poi, en-ni-mi
when she comes, then the marraige feast. You (others)

pak-cä äya' cia-cä äya', ämi sang-ra en sak
can (kill) a little pig, a little cow, whatever pleases your heart,

khu-si en ya ä khä[=khai]-mi, tong cö sat ngar lök dia.
you can consider (?), which one animal.(you should) offer.

khöt[=khök]-nö n'mi hön dä, thak poi pee poi dia, khöt[= khök]-nö
Afterwards with this give the "thak" festival (distribute it), and after that

m'si wüa k'cak wüa 2 [pre] ya' kö dä nö khök khee dä,
the two are (called) wife and husband, and when finished

nö cang ä cum-lang ngia [mia?, s. below!] phak, nö bong ä köm [kom?]-mi ämia
make (a ceremony for welcoming the new wife). And for the bongkom ceremony,

yong-mi long-tang mä-cö, nö wong [wang] lup äri mi n'mi p'dö,
whatever kind of an elder performs the rule,. drop this.

k'yüa 2 [pre]-ya' köi, cang pee äri köm-mi dä, p'dö
For the two people, make (give) the future (following) rule: drop

älai kö süng-nam ri, bong-kom ri eng sap, k'yüa 2 [pre]-ya'
the former spirit's rule, the bongkom rule, distilling (?); to the two people

kö dä, nö khin ä-nö ä-ni hön-ya nö, tä-cum p'mi[= pai-mi] won[= wan] köi
It is time., now from today onward, the couple, in these things,

cöng kä-e, pä-pho lä, pä-plet lä, ämat cing poo
in life let hem succeed (develop?), let them thrive, they themselves too

khök ran do-sa wöi dö[i] köi, mru-mi ya lim-ma dö[i] köi.
shall finish (nocuous?) quarrelling, men shall no longer be able be treacherous/deceptive,

ya kwai-cai dö[i] köi, cöng yung hä[=hai], kun yung hä[=hai], ko yung hä[=hai],
be able no longer to be malicious (?); life shall be good, and the "stomach" and "throat" [see p 2] shall be good

tä-lek hä[=hai] tä-sang hä[=hai], p'dä p'ya wöi lä lö'.
compatible and agreeing with each other, (so let them be able to stay as they want?)

cu yung ca yung poo, pä-thon lä lö, p'mi dä n'mi pra-si,
children and grandchildren are good, let them educate (as they want?). That there, this prayer,

n'mi pia-ti yat dä, nö tek pee ärui nö n'mi cur nö,
this long blessing, say them (by preference) and that's the end.

tong poo älup dä wöi döi, khöt[=khöt]-nö n'mi ni kö {~~dä~~} äkom-mä.
there is no more work. After this day.there will be many.

{en-ni nö tä-yung wöi cä-cee dä n'mi ni dä} en-ni
[repetition]

nö tä-yung wöi cä-cee dä, wong[=wang] po lä khä-böt nö,
Then, in order to stay well, after you have brought (your wife),

1. yak e nö khök khee ho pee pra-si äplau dia.
   at the end of the evening say a prayer once

2. nö wor [war] thöi khök khee ho pee äplau dia.
   then next morning say it once

3. en-ni ca hom böt pä-khee ho pee äplau dia.

before you eat rice say it oncc

4. ki ca hom khee pee plau dia.
   after eating (say it) once

5. sat köi khee pee äplau dia.
   at sunset (say it) once

nö mä-khöt nö, nö na köm tai pra-si pek kröng-cä,
so much, and a little bit prayer giving will become real.

tong poo ä-krat-mi na döi thurai ri dä
There is no difficulty (in) this god's rule

äteng-mä cü-mü, nö en-ni mi dä, mi äri köi,
but it is straight, and (when) you follow (do according to) this rule

cang ä nö, tong pö[= poo] ön döi, yung köm tai, nö mä-khöt
there is nothing to detest, it will become good, and the amount is

kh'ti nö, na köm tai. nö mä-mi dä m'si pok won [wan] en-ni kö[i].
correct and will become real..And this is the wive-taking stuff (ceremony) for you.

.

khöt[=khök]-nö nö cang cum-lang mia[= ngia ? see above] phak, nö tä-cong nö tä-
After this you make (the blessing given when the new wife enters the husband's house), then

lang, cöng köi nö ka cu' nö ka ca, khöt-nö {long-tang,} long-tang khin
sleep with each other and conduct a married life In life get children and grandchildren. After that

long-tang nö, ro-ka nö wong[=wang] töm, ro-ka' wong[=wang] töm nö köng.
aged people time: When aged, illnes will arrive, and (when) illness arrives (you will) die.

khöng [= köng] khök khee dä p'dä, mä t'[ong] yong cö äri krang dä.
When death has occurred, which kind of rules (are to be followed)?

nö lup ä-köm nö khök khee dä, nö köng, nö kong pee
When work to be done has ended in death, plait a

pang, {köng nö kong pee pang nö}, ceng ä mai köi. nö n'yong
coffin [repetition] burn it in the fire. If that way

na döi c' poo {dup ä c'poo} yung dup ä, c'poo cöng bon
is not (possible), [repetition] burying it is also good; it will also become earth.

köm, tä-ceng ä c'poo cong[=cöng] bon köm tai, köng khä-böt nö
Also when burned it will become earth. When definitely dead

sak that nö, älen-mä thurai phung köi lüm la' dä
the heart (stands still?) and the soul climbs quickly to God's

pün khai, nö yong ra-e ärau poo tuk döi, ä-bau poo
world. .In such a place it doesn't know neither illness (=pain) nor unhappiness

tuk döi, mai nö tik ä khä[=khai] pä-cee rau dö[i], dup ä khä[=khai]
When the fire burns it, no (physical) pain, when you bury it,

pä-cee ng'ree dö[i], bon cä-tai buk dä, lüm-la' dä
no tribulation (spiritual pain),. the body will become soil, the soul has

lot khä-böt, köng khä-böt nö, i lu-sa buk dä bon tai.
been released. When (definitly) dead, our human body becomes earth.

cä-cee dä, köng khä-böt nö, en-ni sak khu-si.
Therefore, when dead, your heart is glad.

ä'ko dä to rong köi ä-i wöi reng tai cä-cee dä,
Because now we will be staying in the jungle,

kong pang c'poo ya', cü-mü, mrö thee kö,
we also can plait a coffin. But when we must put up in town

p'ni khin i wöi khä-böt khee dä, pang mia i ka' döi,
one (future) day, we cannot get bamboo strips for the coffin.

ra-e, i dup ä tö, nö khee dä, nö khök khee dä
Here, this time, we depose it by burial, and when this time has ended,

kröi mü reng c'poo sop poo ka' dö[i], sing pö[=poo] ka' dö[i].
(we) continue to try to find; nevertheless, we can not get fire-wood and trees..

cä-cee dä, ädup tö tä-kla köm, äcap t'ra kwar wüa
Therefore depositing by burial will be necessary, like the ruling people, the Bangali,

yong; kä-büa poo, älai kö süng mung
they too formerly had hilly bush

'o ta', mi-khin khök khee dä ceng ä. ko-mat cee
..(?). ., in this passed time they also cremated (the corpse), now

ämat-ni dup ä nö cü-mü, tong plong, mrö thee kö,
however they bury it. Why? Staying in town,

ämat-ni reng ri, mang{ng} ri wöi ka' sing döi ka' kau dö[i].
their continuous kingly rule is, that they don't get trees nor bamboo.

ra köi, mat-ing dup ä nö, n'mi yong dä, i' poo,

Therefore they bury them, and in this way we too (will do it).

i poo ko-mat dä, ka' sing ka' kau reng tak cä-cee,
Also we now still get trees and bamboo, therefore

ceng ä c'poo yung tai, nö en-ni sak khu-si khä[=khai] nö, en-ni
also burning will be good, and when your heart is glad (= when you like), you

dup ä, c'poo nöm-nöm yung, tup [= dup] ä c'poo n'yong tö',
bury them, and (this is) truly good too. Depositing them by burial is as

äyung krang dä, ceng c'poo n'yong tö, äyung krang dä,
a good way as burning is a good way of depositing.

ceng ri ang tek lä en-ni, köng khä-böt, sak that khä-böt
I tell you about the burning rule: when dead, the heart has stopped,

nö en-ni mi äprüa hä[=hai], äyia hä[=hai], {äyia hä[=hai],
you are (filled) with grief and mourning [repetition]

~~h?~~} kar hä[=hai] ho' pee thurai dit-cä dia, u' thurai{y}ö
and weeping, please call to god. O God!

sak that yan böt cä-cee m'yong khöt[=khök] tä-böt äsing ...
because the heart has completely stopped, so life has defnitely ended

nö cä-poo en lot ä dia, cöng kö[i] cang lä ng'ree khä[=khai]-mi
now, however, release him, who in life had tribulations,

wong [wang] tälen-mi, cöng dä ru yung phua yung, po wong[= pä-wang] ru lä.
who comes quickly, (in) life gets up well, is born well, let him come and get up,

p'mi dä, äprüa hä[=hai], äyia hä[=hai] kar hä[=hai] yuk hä[=hai]
that one. -- With grief and mourming and great (?) weeping

en-ni ho pee thurai dia; hä[=hai]-nö won [= wan] hön dua ä'
you should call to God, and then wrap him (he dead body) into a cloth

dua khök nö, kong pee pang dia, pang bo, pang hee -
after wrapping, plait a coffin (beautifully decorated)

en-ni ya ä, pöm yong kom[=kong] pee pang nö, pang lak kö
in the way you can (do) it, plait the coffin, and into the coffin

cer ä dia, pang lak kö cer ä nö, en-ni-mi
place (the corps), when you have placed it into the coffin,

tö ä sat {~~t~~'}döng c'poo yung tö ä ä-ni c'poo yung.
to keep it till noon (half a day) is as good as one day,

tö ä äwor [äwar] c'poo yung, tö ä 2 [pre] ni c'poo yung.
to keep it a night is as good as to keep it two days.

nö 2 [pre] ni klen kö dä, mün khöt prüa yia c'poo,
more than two days, how much sorrow and grief (you may have),

hang kö tö büa, lak dä su khä[= khai] bun khä-böt nö,
don't keep it, it will instead get rotten, eaten by vermin,

mru ä ng'ree tang, nö n'yong ra-köi, 2 [pre] ni tai, 2 [pre] ni
and men will get tribulations, therefore two days,

klen kö tö khä[=khai] nö, mru äkom ron-lai cee dä
keeping it more than two days, is too much trouble for men, therefore

äprek kla khai tä-pün döi, tia-mä köi
parting is necessary, to climb up together is not.. On the way

sing dung kau pai ree s'rai, hot tä-than khä[=khai],
tree-stumps, bamboo baskets, spears and daos, (? remain when crossing?)

pom[= pöm] yong tä-mang tä-kat dö[i] ämat-ni thurai
like this do not go along with at the same time. They, who climb

pren kö pün-mi wüa poo, yong ra-e khöt pat-mä
to God's country, in such a place, they too do not want the whole amount

mru kä-e äya' poo pee lä ng'ree köm-mi dä nau dö[i]
(of what) will bring tribulations to men.
.
ra-e äsu nam, ähang nam, mru nö ta hüp köm mi;
Here, when men will stay breathing in rotten smell, offensive smell,
.
äkom äpai-mä, 2 [pre] ni klen kö dä hang kö tö büa nö
distress is spreading, so.do not keep it more than 2 days
.
lak ngam dä en-ni ya' lup äri dö[i] dä, ngam dä
(if you should, contrary to all expectations, not be able to do the work)

äsat döng tö, tö c'poo yung köm tai, tö ä-ni c'poo yung, tö
keep it half a day, this will be good too, to keep it a day is as good

2 [pre] ni c'poo yung, 2 [pre] ni dä, nau khä[=khai] lai tak kä-ca-lö,
to keep two days, in case you want more than two days

prüa khä[=khai] lai tak khä[=kä]-ca-lö, khöt hön-ya nö tö khee
grief will be more from the amount when keeping

dök pö [=poo] cä-cee dä, tö ä büa tai, nö cong wok[=wak]
because you also are seized,. keep it, and those who must guard the dead body

khä-böt mi khin khök khee dä, en-ni, en-ni-mi,
when their time is ended, you, you (others)

wok [wak] pang kö cer tö khä-böt nö, en-ni nö cong wok[=wak]
must place the dead body in the coffin, when your guarding the dead body

nö khök khee dä, nö n'mi, {en-ni nö cong wok[=wak] khök khee dä,
has come to an end. [repetition; should read: Before the time of guarding the dead body has

~~khlö hä~~} en-ni-mi, khlö[=klö] hä[=hai] ren-si hä[=hai],
come to an end] you others, making noise and joking,

tä-ngok hä[=hai] tä-krau hä[=hai], m'la-cä-wüa klang-wüa-wüa.
playing together and teasing (?)each other, young girls and boys,

kim reng-cä-wüa long-tang-wüa {~~n'~~}wong [=wang]-mi-wüa poo,
married people, aged people, coming people (= visitors) too

hang küa ren-si, tor-lor dia, ärö wöi döi.
do not make (offensive?) joking and amusement (?), it is not custom

thurai ri nö dä köng p'cee, n'yong khöt[=khök]-nö kröi
in God's rule for death. After this has happened, those who will look for

hä[?=hai] tor-lor, kröi küa ren-si, kröi cang khlö[=klö]
amusement (?), look for (offensive) joking, for noise-making,

köm-mi, khlö[=klö] hön dä hang wöi dia, ren-si hön dä,
do not stay with noise, do not stay with joking,

hang wöi dia, tängok täkrau hön dä hang wöi dia,
do not stay with playing and teasing (?),

kong mu hön t'ra hön dä wöi dia, sek-ca hö
stay with religion, with justice, stay with humility and

kh'ti hön dä wöi dia, kong-mu' t'ra lai klö dä
correctness, speak words of religious justice

tek dia, nö äsu pra-si hön dä kröi tang pee
and with praying for blessing look for dedicating

phung dia, kröi tang pee lüm-la dia, hä[=hai]-nö,
(the person), for dedicating the soul, and then

en-ni-mi nö cong ä, nö cong ä khee dä, t[ong]-yong cö,
you others wait for it, and while waiting for it, in which way ever

en-ni sak lö rom ä khä[=khai] c'poo, süng-nam ri
your heart wants to make merry too, do not perform

kö dä hang kö cang lä, nö ärom poo süng-nam ri
(hill) spirit rules. Because making merry (dancing with the coffin before cremation)

na dö[i] cä-cee, en-ni sak lö n'mi poo.
is not a (hill) spirit rule, your heart wants this too.

hä[=hai]-nö lü luk hom en me see lö-ra älai kö,
and then "lü luk hom" [rice cooked by the tutma and put near the head of the corpse] you ...(?) anyway before

äri prim-mi dä tong pö[= poo] wöi dö-böt, ärang köi
the old rules do not exist anymore. Actually plait a coffin

kong pang ki kong pang nö won[= wan] hön dua tö nö
and after you plaited the coffin, wrap the body with cloth and

pang kö cer tö; ni khin{g} kh'ti tö ci nö, tö ä 2 [pre] ni mi
put it in the coffin, keep it the correct number of days only, those who keep it 2 days,

äni köi 3 [sum] kat ci ho pee thurai, büa.
on each day call to God three times.

en-ni nö cong wok [wak]-mi yak-e poo ho pee büa.
You, who guard ithe dead body, in the evening also call to him

rau-e poo ho pee büa, khöt[=khöt]-nö kh'ti ni tö ci nö
in the morning too call to him. After keeping it each of the correct days,

en-ni-mi nö ceng ä, nä nö po lä, nö kla lä
you (others) cremate it. For this you take and bring down(stairs)

wok[=wak], nö kla lä wok[=wak] köm cä-khök khee, ho lä pee
the corpse. When it is time to bring the dead body down, you call to

thurai. nö yu kö, nö kö ceng ä tom[= tong] pang
God. Then down there you bring the coffin

nö tik ä köm mai kö, mai kö tik tö khök nö,
to burn it in the fire. After burning it in the fire,

n'mi ria-e dä, cang pee pröng kim ca-la cö
near this place, make a small house on a platform

lök dia. n'mi hön dä, long-tang-wüa mätang-wüa
From there aged men and women,

sak khu-si-mi änau änau älök älök,
whose heart is happy, whoever wants one by one

en-ni-mi kö ho pee thurai, kö lü pee,
you go and call to God. Ask him for

älot dia, sing khök khee kö, tong cö cam cang lä
deliverance. When life is ended, what to do to achieve

äprek ng'ree khä[=khai]-mi {khä-mi}, äpa thurai{y}ö,
division from those with tribulations. Oh God Father

en lot ä dit, äni hön-ya dä, en wong[=wang] yu tö ämi
you release him please. From today, the way to climb up

äpün-ra tia-mä dä, pä-ka böt lä, pä-mi dä,
which you let come down, let him finally receive it, that one

'or lü pee älot dia, tang pee lüm-la'
up there to ask for deliverance., to dedicate/bestow a

sai dia, nö n'yong dä cang ä köm, n'yong en-ni mi
pure soul.. Like this it will be done, (thererfore) do it that

cang ä büa dia, nö n'mi dä sak hön hang kö tä-ma
way..and don't forget it in your heart,

tä-khä[=khai] büa, nö wong[=wang] mang khök; nä nö wok[=wak] dong köi
consider (?) it, and it is time to go away. After this: the "mourning time:"(when the dead body is still in the house)

en-ni cang wok[=wak] ngar c'poo yung tai, tong na dö[i],
You offer an animal for the death ceremony, but this will be good. What is not?.

su nam hang nam mä tö ca tö kham döi, wok[=wak] dong köi
do not eat nor drimk together with rotten and offensive smell, when the corpse is in the house,

en-ni cang dö[i] kä-ca-lö, ki ceng tö wok[=wak] khök khee phai.
you don't do it under these circumstances, sit together after burning the dead body...

ki ceng lä wok[=wak] wong[=wang] mang khök khee phai, sau tö kim, sau tö
After burning the dead body and coming back sit together, keep the house washed,

üa pröng nö, en-ni mi nö khök khee phai phet ngar yang.
the platform (?) washed, and when you finsished, sit together to eat the sacrificial meat

phet wok[=wak] ngar yang, köng poi. mä-köng poi nö khök khee cee

Eating together the sacrifial meat is the death festival.. When the festival for the dead is finished

lang yung röt tak. nä nö khök khee dä bök-bök mru
(eating together) is still better.. When this is finished (like this), all people

äsui äring wöi dö-tham, nä nö kham köm tai.
have no disgust and discomfort at all. After this (you) will drink.

nö mä-köng ri dä mä-khöt nö mä-khöt, mä-khöt, cü,
So much about the rules for the dead, so much, when departing,

cü-mü, mä-köng ri dä, hang kö {hang kö ta tö [b]üa}
however, (another) rule for the dead: don't depart and leave behind

{hang kö ta tö} preng, hang kö ta tö nang lä büa.
[repetition] quarreling with each other..

nö kh'ti krang cöng köi, nö tang phung kramma kong mu'
Now, the correct behaviour in life: to dedicate (the person) to Kramma religion.

kö, öm khä-böt nö, p'dä, pöing-ya hä[=hai] preng-ya hä[=hai], tä-an hä[=hai]
you must believe (To do this) with ability, with knowledge, with conscience. Are

3 [sum] kh'ti, yu, ärak, bing, bang, kuk lik, dam-mra,
three correct ways.: Beer, arak, opium, ganja, stealing, robberies,

da kaik, yot yang, bing kham bang kham, yu' kham ärak kham.
dacoity, extramarital intercourse, smoking opium and ganja, drinking beer and arak:

n'mi mi dä tä-tö'. nö äni kö 3 [sum] krim ci dä
stop all this. And each day three times,

lü su yung lü pra-si yung. -
ask for good blessings, say good prayers.

m'la-cä-wüa-wüa, klang-wüa-wüa kä-büa kö dä,
To young girls and young boys,

ämat-ni tek pee c'poo ämat-ni man ni [?] dö-tham.
whatsoever you tell them they don't (listen?) at all

nau dö-tham, nö nau khä[=khai]-mi dä nöm-nöm ya cang
they don't want at all. (But) those who want, they really could do

khä[=khai] pä-nü yung nö, ä-mat-ni na lä dö[i] cä-cee,
good (things), but because they do it wrong

na c'poo mat-ing ma po, toi lü su dä, toi lü

they even also forget it and spoil asking for blessings, spoil saying

pra-si dä c'po t'lang-ku ni hä[=hai], kras'ti ni töng dä,
prayers, however, they should pray every Sunday and Thursday.

kä-büa nö tä-kla, nö nä-khöt dä poo lü lä,
thus, how much they ever pray for,

nä-khöt cö lü dä lai nö sek-ca tee dö-tham buk kö,
how many prayers they speak, humility will not stay in the body,.

nö nä-khöt cee, lü kh'lai [?] nö, sek-ca dä, nö tee. -
if (you) pray like this, humility will remain..

nö äcup krang, en-ni mi, nö cup nö yia
Now saluting: you should salute (praying in joining hands)

.

nö lü ä su khä-böt khee dä, m'si-wüa dä prek[=! pre] khok kung tua,
and when it is time to ask for blessings, women should put their two legs on the knees

nö 'or lü su yung köm, pree bong rut kop, sak nau
and ask for good blessing with the two hands joined,

äsu yung nö 'or lü köm. - .
ask above for heart-felt good blessings.

nö k'cak-wüa dä, töm phai tlang phet phai cöm, pree bong
Now the men: they sit cross-legged and with their two

mä-pra pree kop rut sua nö 'or lü su yung.
hands joined and stretched out, ask above for good blessings,

{nö 'or lü pra-si yung köm}, hä[=hai]-nö n'yong
[repetition], and then like this,

p'dä, 'or lü pra-si yung, ämat-ni khu-si sak kla nö. -
asking above for good blessings, in what they like, they are comforted

.

nö thurai tä-kua-ra kim köi, äkat nö,
Now: in god's meeting house, one time,

nö cup, nö lü pra-si khök khee dä, thurai
then salute and when having asked for a blessing, god's

tä-kua-ra kim, {thurai tä-kua-ra kim} kim cang
meeting house, [repetition] In the house area

ra-e dä äsai-mä, änai tlang äsai-mä
it is clean, flat, neat and clean.

nö hen ä köm, n'mi-e cä-cee dä kua dak kim dak köi,
it has to be cleared (kept clean). Therefore don't build it below the village or lelow

hang kö cang büa, kua kö yong khöt kui ki pak ki',
houses. As in the village there is much defecation of dogs and pigs',

kim cor tui car cor tui kö kla dö[i] pöm yong n'mi
this way, water flowing down from houses and verandas does not fall into

thurai kim cang ra-e dä, äcap äkom kui ki',
God's house area, many things like dogs' and pigs' defecations

pak ki kö tö nöng dö[i] pöm yong, älöng-cä ra-e,
can not deposit this way.. Build it [the temple] in an upper place,

cang ä büa, sai-mä nai-mä, n'mi-e klom hen ä.
clean and flat, in this, keep the area cleared..

P6c) Some remarks on the Riyen alphabet

Since in the meantime the Kramma signs had been allotted by Menrum to the English keybord, I made some proposals for a change:

Comment: Underlined = re-allotted (my proposal: 7, Riyen: 12); ⌗ = glottal; ⑦ = written ng.

Of the Riyen allotments 3 are useless changes and really misleading:

U > ⌗ (instead of u = o), O > u (instead of o = z), W > ❻ (instead of w = v)

Tell menrom (Menrum) that the primary (Latin) value of U is now written oo in English, but sometimes still u as in "bush", while the primary value of O is still to be found in words like "long". For ❻ Latin writing has no special sign, the sound may be written differently according to whether short or long, as in English "the shirt", "a turn" (French: "me", "soeur"). Missionary Mru script either "a" or "ow", my proposal: put it above A and allot it to Q!, but k = q to K, kh = k to X, since the latter is originally a Greek letter (where it was pronounced kh).

At first I arranged the Kramma letters (as far as I had come to know them from a Kramma text) according to the South Asian order. By courtesy of Céline Mouchet I received a „Primer of Mro" in Kramma letters, written by a Mru who spells his name (on the title page) in Roman letters as „Chawng Now Manyang Mro". A few pages later, it appears also in Kramma letters which I would transliterate Cöngnau Menyang Mru „cögnau Menyang Mruö. Menyang's transliteration of Mru words and sounds into Roman letters does not follow the system as used by the Christians. As a result it shows the enormous difficulties which a Mru has to face when he starts from the English value of the vowels.

I am not going to comment on them[i], but on Menyang's table of Kramma letters. Since it appears repeatedly, the order given cannot be haphazard. At first sight it seems to follow no conventional sequence at all: t, ng, y, m, b, d, a, ph, kh, h, i, c, etc. A closer look, however, reveals that the order basically is the order of the Roman alphabet. The following list will show this:

Latin	English		Kramma	Mru
t	tee		ta	t
	g		ngi	g
	y		yu	y
	*M		mim	M
a	ei		a	a
b	bi		ba	t
c	si			t
d	di		da	t
e	i	a	a	e
f	ef	P	phi	t
g	ji	k	khoi	g
h	eic	h	hau	t
i	ai	i	döi	t
j	jei	c	cu	t
k	k‘ei	k	kou	t
l	el	l	ol	t
m	em	m	mum	t
n	en	n	nin	n
o	ou			t
p	p‘i	u	u	t
q	k‘yu	o	o	t
r	ar	r	ru	r
s	es	s	si	t
t	t‘i	T	tho	t
u	yu	ö	ö	u
v	vi			t
w	dabl yu		wa	t
x	egz			t
y	wai			y
z	zet			t
		e	e	e
		*k	ku	t
		*l	lan	t
		*l	la	t
`		*öi	höi	"
‘		o	ri	t
		*T	tek	t

Explanation: „Latin“ here stands for our classical set of letters, „English“ for the English name of the letter, „Mru“ for the Kramma letter transliterated in „simplified Latin“, „Kramma“ for the name of the Kramma letter; „simplified Latin“ implies: e and o = open vowels, ö = o barbu, ng = velar nasal, c = palatal affricate. The asterisk (*) characterises letters not really required for a consistent system in which a single letter will be allotted to any Mru phoneme (see below). I assume the alphabetical order to have been devised by the prophet. So let us try to understand what induced him to introduce the deviations from the classical order.

The value of /c/ (as used here) (unless in the digraph |ch|) is unknown to English speakers, while its actual English pronunciation („s“ or „k“)[ii] shows no obvious common basis. Hence it received no equivalent in Kramma; v, x, and z are not used in Mru; they received no equivalent letter as well. On the other hand, there are Mru consonants, for which Latin has no single graph. As far as possible, the Kramma order allots the slots of English consonants not needed for Mru to similarly sounding consonants in Mru. Thus f > ph, g > kh, j > c, t > th. If we do not classify the latter as a replacement (the English /t/ is pronounced aspirated), we'll have to treat p > p and k > k as replacements, and this looks a little bit curious. |ph| may be pronounced „f" both in English and in Bangla, English |th| tends to be replaced by dental /t/ in Bangla, while English |t|, to keep it distinct, is pronounced more retroflex. Thus a Mru, having learned English from a Bangla teacher, may assume that the Latin alphabet has no slot for Mru (unaspirated) /t/. Also for /ng/ an appropriate slot is missing. Why Kramma did not identify English and Mru /y/, I don't know. Anyhow, it became the third consonant to be put in front of the alphabet. Christian Mru spelling, since based on Lushai which uses no /y/, had missed to introduce /y/. It used „i" instead, thereby causing possible confusion and imprecise spelling: what will you have to write to differentiate between /ya/, /ia/ and /yia/?

A fourth sign was added in front: a second /m/. It would not have been required, unless you wanted an equivalent of the Bangla and Marma sign for a nasalising /m/. But there is no such sound in Mru.

At the end of the alphabet (which now could be called „ta-ngi-yu-mim") more of these unrequired signs were added: a second /l/, and even a third /l/ to be used in combination with a second /k/ to write the Mru initial cluster /kl/, written in the Christian Mru spelling system „tl." As such it was imported form the Lushai who (phonologically, but not phonetically, correct) identified both. Hence /kl/ became marked as something „special." As a matter of fact, the Kramma signs for this cluster are nothing but minor writing variants of the „primary" /k/ and /l/, in principle quite similar to the combinations used in Burmese and (especially) Bangla script in abundance. Burmese even has special signs for whole words (not unlike English |&| for „and"). Following these examples, Kramma „needed" such a sign as well: the last letter of the „ta-ngi-yu-mim" reads „tek" (= to say), but it can also be written by single letters. Also the second /l/ may have been instigated by the example of the Bangla and Burmese alphabets – they contain an unrequired second /l /as well. In Kramma it is called „la" and can be used as a shortcut for „la" (= month).

Finally, there is even an additional sign for a diphthong which also might be written by a combination of its components (öi), while on the other side Kramma lacks a special sign for its seventh vowel, the „u barbu". This has to be written by a combination of Kramma |u| and |ö|. Compare the English speciality, also imported into the Christian spelling for the Mru: the monophthong (open) /o/ is written |aw| while the diphthong /ou/ is written |o|. Otherwise the Latin vowels have been imported with their Latin value – but the deviant English names for them must remain confusing for a Mru. In Kramma they mostly are called by the vowel they represent, but not always.

When we take a closer look at them. the first remarkable phenomenon is that they had to shift their slot. /a/, called „a" in Kramma, but „ei" in English, moves to the slot of the Latin sign pronounced „e" when reading a Christian Mru text, but called „i" in English. /i/, however, keeps its slot. On the other hand, /o/ is thrown out and replaced by /u/. This /u/ is the same as in „Mru", but (by Menyang) transliterated as closed „o" when written in Bangla or English. This kind of spelling was induced by the Marma in whose language, due to a Burmese sound change more than 1000 years ago, the Mru are called „Mrô". The first ones to use this Marma term too were, in colonial time, the British. By now the Bengali adopts it as well. As a consequence, when Menyang writes Mru with Bangla letters, he uses Bangla /ô/ once more. What a muddle when seen from outside, but seen from the Kramma side, these changing of slots has a remarkable effect: the vowels (though interspersed with consonants) regain their old South Asian (Bangla and Marma) order: /a/, /i/, /u/, /e/, /o/.

But Mru has two more vowels. One of them is (as mentioned) represented by a digraph. Remains the question where to insert the other one: Burmese put its equivalent behind /o/, Kramma chose the vacated slot for /u/. This is quite ingenious for two reasons. 1) The English pronunciation of the /u/ more than often (be it in words like „but,“ „cut,“ or like „bury,“ „incur“) approximates Mru /ö/; 2) the Burmese equivalent, the actual pronunciation of which is different, but in the Mru pronunciation of Marma replaced by Mru „ö“, is written by a combination of the graphs for /i/ above and /u/ below the consonant which brings us back to the English name of /u/, „yu“.

Still, South Asian scripts have another advantage over European scripts: they have a special sign for the glottal preceding every vowel when pronounced singly or at the beginning of a syllable. When used without further additions to indicate the vowel, it represents the short /a/. The latter in Bangla has changed its sound to (short, open) /o/. As Kramma thought it necessary to have a similar sign too, it had to be called a name, and it was called as in Bangla. But this meant forgoing the possibility to call the Mru /o/ as pronounced, a new name had to be invented, and it is now called like another Bangla (somewhat enigmatic) „vowel“ sign: „ri“ To add to the confusion, the Kramma order of the alphabet moved the /p/ behind the /m/ and added the glottal sign called „o“ behind the /u/ replacing the |o|, so that by now it takes the slot of /q/. Is it an incident that some Euro-American linguists may also use the |q| as a simple transcription either for the glottal constriction or for the soft velar fricative? Probably. But it may not be an incident that the Kramma sign for the glottal looks very similar to a Bangla |h|: In Lushai the final glottal constriction is written by an |h| (which therefore in initial and final position has a quite different value!). This usage was adopted for the Christian Mru spelling. But the soft Mru final glottals have a usage and function quite different from the hard ones of Lushai. As a result, these final |h| appear in Christian Mru script rather haphazardly. The Kramma sign, when used as a final, appears even more haphazardly (but as a rule not in the same positions as the „Christian“ |h|). As an initial it is not really needed, since in Kramma syllables are written apart Thus, instead of being prefixed to every initial vowel, it in principle (probably because of it’s name) is used before /o/ only. In this way becoming deprived of all proper sense, it may even precede medial /o/.

As a result, the Kramma script is loaded with futile elements of other scripts. We might call this useless, but the inventor of the script apparently attached to them some significance; by their very superfluousness Kramma shows the same characteristics as the established scripts. Most probably for the same reason, the spelling system had to be made more complicated than reasonable. I already mentioned the digraph for a monophthong and the monograph for a diphthong. But more could be done. In English, |y| and |w| can be used as consonants and vowels, |y| by itself, |w| in combiatiom with other vowels, whereby it can totally change its quality. Kramma achieved a similar effect by using |w| as a vowel of its own. If you write |kaw| it is to be read „kau“ (a spelling even linguists will approve). But if you write |kw|? „Kou“! But don’t think that |w| as a vowel is always to be read as „ou“ (since |kaw| might be reinterpreted as „kaou“ and still stand for /kau/). As soon as it follows /u/ it changes its character once more and becomes „a“. That is |kuw| is to be read „kua“. Correspondingly |kiy| will stand for |kia|, but there is no |ky| as Mru has no „kei“, and „kai“ has still to be written |kai|, not |kay|. But if there is a |kay|, how to pronounce it? I don’t know (maybe „kaia“). In order to understand how |kiy| and |kuw| may come to be read „kia“ and „kua“, try to write these syllables with Burmese letters – you’ll have to write them exactly „the Kramma way“. But if you want to transliterate „kou“ into Burmese, you’ll have to proceed differently. You cannot turn to Bangla either – |kw| is a genuine Kramma invention.

Finally, let us have a short look on the Kramma names for the letters. They are far more colourful than those of English. English in a few instances uses a quite deviant name – Kramma as well, but normally in connection with letters quite different from English. A real exception is the name for /l/ (standard version): „ol", and it is the more exceptional as Mru syllables (unless recent loans) never end in final /l/. One may assume this name to be acceptable at best for the second /l/ imitating the unused second /l/ in Bangla and Marma, to be derived from a Sanskrit „syllabic l". But this variant is called by a name as conventional as possible: „la". To call a relic from the past by the conventional name and the normal Mru usage by a name imported from English (but with the basic vowel converted from the English („e") into the Bangla („o") minimal vowel based on /a/) and to impose it on the Mru who traditionally did not use it, is a most remarkable piece of symbolic transformation.

Perhaps there is still more behind it. Let me recall that the British introduced their names for the tones of the gamut; the basis is „dou", six steps up we reach „la" – the prophet is said to have reached the sixth grade in school, and the symbols he invented for /d/ and /l/ are nothing but each other's inversion: Kramma |d| corresponds to the classical Greek graph for |l| (lambda majuscule) with the opening turned down, while Kramma |l| has the opening turned up. One may object that Kramma |d| is not called „dou", but „da". However that may be, the message points in one direction: leave classical „da-la" (by the way the name of the Lushai missionary who tried to bring Christianity to the Mru) and accept modern „ol" (meaning „easy" in Lushai) to represent the common standard.

There are two more consonants the Kramma names of which are closed by consonants: „mum" and „nin". Why not simply „mu" and „ni"? Perhaps because in Arabic script (to be associated with the Islam) they are called „mim" and „nun". Moreover Kramma contains a (rather useless) „mim" too. The sign looks a little bit like a mirror image of the Arabic „mim." If we don't write Arabic from right to left, but produce a mirrored image by writing it from left to right, there appear two more signs written similarly in Arabic and Kramma: |h| and |w|. In Arab they are called „ha" and „waw" (= wau), in Kramma „hau" and „wa". Nothing but a coincidence? Or a systematic reversal, containing once more a symbolic message? Both with the Bengali and the Mru right and left (for hygienic reasons) are associated with the ideas of pure and impure. For the prophet, to reach the „pure" was a major aim. If you ponder about the truth behind the surface, you may find out that Muslim writing starts from the pure side to proceed to the impure side, English like the Sanskrit-derived scripts of Bangla and Marma, start from the impure side to proceed to the pure side. Initially Islam was a pure and good idea, but finally the Muslim behaviour, as experienced by the Mru, is impure and bad. To approach the pureness of Allah (to be equated with the Mru „Thura Turaii" – a Kramma believer should pray to him five times a day), you'll have to reverse the direction taken by the Muslim Bengalis.

However, even if the prophet had similar ideas in his mind, he cannot have expected that he could convey them to the Mru people by means of the new script. Due to their limited knowledge, they would be unable to understand them. Therefore, „symbolic messages" is amisnomer. But what then are they meant for? Most probably for nothing. We have to reverse our perspective: The messages were not meant to issue from the script, but the script (in part) was derived from the messages. If you want to create a new script, you must have some ideas to guide your endeavours. It may be rare that they crystallize in signs and names of letters, but it does not seem impossible that a few of them do surface and subsequently can be retrieved.

Q Death

Q1 The normal burning ceremonial (*romma)*

Q1a) Laying out the body

The Mru acknowledge that someone has died when respiration, pulse and eye movements cease, the body cools, and rigor mortis sets in. The women then start to weep and wail loudly, while the men fall silent. There are no special rituals that must be performed immediately before or after a death, but on the day it occurs, the members of the household may not perform their normal tasks (*chak-dük khang,* "heart-dying ban"). Provided that the death is from natural causes (*char,* see Q2a), and there are no exceptional circumstances (see Q2b), preparations are made for the customary death ceremonies (*romma).*

The body is washed with warm water and dressed in new clothes. These tasks fall to the oldest close relative of the deceased; this may be a parent, a brother, a sister, or an older child.

The women put on their jewellery and the men don new white turbans. A small basket (*kho*) is made, and a chicken is killed; any method of slaughter may be used. Exactly one *ser* of rice is cooked. This is measured with the aid of a bamboo internode calibrated in units of one *pwa*; a quarter of a *ser* is about half a pound. The cooked rice and the legs, stomach and head of the chicken are put into the basket (*hom-kho*) and placed near the head of the body, so the rice is called *laluk-hom* ("head-rice"). A second basket (*em,* see D2a) is put over the *hom-kho* to protect the contents from vermin.

The relatives are informed with all possible speed. All members of the deceased's own sib and sibs related by marriage must be given the opportunity to view the body if at all feasible. It must therefore be kept in the house for two to three days, even though decomposition proceeds quickly. A coffin (*pang)* is made as soon as possible. It is usually placed exactly in the centre of the main room (*kim-tom*) with the foot pointing towards the front of the house (*kimma)*, though the Chüngma Mru put it diagonally across the middle of the *kim-tom.* The body is shrouded with a cloth so that only the head can be seen.

A large pig is then killed (*ceng-to-ngar,* "burning-dead-person-sacrifice") and a small amount of its blood is dripped into the coffin. Some of the meat is cooked and eaten, and the rest is dried. So long as the body remains in the house, it is given a portion of this meat three times a day; the meat is put into the *hom-kho* together with a handful of rice from the three daily meals of the family. At one time, the rule was to slaughter one pig – or if this was not possible, at least a chicken – each day that the body stayed in the house.

In addition to the pig, a cow may also be killed (*ceng-to-ngar-cia*), provided that it already belonged to the family and was especially valued by the deceased. The cow can be tethered almost anywhere, and its throat is cut with a chopper. It may not, however, be tied to the *ling-pu* (sacrificial pole), and there can be no dancing round it when killing it with a spear. Its meat may not be given away to other households; any meat that is not eaten in the house during the days when the body is there must be thrown away at the end of that time.

During the period when the body lies in the house, scraps from family meals and water used for washing dishes must be thrown down the stairs; they may not be thrown from the platform of the house. This custom is unique to the funeral ceremony: if someone throws waste down the stairs at a cattle feast, for example, he or she must pay a fine to the feast-giver (MK, 18.08.1956; Kangku Catumma, 7.09.1956).

The relatives of the deceased bring pots of rice-beer for the funeral celebration. The marker (*pleu*) for measuring drinks is not the ordinary vertical one (see K2b); at funerals, a thin bamboo stick with a notch in the middle is placed across the top of the pot and bent downwards to form a right angle. This is the marking device used by Khumi and Twipra for measuring drinks, but it is forbidden amongst the Mru at times other than funerals (Langcong Changkan, Chamklaa-KP, 25.01.1957). Any remnants of beer are poured away on the same day. On all other occasions, for example when relatives come to visit, the remnants of the beer are not poured away until the next day: there is a belief that if it were poured away at once, the guests would die after departing (MK, 18.08.1956).

The beer pots are placed in a row by the side wall. A bundle of items is fastened into the roof over them; it contains a bracelet (*kwai-che*), a silver hip chain (*keng-ko*) a "vampire ear decoration" (*ui-mukma ram cheng*), a hoe (*tim*), a roast chicken (*wa-kreng*), and a thread square (*chaling-chalap*). The bundle is tied up with red and black cotton threads; I was not able to discover its significance. The same bundle must be made up for every person who dies, whether man or woman, even though the bracelet, the hip chain and the ear decoration are used only by women. The ear decoration is an imitation of the ordinary silver *param cheng* worn by women (see F2b), but made of bamboo; it is also called *li-cong* ("wind-fish-trap") and does not resemble any of the bamboo items made for other sacrifices. It consists of a piece of bamboo about 4 cm long, with a node in the middle; the outer surface is shaved in broad strips from the end towards the node using a chopper (Chamklaa-KP, 25.01.1957). The bundle probably has an apotropaic significance, and is thought to offer protection, for example, against an *ui-mukma* (see P1j).

Q1b) The coffin

The bottom of the coffin is made from bamboo canes split in half lengthways. Two pieces about 1.80 metres long are placed half a metre apart to form the long sides. Three more about 60 centimetres long are laid across them, one at each end of the coffin and one in the middle. Where they overlap the long canes, holes are made through both with a chopper, and bamboo sticks about half a metre long (six in all) are inserted vertically through these holes from below. The lower ends of the vertical sticks end in a node, so they cannot slip through the holes. More bamboo sticks (*chek-rek*) are laid lengthways on the three shorter lengths of bamboo, five for a man and six for a woman; the reason for this is explained in Q1i. Two mats are then tied to the vertical bamboo sticks, one at the head end and one at the foot end. They lie along the base of the coffin and hang over the edges.

The two mats are woven in particular patterns. The mat at the foot end must be woven in *kan-kok* for both sexes. The mat at the head end is woven in *lum-chi* for a woman and in *la-ke* for a man (see drawings). *Kan-kok* is the name of a fern; when people travel, they push stems from the plant into the sand at the edge of a stream, near its mouth, for *o-reng-nam* (see P1f). The term *kan-kok* could also be taken literally, to mean "crossed wrapping". I was not able to discover the meanings of the other two terms. *Lum-chi* is the name of the patterns often used on the embroidered stripe in the women's skirts. When the mat is fastened to the coffin, the stripes running around it have the green side outwards so that the patterns show clearly; the vertical ones are coloured beforehand. In the home of Langcong Changkan, they were bright red, and left as loose threads because no more cross stripes were passed through on the floor side. To fasten the mats to the vertical sticks, bamboo slats are placed along the four sides of the coffin and fastened through the mat to a slat on the opposite side. Along the lower edges, a narrow bamboo slat is fixed on the outside through the mat to the bamboos that constitute the floor of the coffin on the inside. Uncoloured strips are wound diagonally round the upper and lower slats; this results in a rhomboid pattern which fixes the mats firmly to the six vertical sticks.

The coffin, the widower and the drummers

The coffin is then lined with a blanket and the body is laid in it. More bamboo slats are placed diagonally over the coffin and a second blanket is spread over them. So long as the body is in the house, the blanket does not cover the whole of the coffin, but hangs down from a bar placed across the coffin over the chest

of the deceased, so that the head is still visible. More blankets and cloths can be spread on top of the first coffin cover. A final blanket is hung on the central post.

A fly whisk is also made in the red-green woven pattern described above, and attached to a handle. The ends of the red woven stripes hang as loose threads at the lower edge. Another fly whisk is placed on the coffin blankets, and used when necessary. All other gifts for the deceased are put at the head of the coffin.

In Chamklaa-KP (26.01.1957), the two vertical sticks at the head of Langcong Changkan's wife's coffin were extended by small bamboo sticks, the upper ends of which were scraped to form tassels. On these hung a mirror and chains on the left side of the coffin, and earrings, bracelets and another red chain on the right. On the floor to the right of the coffin stood a bottle, a betel box, a bundle of cotton threads, a small basket of the kind used to hold jewellery, a bunch of green bananas, and a small pile of tobacco leaves to which visitors helped themselves; to the left stood a drinking bowl and a bamboo tube for water, a small round pot containing coins, and some banana leaves.

Weaving patterns: above, *lum-chi*; middle, *la-ke*; below, *kan-kok*

Q1c) The death dance

When the body has been laid out and the first animals have been slaughtered, the death dance begins. It is performed three times a day, before or after the meal. The accompaniment is played on three *tu* (gourd-pipes), three *ner* (plate gongs), and three *tömma* (drums). A *tu-charaa* (*tu* master) is needed to play the *tu*. His payment for this service varies according to his relationship to the head of the household (see L4). If he is a *tutma* (member of the father-in-law sibs), he receives a *re* (spear); if he is a *pen* (member of the son-in-law sibs), he receives a *lapong* (turban) and one to two rupees. If he is not related in either of these ways, he is paid three to four rupees. However, for compassionate reasons, the *charaa* does not normally accept the full amount, but settles for one rupee (MK and Kangku Catumma, 7.08.1956 and 7.09.1956).

The musicians who play the *tu* and the *ner* walk around the body, but the drums remain in one place, either at the feet of the body or on the outdoor platform (*car*). *Tür-ram* (fish palm) leaves are tied to all the instruments. They must be of a special yellowish kind (*Didymospermum porphyrocarpon)* that is used only at funerals; the green *tür ram* used for cattle feasts is forbidden at this time. The *tu charaa* indicates to his assistants which verses they should play. If possible they should have some previous knowledge of them, though the words are not sung. The note corresponding to each syllable is at one of three pitches, roughly C, E and F. Each note can be played on one of the *tu,* which are tuned to different pitches. Anyone can play the *ner,* though they should follow the correct tonal sequence if possible.

In Chamklaa-KP (26.01.1957), the three *tu* and three *ner* were not accompanied by the drums; the latter were beaten beforehand in the house. This was done three times a day, as on the previous days. Three people beat the two drums. The first person beat the larger drum slowly with a stick, producing bass notes; the second beat the front part of the other drum with his hand or a stick, producing notes at an intermediate pitch; and the third one squatted at the opposite side and used two sticks to play a fast beat at a higher pitch. Listening to the performance, I did not see how it could be made to accompany the words of a song. The drummers were young boys.

There are, however, *tömma-long (*drum verses). According to Kangku Catumma (7.09.1956), the first one is: *ku-di ca-paa köng ding-dök*, *caa-ta tui-ku-löng pang-üm* ("Kudi, the son, is dead, is no more; eaten up by cockroaches, with swollen bellies"). The drum accompaniment is: low, low, middle, middle, high, high, high, low, low, middle, middle, high, high, high.

The rule is that the *tu* and the *ner* must be played and the drums beaten during each day that the body lies in the house. I wondered, therefore, whether this *tömma-long* might have been in fact a *tu-long* (*tu* verse). However, my informant Mita Tang, a *tu-charaa* from Langpoi-KP (Horinjhiri-Mouza), did not know the verse.

According to Kangku Catumma (7.09.1956), sixty *tu-long* are played at funerals, though it is difficult to know what counts as a unit. The number of times the players walk around the body does not depend on the number of verses played, though in Inglai-KP, at a cattle feast held on 21.02.1956, there did seem to be a correspondence. In the following sections I therefore give two numbers for each verse, the first according to the names of the *long* as supplied by Mita Tang, and in brackets, the number one obtains if one omits repetitions of verses.

Q1d) The *tu-long*

Only the *tu-charaa* know all the words. The masters are few in number, and not all of them are willing to share their knowledge. I could not always arrive at an interpretation of the individual verses that made sense. Interpretations of which I cannot be sure are given in brackets, the informant's suggestions in round ones and my own in square.

1) *taröng*
Ng ho

2) *taröng plang*
Ng ho

These are the first "circuit" (*taröng*) and its variation; they do not have words of their own.

3) *bung-ku*
rüm bung-ku ngau-chau ng, mü ni-ko rung, khi ni-ko rung döi ö.
The forest owl (feathers erect) does not look at the sun rising,
does not see the rising sun (i.e., the dead person can see the daylight no more).

4) *wa-ma ko thak* (also called *uram thak wan*)
wa-ma ko thak thak ng, thak taruk rau chong-a.
The chicken (i.e., the stepmother: *u ram)* curses, curses six mornings (in a row).
[Perhaps instead of *wa-ma,* it was originally *tua-ma*, which means sister: death began because of the curse upon her. See Q1e.]

5) *wa-kheng*
kheng ö parüu parüu ü.
The *kheng* birds dance in a row one behind the other.
[At first this line was placed following the first two *taröng*, as a kind of heading for the following *long*, each of which is dedicated to a bird.]

6) *ting-ra kui*
ting-ra-kui wa-ca ü, paköng ching hai khai ö.
The little *ting-ru-kui* bird "let-die with-life is".
[I.e., a living body has been made into a dead one. The story explains the origin on earth of death and of the *tu* songs. See Q1e.]

7) *padai*
klang-ca padük padai ö, paköng ching hai khai ö.
The man is rigid and lifeless; a dead body has been made from a living one.

8) *khang-cing khang cöng*
khang-cing khang cöng wa-ca ö, paköng ching hai khai ö.
[Cf. 6 and 7. The bird names probably also have a meaning; a *khang* is a ceremonial ban on something.]

9) *wa-hui*
Hui hai ng ng, hai ng ng. (3 x)
(*hui hai* = *wa-hui*, green pigeon)

10) *wa-har*
hor har ng ng, har ng
(*hor-har* = *wa-har*, thrush, *Garrulax leucholophus)*

11) *pong-li*
ching döm pong-li pong-nam, tön kang ang khai,
kaa döm döi u ö, kaa döm döi paa ü.
[The Ficus spirit, controlling the soul deep down] holds me back, [I, the dead man] cannot come down, o mother, I am not allowed to descend, o father.
(The following verses are also words spoken by the deceased.)

12) *plong-cet* (also called *pik-rau-mi tarok cong long*, the "sick person's sacrifice expectation verse")
rümma plong cet lu tu ng, cong rüm tui tang, kaa rüm tui tang döi ö.
[The forest liver inspector with the hair tuft] waits to try forest water, does not get forest water. (The dead man waited for a sacrifice, but did not receive it.)

13) *wa-tur*
pongma tur lip ö, chaima tur lip ö, ang nguk ui ting, ra nging ngüng.
(Like) the Ficus pigeon, the Banyan pigeon, I coo [plaintive cry?].
(The deceased worries about his children.)

14) *wa-ak*
ng au ak, ng au ak, au ak, ng au ak
(The dead man croaks like a crow.)

15) *wa-wang*
tui-boi wa-wang, hor leng ce plai ö, ce plai ö.
The wagtail runs dancing around, dancing around (the dead body does not find rest).
[Reference to the death dance which is represented phonetically in the following?]

16) *wa-wia*
ng ho, ho ng
(plaintive cry)

17) *poi-coi*
coi maling mange ö.
Pull out [the neck feathers and] the tail feathers. (The deceased feels like a pigeon from which all the feathers have been plucked.) [Probably a demand by the deceased to offer a chicken and, in the following *long,* to kill a pig.]

18) *ria-kwek*
yoi-ca kek kek ria, kek ria.
Empty the entrails, empty the intestines. Empty the intestines (by pressing them with hand or stick).

19) *wa-ham*
pu wa-ham klung ram cia ton, cia ton a.
(The cows kept by Grandfather Wa-ham eat *klung*-leaves. How fortunate they are, compared with me!)

[I am inclined to translate *klung ram as* "to tether (or lock in) and enclose" (that is, at the sacrificial post), and to render the passage as: "Grandfather Wa-ham, tether the cattle, enclose the cattle". Now, however, when a cow is killed for the deceased, it is forbidden to carry out the sacrifice in the manner used at feasts of merit.]

20) *wa-klöt*
kimma wa klöt ö, yan chom reng a.
(The house falls to pieces completely???)

This is where the first part ends, as does the beating of the *ner-cham* (slow gong). The *ner-khöng* (rapid beating of the loud gong) begins.

21) *plai khöng taröng*
ho, ho

22) *plai khöng taröng plang*
ho, ho

23) *dök ching dök kau*
dök ching tahöm tahöm, dök kau tahöm tahöm.
Seize a beautiful and well-formed tree, bring bamboo beautiful and well-formed. (This and the next verses describe the making of the coffin and the placing of the body in it.)

24) *rin*
pön ce rin, pön ce büa, pön ce rin.
Cut the bamboo tubes to size, cut the bamboo tubes, cut the bamboo tubes to size.

25) *chek* (slats)
chön-a pe chek ö, lang-a pe chek ö.
Split the bamboo slats, split the bamboo slats lengthways.

26) *a-chüa* (colouring)
chüa pong kon a yü ri yau, tam pong kon a yü ri yau.
What will the turmeric mat be? Make it wet! What is the ginger mat becoming? Make it wet! (Colour the strips.)

27) *a-kong* (weaving)
kwai lin kuai len, kwai lin len.
Put it lengthways, put it crossways, put it all over.

28) *kep* (border slats)
lang pin thö röi, lang pan thö röi.
(Bind the slats neat and tight.)

29) *dak-kep* (the lower border slats)
langma röi ü, pang chang kwai.
The long good strips, put them through the coffin lattice.

30) *cöng-cöi*
cöng-cöi dom lo, cöng-cöi ö, cöng-cöi pai lo.
Cöng-cöi, take hold with both arms, Cöng-cöi, o Cöng cöi, carry in both arms. (Cöng-cöi is a spirit [see P1c]. Here, however, the name is used of the person who puts the body into the coffin. Why he is called Cöng-cöi is not known.)

31) *hom kho ton*
chüngma kan pen khu ö, re dam pachaa kan ö.
Large pieces of mountain pumpkin and a side dish made from two kinds of fish. (The fish names stand for chicken and pig; a small amount of meat from each is put into the rice basket, or *hom kho.*)

32) *kan-chur kan leng*
kan-chur ka- leng kan ö, (repeat) *dam pacha kan ö.*
(two vegetable kinds) side dish (two species of fish) [see above]

33) *hü (*also called *kar wan,* lament)
hü ha,hü ha,hü ha ...
(lamenting like a hanuman monkey = *hü*)

34) *rak cam tui*
kimma rak cam tui, wang-a prik prük.
From edge of the house roof the water, the rain drips and drips.
(The tears drip from the eye.)

35) *chön pu malaa*
chön pi, chön pu, chön pu malaa,
chön pu malaa ko chaima, chön pu malaa ko chüama.
The rattan girl white and beige, the rattan girl white and yellow. [?referring to the story of Chong-pau malaa, see S4a.] (White and yellow flowers grow over the dead body wrapped in woven mats, *chön pi* and *chön-pu.*)

36) *than caa*
than caa taa, dun pen pon.
(Corpse eater), (decay).
[This verse is closely linked to the next one. Both should be understood as signifying a burial. The deceased himself speaks once more. The corpse-eater, introduced as a spirit, has devoured him – *caa taa –,* and wraps him up – *dun*.]

37) *klung*
klung-a cai ang ö, than-a cai ang ö.
(He) crumbles me to bits of earth, (he) reduces me to termite droppings.

38) *wa-taa*
ang toi cöng-ce rum-ce, wa-ta long-leng ce-plai khai ö,
kaa döm döi u ö, kaa döm döi pa ö.
I am changed (into a king pigeon?), as a bird I fly around and dance. I cannot come down (o mother), cannot descend (o father).

39) *yom*
klang-ca angum-angia ö, chot karek laa-ma yong.
Like Angum-angia pierces stars and moon (so I am hunted).
[*Yom* who causes the eclipses, is Yama, the god of death, with whom the man *Angum-angia* is equated here, see P2b.]

40) *leng*
leng bü, leng yoi.
(Meaning unknown, probably refers to the cremation.)
[*Bü* and *yoi* in songs are used as parallels to *plong* and *mu,* which mean "liver" and "stomach". *Plong* also means "the centre of life"; *leng* could

mean "to stroll" but also "to move to and fro". If the verse is taken together with the previous one, it could mean that the god of death checks the "heart and kidneys" of the deceased. Alternatively, it could mean that the soul moves around restlessly until the end of the year, as indicated in the following verses.]

41) *rüm-rup*
o ko rüm rup pau ö, khüm ning tahöm khai ö.
By the brook the *rümrup* flower is beautiful, come the end of the year.

42) *yüm-yüa*
ning khüm yüa ri pau ö, khüm ning tahöm khai ö.
The real year's end flower, by the end of the year it is beautiful.

43) *pur cen*
ning ria cen long pau ö, khüm ning tahöm khai ö.
At the turn of the year, the *Gardenia,* by the end of the year it is beautiful.

44) *pur cin (*1)
o ko cin-ca, hai mum ö.
By the brook the Cinjwe starts to build a nest.

45) *pur cin (*2)
o ko cin-ca, hai thar ö. .
By the brook the Cinjwe destroys the nest.
[The meaning of this probably depends on the following two verses. I feel inclined to translate *hai mum* and *hai thar* as "starts to bud" and "starts to open". Since this can only refer to a flower, I would then assume that for *pur cin* one should read *pur cen* as far as verse 46.]

46) *pur cin (*3)
o ko cin-ca hai bai.
By the brook the Cinjwe starts to build a nest.

47) *pur cin (*4)
o ko cin-ca. bai kwak.
By the brook the Cinjwe throws the nest away.
(Just as the bird discards his nest, the deceased now leaves his home.)
[I am not aware of any Mru stories about the Cinjwe bird that might have a bearing on this image. The Cinjwe bird is distinguished by its plaintive call, which consists of a short note followed by a long falling note; the sound is well represented by the Marma name Cinjwê, if one pronounces the final e with a falling intonation.

There is a Marma story about two brothers who had used up all their food, and set off to look for something to eat. The older brother found nothing; the younger brother found just one grain of rice, and ate it at once. When the brothers met again, the younger one confessed that he had eaten the only grain of rice that he could find. The older brother was angry. In his rage he killed the younger one, cut him open, and ate the grain of rice from his stomach. He was then overwhelmed by grief, and thought how beautiful it would be if he had his brother back again. He began to mourn, and called "*cinjwê, cinjwê*!" (Kyo Thwân Ong, 19.04.1956).]

48) *long hin*
pri po long hin ö.

The tiger takes (domestic animals?) (as the spirit drags away the deceased.)

49) *pri-kung*
prima kung pang pang.
The tiger's back is clearly visible (everywhere?).

50) *tui-pan*
oma tui pan lip-lop-lap
The green splashes on the waters of the river.

51) *kan-car*
tui hu kan car, ching klöng han raa.
The flood rises over the shores, tree-trunks are washed away.
[Fauna and flora are carried away by tigers and floods. In the same way, the body is carried off now; or must the deceased overcome these dangers on his way to the hereafter?]

52) *kau-ting*
kau ting klong köi nia pün khai, nia mo,
lök-po dön-po kalüm döi, nia mo,
lök-po dön-po kahau döi, nia döi.
The penis has climbed to the bamboo roof ridge, the erect penis; nobody, no one, lifts him up, nobody, no one, lowers him, the penis, the erect penis.
(53) *kan-chur bia köi kai klaa khai, kai mo, kai mo,*
lök-po dön-po kalüm döi, kai mo.
The vagina has fallen into the bowl of sour vegetables, the engorged vagina; no one nobody lifts it up, the engorged vagina.
(The spirit talks to the dead man: Who copulates with you now? No one but me.)

53 (54) *wa-tur*
wa-tur tahau yung klai döm he hot.
The pigeons mate, they ejaculate and receive.
[Bodies apparently have something to do with fertility. The hut for the deceased is called *tur-kim* (pigeon? house), and the most important place at the foot of a tree in the field is called *tur-tut* (see J2e). This may be a link to the next verse.]

54 (55) *ching chüm*
chüm ching tut, lu thur chüm chua khai ö.
At the end of the tree, under the tree, it is put away (by God?).
[The meaning of *chua* is "put in below and pull out above". My informant therefore thought that the line meant "God has taken the dead man with him", since in the hereafter, the deceased person sits under a tree. However, I have my doubts as to whether *lu thur* should be equated with *thurai* (which in songs is often shortened to *thur*). *Lu* is the head, and *lu t(h)ur* could be an "ancestor skull." Today skulls have no special significance – although there still is a "skull day", see below – so the verse is no longer meaningful.]

55 (56) *tö hüa*
tö hüa lö, tö hüa lö, tö hüa di köi, tö hi hüa.
Must leave [leave behind and go away].
(The living remain behind, the deceased goes away.)

The following is an utterance made by the *tu-charaa*:

Pai-a tahüa-taa chum pat cekhai; chum kat tai ang tek kom-baa. Mi köda chum pat pat kom mi köda, ma-khöt tai kömi wöi döi. Yong kh'e chum pat khü cayang, chum pat nö tö hüa kön tai; chum ö then hom kön tai-a. Tang hom nö, plan won yok kön tai. Pai-mi köka ani chum kat, chum kat ci plai kom-baa.

(The "leave behind for good" lines mean I must walk three times around; three times I must say it. When the three times around are finished, there are no more. If one goes three times around at that time, after the three times around, one must leave it behind. Rice must then be offered. After the offering of the rice, one must start again. So it happens three times a day, three times we will dance. (Finally follows the *krong döm* [s. Q1g]:)

56) *taröng*

57) *taröng plang*

58 (57) *tür-ram*

59 (58) *chang-ku*

60 (59) *wa ce-ca*

Tür-ram and *chang-ku* are the same as the *tu long* for cattle sacrifices (see R3g); for *wa-ce-ca* there are no words. Rice is eaten, then the verses continue as follows:

61) *taröng*

62) *taröng plang*

63) *tür-ram*

64) *chang-ku*

65) *wa-ce-ca*

66 (60) *kwai hiu kwai tam*
kwai hiu bu ü, chöng lak ng lak,
kwai Tam bu ü, chöng lak ng lak.
The earth-bee nest whirrs and buzzes,
the honeybee nest whirrs and buzzes

(During these verses the performers dance around three times to the right and three times to the left, then strike the house wall with a hewing knife.)

67) *taröng*

68) *taröng plang*

69) *tür-ram*

70) *chang-ku*

71) *wa-ce-ca*

At this point the musicians return to the house.

Qle) The origins of death and the *tu*

Mita Tang told me the following story about the origins of the death dance (29.06.1957):

Acük kh'e mami pren da.köng döi pe, cük kh'e baa.

In former times this country did not die, it is said, in former times.

Or-mi pren pe, or köng-mi nö, mi pren mru or-mi pren köi, katö-taa tuama khai.

In the upper country, it is said, people died. The people of our country gave their sister into that country.

Katö-taa tuama khai nö or köi a tuama caa köng khai, pe.

They gave the sister away, and the sister's son died there, it is said.

A tuama caa köng khai nö, wang po apin tai-a.

The sister's son died, and those people had to invite them as guests.

Wang po pin hön-yaa nö, katoi mü, katoi mü!

The hosts came and said: Come along and see!

(*Köng nö, kapo pin nö, katoi mü köka ang tek yong röp-röp ari ayang.*) (When somebody dies, an invitation is issued: "come along and see!" as we say. The story therefore is true to the rules and customs.)

Bök-bök cang, tum tu mi, thep ner mi, toi mat-la reng tai-a.

All that they did there, blowing the *tu*, beating the flat gongs, the people were told to remember carefully

Toi mat-la reng panö, ani kh'e mat-ya ce wang döm-a hanö,

and they remembered everything. One day after they came back here,

ping-pang bök bök kong-a hanö, ting-ru-kui wa ce tep-a köt pe ba nö,

they prepared a coffin and everything else, and killed a little Tingrukui bird

or tuama köi kapo ape kapo apin, ce ang caa köng khai.

and to the same sister they sent a message and invited her, saying: "Come, our son has died,

Ang caa köng khai, toi mü ang caa dit, toi mü ang caa wak dia.

our son has died, come along to see our son, come along to see the body of our son,

Pa pe atuama köi. Nö a tuama tek pe da,

thus they spoke to the sister. Then the sister said:

Tong pacö maköng cö, tong pacö maköng cö. Yu köi i wöi mi da

"Why has he died, why is he dead? Down where we live

maching pren, ma köda maköng pren, ma köi tai kaköng.

is the country of life. Here is the country of dying, here one must die.

Yu köda köng döi kabö. Köng döi kabö kat po, tong pacö maköng cö

Down there, people do not die. Since they never die, why has he died?

dit po öm döi pe tuama. Köng khai, köng khai,

I do not believe it, not at all", said the sister. "He is dead, he is dead.

toi mü ang dia, toi mü ang dia. Pa pe ka a tuama toi mü.

Come along and see, come with us and see!" When they said this, the sister came.

Kar khai t'i-a li, a talung köng cace kar-la rom-rom,

She cannot help crying because her brother's son has died. She cries and laments;

kar-la rom-rom che pe, wang toi mü.

crying and lamenting, she goes to see for herself.

Wang toi mü pace, mü nöm kh'e da,

She comes to see, and when she really looks,

pai ting ru kui wak cayan khi-taa khai. Mru wak khi döi, ting ru kui wak cekhai.

she sees the dead Tingrukui instead. A man's body she does not see; the Tingrukui is the dead body.

(*Na-ka ce pai-a hoi-a tu long.*)

(The *tu* verse tells of this.)

Hai la köng khai, pada ca pai ting ru kui wak ka-e,

With this the dying started. It began with the dead bird.

paköng ching hai, hai-a pala cayia dabö.

The living died, so things became exactly as they are now.

Hai-taa köng khai pai-a ting ru kui wak. Na-yong ka, uh,

It started with the death of that bird. In this way, oh,

en-nit. po, ang-ing yong köng pren cekhai madök en-ni nau khaï?

would you want your country to be a country of death, like ours?

Na-mi ra-köi, mang kh'e, a tuama tui-la pe dong taruk kat pe.

So when she left, the sister struck the tree-ladder six times with her foot.

Ö, na-mi hön-yaa taplan lut nö, ma-mi cöng maköng pren khai,

Yes, since then, it has all turned around: this became the country of death,

or-mi cöng maching pren khai, ö na mi ri ko mat talup a,

and the upper country became the country of life. Their rules are copied now,

ma-hön-yaa ni tu ri po wang, ma-hön-yaa tak tu ri po,

the *tu* rule also comes from there; we have the *tu* rule therefore.

tonma tak, tu tum ma-hön-yaa po.

In recent times, blowing the *tu* is also copied from there.

The following version of the story was first told to me by Kangku Catumma (7.09.1956):

Once there were seven brothers and a sister. The brothers lived here on earth, where at that time there was no such thing as death. The sister, however, married and went to live in a place where people died. When one of her children died, she sent for her brothers to come to the death ceremonies. They found the child laid out in exactly the same way as is now customary amongst the Mru.

The brothers liked the custom very much, and when they returned home, they tried to copy the ceremony. One brother caught a Tingrukui in the forest. (The Tingrukui is a bird the size of a sparrow, with a yellow tail and green wing feathers). The brothers laid it out as if it were a dead man, and sent a message to their sister saying that one of the brothers had died. The sister was deeply grieved. When she arrived, she was astonished to find all her brothers alive, and in the coffin, a dead bird. Her grief gave way to fury, and as she left, she put a curse upon them, that they too would die, and feel the pain that she had felt.

On her way back, she passed two ficus trees, a *pong-ching* that had few leaves, and a *chai-ching* that had many small leaves. She kicked the two trees, saying: As your leaves fall, the people shall die. Now, they are like the *chai*, which has many leaves; but they will become like the *pong*, which has few.

The sister returned to her country, and died there. Ever since that time, the people on earth die too. (Nowadays, if someone catches a Tingrukui, brothers are not allowed to share the meat with their sister.)

It is said that at one time, the *tu* were not played at funerals. The people of the Tamtu sib were the first to make *tu*, and to make rules for playing them. The death dance with the *tu* is therefore called *Tam-tu-ca-ri* ("rule of the Tamtu people"). The Chüngma Mru do not follow this custom.

Q1f) The role of the *tutma*

There is a second type of death dance (*wak-plai*), mainly performed when the deceased is a woman who was mistreated by her husband. When the man's *tutma* (i.e., the sib relatives of the wife) are informed of the death, they give advance warning of their intentions: they may come peacefully, or they may perform the *wak-plai* in anger.

If the *tutma* come in anger, the husband must always present a spear to any member of the sib who arrives. Whatever the visitors' intentions, they must be given about 30 rupees in total. They bring pork and cooked rice, and in return they are entertained with chicken and with rice. Before they go up the tree-ladder and into the house, they must be served as much liquor as they want.

While this is happening, the *wak-plai-long* (= *tömma long,* see Qlc) is beaten on the drum. The members of the *tutma* then enter the house, climbing over each other, and perform a wild dance around the woman's body. Sometimes they destroy the whole house in the process: they trample the stove, tear down the racks, knock over the baskets, and break everything in them, spoil the utensils, and chop holes in the walls (MK and Menching Atwang, 8.09.1956).

When the *tutma* come in peace, they must be entertained in the same way. They must be served rice-beer on arrival. When the cremation takes place, the brother of the dead woman is entitled to receive from her husband or son the *cham-kam* ("hair-burn") money (MK, 11.07.1956). Amongst the Anok, this is between five and seven rupees. The hair money is the final symbol of the woman's connection to her sib of origin; I was told that a woman's hair is regarded as a property of her brother.

The *wak-plai* demonstrates the right of the relatives to watch over the welfare of women even after they have married into other families. The *tutma* must be invited and are free to express their final verdict on the woman's treatment. They have the right to come even if they are not invited, to make sure that their relative's funeral is carried out with the proper ceremonies. Their consent is needed if for some reason the woman's body is to be buried rather than cremated (see Q2b). If a woman is buried without the consent of the *tutma,* the husband must pay a fine of 30 rupees to the brother of the deceased; this is called *wak-pong-dong* ("corpse-decay-mourn"). If consent is given, the amount is reduced by half. If the woman has died a "bad death" (*char,* see Q2a) neither the *cham-kam* nor the *wak-pong-dong* is paid (MK, 14.07.1956).

Qlg) *döm krong*

The *döm krong* ("coming down to the earth") takes place the evening before the cremation. First, the bundle wrapped in red and black threads (see Qla) is taken down from its place under the roof and put at the foot of the coffin, while it is still in the house. In front of the house, a miniature altar or *pak-plai-kimca* ("pig-dance-hut", see J4c and P1g) is built on the *wia-dong* (the earthen foundation of the house). Two chickens are killed and rice is cooked. One of the two chickens, the *krong-döm-waa,* is cut in half. One half is cooked with flour and the other one without, though both halves are seasoned with salt and pepper. They are simmered in a pot, then the two halves are placed on a winnow (*per*) with some rice. A hoe (*tim*) is put in the rice. The other chicken is prepared as *wa-kreng* ("chicken-roast") and placed on the *pak-plai-kimca.* A chick or part of a fully grown chicken is enough for the *wa-kreng,* which is not intended for eating. A bracelet (*kwai-che*) and a woman's silver hip chain (*keng-ko*) are placed with the *wa-kreng* on the *pak-plai-kimca.*

Two *teng* (thin bamboo tubes with forked tails) are stuck into the ground next to the *pak-plai-kimca* and filled with *hom-noi* (cooked rice with water). Unlike the *teng* used at other ceremonies, the ones used at death ceremonies are cut from a top internode, so the bamboo stands upside down relative to the direction of growth (MK, 20.07.1956). The *hom-noi* in the *teng* is not used during the ceremonies. An additional supply is put in a small round pot (*ci-ö-ca*), and is called *ci-ö hom-noi.* This is used for spitting (*phyöi cho*) on the *tu* and *ner* (plate gongs) during the circuits made by the players round the *pak-plai-kimca.*

The chicken and the rice on the winnow are eaten by the players of the *tu*, *ner*, and *tömma*. The hoe has no further function, and is set aside. The players walk three times (*wir* = circuit) around the *pak-plai-kimca* and the house of the deceased, playing the *tu-long* set out in section Qld above. The order in which the circuits are made can vary. According to Kangku Catumma (7.09.1956), the players go three times round the house of the deceased in the usual direction, that is, clockwise, then three times around the *pak-plai-kimca.* They then go three times round the *pak-plai-kimca* in the opposite direction, that is, counter-clockwise. Finally they go up into the house, where they play the rest of the *tu-long.*

In Chamklaa-KP, however, on 26.01.1957, I saw a funeral where the players made six circuits round the *pak-plai-kimca,* which stood in front of the *kimma* (front house), then moved to a corner to the left of the *pak-plai-kimca,* between the house and the platform. The position of the *pak-plai-kimca* prevented them from walking right round the house. They then made three clockwise circuits, then danced three double circuits in the opposite direction and back again. After the last clockwise circuit, they returned to the *pak-plai-kimca.* During each anticlockwise circuit they struck their knives against the house wall. Once the players have re-entered the house, no more dancing is allowed during the death ceremonies.

During the *döm krong,* a bundle of wood is brought from every house in the village and placed on the front wall of the *kimma.* After the *döm krong,* the young people begin a game in the house of the deceased. They take coins – old *paisa* pieces – from the small round pot (*ci-ö-ca*) at the head of the coffin, throw them in the air, and catch them on the ground under a "cup" (a thick piece of bamboo internode about 4 cm long). The others are then asked to guess *yuk* or *pau*, or, using the Bangla words, *bandor* or *phur* [< **phul*]); the words mean "monkey" (the side with the king's head) or "flower" (the side with the number). The players make bets, and the pot serves as a bank. After the game, the pot goes to the feast-giver, who does not take part in the game himself. The money is meant as a gift for the deceased, but the amount is kept small because of "grave robbers". The old coins are especially sought after by Bengalis, who sell them in Burma (Chamklaa-KP, 26.01.1957).

Q1h) The last morning

Cremations never take place on Wednesdays or Saturdays; it is believed that the fire would get too hot on those days. Before daybreak, a few wood shavings are placed at the head of the coffin, and a cotton thread leading from them is tied to a thin bamboo stick. The pile of shavings is ignited with a match and the thread catches fire. The light – which is called *tem-cong-mai* ("water-snail-fire") – is meant to illuminate the dark path that the deceased will take to the next world. Each member of the household then twists *ting-ai-ram* (palm leaves) to make *yia-rung* (calabash stoppers); a long handle is left protruding from the small, cone-shaped stoppers, and the whole thing looks like a pipe or small scoop. These *yia-rung* are put into the coffin near the head of the body, as mementoes of all the members of the household. The fire in the hearth is then extinguished, and everything that has been in contact with the body and is not to be taken for burning with the coffin is thrown out over the tree-ladder.

The blanket covering the coffin is removed for short time. A strong bamboo carrying pole (*kot-long*) about three metres long is fixed lengthways over the coffin by means of bamboo strips that pass under the corners of the coffin base. The pole is decorated with a simple pattern consisting of alternate straight and wavy lines; these are made by scraping off the bark, and they run the length of the pole. Holes are made in the ends of the bar and a bamboo tassel on a short mount is pushed into each hole. The coffin is then covered with the blankets again and bound with strips of bamboo that go all the way round it. The body is carried out feet first through the door and down the tree-ladder. The relatively large number of items that must also be taken to the cremation are put into one or more baskets and carried by those who follow the coffin.

The wood given on the previous day by the other families will by now have been packed into bundles, and more wood – the larger part of the wood that will be needed – is taken from the store at the home of the deceased. The bundles are also attached to a carrying pole borne by two people, one at the front and one at the back. It does not matter who performs these tasks; the helpers need not be members of the household. Only the *hom-kho* – the rice basket into which the food for the dead person has been put until this point (see Q1a) – must be carried by a member of the deceased's household. If a stranger were to carry the *hom-kho*, his life would be in serious danger, and the risk could only be mitigated by the sacrifice of a *met-pak* ("atonement pig").

When the bearers reach the village exit (*plon*), they take a short rest. They put the coffin down a short way to the side of the path, at the *hom-chok-raa* ("rice-feeding-place"), and push four bamboo poles into the earth at its four corners, pointing forwards and backwards away from the coffin and in line with the sides. A member of the party – anyone who wishes to perform the task – stands by the coffin and takes from the person carrying the *hom-kho* the accumulated food that has been given to the body. He receives rice and some meat from the *ceng-to-ngar-waa* (the first chicken sacrificed after the death, see Q1a). This he lifts from his lower right side to his mouth, and then throws it away over his left shoulder. The next piece he lifts from his lower left side and across his body to his right shoulder. The actions are then repeated from the right shoulder to the lower left side, and from the left shoulder to the lower right side. The whole *hom-kho* is then thrown away, together with a winnowing plate (*per*), a pair of fire tongs (*kep*), a mixing spoon (*kou*), a calabash (*tui-yia*), the bundle wrapped in red and black threads, and finally some bamboo drinking tubes (*röng*) that were made while the coffin was being prepared and were used for drinking liquor. (The liquor can also be drunk from glasses, but then the glasses would have to be thrown away, and this would be costly.)

The coffin is carried to the burning place

This part of the ceremony does not take long. The party then continues towards the place where the body will be cremated. The coffin bearers walk in front, followed by three people beating plate gongs, and finally the helpers (MK and Kangku Catumma, 7.09.1956; Chamklaa-KP, 27.01.1957).

The coffin is carried over the brooklet and disappears in the dark

Q1i) The cremation

The place where cremations are carried out (*ceng-rong*) is always near a stream or river, below the village if possible. Mills (1931:521) writes: "Every village has its burning place near a stream and usually close to a pipal tree [*Ficus religiosa*]." However, the presence of this tree – or, indeed, of any tree at all – is not regarded as essential.

There are two places that are important. One is the spot where the cremation takes place; this only remains recognisable for a short time after use. The other is the place where miniature houses are built for dead persons. As soon as the members of the funeral party arrive, they begin to prepare both places. Some of them cut materials to make a house for the deceased, while others build the woodpile. As is customary amongst the Marma, the funeral pyre is built with five layers of wood for a man and six for a woman. Both peoples give the same reason for this: the women are responsible for the supply of firewood, and this is recognised and honoured by the addition of an extra layer on the funeral pyre (Kangku Catumma, 7.09.1956). We have already noted that an extra slat is added to the base of a woman's coffin (see Q1b); this is also a mark of recognition of her role in providing wood.

A study of Mru songs (see S2a) suggested to me that certain numbers have a symbolic value. The number 5 seems to symbolise the liver, and thus the centre of life, while the number 6 indicates the indestructible spiritual force deep inside the individual. However, I cannot see any evidence of such symbolism in the death ceremonies, so I am inclined to believe that it derives from Marma customs.

The pyre has been set afire

All the layers in the funeral pyre are laid from east to west. To keep the layers distinct, a few logs are placed at right angles between them. Two small wooden posts or lengths of strong bamboo are pushed into the ground on all four sides to make sure that the pyre does not fall apart. It is supposed to be at least a metre high, so at least a cubic metre of wood is needed. Unlike the Marma, the Mru place no restrictions on the kinds of wood that may be used (MK, 30.09.1956).

The coffin is then placed on top. A piece of the carrying pole (*kot-long*) is cut off and used after the cremation (see Q1k). A man's body is laid with the head to the west, so that he can see the sky brightening and get up to work. A woman is laid with her head to the east, so that she can see when evening is approaching and it is time to cook the (main) evening meal for her family.

Just before the cremation, a ritual is carried out to determine the cause of death. A *ting-ai-ram* (palm leaf) is cut into small pieces measuring about 2 x 3 cm. Between three and five of these pieces are cleaned carefully and placed one on top of the other to form a "book". This is put on the coffin, and the deceased is asked to tell which spirit is to blame for his or her death. The pyre is then lit with a bamboo torch about half way up, first at the head end and then at the foot. Any member of the funeral party may light the pyre. At the moment when it is lit, a small chicken is killed by wringing its neck, and it is thrown into the flames (*dung-cot-waa*, see Q1o; for details of other sacrifices made during the cremation, see Q1m).

As soon as the flames start to rise, the *ting-ai-ram* leaves are taken down and examined. If there is earth between them, a field spirit was to blame for the death. If there are hairs of a wild cat, the deceased was bitten by a *chit-ci* (an evil spirit, see P1k). If some kind of writing appears on the leaves, the death was brought on by sorcery (see P1k). If there is water between the leaves, the death was caused by a *tui-ching-nam,* or snake spirit, that lives in a *rua-lim-kua* ("king-cobra-hole") (see P1e and J1d; if a man makes his field where there is a *rua-lim-kua,* he puts himself in serious danger).

The Mru have complete faith in this oracle; signs of this kind can only be made by spirits. However, the result has no practical implications for the funeral, since no additional sacrifices are required in response to particular outcomes (MK, 8.09.1956).

Q1j) "Spirit flesh"

I witnessed a cremation on 27.01.1957 in Chamklaa-KP. Either there was not enough wood to burn the corpse completely, or the fire was not hot enough. The two people who were tending the fire with long bamboo poles had more wood brought; they put some of it on the top of the pyre, and pushed the rest in below the body. That did not work either. They therefore removed what was left of the body from the pyre and built a new fire, cutting down fresh bamboo and slitting its internodes so that they would not explode. The second fire also proved ineffective, even though they poured petrol over it.

Part of the back of the corpse remained intact. This was taken out of the embers, examined, and cut out with a hewing knife (chopper). Parts of the flesh looked completely unburned, but they were hard, like wood. The unburned parts were declared to be "buffalo flesh", which spirits had caused to grow in the woman's body and which had caused her death. This may have been linked to the fact that her husband had recently given a buffalo feast. The hard pieces of flesh that the spirits had caused to grow were cut out, put into a small pot (*ci-ö-ca*), and buried. The remaining pieces of normal flesh were thrown back into the fire, and this time they burned. Langcong Changkan, the woman's husband, was worried at first that the burning of the body might be causing his wife some pain; the others reassured him repeatedly.

Not all the villagers agreed that the hard parts of the body were "buffalo flesh". Kröilö Prenchü, for example, insisted that the fire was not hot enough because the original pile of wood was too small. My Marma boy Kyo Thwân Ong on the other hand said it happened at Marma funerals too that parts of the body would not burn because they were "spirit flesh". He said it was usually pig flesh, and there were even people who would eat it, though usually it was cut out and thrown back into the fire. Their custom was to burn the body twice: the pyre would be lit in the evening and allowed to burn down, then, if there was anything left next morning, the remains would be burned a second time (Kyo Thwân Ong, 27.01.1957). The "pork" would often be found in the chest, which would then be cut open with a hewing knife, and blood would spurt out.

If the body rises on the pyre or is thrown off it, spirits are held responsible. In one village, a child's body that had been placed on a pyre was found undamaged an hour later, at some distance from the fire. It had to be burned a second time (Kyo Thwân Ong, 8.05.1956).

I spoke with a Bengali Muslim, who declared that only people who ate pork would burn. The body of a Muslim would not burn at all.

Q1k) The *taham*

When the pyre has burned down, the embers are extinguished by pouring water over them. The piece of bamboo internode that was cut from the carrying pole (see Q1i) is made into a kind of long trough: cuts are made along both sides and the middle part is pulled out. Water is poured into the trough, and rice pressed into lumps by hand (*hom cem*) is put in the water. A piece of bone is added from the remains of the pyre. The trough is then placed on the ashes, and beside it other pieces of bone, a hewing knife (*charai),* a hoe (*tim*) and any jewellery and coins that may be found in the ashes. The bones and utensils are covered with a cloth, which is fixed to the ground along its sides with small bamboo pegs. Of the six vertical posts that held the pyre together, four are pulled out and the unburned ends are placed along the edges of the cloth to form a square. The trough is placed in front of the cloth bundle, lying from left to right. In front of the trough is the pot in which the water for washing the body was heated (*tui-chaa-pu*) and the bamboo fire tongs (*kep*) that were used at the time. A fly whisk is made from thin bamboo strips, similar to the one made before, but in one colour only. Slits are made in the handle and horizontal stripes are woven through them. The handle of the fly whisk is then stuck at an angle into the ground.

All these items are arranged in a long rectangle. Small bamboo sticks are stuck at an angle at the corners, in line with the sides, like those used for the *hom-chok-raa,* but smaller (see Q1h). A flat structure is made on the ground and then fixed above the articles on the ground to form a sloping roof (*taham*). Four lengths of split bamboo are tied across three bamboo poles about a metre long. A cloth is spread over them, followed by a layer of bamboo leaves; then another four split bamboos at right-angles to the first; then three more unsplit poles, at right-angles to the previous layer (MK, 7.09.1956).

In Chamklaa-KP, the roof was simpler, and was made with less effort. The first three bamboo poles were fastened to a pole at one end. Five narrow bamboo canes were fixed across them and covered with bamboo leaves. The roof was set at an angle over the items on the ashes, with its shorter side resting on the ground and the opposite side supported by two vertical posts about half a metre high and forked at the top. The roof does not have to face any particular way; in Chamklaa-KP, it stood at right angles to the position in which the coffin had been placed and with its open side facing south towards the stream. Mills (1931:521) offers the following details, which differ from my own findings: "The finer ashes, left where the man was buried [burned?], are also treated as if the spirit of the dead man were there too. Over them is laid a piece of cloth. This is pegged down at the edge and on it are laid a *dao* [chopper], a hoe, etc. Over this another cloth is laid and by them are placed pots of food and drink. Finally a little lean-to shelter is built over of the heap, with the open side towards the east."

The *taham* with the fly whisk and part of the bones wrapped in a towel

My own observation was that *taham* are not treated with respect. In Chamklaa-KP, while the wood was still burning, boys who had nothing to do searched amongst the remains of old funeral pyres for any jewellery or coins that might still be there. They found one eight anna piece, but nothing more.

Q11) The house of the dead

While two men were busy with the fire, the others built a small house (*tur-kim*) nearby for the deceased. The *tur-kim* looks like a miniature version of a normal dwelling house. It is interesting to note that the Anok do not build a copy of a normal Mru house, but of a modest Khumi house, which the southern Chüngma have also adopted as their preferred style of dwelling. As a result, there is little difference between the Anok and Khumi miniature houses for the dead, except that the Khumi ones are distinctly larger. The Dopreng Mru, however – unlike the Anok – build their *tur-kim* (houses of the dead) to look like the *kimma* (front part) of a Mru house. The gabled roof has the distinctive ridge, and slats on the sides. The walls consist of bamboo strips five to ten centimetres wide, woven between upright canes (Rüm and Rümputma-Kua, December 1956).

Preparing the *tur-kim*

The Anok *tur-kim* is about one metre long. The roof projects about 25 cm at the front and back, so the total length is about 1.50 m. The platform is about 25 cm above the ground. The house is just over half a metre wide; the roof ridge is about one metre above the ground, and the top of the side walls is about 30 cm lower than the roof ridge. The house stands on nine vertical stilts (*cüng)*, three at the front, three in the middle, and three at the back. Bamboo poles (*hong*) are tied across the house at floor level in front of the first and middle rows of *cüng* and behind the back row. In Chamklaa-KP, these bamboo canes were left sticking out about 40 cm at both sides of the house. Further canes were tied on to these, parallel to the sides of the house, and the points where they crossed were supported by forked branches stuck in the earth.

Six more canes (*hong-klang)* are placed to the right and to the left of the stilts (*cüng*). These canes project about 30 cm in front of the house. Another layer (*kan*) of closely spaced strips is laid across them, followed by more strips or slats (*chek-rek*) to make a floor (*phai*), which is held in position by slats that cross it a the front and back and in the middle.

The top ends of the stilts (*cüng*) are sharpened to a point, which is used to fix the canes (*pang*) for the roof ridge (*klong*) and the eaves. The upper canes that run across the house (*klöng*), like the lowest layer of the platform (*hong*), are fixed in front of the first and middle rows of stilts and behind the back row. This means that the *klöng* are over the *pang* of the side-walls.

Next, the walls are built, two long walls at the sides, one at the back, and one in the middle. The middle one has an opening for a door towards the left. The front of the house remains open. The back and side walls are richly adorned with bamboo tassels. When the walls are woven according to the old rule from narrow strips, they must be woven in certain patterns. The back wall of a house for a woman is woven in the *lum-chi* pattern. For a man, the *la-ke* pattern is used. For both sexes, the wall in the middle is woven in the *kan-kok* pattern (see Q1b). The patterns correspond to those used for the coffin. The back wall of the house of the dead is the "head end"; the strips are not coloured, but the pattern is made by using white strips from the inside of the bamboo and green strips from the outside. To make the side walls, horizontal strips are woven through the vertical strips as follows:

1) over three vertical strips, under four, over 3, under 2. Repeat.
2) over 3, under 2, over 3, under 4. Repeat.
3) and 4) the same steps, but two columns to the right.
5) and 6) the same steps but two more columns to the right and so on.

This creates a pattern of blocks three strips wide and two strips long, rising diagonally across the wall.

In Chamklaa-KP, people were still making the walls of the *tur-kim* in this traditional pattern. In Tapwüa-Kua, a new and simpler method is used, so the whole process is less time-consuming. The vertical strips are replaced by thin bamboo canes about 10 cm apart. Thin bamboo canes and bamboo strips are woven through the verticals, first 5 to 7 canes at the top and bottom, then a bamboo strip 5 cm wide, then two more canes. They are woven through the vertical canes in the normal way, i.e., over one cane and under the next on the first row, then under and over on the following row. A 10-cm gap is left between the woven part at the top of the wall and that at the bottom; since the upright canes are also 10 cm apart, this forms a row of open squares. Thin, flexible pieces of cane are fixed to the top and bottom edges of the holes, to form overlapping arches that fill each square. The effect is that of a circle inside each square. Finally, a cane is placed horizontally across the middle of the circles.

After the walls are added, the front half of the house remains open. Three or four canes of decreasing length are placed in the gable area in front of the *cüng* over the front *klöng*. The back wall projects about 3 cm beyond the side walls (Tapwüa-Kua, 30.09.1956).

When the *tur-kim* is finished (see Q1m), a roof is put over it. Bamboo slats about 1.80 m long are woven together using bamboo strips. This layer is covered with bamboo leaves, then a second layer of woven slats and strips is added. The whole structure is held together by bamboo canes placed over and under the layers and tied together using strips threaded through them. The roof is made on the ground, then placed over the house forming an arch that is attached to the eaves at both sides. In Chamklaa-KP, the eaves then were fastened by means of bamboo strips to the projecting sides of the floor.

The front platform is added last. In Chamklaa-KP, it was a modest affair made of thin canes. In Tapwüa-Kua, it was woven from bamboo canes in the manner of the shutters of a normal house, and laid on the projecting ends of the bamboo canes that form the floor (*hong-klang*). To make a "tree-ladder" (*dong*), a single piece of bamboo is placed at an angle against the platform, near the front door. The ends of all the unsplit bamboos (*hong, pang,* etc.) are decorated with a simple stripe pattern (*thau-rin-bong*), which is made by scraping the outer skin from the bamboo.

Further decorations are set around the house, namely, tassel poles about 3 metres high, and "flags" that are even taller. The tassels are not bushy, but consist of thin shavings made by scraping upwards along the two top internodes of the poles. The "flags" are flexible bamboo poles more than two metres high, with strips of cloth hanging at the top. A piece of old cloth (*pahaa*) is rolled up and cut across one end at an angle, then the cloth is folded over and cut at an angle once more. When it is unfolded, there should be a row of connected diamond shapes; if it does not quite work out, there will be strips of cloth cut at an angle at the sides.

Two rows of three tassel poles are set into the overhanging roof of the house on the two long sides. A seventh, thin, tasselled bamboo two internodes long, with extra *chit-wai* pushed into the internodes, and an eighth one about 2 m long, with its top internode scraped white on the outside and split to form a funnel shape, are placed at the left front corner of the house and tied to the projecting *pang*. Three flags are pushed into the roof along the front and back of the house. This means that a tassel stick and a flag are placed together at each corner. A seventh flag is erected next to the *taham.*

The houses for the dead were not so richly decorated in Tapwüa-Kua as in Chamklaa-KP. In Tapwüa-Kua, the best preserved house that I found had only the remnants of a single tassel stick about 2 m long in the middle of the back wall. Its top internode was splintered to make a funnel shape, and the next internode was scraped up towards the node and adorned with further *chit-wai*, in the manner of a *caroca* (see P5b). I found another variant on 3.04.1957 in Borpara, a hamlet near Tapwüa-Kua. Six *caroca* stood round the *tur-kim.* There were no flags, but by the tree-ladder was a *dong-lu-kau* (a mast used for cattle feasts) bearing *laa-ui* ("cotton fruits"). In these respects it resembled the house of a Khumi feast-giver (see R5c).

Preparing the *tur-kim* with tassels and a red *be* for a woman

Q1m) The final ceremonies

Before the roof is put on the *tur-kim*, everything must be in place inside it. The back room has a round hearth (a *taping,* without hearth stones). A cloth "cradle" (*be*) is hung between the back of the roof and the central roof ridge support. Bone remnants are collected in another piece of the same cloth from which the flags were made; in Chamklaa-KP, this was a kind of cape, red with yellow flowers that had been bought in the market. There is no rule about how the bones are divided between the *taham* (see Q1k) and the *tur-kim*, but pieces of bone from the skull should be left in both places. In addition to the bundle of bones (*hut-cur*), if a woman has been cremated, other items must be placed in the back room; these are bracelets and chains, or remnants of them recovered from the ashes; and a new skirt (*wan-klai*), in which a cut is made. If the deceased was a man, a loincloth (*dong*) is put in the back room instead. Larger articles are placed in the open front area. One of these is a basket (*töm*), which is made on the spot, and resembles a *hom-kho*, though it does not have to be of any particular shape. The others are a drinking bowl (*bia*); a bottle containing some liquor or rice-beer; and some tobacco leaves. A hewing knife for the deceased is put under the *taham,* not in the house. A small *pak-plai-kimca* (sacrificial altar) is built immediately in front of the house on the left. Two *teng* (narrow bamboo canes with forked tails) are placed next to it. When the *caroca* (tassel stick) is in place, everything is ready for dedicating the house.

The *tur-kim* with gifts (pots and baskets)

Soon after the pyre is lit, a pig and a chicken are killed as *ceng-to-ngar*. They are butchered and cooked at the spot where the *tur-kim* will stand. Fresh rice is cooked at the same time. If no animals are killed at this point, some meat from the pig and the chicken killed for the first *ceng-to-ngar* (see Q1a) is kept for the ceremony. The deceased receives a small portion of all the foods (see below); the rest is eaten by the persons present at the cremation, though the meat from the chicken may only be eaten by members of the dead person's sib (MK, 8.09.1956). In Chamklaa-KP, a dog was taken to the pyre, killed, placed briefly on some of the burning embers, then dragged away. The dog (*tiama-tüng-mi,* see Qln) is not normally used as food. On the rare occasions when a dog is eaten, none of its meat may be given to the deceased.

When the men start to build the funeral pyre, boys are sent to the stream to catch prawns (*chong*) and small crabs (*röp*), which are cooked in a bamboo pipe (*röng*) to make *chur-ngar*. Small portions of the *chur-ngar*, the *ceng-to-ngar* (see above), and the freshly cooked rice are wrapped in leaves to make six parcels. The first is put into the *tur-kim*, which is then closed and opened again, and the portion thrown away. This *chör-köm* ("offering a mouthful", cf. *köm pot,* P2h etc.) is repeated six times, and constitutes the ceremony by means of which the deceased takes up residence in his new house.

Two earthenware bowls are then placed in the entrance, one containing rice and the other meat. A small portion of cooked rice (*tur-bo-hom*) is also put into the newly-made basket. The baskets in which the articles needed for the cremation were brought from the hamlet are placed on the roof of the house, and small quantities of chopped food are placed in them. A banana tree or a small lemon tree may be planted in front of the house in memory of the dead person. Finally, a small amount of husked but uncooked rice (*rui-khaa-mi,* "scatter-enclosure-rice") is thrown around the house. In areas where the Buddhist influence is stronger, a pot is also placed there containing water that people have poured over their hair beforehand in the house of the deceased (*klai-ri-chak*). This is thought to help the dead person to find his or her way to Buddha (MK, 10.09.1956).

Some knives, cooking spoons and a round beer pot (*yu-kong*) are left in front of the miniature house. A length of bamboo internode is left as a drinking tube, and the drinking measure (*pleu*) is laid on the ground (see Q1a). Before leaving the site of the cremation, all those who took part in the ceremony must drink from the pot and spit (*cho)* the liquid out again. A small piece of *ai-dam* (wild ginger) is skewered on a stick and dipped into the water in the pot. As each person takes liquid in his mouth and spits, his forehead is touched with the ginger. It is believed that if this were not done, he would suffer from headaches afterwards.

On the subject of the miniature houses, Mills writes (1931:521): "The body is burned and the calcined bones are removed and placed in a little house on piles a few yards further back from the stream. To the sides of the house are fastened bamboos festooned with tassels of bamboo shavings, and in the house are put with the bones food, drink, pots, and bits of rag for clothes. All vessels, whether of pottery or brass, are broken."

In my experience, the pots were not broken. In Chamklaa-KP, the only item to be broken was a small phial that had contained oil which the persons present had rubbed into their hair. The phial was cracked by means of a blow from a hewing knife before it was put into the miniature house.

Once the body has been cremated, a new fire can be lit in the home of the deceased. The work at the cremation site usually continues into the afternoon. When the members of the funeral party return to the village and climb the tree-ladder to their houses, those who stayed at home – women do not go to the cremation – throw a small amount of ash over them. This is to prevent evil spirits from entering the house.

In the afternoon, a *cari-yöng* (see P5b) is held in the house of the deceased. The next day there is a *khang* (work ban) that applies to everyone in the village. That day is called *lu-khaa-ni*; the literal meaning of the term, and its significance in this context, are unclear, but it might mean "head-enclosure-day" (cf. Q1d, *tu-long* No. 54).

The miniature house built for the dead person does not receive any further attention. It slowly disintegrates, and its remains can be moved aside if necessary, though some people are afraid to go to the cemetery, especially at night, when it is said that voices can be heard. Rich people may hold another feast soon afterwards in memory of the deceased; this is called a *hom-hü-poi* ("rice-call-feast"). The ceremonies resemble those of a small cattle feast (see R1). An invitation to a meal is addressed to the deceased, while a small amount of meat from the sacrificed animal is thrown down the tree-ladder, together with some rice spirit, some beer, and some sticky rice, coconut, sugar cane and betel (MK, 8.09.1956). Any feast may cause people to remember their dead: at a feast in Rengyong-KP, a woman who was not related to the feast-giver started a loud wailing, calling to her son who had died about 14 days previously, and lamenting because he could not be there to enjoy the festivities (5.04.1956).

Q1n) Notes on a Chüngma funeral

Céline Mouchet kindly sent me the following notes in 2005. They contain some extra information.

"Regarding the death ceremony, I saw one in Thanchi area with some Chüngma Mru; three Mru were playing each 1 *ner*, 1 *tömma*, and 1 *mong*, when the corpse was brought to the cremation place. After the cremation, and before the Mru left the place, they very quickly made 6 pipes, very small, and then 6 men played these pipes with the 3 others playing the *ner*, the *tömma* and the *mong*. The pipes were made of bamboo (taken from the cremation spot), looking like a small flute, I would say. Six men made one flute each very quickly with their dao (I didn't get the time to watch them doing it). Then the men blew into them – only a small sound appeared and it was not really made to play 'music'. After playing the pipes during less than a minute, the men installed them above the door of the small replica of a Mru house. Then, quite quickly, everybody started to leave the place, going back to the village. I was staying a bit behind in order to finish drawing some things and taking notes and pictures, but the men wanted me to leave as well with the group. So we all left towards the village …

During this cremation, there were over 15 men, each one of them knowing exactly what he had to do, and during more than 3 hours they were all busy doing something, like preparing the pyre, building the replica of a small Mru house, etc. It was hard to follow everybody in all their activities. It was also hard to watch, seeing a 4 years girl being cremated; and moreover, to take my notes, coming closer and taking pictures, I had quite a lot of reserve in front of the people crying for their child. A Mru relative of the girl that we knew allowed us to watch the ceremony. Following it, I didn't manage to get straight away an explanation of the things I saw, and when I could ask later on, I got only vague explanations. It was as if they knew exactly what they had to do during the cremation and were executing task after task without any hesitation, but repeating some old customs, the exact meaning of which was mainly forgotten. I think also that the people maybe were feeling not comfortable to talk about the death ritual … I remember once I went with a group of Mru for fishing in the Remakri river. At one stage, I noticed at one place of the river a spot where I could see lot of fishes. I called the Mru to come but, strangely, everybody was completely ignoring me, keeping some distance, not even coming closer to have a look, and continuing to search in other places of the river. I was not understanding this sudden attitude, and a man, seeing my confusion, finally came to me and told me quietly that the cremation place was very close to here and that there were always lots of *chüng-nam* surrounding it, so it was not good to remain here …"

Q1o) The journey of the dead

Hutchinson writes (1909:43) that the Mru "believe in no here-after, and that complete annihilation follows death". However, the Mru funeral ceremonies clearly indicate a belief in life after death. I did not meet anyone who claimed to know exactly what happens, but there seemed to be a general belief that before reaching the next world, the deceased must make a dark, gloomy and "bad" journey. The *tem-cong-mai* (see Q1h) is burned to give him some light. The chick that is burned with the body (see Q1i) is meant to go ahead of him and pick up all the vermin; it is therefore called the *dung-cot-waa* ("vermin-peck-chicken"). A dog is also dispatched with the deceased; it is called the *tiama-tüng-mi* ("way-showing-one"). The dog is not burned, just laid on the embers for a short time and then discarded (see Q1m).

Once the deceased has made the "bad" journey, he comes to a fork in the road. There, someone is waiting to tell him which way he must go. This being is called *cakraa-mang*, which is the Marma name for the lord of heaven. A bad person is sent to a bad place, from which there is no return. A good person goes to a heaven in the sky. There it is clean and beautiful, and even if someone was old and frail when he died, he arrives as a child.

In former times, when there were still *rok-mi* (shamans, see P3a) among the Mru, one of them would accompany the deceased on his journey. When they arrived at the place in the sky, the shaman would take the traveller, who had now become a child again, to its parents. The child would feel ashamed at first, thinking itself to be dirty and smelly, and unable to believe that these people really were its parents. After a while, however, it would settle down and feel at home, and start to play with its new comrades.

A favourite pastime would be to sit in the shade of a big banyan tree and play with sword beans (*ching-kling-ko, Entada scandens).* This game, called *ap ching-kling* ("shooting sword beans") was still commonly played by Mru children at the beginning of the nineteenth century. The rules of the game now seem to have been forgotten, at least around Tapwüa-Kua. The children in heaven are also said to play with bows (*chalang-baa*) and arrows. At one time, the *chalang-baa* was a weapon that the Mru used in battle; it is not used now.

The soul of the dead person does not have to stay in heaven. If it wishes, it can come down to earth again and be reborn as a child (*wang-ru*, see N1d).

Even those who died a "bad death" (see Q2a) go to the home in the sky, and they too have a clean and pleasant place to live. When someone dies and goes up to heaven, the ladder is visible as a rainbow (*char-dong*, see

P5a) rising vertically over the sea (MK and Kangku Catumma, 7.09.1956). This heaven in the sky (*kwang-köi*), where all the souls (*lüm-laa*) live, is called the *ching-raa* ("life-place"). It is uncertain whether anything more than the dead body remains on earth. Cremation grounds (*ceng-rong*) and graves (*dup-raa*, see Q2b) are frequented by many evil spirits, which is why people take protective measures when they return from a cremation, and why they fear burial grounds (Q1m). The spirits (*nam*) are not identified by any particular name, but they are thought to be capable of causing illnesses (see P1i).

It is not impossible that the spirits have some connection with the dead bodies, because the Mru believe that both men and women have seven souls. These seven souls (*lüm-laa*) are like brothers. One of them – the youngest and smallest – is good, but there is another one that is bad. The Mru language does not normally distinguish between singular and plural, so all we know is that "*lüm-laa*" go(es) to the sky. However, it is believed that the person can only be reborn when all seven souls are reunited (MK, 10.09.1956).

Q2 Funerals without *romma*

Q2a) "Bad death"

People who drown, burn to death, fall from a tree, are killed by tigers, or are pierced in a spear trap, die a "bad death" (*char*). Death during childbirth is also considered *char*, though there is a difference between *kak-köng* ("death in receiving": the mother dies while actually giving birth) and *rui-taa* (the mother dies after the umbilical cord has been cut). This second form of "bad death" applies if the mother dies at any time after the child is born and before it reaches the age of three. A. S. Hands (in Mills, 1931:520) seems to imply the same thing: "The Mros do not do puja at the funeral house of a woman who died before her child is three years old." In the case of *rui-taa,* the cause of death can be an ordinary illness, so this form of "bad death" is not feared as much as *kak-köng*.

Whenever a *char* takes place, the local people take special measures to protect themselves. All hamlets that receive the message impose *khang* for a day, that is, the villagers do only such work as is unavoidable. The hamlet may be closed to outsiders, though this is not regarded as essential.

When any death takes place, the relatives (*tutma* and *pen)* must be invited immediately. In the case of a *char,* however, when they reach the entrance to the village (*plon)* they must take in their mouths a bitter berry (*kür-khaa*), chew it a little, and spit it out again (*cho*). This *kür-khaa-cho* must be repeated when someone returns home from a hamlet where there has been a *char*, and he must also perform it before being allowed to pass through any other hamlet that lies on his way (MK, 8.09.1956).

The coffin (*pang)* is made in the usual manner, but all the bamboo parts must be white. The mats that are draped around it are woven not in a pattern (*bong*), but in a simple *klik-chumma* weave (over three strands, under the next three). A pig is killed; this is the *met cing* (cf. P4x). A few drops of its blood are sprinkled into the coffin, but the pig itself must not be eaten. Initially, no meat of any kind may be consumed (see below). The pig's carcass is therefore discarded outside the village. No *hom-kho* (rice basket, see Q1a) is prepared; the rice is cooked in a pot, which is then placed near the head of the deceased (*laluk-hom*). The death dance is not performed (Khamcöng Khongtör, 10.09.1956).

It is said that the body of someone who has died a "bad death" decomposes particularly quickly. Nevertheless, it must be kept until all the relatives who can be reached have seen the dead person once more. All members of the deceased's sib who are aware of the *char* must keep a *kan-khang* (ban on side dishes) until the body has been burned. They may eat only cooked rice, without spices, and they may drink nothing but water. Outsiders may not eat anything at all in a house where a *char* has occurred, although they are not forbidden to enter it. This restriction applies to everyone until the next new moon: until then, there is a state of *laa chür* ("moon dirty"). After the new moon, a state of *laa chai* ("moon clean") prevails, and there is nothing more to fear (MK, 8.09.1956).

When the coffin is carried to the cremation ground (*ceng-rong*), a *hom-chok* (see Qlh) is performed at the village exit. *Kür-khaa* berries are used instead of meat. No animals are killed during the cremation. *Taham* and *tur-kim* (see Q1k–1) are built, but the wall of the miniature house for the dead is woven as for a *bok* (field hut), that is, in the Dopreng manner (see Q1l). If a pregnant woman dies, a plant (*kir-kum*), identified as *Curcuma zeodaria*, must be set in front of the *tur-kim*; this is not done under any other circumstances (MK, 15.04.1957). On their return, the men perform another *kür-khaa-cho* at the village entrance. When they arrive home, a second *met-cing* is killed and discarded. Until fairly recently, it would have been eaten, but only by members of the sib.

A third pig is killed on an old mat (*bai-klep)* on which lie two kinds of flowers (*chong-prem* and *chapan*)*,* pieces of two types of pumpkin (*chang-pho* and *kan-pen*), and a cucumber (*ngui*). This pig is the *met-ngat* (*ngat* = to cook thoroughly). Parts of the offal are boiled in a *pak-kia-pak-chak-röng* ("pig-kidney-pig-heart-tube") and poured out with an invocation (*phyöi-tamma*) on to the old mat. If these two pigs were not sacrificed, any visitors would also would die a "bad death".

After the sacrifice of the *met-ngat*, the relatives are permitted to eat meat again, and to wear flowers in their hair, which is forbidden up to this point in the funeral. The pig is cooked, and eaten by all those present. In the afternoon of the day of the cremation, a *pak-tan* (see P5d) is carried out. This is a pig sacrifice that in theory should be made at the end of every funeral, but in fact it is normally replaced by a *cari-yöng* (Khamcöng Khongtör, 10.09.1956).

If someone dies a *char* death, *cham-kam* and *wak-pong-dong* (see Q1f) are not paid. Any part of a dowry still owing is cancelled (dowry = *mang-tang*, see M4e). If the husband were to demand payment, he himself would die a "bad death" (MK, 10.09.1956). If a married woman dies a "bad death", however, her husband must make a special sacrifice of a pig, called the *char-met*, for her relatives (*tutma*). Until this is done, the *tutma* may not eat chicken in the house of a member of the husband's sib, although the sib relationship continues (MK, 30.11.1956).

Q2b) Burial

The only reference to burial amongst the Mru is to be found in Lewin: The "Mru bury their dead" (1870:237). It seems safe to assume, however, that "burn" was intended, not "bury" (see Q4a). Bodies are buried occasionally, but in theory this only happens if a child dies under the age of three months, in which case *pöng ching pöng kau döi* applies ("wood and bamboo not to be used"). If a child dies over the age of three months, the same rules apply as for adults, so the body can be burned with the full ceremonies (*romma)*. In practice, however, the bodies of children up to the age of two are normally buried. Burial (*dup*) is also usual during epidemics. Death from cholera or smallpox is not regarded as *char* ("bad death"), but the people who remain healthy tend to leave and found a new hamlet, and their relatives will stay away from the old one. Many helpers are needed to carry out a full *romma*, and burial is much simpler. As in the case of a "bad death", the relatives of a married woman do not press for payment of any part of the marriage settlement that is still outstanding.

Burial place with basket and rain-coat near Mangking-Para

If it proves possible to cremate a victim of an epidemic in spite of all the practical difficulties, no animals are sacrificed. Even if they were, other people would not be allowed to eat the meat.

Adults who die a normal death are seldom buried; this would only happen if the relatives were too poor to afford the full funeral ceremony. In fact it does not usually come to that, because relatives and the other villagers will normally help, and will not expect payment. The *tutma* may also help, because their consent is needed for a burial (see Q1f). If they insist on a *romma,* they commit themselves to paying the costs and providing assistance.

If a body is buried, it is simply wrapped in cloth; there is no coffin. As in the case of a *char*, no animals are killed and no *hom-kho* (rice basket) is prepared; all that is necessary is to place a pot of rice near the head of the deceased (*laluk-hom*).

The grave (*kwam)* is dug with a hoe (*tim*). The body is lowered into the grave, and buried together with some clothes, and any jewellery that the person might have possessed. A number of items are placed on the grave and left there. These are the *la-luk-hom-pu* (the rice pot that was placed near the head of the deceased) with a cover (*hom-pu-khüm-bia,* "rice-pot-cover-bowl"); the *tui-chaa-pu* ("water-hot-pot") in which water was heated for washing the body; the calabashes (*tui-yia*) used for pouring the water into the pots; the pair of fire tongs (*kep*) used for moving the hot pots; the hewing knife (*charai*) used to cut the materials for the fences around the grave; the hoe (*tim*) used to dig the grave; the basket (*töm*), used to carry the tools to the grave, together with its carrying strap (*nam*); and the carrying pole (*kot-long*) used to move the body. None of these things may be used again, but they are not destroyed (MK, 10.09.1956).

Two rows of *kung-kaa* (bamboo sticks bent into arches) are set in an oval around the grave. The outer row of arches is a little lower than the inner one; it stands about 40 cm high. On the burial ground at Mangking-Para (20.06.1956) I saw that on two graves, a fence had been added between the outer and inner rows of *kung-kaa.* It was made of horizontal bamboo slats and stood about a metre high; it would have kept out wandering animals such as pigs. In addition to the pots that stood at the head of the grave, an assortment of blankets, loincloths and smaller baskets lay on small mounds. Large carrying baskets and a woven raincoat hung on the poles by the gate.

Q3 Funeral customs amongst the Khumi of the Chittagong Hill Tracts

Q3a) My own observations of Khumi and Longhu funerals

Like the Mru, the Khumi and the Longhu (or southern Chüngma Mru) dispose of bodies by burning them. A bamboo coffin is made; the mats, however, unlike those of the Anok Mru, are not woven to any particular pattern (*bong*). A pig, which the Khumi call the *tang-kadam,* is killed and a few drops of its blood are sprinkled on the coffin. A chicken is also killed and some rice is cooked. On the first day, the body is offered one meal of chicken and rice; on the following days, it is given three meals of chicken. If the family are too poor to do this, the chicken can be replaced by an egg. The drums and plate gongs are beaten every evening.

The house for the dead person is made on the day before the cremation; unlike the small *tur-kim* built by the Mru, it is big enough for a living person to use. Before it is finished, another pig and a dog (*tur-ca met* in Longhu) are killed in the family house. The Khumi do not follow the customs of putting a bundle under the roof, playing *tu*, using particular palm leaves such as *tür-ram* and *ting-ai-ram,* and performing a *döm-krong* (see Q1g).

On the morning of the cremation, another chicken is killed. At the entrance to the hamlet, one of its legs is put into a bamboo funnel with some cooked rice. Some cuts are made in a winnow and it is thrown away, also at the hamlet entrance. The drum and plate gongs are played on the way to the cremation ground. No animals are killed there.

A *taham* (see Q1k) is set over the ashes, but nothing is put on it. Pieces of the bones are placed in the house made for the dead person. Three days later, the remaining bones are collected and are also brought there. A hewing knife, a hoe, some clothes and jewellery, and all other items that are needed for daily life are left behind for the deceased.

On the first day, while the body is still in the family house, it is offered rice and meat from the first pig to be sacrificed. The meal is proffered just once, not repeatedly. Later, the body will be offered a little rice with prawns every third day. This continues until the beginning of the next harvest; the deceased is not offered any more food after that.

Unlike the Mru, the Khumi do not perform ceremonial spitting (the Mru *cho*) when they leave the cremation ground, to keep evil spirits at bay. However, when they arrive at home, a small amount of ash is thrown on to their heads while they climb the stairs. They have no ceremony corresponding to the Mru *cari-yöng.* At the end of the year – that is, when work stops in the fields – the deceased is honoured one last time with a farewell celebration (Longhu: *tur-wak*). As many people as possible go from the village to the cemetery, taking rice, fish, spiced meat, rice-beer and spirits. No animals are killed there. The celebration consists in a shared meal and drinking, both inside and in front of the house that was made for the dead person (K Müilün Tamchaa, 26.03.1957).

The ceremonies are a little different if the person who has died has given great feasts of merit during his lifetime. If he has fallen on hard times since then, and his relatives are also poor, a pig is sacrificed, otherwise it should be a cow (Tindu-Para, 24.03.1957). If the deceased is a man who has performed a *tang-thau* (see R5g), then as long as the body stays in the house – which can be more than ten days – a "snail fire" (Longhu: *tem-cong-mai,* see Q1h and R6h) is lit at the four corners of the house, and the drum and plate gongs are beaten outside (K Müilün Tamchaa, 26.03.1957). The house of a dead feast-giver must be decorated as it would have been during a feast. If he has given a *tang-iüng*, a roofed sacrificial area must be prepared, complete with sacrificial post, "rest-house" and "horse" (see R5g). During the time he lies in the house, there is dancing in the house three times a day at meal times, to the music of *tang-plung* (festive mouth organs; see R6j) (Khuchi Patlaica, Wöila-KP, 28.03.1957).

If a married woman dies, five rupees must be paid to her brother as "hair money" (*cham-kam,* see Q1f: the Mru also follow this custom). If it was a "bad death" (Khumi: *cha-di na*), a normal funeral is carried out and the body cremated. This is done not by her own household, but by members of the father-in-law sib (*pakiüng*), who must be given between 50 and 80 rupees and a large pig for doing so. Two more pigs must be killed, one for all the villagers and one for the relatives from other villages. If there is a "bad death", the nearest police station must be informed; if somebody from the police comes, a third pig must be slaughtered and eaten in the village. Finally, a large sow, that is, one that has farrowed at least once, must be killed at the beginning of the new moon. Until this is done, outsiders cannot stay in the village or be fed there. To the Mru, all these slaughtered pigs would count as *met-pak.*

The Awa sibs of Nemlau and Kiaco impose an additional restriction in cases of "bad death". When they return from the cremation, the people of the dead person's household may not enter any other house in the village or receive visitors themselves. This lasts until the appearance of the new moon (H. Renglün Amchang, 26.06.1957).

Q3b) Accounts in the literature

Lewin writes about the Khumi (1870:230): "They burn their dead, first filling the mouth of the corpse with rice and rice-beer. The ashes are afterwards placed in a small hut built near the place of cremation. Here are also deposited the every-day clothes, the eating utensils, and the sleeping mat of the deceased." Hutchinson reports (1909:45): "The dead are burned, and a curious procedure is introduced in the preservation of the calcined remains of the bones. These are collected and wrapped up in a new piece of cloth, and are then placed in a small bamboo house erected for the purpose which partakes of the nature of a family vault, for only the remains of the members of one family may occupy the same house. No repairs are made to this house and a new one is erected when required. Once in every seven days, for a period extending to one year, a full meal is placed in the house containing the ashes. The same ceremony is followed for both sexes. There is no particular way of laying the corpse on the funeral pyre, and no subsequent period of mourning."

My own observations are at odds with many of these statements. The cloth need not be new; the houses for the dead are not family vaults; and some things are in fact done differently depending on whether the deceased is a man or a woman. A new house is built for each dead person, and the deceased is cared for until the new harvest; this could be considered a period of mourning. Other points are also arguable.

In the field notes of A. S. Hands (1928), we find the following passage about "Mro cremation"; the source of the information was Headman Rengkeng (probably Rengkhen) of Renikhyong-Mouza, near Banderban: "Bodies are all burnt in the same place and a small sloping machan is erected over the ashes, and on the machan [= ? below the *taham*] is placed the belongings of the deceased. After that a bamboo house is erected and the remains placed in that. Sometimes a cow or bullock is sacrificed at the house and subsequently an offering of cooked rice is put near the house once a month for twelve months. No particular date in the month is chosen for this. The houses are all arranged in a group a short distance away from the cremation ground. There was one house less elaborate and separate from all others, which I was told contained the remains of a woman who had died before her child was three years old."

The mention of "sometimes" sacrificing a cow or a bullock probably relates to the Khumi; the bringing of food reported here is also untypical of the Mru. Food is put out not for twelve months, but until the end of the year, that is, until the new harvest. The other details as given apply both to the Mru and to the Khumi.

Q4 Death customs in Arakan

Q4a) Early reports on burial grounds and cremation

An unusual note on the Mru of Arakan is to be found in Tickell (1854:99): "These people burn their dead, and place the bones and ashes in little toy-houses adorned with flags; carefully removing the relics from place to place, as they shift their habitations. Two forked sticks are placed in honour of the Nat, or tutelary spirit of the spot. These sticks, after elaborate carving and painting with black marks, are sanctified by sacrificing buffaloes to them, and then placed in the spot to consecrate it for their dead: the Nat being supposed to take up her abode in the tree against which they are propped. A little stream ran under the spot and a steep wooded hill rose immediately behind it." On p. 103 of his report, in a description of a village, Tickell says: "I remarked several of the sacrificing posts before described."

This report, the content of which is not supported in other sources, may be based on a custom I noticed amongst the Khumi and the Longhu (south

Chüngma Mru) of the Chittagong area. If the deceased was a great feast-giver, replicas are made of the festive decorations that adorn a house during feasts. Cattle may also be slaughtered (see Q3a). The greatest feasts of merit are no longer held, but until the beginning of the 20th century, the painted forked posts were also erected for the funeral (see R7a). The statement that people take their "relics" with them when they move to set up a new hamlet suggests a close bond with the dead that is not in keeping with the present attitudes of the Mru of the Chittagong Hills.

Stilson (1864:218) writes as follows about a visit to a cremation ground of the Khami: "By a short and winding steep ascent through the tall grass, we soon gained an eminence overlooking the stream, where we found the ashes and bleaching bones of many human bodies. Near the ashes of several we observed neatly constructed dwellings in miniature, resembling in form those in which the people reside. In these were placed the identical implements used by the deceased in their industrial pursuits, such as the heavy knife, the spinning wheel or loom, etc. By the side of each miniature house was suspended a basket-like cage, in which was placed a fowl, with a little rice for its food. Only one of these animals did we find alive; the rest had all starved in their cages. The main idea we gather from these relics, as confirmed by the people's testimony, is, that the deceased go to a place where they will need to use these implements, and will likewise need food; hence the fowl is provided for the sustenance of the departed one."

The locating of the burial ground near a watercourse is repeatedly confirmed. There is also general agreement that the people provide their dead with the implements needed for work. It seems strange, however, that the sacrificial chickens were not killed. One wonders whether they were meant only as food for the dead, or whether they had some other significance.

Details of the death customs of the Khumi may be found in St. Andrew St. John (1872:240), who writes: "When a person dies the body is laid out in the house, and a feast made; food is set apart for the ghost, which is supposed to remain over the house as long as the body is there. Seven packets of rice for a man, and six for a woman, are left at the place of cremation for the ghost to eat, and neglect of this custom is a bar to inheritance." "When a person dies, and has been burnt, the ashes are collected and placed in a small house in the forest, together with his spear or gun, which has first been broken in pieces. These small houses are generally placed in groups near a village, and are sometimes large enough to be mistaken for one. After harvest the whole of the deceased's relatives cook various kinds of food, and take them, with pots of 'a-moo', or liquor made from fermented rice, to the village of the departed; the doors of the houses are opened and food having been placed inside, are then re-closed; the relatives then weep, eat, drink, and return home." This ceremony is called "*Ta-proung Pa-oung*" or "the opening of the house of the dead" (*taprong*: to open, *pa'ong:* burial place), and it is held in honour of the "departed spirits called 'hpa-law'". The custom is not known amongst the Mru. "The Chyaung-thahs [i.e., the Marma] perform this ceremony thrice a year, but it simply consists in setting aside food and drink for the deceased for a short time, and then throwing it away" (St. Andrew St. John 1872:238).

The period for which the body was kept in the house varied according to the importance of the deceased, as it still does. This is confirmed by Hughes

(1881:22): "In the case of the death of a chief the corpse is kept for several days, the length of time varying according to his status."

Q4b) The Mru

According to a report from the Ponnagyun Hills, "... when a Mro dies a natural death, his body is placed in a coffin made of split bamboos woven into a basket as long as the length of the body. Then a pig is killed and its blood is poured over the coffin. The pork is offered to the visitors. The dead body is then carried to and burnt at the grave. The pieces of bone, which remain unburnt, are gathered and placed on a platform in the cemetery with *khaung* [rice-beer] and other eatables for the deceased to drink and eat" (U Ohn Pe 1931:260).

This "platform" probably refers to that of the miniature house for the dead person, as is indicated in the next and more detailed report: "When a person dies, a bamboo-coffin is made immediately. A fowl is killed and boiled. Rice is also boiled. The boiled rice and fowl are put in a plate which is then placed near the dead man with the saying 'eat'. A pig is killed and those who made the coffin touch the blood with their hands. The idea is to ward off ill-luck which might come on to them as the result of making the coffin. The pork is then cooked and eaten by those present. The corpse is usually kept for one day. Rich or well-to-do persons, however, sometimes keep the corpse in the house for 8 or 9 days during which feasting and drinking 'khaung' are done. Early in the morning one day the corpse is carried to the place of cremation. While the corpse is being slowly burnt, a small house for keeping the bone-ash is constructed. When the dead body has been burned down to ashes, the bone-ash is collected, placed on the blanket which has been used to cover the corpse and put on the 'bone-house'. The plate containing the rice and curry, the water-pot, the pots for cooking rice and curry are all brought and kept in the bone-house. The people then come away. On arrival in the village, a young pig is killed and the people who have taken part in the cremation of the dead body touch the blood with their hands. No food is taken to the bone-house afterwards as in the case of Kami (Ayaing)" (U Ba Myaing 1934:136). Washing with blood has not been reported amongst the Mru of the Chittagong Hill Tracts. However, it seems most likely that this Arakanese final ceremony is also a kind of *cari-yöng* (see Q1m, P5b).

U Ba Thin (1931:250), writing about the Saingdin area, reports: "All three tribes [Mru, Areng Khumi and Awa Khami] cremate the corpses of those who have died of ordinary diseases and, after the burning ceremony, they collect the bones and ashes and store them up in a small hut over the burning place, but certain clans of the Mro build the small hut, not over the burning place, but at the village landing-stage and store the bones and ashes for about two or three months. After this period, the bones and ashes are carried back to the original burning place where a fresh hut is erected for keeping the bones, etc."

I wonder whether there could be some connection between this and the habit described by Tickell of taking the bones when moving to live in another place (see Q4a). The custom described here is particularly remarkable because in the Chittagong and Ponnagyun Hills, links with the dead are broken immediately after the day of the cremation. The end of the year is mentioned repeatedly in the *tu-long* of the Anok Mru (see Q1d), but it has only practical significance amongst the Khumi.

Q4c) The Areng (Khumi)

Writing about the Ponnagyun Hills, U Ohn Pe (1931:262) reports: "When a person dies a natural death, the body is bathed with cold water and kept on a mat after being covered with clothes. It is kept in this way for at least 5 days, and at the most 8 days, according to the means of the householder to entertain visitors. No action is taken to prevent the dead body from becoming putrid. During the above mentioned days, pigs, cows and fowls are killed and offered with *khaung* [rice-beer] to the visitors. On the funeral day a coffin is made from split bamboos, into which the corpse is put. On that day the brother of the deceased, if a woman, asks his brother-in-law to give him compensation, for his sister's hair will be lost by being burnt at the cemetery. At least Rs. 5 and at most Rs. 15 is then given by the brother-in-law. No such demand is made in the case of the death of any other relation. New clothes are put into the coffin on the funeral day so that the deceased may wear them. The dead body is then taken to the cemetery and burnt there. The unburnt bones are gathered and kept in a small hut built for the purpose at the cemetery."

The payment of the hair money is general throughout the Chittagong Hill Tracts. The only real difference from the customs there is that the body is put in the coffin on a different day. This is confirmed by a second source, U Ba Myaing (1934:135–36):

"When a person dies, the dead body is put on a blanket under which a bamboo mat has been placed. Two or three blankets are then put over the dead body which is kept in the middle of the hall. The next day a bamboo coffin – like a basket – is made and placed on the shelf near the roof of the house. A fowl or pig or cow, according to the means of the house is killed. Some blood is put on the dead body. The fowl or pork is cooked and eaten by all. The next day a small house is built in the jungle not far from the place where the body is to be cremated. On this day, a pig and a dog are killed and the people who build the small house which is to be used later for keeping the bone-ash are fed with the pork curry and the curry of the dog's flesh. In the evening the 'khaung' is freely drunk and the visitors are fed with all kinds of curry such as can be obtained. This drinking and feasting will continue for at least five days but never more than 10 days. On the last day early in the morning the dead body is placed in the coffin brought down from the shelf and the dead body is taken to the place of cremation and cremated. When the dead body has been well burnt down to ashes, the bone-ash is collected and placed in a piece of blanket and put on the small house. The boiled rice which is brought in a cooking pot is taken out and placed in an earthen plate which is then put on the small house near the bone-ash with the word 'eat'. All then return home. No one goes to the graveyard a day after the cremation. On the morning of the second day some rice and curry is put on a leaf and placed on the earthen plate at the small house. Water in a small pot is also kept there. The man returns home. No one goes there in the evening. On the 6th day in case of a woman, and on the 7th day in case of a man, the morning meal is again placed at the small house.

Later on, once a month, a morning meal is placed at the small house until the time arrives when people start enjoying the vegetables grown in the

Taungya [swidden], i.e., until the month of *Waso* [July]. After this month no morning meal is taken to the small house until the paddy in the Taungya has ripened. At this time a pig is killed and cooked. Taking 'khaung', pork curry, and rice the people proceed to the bone-house. Some rice and curry together with a bottle of 'khaung' are placed there with the saying 'eat and drink'. The people enjoy their meal, drink the 'khaung' and return home.

It is to be noted that the members of the house where death has visited do not eat salt the next day after the cremation of the dead body. On the second day they eat salt, but no vegetable, fish and meat. They must not travel. On the 11th day after the cremation they may eat vegetable, fish and meat but they may not still go out of their village. On the 12th day they may go anywhere they like."

These rules apparently apply no matter what kind of death has taken place, since in case of a "bad death" (in childbirth) the restrictions are even stricter (see Q4e). It is interesting to note the numbers: 5 (days) for a man and 6 for a woman. I assumed that the Anok Mru had taken these numbers over from the Marma. The Khumi of Arakan – as seen in the report by St. Andrew St. John (Q4a) – replace them with the numbers 7 (packets of rice) for the man and 6 for the woman.

There are no detailed data on the Areng of the Saingdin area. The only information I could locate was U Ba Thin's description of the coffin (1931:250): "All the tribes make the coffin with split coloured bamboos. It is rectangular in shape with a separate lid of the same materials, but the Mro and certain Awa Khami, instead of a bamboo lid, use cotton rugs or blankets."

Q4d) The Awa (Khami)

The only more detailed account is that of U Ba Myaing (1934:136/37): "When a man dies, the dead man's face and feet are washed with warm water. A bamboo coffin is made, a pig killed and cooked. Rice for the dead man is boiled in a separate pot and this rice together with a little of the pork curry is put in an earthen plate and placed near the corpse. Water is also kept there. When the bamboo coffin has been made, the corpse is placed in it, and the people partake of rice and curry. Meantime, a chicken is boiled and put on the coffin. After a while it is taken out and kept on the shelf above the fire-place. If the person who dies be a man, his sister's husband comes with a pig and 'khaung' and together with his friends and companions dances round the corpse with gong, 'Han Ra' [plate gongs] and cymbal for a day. [If 'friends and companions' means 'sib members', this would make it the duty of the *pen* members to perform the death dance for a man.] If the person who dies be a woman, her brother would come and do the same. He would also demand from his brother-in-law the money – usually Rs. 10 – as the price of the sister's hair being burnt. On this day, a fowl is killed. The two wings are tied with a string and cooked together with the remaining flesh. When the curry is cooked the wings are taken out. One wing is placed with rice in a plate and thrown down the house. The boiled chicken from the shelf [the rack over the fire place] is taken down and kept in [?'on', see below] the coffin. The people then have their meals. When all have had their food, the boiled chicken is again taken up and placed on the shelf above the fireplace. The next day in the morning a pig is killed and the two joints of the two front legs are taken, tied with a string and cooked with the rest of the pork. One joint as in the case of [the] fowl's wing is put on the coffin with some rice and the other joint with some rice is thrown down the house. The boiled chicken is again taken down from the shelf and kept on the coffin and then taken back and placed on the shelf where [?when] the people finish their meal. In the evening rice is pounded and small cakes of rice are made. Seven cakes are required in the case of a man and six in the case of a woman. In the morning one fowl is killed and the wings are disposed of as in the previous case. The corpse is carried. At every place where it is put down by those carrying it a rice-cake must be left behind. On arrival at the grave-yard the corpse is put on the pyre. The boiled chicken from the shelf is also brought and put there. A special man lights the pyre. Others may light it after him. While the corpse is being burnt the man who first lights the pyre cooks some rice and fish curry – fish is necessary. Some people make a small platform. When the dead body has been burnt to ashes, the ash is collected. Half of the rice and fish curry are put on ash-heap. The platform is brought and placed over it. On the platform the remaining rice and fish curry are put on."

It is not clear whether this platform is a kind of *taham* (see Q1k). There is no mention of a miniature house for the deceased.

U Ohn Pe (1931:258) does not offer any further detail. "The people then come away. As they return each of them present at the funeral rubs their joints on the way with a boiled egg which has been brought from home in the morning. On arrival at home the members of the house where the man died kill a fowl and besmear all the utensils in the house with the blood. A leg of the fowl is roasted in the fire and kept on the side wall of the house."

Q4e) "Bad death" and burial

The first mention of "bad death" appears in St. Andrew St. John. Of the Khumi he writes (1872:240): "The ceremony of 'ya', or taboo, is strictly enforced on the following occasions: ... when any person belonging to the village is killed by a tiger, alligator or other animal, or when any woman of the village dies in child-birth, or when the body of any person who has died as above has been brought into the village, all intercourse with that village is cut off until the appearance of the next new moon; ... when a villager dies by accident, intercourse is forbidden for a day."

According to Hughes (1881:22), people who had died a "bad death" were not cremated: "The explanation of this custom given by the people is that hill lore abounds in legends and stories, tending to prove that when for those who have departed this life by other than natural causes obsequies are performed similar to those which are bestowed on those who receive their quietus in the ordinary course of nature, the rulers of the same family have again fallen victims to an alligator's hug or a tiger's embrace."

It is questionable whether these details are true of all the ethnic groups. In the Census reports of 1931 on the Ponnagyun Hills area, burial following a "bad death" is reported only in connection with the Areng Khumi and the Mru. U Ohn Pe (1931:262) writes: "In the case of death from an unnatural cause, the dead body is buried in the cemetery and never disinterred. In the case of death from inability to give birth or from child-birth, the whole house, including all possessions, is abandoned by relatives and strangers alike. The husband will refrain from eating flesh and vegetables for 40 days and lives in a hut erected specially for the purpose. He will live on only rice and water. He is not allowed to sit with others but is allowed to speak with them."

These Khumi customs appear similar to those of the Mru: "In the case of unnatural death, the body is buried without any food or the killing of a pig ... the husband or wife, as the case may be, will refrain from eating flesh or vegetables for 40 days after abandoning the house and all belongings for ever. He or she, as the case may be, will live on only rice and water in a hut specially built for the purpose. He or she, as the case may be, is not allowed to sit together with others, but may converse with them" (U Ohn Pe 1931:260).

U Ohn Pe does not provide the corresponding information about the Awa (Khami), but U Ba Myaing, who visited the same area ten years earlier (1921), writes (1934:135): "If a woman of the house dies of child-birth, all the utensils and properties in the house together with the house have to be abandoned and a fresh house has to be built and fresh utensils obtained. For this reason, valuable and costly articles are removed before the woman dies. As soon as the woman dies all things in the house cannot be used after the cremation of the dead body. This custom holds good for the Kami (Ayaing), Mros and Kami (Awa)."

The information provided by U Ohn Pe suggests that when U Ba Myaing speaks of "cremation", he may be mistaken. There is no available information about the Saingdin area. In the Chittagong Hills, even those who die a "bad death" are normally cremated. However, there are still strict measures to be taken in case of women who died in childbirth. These details are not mentioned in U Ba Myaing's account.

With regard to rules about food, U Ba Myaing writes (1934:135): "In the case of the Kamis (Ayaing) the husband – if the husband be dead, then the head of the house where the woman lives – refrains from eating salt for 9 days and from eating meat and fish and vegetables for 30 days. The Mros also follow this custom, though a very small percentage of them has dropped off this practice. The Kamis (Awa) have no such custom."

There is only one report on the Awa Khamis of the Ponnagyun Hills about burial practices during epidemics: "In the case of death from contagious disease, the head and the body are buried; after one month the remains are disinterred and burned to ashes" (U Ohn Pe 1931:258). Except for the moving of the bones by some Mru in the Saingdin area (see Q4b), this is the only report of a second burial. The only other comment on the ethnic groups of the Saingdin Hills is to the effect that "... all persons who die of contagious or infectious diseases, particularly small-pox and cholera, are buried immediately and no form of shelter is erected over the grave" (U Ba Thin 1931:250).

The same writer also reports: "There is no practice among any of the tribes of erecting a stone cairn over the grave of the dead, and no custom prevails among them of disposing the dead in trees, cliffs or *machans* [i.e., on platforms]. No tribe separates the head from the body." The only exception to these general rules is found in the work of U Ohn Pe (1931:262). Writing about the Areng Khumi of the Ponnagyun Hills, he says: "In the case of the death of a baby soon after birth, its body is wrapped with a piece of cloth and kept in a bamboo basket hung up in a tree till it decays to nothing."

Amongst the Mru of this area, children under the age of three are also cremated. U Ohn Pe reports a special feature of these cremations (1931:260): "In the case of the death of a young person up to the age of three years, a dog is killed to show the way to the young deceased, and carried in the coffin with the child's body to the grave, where the corpse is burned without the dead body of the dog. The pieces of bone, which remain unburned, are then placed on a platform without *khaung* [rice-beer] but with eatables. The dead body of the dog is thrown away in the cemetery."

The custom of the Anok Mru of providing the dead with a "guide to show the way" (*tiama-tüng-mi,* see Q1n) seems here to be observed only when a child has died.

Q4f) Survival of the soul

The oldest information on Khumi notions of life after death is to be found in the writings of Latter (1846:63): "They reverence also the spirits of the dead; these, they say, at times flit over their ancient haunts, at others wing their way like birds over mountain and vale. The spirits of the good they think ever happy, those of the wicked miserable."

These ideas are not confirmed in later reports. However, they are consistent with the Mru *tu-long* (see Q1d) in which the deceased appears in the shape of different birds. These birds must fly around in the sky because they cannot return to earth.

Apart from Stilson's remarks (see Q4a), there are no more early discussions of life after death, though the question is taken up in census reports. Some groups were thought to have been agnostic: of the Mru in the Ponnagyun Hills, it is reported that "they have no ideas as to the ultimate abode of the dead" (U Ohn Pe 1931:260). With regard to the Areng of the area, the same author (1931:262) writes: "They believe that the deceased disappear for ever and do not return to this earth in a future existence." The Awa "believe that a man, after death, will enjoy the same kind of life in the next existence, but do not know in what form or shape that existence will occur" (U Ohn Pe 1931:258). U Ba Thin reports in slightly more detail from the Saingdin Hills (1931:250): "There is no believe among the Awa Khami regarding transmigration, the soul disappearing after death, but the Ahraing Khami and certain clans of the Mro believe in the transmigration of the souls of the dead into living beings, but certain Mros believe that there is no transmigration and that the souls of the dead disappear after death. There is no believe among any tribe here as to the transmigration of the dead into butterflies or other insects."

Neither U Ba Thin nor U Ohn Pe draws any conclusion from funeral customs about the destiny of the dead. Only U Ba Myaing explains a funeral rite in these terms (1934:137–38): "Kami (Ayaing), Mro, and Kami (Awa) all alike put two quills of a porcupine ... in the coffin. The idea is this: When a man leaves this world he is met on the way by an ogress whose hair is as sharp as the blade of a sword. She always falls into sleep when one searches for lice among her hairs. If the man has nothing with which he can touch the hairs, he must touch her hairs with his hands and receiving cuts in his hands he will bleed and become exhausted. When the ogress wakes up she eats him up. On the other hand if he has these 2 quills of a porcupine he can touch her hairs with them and when the ogress goes to sleep he can flee after pushing her down and thus reach the happy abode."

Clearly this ogress has nothing to do with the judge of men identified by the Anok Mru as the lord of heaven, and called by the Marma term *cakraa-mang*. However, there may be some connection with the ogre as a result of whose death the first great feasts of merit were introduced amongst the Mru (see R8a).

R Cattle feasts

R1 The lesser Anok feasts

R1a) Types of feasts

The term "cattle feasts" here means all feasts that centre upon the ceremonial killing of cattle. These may be domestic cattle (*cia*), which are of two kinds, *mang-chang* and *ratca* (see G1a); they can also be gayal (*cia-nöm*) or buffalo (*naa*). In principle, there is no reason why cattle cannot be slaughtered simply for meat, like pigs. In practice, however, this almost never happens, because the people are loath to miss the chance of a small feast, and tend to hold one even when a domestic animal has to be put down for some reason.

For ceremonial killings, the proper instruments must be used. A pig must be killed with either a bamboo dart or a spear; cattle must be killed with either a spear or a hewing knife. When a pig is sacrificed, a spear can be substituted for the dart without necessitating any change in the ceremonies, since in either case, the pig is killed by stabbing. When cattle are sacrificed, however, the manner of death is important because stabbing with a spear is part of one kind of feast, and cutting the throat with a hewing knife is part of another kind. There are various differences between the *chot-poi* ("stabbing-feasts") and *ngen-poi* ("[throat-]cutting-feasts"). The *chot-poi* (MK, 26.06.1956) are particularly important.

The lesser cattle feasts are generally called *chot-cia* or *chot-cia-poi* ("stabbing-cattle-[feast]"). The larger cattle and gayal feasts are in fact also *chot-cia*, but they are usually called *caa-cia ning* ("eating-cattle several"), while buffalo feasts are called *caa-naa* ("eating-buffalo"). There are several variants of all these large feasts, each type having its own ceremonial equipment. The different feasts are often called by the name of some piece of equipment rather than the more generic terms *caa-cia ning* and *caa-naa;* thus we have the *riyang* (called after the bamboo rack of same name), the *cang-chuk-calaa* ("making-roof-over") and the *tung-taing-cawaa* ("spreading-out-awning"). The simplest stabbing feast has a name of its own: it is known as the *ling-lang-cia-poi*, or the "go-and-return-cattle-feast", since the animal is tied up and killed on the same day. There is no dancing (see R2) at a feast of this kind; the sequence of events resembles that of the *cari-yöng* (see P5c) (Kangku Catumma feast, 18.04.1957).

There are two reasons why someone might decide to hold a feast. The first is as a sickness offering, when a member of the feast-giver's family is ill; the second is simply the desire to hold a feast, and to acquire a reputation as a feast-giver. However, one cannot assume that large feasts are given simply to enhance the giver's reputation, while small ones are always sickness offerings. A feast may be given for both reasons: a large feast can be held as a sacrifice on behalf of a sick person (*nat-cang*), but of course it will also enhance the feast-giver's standing. On the other hand, even small feasts can be given just for their own sake, and to demonstrate the feast-giver's hospitality.

The reason for giving the feast makes no difference to the actual ceremonies. Guests are almost always allowed, though there are some small villages which close themselves off for special cattle sacrifices. This is not a general rule, but a custom which a particular village may adopt. When it happens, people in the neighbouring villages infer that the people in the closed village are afraid of outsiders coming in and eating too much of their food. The shutdown (*khang*) is considered improper, but outsiders must respect it because anyone who breaks the *khang* must pay a fine (see C1b; MK, 10.09.1956).

If there is sickness in the family but the householder does not have the means to hold a *nat-cang* cattle feast, he can sacrifice a pig, with the promise that he will hold a cattle feast later (*tang-pük*, see P5b). The cattle feast is then held as soon as means permit to give thanks for the sick person's recovery. There is a third reason why someone might wish to hold a feast, namely, to avert misfortune. If a householder acquires wealth – for example, if he has a particularly good harvest – and he then fails to hold a feast, he risks suffering material or social hardship later. One of the functions of the ear-piercing feast (see N3) is to hold off bad luck.

These are only preliminary notes. The reason why the feast is held does not tell us much about it. In the following sections, the feasts themselves will be described, then some interpretation will be attempted. The smaller cattle feasts, the simple *chot-cia,* will be described first. The large feasts, which form a distinct group because they share a number of additional ceremonies, are described in section R3. All the rules for the smaller feasts apply to the larger ones, but not vice versa.

R1b) Preparations

The size of the feast determines the extent of the preparations. If a family has some rice-beer and spirits, or can easily obtain some, a small feast can be organised from one day to the next. The preparations for a big feast, however, can take weeks, or even months. Almost all feasts are held between October and April; a feast is only held outside this period if someone is ill. The local view is that December and February are the main months for feasting; in practice, however, feasts are often left until March or April. I myself attended feasts on the following dates: December 3–4 and 24–25; January 30–31; February 2–3 and 20–21; March 8–10, 15–16, 19–20, 25–26; April 1–2, 5–6, 13, 29–30; July 12–13. Approximate dates for the larger feasts are usually announced soon after the end of the harvest. However, it may be several weeks before the festivities begin. Preparations must be made, and they may take longer than expected. The most important guests may be away at other feasts, or the animals may not be delivered in time. A waxing moon or a full moon is considered most favourable for a feast, but because of the various delays that are possible, most feasts tend to fall during the period when the moon is waning.

The most important preparations are making the rice-beer and spirits, procuring the animals, and inviting the guests. For big feasts, large quantities of rice must be hulled, and this can take days. It must also be done before the beer and spirits can be made.

The sacrificial animals, whether male or female, are almost always bought. It is not forbidden to sacrifice one's own cattle, in fact it is considered especially good. However, the feast-giver and head of the household usually feels too close to the animals he has raised himself to use them for his own feast, unless there is some reason why they have to be killed in any case. There are similar difficulties about buying cattle from other people in the village, or from people who will come to the feast. The cattle to be sacrificed are therefore bought from outside, often from the

Marma or the Bengali. The Mru hardly ever keep buffaloes, so these are usually bought from a dealer (MK, 26.06.1956).

For large feasts, procuring the cattle is less difficult than obtaining the many chickens that are needed to feed the father-in-law's people (*tutma*). The Mru themselves do not keep chickens in large numbers (see G5a), so once again they must depend upon a dealer. In hamlets that are remote from the bazaars of the plains, weeks may pass between the placing of the order and the delivery of the chickens (K Kangnong Kanbeo', 4.03.1956).

An old *lingpu*, one stem has taken root and is regrowing

Before the feast can start, there are festive decorations to be made, which involves various kinds of bamboo work. However, even for a large feast, this rarely takes more than two days, because all the men of the village help. For a small feast, a few hours are usually enough.

At least two kinds of ornamental bamboo poles are needed for any feast: the *dong-lu-kau* ("stairs-head-bamboo") and the *chit-cüng* (tassel-pole). In addition, for each head of cattle to be sacrificed, a post is fetched and erected in the customary place in the village, alongside posts that are left there from previous feasts. The animal to be sacrificed is tethered to the post on the first day of the feast.

This post is called the *ling-pu,* often pronounced *lim-pu.* I was not able to discover the meaning of either part of this term. The *ling-pu* is simply a length of tree trunk about the height of a man; it is not decorated. The species of tree that may be used are: *catunma*, *cangma*, *klo*, *mam* and *um-pang-ching*. For small feasts, there are no special rules about the way in which the *ling-pu* is cut and erected (MK, 15.07.1956; but cf. R3i).

R1c) The *dong-lu-kau*

The *dong-lu-kau* of Yöngtu-HP

The *dong-lu-kau* ("stairs-head-bamboo") is the public sign that the householder is giving a feast. For each gayal, buffalo or head of cattle to be sacrificed, whatever the size of the animal, one *dong-lu-kau* is prepared and set up at the feast-giver's house so that it stands next to the head of the staircase. The *dong-lu-kau* is more than ten metres tall, and can be seen towering above the house from far away. There is little variation in design; the *dong-lu-kau* described below is one that I saw made in Yöngtu-HP (30.01.1956).

First, a support is made by crossing two lengths of bamboo, each about 1.80 metres long, and tying them together. They are held in an upright position by means of two further sloping poles tied to their upper ends. The first two bamboo poles form a fork in which the upper end of the *dong-lu-kau* pole rests.

Preparation of *chit-wai*

The bottom 3.50 m of the *dong-lu-kau* is not decorated. In the next internode, the outer layer of about 20 cm of the pole is scraped off; this section is painted, using soot, with black criss-cross lines that form rhomboids. Alternatively, the 20 cm length is scraped in lengthwise stripes, leaving a pattern of alternating light green and dark green parallel lines (*thau-rin-bong*). The next one and a half metres are left plain. Above this, a similar length is decorated with *chit-pap* (tassel rings), which are made as follows:

Bamboo strips are cut to a length of about one and a half internodes (30–40 cm) and a width of half a centimetre. They are cut just above a node; the inner surface is then scraped with a hewing knife as far as the remaining node, thus forming a tassel 10–20 cm long. The edges of these tassel sticks (*chit-wai*) are then scraped a little on the right and left alternately to make small "tassel buds".

Three strips of bamboo are then twisted together to make a ring about 20 cm in diameter. Between ten and twenty *chit-wai* are put into the ring so that the tassels are just above the ring and hang down over it. The inside surface of each stick faces outwards. The lower ends of the *chit-wai* are tied to the main pole. Nine of these tassel rings are placed along the pole at intervals of 10–20 cm; they look rather like a series of funnels, one above the other.

Extra tassel sticks are placed in the top and bottom rings. These differ from the previous ones in that they are a little more than two internodes long, and each bears two tassels, one under each of their two nodes. Nine of these extra sticks are put into the top ring and four into the bottom one. They are placed so that the second (extra) internode emerges from the ring next to the long tassel and the tassel buds.

For about four metres above the top tassel ring (*chit-pap*), the branches – of which there are two to four at each of nine to eleven nodes – are cut back to a length of about 15–20 cm. Tassel sticks one and a half internodes long are tied to the remaining part of each branch. These sticks have a tassel 15–25 cm long at the end, but no buds on the sides.

Above the branches is attached a decoration called *dam-khai* ("fish-weave"). This decoration is two and a half to three metres high, and is fixed to the main pole so that half of its length projects above it. The "fish" is made as follows:

One starts with a strip of bamboo about 1 cm wide, which will run lengthways through the middle of the decoration. A second strip the same width but only one internode long is placed at right angles over the first strip. About 4 cm to the right of the first strip, the second one is folded so that it faces diagonally downwards, and it is threaded underneath the central strip. The same happens to the left hand end. This whole process is then repeated with more strips, at intervals of about 8 cm. Each one is placed over the central strip and right angles to it. The ends are folded diagonally downwards and threaded under the central strip. The result is shown in the diagram.

The flexible top of the *dong-lu-kau* pole is then bent over and the "fish" is attached to the top by a narrow bamboo strip, and again at a point two internodes below the lower end of the "fish". When the mast is set up, the fish is stretched and points down towards the earth.

A crown (*kruk*) is tied to the upper end. To make this *kruk*, a wheel about 15 cm in diameter is made from a strip of bamboo. Four pairs of small sticks, curled into tassels at opposite ends, are attached to the wheel to make eight spokes. Between the tassels, four small decorations about 4 cm long are suspended from strips of bark. They are bell-shaped, made from bits of banana leaf held in place by small bamboo spikes. The bark strip passes into the bell through the top, and a small cotton ball is attached to the bottom end to make a "clapper". In addition to these four small bells, a large clapper is fixed to the centre of the crown by means of a strip of bark. This large clapper is made from banana leaves and filled with bamboo scrapings (*löng-chur-baa*, ants' nest). The crown is attached to the end of the "fish" by means of another strip of bark fixed at the point where the "spokes" cross. It is heavy enough to pull the strip downwards and to hang freely once the pole is in position.

I only saw a crown made in this way in Yöngtu-HP (north Chüngma area). In Menring-KP (2.02.1956) and Inglai-KP (20.02.1956) they used a different design, but when I saw it, it was already hanging from the completed *dong-lu-kau,* so I cannot describe in detail how it was made. Instead of the *kruk,* there was a ring about 20 cm in diameter, made from a variety of tassels (*klang-ui-chu*, see R1d). Instead of the *löng-chur-baa,* a longer tassel hung down from the middle of the ring. The outer edge of the wheel from which the tassels hung was studded with thick cotton balls.

A second and simpler form was also used in these villages. A cluster of five *klang-ui-chu* tassels was suspended so that one hung down the middle, while the other four were spread apart by means of a ring about 10–15 cm across. Above and below the ring, the strips which held the tassels passed through one thin bamboo cane.

In Menring, when two buffalo and one head of cattle were sacrificed, the villagers made two of the large crowns and one of the smaller ones. In Chikcaron-KP (2.03.1957), where they sacrificed a buffalo and two head of cattle, they made one large crown and two small ones. In Inglai-KP, two head of cattle were sacrificed. An enclosure was built for one of these, but the other was simply tethered. One large crown and one small crown were made. The rule here appears to be that a small crown is made for the animal that is tethered, and a large crown for one that is enclosed. This applies whether the animals are cattle or buffalo.

R1d) The hornbill

The Anok version of the *dong-lu-kau* bears one further decoration, which is not used by the northern Chüngma. The Anok add a bird, the *rang-pangma* or hornbill. This is made from bamboo, and fixed to the highest point of the *dam-khai* ("fish-weave"). Nothing is known about the significance of this bird – or indeed about the meaning of the other decorations – but in the story of the "white bamboo rat" (see R8b), the hornbill appears as a brother of the first feast-giver.

Sometimes there was no hornbill on the *dong-lu-kau.* When I asked why, I was usually told that nobody knew how to make one. I will therefore describe how it was done by a man from Rengyong-KP, who made one for me (10.07.1956):

1. Neck and beak. A strip of bamboo about 50 cm long and 1 cm wide is bent into a loop and the ends placed together. When the loop is held up with the ends pointing forwards like a bird's beak, it does not hang straight down, but bulges out backwards. The bulge forms the bird's breast and belly. The last 6 cm of the ends of the loop are folded forwards and downwards at an angle of 80 °; these ends form the head. The last 2 cm of the ends are tapered along the top edge, sloping the opposite way from the neck fold; they then look like a beak. The ends that now form the neck and the beak are tied together using a strip of bamboo about 3 mm wide; this holds them in position.

2. Back. A second strip of bamboo the same length and width as the first is placed crosswise with its centre on the nape of the bird's neck, i.e., just over the point where the ends of the neck strip are folded. The two ends are folded forward, then threaded back between the two neck strips, without crossing. The last 15 cm or so of each end are split in two; these split ends form the middle tail feathers.

A *dong-lu-kau* with *rang-pangma*

3. Legs. A third strip about 1 cm wide but only 30 cm long is placed across the front of the neck, at the fold in strip No. 1. The ends are threaded under strip No. 2, near the nape of the neck, then folded back down and passed over the outside of strip No. 2 and through the loop made by strip No. 1. They thus form a double layer which makes the neck. A strong, rigid stick about 1 cm wide and 30 cm long, is inserted between the layers and tied with a strip 3 mm wide. The stick protrudes between the legs and is used to fix the bird into the fish weave at the top of the *dong-lu-kau*.

4. Tail. A strip somewhat shorter than Nos. 1 and 2 is passed through the neck, behind No. 1 and under No. 2. Its two ends are folded backwards at an angle of about 60 °, at points 3 cm to the right and 3 cm to the left of the centre. The ends pass backwards over the sides of No. 2, cross in the middle, are threaded under the ends of No. 2, and split lengthways. The split ends of Nos. 2 and 4 form the tail, the ends of No. 2 being the inner feathers and those of No. 4 being the outer feathers. These eight feathers are held in position by a strip about 4 mm wide, which is woven to and fro between them five times. The tail is 14 cm wide.

5. Wings. Another strip like Nos. 1 and 2 is placed across the front of the neck, over No. 2 and just under the place where the ends of No. 1 meet to form the head. This strip is bent towards the back, and its ends cross just above the middle section of No. 4, which runs diagonally, so that they stick out sideways over No. 2 at the point where Nos. 2 and 4 first cross. The ends of No. 5 are then bent downwards and are pulled around the cross, up

again inside Nos. 2 and 4, then backwards in the direction of the neck, parallel to No. 4. No. 5 is crossed back over itself, then immediately threaded under the diagonal middle section of No. 4, folded upwards in front of it, and passed back on the outside at an angle of about 60 °. It fits into the angle formed by No. 4 and passes under it again. This leaves about 10 cm free at each of strip No. 5. These ends are split into four lengthwise. A narrow strip is woven amongst them 8 to 10 times and holds them in position as wings.

6. Crop. A ball about 2 cm in diameter is made by twisting together two strips 3–4 mm wide. This ball is fastened by the end of one of the strips to the front of the neck between No. 2 (below it) and No. 4 (above it). The strip passes between the two ends of No. 1 and is pulled under the point where No. 5 crosses over. The *rang-pangma* is now ready.

R1e) *chit-cüng*

The *dong-lu-kau* (see R1c) is not the only kind of decorative pole. There is a second kind, called *chit-cüng* (tassel pole), which is placed next to the sacrificial post (*ling-pu*). The *chit-cüng* is usually two and a half to three metres high, although in Inglai-KP, poles of this kind could be up to four metres.

Either one or two *chit-cüng* can be placed by each *ling-pu.* Poles of the same, or a somewhat simpler, design are also used to adorn the enclosure (*rum,* see R3e) and the gong stand (see R1f).

The upper end of the *chit-cüng* is half an internode long, and can be decorated with *thau-rin-bong* (longitudinal stripes made by scraping). The next three internodes below it are scraped upwards to make hanging tassels at the upper nodes. Tassels may be either on opposite sides of the pole, or all around it. Additional special tassels, called *klang-ui-chu,* are inserted into three holes cut in the top internode.

A *klang-ui-chu* is made from a bamboo tube one internode in length. It is scraped upwards on all sides to make a tassel at the top node. The rest of the bamboo cane is then cut off, leaving only the node and a thick tassel 20–30 cm long. The node is cut open in the middle and a flexible strip one internode long and about half a centimetre wide, with a node at one end, is inserted. The tassel hangs from the thickened (node) end. The other end is put through one of the holes in the *chit-cüng*.

Three *klang-ui-chu* tassels may also be attached in each of the next two internodes of the *chit-cüng,* in which case the *cüng* will have nine *klang-ui-chu* rather than three. The ends of all the tassels, like those of the *dong-lu-kau,* may be coloured red – or, less frequently, green – with dyes bought from the bazaar (Menring-KP, Kingdöi-KP, Inglai-KP, Tapwüa-Kua). This gives them an even more striking appearance.

In Yöngtu-HP (30.01.1956) I saw a variation on the *klang-ui-chu* on a single *chit-cüng* which stood in the place reserved for the beer pot. In this instance, a fourth set of three *klang-ui-chu* tassels was placed under the three sets just described. This fourth set differed from the others in that about half of the bamboo tube from which the tassel had been scraped was left attached to it. This tube was then fixed to a small strip, which was inserted into the *cüng*.

The *chit-cüng* is similar to the *caro-ca* set up for pig sacrifices (see P5b), but differs from it in that the fatter *klang-ui-chu* tassels are used instead of *chit-wai* (tassels scraped on small bamboo sticks). Unlike the *caro-ca,* the top half internode of the *chit-cüng* is not split.

A *caro-ca* is needed for cattle feasts as well as for pig feasts, but for a cattle feast, it is smaller, being little more than a metre long. It is attached to the front end of the roof ridge of the feast-giver's house. In Inglai-KP, where two head of cattle were sacrificed, two *caro-ca* were placed at each end of the roof ridge. They were all about two and a half internodes long, but one of each pair was a little shorter and protruded outwards a little further. Only the longer one, which stood vertically, was split into a funnel shape. Both kinds had tassels at their two nodes, made by scraping upwards from below. Four simple tassels (*chit-wai*) were then inserted into the lower internode (21.02.1956).

R1f) The gong stand

The gong stand (*mong-töng*) is not an essential item: if the household does not possess a gong (*mong*), and if the feast is a small one, then there is no need to make a stand. However, if the family does possess a gong, then a gong stand is made even for the smallest of feasts. *Chit-cüng* are needed to decorate it.

The *mong-töng* is a gate-like structure. In Yöngtu-HP (30.01.1956) it was made as follows:

Two *chit-cüng* about 1.80 metres high were pushed into the ground so that they stood vertically about 1.80 metres apart. The top half internode of each was decorated with three tassels (*klang-ui-chu*). The next internode down was scraped upwards all around to form a tassel at the upper node, while the following internode was left unadorned. The next one was scraped upwards to form tassels on two sides. The rest of the pole was left plain.

Between these two *chit-cüng,* two bamboo poles were bound horizontally, one above each of the nodes bearing the tassels, i.e., nodes 1 and 3 counting from the top. They were attached to the side of the *cüng* facing the place of sacrifice. The two cross pieces were also decorated with tassels, made by scraping two internodes from the middle out towards the sides. A bamboo pole about 1 metre long was then tied to the middle of the lower bar by means of the usual lashing, so that it stood vertically behind the upper bar. It stood higher than the uprights at the sides, but was decorated in the same way, with three *klang-ui-chu* at the top. Its two remaining internodes were scraped towards the upper node to form tassels. The gong was tied to the middle of the lower bar.

The *mong-töng* in Kingdöi-KP (14.02.1956; feast on 4. and 5.02.1956) differed from the one in Yöngtu-HP only in that there were two of each of the bamboo poles, and a red wooden point (*deng-bong-tong*, see R3c) was fitted into the tops of the vertical poles that stood closer to the feast area. In Inglai-KP (21.02.1956) and in Tapwüa-Kua (13.04.1957), the gong stands were about 1.10 metres narrower, but about 2.50 metres taller. In Tapwüa-Kua, the lower horizontal bar, from which the gong hung, was attached at a height of about 1.80 m and supported from below by two plain bamboo poles, pushed into the ground directly in front of the end *cüng* on the side facing the feast area. There was no vertical pole in the middle. The side poles were made in the same way as the *chit-cüng* at the *ling-pu,* i.e., with three internodes scraped to form tassels, plus three added *klang-ui-chu* tassels. There is one place for beer pots next to the gong stand; more pots can be placed on the opposite side of the feast area. For larger feasts, both these places (see R1h) are decorated with a single *chit-cüng* (Kingdöi-KP, 14.02.1956).

R1g) The beginning of the feast

A useful way to gain an overview of the ceremonies performed at small feasts is to study the *ling-lang-cia-poi,* which last for one day only, except for any time spent dancing. The number of the visitors is not large, so the feast-giver (*poi-yaa*) is able to carry out the ceremonies himself. At larger feasts, many of the tasks are delegated to helpers; since a single observer cannot follow everyone at the same time, I would not have been able to gather all the information.

The next two chapters contain accounts of the observations I made at a *ling-lang-cia-poi* held by Karbari Kangku Catumma in Tapwüa-Kua (13.04.1957), and information I gathered afterwards (14.04.1957). The immediate reason for the feast was that a small cow appeared to be ill, and the karbari thought it best to kill it before it died anyway. This needed to be done quickly, and since the karbari lacked the resources to give a large feast, he dispensed with lengthy preparations and did not invite many people. On the 12.04., the date was fixed for the morrow.

On the morning of the 13th, I arrived at the feast area to find that a tree trunk and bamboo canes had already been fetched. *Ling-pu* posts from previous feasts still stood on the sacrificial area. A hole was dug with a hewing knife close beside the old posts, and a banana leaf was thrown into the hole before the new *ling-pu* was placed in it. Nobody knew the reason for the banana leaf. The animal was tethered by means of a rope round its neck, long enough to allow it to move its head freely, but not long enough to let it lie down. The *dong-lu-kau, chit-cüng* and *mong-töng* (see R1c, e, f) were prepared and set up. A *chit-cüng* was put in the hole together with the *ling-pu;* two bamboo canes about a metre long were tied together, filled with water, and placed upright, leaning against the *ling-pu*. These are called *tui-lam* ("water-bamboo"). All this work was done by the men of the feast-giver's household, together with members of the Atwang and Khongtör sibs in the village.

While the *chit-khai* (tassel decorations) were being prepared, the karbari killed a chicken in his house. This was the *krong-döm-waa* ("earth-descending-chicken", see Q1g) or the *taröng-waa* (for the *taröng,* see R1h). It can be killed by any method.

The chicken was then divided into two halves. One half was cooked as *wa-pan,* with ginger, salt and rice flour, while the other one was cooked as *yong-kang,* with no added ingredients. A *hom-noi pu* was also prepared. This is a "boiled rice-mixture-pot", and contains a little cooked rice, some crushed ginger and turmeric, and some water. Near the *mong-töng* (gong-stand) a *pak-plai-kimca* ("pig-dance-house", see J4c) was built. As is customary, a piece of banana leaf with *mi-tut-ai-dam,* "rice with wild ginger" was placed upon it, and next to this a *hom-noi pu,* and a calabash (*tui-yia*) containing water for filling it. The karbari brought from his house a basket or *em* (see D2a); its contents included eight *teng* (swallow-tailed bamboo canes), but these were set up later, during the *taröng.* A second basket, a *klai-puk* (see D2c), was placed upside down and served as a *pui-tang* (small table), on which were placed an iron arrow (*chia*), some rice wrapped in a banana leaf, and the *krong-döm-waa* (the two halves of the cooked chicken). Finally, a winnowing tray was added; this would later be used as a plate. The iron arrow is not obligatory, and does not serve any practical purpose; however, it is customary to add one, and it is regarded as a good thing to have (K Kangku Catumma, 19.06.1957).

A small part of the wing is taken from the half chicken cooked as a *yong-kang,* and is wiped over the head of the sacrificial animal and the *dong-lu-kau* (stairs-head bamboo). The wing is then thrown away at the *dong-lu-kau.* The rest of the *yong-kang* is left with the *wa-pan* on the *klai-puk*. After the *taröng* ceremony (see R1h), the karbari's grandchildren, Menrau and Langrung, ate the rice and the *wa-pan,* using the winnowing tray as a plate. This meal is called *ngiawüa-hom* ("child-rice"), though any member of the feast-giver's family who wishes to eat some of the chicken may do so. This includes the *charaa* (master of ceremonies), if he is a *tutma* of the feast-giver. The *yong-kang* may be eaten straight away, or it may be kept till evening for the *pang-cang* (see R1i), unless there are plans to sacrifice another chicken for this purpose.

ngiawüa caa hom

R1h) The sacrifice

Before the cow is killed, the *taröng* and *cho* ceremonies must be performed. Three plate gongs (*ner*), a cymbal (*chokca*) and a drum (*tömma*) are used for the *taröng.* The gong (*mong*) must also be struck, though it does not have to be beaten regularly (MK, 26.06.1956).

After a rehearsal in the karbari's house, the players – Netnöng, Menching, Retyaa, Menlong and Nia'ok, who also helped with the preparations – came down to the place of sacrifice. The karbari took a sip of *hom-noi* (rice mixture) from the pot (*hom-noi pu*) and spat a little at each instrument, touching it at the same time with a piece of ginger.

Karbari Kangku touches the instruments

By now, it was about 9.30 a.m. The players started to beat their instruments and circle clockwise around the sacrificial animal. They played in two different tempos, the *ner khöngma* (strong rapid beats) or the *ner acham* (more slowly and gently). They made fixed numbers of circuits (*wir*): first eight *khöng wir,* while rapidly beating (*akhöngma*) the plate gong (*ner);* then four *cham wir* (these are optional); and then eighteen circuits of *khöng.* At large feasts, the first eight *khöng* can be performed on the evening of the first day; *cham* can be played as desired throughout the night; and the last eighteen *khöng* can follow the next morning.

The *taröng*

During the first and second *khöng wir,* the karbari pushed the *teng* (swallowtail bamboos) into the ground, two beside each *pak-plai-kimca* (see R1g), which were at the eastern and western sides of the feast area. He then filled them with *hom-noi.* The *teng* near the *pak-plai-kimca* and *mong-töng* are said to be filled with *kacakwüa-noi* ("men-mixture"), while those on the opposite side are said to be filled with *machiwüa-noi* ("women-mixture"). When there is dancing at night, beer pots are set out in these two places. They are called *kacakwüa-yu,* or *klangwüa-yu* ("young-men-beer") and *machiwüa-yu* or *malaawüa-yu* ("young-women-beer").

During the *taröng,* all the members of the karbari's family came out, followed by the family of his son Mowai, who lived in the same house. Using a small piece of bamboo as a sucking pipe, they all took a mouthful of *hom-noi* from the *hom-noi pu* and spat it at the *ling-pu*. This ceremony is called *cho*.

The family members make their *cho*

On the 13. and 14.07.1956, I attended a feast given by Khamcöng Khongtör in Tapwüa-Kua. His wife was too ill to come out to the feast area, and the two children were still too small. Khamcöng therefore performed four *cho* himself. For each *cho,* he took a mouthful of rice-beer from the beer pot *(kacakwüa-yu)* at the edge of the feast ground. He then spat a little of it to the right and left of the *ling-pu* and on to the *dong-lu-kau*. When he made the fourth *cho*, he left out the *dong-lu-kau*.

While the others ate the *ngiawüa-hom* (see R1g), the karbari now fetched a spear made entirely of iron (*chin re*), though any type of spear would have been permissible. He held it in his right hand, together with a banana leaf in which *mi tut ai-dam* (rice and wild ginger) were wrapped. Every member of the family had to stand on the sacrificial area and grasp the spear. The karbari then went to the right side of the cow and threw *mi tut ai-dam* on to it six times. The sixth time, he threw the banana leaf too. He then uttered a *phyöi-tamma* (invocation of the evil spirits).

The karbari speaks the *tamma*

Up to this point, he had held the spear in his right hand, pointing upwards. Now he raised it, gripped it with both hands, and thrust it into the cow's right side, between its ribs. The cow collapsed immediately. The karbari pulled its tongue out of its mouth, pierced it with the spear, took the hewing knife from his belt, and cut off the front part of the tongue. While he did this, a helper was pouring water over the tongue from the *tui-lam* (these were the bamboo tubes filled with water that had been kept by the *ling-pu*). The karbari smeared a little blood from the spear on to the *ling-pu*. Retyaa used a bamboo awl to nail the tip of the tongue on the top of the *ling-pu*.

The tongue is "nailed" on the *ling-pu*

The cow is pulled away

The rope was then untied from the post and some helpers quickly began to pull the cow away, while Menching and two boys hung on to the tail and tried – without success – to pull it in the opposite direction. They were playing *tön-cia-mai* ("pull-cattle-tail"). We shall meet the cattle tail pullers again when we look at the larger feasts (see R3k).

The cow was killed towards 10 o'clock. It was then dragged up the tree ladder to the karbari's house, and on to the open platform (*car*), where it was cut up. A small amount of dung from the entrails was thrown on to the *ling-pu* and the front wall of the *kimma* of the feast-giver's house.

The Mru, except those from the southern Chüngma, do not save the skulls of sacrificed animals. Particularly large and fine gayal or buffalo heads may be cleaned and dried on the open platform, then taken into the *kim-tom*, the large room of the house. They may be placed by the front or the back wall, or fastened to the wall, or tied to a main post. No further attention is paid to them. Cattle skulls are usually broken up and the bones given to the dogs.

R1i) The last part of the feast

After the animal has been butchered, and before the meal is eaten, the *cak-kwak*, *bong-kom* and *köm-pot* ceremonies (see P3d and 5b) are performed. These ceremonies also feature in pig feasts. During the butchering, the cuts of meat needed for the ceremonies are set aside, and boiled or roasted separately.

In the karbari's house, two beer pots were set out. One was the *bong-kom-yu* ("wrist-tying-beer"); this bore a *cam-pau-pleu,* which is a handsomely decorated sign marking a drink measure (see K2b). When the food is thrown out for the spirits *(cak-kwak),* some beer is siphoned (*üng*) out of this pot for the *bong-kom* ceremonies.

Kham yu in the karbari's house

Shortly before 1 p.m., Karbari Elai Atwang, in his capacity as the feast-giver's *tutma* (see L4a), gave out the *bong-kom* threads to all the members of the feast-giver's household, starting with youngest. Following this ceremony, the *köm-pot* ("spirit-morsels") are distributed amongst various places and pieces of equipment, including the *ling-pu* (the sacrificial post) and other items that had been used on the feast ground.

All the villagers partake of the meal which follows. The men eat first, then the women and children. After the meal, the door to the feast-giver's house is locked, and all the women in the house are fetched one by one to drink from the beer pots. Those who have drunk a *pleu* measure are marked with a line of pot black on their foreheads. Anyone who does not bear this mark is not allowed to leave the room.

Thithop receives the *bong-kom* from Elai

The final ceremonies follow in the evening. These are called *cang-pang-yöng* ("make-invitation-stand") or *pang-cöng* ("invitation-prepare") or *pang-cang* ("-make"). If there was dancing in the afternoon and early evening, the dancers now enter the feast-giver's house (*kim-pün*, "house-climb"), dance for a while in a circle, and then sit down. The *taröng* players go on to the open platform (*car*). If there was no dancing earlier, and if the instruments are still outside, they are now taken back into the house and then brought on to the platform. The *cam-pau-pleu* on the *bong-kom yu* is now cut down, and the *bong-kom yu* is brought on to the platform, as is the *yong-kang*, the chicken cooked without seasoning (see R1g). One by one, any four of the feast helpers take a piece of the *yong-kang* in their right hand, suck rice-beer from the pot, and then throw the chicken pieces and spit the beer on all the instruments, while the players continue to play. They then take a second mouthful of beer and spit it to the left and to the right of the beer pot. Four more *pang-kan* ("hospitality side-dishes") of the same kind are brought. Finally, with the fifth *pang,* the head of the household makes a last *tamma* (invocation), then he too throws the rest of the chicken and spits beer at the instruments, and ends with a *cho* (spitting) to the left and to the right of the beer pot. Water is then poured over the head or neck of everyone who has performed the *cho*. This ceremony is believed to drive away the spirits (*chüng-nam*). The instruments are now put away and may not be played any more until the next feast. At larger feasts, the four helpers are entitled to a payment which varies according to their relationship to the feast-giver (see L3). If they are *tutma* ("father-in-law people"), they receive a hewing knife; if they are *pen* ("son-in-law people"), they receive a turban (Khamcöng Khongtör, 14.07.1956).

R1j) *Ngen-poi*

As mentioned earlier, the *ngen-poi* (slaughtering feasts in which cattle are killed by having their throats cut) are not of especial interest. I did not attend a feast of this kind. No particular invitations are issued, and hardly anyone is informed outside the village, so a stranger would only come upon one by chance.

There is even less preparation for the *ngen-poi* than for the *ling-lang*; nevertheless, I was told that there are three forms of *ngen-poi*, namely, the *cia-poi* (cattle feast), the *naa-poi* (buffalo feast) and the *ngen-poi* (throat-cutting feast). These feasts can be held independently, or alternatively, a *ngen-cia* ("cattle-slaughter") can be part of a large *chot-poi* ("stabbing-feast") such as a *chukcalaa*. In the latter case, however, the animal is not slaughtered on the feast area, but at the watering place (see R3j), during the afternoon before the first *taröng*. For an independent *ngen-poi,* the animal is tethered not to the sacrificial post, but to any convenient tree. Slaughter by throat cutting is not permitted at the *ling-pu*. There is no *dong-lu-kau* (see R1c), no playing of *taröng* (see R1h) and no dancing.

At feasts where the *taröng* is played and danced, slaughter by throat-cutting is not allowed. For a *naa-ngen-poi,* there is no need of a *naa-lu-ngua,* which is a "buffalo-pillow", a small cow (see R3j). A small dog can be killed on a buffalo, but not on cattle (cf. *klan-ria-pak*, R3j). The dog is clubbed on the head; like the *klan-ria-pak,* it must not be killed with a spear or bamboo spear (MK, 26.06.1956).

Although they involve few ceremonies, the *ngen-poi* are not just feasts that are held when an animal is slaughtered. They are also sacrifices, and there must be *cho* (ceremonial spitting) and *tamma* (invocation of the spirits). If *cho* and *tamma* are performed over a cow, it must be killed by stabbing or by having its throat cut. However, cattle can be killed with a spear or by throat-cutting even when the killing is not intended as a sacrifice, so in theory, any weapon – for example a hewing knife, or an iron or bamboo spear – can be used for non-religious slaughter. The Marma (Polika, 26.12.1955) and Twipra (Noliram-RP, 17.01.1957) usually shoot buffaloes with a gun, as do the Bawm, who have been converted by missionaries (Munnuam-Khua, 23.12.1964, on the shooting of a gayal). Amongst the Mru, shooting is not usual, though it is not forbidden as a way of slaughtering animals for consumption. Guns are used to kill chickens that are hard to catch, or pigs that have been reared under the house and are so large and dangerous that there is no other way.

R2 Dance

R2a) The significance of the dancing

Very few *chot-poi* ("stabbing-feasts") are over in one day. Even when there is only one animal to be sacrificed, the feast will normally last two days. The ceremonies are the same as those described in the last section, but the first *ner-khöngma* (see R1h) does not take place until the evening of the first day, and the second *ner-khöngma* and the killing of the animal are left until the next morning. The night of the first day and the whole of the second day are devoted to dancing, which is the real attraction of the cattle feasts. The other ceremonies can pass more or less unnoticed in the general hurly-burly. Eating, drinking, and, most importantly, dancing, are the real business. Dancing is more than a social pastime: it is an integral part of the occasion. There is no dancing at any other time.

The Mru enthusiasm for feasts begins in childhood. *Plai* (dancing) is a normal part of children's games, and in September 1956 in Tapwüa-Kua, my own interest in dancing made *plai* the game of the moment. The older children helped: they set up a *mong-töng* (see R1f) to serve as a gong stand, and they made "cattle" from pieces of banana tree and bamboo sticks, and then tethered them to the *ling-pu*.

After a while, Mowai Catumma started to feel uncomfortable. On 25.09.1956 he declared that if the children did not stop, somebody would have to give a real feast, because all that dancing round the sacrificial post would call up the spirits, who would seize one of their parents. Eventually, a solution was found: the children set up a piece of wood to serve as their own "sacrificial post", well away from the real one. Soon after that, however, they started to lose interest in *plai*. Part of the reason was that I went off on another trip.

This story suggests that if the young people really wanted to dance, they could do so, provided they kept away from the sacrificial posts. I was told, however, that they would be ashamed to do this (MK, 2.09.1956). The young men are fond of meeting in someone's house to "rehearse" with their instruments (*plung*). If they "rehearse" night after night even though no feast is planned in any of the surrounding villages, this exerts a certain social pressure on their elders. The young people badger the adults, who eventually start to look around for someone who might be persuaded to hold a feast. If a particular householder looks as though he might be in a position to do so, he soon comes under pressure from the entire village. Meanwhile, every evening, the pipes sound in his ears; and he might even wonder whether the spirits are hearing them too ...

The social importance of the dancing is more apparent than any religious significance. It is true that the *cam-pua* (see M1b) gives young people from different villages a chance to get to know each other better. For the girls, however, who do not leave their villages for the *cam-pua*, the dance feasts are almost their only chance to visit other villages, and broaden their horizons a little before marriage. Their only other opportunity is when they accompany their family to wedding feasts.

A feast is a much more exciting prospect than young men getting together in the "spinning room". During a feast, the normal weekday routine is overturned, and for the men – more so than for the women, who generally do not drink much – it brings a supply of alcohol, which creates a certain mood, and heightens their sense of adventure. The dancing is the main reason for attending a feast. It is in fact a medium of social communication, the significance of which goes far beyond the dancing itself.

As early as the 1950s, Mru dancing took on an additional function amongst the Mru living around Banderban, the headquarters of the administration. People no longer danced only at cattle feasts. Around Banderban, people now danced at other times, not because they wanted to do so, but under pressure from foreign authorities. Pierre Bessaignet (1957), for example, visited the village of headman Dewai, accompanied by a male member of the Bohmong family. (I was told that Dewai usually spent his time in Banderban, sucking up to the authorities, with whose backing he harassed his own people.) Bessaignet wrote that at the end of their visit "young boys and girls performed a dance in our honour" (1958:54). We ourselves – Dr. Kauffmann and I – saw our first Mru dancers on 12.12.1955 in Banderban, where they danced a few dispirited rounds in front of the Bohmong's palace, in honour of his Punyah day. In January 1956, the Public Relations Officer of the Banderban Circle brought a Mru dance group to Dacca, the capital of East Pakistan, where it was to appear as a "folk dance group". In 1964, there were pictures of Mru dancers in travel brochures for the Chittagong Hill Tracts, and on postcards sold in Chittagong. This did not necessarily mean that the old customs had broken down; these performers may have broken their own rules under outside pressure.

In 1964, the area was closed to foreigners. After it was re-opened in 2000, the Mru were still dancing, as documented by Céline Mouchet.

R2b) The dancers

When the harvest is poor and times are hard, a whole year can pass without any dancing in a neighbourhood. In better times, however, all the larger villages can afford a feast every year, and there are times when one can go straight from one feast to another. In January and February 1956 in the Pantola-Mouza, the following feasts were held: 27.01. and 28.01., Longrau-KP, one cow; 29.01. to 1.02., Yöngtu-HP, two buffaloes and three cattle; 2.02. and 3.02., Menring-KP, one buffalo and two cattle; 4.02. and 5.02., Kingdöi-KP, two buffaloes and one head of cattle.

When so many feasts are held over a short period, there is competition to attract the dancers. The more visitors one has, the greater the honour. Unfortunately, the large feast given by headman Yöngtu attracted no more dancers than an "ordinary" feast. In Menring-KP, however, the feast given by Duitong, the karbari's older brother, drew more dancers than I ever saw at any other feast. Duitong apparently stood higher in the visitors' favour than Yöngtu.

Sometimes more visitors arrive than the feast-giver expected or wanted. On 13.07. and 14.07.1956, Khamcöng Khongtör held a feast that caused him serious hardship and obliged him to borrow money. The feast was held at a time when not a single other feast was planned in the neighbourhood, not even a small one (see R1b). Large numbers of young people came from the neighbouring villages and ate up the small amount that the host had to offer, given that times were hard. Khamcöng received them half laughing, half crying.

The Khumi of Betchora-Mouza had only limited contact with larger Khumi groups, and in consequence they forgot their own dances. They were not yet sufficiently assimilated to the Mru to have taken over the Mru style of dancing. Since they too believed that there is no good feast without dancers, they invited Mru from the neighbouring villages to dance for them (K Lenten Camthang, 17.02.1957).

Young Mru who go to dances do it for their own pleasure, but it benefits the feast-giver too because their coming brings him honour. If only a few dancers turn up at a feast where dancing is to be expected, this brings disgrace on the feast-giver.

Whether the feast is large or small, the feast-giver must give presents to the young people to thank them for coming. If boys from another village come to the feast *cam-hom-plai* ("seeking-gang-dance", that is, they come as a group), and if there are at least three to five of them, without instruments, they must be given a bottle of rice spirits (*plai-hom-arak*, "dance-gang-spirits"). If the boys bring their instruments but come without girls, they are given a pot of rice-beer (*plai-hom-yu*). (They may bring gourd-pipes, *plung*, see E1a; drums, cymbals, and so on are not permitted.)

If they bring girls too, they are given an extra bottle of *plai-hom-arak.* The visitors from each village are rewarded separately: if visitors from another village are not given their arak or beer, the people of that village will bear a long-lasting grudge against the feast-giver and all his own villagers. If dancers from the feast-giver's village subsequently attend a feast in the village whose people were not properly rewarded, these dancers will be treated in kind (MK, 13.09.1956).

Dancing in Menring-KP

The feast-giver obviously cannot accommodate all the young people and other visitors in his own house. They therefore stay with relatives, taking with them any friends who have no relations in the village. This means that the visitors stay in family groups, or even in village groups, and the players of the dance pipes must rely on being able to meet in someone's house, for example the feast- giver's house, or the karbari's, to tune their pipes. They spend the whole afternoon of the first day in these tuning sessions, and they re-tune their instruments before every appearance. The girls spend the time combing their hair and making themselves beautiful. When evening comes, everyone waits for the first eight *taröng* rounds, which are the signal for the dancing to start.

R2c) The rules of the dance

Dancing to the gourd-pipes (*plung-plai*) starts at the end of the first day, usually just before midnight. By 3 a.m. at the latest, the players and the dancers tire, and usually go away and sleep. If the feast is a large and successful one, however, they may keep going until sunrise, then leave to eat and have a short sleep until the dancing starts again between 10:00 and 11:00 on the second day. After a break for lunch, the dancing recommences between 15:00 and 16:00. There is a further break for dinner towards 18:00. The dancing ends between 20:00 and 21:00 with the *kim-pün* (see R2d).

Very few young people take part in all the 12–15 hours of dancing. The young men partake of the alcohol quite freely, and are often sleepy on the second day, so there can be delays while a band is mustered that is fit to play. Some of the permitted spells of dancing may even be left out altogether. The greater the number of the young men present, the better the chances of long bouts of dancing.

Sometimes the *plung-plai* starts straight after the *taröng* (the ceremonial circling, to the music of plate gongs, cymbals and drums). More often, hours elapse before it starts. A kind of hooting sound made on the gourd-pipes (*plung*) announces the arrival of the dancers, who move into position as they approach. They are then ready to start dancing in their circles at the *ling-pu* without having to stop to regroup. The circles always move clockwise, as in the *taröng.* The blowing of the *plung* then becomes rhythmic, and the *taröng* instruments, played in *ner-cham* (see R1h), join in to complete the band.

After a few minutes, the "hooting" starts again. When it does, the regular, rhythmic steps give way to small running steps, until the strong beat starts again about half a minute later.

The boys and the girls dance in two groups facing each other. The girls always move backwards and the boys forwards. The girls dance in a row with the older ones on the left, i.e., closest to the *ling-pu*; the younger ones, who may be less than ten years old, are on the outside. This means that the little ones have the longest distance to cover, so even if they know the dance steps, they usually have to run to stay in the row. The girls hold hands, elbows slightly bent, and clasping each others' fingers (*taprak*).

The boys follow. They move in no particular formation, though the players of small pipes usually go in front, while those with larger pipes follow. The boys who are playing plate gongs, cymbals and drums may follow behind, or go on ahead, or run around amongst the others. They are allowed to play their instruments whenever they like. The players do not need to be especially skilled, but there must be the proper number of them. They change frequently; the children can take turns with the instruments, and the older men, who are often a little drunk, like to relive their dancing days. Men almost never dance without an instrument, and the gourd-pipes are only played by *klangwüa*, or unmarried men. Except for the children, those who do not have pipes must wait until others hand theirs on. In the meantime, they may join in by whistling through their fingers (*rai-wik*).

The girls do have some "instruments" of their own: they use the rattles on their legs to set the beat, and they wear chains with coins that rattle in time with their body movements. The women are subject to the same rule as the men, namely that married people cannot join in the main part of the dancing. However, the older men may join in at any time if they are playing *taröng* instruments, and the married women have their chance on the second day, when they have their own special dance.

The dance for the married women is held at all the larger feasts, and may be held at smaller ones. It takes place during the *plung-plai,* and only the old women take part. They do not join the girls' line. In Menring-KP, they formed a row of their own and danced forwards, while the girls and the women danced backwards behind them. In Rümpöng-KP, they danced backwards in the middle of the whole formation, holding on to the enclosure surrounding the sacrificial animals. They danced slowly, and were overtaken by the other dancers. When older men played the *taröng* instruments during the dance of the old women, they danced facing them, just as the young men dance facing the girls. I felt that the dance of the elders must have some special significance, but if this was the case, I was not able to discover what it was.

R2d) Dance steps

The dancing is in 4/4 time. The first beat is emphasised and is extended slightly, which seems odd at first because it is not accompanied by a dance step. On the second beat, the girls step backwards on one foot, and the boys step forwards. On the third beat, the girls bring the second foot level with the first; the boys do the same on the fourth beat. These steps are then repeated, starting with the other foot. The whole sequence occupies two bars of the music.

The girls bend their knees on every beat, emphasising the first and the fourth. Theoretically, on the second beat, the boys move forwards on to the right foot when the girls move backwards on to the left, and vice versa. In practice, however, this often breaks down because the boys are not in an ordered group, and are not tied to a set sequence of steps. The boys can also step sideways instead of forwards. Those on the inside may be obliged to do so, when there is not enough space in front. They can step to either side, depending on where there is most room. While the girls bend their knees on each step, the boys bow on the forward steps, so that the tubes of the gourd-pipes move up and down. This is the basic sequence of steps for the rhythmic dancing. Variations can be introduced, depending on the skill of the dancers.

Dancing in Tapwüa-Kua

During the periods when the rhythmic playing stops and is replaced by the hooting on the gourd-pipes, the girls abandon their step sequence and move backwards with tiny running steps. When they do this, they cover a larger distance than when they dance the step sequence. They then move forwards towards the boys, who retreat a little. This surging to and fro is repeated several times, until finally the girls run rapidly backwards, with the boys following close in front of them. The youngest girls, who are at the outer end of the line and have the greatest distance to cover, change places with the boys, who would otherwise stride along in the centre of the circle and move forward too quickly.

The rhythmic playing of the *plung* and *taröng* instruments then starts again, the pace slows down, and the girls re-form their row, with the oldest in the middle. The smaller ones eventually find their places again at the outer end of the line. During the hooting on the gourd-pipes, the tall girls in the middle of the circle may jump up and down on both feet instead of running in small steps. This means that the little ones on the outside do not have to move so far.

The girls can vary the rhythmic part of the dance by taking small steps forwards and backwards instead of doing the knee bends. They take four small steps forwards on beats 4 and 1, then four larger steps backwards on beats 2 and 3. The change of direction is accomplished by moving the foot that has just taken a step forwards on the extended first beat back again on beat 2, after a short knee bend. The girls move backwards at some speed. This is too much for the men to keep up their step sequence, so they simply use walking steps (Tapwüa-Kua, 29.05.1956 and 14.07.1956).

The sequence just described can be doubled to eight steps forwards and eight steps back. When the girls take eight steps forward, the boys have to stop and are pushed to one side.

The girls take the initiative in varying the steps. The young men, especially those playing the large gourd-pipes, need to stay in their less vigorous sequence, because if they move too quickly, the wax holding the gourd-pipes together may fall out, creating additional air holes in the instrument. There is always one boy or another standing outside the dance, busy resealing his pipes.

In Nongneng-KP (Rümma area, 12.03.1957) I saw the girls performing a variety of dance steps, including the following:

1) jumping backwards with both feet on every beat.

2) the same, but turning alternately to the left and to the right. Adjacent pairs of girls turn in opposite directions, so that they face each other, then turn their backs to each other.

3) putting one foot in front of the other on every beat, changing foot at each step. This is strenuous and tiring, because they are moving backwards at the same time.

4) one foot stays still, while the other moves forward and strikes the ground three times. On the fourth beat, this foot is drawn back with a knee-bend, and the dancers change feet on a backward jump.

5) on a sequence of four beats, the dancers step back with the left foot, stamp with the right foot, step back with right foot, and stamp with the left foot. This is a slow and leisurely step.

Before the guests go into the feast-giver's house (*kim-pün*), the dancing may end with a circle dance. The big girls at the inner end of the line stop dancing backwards and move suddenly forwards, between the sacrificial post and the boys. They break through the group of gourd players and move towards the younger girls at the outer end of the line. Keeping hold of each others' hands, they form a ring beside the sacrificial post, thus enclosing some of the players. The other young men continue playing outside the ring. The whole thing happens so quickly that the boys really are "caught".

The ring now approaches the feast-giver's house, circling slowly, and breaking up at the staircase so that dancers and players can climb up into the house. The circle is re-established inside. The girls face the centre of the room and form a circle, and as many of the players as possible form a second circle inside the first. As usual, everyone dances clockwise. The remaining players stay outside the circle and are pushed into the corners. There are no dance steps now: the dancers simply stamp as hard as they can, which makes the floor vibrate and the whole house shake. The floor of a good Mru house must be strong enough to withstand this treatment. If the feast-giver has any doubts about his floor, he would be well advised to have it renewed before the feast. Another tactic is to set out a large number of beer pots and assemble a good number of guests in the house. This leaves less space for the young people to indulge their high spirits.

R3 The big feasts of the Anok

R3a) Classification

It is difficult to draw a clear distinction between the big Anok feasts and the lesser ones. The names of the feasts do not provide clear guidance. The *caa-cia-ning* and the *caa-naa* certainly count as big feasts (see R1a); but feasts such as the *chot-cia-poi,* which cannot claim either of these titles, can also be held in the style of a large feast. It is also impossible to make the distinction on the basis of the number of animals sacrificed, because a feast for which two cattle are killed can be held as a large feast or a small one. The reason for holding the feast is also not definitive: if, for example, a feast is held as a sickness offering (*nat-cang*), only one animal is needed, but the feast-giver may sacrifice an additional animal to maintain his good name. Conversely, a feast at which a single animal is sacrificed (*cia-lök*) is not necessarily held as an offering for a sick person. If the feast is a *cia-phir* ("cattle-double"), this indicates that the feast-giver is making the sacrifice not only as a duty, but also as a social event, so guests are always welcome.

The above statements include elements of interpretation, but these were made by the Mru themselves. All that can be said with certainty is that there are circumstances under which at least one head of cattle must always be sacrificed; whether additional animals are sacrificed is at the feast-giver's discretion. If two cattle are sacrificed (*cia-phir*), then the feast is also a higher-level offering to the spirits, and the feast-giver is entitled to include some ceremonies which go beyond those already described. He is not obliged to hold these ceremonies, but he would normally do so, because the *cia-phir* offers greater opportunities of connecting with the spirit world than the usual *nat-cang*.

Sacrificing more than two animals may increase the feast-giver's social prestige, but it does not entitle him to expand the ceremonies any further. Feasts with more than two animals are called *caa-cia-ning* ("big beef meals") or *caa-naa* (buffalo meals).

All feasts at which buffaloes or gayal are sacrificed are prestigious. First, these animals cannot be sacrificed alone: one head of cattle must also be sacrificed, even if the animal is only a small one, as a *lu-ngua* (see R3j). Second, these animals are more expensive than cattle, wich would entitle one to the same ceremonies.

The first time a man gives a feast, he is only permitted to sacrifice cattle, though he can sacrifice as many as he likes. If there are two or more, he is entitled to hold the additional ceremonies. A gayal or buffalo feast may only be given by someone who has previously held at least a *cia-phir*. These are the only rules governing the order in which an individual may hold feasts (MK, 26.06.1956). The amount a man may spend on a *caa-cia-ning* or a *caa-naa* is not affected by these rules. He is therefore free to hold a *riyang* or a *chuk-calaa* (see R3b and d) as a big feast. However, while a *riyang* can be erected without the sacrifice of a buffalo, e.g., for a *cia-phir*, there can be no *chuk-calaa* unless at least one buffalo and two cattle are sacrificed.

A feast can be identified unambiguously by stating both the number of animals offered and the nature of the festive decorations. A *naa-phir riyang*, for example, is a feast at which a *riyang* is erected over two buffaloes plus, of course, the obligatory head of cattle.

R3b) The *riyang*

The *riyang* is a flat, square canopy made of bamboo. When a feast meets certain conditions (see last section), a *riyang* may be erected over the sacrificial posts and the animals. The *riyang* gives its name to the type of feast in which it features.

In the following paragraphs, I describe the structure of a *riyang* that I saw in Menring-KP on 3.02.1956. The terms front, right, and so on are from the point of view of the animals tethered underneath it. The *riyang* consisted of two flat, horizontal bamboo layers held some 40 cm apart by a system of bamboo supports. The lower layer stood about 2.10 m above the ground, so the top layer was about 2.50 m above ground level. The whole structure was about 6 m^2. Each layer was a kind of lattice made from two sets of parallel bamboo poles placed at right angles to each other and fastened together at the points where the poles crossed. The support structures which kept the layers apart were (1) vertical bamboo poles about 60 cm long, bound to the lower layer and fastened with bamboo strips (*mia-kom*) to the outer edges of the upper layer, and (2) four diagonal poles per side. These diagonals connected the corners of the upper layer and the central points of its sides to the quarter and three-quarter points on the sides of the lower layer, forming two shallow V-shapes along each side of the structure. The top layer consisted of 20 poles from front to back and 17 from left to right. The lower layer had 32 poles from front to back, and 30 from left to right. There were 17 short vertical poles at the front, 18 on the left side, 20 on the right side, and 19 at the back. These numbers are all close to 18, which is the number of ritual circuits made around the animal, but no specific number was actually required. Unspecified numbers of short verticals are placed in the middle of the *riyang*.

The short bamboo poles were about two internodes long (½+1+½). They were scraped upwards at the sides to form tassels at each of the two nodes. Three further tassels (*klang-ui-chu*, see R1e) were inserted into the side of the top half internode and a red wooden ornament was pushed into the top.

The wooden tops are the only wood carvings produced by the Mru. They are called *deng-bong-tong* ("banana-fruit-piston"). The part of the wooden top that shows above the bamboo pole is about 8–10 cm long, and consists of a curved cone shape on a shorter and narrower base. Below the base is a plug several centimetres long which is inserted into the opening of the bamboo pole. All the tops are dyed red after carving.

riyang in Kingdöi-KP

The *riyang* is also decorated with so-called *te-pau,* which is the name of a kind of red flower. These decorations are thin bamboo strips scraped to form "little buds"; the whole strip is then dyed red. The strips are hung in overlapping semicircles from the ends of the bamboo poles forming the lower layer of the *riyang*, each strip spanning five poles, i.e., from pole 1 to pole 5, from 2 to 6, from 3 to 7, and so on.

The whole structure rests on four posts placed around the *ling-pu*. The posts are forked at the top; two thick bamboo poles lie across them at the front and back of the area. Two more strong poles are tied to these, just inside the forks, so that they extend from front to back. These four "beams" support the double platform of the *riyang,* which is tied on top of them. They reach almost to its outer edges.

The *riyang* is not made on the ground and then placed on the frame, but built *in situ*. The only preparation is the scraping of the tassels on the individual poles. In Menring-KP, the branches of a sacrificial post had grown into a tree, which had pushed up through the *riyang*. In Kingdöi-KP (seen 14.02.1956), a kind of bamboo tassel umbrella stood in the corresponding position on the *riyang*. In other cases, however, the *riyang* was flat on top (Rümpöng-KP, Nongneng-KP, Chikcharon-KP).

The most elaborate *riyang* I saw was in Kingdöi-KP. Its supporting beams, which lay on the four forked posts, were about 7 cm in diameter. The bamboo poles in the lower layer of the *riyang* were much closer together than in the Menring-KP example – there were more than 40 of them in each direction – and the internodes of every pole were scraped to form thick tassels, not just at some of the nodes, but at all of them.

The distance between the upper and lower layers was about 60 cm. The short vertical bamboo poles were about 80 cm long, and they all had tassels and red wooden tops. There were 22 of these on each side (84 in total, since there was only one at each corner). In the middle stood a second set of poles, about 9 x 9 in all. Many of them had six *klang-ui-chu* tassels rather than three, because two were pushed into each hole. The *te-pau* spanned 7 bamboo poles each, i.e., skipping six, because the poles were closer together. Unlike the *riyang* in Menring-KP, this one had *te-pau* looped around its corners. The edges of the upper layer of the *riyang* were also adorned with *te-pau*.

Instead of the simple red wooden tops, the four corner bamboo poles of the *riyang* had double spiky tops. These were made by inserting large tops, over 20 cm long, into the ends of the poles, then putting smaller ones about 15 cm long on top of them. Four holes were bored round the base of the smaller tops; into these holes were set small wooden slats 5 cm long, which pointed upwards and outwards. These little slats also bore wooden tops 5 cm long.

Four additional "spokes" approximately 15 cm long were also set into the base of each of the largest wooden tops. About 5 cm from their outer end, each bore a crosspiece about 10 cm long, pinned on with a wooden spike. Wooden tops about 8 cm long were inserted into the ends of all the crosspieces and into the first third of the main "spoke". This meant that the four double tops at the corners of the *riyang* each bore l+4 additional small tops, making 20 altogether.

R3c) The *chit-cüng* and the *cakku-thi*

In Kingdöi-KP on 5.02.1956, two white buffaloes and two heads of cattle were sacrificed. To mark this special occasion, two special decorations were made in addition to the *riyang*. One was a bamboo *cakku-thi,* which was erected next to the *mong-töng* (gong stand, see R1f). *Cakku-thi* is a Marma word which means "paper-umbrella"; these "umbrellas" are used by the monks at Buddhist festivities. Secondly, the *riyang* was adorned with two additional decorative poles, which rose above it. Both these poles were called *chit-cüng,* though they were not at all alike, and they bore no resemblance to the usual *chit-cüng* (see R1e). They stood in line with the sacrificial posts (*ling-pu*). The one on the right was a little more than 3 metres high and was made entirely of wood. The one on the left was a bamboo pole 4 metres high, and similar in form to the *cakku-thi*. In the following paragraphs, I shall describe these items as I saw them on 14.02.1956.

The wooden *chit-cüng* was about 8 cm in diameter, and was painted red from top to bottom, including the decoration. About 50 cm from the top, four "spokes" had been inserted, each about 30 cm long. At a distance of 10 cm from the post, three wooden tops, each about 15 cm long, were fixed in each spoke. A further spoke had been inserted into the base of each of the outermost tops on the right and on the left. These spokes had no tops, at least when I saw them.

mong-töng (without gong) and *cakku-thi*

The post itself ended in a large wooden top, about 30 cm long. The base of the top bore four spokes, each approximately 25 cm long. These radial spokes supported four narrow slats, each about 40 cm long, two lying from left to right and two from front to back. The result was a grid consisting of 16 squares, each about 10 cm by 10 cm. The four main spokes, which protruded a short way beyond the grid, bore bars about 50 cm long at their ends. These bars formed a square frame around the grid, and about 5 cm away from it. A wooden top was fixed on each of the 24 intersections of the grid.

The other *chit-cüng* had only a simple red top, below which three large *klang-ui-chu* tassels had been inserted. The next node below this bore a tassel that had been scraped upwards from the bamboo. About 30 cm below the tassel was a tassel wheel approximately 60 cm in diameter. The ends of its tassels were coloured green.

The structure of the wheel is easier to see in a "real" *cakku-thi,* which is about 1.80 m high. At its upper end is a wooden top about 20 cm long; into the base of this top are fixed five spokes (not four, as elsewhere). Each of these five spokes bears a small wooden top in the middle and another one at the end. The tops at the ends of the spokes have three smaller spokes fixed into their bases, one pointing outwards, and the other two at right angles to it. These three small spokes end in small tops about 5 cm long. This means that each spoke branching out from the main top bears five smaller tops. This main top is plugged into the uppermost half internode, into the sides of which four *klang-ui-chu* tassels with green ends are inserted. The next internode below this is scraped upwards to form four tassels at the upper node. In the middle of this internode, four holes are made into which two bamboo sticks are inserted, forming four spokes. Nine long bamboo strips are tied in concentric circles around these spokes, the two outer ones being tied together by thinner strips. More than 50 sticks bearing shaved buds are interlaced through these rings from the outside. The ends of the sticks bear *klang-ui-chu* tassels with green ends. The wheel, with its border of tassels, looks like an umbrella.

The next internode below the umbrella is scraped up on two sides to form tassels. About 30 cm above the ground, the central bamboo tube slides over a thinner one that acts as a stand; the umbrella can be placed on the stand or taken off at will. I did not ask whether the umbrella was carried around during the feast.

I have no information about the time it takes to make a decoration of this kind. In Menring-KP, I watched a man cut a wooden top into shape and finish it with a hewing knife. It took him almost ten minutes. In Kingdöi-KP, there were almost 320 wooden tops. About 50 man-hours would therefore have been needed simply to make the tops.

R3d) The *chuk-calaa* and the *caing-cawaa*

At the time of my visits, the only author who had written previously about the Mru of the Chittagong Hill Tracts, and who had mentioned the cattle feasts, was Mills (1931:520). He wrote: "A man who has given the full series of feasts of merit may build an extra large house and may set up by the side of it four or five long bamboos. The series of feasts is (1) fowls, (2) a pig, (3) a dog and (4) bulls, buffaloes or (goyal) up to three in number."

There are some obvious misunderstandings here. The animals listed do not have to be sacrificed in that order at a series of feasts; indeed, on the occasion of a big feast, they must all be sacrificed. Three is not an upper limit for the cattle, etc.: the "four or five long bamboos" mentioned by Mills are the *dong-lu-kau*, of which one is set up for every animal. The fact that there were more than three of them indicates that more than three cattle had been sacrificed. Anyone who wants a large house can build one; this has nothing to do with the feasts he has given.

The account given by Mills contains a number of further misapprehensions. He writes: "A plain upright post in the middle of the hamlet marks the place of sacrifice … In some hamlets, but not in all, a small pointed stone is set up at the foot of the post for every animal sacrificed … In one hamlet I passed through the sacrificial post was very high, and was surrounded at a distance of some feet by a circle of lower posts notched at the top. When the feast is to be given rafters are fitted from these side posts to the centre post and the whole roofed over, so that the ceremonies can take place under cover."

In fact, there may be more than one post marking the place of sacrifice: there can be a good many *ling-pu* posts, depending on how many feasts have been held. Stones are not placed at all feasts (see R3i). Mills' account suggests that the "roofing over" of the sacrificial area was a durable construction; this cannot have been the case, because all festive decorations are used only once, though the rule is that they are left standing afterwards. The posts that Mills saw might therefore have been left from a previous feast, and were probably the remains of a *cang-chuk-calaa* ("making-shelter").

The *chuk-calaa* that I shall now describe is the one that was built for headman Yöngtu (31.01.1956). Between the sacrificial posts (*ling-pu*), a tree trunk about 5 m long and more than 10 cm in diameter was set into the earth so that it rose about 4 m above ground level. This central post was forked just below the top, but the fork was of no significance: it might just as well have been completely straight. A beam about one and a half metres long passed over the two forking branches. This would form the roof ridge. Four corner posts were set around the central post, marking out a square whose sides were about 5 m long. Further posts were set in the middle of each side; these were about 2 m high, and supported the roof edge beams.

The *chuk-calaa* seen from within

This was therefore to be a hipped roof, unlike the usual Mru house roofs. Bamboo poles connected the ends of the roof ridge with the corner posts:

two bamboo poles were placed at each corner with their thicker ends on the corner posts, then bent over the roof ridge beam to the other side. This meant that there were four poles connecting each corner with the roof ridge. The bundles of four poles were bound together. The rafters were fixed at right angles as roof rails (see C3d) over these thinner bamboo poles. The rafters did not all converge upon the roof ridge, but remained parallel: the outer ones led up to the bundles of bamboo poles, thus becoming progressively shorter. Over the rafters were placed horizontal laths, which were tied to the rafters and the rails below them. Finally, the roof was thatched with grass in the usual manner. A white cloth and a red cloth were wrapped clockwise around the central post, forming a striped spiral starting about 30 cm below the forked part at the top and ending just above the ground. The post in the centre of the side that would be to the right of the tethered animals was wrapped in the same way. That side is usually the one through which one would enter from the village square, though the position would depend on local conditions.

In addition to the *riyang* and *chuk-calaa,* a third type of "roof" can be made by "spreading" (*tung*) a large canopy (*caing-cawaa*) of material with a coloured print. This canopy covers the dance floor and the sacrificial area. The *tung-caingcawaa* is a sign that the feast is a very big one – indeed, that it is on a scale rarely experienced. A *caing-cawaa* canopy costs about 500 rupees, and nobody knew any Mru people who owned one. The word itself is of Arakanese origin. The canopy would have to be hired for the feast from a Bengali, who would charge between 50 and 100 rupees (MK, 15.07.1956).

Finally, there is yet another structure that can be set up for small feasts. It is the *watur-kim* (pigeon-house). The *watur-kim* is a miniature version of a *riyang,* and consists of two horizontal layers, the one about 30 cm above the other. Each layer consists of two sets of seven parallel bamboo poles about 1.5 m long, placed at right angles to each other. The whole structure stands directly over the sacrificial post, about 2.5 m above the ground. It is decorated with the usual short tassel sticks placed vertically (Tapwüa-Kua, 14.07.1956).

R3e) The enclosure

Buffalo and gayal are tethered to the *ling-pu* and surrounded by an enclosure (*rum*). Enclosures may also be built around cattle, but this is not usually done if there is only one animal. When there are several animals, separate enclosures are made by building partitions between them.

Although the enclosure may be decorated, it has no ceremonial function of its own, serving mainly to restrain the animals. One danger is that the animal may become restless while it is tethered and the young people are dancing around it, and it may try to break away. More importantly, when the time comes to kill it, the spear may hit it in the wrong place, and it may try to defend itself. The Mru are particularly careful with buffaloes, which are big, strong animals. They do not normally keep buffaloes themselves, and are not familiar with their disposition, so they have no faith in the animals' good nature. The Mru men therefore set up the sacrificial posts before the arrival of the Bengali, who will then bring the animals into the village and tether them. While this is happening, both men and women maintain a safe distance (Yöngtu-HP, 30.01.1956). If the men have to tether the buffaloes themselves, they will not do it until the animals have been rendered largely immobile by means of a wooden beam, which is fastened with a rope round the buffalo's neck and then pulled between its legs (Rümpöng-KP, 24.12.1956).

The buffalo's head is tied to the post as low down as possible by means of a thick rope wound around its horns, so that its forehead presses against the post. A second rope is wrapped around the animal's hind legs and tied to a peg pushed into the ground at an angle behind it. Posts are then set into the earth on both sides of the animal. In Yöngtu-HP, there were three posts at each side, each about 5 cm across, and rising to a height of 1.50 metres. A tassel bar (*chit-cüng*) is set beside every post. The posts are then joined by five horizontal bamboo canes which are fixed around all of them (in Menring-KP, only four bamboo canes were used). Notches were made on the inner surfaces of the bamboos at the points where they had to bend. The canes were fastened to the posts in the usual way, using bamboo strips. Where there is a double enclosure for two buffaloes, bamboo canes are pushed underneath the animals, just behind the front and hind legs, so that they lie in the backs of the knees and prevent the animals from lying down. These canes are then tied to the side posts, but they normally break under the weight of the animals.

The enclosure may be further decorated by the addition of smaller tassel poles, about 1.20 m long. These are adorned at the top with *thau-rin-bong* (simple strip shaving), while the sides of the other internodes are partly scraped upwards to form tassels. They are placed between the posts. In Yöngtu-HP, the horizontal bars of the enclosure were also decorated with simple scraped tassels. In Menring-KP, only the bamboos that stood beside the six enclosure posts bore tassels; the remaining vertical bamboos were tied together in pairs, and red strips bearing little scraped buds (*te-pau*) were fixed into the upper ends and draped over the buffaloes from side to side. In Menring-KP, the cattle were not enclosed at all. In Yöngtu-HP, however, extra enclosures were added to the right and to the left of the buffalo pen, and one head of cattle was tethered in each. These extra enclosures were not so strong and bore less decoration. A third head of cattle was tethered, unenclosed, facing the others. The arrangement is shown in the illustration.

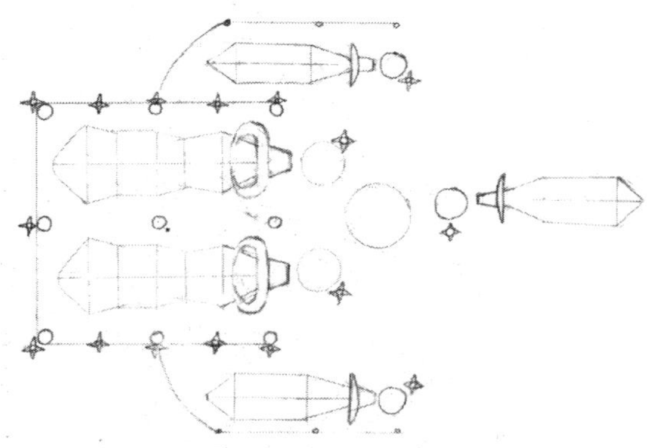

In fact the layout was less symmetrical than shown in the drawing, because four thick posts and two thinner ones still stood beside the central post of the *chukcalaa* from earlier feasts. The enclosure was built so that the heads of the buffaloes pointed roughly eastwards. This, however, is not a general rule. In Menring-KP, Rengyong-KP and Kingdöi-KP, the enclosure lay east to west, but the heads of the animals pointed westwards. A cow that is tethered without an enclosure can stand in any position unless it is tied too closely to the post.

R3f The role of the *tu*

The *tum-tu* (blowing of three single-pipe gourds) is a feature of large Anok feasts. The Northern Chüngma do not use *tu*, otherwise their ceremonies are almost the same as the Anok ones.

The *tu* should not be played by one man alone; there must always be three instruments, tuned to three different pitches (see E1d). *Tu* cannot be used at feasts for which only one head of cattle is sacrificed. Blowing the *tu* calls for special skill, so a *tu-charaa* (*tu* master) must be present. The other players can be any two men, married or not, but if at all possible they should have some previous experience. They are instructed by the *tu-charaa.* Anyone can train as a *tu-charaa,* but there are not many of them, perhaps one per thirty villages. The *tu* master receives as his pay one to two rupees, a bottle of rice spirit, and other items depending on his relationship to the family (see L3). If he is a *pen* (member of a son-in-law's sib), he receives meat from the sacrificed animal and a turban (*ngapong*). If he is a *tutma* (member of a father-in-law's sib), he receives a chicken and a spear (*re*). If he is a *tainau* (a member of one's own or a brother's sib), he receives only the meat from the sacrificed animal; a payment in money is not required, though he is usually given one rupee. The two other players do not receive any particular reward. The *tu-charaa* is not obliged to take over any additional tasks at the feast, but he may do so. He may, for instance, undertake the collection of the *chi-khang* plants (see R3i), and – unless he is a *pen* relative – he may distribute the *bong-kom* (see R1i) (MK, 26.06.1956).

The *tu* are believed to be the *plung* (gourd-pipes) of Cöng-cöi (see P1b). Cöng-cöi is a spirit, and if he takes possession of a man, he can only be appeased by a cattle sacrifice. He comes as soon as he sees the blood from cattle; he can be seen by people who are insane. The *tu* are his pipes, and they are blown during the *taröng* (see R1h). Each circuit around the cattle has a corresponding *tu-long* (*tu* verse) (Kangku Catumma, 26.06.1956).

At the feast in Inglai-KP, the circuits started in the morning at about 7.45, after a *param-rui-poi* (see N3a). The *tu* had been left on the sacrificial post overnight, and the *tu-charaa* had tried them out again at about 6.15. Nine men then arrived to play the *taröng,* three to play the *tu*, three carrying *ner*, two with cymbals, and one with a drum. At *tu* feasts, a *klingca* (small cymbal) is needed for the *taröng* in addition to the *chokca* (large cymbal). During the circuits (*wir*), the *tu* were carried but not played. After each circuit, there was a break in which the *tu-charaa* sang the notes to be blown by his assistants, who were two old men. Each player then blew a double note, first blowing and then sucking. If one of the assistants proved incompetent, he was replaced by someone else.

In Yöngtu-HP (30.01.1956), the *taröng* were performed towards 10 p.m. During the first three circuits, two *tu* were carried, but were not blown at all, although there was a break after each circuit here also. Yöngtu-HP is part of the North Chüngma area, where *tu* had not originally been used. They were only introduced there for a while after a *tu-charaa* from the Anok area had moved in, but the knowledge appeared to have died out again (MK, 14.04.1957).

The situation seems to have been similar in the Dopreng area, where Céline Mouchet saw the *tu* used at a small *cia-chot-poi*. In what follows, I have made some changes to her transcription of Mru words:

"Après que les assistants aient bu, le *wia-ma* frotte les *tu* avec un peu du mélange (*mi tut ai-dam chür,* composé de riz non cuit, d'une variété de gingembre jaune, d'un peu d'eau, tout posé sur une feuille de bananier [cf. R1g and R3i]) ... Les *tu* sont toujours par trois; il sont rarement joués ... Ensuite, les instruments, tambour, gong à bosse, gongs plats (seulement deux ici), et cymbales sont rapidement testés, puis le *wia-ma* pose un peu de *mi tut ai-dam* sur chacun d'eux, boit un peu de *yu* ... et crache la bière sur les instruments. Assistants (*plai-ria*) et *wia-ma* soufflent alors dans les *tu* pendant un temps assez court, puis les trois hommes boivent encore du *yu* ... et viennent le cracher sur le veau ... Le *wia-ma*, ses deux *plai-ria* et d'autres hommes s'emparent des instruments, et, tout en jouant, effectuent 6 tours autour de l'enclos, sur un rythme rapide et dans le sens contraire des aiguilles d'un montre. Ensuite, *wia-ma* et assistants soufflent encore un peu dans les *tu*, puis reprennent les instruments et repartent avec les autres Mru pour 18 tours autour du veau dans le sens des aiguilles d'une montre, tout en jouent. ... les *tu* ici ne sont pas transportés et ne bougent pas de leur droit endroit d'origine." (Mouchet 2001:64–65)

The *tu* and the *tu-long* also feature in the Anok death ceremonies (see Q1c–e). Both at the death ceremonies and at the big cattle feasts, pieces of *tür-ram* (*Didymospermum spec.*) leaves are tied to the pipes of the *tu* and to the carrying cords of the plate gongs (*ner*) and the drum. For cattle feasts, however, leaves from the green species and not the yellow one must be used (Kangku Catumma, 7.08.1956).

R3g) The *tu-long*

The number of *tu-long* played during cattle feasts is determined by the fixed number of *taröng* circuits (see R1h). This means that eight verses are played the first evening and eighteen the next morning. The verses played in the evening are numbers 1–7 and 17 of the full sequence that will be played in the morning. During a feast, these verses are only played, not sung; however, the words are as follows (Mita Tang, 28.06.1957):

1) taröng long: ng ho

2) taröng plang: ng ho ... (1 and 2 without additional text)

3) tür-ram:
tür-ram talek, ram talek ü.
(Wave the fish palm leaves, wave the leaves, oh)
Wave the fish palm leaves! (See R3f.)

4) *chang-ku:*
marüm chang-ku bür ü, hang pe cim baa.
(Woods [tree name] pipe tongue spread give chalk well)
Spread chalk on the *chang-ku* pipe tongues!

5) *plai-chet* (dance-sweep):
waa-di chai ce ü, chai ret ret, waa-da chai-ce ü, chai ret ret.
(Little bird oh, clean scratch [with your feet])

The *chai-ca* birds scratch (the place) clean.

(The young men who with their gourd-pipes dance around the tethered animals appear [seen by the spirits] as birds.)

6) *cia-leng* (cattle-rope):
rui-ching kong ü, nam-ching kong ü.
(Liana tree stem oh, strap tree stem oh)
The liana, the strap be strong like a tree-trunk.

(It should be noted that *rui-ching* and *nam-pramca* (here *nam-ching*) are also the names of "medicine" plants used for the feast (see R3i).

7) *cia-klik* (cattle-tether):
o-ma dam ti kon ö, o-ma dam long kon ö, paprop chöi löi.
(River fish the one, river fish the other, beside each other lie)
Two fishes lying beside each other in the river (are the cattle).

The first evening, 17) follows at this point.

8) *chüng-ku hua mang:*
chüng-ku ü hua mang ö, pacöng chüng ku hua mang chö.
(Hill country? oh stone lord, may become hill country? stone lord.)
The informant explained: "The man shall be like a big stone." The *hua-mang* looks like a big stone block, but it can also represent the embodiment of the mountain lord. The verse probably refers to the setting of a stone (see R3i); I would translate it as: "Oh a boulder of the hill country shall become a stone lord of the hills!"

9) *ching-klong wia chang:*
ching-klong ü wia-chang ö, pacöng ching-klong wia-chang chö.
(Tree-ridge oh all around? famous? let ... become tree-top ...)
This verse is also called *bong-kom long*, i.e., the verse referring to the *bong-kom* ceremony (see Rlh); the informant, however, could not give a reason for the name, or even the reference. He suggested instead that the verse might be about a dead tree of hard wood (*ching-chang klong)* which remains standing for a long time, that is, something which lasts for a long time after its own death. It might also be an allusion to a phenomenon that is seen occasionally, namely, the sacrificial post putting down roots again and growing into a tree. However, usually only soft woods are used for the sacrificial post.

10) *rüm-rup*:
o ko rü- rup pau ü, khüm ning tahöm khai ö.
(Brook (name) flower oh, seize year itself adorning is.)
At the brook the *rüm-rup* flower blossoms by the end of the year.
(For the verses 10–14 cf. the identical verses 41–45 of the *tu-long* for the death ceremonies, see Qld.)

11) *pur-cen:*
ning-ria cen long pau, khüm ning tahöm khai ö.
(Year end (name) flower, seize year itself adorning is.)
The year end *Gardenia* flower blossoms by the end of the year.

12) *yüm-yüa*:
ning khüm yüa ri pau, khüm ning tahöm khai ö.
(Year seize true rule flower ...)
The true year's end flower blossoms by the end of the year.

13) *hai mum:*
o ko cin ce hai mum ö.
At the brook the Cinjwe builds its nest.

14) *hai thar:*
o ko cin ce hai thar ö.
At the brook the Cinjwe destroys its nest.

13) *hai mum* (= 15*)*

14) *hai thar* (= 16)

According to my informant, the bird's nest symbolises the meeting of all the relatives, *tutma* and *pen,* for the feast, after which they all part again. The three previous flower verses may express the wish that such a beautiful feast might happen again every end of the year. This explanation, however, is incongruent with the fact that the same verses are played for death ceremonies. They may not be in their proper setting here.

17) *krong-chari* (lord of the earth):
krong-chari ü wöi tam baa, nam-chari ü wöi tam baa.
(Earth lord oh can we reach you?, spirit lord oh can we reach you?)
Oh lord of the earth, are you there, oh lord of the spirits, are you there?

18) *cia-köng* (cattle dead):
cia maköng ü, baa ting tüa ü, naa maköng ü, baa tang tüa ü.
(Cattle dead oh, bow? bounce? buffalo dead oh, bow? make tight? oh)

(The informant did not know how to translate the second half-lines, nor could he offer a meaning. I do not feel confident in expressing an opinion as to whether the talk here really is of a bow: nobody shoots cattle with bows, but an iron arrow (*chia*) is part of the equipment used in sacrificing cattle [see R1g].)

R3h) The *riaca*

The *tu-charaa* is not the only helper at a large feast. There are many other tasks that the feast-giver (*poi-yaa*) does not perform personally, but assigns to special feast helpers called *riaca.* Amongst the Anok, orders can only be given to the *riaca* when the *tu* are blown. The *tu* players themselves do need not to be *riaca.*

Any man from the village may volunteer to become a *riaca.* His suitability is ascertained by the *khüm-ria* ("grabbing the *riaca*") test. On the day before the first festival day – the first day being that on which the cattle are tethered to the *ling-pu* – this test may be performed by any man. He cannot be one of the volunteers, but he could, for instance, be the *tu-charaa.* He takes a flat piece of bamboo (*chek*), places a small amount of hulled rice on it, holds it over the fire, and asks whether the candidate is fit to become a *riaca.* If the rice remains as it lay, the man is fit; if the grains slip or fall, he is unsuitable. The number of *riaca* to be chosen is determined by the number of sacrificial animals: two *riaca* are needed per head of cattle or buffalo.

For the duration of the feast, the *riaca* are not allowed to eat certain foods. These taboos are called *kan-khang,* and two people from the feast-giver's house must also submit to them. There is no special designation for these two people but they can be described as *ria.* Unlike the *riaca,* who come from the village and must always be men, the people who fast in the feast-giver's house can be women.

As soon as a *riaca* is chosen, he receives from the feast-giver a bottle of rice spirits, which he must drink there and then with the people of the feast-giver's household. Each time a glass is filled, it is passed around the group. After this ceremony, the *riaca* must abstain from rice spirits for five days, but they are permitted to drink rice-beer. They are not allowed to eat anything cooked with oil or masala spice, or have any kind of accompaniment to the rice other than meat from the sacrificial animal, salt and ginger. During the same period, the *riaca* must also abstain from sexual intercourse. These restrictions do not apply to the feast-giver (*poi-yaa*) himself. If there is to be no blowing of *tu,* no food restrictions need be observed even in the feast-giver's house (MK, 16.08.1956). In Rengyong-KP (15.04.1956), however, where a householder held a *cia-phir* ("cattle-double-feast") because of the illness of his sister, the feast-giver and his family abstained from oil and spices, though they the drank rice spirits.

The food taboo (*khang*) starts with the *khüm-ria* ceremony and ends five days later, though the first time a man gives a feast, the *khang* lasts for nine days. It ends with a special ceremony called the *kan-khang-phöt* ("side-dish-taboo-lifting"). For this ceremony, one or two prawns (*chong*) are crushed with *wia-chur* leaves (*kan-chur-wia*) and cooked in a small piece of bamboo internode (*röng*). This mixture, called *chur-ngar*, is spread on the fingernails and toenails of the *riaca.* With this, the *khang* is lifted. Finally, the *riaca* take meat from the sacrificial animal (*cia-chu*) which was set aside for this purpose, put it into a pot and cook it, distribute it amongst the people of the feast-giver's house, and partake of it themselves (MK, 26.–27.06.1956, 16.08.1956).

R3i) "Medicines" and the placing of stones

The duties of the *riaca* include caring for the guests and generally helping, mainly in the feast-giver's house. During their period of office, they are permitted to enter the *kimma* (private room) of the feast-giver's house without ill consequence (see C3b), but only when the feast-giver instructs them to do so, for instance to fetch spears or cloths which are needed as gifts, or to distribute *köm-pot* (see P3d and P5b) (MK, 13.09.1956). The *riaca* also help at all the major ceremonies.

Preparing and distributing *chi-khang* (medicine) is one of their first duties. The plants needed for the "medicine" can be collected by anyone who knows the plants; the *tu-charaa* usually performs this task (see R3f). The following plants are needed: *waama-ki-or* (*Clausena heptaphylla*), *la-ia* (*Aglaonema hookeria-num*), *la-bang* (*Homalomena aromatica*), *rui-chingma* (not identified), *kom* (not identified), *re* (*Amomum dealbatum*), *yotma* (not identified), *ching-thiuma* (not identified), *nam-pramca* (not identified) and *ngetca* (not identified), and *kau-dan* and *ching-dan* (fused bamboo and fused wood). The Dengwüa also use leaves from the *ching-cu* (not identified), *chöng-cöi-la-ia* (not identified) and *ruica-dong-bök* (a liana, not identified).

These plants are pounded with water in a bamboo tube. The feast-giver must offer a pig, which is killed at this point by any man present and without any special ceremony. A small amount of its blood is added to the medicine plant mixture, which is then called *chi-khang-chai* "clean medicine"). A small pig would be enough to make the "medicine", but it is usual to slaughter a large one, to provide more meat to feed the guests.

A small dog is also killed by clubbing it on the head and cutting its throat. Blood from the dog is then mixed in a bamboo tube with the medicine plants, water, and some pig's blood; the result is called *chi-khang-chür* ("dirty medicine"). Cotton is wrapped around the ends of two small bamboo sticks, one each for the tubes containing the clean and dirty "medicine". These then serve as brushes. It is the duty of the *riaca* to distribute the "medicine"; no one else may do it.

During the first night of dancing, the *chi-khang-chai* is painted first on the foreheads of the sacrificial animals and then on the foreheads of all the people taking part in the feast, even if they are asleep. This is done three times. In Inglai-KP (21.02.1956), one *riaca* started painting at about 1 a.m., after the beginning the *plung-plai* (dance with the gourd-pipes), and he made his third round towards 6 a.m. In Yöngtu-HP (30.01.1956), two *riaca* performed the task. Both of them painted the medicine on the foreheads of the people first, and of the animals second. They made their first round at about 10 p.m., after the beginning of the *taröng,* and their second and third rounds about half an hour later. The *plung-plai* did not start until they had finished. In Menring-KP, there was no playing of the *tu,* therefore no painting with *chi-khang*.

Chi-khang-chür is used throughout the feast. The first time is when the *ling-pu* (sacrificial post) is fetched. Anyone may cut down the tree, which must be of a suitable kind (see R1b). If the post is intended for a *tu* feast, then once the trunk is cut, a chicken is killed by wringing its neck and the intended post is moistened both with chicken blood and with *chi-khang*. Once the post (*ling-pu*) is erected, it is moistened with *chi-khang* once more. The *dong-lu-kau* ("stairs-head-bamboo", see R1c) and the *teng* ("swallow-tailed-bamboo", see R1g), are sprayed with *chi-khang*. Finally, *chi-khang-chür* is used at the brook when stones are fetched on the second day of the feast, after all the animals have been killed.

Fetching the stones is one of the duties of the *riaca,* although other men may help him. In addition to the "medicine", they must take to the brook *hom-klep* (cooked rice wrapped in a banana leaf); *cia-lu-klok* (brain from one of the cattle sacrificed at the *ling-pu*); *yu-lon-tui* (beer siphoned from a jug with the aid of a tube); and *arak* (rice spirits). These items are placed by the brook at the place from which the stones are taken. No invocation is made. One stone – and one only – must be fetched for each head of cattle and each buffalo. The size of the stones is not important: very large stones are chosen only when the distance over which they have to be carried is short. The stones are placed in small holes (*dup-hua*, "digging-in-stone") which are dug in front of each *ling-pu* to which an animal has been tethered. The lower jaws of the sacrificed animals are buried to the right and to the left of each stone. After a short time, only the stones remain, because the dogs dig up the bones and take them away (MK, 24.06. and 26.06.1956).

R3j) Sacrificial ceremonies

During the afternoon of the first day, before the dancing begins, a chicken is sacrificed – usually by a *riaca* – a short distance below the place from which water is fetched. The only decoration is a *chit-wai,* which is a bamboo stick a little more than two internodes long, with tassels scraped up against the insides of the two nodes. The *chit-wai* is set into the ground next to the brook so that the blood of the chicken can be dripped over it. For the *wi-klak* ("blood-shedding"), the chicken's throat is cut with a hewing knife, and an invocation is made to the *o-reng-nam* (the river spirit, see P2f).

The *klaa-wi* ("shed-blood") is followed immediately by a *cok-malong;* this term probably derives from the Arakanese *kyok hma' 1ong,* "pouring on the stone*"*. Above the place where the *wi-klak* was performed, a small *tong-mang-kim* ("hill-lord's-house") is built for the *cok-malong*; for the method of construction see P2f. The *tong-mang-kim* has four *chit-wai* (tassel sticks), on which are scattered *pok-pok-kom* (puffed rice and bits of flat dough-cakes made of rice flour). A cow is killed by cutting its throat over this little house. It is sacrificed to *reng-ma* – also called by the Marma name *tong-mang,* "hill-lord" – while water is poured over it six times from a *ri-chang-pön* ("water-pouring-tube") (MK, 22.08.1956). All the equipment used for this and for the *klaa-wi* ceremony is the same for a goat sacrifice at the river (see P2f and P3b); at most *tu* feasts, only a goat is sacrificed rather than a cow.

The sacrificial animals are tethered to the *ling-pu* without any ceremonies. From this point until they are sacrificed, they are not fed, but some attention is paid to their welfare: during the day, they are given a bowl of water, and water is poured over them until they are thoroughly wet (Menring-KP, 4.02.1956). At night, blankets are hung over them (Inglai-KP, 21.02.1956). This warming of the animals seemed little more than symbolic: in Menring-KP, only one blanket was brought, for three animals. Towards 9 p.m., an old man "warmed" all three posts for a short time, then "warmed" the post of the *lu-ngua* calf (see below) for a longer time. He then put the blanket over the back of the cow for a while. Finally he took the blanket away, without paying any attention to the buffalo.

The *lu-ngua* (pillow) is a small cow or a calf without which buffaloes or gayal cannot be sacrificed; if only cattle are sacrificed, they do not need a "pillow". Only one *lu-ngua* is needed, regardless of the number of sacrificial animals. Of all the animals tethered at the *ling-pu,* the *lu-ngua* is killed first; after this, the animals can be sacrificed in any order.

The killing of the cattle begins on the second day at some time after sunrise. It can also follow the singing of the eighteen *tu* verses and the *taröng* circuits of the animals, or even take place while these are in progress. If the feast is a *param-rui-poi* (see N3a), the ear-piercing ceremony precedes everything else. It is usual to kill only two animals after sunrise, or even just one. Their fate does not overtake the others until the afternoon; when the later ones are killed, the blowing of the *tu,* like the *taröng* and the attendant ceremonies (see R1h), are not repeated. The ceremonies appropriate to a small feast are thus performed only once. It is thought desirable that the feast-giver (*poi-yaa*) should kill the animals himself, but in fact he usually leaves this to the *riaca*, while he stays in the house and attends to the guests. The *riaca* are also not obliged to complete the killing: one of them must bring the spear, but then anyone else may deliver the death blow. The person who does the killing incurs neither rights nor duties (MK, 26.06.1956).

At all big feasts where *tu* are blown, a piglet must be killed with a spear or a bamboo spear on top of the largest of the animals sacrificed in the morning. The piglet must already have been castrated. It is called the *klan-*

ria-pak ("help-*ria*-pig"). "*Ria*" may mean "beside" in this context. It has no special reference to the *riaca:* one of the *riaca* may kill it, but so may anyone else, and it is not prepared separately for the *riaca* to eat. It is placed to the right of the cow near the stab wound, lying in the same direction as the animal. The man who is to kill it holds it down with one foot and his left hand. With his right hand he throws *mi-tut-ai-dam* (rice with wild ginger) over it from a banana leaf, and then stabs it to death from the right (Inglai-KP, 21.02.1956). Only one *klan-ria-pak* is needed for a *tu* feast, but all the other sacrificed cattle must be touched with the dead piglet.

When a *riyang* was held in Chikcaron-KP, a dog was used instead of a piglet. First it was clubbed to death, then its throat was cut over the main sacrificial animal, which was a buffalo. Although the feast-giver was a member of the Chikcaron Klenchangnau, which is an Anok group, the ceremonies were largely those practised by the Chüngma. There was no blowing of *tu* or preparation of *chi-khang*. In the rest of the North Chüngma area, for example in Menring-KP, I saw no *klan-ria-pak* sacrifice. Instead, there was a special ceremony at feasts for which two or more cattle were sacrificed. After the sacrificial animal had been dragged away, and just before it was butchered – that is, when only the first cuts had been made – some turmeric was wrapped in a blanket and placed between its front hooves. The Tamtuca branch of the Anok also performs this ceremony. I was not able to discover the reasons for it (Chikcaron-KP, 2.04.1957; MK, 14.04.1957).

All cattle and pig sacrifices are accompanied by invocations (*tamma*), which begin with *phyöi.* All *phyöi-tamma* refer to *nat-cang*, that is, the sacrifices are directed towards the *chüng-nam*, or evil spirits. None of the spilled blood may be brought into contact with the rice of the *ca-pam* or the *khong-rau* (see D2h). However, the good spirits are also invited to the cattle feasts, so while the cattle are being sacrificed, a chicken is killed in the house, in front of the rice store in the entrance to the *kimma.* The killing is accompanied by *phyok-carok tamma*, that is, an invocation of the good spirits (MK, 18.08.1956). For the sacrifice of the chicken a special beer pot is placed beside the rice store; this is called the *pam-ria-yu* ("store-beside-beer"). The guests may also drink this beer. Normally, strangers are not allowed to enter the *kimma,* or private room, but on this occasion, they may go in as far as the pot. Anyone who has ever been bitten by a tiger is not allowed to drink the *pam-ria-yu.* Should such a person exist, he or she would also be forbidden to drink the *bong-kom-yu* ("wrist-tying-beer"), with which the *phyöi-tamma* are made at *bong-kom* time (see R1i) (MK, 13.09.1956).

R3k) Social relations

As previously noted, all the feast-giver's closer relatives are invited to the feast; where the line is drawn depends on the intended scale of the feast. All relatives are entitled to take part in the feast, whether or not they are known to the feast-giver. All of them may enter the feast-giver's house and be served, or help themselves from the beer pots.

In addition to these informal rules, there are some more formal ones which the feast-giver must observe. The rules for welcoming the players and dancers have already been described (see R2b). The feast-giver is not allowed to use the meat from the sacrificial animals as soon as they have been killed: first he must "ransom" them from the people of his brother-in-law's sib. Before the slaughtered animals can be hauled away and butchered, either in or near the feast-giver's house, one or two men must sit down on them. These men are called *tön-cia-mai-mi* ("pull-cattle-tail-ers"), and their only duty is to cause the feast-giver difficulty or expense. Any member of the feast-giver's *tutma* or *pen* (brother-in-law's family, see L3b) may act as a "cattle tail puller". Even at small feasts, where no *tu* are blown, one *tön-cia-mai-mi* is allowed. However, this is not usual, even though the animal's tail really is pulled on those occasions (see R1h), which is not the function of a "proper" *tön-cia-mai-mi*. His real duty is to declare that if the feast-giver does not give him this or that, he will not release the buffalo or the cow. These demands must be met, either by the *riaca*, or, at a small feast, by the host himself, or someone from his house. If the *tön-cia-mai-mi* is a member of the *pen*, he can claim to a new turban (*ngapong*); if he belongs to the feast-giver's *tutma,* he claims a spear (*re*) (MK, 27.06.1956). He also receives a bowl of rice spirits after the *riaca* has taken a mouthful of it and spat it (*cho*) over the head of the *tön-cia-mai-mi* (this happened only in Inglai-KP, not in Menring-KP).

The *tön-cia-mai-mi* sitting on the sacrificed cow waiting for new turbans

In the evening of the day of sacrifice, or at another suitable time, a *tainau* and *tutma-pang* is held (cf. M4c). Anyone who bears a *tainau* or *tutma* relationship to the feast-giver – that is, anyone who belongs to his brothers' sibs or father-in-law's sib – may take part in this *pang,* provided that he is willing to make the necessary commitments. These men all are people with whom the feast-giver would wish to cultivate good relations, and the *pang* can be arranged when they are invited. A *tainau* who is willing to participate will have brought a pot of beer for the feast. When the *pang* takes place, the feast-giver presents him with a *yu-kong* (beer pot) and 5 to 10 rupees; the *tainau* does not have to perform any immediate service in return for this payment. However, when he himself holds a feast, he will make a special point of inviting the feast-giver, and will give the money back to him at the feast.

A special honour is conferred upon members of the *tutma,* especially in the narrower sense of the mother's brother. Not only is he invited: he is met and accompanied by an emissary, if possible a member of the feast-giver's household. The *tutma* stops at the entrance to the feast-giver's village; a messenger runs to the village, and announces the arrival of the guest to the host. If possible, the feast-giver then goes in person to the village entrance to welcome the *tutma,* taking with him one or more bottles of rice spirits. Guns are fired or firecrackers lit to mark the *tutma's* arrival: the *tutma* uses this means to make his arrival known before he enters the village, and the feast-giver has the salute answered. (The fireworks, which are produced by the Bengali, can be extremely dangerous: in Yöngtu-HP I met a man whose forearm had been blown off by a cracker that exploded at the wrong time.)

The *tutma* and his companions stay just outside the village until the drinks offered by the feast-giver have been consumed. In return, the *tutma* brings a piglet as a gift. They then proceed in cheerful spirits to the feast-giver's house, where a meal awaits them, no matter what the time of day. The meal may consist of a chicken sacrificed for the *wi-klak* at the river, or meat from the first head of cattle to be sacrificed. An especially large beer pot is brought in and broached in the presence of the *tutma.* Once the guest of honour has drunk from this *tutma-yu*, he invites other men to whom he feels particularly obliged to drink also. Later, anyone can drink from the pot.

The members of the feast-giver's sib will help him to care for the guest of honour, since he is their *tutma* too. From the time he arrives at the hamlet in the early afternoon, he is entertained so well that he soon feels tired, and will probably go to sleep before evening. He is at the centre of attention the next day too. He is asked to come to the feast area for the killing of the main sacrificial animal; he is the first one entitled to take the role of *tön-cia-mai-mi*. After that, he must rouse himself one last time for the *pang*, before settling down to sleep through the night. Next morning, if someone feels equal to accompanying him, he can return home, though others will try to persuade him to stay longer (K Kangku Catumma, Chikcaron-KP, 1.–3.04.1957).

In addition to the beer pot, each of the *tutma* men taking part in the *pang* receives a spear, for which he has to pay with cloth, depending on the value of the weapon. The *tainau* and the *tutma* are also presented with special cuts of the beef and the buffalo meat to take home. The meat must be taken from the neck, ribs or loin; leg cuts are not permitted (MK, 29.08.1956). The *tutma* must also receive chickens. The feast-giver must acquire a sufficient number beforehand; his own *pen* relatives will also bring along chickens. If a member of the *pen*, the son-in-law sib, wishes to indicate special attachment to the family, he may bring a spear; in return for this, at the *pang,* the host gives him the equivalent value in cloth in form of a new turban. The distribution of the spears and the turbans is generally one of the duties of the *riaca.*

Where *pang* are held, they obviously add to the feast-giver's costs. The North Chüngma people, however, have taken over a rather different custom from the neighbouring Marma: the feast-giver may send the *riaca* to invite the more prestigious guests into his house, where he will ask them to drink from a tall *yu-kong* (beer pot) and then ask them for a financial contribution to his feast. If a literate man is available, the feast-giver asks him to make a list of donors' names, and the amount that each will give. This system was said to be the main reason why a large feast – a *cang-chuk-calaa,* with two buffalo and three cattle – held by Headman Yöngtu was relatively badly attended: people were afraid that the headman might tax them too heavily (1.02.1956).

R4 The feasts of the other Mru groups

The following notes are not intended to provide a comprehensive guide to the kinds of feast held by Mru other than those belonging to the Anok and the North Chüngma groups. They are simply a set of observations about feasts that I visited, supplemented by the answers to questions I asked. Rather than summarising the material, therefore, I will describe it in the context of the individual feasts.

R4a) A Dopreng *chot-cia*

This feast was held at Taa-uk-Para, Prumma-Mouza, on 3. and 4.12.1956. The feast-giver was Karbari Monglung, who had fallen ill. I did not reach the village until the evening of 3.12., by which time the feast preparations were already complete and the cattle tethered. Four *chit-cüng* (bamboo masts with tassels) had been erected. They were more than three metres long. Three *klang-ui-chu* were fixed into the top half internode; the next four internodes had been scraped upwards to form tassels at the upper nodes. Three of these *chit-cüng* stood by the *ling-pu* (sacrificial post), and the fourth at the *dong-lu* (stairs-head), where it took the place of the Anok *dong-lu-kau* (see R1c). Beside it and next to the staircase stood a *pön-cüng* (small bamboo cane post), about 1.20 m high. All its internodes were scraped upwards to form tassels, and the top half internode was decorated with *thau-rin-bong* (stripes made by scraping in vertical bands). The three internodes below this bore holes that seemed randomly distributed, and into which simple little bamboo sticks were inserted, rather than *chit-wai.* Almost half of these bore small bamboo tubes ending with a node at the top; these stood out at an angle, like branches.

The feast-giver's house was decorated with three bamboo poles of which all the internodes were scraped upwards to form tassels. One of these poles, called the *rai-cün-cüng*, ran parallel to the stair rail (*rai-cün*); a second one, called the *dong-lu-mom* ("stairs-head-lining"), ran parallel to the ground at the front of the outer platform (*car*), which one crosses to reach the entrance to the house (see C3f); and the third one, called the *yang-pang-mom*, ran parallel to the *dong-lu-mom* at the *yang-pang*, with the top bar of the platform rail lying over it. Under the platform (*car*) stood four shorter bamboo poles, decorated with scraped tassels. These were fixed to the bamboo supports of the platform in the square, and were called *cia-yu-pai* ("cattle spleen"); no one could tell me the significance of this name. The *pön-cüng*, the *rai-cün-cüng* and the *dong-lu-mom* each bore a square decoration made of threads, and called a *chalik-mai-lap* (Anok: *chaling-chalap*, butterfly). These decorations consisted of a cross made from two narrow bamboo sticks. Red and black threads were wrapped around the sticks in five rounds, working from the inside to the outside. The effect was that of a spider's web. The end of the thread was used to hang the square from one of the bamboo tassels. I could not discover any meaning behind the *chalik-mai-lap*.

The place where the beer pots stood was at the western side of the feast area. There were two *yu-kong* (beer pots) for the dancers, four *ci-ö-ca* (small round clay pots) containing *hom-noi* (rice water), and a *pui-tang* (small woven table) on which lay a *tu* (a single-pipe gourd, see E1d). A simple *mong-töng* (gong stand) stood nearby; it consisted of two vertical bamboo tubes and one horizontal crosspiece, with no decorations.

Unlike the Anok, the Dopreng taken the gong down from its stand during the *taröng* circuits, and the player carries it around and beats it with his fist. Before the *taröng,* which took place at about 9 in the evening and 8 in the morning, all the instruments were spat upon (*cho*) with *hom-noi* (rice water) from one of the *ci-ö-ca,* to an accompaniment of *tu* blowing. The instruments were one drum, two plate gongs, one large gong, and one cymbal. *Tür-ram* (*Didymospermum*) leaves are tied to the *tömma* (drum) and the *tu* (gourd-pipe). The *tu* is also carried around and blown during the *taröng*.

The young people's dance (*plung-plai*) started toward 10 p.m. The girls danced enthusiastically and well, but the boys were lazy, they blew the *tu* badly, and were generally more interested in the rice-beer, though occasionally one would fetch a girl from the row of the dancers and invite her to drink from the beer pot. There were virtually no visitors or onlookers, just one old woman, who warmed herself at the open fire at the edge of the feast area and tried to encourage the dancers. The dancing continued till dawn, though there were often very few boys to blow the *plung*. The girls went on dancing to keep warm.

After sunrise, the square was deserted. At about 8 a.m. a *taröng* took place, then things became quiet again. The cow was stabbed to death only after the *plung-plai* had recommenced, at about 11. It was killed not with a spear, but with a simple iron arrow (*chia*). The tongue was cut off and fastened on the *ling-pu* with a spike, while water was poured over it; this was done in the same way as amongst the Anok. The dancing ended at about 1 p.m.; it was not resumed, and a *kim-pün* (see R2d) is not usual. However, like the Anok, the Dopreng held a *pang-cang* in the evening (see R1i).

R4b) The Dopreng *riyang* (*naa-phir*)

A feast took place in Prumma-Mouza, Rümpöng-KP, on 23.–25.12.1956; my main informant was the *riaca* Rengnap. The feast-giver, Irtu, had had a particularly good cotton harvest. From the proceeds he had bought two buffaloes from the Marma, a large buffalo ox for 300 rupees, and a small buffalo bull for 200 rupees.

During the morning of 23.12., the bamboo ornaments were made, and the feast-giver's house was decorated with them. The decorations were generally the same as those at Taa-uk-Para, though there were no thread squares, and the whole of the outer platform (*car*) was rebuilt and covered by a roof. The "cattle spleen" (*cia-yu-pai*) here consisted only of three sticks joined together to form a triangle, and another one hanging down from it to the ground. At the *dong-lu* ("stairs-head") two taller *chit-cüng* (tassel posts) and one shorter one were erected, a large one for each of the two buffaloes to be sacrificed, and one for the cow. The upper half internode of each was decorated with *thau-rin-bong* stripes, and two *klang-ui-chu* tassels were inserted. On top of each of these *chit-cüng* sat a *rang-pang* (hornbill, see R1d), woven from bamboo strips. Nobody could tell me its significance.

Towards noon, there was a *ner-thep* (beating of plate-gongs), accompanied by the *taröng* instruments, in the feast-giver's house. At noon, a pig was slaughtered in front of the house; its killing was not marked by any special ceremonies. This pig was the *chit-khai pak* ("tassel-ornaments pig"), and it must be slaughtered even if only one head of cattle is sacrificed; it is intended for the workers who prepare the festive decorations. The size of the pig depends on the amount of the work to be done. In Rümpöng-KP, the large pig that had been intended as the *chit-khai pak* could not be caught, so the housewife caught two smaller ones instead. Unlike the Anok, the Dopreng do not collect the pig's blood in special bamboo tubes to make *chi-khang* (medicine). They do not prepare *chi-khang* for any feasts, whether large or small, nor do they paint the foreheads of the participants with any other substance.

In the afternoon the *ling-pu* posts were erected. Holes were dug for them, and a *ram-nget* leaf (not identified) was thrown into each. I was not able to discover the significance of this custom. Two *chit-cüng* were set next to each *ling-pu*; they were of the same type as the ones at the *dong-lu.* Three more *chit-cüng* were placed to each side of the buffaloes, and a final one behind the animals, in the centre. The *riyang* (see R3b) was tied to these *chit-cüng*, not to specially made supports. It was about four metres square; each of its two canopies was a grid consisting of 15 bamboo tubes in each direction. Its edges were decorated with tassel sticks about 1.50 m long, inserted vertically at irregular intervals; there were about ten per side. To support the *riyang* on the fourth side, that is, in front of the buffalo, crossed bamboo canes were placed underneath it at the front of the buffalo enclosure. The various ornamental tassels were not coloured.

To limit the buffaloes' freedom of movement, their hind legs were tied with a rope which was then fastened to a short piece of bamboo pushed into the ground behind them. A length of wood was placed underneath the animals from side to side, and supported on two pairs of crossed bamboo poles. During the night, the large buffalo collapsed, presumably because the animals had not been given any food for three days. On the morning of 24th, the buffalo bull also lay down. The animals were then offered a few banana leaves, but the large buffalo refused to eat, and it was stabbed while lying down. Some people declared that it had been dead already.

A small cow, the *naa-lu-ngua* ("buffalo-pillow", see R3j), was tethered so that it stood outside the enclosure, next to the buffalo on the right. The heads of all the animals pointed north-east. The enclosure itself was completed somewhat untidily on the evening of the 23rd, since it was growing dark; the small *chit-cüng* that decorated it were not spaced evenly.

The same evening, before the dancing started, a chicken was sacrificed (*wi-klak*, see R3j) at the place where water was drawn. This was the only sacrifice offered at the stream during the feast. After the *wi-klak,* the *taröng* instruments were brought down from the feast-giver's house to the feast area. First they were ceremonially spat upon (*cho*) with *hom-noi* (rice water). Unlike the people of neighbouring hamlet Taa-uk in Rümpöng-KP, the people of Prumma-Mouza did not blow *tu* (single-pipe gourds). However, three *tu* were brought out and, like the drum, they were adorned with *tür-ram* (*Didymospermum*) leaves. The three *tu* were brought out at all feasts, even if only one cow was sacrificed, but they were never blown, because the hamlet had no *tu-charaa* (*tu* master). They were simply placed near the beer pots, and carried for a while during the *taröng* circuits. The dancing (*plung-plai*) started at about 10 p.m., and lasted till sunrise. Numerous visitors arrived from neighbouring villages. All the girls danced together. The boys split into three groups and spent the day tuning and coordinating their *plung* pipes. During the dancing, the groups relieved each other as necessary.

The next morning, the festivities began at about nine o'clock with the *ner-tep*, or playing of the *taröng* instruments. During the circuits, the members of the feast-giver's family – three girls, an old woman, and two men – performed a *cho* (ceremonial spitting) at the *ling-pu*. One after another, in the order given, they sucked some rice water from four small *hom-noi-pu* (rice-mixture-pots) and spat it on to the sacrificial posts. Whatever the type of feast, there must be four of these pots.

After the *taröng,* two more men performed a *cho*; these were the two *riaca* (see R3h). Whatever the number of animals sacrificed, the Dopreng appoint two *riaca* and a woman, who is called a *riama*. They are chosen (*khüm-ria*) in the same way as the Anok riaca, but they are not subject to any food taboos (*khang*). The *poi-yaa* (feast-giver) is not required to abstain from meat or alcohol, so there is no *kan-khang-phöt* ceremony (see R3h).

After the *taröng,* one of the *riaca* put a blanket over the animals, first on the buffaloes, then after a while on the small cow too. The Anok practice of "warming" the animals at night is not followed. While placing the blanket over the animals, the *riaca* pricked them with an iron arrow (*chia*) and threw *mi tut ai-dam* (rice with wild ginger) at them. However, the animals were not killed yet: first there was another *ner-tep* (though the spitting ceremony was not repeated), following which the dance (*plung-plai*) started again straight away. By now about two hours had elapsed, so it was about midday.

First the *riaca* stabbed the *lu-ngua* cow to death, then the large buffalo, both with a spear. The wounds were administered from the right side. Although the *riaca* did the killing on this occasion, any other person could have done it instead. As usual, the tongues were pulled out and cut off; only now was water poured over the animal's muzzle. The water was poured from a calabash, but not with a *tui-lam,* as is the Anok custom (see R1h). The pieces of tongue were then fixed to the *ling-pu* posts, the buffalo enclosure was broken open, and the animals untied. They were then dragged to the feast-giver's house, accompanied by the dancers. At the house, they were hauled on to the outer platform and cut up.

It is not a general Dopreng rule to kill animals at noon. The previous evening, the *riaca* had been expecting to kill them during the morning. The long break between the *taröng* with the *cho* and the killing could have been due to a misunderstanding between myself and the *riaca*: I had asked him to wait for a moment while I changed my film, and he was perhaps trying to comply with my wishes.

During the second playing of the *taröng* instruments and the *plung* dancing, and right up to the time when the buffalo was dragged away, some old women also made circuits of the enclosure. They held on to the edges of the enclosure and followed the *taröng* players, rather than dancing backwards like the girls. One after another, they were given a glass of rice spirits by men who danced around with bottles in their hands and took an occasional pull to reward themselves for their efforts.

After the cow was dragged away, the buffalo was left lying in the enclosure for a while. A *riaca* distributed rice spirits among some of the older men, who were *pen* of the feast-giver (people of the son-in-law sibs) and who had brought chickens and spears. He wrapped new turbans around their heads; thus adorned – and more than a little tipsy – they too joined in the dancing. Anyone who was still thirsty could drink his fill from the beer pots that stood in the appointed place, or from the four *hom-noi-pu* (small pots containing rice water). The latter were refilled with the aid of a small bamboo tube with a red tassel at the end.

Until the second buffalo was killed, at about 3 p.m., girls from the row of dancers were brought to the pots to drink beer. The second buffalo was eventually dragged away by its rope to the house of the feast-giver. Again the animal was accompanied by young men playing the *plung* and by girls dancing; *tön-cia-mai-mi* (see R3k) and other ceremonies are not usual. The dancing group then broke up; the dancers did not go up into the house, and there was no more dancing that evening. However, an informant from a Dopreng hamlet on the Lama river said that in his community, it was usual to have a *kim-pün* (see R2d). After the second buffalo had been butchered in the feast-giver's house, the *bong-kom* (see P2e, R1i) were given out. A *pang-cang* (see R1i) followed after dusk.

On the third day of the feast – which I did not spend in the village – stones were to be fetched and set in position. I was told that this part of the proceedings was not accompanied by any further ceremonies. One stone would be set for every animal sacrificed; this is done even if only one head of cattle is killed. Red and black cotton threads (*ing-pring*) are wrapped around the stones (the Anok do not do this). The lower jaws of the animals are buried to the right and to the left of the stones. Like the Anok, the Dopreng only keep the skulls of particularly large animals. Even when they are kept, the skulls have no further use during the feast.

In summary, my impression was that the Dopreng in Prumma-Mouza used some decorations that were not usual amongst the Anok, for example the *cia-yu-pai, pön-cüng*, and *ing-pring*. The ceremonies, however, were fewer in number and less elaborate, some being no more than rudimentary. There were no essential differences between large and small feasts.

R4c) The feasts of the Rümma

A feast took place in Songkhö-Mouza, Nongneng-KP, on 11.–12.03.1957. My main informant for the material presented in this section and the next was Karbari Nongneng Krimchang.

Like the Dopreng, the Rümma follow essentially the same procedures whether they are sacrificing one animal only or several heads of cattle, buffalo or gayal. Every feast consists of four distinct days: on the first, the ornaments are prepared and the cattle are tethered, and the dancing starts in the evening; on the second day, the cattle are killed, and the *cia-lu-ting* ceremony (see below) is performed; on the third day, stones are placed and the *bong-kom* are distributed; on the fourth and last day, the feast ends with the *pang-cang* (see R1i).

At this feast, two cattle were killed: one was given by the son of the karbari, and the other by another household belonging to the Krimchang sib in the village. During the morning of the first day, the houses of the feast-givers were renovated. The *cang-kim* ("making-house") consists mainly in replacing some of the bamboo canes under the outer platform, which are supported at an angle from the left and the right. A dog and a pig are killed as *kim-cang-kan* ("house-making-side-dish"). Later, some cooked meat from these animals must be wrapped in a banana leaf and set down at the *ling-pu*. The *kim-cang-pak* is the only pig that *must* be killed for the feast, but as many more as desired can be killed to provide additional food.

Bamboo canes decorated with scraped tassels were drawn over the outer platform and covered with a mat. At the stairs head one *dong-lu-kau* was erected for each of the cattle sacrificed. Like the Anok equivalent, these *dong-lu-kau* have nine *chit-pap* (tassel rings, see R1c) at approximately the height of the roof. *Chit-wai* (tassel sticks) are tied to the small bamboo branches above the *chit-pap,* but there is no "fish weave" (*dam-khai,* see R1c) at the top, and no hornbill (*rang-pang*). Like the smaller Anok version, the top of the *dong-lu-kau* consists only of five hanging tassels, four of which are kept separate by a ring. There is, however, a *dam-khai,* which hangs down from the gable over the front wall of the house (*kimma*). At the end is a banana leaf which contains the waste material from the tassel making; this is called *löng-chur-baa* ("ants-stinging-nest"). Its significance was unknown. The Rümma do not hang thread squares (*chaling-chalap*) as decorations.

No ceremonies were performed when the pieces of wood were brought for the *ling-pu* posts. Before they were erected, some water was poured from a *tui-lam* bamboo cane (see R1g) and a *kwai-chek* (bamboo stick) was thrown into the hole that had been dug to receive the post. It is said that once upon a time, there were two children whose mother had died. Their wicked stepmother blindfolded them and led them into the jungle. However, the children did not perish, because they followed the call of a pigeon, which led them to a place where they grew up safely, and gained great wealth. They then organized a big feast. The wicked stepmother was invited, and was crushed to a pulp under the sacrificial post. Even today, people throw a *kwai-chek* into the hole, so that she will feel pain.

I heard the full version of this story under the title "*büi-ko chang-cia"* (see R8b). It was told to me by Aroi Chüngprim.

Two tassel masts (*chit-cüng*) three to four metres high were erected at every *ling-pu* post. Only one enclosure is built, even if there are several animals. No enclosure is thought necessary for smaller animals. For any large feast, a special ornamental *riyang* (see R3b) can be erected. A small tassel mast 1.4 to 2 metres high was placed at each of the two places where the beer pots were set out. At one of these, the gong stand (*mong-töng*) was erected. It had no tassels. Unlike the other Mru groups, who indicate the second drinking place only symbolically by means of two swallowtail bamboo (*teng)* filled with *hom-noi*, the Rümma put beer pots at both places (*kacakwüa-yu* and *machiwüa-yu*). The dancers are supposed to call each other to drink, though in practice, it is always the boys who call the girls.

Two further *yu-kong* (beer pots) must be set out, one on the outer platform near the *dong-lu-kau* and one in the house by the *ca-pam* (rice store in the entrance to the *kimma*). The Rümma, unlike other Mru, do not generally place beer pots in the house, in the middle of the *kim-tom*. This will be done only if the feast is a large one; the beer is then intended primarily for the relatives from the brother-in-law's family (*can*, that is, *tutma* and *pen*).

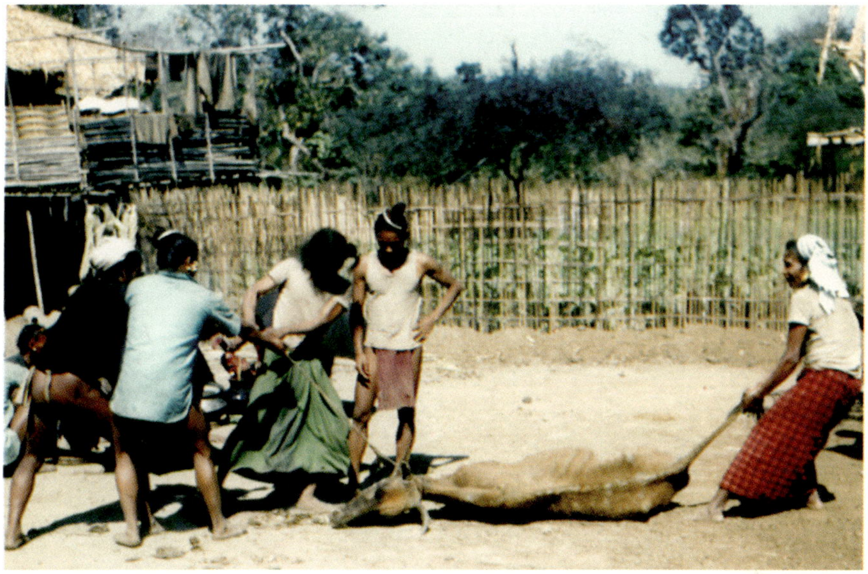

tön-cia-mai with the Rümma

Girls are fetched for drinking *yu*

Whatever the size of the feast, two men are needed to act as *riaca* (see R3h). They are chosen by the feast-giver, not by means of a *khüm-ria* ceremony, as is customary amongst the Anok. They receive their instructions from the *wiama* (master of ceremonies), whose role corresponds to that of the Anok *charaa.* The main duty of the *riaca* is to care for the guests, though the *riaca* also take over all the work in the feast-giver's house, such as cooking and distributing the meat. The rice, however, must be cooked and distributed by the occupants of the house themselves. The presence of a *riaca* is essential when the animals are sacrificed, since he must inflict the first wound. After that, if the animal is still alive, anyone else may finish the task. A *riaca* may also distribute the *bong-kom* (see R1i), though this is normally done by the *wiama*, even if he is not a member of the feast-giver's sib (cf. R3f: at an Anok feast, the *tu-charaa* can distribute the *bong-kom* instead of a *riaca).*

The *wiama* and the *riaca* are not paid in money. If they are *tutma*, that is, if they belong to the father-in-law sibs, they receive a spear; if they are *pen*, that is, members of the son-in-law sibs, they receive a turban. For the four days of the feast, the *riaca* and everyone in the feast-giver's house must observe a *kan-khang* (side-dish-ban) on *arak*, fish and *chur-ngar* (sour prawns). Beef and beer are permitted. The *kan-khang-phöt* resembles that of the Anok (see R3h), and ends with the spreading of *chur-ngar* juice on the fingernails. A *pang-cang* (see R1i) is held for this purpose; it can only take place after the *riaca* have received their turbans or spears.

R4d) The Rümma *param-rui-poi*

At 5.30 in the afternoon of 11.03.1957, the cattle were tethered. There were no more ceremonies until about 8 p.m., when the *taröng* instruments were beaten for the first time in the karbari's house for the *rik-plai* (initial dance). At about 10 p.m., the instruments – three plate gongs, one drum, one cymbal, and one large gong – were carried to the festival ground. There the *wiama* spat rice-beer upon them (*cho-yu*, "spitting-beer") from one of the pots by the gong stand. While doing so, he uttered a *tamma* (invocation). All players gave six loud *hu* cries, on a tune that rose sharply then dropped briefly, after which they started to move around the *ling-pu.* Each single *wir*, or circuit – of which there were eighteen – was counted loudly. The gong was not left on the stand, but carried around in the circle. One *tu* (single-pipe gourd) had to be there, but it was not played because the rules governing its playing were not known. (The Rümma do not play *tu* at death ceremonies either.) However, I was told that the members of the Kung sib, which is classed as Rümma, blow three *tu* at feasts.

At about 11 p.m., the *plung-plai* began. This is the dance of the young girls and the men who play the dance-pipes, which are gourd instruments with several pipes. The dancing became less energetic by about 1 a.m., and stopped almost completely at about 3 a.m., when some people felt inclined to give up. However, they decided to continue, and held out till dawn. The players had been somewhat inebriated ever since the previous evening, and the beer pots had been replenished throughout the night.

The following day, the dancing (*plung-plai*) started again at about 10 a.m., and the *taröng* players appeared about half an hour later. The *cho-yu* ceremony (spitting on the instruments) was performed again before they were played, and the players again gave six *hu* cries. They then made circuits of the *ling-pu* in *ner-khöngma* ("fast-gong" time; see R1h), overtaking the *plung* players several times at speed. Eighteen *wir* (circuits) are the rule, as in the evening. Meanwhile the *wiama* brought the spear. As with the Anok, a banana leaf containing *mi tut ai-dam* (rice with wild ginger) was fastened under its point. A little of this mixture was thrown over the animal six times before it was killed.

No one sat on the slaughtered animal, but when it was dragged away, some of the young people held on to its tail and tried to pull it back (see R3k, *tön-cia-mai*). In Cöngpat-KP, which is in an area north of Alikodong where the Rümma and the Dopreng people mix, I was present at a feast on 19.03.1957. Only the tail of the last animal to be sacrificed, a buffalo, was pulled. When it was about to be dragged up to the outer platform of the feast-giver's house, several young men hung on to it so that it had to be let down again several times. While this was going on, the dancers who had accompanied the animal to the house in the Dopreng manner continued to dance in front of the house.

In Nongneng-KP, the procedure was the same as amongst the Anok, in that the animals were not accompanied by dancers and *plung* players when they were dragged to the house. No one hindered the people who were trying to raise the animal on to the platform.

As soon as this had been accomplished, the *param-rui* ceremonies took place (ear-piercing; see R3j for a different sequence of events). This ceremony was held on the outer platform, unlike the Anok practice (see N3a). The *wiama,* who takes the role of *param-rui-charaa* ("ear-piercing-master") pierces first the child's right ear, then its left, then touches them briefly with a taro bulb. The Anok custom of performing a ceremonial spitting (*cho*) is not usual here, nor is it thought necessary that a dog should share the pain. Instead, the dead animal is pricked in the left ear, and a small piece is cut off the edge of the same ear. This is later used for the *bong-kom* (see below).

The ear-piercing is followed by the *cia-lu-ting* ("cattle-head-arrive"); at feasts other than the *param-rui,* it follows the raising of the animals on to the platform. During this ceremony, the *wiama* takes into his mouth some hulled rice and rice-beer from a bowl, then spits on the head of every member of the feast-giver's family. In Nongneng-KP, he started with the child whose ears had just been pierced. He is said to spit "over" the head of the animal, but in fact he spat beside it. He then blew six times over a *laa-rik-yia* (a small calabash containing grains of rice, see P5b) and shook it a little while making a *tamma* (invocation). The animal was then butchered and cooked. None of its meat could be eaten, however, until the *krongma-namma cak-kwak* ("earth-spirit meal-throw-away") had been performed (cf. R4e). The Anok hold this ceremony before the *bong-kom* (see R1i).

R4e) Dömrong feasts

According to statements from Kukcong-HP (Kwainjhiri-Mouza, 15.03.1957), in the Dömrong district, there are no generally applicable rules for cattle feasts, rather each clan, e.g., Hualing and Pongmi, has its own style. In Mangpung-KP (Kamichora-Mouza), on the other hand, a number of informants said that among the Dömrong there were two rules, the *chüngma-ri* and the genuine *dömrong-ri*, but they did not know much about the latter. The Chüngma rule has (at least nowadays) nothing to do with the rules that the Chüngma follow: to a large extent, the northern Chüngma are allied with the Anok, whereas the southern Chüngma (see R6) are allied with the Khumi (see R5).

I had no opportunity to take part in a feast according to the (or a) Dömrong rule; I arrived at a feast according to the Chüngma rule, held from 14. to 16.03.1957 in Mangpung-KP, only on the evening of the second day of the feast, so that for many points of the description below I have to rely on additional information obtained from a group (of changing composition) of participants in the feast on 16.03.1957.

On the first day, the festive decoration is produced; for each animal a *dong-lu-kau* (step head bamboo) is erected, with (as among the Anok) nine *chit-pap* (tassel rings) and a *dam-khai* (fish weave) at the upper end. A *rang-pang* (hornbill) is placed, not on the tip of the bamboo mast but on the uppermost tassel ring. This bird stands on a small rectangular woven structure (and thus is similar to the *waa-taa* of the southern Chüngma, see R6). On the lowest tassel ring, a banana leaf is suspended, with some cotton wool wrapped up in it: *löng-chur-baa* (see R4c and R1c). Nothing could be discovered about the significance of *rang-pang* and *löng-chur-baa*. The outer platform of the host's house is covered with transverse poles and roofed over with a mat; the gable of the roof is decorated with a *caro-ca* (see R1e), under which, in the front wall of the house (*kimma*), three *chit-wai* (tassel sticks), called *cong-chit* (sticking tassels), are stuck. It is also possible for several *cong-chit* to be stuck in the wall (as in Mangpung-KP), but three are required. On each *ling-pu,* two *chit-cüng* (see R1e) are placed, and in addition the animal is framed with six *chit-cüng*. Two more *chit-cüng* (transversely fastened with a tassel stick) form the *mong-töng* (gong stand). Rice-beer pitchers are placed only on the *mong-töng*; there are not two places, as among the Rümma. A small red coloured bamboo case (*röng*) serves to pour the water (from a round metal pot) into the beer pitchers with tassels. At the beer-pitcher place, four *teng* (swallow tail bamboo) are knocked in, in a square around a *ci-ö-ca* (pot-bellied earthenware pot) with *hom-noi* (rice mix water), from which all members of the host's household must be spat upon before the sacrificial beast is killed. After the killing, copious amounts of cattle dung are thrown at the house wall and the *ling-pu*.

No *tu* (single tubed gourd-pipes) are played for the feast, no *chi-khang* (medicine) is prepared and no stones set; not even at big feasts, such as a *riyang*, for instance, which is called *tui-ko-kröng* (water bottle stand) among the Dömrong. Also the outer platform of the host's house is never renovated. In the case of pigs, only one animal has to be killed: the *chit-khai pak* (tassel decoration pig), namely at the time when the *dong-lu-kau* is placed. The Dömrong rule, on the other hand, requires a completely new outer platform (on which a beer pitcher has to be placed) as well as the blowing of *tu* and the setting of stones at each feast, that is to say, in the case of a head of cattle as well.

According to both rules, two *riaca* (feast helpers) are necessary at each feast; according to their choice (by *khüm-ria*) they drink some of the *capam-yu* (beer on the rice container in the entrance to the *kimma*), then some uncooked rice that has been steeped in water is daubed on their foreheads. According to the Chüngma rule, they are not subject to any *kan-khang* (ban on side dishes) and can also drink rice spirit like all other partakers of the feast. Apart from the *riaca,* a *charaa* (master) has to be engaged who oversees the work and is rewarded for his efforts with a house knife (if he is one of the host's *tutma*) or a turban (if he is one of the *pen*). In addition, a so-called *pröi-cöi is also necessary*, whose only task is to consume meat and rice spirit together with the *charaa* and the *riaca*. Afterwards, the *riaca* escort the *charaa* home (*chaa-charaa*, to escort the master) and are given beer and spirit in his house as a reward. It is also one of the tasks of the *riaca* to stab the sacrificial beast to death (or at least to make the first wound, after which anybody can kill it completely). A *riaca* or the *charaa* also gives the *bong-kom* (cf. R4c), although not for all people from the host's household, but also to all other members of the clan and also their *pen* (male members of the son-in-law's clan), namely both according to the Chüngma and Dömrong rule.

On the morning of the third feast day (16.03.1957), I observed the *cak-kwak* (throwing away food). Already on the first day of the feast (the *chit-khai-ni*) a *khö* (a round basket) is woven. Cooked rice and meat from the sacrifice (parts of the intestine and the liver) are put into it. A *teng* (a small bamboo case with a swallowtail) is filled with rice-beer (*khong-tui*, i.e., sucked-up beer). The entire family gathers on the outer platform. A (random) family member takes rice and meat out of the *khö* six times and divides up the meal among his relations. Each first receives a portion in his left hand, then takes some beer in his mouth and throws and spits both away with a "*phyö*" cry. Then each receives a second portion, also in the left hand, and the same action takes place. The third portion is handed to the right hand and again thrown away with *phyöi*. Then everybody grips the basket with the right hand and throws it over the platform and down the staircase, the rest of the beer is emptied after it, and in addition a certain amount of spirit is poured out. Finally more raw meat, cooked meat and ashes follow, having been wrapped in a banana leaf for this purpose earlier. Under the staircase, the dogs scramble for the scraps of the feast – the people withdraw to their houses; it would not be good to stay outside now, since the spirits are coming for their meal.

R5 The feasts of the Khumi

a) Overview

I took part in two cattle feasts among the Khumi, a *chi-hare* of the Karbari Müilün Tamchaa on 26.–27.03.1957 and a *ku-ram* of the Karbari Kangnong Kanbeo' on 8.–9.03.1956. I searched for more information in Kangnong-KP during a second visit (23.03.1957, in the company of Menkröi), when apart from the karbari himself, among others, his father and Karbari Rennoi (who gave a *reng-tang* feast in 1957, which I learned of too late) provided useful information. According to the last-named informant, there are three kinds of feasts of merit among the Khumi: *ku-ram*, *tang-thau* and *tang-iüng*. The last two are also jointly known as *reng-tang*. A *tang-iüng* presupposes a *tang-thau*, a *tang-thau* presupposes a *ku-ram*. According to K Müilün, a *ku-ram* also cannot be given freely, rather the giver of the feast must previously have held a *chi-hare*. According to my informants from Kangnong-KP, all cattle sacrifices below the *ku-ram* are to be regarded as *nat-cang*, i.e., mere sickness offerings. They are finished on the last day of the feast; the feasts of merit on the other hand require as a follow-up (during the rainy season) a field feast, known as *lau-reng*, during which stones are set. Only married people can give such feasts. The number of sacrificial beasts is determined by the kind of feast. The following are necessary:

	Chi-hare	*ku-ram*	*tang-thau*	*tang-iüng*
Cattle	1	1, 3, 5	5, 7, 9	10, 15, 30
Pigs	1	7	15	ca.30
Dogs	–	2	4	ca. 10
Goats	–	–	2	2
Ducks	–	–	1	1

The number of chickens is not determined.

For the *lau-reng* that follows, in the case of *ku-ram,* another goat and four pigs should be sacrificed, and according to the *reng-tang,* cattle as well (see R5h). Instead of cattle, buffaloes or goyals can be sacrificed, but in the case of *ku-ram* and *tang-thau,* their number must always be uneven. It makes no difference to the character of the feast, however, whether 1, 3 or 5 cattle (or buffalo etc.) are now sacrificed. For example, for the *ku-ram*: with one head of cattle, a person can give both a *chi-hare* and a *ku-ram*, with five beasts, both a *ku-ram* and a *tang-thau* as well. The stated number of pigs represents the lowest number for the particular type of feast in each case. Other animals can be killed for eating, and, in addition, certain procedures associated with pig sacrifices can also be adopted from higher-ranking feasts (see R5b: *chi-hare* with three pigs). Almost all large feasts that are given nowadays are *ku-ram*; *tang-thau* are very rare and *tang-iüng* now exist only in hearsay. The *tang-iüng* include, as far as is known, all works of the *tang-thau*, for which a large roof (probably corresponding to the *chuk-calaa* of the Anok, see R3d) is erected above the place of sacrifice; the length of the feast was given as thirteen days. It was not possible to obtain more precise details about this big feast that went beyond mere opinions from the Khumi that were asked. However, see section R6: the *tang-tum* of the Longhu described there should correspond to the *tang-iüng* of the Khumi. According to this, the number of cattle given above should be corrected from ten to nine, so that at all feasts the number of cattle to be sacrificed would be uneven. As in the sections above, I also use Mru terms in the following descriptions, wherever an English equivalent of the specialist expressions exists only to a limited extent. It would doubtless be more correct to use the relevant Khumi terms in this case, but I have three reasons for staying with the Mru: 1) I thus spare the reader from having to learn more new vocabulary; 2) when recording I often dispensed with asking for the Khumi expressions myself, while I was receiving my information in the Mru language; and 3) in this way the comparisons between the Khumi feasts and the corresponding feasts of the Longhu (southern Chüngma Mru, see R6) are made considerably easier. Wherever it is not clear from the context, I identify Khumi words with (K).

R5b) *chi-hare* (K)

Literally translated, *chi-hare* means "one head of cattle", "a cow", but at the same time it stands for the simplest type of feast. In the simplest version, apart from one head of cattle, only a pig (as *bong-kom pak*, see below) and a chicken are necessary as sacrificial animals. This chicken is called *chi-lu-a* (K; Mru: *cia-lu-wa*), that is to say "cattle-head-chicken"; at the same time, however, *chi-lu* is the general designation for the place (characteristic for each Khumi house) where the skulls of previous sacrificial animals are suspended above and alongside each other. In front of this skull wall, a *hom-noi-pu* (rice-mixing-pot, see R1h) with sucking tubes is placed. The giver of the feast takes some liquid in the mouth out of this three times and spits it with "*phyöi*" and "*tamma*" (that is to say, a call to the evil spirits, see R3j) over the chicken (that is to say, unlike with the Mru, there is no counting to six with a single *tamma*, but a threefold repetition). Then he cuts through the chicken's neck with the hewing knife, which he lays on the floor for this purpose with the cutting edge facing upwards, holding it in place with his foot. He drips some of the blood on the *mong* (large gong) and *ner* (plate gong) hanging on the *chi-lu*. Then the gong is struck (in the observed case by the karbari's grandson). The chicken is cooked with salt and pepper; meanwhile the sacrificial beast is tied up in front of the house. Some of the meat of the chicken is wrapped with cooked rice in three small portions in pieces of banana leaf, that is to say *köm-pot* are prepared. After another threefold *cho* (spitting) with *hom-noi* (rice water) before the *chi-lu* stand, one of the *köm-pot* is made fast behind a small bamboo cane on the central house post (beside the *chi-lu*), the other two are thrown away on the staircase with *tamma* (calling, but without *cho*). Then the giver of the feast dips a piece of ginger into the chicken broth (the water with salt and pepper in which the chicken was cooked) and dabs the foreheads of all members of his household with it. The piece of ginger is then skewered on a small stick and also stuck on the central post of the house. The drum, plate gong and large gong are daubed with some chicken meat; then the giver of the feast picks out the tongue from the chicken meat and sticks it with another little stick to the central post. The rest of the meat and rice is eaten by the grandchildren along with the giver of the feast and his wife. To conclude, the plate gong is struck three times and the large gong is also struck three times.

After that the karbari takes the plate gong, goes to the feasting-place and makes a *ner-tep* (beating the plate gong), stands beside the animal and calls "*ho"* twice, beats for a short while in two different rhythms and then goes back into the house. There is no *taröng* in the Mru style; also only one plate gong is struck on ceremonial occasions. For the dance at larger feasts, however, as many other plate gongs can be beaten as desired. There is no need for dancing for a simple *chi-hare*; however, if people want to dance too (as in the case of the observed feast), the dancing begins on the night after the first day (on instruments and dance forms see R5e).

The next morning, the bamboo decoration is prepared (this generally also takes place on the first day, but because of my visit the host advanced the date of his feast, without giving the men of the hamlet who helped with preparation of the decoration enough time to change their plans).

Beside the staircase (which leads to an outer platform that connects the main house with the side house), a *dong-lu-kau* (K: *chamthing-kaleong*) is set up; it consists of a bamboo mast over 10 m long without leaves, the only adornment of which consists of a *dam-khai* weave (K: *catam-tamai*, crocodile tail) tied on at the tip, at the end of which a simple crown hangs. The crown consists of four long bamboo tubes somewhat over 10 cm in length, which lead to a ring of approximately 10 cm diameter from which three white paper snakes approximately 50 cm long hang down. Directly beside the *dong-lu-kau* a *caro-ca* (K: *careo'*) (approximately 5 m in length) is set up: a tassel pole, the uppermost half internode of which is decorticated and split up in the shape of a funnel, while the two internodes that follow below, which are scraped up on two sides each to form tassels, each have three chit-*wai* (tassel sticks) inserted. These *chit-wai* (K: *capon*) are one and a half internodes long and carry two tassels; one hangs at the upper end of the stick, the other, at the central node, terminates just before the insertion hole in the tube of the *caro-ca*. Between both tassels, the sticks have some scraped buds. (I did not find among the Khumi the tassel that corresponds to the *klang-ui-chu* of the Mru.)

Three more *caro-ca* of approximately the same size are placed (right, left and centre) in front of the main house, a fourth is placed in front of the centre of the side house. A fifth duplicates and extends beyond the above mentioned *caro-ca* in front of the centre of the main house by somewhat more than one metre. A sixth *caro-ca* is finally placed on the *ling-pu*. On the platform between the staircase and side house, a porch is joined on (at the same height), on which a beer pitcher is placed. The corner posts of this little platform, which are about three metres in length and extend about 1.20 m above the platform, are also given the form of *caro-ca*. A third central pole has tassels inserted like the *caro-ca*, but no split top (*lu-kut*). A wall that consists of six bamboo strips is attached to these three poles. The strips are attached in such a way that one green and one white side face outwards alternately.

Above this wall a red cloth with a printed pattern of yellow flowers (of Bengali origin, used normally as a drape) is spread. In addition, the wall is decorated with *tür-ram* fronds (K: *ku-ram*, *Didymospermum spec.*); two other *tür-ram* are tied to both stems of the *ling-pu*. For each head of cattle, two stakes have to be set; the tether is wound round both. *Tür-ram* and drape are not necessary for a simple *chi-hare*; if they are employed, then apart from the pig (*bong-kom pak*) that is killed in the house when the *ling-pu* is set, two more small pigs and a dog have to be sacrificed for the *tür-ram* decoration and the hanging of the drape. Before their meat is eaten, for a short time, the plate gong and large gong are struck in the house. Towards 10:30, the young people start the dance again, during which time the foreheads of the skulls on the *chi-lu*, that of the sacrificial beast, the dancers and finally the remaining participants in the feast are daubed with steeped rice meal (K: *takô*). The band consists of a drum, a plate gong and a cymbal; the large gong hangs on a completely unadorned gong stand and is struck with a drumstick. Neither beside the gong stand nor on the feasting place are there any ceremonial places at all, with beer pitchers or even only swallowtail bamboo tubes.

Towards 11 o'clock, without any special invocation or the like, the ox is killed with a spear. Since it is tied up with a long rope so as to give it great freedom of movement, and the young people who are trying their hand at spearing are not very skilled, they repeatedly strike the beast either on a rib or a shoulder blade. A spear goes into the belly, the spear point gets stuck and a new spear is brought. The cow keeps falling to its knees, but since it is not considered permissible to kill the beast while it is lying down, water is poured over its snout so that it gets up again. Only after about nine attempts, which provides a great deal of amusement for the watching children, the cow collapses completely; although it seems not to be completely dead, its tongue is pulled out, stabbed with the spear (without sprinkling of water), cut off with a hewing knife and placed on the *ling-pu* post (not pinned in place). The cow is then untied, dragged to the house of the feast-giver and heaved up on to the platform. The dancers follow, dance first in front of the house on the ground, then on the platform around the dead animal. Meanwhile they get *khong-tui* (sucked up beer) to drink.

Before the beast is cut up, the karbari's son brings his wife. She is pulled from behind under the right hind leg of the beast (the animal is lying, as usual, on its left side, with the entry wound facing up). As she stoops, this hind leg is pulled away over her six times. This ceremony is intended to help her to be fruitful, since she has had no children so far. A second ceremony follows: on the place between the beast and the house the *tutma* (mother's brother) of the karbari's son throws two cowry shells on to the shovel shaped winnowing fan (*per-ching*), in order to ask whether the karbari's grandson is to be given the name of the karbari's (dead) elder brother; the child's mother had had a dream to that effect. If one cowry falls on the front side and one on the rear side, the name counts as good and is accepted, whereas if both the cowries come to rest on the same side, the name is rejected. The same ceremony is also carried out at the birth; after the name giving, the two cowries are given to the child tied round its right wrist with a cotton thread.

As a third ceremony, at the same place between beast and house, the *kom-bong* (wrist-tying) follows. The one officiating in this case is again the *tutma*; this is a rule for all cattle feasts of the Khumi, only for pig sacrifices does the householder himself give the *bong-kom*. The *tutma* is to be rewarded for his services with a bottle of rice spirit and a spear (other helpers at the feast are not necessary in the case of a *chi-hare*). The *bong-kom pak* (wrist-tying pig) has already been killed (together with the other pigs) in the morning; some of its blood is sprinkled on the prepared *bong-kom* threads. Now the *tutma* ties one of these threads around the right wrist of each family member, then bites some ginger from a stem, takes a mouthful of spirit and spits both with a *tamma* over the arm of the person in question, blows then six times over a small calabash, in which some husked rice has been placed (*la-rik-yia*), and finally dabs some rice meal paste (K: *takô*) on the forehead of the person in question, and the back and breast. Not only the people of the feast-giver's house are so treated, but also (provided they are present) married sons of the feast-giver who have already made their own homes, and younger brothers, provided they have the same father and mother as the feast-giver.

Now the animal is cut up and the meat cooked. Small pieces of the cooked meat are wrapped with rice in fragments of banana leaf; these spirit morsels are distributed in the house (the *chi-lu* stand being the chief place for this observance), the remnants are thrown away down the staircase. Four bowls with rice and cooked meat are placed on the outer platform and people dance around them to the sound of the beating of the plate gong and drums and the playing of a *tang-plung* pipe (see R5e); then these dishes can be eaten (by anyone who wants to), and shortly afterwards (at about 15:00) the general meal follows. Late in the evening more dancing takes place, this time inside the house.

R6 The feasts of the Longhu

R6a) *klik-lök*

Historically, the Longhu (Southern Chüngma Mru) have adopted the feasts of the Khumi, though some variations have arisen with the passage of time. The feasts take place in the following sequence: *klik-lök*, *ku-ram-calang*, *ku-ram-bung*, *catang-thau*, *tang-tum*. Before any of the feasts in this series can be held, the preceding one must already have taken place. A *klik-lök* ("tethering-one") can be given by anyone who is married. One head of cattle and one chicken are needed. First, the chicken (*cia-lu-waa*, "cow-head-chicken") is killed; this is done at the place in the house where the skull of the cow will be hung later. If there are already skulls hanging there from previous feasts, some of the chicken's blood is dripped on them. The chicken is then cooked, and some of its flesh, together with some boiled rice, is put down in front of the *cia-lu* stand (the equivalent of the *chi-lu* of the Khumi, see R5b). This offering is called the *kan-köm* (*kan* = side dish). A *cia-lu-waa* is needed for any cattle feast, whatever its size (Phungkri Kwai, Uikük-HP, 14.01.1957).

Further pieces of the chicken flesh are fastened to the upper ends of two long, thin *teng* (swallowtail bamboos), which are placed in front of the *hungma-cüng* (central post of the house) for all feasts, even for pig sacrifices. Nothing is put into the *teng* themselves. In the afternoon or evening of the first day, the sacrificial cow is tethered on the feast area. Dancing is not customary at a *klik-lök*. Before the animal is killed the next morning, *kom* (rice flour paste) is prepared. Some of this paste is dabbed first on the old skulls that are kept in the house, then on the forehead of the cow and on the brows of the people taking part in the feast. For a *klik-lök,* no turmeric may be added to this paste (K Laichia Rongdim, 28.03.1957).

As is customary amongst the Khumi, two poles must be erected per animal as *ling-pu*. These can be very small, and need only be about the height of a man if there are no longer enough old posts on the *ling-pu* area. Khuchi Patlaica, for example, in Tuktöng-Para, set only two 50-cm branches from a *ram-nget* tree instead of posts. One of these branches was forked. Amongst the Anok, the sacrificial posts are always unforked; with the Longhu, forked posts are not usual, but they are not against the rules. On the day of the tethering, only one of the posts is put in position; the second follows the next morning, together with two decorative bamboos, a *caro-ca* and a *mang-kut*. The *caro-ca* is about three metres long, and takes the same form as with the Anok and Khumi (see R5b). The *mang-kut* resembles the *caro-ca* in that it also has *chit-wai* (tassel twigs) set into it; however, the top half internode is not split open, but only decorated with strips of shaving (*thau-rin-bong*). The two tubes are tied together with a bamboo thong. They must be erected at all cattle feasts, no matter how many beasts are being sacrificed, and therefore regardless of how many *ling-pu* posts there are (Khuchi Patlaica, Tuktöng-Para, 15.01.1957).

The *dong-lu-kau* ("stairs-head-bamboo") resembles that of the Khumi in its simplicity, although it bears not one, but three *dam-mai* ("fish-tail"; also called *dam-lip*, "fish-scales", which is the equivalent of the Anok *dam-khai*; see R1c). One *dam-mai* is fixed to the top of the *dong-lu-kau* and the other two are fixed to lower branches. At the feast of the Karbari Wöilaa (Tuktöng-Para, 26.03.1957), the *dong-lu-kau* had only two *dam-mai*; the third one hung from the top of a *caro-ca* which projected from one of the front gables of the feast-giver's house. A *klang-chu* ("?tree name?-grape") is suspended from the lower end of the *dam-mai.* It consists of a bamboo ring from which hang four twigs (*chit*). Some cotton is hung from one of these twigs, and some tobacco from a second. I was not able to discover the reason for this. The Khumi use only cotton (H. Renglün Amchang, 26.06.1957). A *caro-ca* is placed immediately next to the *dong-lu-kau*; another *caro-ca*, which hangs over the gable, stands centrally in front of the house. A further *caro-ca,* together with a *mang-kut,* is placed on one side of the platform. A *hom-noi-pu* (a pot holding mixture of water and cooked rice) must also be placed there, but is not used. At larger feasts, a beer jug from which people can drink stands in the same place (K Wöilaa, Tuktöng-

Para, 28.03.1957). As is customary amongst the Khumi, the sacrificial beast is hauled on to the platform and cut up there. The head is set aside and is not cut any more until the third day of the feast. The lower jaw is then removed, together with the skin and the remaining flesh, except for of a piece of skin over the forehead. The prepared skull is then put on to a support (*tui-krök*) over the stove to dry. On the third day of the feast, there is also an expedition to the river or stream to catch small fish and crabs, which are cooked in a pot with bitter leaves to make a *chur-ngar*. In the evening, the plate gongs in the feast-giver's house are struck. In common with other Mru groups, and unlike the Khumi, they have three of these gongs. The cymbal is also struck; this is the only time it is used at the feast. On the morning of the fourth and last day of the feast (every Longhu cattle feast thus lasts at least four days), the *chur-ngar* ("sour- meat"?) is eaten; if a *kan-khang* ("side-dish-ban") is to be observed, only the meat is eaten. At a *klik-lök*, a *kan-khang* is not obligatory. The final ceremony is the *cia-lu-kop*, the fixing of the animal's skull to the *cia-lu* stand; no new sacrifices or invocations with *cho* (spitting etc.) are needed; however plate gongs and drums are beaten during the ceremony, and also a gong if available. The instruments are put away afterwards, and may not be beaten again before the next sacrifice (K Wöilaa, Tuktöng-Para, 28.03.1957).

A *chüng-nam-prüt* ("spirits-chasing out"), such as the *cang-pang-yöng* (see R1i) practised by the Anok, is not known amongst the Longhu (Phungkri Kwai, Uikük-HP, 16.01.1957). At least at larger feasts, the Mru and the Khumi carry out a *wi-klak* ("blood-shedding"), by sacrificing a hen or a goat at the watering place or the bank of a stream. The Longhu do not regard this as necessary, even if the feast is a large one, though anyone who wishes to do so may kill a chicken (Khuchi Patlaica, Tuktöng-Para, 28.03.1957). Cattle can be sacrificed on the bank of the stream, but not in the context of the feasts described here. The practice of *khyong-malong* (Burmese: "pouring into the river", cf. the Anok *cok-malong*, R3j) is a sacrifice (*nat-cang*) to guard against illness, such as fevers and dysentery. In that case, the cow is slaughtered with the hewing knife; a hen and an egg are also sacrificed. The offering is made to the spirit of the mountains (*tong-mang*, Burmese "mountain-ruler"), even though the sacrifice must be carried out on the bank of the stream (Phungkri Kwai, Uikük-HP, 16.01.1957).

R6b) *ku-ram*

The Longhu have two *ku-ram* feasts, the *ku-ram-calang* and the *ku-ram-bung*. These differ according to the number of pigs and cattle required. For a *ku-ram-calang,* one to three head of cattle[483] and at least three pigs are needed (*pün-töm-ngar*, *ku-ram pak* und *bong-kom pak*; see below). For a *ku-ram-bung,* however, there must be 3, 5 or 7 head of cattle (even numbers are

[483] At the feast which H. E. Kauffmann in a manuscript calls *reng*, and which is identifiable as a *ku-ram-calang* because of the use of turmeric paste, a calf and a buffalo were sacrificed. This is reminiscent of the Dömrong combination of buffalo and *lu-ngua*.
On the evening of the first day, the first pig is stabbed with a bamboo spear on the platform of the feast-giver's house, at the post that supports the front end of the roof ridge. This pig is called the *kung-prek pak* (also the *kung-takek pak*, or "helper-choosing-pig"). As with the Khumi, there is no special ceremony – such as the *riaca-khüm* of the Anok – for choosing the feast helpers (the *kung*, see below). The same night, there is dancing in the house of the feast-giver; at this *bök-bök-plai* ("everything-dances") plate gongs, drums, cymbals and single-pipe gourd-pipes are played.

forbidden) and 9 pigs. The scale of the *ku-ram-calang* is similar to that of the *chi-hare* of the Karbari Muilün Tamchaa (see R5b). Here too, the typical festive decoration consists of the *tür-ram* leaves (*Didymospermum sp.*), which are tied to the platform of the feast-giver's house and to the sacrificial posts. The Longhu and the Khumi call this tree the *ku-ram*, and it gives its name to the feast. Although there must be at least three head of cattle for the *ku-ram-bung,* its scale does not go much beyond that of the *ku-ram* feast of the Karbari Kangnong Kanbeo' (see R5c–f). Buffaloes or gayal may be sacrificed instead of cattle; this does not entail any change to the way the feast is conducted. With regard to the cattle, the sex of the sacrificial animals is not important; in the case of the pigs, they must be castrated animals. The number of chickens to be sacrificed is not prescribed. However, before the first cow is killed and butchered, there must be enough chickens to feed all the helpers and – most importantly – all the guests who have a *tutma* ("wife-giver") relationship to the person giving the feast.

On the second day, the platform of the feast-giver's house is renewed, and the necessary decorations are put up. A pig and a cow are then killed on the rear platform (*kung-car*) and a further pig and cow on the front platform (*pur-car*). These are called the *pün-töm-ngar* ("climbing-place-sacrifices"). Large quantities of *ku-ram* leaves are brought to decorate the front platform and the *ling-pu*; when they arrive, a *ku-ram pak* is killed as an offering for them. No leaves other than those of the *ku-ram* are used as decoration for the feast. At a *ku-ram-bung*, a further pig is killed for all those who help to decorate the house; this is the *pau-dim pak* (the "bamboo-'flower'-pig"). The water for washing the pigs and filling the beer jugs is fetched not in calabashes, but in lengths of bamboo (*tui-lam*). The water carriers have their own beer jug, which must be placed on the new platform. In addition, a further pig is slaughtered; this is the *tui-lam pak* (the "water-tube pig"). Some of the blood of this pig is painted on the bamboo tubes.

There is no particular ceremony attached to the felling of the sacrificial posts for the cattle. Those who bring the posts into the village stop at every house on the way to the feast area and are entertained with rice-beer and rice-liquor. Before the posts are erected at the *ling-pu* area*,* a pig (*ling-pak*) is killed; some of its blood is painted on to the new posts before they are set into the holes. Nothing else is thrown into the holes before the posts are placed in them. Before the cattle are tethered to the posts, another goat is sacrificed at the *dong-lu-kau* ("stairs-head-bamboo") of the feast-giver's house. It is killed by cutting its head off with the hewing knife (*ngen rua* = "ritually slaughtered goat").

When the cattle have been tethered, the people dance round them with plate gongs, drums and cymbals. This *cük-cük-rik plai* (very first dance) corresponds to the *taröng* of the Anok (see R1h), but the number of circuits is not prescribed, and they are not counted. For the *ku-ram-bung* dance, three *tu* (single-pipe gourd-pipes) are also needed. Unlike the Anok custom, the *tu* and the drums are not decorated with leaves, and no particular *tu* songs are known. (Neither the Longhu nor the Khumi use *tu* at funerals.) A gong is hung at both *ku-ram* feasts. The *mong-töng* (gong stand) consists of a "gate" made of two upright bamboos, with a third one bound across them; these are not decorated with bamboo twig tassels. After the *cük-cük-rik-plai*, gourd-pipes (*plung*) are added to the other instruments and the general dancing (*bök-bök-plai*) begins. If the dancers have the stamina, it lasts until morning.

The dance starts again at midday on the next (third) day. From time to time, the older people join the younger ones and dance the *long-tang-plai* ("old people-dance"). Before the older people of the feast-giver's house come down, a further pig – the *preng-kom pak* ("turmeric-paste pig") – is killed on the front platform of the house. Drops of its blood are mixed with water, cooked rice and ground turmeric in a bamboo receptacle. Some of the resulting paste (*kom*) is smeared first on the foreheads of the sacrificial animals and then on the brows of everyone taking part in the feast. This *kom* is not the same as the *chi-khang* of the Anok, who only make it for *tang* feasts (see R6d).

The ninth and final pig is only killed after the sacrifice of the last head of cattle. This pig is the *bong-kom pak,* ("wrist-tying pig"); some of its blood is dripped on to the *bong-kom* threads. All the cattle can be sacrificed in one day, spread across the morning and afternoon.[484] They are hauled on to the new platform of the feast-giver's house to be cut up, and the *bong-kom pak* is killed on one of these animals. Before the *bong-kom,* the feast-giver gives a *pang-cöng,* an event at which not only the *tutma* (members of the father-in-law's sib), but also the *pen* (members of the son-in-law's sib) are honoured – and especially the husband of the feast-giver's sister. The sister's husband must be offered a jug of rice-beer and at least one bottle of rice liquor, plus an upper leg of beef and one of pork. He also has a new turban wound around his head. In return, the feast-giver receives from his brother-in-law between five and fifteen rupees in cash. The day ends with a *kim-pün-plai* ("house-climbing-dance"), which is customary at all cattle feasts except for the simple *klik-lök*, where there is no dancing. This dance takes place in the feast-giver's house, and once again all the instruments are played, including the *tu* (single-pipe gourd-pipes). Finally, the dancers may quench their thirst from the beer jugs set out for them in the house.

During the feast days, the feast-giver has helpers at his side for all the important tasks. These correspond to the *riaca* of the Anok (see R3h) and are called *kung-kham-mi* ("afterwards-drinkers") or simply *kung*. As with the sacrificial cattle, the number of the helpers must never be even, i.e., there must be, for example, 5, 7 or 9 *kung*. A group of seven men is regarded as particularly good. The total number of *kung* is independent of the number of sacrificial beasts. There are no particular ceremonies for choosing the *kung*; anyone who wishes to do it and is acceptable to the feast-giver can undertake this role. The most important task of the *kung* is to make the guests comfortable, serving them food and drink, and taking care of any who have overindulged. The feast helpers themselves, like the feast-giver, may drink neither beer nor spirits, and they are also subject to a *kan-khang* ("side-dish-ban") since they must avoid *ngaa-pi* ("rotten fish"), all pickled food, seasonings other than salt, and oil and taro tubers. However, they can eat unseasoned meat. The feast-giver and his family, who are subject to the same *khang,* may eat meat from the sacrificed pigs and cattle. In return for their services, the *kung* receive five rupees each; rice liquor,

[484] According to Kauffman's report from Changyung-Para, the calf and the buffalo were killed in the morning, hauled *on* to the *car*, and then the *bong-kom pak* was sacrificed there. The preparation of the turmeric paste and the *long-tang-plai* did not take place until the afternoon.

beer and meat are put aside for them until the last (fourth) day of the feast; and when the guests have left and the *chur-ngar* (see R6a) has been eaten, they make up as well as they can for what they have missed. However, this does not mean that the *kan-khang* placed on the helpers and the feast-giver is completely lifted. For a further year they are forbidden to eat *ngaa-pi*, curry spice, onions, taro, game, and the meat of wild pigs. Domestic pigs, however, can be eaten seasoned with pepper, salt, turmeric, ginger, etc. When the year is over, a new feast must be held for the lifting of the food restrictions (*kan-khang-phöt*). This new feast is a small *ku-ram-calang* with one head of cattle and two pigs. It is called the *pün-töm-lat* ("climbing-place-unbinding"). At this event, the existing platform of the feast-giver's house is dismantled and then restored to the form that is usual before a *ku-ram-bung*. For this feast, as for *ku-ram-calang* in general, three *kung* are appointed. These *kung* only hold office for three days; they are allowed to drink alcohol during the feast; and they are not paid. No further *kan-khang-phöt* are needed for these *kung-kham-mi* at the *ku-ram-calang* (Phungkri Kwai, Uikük-HP, 14. and 16.01.1957).

R6c) *üa-reng* (*ku-ram-pahok*)

Like the Khumi *lau-reng* (see R5h), the Longhu *üa-reng* (field feast) can be held without a preceding cattle feast. When there is illness, for example, it can be pledged as a *nat-cang* (sickness offering). However, when a cattle feast has been held, it must be followed by an *üa-reng* (which is also called a *ku-ram-pahok,* though *ku-ram* leaves are not used for this feast). The *üa-reng* is held after the first weeding; unlike the *chia-dong-khang* (see P3e), it is a private celebration, and does not affect the everyday life of the other people in the village. For lesser feasts, up to and including the *ku-ram,* it is enough to sacrifice one goat; from a *ku-ram-calang* upwards, at least one large goat must be offered. A *ku-ram-bung* requires a cow. In addition, two pigs, a small dog, a chicken and an egg must be sacrificed in each case.

On the day before the sacrifice, *waa-taa* and *waa-tur* decorations are woven at home. The *waa-tur* is a (wild) dove; the *waa-tur* decoration is the counterpart of the *rang-pangma* (hornbill) weaving made by the Anok (see R1d). *Waa-taa* in fact means black hornbill (*Aceros spec.*), but the *waa-taa* decoration does not look like a bird. It is a simple, flat mat, measuring about 10 x 20 cm, and made by weaving bamboo strips. The plaits are then pulled apart in the middle so that the mat can be clamped horizontally on to the end of a plain bamboo pole. The "dove" is then stuck into the top opening of the bamboo pole, so that it more or less sits on the *waa-taa* mat. Finally, a *chaling-chalap* (butterfly) is hung from the pole; this consists of two crossed bamboo strips, around which red and black threads are wound, working from the centre outwards. (The Khumi do not use thread crosses for the *lau-reng;* see H. Renglün Amchang, 26.06.1957.)

Eight of these woven decorations with their with thread crosses are placed on bamboo poles around the roof, one on each of the four corner posts, one at each end of the roof ridge, and one on each side of the roof in the middle. Inside the house, a string is stretched from the front end to the back end of the roof ridge beam (*klong*); from this thread, *waa-tur* and *chaling-chalap* are suspended alternately, eight of each as before. An additional *waa-taa* – normally without *waa-tur* and *chaling-chalap* – is set on a bamboo pole about 2.50 m long and placed at the *ling-pu* (sacrificial post). Another one is placed at the *plon* (entrance to the village). After a *ku-ram-bung,* a *waa-tur* is set on each of these *waa-taa,* but no thread crosses are hung from them. Two further *waa-taa* are placed at the entrance to the village. Finally, these woven plaits are also needed in the field. Eight *waa-taa* are placed on the field house, in the same positions as on the village house, though only the *waa-taa* on the front gable has a *waa-tur* and *chaling-chalap.* Six more *waa-taa* are set at the *tur-tut* (see J2e), that is, the field sacrificial place (Angkaa Atwang, Laichia-KP, 27.03.1957).

Before the trip to the fields, some of the sacrificial animals – a pig, the small dog and the chicken – are killed on the platform of the village house. The pig is stabbed with a spear, the dog is beaten to death with a club, and the chicken is killed by cutting its head off through its open beak with a sickle. The flesh of these animals is cooked, taken to the field, and eaten there. The cow (or the goat) is led out alive to the *tur-tut* area, tied to a post (called the *muk-ma)*, and the people circle around it to the beating of *ner* and *mong* (plate gongs and a large gong), but without drum accompaniment (Phungkri Kwai, Uikük-HP, 14.01.1957). Animals are not killed in the field. However, a narrow strip of the cow's ear is cut off and thrown away at the *tur-tut*. There too, the egg, which the villagers have brought with them, is broken and its contents are allowed to drip on to the ground. Rice leaves and a small branch from a cotton bush are then fastened to the cow's tether and the animal is led back home. Some of the egg-soaked earth from the *tur-tut* is also brought back. The returning villagers may not enter the house straight away; they stay at the *ling-pu*, where the cow is tethered to *ling-pu* posts (with *caro-ca*) that have been newly erected for this purpose. The people then circle around it to the beating of *ner* and *mong*. (The same happens with a goat.)

When they return from the field, the villagers bring with them a stone from the stream. This stone is now set up (Angkaa Atwang, 27.03.1957). For a large cow, the stone is a large one; for a small cow, it is smaller; for a goat, it is a very small one. Before the stone is set up, a pig is placed upon it at the *ling-pu* and stabbed to death (Phungkri Kwai, 14.01.1957). The foreheads of the participants are smeared with the egg-soaked *tur-tut* earth; at the same time, they are given beer into which pig's blood has been dripped (*caloima-yu*). The beer is served from a bamboo receptacle (*röng*).

When all the participants have drunk, they may go back into the house. The cow is then untethered, and hauled on to the platform of the feast-giver's house, where it is killed with the hewing knife (Angkaa Atwang, 27.03.1957). While this is happening, the plate gongs are struck, and now for the first time there is dancing, with sword and shield. It is only at an *üa-reng* following a *ku-ram-bung* that there is a big dance (*bök-bök-plai*) at which gourd-pipes and so on are played, and which lasts all night (Phungkri Kwai, 14.01.1957). Guests are welcome at this feast – the village is not closed – but the next day, the people who live in the feast-giver's house must observe a *khang,* during which they may not work in the field. On the following day, the rice-beer left over from the feast is poured away, and then another day's *khang* must be observed; this is called the *yu-kwai-khang*, "beer-emptying-ban" (Angkaa Atwang, 27.03.1957).

R6d) *catang-thau*: sacrificial animals

When a *catang-thau* is held, 5, 7 or 9 head of cattle must be sacrificed, as well as 15 pigs, 5 dogs and 1 goat, and any number of chickens. The pigs and the dogs are killed to mark particular stages or for particular purposes, as follows:

1) a *kau-bong met,* for the cutting of the bamboo to make the bamboo decorations (*kau-bong*);

2) a *kung-prek pak*, as advance payment for the helpers (*kung*), who begin their duties after eating this meal;

3) a *tung-chöng pak*, after the lower part of the house (*tung*) has been renovated and surrounded by a border (*chöng*);

4) a *carai'-ngar*: a pig and a dog are sacrificed when a "rest house" (*carai'*) is erected;

5) a *kong-no met*, a pig and a dog, when a decoration in the shape of a horse's head is woven (*kong-no = horse*);

6) a *kim-kung-car pak*, for the rear platform (*kung-car*) of the house (*kim*); a dog is also killed;

7) a *kim-pur-car pak*, for the front platform *(pur-car*), again in addition to a dog. (6) and (7) together make up the *pün-töm-ngar*, i.e., the sacrifice for the places via which one ascends into the house (*pün-töm*, see R6b);

8) an *anum-röng pak*, for the covering (*anum-röng*) of the front platform with pieces of cloth, to create shade (see R5c);

9) a *tui-lam pak*, for the bringing of the necessary water to the feast-giver's house. The water is brought in bamboo pipes (*tui-lam*);

10) a *ku-ram pak*, when the palm leaves (*ku-ram*) are brought for decorating the platform and the sacrificial posts;

11) a *dong-lu-kau ngar*: a pig, a dog, and a goat, which are killed on the staircase (*dong*) of the front platform, together with the

12) *dong pak*. Some blood from the dog must be dripped on to the stairs head bamboo (*dong-lu-kau*), the new stairway (*dong*), the platform (*car*) and the sacrificial posts (*ling-pu*). Some goat's blood is dripped on to the *dong-lu-kau* and the *dong;*

13) a *preng-kom pak*, for the making of preparations like *chi-khang* (a medicine, see below) and *preng-kom* (turmeric paste);

14) a *li-tong-khep pak*, on the occasion of a dance with rice pestles (*li*), which is described at the end of the next paragraph; and finally

15) a *bong-kom pak*, for the closing ceremony, in which threads are tied (*bong-kom*) around the wrists of the members of the feast-giver's family (Khuchi Patlaica, Tuktöng-Para, 15.01.1957).

In general it is not clear why particular animals are to be killed in connection with particular tasks. There is a general view that the animals are killed for the people who carry out those tasks, since they receive no other payment. However, the workers do not eat the animals themselves, but share them with other people at the feast. "For the workers" means not so much "for the workers to eat" as "for their welfare". The avoidance of accidents is part of this; the sacrifice can therefore contribute to the success of the work. A large feast may arouse envy, which must be assuaged. The dangers are conceived in the form of spirits. This is suggested by the use of the word *met* for some of the pigs: this term carries implications of reconciliation and atonement. Dogs have a certain role in the sacrifices because they are guardian animals. The term *ngar*, which is used in this context, is often used to denote animals of various kinds that are killed in connection with a ceremony.

The links between the pigs and particular tasks or objects are fairly loose. One does not have to specify which pig is being sacrificed for what purpose. It is therefore not surprising that the list given above does not agree exactly with that given for the Khumi *tang-thau* (see R5g). The Khumi, for example, only specify one pig for a *pün-töm* sacrifice, whereas here we have two. For the Longhu, the number is determined by the fact that two pigs must already have been sacrificed as *pün-töm-ngar* at the *ku-ram-bung.* However, instructions for Longhu feasts also differ amongst themselves, for example, for the *ku-ram-bung* – as for the *tang-thau* of the Khumi – a *pau-dim-pak* and a *ling-pak* are indicated. Equivalent sacrifices are therefore to be expected at the *catang-thau.* Furthermore, the above list contains a *li-tong-khep pak,* which the Khumi do not include. The Khumi on the other hand have a *therca pak* and a *tem-cong-mai pak*, which are not listed for the corresponding Longhu feast. However, the largest feast, the *tang-tum*, includes all the pig sacrifices listed here.

This suggests that certain ceremonies belonging to the higher feasts can also be carried out in part at lower feasts (cf. the numbers of cattle, which can be the same at different feasts), so long as a pig is sacrificed as appropriate. However, it also seems that some tasks and services that are needed even at smaller feasts are only rewarded with a pig at larger feasts. For a *tang-tum* (see R6i), a *deng-hap-mün-mi pak* must be killed; this is a pig (*pak*) for those people (*mi*) who cut up (*mün*) the banana leaves (*deng-hap*). A *büi-pak* is also killed; this is a bamboo-rat pig, for catching the bamboo rat that is needed also at a *tang-thau.* In general it seems clear that as long as at least 15 pigs are killed, they do not have to be assigned individually to designated tasks or purposes, except where a particular time and place are specified for the sacrifice, as for example in the case of the *bong-kom pak.*

The feast-giver has to convey some cooked meat from the sacrificial beast, together with rice, to every household in the village. This rule is known as *kim-leng* (house-visiting). After dividing up, comes the *chup* (honouring): Anybody who so desires eats some raw meat with salt and pepper and throws a bit of it over the outer platform for the spirits. This is the only occasion among the Mru when raw meat is eaten (*pri-yong*, like the tiger). Meanwhile the young people's dance continues past mid-day, although there are now sizeable gaps in the ranks because of the amount that has been drunk. Everybody that the host esteems highly has a bottle of *khong-tui* (sucked-off water; pale beer sucked out of the pitcher) hung around his neck on a strap made of a strip of bamboo and is invited to join the dance; to do this the dancers (like the host's house previously) are pelted with dung from the entrails of the sacrificial beasts. The dance ends when the banquet begins, early in the afternoon; a *kim-pün* (see R2d) is not usual.

On the third day of the feast, which I didn't attend in Nongneng-KP, stones are set, a stone for each beast, even if only one beast has been sacrificed. Some blood and dung from the animal is left at the place in the brook where the stone lay. Some of the stone is ground off: together with pieces of the animal's ear, which was cut off the previous day, the grit from the stone is twisted into the threads that are finally used for the *bong-kom* (tying around the wrists of the family members). The *laa-rik-yia* is blown again (see above) and the beer jug on the *dong-lu-kau* serves as a *bong-kom yu* (wrist-tying beer). – As usual, the stone is buried together with the lower jawbones; afterwards it receives a *köm-pot* (spirit portion), as does the *ling-pu*. More *köm-pot* is distributed in the house. Then all those who have taken part in the dance have to be fed by the host. On the fourth day, as mentioned at the beginning, *kan-khang-phöt* and *pang-cang* then take place. For the *pang-cang* (according to Karbari Rengtan Rümthu, 14.03.1957), four *pang-kan* (see R1i) have to be given: one for the *wiama*, one each for both *riaca* and one for one random person from the host's house who is present and wishes to partake.

R6i) *tang-tum*

This feast lasts fifteen days. The fifteen days do not include the preparation period, during which the necessary animals and gifts are bought, rice is winnowed, beer and liquor prepared, and the guests informed. As usual, the number of cattle to be sacrificed must be uneven. There must be at least nine, though the number can be anything up to twenty-nine, in addition to which one extra animal may be sacrificed. This means that thirty head of cattle is the maximum; it is also the only even number that is permitted. The *tang-tum* is the largest feast possible. (Kuchi had never heard of a *long-plai*, see R6h; he thought the largest feast was probably a *bong-wai*, see R6g.)

More than thirty pigs are also needed. They are killed in connection with the following tasks or items (the list may not be complete):

ping-tung: for closing off the space under the house

kung-prek: "feast helper-choosing"

mung-kau-mi: gathering bamboo: a pig and a dog are sacrificed on the heap of bamboo

büi: the bamboo rat

kim-kung-car: the rear platform; a dog is also killed

riyang: the side platform; a dog is also killed

carai': the "rest house"; a dog is also killed

kong-no: the woven horse's head; a dog is also killed

kim-pur-car: the front platform; a dog is also killed

pau-dim: the bamboo-"flowers", i.e., tassel decorations

tui-lam: the pipes in which the water is fetched

ku-ram: the palm leaves

anum-röng: for shading the front platform; a dog is also killed

dong: the staircase; a dog is also killed

dong-lu-kau: the bamboo at the top of the staircase; a dog is also killed

ling-pu: the sacrificial posts

ling-dup: erecting the sacrificial posts

rung: the roof over the feast area; a dog is also killed

ther-kim: the partitioned part of the house

deng-hap-mün-mi: the people who cut the banana leaves

chi-khang: the medicine, see R6e

li-tong-khep: the pestle dance

tem-cong-mai: the "snail fire", see R6h

leng-ker: making ropes from new cotton for all the cattle

tang-plung and

tang-chaa-mi: see R6j

preng-kom: turmeric paste

bong-kom: threads to wind around people's wrists in the closing ceremony.

The *reng-lu*, who is a brother of the feast-giver, receives a live pig. If there are several brothers, it does not matter whether the *reng-lu* is an older one or a younger one. He is the organizer of the feast, and is responsible for making all the preparations and the business deals. During the feast itself, he allocates tasks to other people. Since he stays in the feast-giver's house, he is subject from the beginning of the feast to the same *khang* as the rest of the household (the *khang* are the bans on certain foods and the ban on leaving the house). He too has his meals brought via the staircase at the back of the house. He is allowed to drink rice liquor.

The feast-giver's *pen* (son-in-law) must provide the people living in the house with chickens for the duration of the *khang*. He too receives a live pig. The *nai* amongst the *pen* relatives – in the first instance this would be the feast-giver's sister's husband – leads the feast-giver's wife into the dance on the occasion of the *krong-döm*, which is the coming down from the house when the ban on leaving it ends for those who have been staying inside. The *nai* receives from the *reng-yaa*, his brother-in-law, a *caa-liu-yöng* ("rice-seed-basket"). This basket is two thirds filled with paddy, and a gourd (*kan-pen*) and a whetstone (*tim-chük*) are laid on top. The *nai* must dance with the filled basket, and may then take it into his own house.

Finally, the *mai-mo-mi* (the "silken" *tutma,* see R6f) receives a live pig as a special honour, plus a second pig that he must sacrifice as a *kwak-bong-kom-met* (an atonement pig for the throwing away of the wrist threads). Like all the members of the feast-giver's household, he has a thread wound around his wrist at the *bong-kom* ceremony. When he goes home after the ceremony, he unties it and throws it away. This is in general forbidden, but if it is done anyway, it must be expiated by the sacrifice of a *met-pak.*

The feast-giver's house is totally renovated during the early days of the feast. The lower part is closed off by a screen of full length bamboos placed upright all around it. A platform, called a *riyang,* is built right around the house. Its walls – unlike the walls of platforms built at smaller feasts – are not made of broad horizontal strips. Instead, at the back and sides of the house, they are made by weaving broad diagonals crosswise over narrower strips. To make the wall of the front platform, broad strips are placed vertically and their upper ends are cut in the shape of swallows' tails. The walls of the platform and the house are decorated all round with *caro-ca* (tassel sticks); this is described in detail in the account of the Khumi *ku-ram*; see R5c. The platform walls on all four sides of the house are hung with *wan-tung* (large printed Burmese cloths), if the feast-giver manages to obtain enough of them. Only the platform at the front of the house is covered at the top.

Two stairways are constructed, as for a *catang-thau.* At the back, there is an ordinary staircase, the upper end of which, however, is carved in the shape of a hornbill (*rang-pang-lu*). But the front of the house must have a proper staircase *ching-phe-dong* ("tree-board-staircase"). The feast-giver must buy it; it is two feet wide, and costs five rupees. Between five and ten silver pieces (rupees) are set into the top, with a mirror beneath them. After the feast, the rupee coins are taken out again, but the mirror is left in place. During the feast, a man keeps permanent watch from a point on the platform near the top of the stairway. (Even for a *tang-thau,* which is a lesser feast, money may be placed in the top of the staircase.)

Next to the front stairway, the trunk of a *klang-kü-ching* (*Meliacea* spec.) is set vertically into the ground to serve as a *rai-cün* (stair rail). This trunk must be forked at the top. The ends of the fork are painted with yellow and black diagonal stripes, using turmeric and soot. The forked ends are called *waa-wang cong-kaa* (*cong-kaa*, forked tail, of the *waa-wang,* or *Enicurus* bird). The forked top ends of the front platform wall are called *waa-wang-chot* (*Enicurus* points). The trunk that serves as a stair rail does not have to be a *klang-kü*: it could equally be a *reng-cang* (*Eugenia*), or a *machat* (unidentified tree). These three, however, are the only woods that may be used. One *reng-cang* trunk and one *machat* must be used for the side platforms: the supporting pole on the right should be a *reng-cang*, and that on the left should be a *machat*. Finally a *pang-ching* (*Bombax* trunk) must be set behind the *dong-lu-kau* ("stairs-head-bamboo"). On this *pang-ching* are hung a shield (*kongma*), a bow and arrow (*chalang-baa*), and a woven bag (*im*) containing betel leaves and nuts.

The trees used as sacrificial posts at the *ling-pu* must include one (and only one) *pang-ching* (*Bombax*), and next to this there should be either a *klang-kü-ching* (*Meliacea* spec.) or a *reng-cang* (*Eugenia*) with a *waa-wang cong-kaa* ("*Enicurus* forked tail"). (According to Menkröi Ngarua', the Anok use *klang-kü* only to make the village *khang* in the "mountain ruler enclosure" [see P2f]. The name *machat* is unknown amongst the Anok, and the *reng-cang* is only used for house building, not for feasts. In his Burmese-English Dictionary, page 1000, Judson notes under *sapre* [*Eugenia*] that the Chin take oaths in court on a twig from this tree, and that Burmese soldiers used to wear *Eugenia* twigs as signs of victory in their hair-knots and their ear lobes.)

Nothing is buried under the *ling-pu* posts. A roof, under which the dancing takes place, is erected over the feast area; this roof corresponds to the Anok *chuk-calaa* (see R3d). It is called the *rung* (Anok *rum;* this is a loan word from the Marma. Burmese *rum:* "meeting hall", "courthouse"). The supporting poles of this *rung* are not wrapped with cloths, as for the *chuk-calaa* (see R3d). The roof, however, is decorated all round with *caro-ca* and *mang-kut* (see R6a).

In addition to the *ther-kim* in the far right corner of the feast-giver's house, a second one is erected on the ground next to the front staircase. A channel made from the leaf sheaths of the banana plant leads from the large beer pot in the inner *ther-kim* (see R6f) to a point over the second *ther-kim;* a pot is placed inside the *ther-kim,* under the end of the channel. Using a straw, the feast-giver sucks beer from the large pot of the inner *ther-kim* and lets it run through the channel into the pot of the outer *ther-kim*. It is a special honour to be the first to drink the *yu-lon-tui* ("beer-drawn-off-water"); the feast-giver may not drink it himself, but offers the honour to someone else. (For other ceremonies that are part of this feast, see the information on the *catang-thau*; unless stated otherwise, this also applies to the *tang-tum.*)

R6j) *tang-plung*

For a *tang-iüng,* the Khumi prepare 6 *tang-plung* and 6 *tu;* the Longhu prepare 6 *tu* and 12 *tang-plung*. (For a *tang-thau* only 3 *tu* are needed.) The person who makes the instruments may make more if he wishes, but only in sets of three. The *tang-plung,* like the *tu*, are single gourd-pipes, which are tuned to three different notes. Unlike the *tu,* the *tang-plung* pipe is not pulled through the bottle gourd so that it sticks out on both sides, but is set in at one side only. *Tang-plung* of this kind may only be made for a *tang-tum* ("*tang*-blowing"). (For an Anok pig feast, called a *tang-pük,* four were made; see P5b.)

The feast-giver himself must make a set of three *tu* and a set of three *tang-plung.* The other sets must be made in a village other than his own. The relationship of the maker of the instruments to the feast-giver is not important. The instrument maker must kill a chicken and a pig (*met-pak*) for his village, and must provide a large pot of beer (*pang-yu puma*). The *reng-yaa* (feast-giver) makes good the man's expenses after the feast. Some of the blood of the sacrificed pig must be dripped on the new *tang-plung.*

On the day when the cattle are tethered in the feast-giver's village, a group of young people from the instrument maker's village bring the *tu* and the *tang-plung* to the village where the feast is being held. This group, which should be as large as possible, consists of those who actually carry the instruments and a number of others who accompany them. Before setting out, they may all drink from the *pang-yu puma,* the large "invitation beer pot" provided by the instrument producer. In addition to the new pipes, they take with them a gong, three plate gongs, a drum, cymbals, and a sword and shield. All the instruments are played along the way.

When the group arrives at the feast-giver's village, the instruments are placed under a specially made protective roof, covered with a mat. Neither the instruments themselves nor the people who bring them may enter the feast-giver's house, or any other house in the village, until later. To mark the arrival of the group, a large pig must be slaughtered. This is the *tang-chaa-mi pak* (pig for those who bring the *tang-plung*). Someone must also wring off the neck of a large cock. Blood from both animals is then dripped upon the instruments. Only then may the instrument bringers enter the houses; the instruments themselves, however, must remain outside.

At this point, a further *ling-pu* post is brought from the village entrance (*plon*), where the feast helpers have left it ready. In front of the post dances a young man wearing the horse's head. (For a *tang-tum,* the horse's head is covered by a cloth, *pan-thin.*) This young man is followed by two people with the sword and shield, and then six more blowing the three *tu* and the three *tang-plung* that were made by the feast-giver himself. The group stops on the way, before the house of a rich villager, and dances in front of his staircase. The house owner must then treat the dancers to a pot of beer. When everyone has drunk, the post is taken to the *ling-pu* and erected there. The people of the feast-giver's village and the *tang-chaa-mi* who brought the *tang-plung* from the other village now dance three times round the *ling-pu,* to an accompaniment played on the drums, plate gongs and cymbals. The dancers with the horse's head and the sword and shield then demand from the feast-giver a pot of beer for their team. After this, the general dancing begins, and lasts all night.

The next morning, the feast-giver must kill for the *tang-chaa-mi* a three year old cow that has not been tethered to the *ling-pu.* The meat from this cow and from the pig slaughtered the previous day on the arrival of the *tang-plung* is put aside and dried; when the feast is over, the *tang-chaa-mi* take it back to their own village, where it is shared out.

A meal is eaten, then the dancing continues. The *mai-mo-mi* (the *tutma* with the silken turban) and the *reng-yaa* (feast-giver) and his wife, who all had to stay in the house until now, come down to the ground (*krong-döm*) to distribute the *preng-kom* (turmeric paste). The *reng-yaa* and his wife each wear a blanket (*wanma*, see F1d) hanging from their shoulders; the two corners of a short side are knotted in front, and the blanket hangs down their back. These blankets sweep along the ground behind them, and are called *krong-cher-wan* ("earth-sweep-blanket"). The *reng-yaa* and his wife each have an attendant, or *kung-lu* ("behind-head"); the *reng-yaa*'s is a young man and his wife's is a girl. The two attendants must step first in cow dung and then on the blanket. They receive 1–2 rupees as an inducement to stop this performance. The *reng-yaa* dances with the *tang-chaa-mi* (who brought the *tang* pipes) and then takes them into his house; once the *preng-kom* has been distributed, the *tang-chaa-mi* may enter the feast-giver's house with their instruments. First, though, they must dance three times round the house on the *riyang* (the platform that runs all the way round it). While they do this, the people of the feast-giver's village stand on the ground and throw bones from the slaughtered animals between the dancers' legs, which means that the dancers have to be careful not to fall. After the third time round, they go into the house, where the feast-giver entertains them lavishly with beer and rice liquor. Meanwhile, fires are built in the two fireplaces that are required for a feast, and peppers are thrown into the flames. The members of the household flee at once; the visitors, however, must dance until, one by one, they are forced to give up. The door is only opened on payment of 5 rupees. This money is returned to the dancers later by the person in their own village who made the *tang-plung*. He in turn is repaid by the feast-giver; the amount will be between 50 and 80 rupees, according to previous agreement.

When the *tang-chaa-mi* go back to their own village at the end of the feast, they leave the *tang-plung* and the *tu* in the village where the feast was held, but they take their plate gongs and so forth home with them. A small house is erected for the pipes that are left behind; it is built next to a *tang-rui-ching* (tree unidentified). On the day of the *cia-lu-kop* ("cow-head-tethering", see R6a) the instruments are placed in the little house with some *yu-lon-tui* (sucked-off beer), *arak* (rice liquor) and *köm-pot* (small portions of meat for the spirits). They are not brought out again until the field feast (*bong-wai*), when they are played at night for the dancing. After that they are put in the *carai'* (the rest house built for the *tang* feast). While they are in the *carai'*, anyone who wishes may play them, so long as they do not take them away. There is a special song (*long*) which is accompanied on the *tang-plung*:

*Yua lö, reng-cang hong po a küm khök ü, küm kha e d*a, *yua lö*

(Yualö, have taken *Eugenia* trunks, was enough, when it was enough, yualö ...)

machat hong po a küm khök ü, küm kha e da, yua lö, [etc.]

tlong-kü ..., ken-ma ..., mun-klü ..., ram-nget ...

"We took *Eugenia* trunks, and when we had enough, we took *machat* trunks, and when we had enough ..." – and so on, through all the kinds of wood that were used to build the feast house.

R7 The cattle feasts in Arakan

R7a) Ceremonies and paraphernalia

The literature on the cattle feasts of the Mru and Khumi is very modest. For the Chittagong Hill Tracts they are mentioned neither by Lewin (1869, 1870) nor by Hutchinson (1906, 1909). Lewin reports from the Khumi of Arakan (1870:229) only: "They have no special festival days, a fortunate war-party, a marriage, or a lucky hunt, are all occasions for merry-making." Latter (1846:63), however, already mentions the signs of the cattle feasts with the Khumi: "Each house ... has suspended from its walls the skulls of the animals it may have killed for food; to these likewise they pay a simple adoration, by placing before each individual a handful of its wanted food, as an acknowledgement of the sustenance it has afforded them in its time. The skulls also of the animals slaughtered by their fathers are in the like manner preserved as much in remembrance of those deceased relatives as a monument of their wealth; frequently will a chief point to them with pride, and tell you how many mountain bulls his father could spear for a marriage feast." And furthermore (p. 65): "The only visible objects of worship of the *Khumis* are the trunks of three or four trees, which have been cut down in clearing a space for the village; and the same number of pillar-like stones. These are fixed in the earth together, in the middle of a large shed, which is also employed as the place of reunion and festivity of the village."

Despite the obvious interpretation faults perhaps it can be gathered from the report, that the skulls (those of cattle, goyal ["mountain bulls"] and pigs, but also hunting animals were hung up) were catered for in the house; probably not with grass etc., but with rice. The sacrificial posts (falsely interpreted as left tree stumps) with one stone each which stood under one roof – probably a kind of *chukcalaa* (see R3d) – would suggest, according to present rules, a Mru hamlet. The detail that marriages also gave reason for cattle feasts may be doubted, but Stilson also brings both in connection. He writes (1862:216–17) on the Khami at the Mi-Khyong:

"The wedding feasts are often preceded by a buffalo fight, as it is called; when the men, all armed with pikes made of sharpened bamboos, try their skill at tormenting the poor animal, by surrounding him, each man standing as guard to prevent his exit, while they pierce him with their cruel weapons. The unfortunate buffalo, after trying his strength for some time, to yield sport, Samson-like, to his savage tormentors, is at length overcome, and falls, to be hastily dressed, cooked, and eaten by the jolly throng. A dance by the light of their fire closes the scene for the night at a very late hour."

The general stabbing of the animal, of which nothing is known in the Chittagong Hill Tracts today, is also reported for the Ku at the Peng Khyong by O'Donel and Reynolds (1864:406). More detailed is the report given by Tickell for the Khumi at the Mi-Khyong. Here he experienced a goyal feast in Tangyong on 24.02.1851. The hamlet lay about 100 metres over the river; at first the sacrificial animal was tied, however, on the riverbank. Tickell writes (1854:110–111):

"The first thing that attracted my attention on the bank, immediately below the little bluff hill on which was situated the village, was a noble bull of the Gayal species, which a number of Koomoois were attaching by heavy ropes to a felled tree. The poor creature was quiet and patient, little conscious of the dreadful fate awaiting it, for it was left tied in this manner for three days and three nights, without food or water, while the dancing and revelry is going on in the village; and, at the end of that time, when the animal is utterly exhausted, and the villagers wrought up to frenzy with drink, it is dragged into the village, tied down to a stake, and then slowly despatched by numberless stabs with spears, the whole crowd dancing round and round and deliberately sticking it in every part of the body." "The village consisted of some ten houses, occupying the entire crown of the hill. They were made entirely of bamboo, exceedingly clean and well put together, and tricked out in such cunning devices of savage finery as were admirable to behold. The roof and sides were stuck all over with bamboos, some split at the top into stars or brushes, others scraped into little streamers, and others adorned with pendants of pith. The fronts of all houses were elaborately painted with black ornamental marks and devices; but on the dwelling of the giver of the feast invention had exhausted all her arts. It was a perfect chevaux de frise of ornamented bamboos, and painted within and without with black and white in elaborate designs, while round the house was a temporary bamboo platform or roofless veranda, on which the dance was performed."

It is remarkable that all houses were adorned and that one painted the whole feast-giver's house. A platform round the house in the CHTs would indicate a *tang-tum*; but Tickell speaks only about two sacrificial animals. That also very big feasts were given at the time, is proved by Hughes (1881:24) who talks about feasts at which up to 90 cattle were killed. On the reason of the feast Tickell writes (1854:112): "These entertainments are not on account of any periodical festival, or on any occasion of rejoicing, or other commemoration, nor at the joint expense of the whole village; but are given by any individual who has amassed the means of doing so; and that these must be ample may be gathered from the fact of the forementioned Gayal alone valuing 80 Rs.; while the pigs, bullock and poultry slaughtered for feeding, and the khon [rice-beer] brewed for inebriating the numerous guests, must cost at least as much more; for these revels last three days, and Koomoois are invited to the spot from the whole country around."

St. Andrew St. John also stresses the merit character of such feasts. He writes, (1872:238–39): "During the dry season numerous feasts are given, at which large numbers of cattle are killed and eaten, and rice-beer and spirits consumed. lt is a mark of distinction to be able to have it said that they have killed so many head of cattle at a feast. The largest number I ever heard of was one hundred and fifty. The gayals, oxen, and buffaloes are tied up to a post and speared, but other animals have their throats cut. Dogs are often castrated when young for use at these feasts. The post used by the Mros is shaped like a Y, and just below the fork carved so as to represent two or more breasts. There is some peculiar signification attached to this symbol, both by the Mros and the Hkamies, and it may often be seen carved on the posts of the headmen's houses, and on the house-ladder. The true meaning I never could get; the usual answer being, 'it is the custom inherited from our forefathers.' The Hkamies and Chins, however, do not carve their posts, but set them up in rough; in the Chin villages I have sometimes seen stones set up on end."

St. Andrew St. John is the only author who reports of carved forked posts. Since at the same time he was the only one who reported about the cattle feasts of the Mru, it cannot be decided whether these forked posts were set up only to very big feasts (as the Longhu material might suggest). With the Mru in the CHTs nowhere forked posts are set up; already Mills wrote with reference to the Anok Mru (1931:520): "Forked post not put up", and also I myself found nobody in the southern Mru groups near the Arakanese border who would have heard something about forked posts at cattle feasts. Regarding the putting up of stones, St. Andrew St. John mentioned in a contribution to the discussion (1872:247) that "when he went into the village all the inhabitants were out in the jungle, and he had no opportunity of finding out what they were erected for. He only observed this rough attempt at a circle in one village, but elsewhere he had occasionally seen single stones, which were said to have been set up in the same manner as a post to show the number of oxen that had been slaughtered, and were then set up simply because they were to hand." This detail might not be quite correct; since he reports only of one stone circle, but otherwise only of single stones, he perhaps refers to Khumi hamlets.

R7b) The feast dance and the instruments

Regarding the goyal feast seen by him, Tickell (1854:111–12) gave the following description of the dance: "... round the house was a temporary bamboo platform or roofless veranda, on which the dance was performed.

The actors in this were some twenty or thirty strapping young fellows, ranged in line facing and encircling the house. The 'music' was composed of about eight or nine of them, sounding the most unique instruments I have ever seen. One was a curious triangular species of metal gong, which when struck whirls round, emitting an undulating sound. [This is a Burmese, not a Koomooi instrument.] There were also four of a singular species of bassoon, formed by a hollow gourd stuck on the side of a thick bamboo, surmounted by a piece of cloth for a flag, and which when blown into gave out a dull blaring note, between a buzz and a snore. The rest of the musicians had cymbals or small drums; and some sprightly youths, unprovided with instruments, put their fore-fingers to their mouths and every now and then gave a whistle which would have done credit to the gods in one of our London theatres. There was no tune to speak of; but a monotonous unchanging succession of thumps, clangs, and blows on each particular instrument ... while ever and anon the blowers of the gourds withdrew their lips and gave vent to a long-protracted bellow, in which all present, male and female, joined till out of breath. The 'step' in this national dance is not complex, or difficult to acquire, being simply that peculiar to the tread-mill, or an alternate genuflexion of either knee, with an occasional shuffle sideways as the circle moves slowly round the house, and a low bow.

The women, who had dived into their houses on my approach, were summoned forth after a while, and made to join the circle – not with much 'courtesy of soft persuasion', but chiefly by being taken by the nape of the neck and shoved headlong into the ring, where, when once fairly established, they entered into the fun with great gusto. Many of our own party, who were also Koomoois, were hauled in, and the mirth grew 'fast and furious', especially as all assembled were copiously regaled with huge pots full of seethed pig's and bullock's flesh, rice, and khon [rice-beer] ... sipping the same through a reed ... The dancing continued for about an hour, when they left off and went about their divers occupations, which (today at least) appeared limited to ornamenting bamboos to an incredible extent."

The description of the Khumi dance given here in all essential points also applies to the form to be found today in the CHTs. Gourd-pipes in the Mru style are not mentioned; the "bassoon" might refer to the *tu* or the *tang-plung* (see R6j) although we should expect three or six and not four pieces. A detail mentioned by Stilson (1862:216), however, proves that also the Khumi had gourd-pipes with several pipes (probably in the form described under R5e): "They have instrumental music ... played upon a sort of organ, formed by the insertion of three or four bamboo tubes, of different length, into a gourd shell. The extremities of the tubes which enter the shell have bamboo reeds attached, which give sounds similar to those of a melodeon; these sounds very nearly form the 'harmonic chord'. Several holes in the tubes enable the performer to give an agreeable variety of sounds."

A more detailed description of the dance forms is found with St. Andrew St. John (1872:239): "The dances which take place at these feasts are very peculiar, but must be seen to be clearly understood. The movement is more of a side-closing step, the body being kept in a position resembling the 'Grecian bend'; the line is headed by players on drums, small gongs, and a wind instrument formed by passing a long bamboo through a hollow gourd [apparently a *tu*], after these come men armed with spears, muskets, choppers, and shields. The young men generally commence the dance, and then drag the girls in between them; the whole line thus formed slowly closing round and round the animal to be speared, whilst the young men make love to the girls by their sides. Before commencing a feast the faces of all are usually smeared with a mixture of saffron and rice-flour [apparently *preng-kom*] which is supposed to keep off the bad effects of drinking." Besides this general feast dance, St. Andrew St. John also describes two special forms: "Sometimes a wild sort of war dance is executed with swords and shields; and there is also a rather clever dance, something like a Scotch sword dance, but between two long heavy rice pestles, which are clapped together by two men to the sound of a drum; if the dancer be not very agile or exact, he is liable to get his leg broken between the two pestles." The second form obviously is the *li-tong-khep* (see R6e) performed in the CHTs at *tang* feasts. The Khumi dance with shield and sword on several occasions, and also Lewin (1870:227) describes a kind of war dance for them:

"Their dance is simple. It is more a species of march than a dance; about twenty young men move round in a circle to measured time; the rhythm distinctly marked, both by the music and the motions of the dancers. The leader, on the occasion I saw them dance, held in his hand a small dâo with a brazen handle, from which streamed a tuft of goat's hair dyed scarlet. The other performers bore, some a shield, some an ordinary dâo or spear; and these weapons they clanked together in time of their movements. The measure went something in this fashion: One step, a pause; two steps, all sink down on their hamps, clank weapons and rise again; another step, then a jump and a shout, and so on. In an adjoining room to that in which this dance took place were the drinkers ..." This dance was performed in the house.

R8 Stories referring to the cattle feasts

More folklore will be reported in the following chapter, but I shall mention already here two stories relating to the coming into existence of the feasts just discussed. The first one which is known in this or that form to every child and which also is told to foreigners, I recorded (on the 2.08.1956) in the shortest and simplest form without any sense for dramatisation from Klingtui Khongtör, an inhabitant of Tapwüa-Kua (Galengya-Mouza), whom I had asked to do so, because I already knew it but now also wanted to record it verbally; the second one, which refers to the *tang* festival (see R6e) and is known only to a few specialists, was told by Aröi Chüngprim, a professiomnal storey-teller from Waling-KP (Galengya-Mouza).

R8a) *chot cia chang-cia*

Cakraa-mang pe a caa Mruca tut-e kachaa pe. Mruca ka mo pa-plau caa chum plau, kwar ka po caa chum plau, mo pa-plau, pa kaa la, pami caa ce-t'ya pe. Chat den lai ka, tui chak-krai lai, hom chak-krai lai raa-köi, cia caa-ta khai pe$_s$ *kwar tut-e yaa ting kröng pe. Nami raa-ka-e, kaplan hanö, cakraa mang hot lün-taa a pe: En kachaa mi caa Mruca tut-e kating. Cia tek pe mi, ting döi, hom chak-krai tui chak-krai lai ra-köi ang caa-taa khai. Nami raa-köi, cakraa-mang kunma-rau khai. Cakraa-mang khöi a pe nor-bong köi. Nayong ka pe, cia ka aluk yüng wöi döi mi. Cakraa-mang pe a keng-tik, pe cia ka-e: mruca mi pacang pe ling-pu hanö, pakhwök en leng hai ling-pu köi, paplai en, pa pe pe. Re hai pachot en, Mruca mi, hanö, en dai da, ling-pu köi, hia a hanö, patak en. Ci-ö-ca tui pathut en nor-bong köi, ang pe en caa mi, mami nor hai en caa-taa khai raa-köi, pa pe pe. Nayong ka pe, cia pen-ca caa alüp alüp mami Mruca caa-böi'.*

The story of stabbing the cow

The sky-lord handed over (to the cow) a writing and sent it to the Mru. The Mru shall weed one time and harvest three times, and also the Bangali shall harvest three times and weed one time, thus said the script. Because the heat was too strong, the cow had too much thirst and too much hunger, it ate up the script; and only the Bangali came to know it. Therefore it returned, and the sky-god asked it: The script, which you brought, did it reach the Mru? The cow said: it did not reach them; because hunger and thirst were too much, I ate it up. Therefore the sky-lord got enraged and kicked it into the muzzle. That way, it is said, the cow has no upper teeth. The sky-lord gave another rule and told the cow: these Mru shall make a tethering pole, bind you with a string to the pole and dance around you, thus he said. With the spear they shall stab you, the Mru, your tongue they shall cut off and pin it on to the tethering pole. From a little pot they shall pour water over your muzzle, because you ate up the script, which I gave you, with this muzzle. Thus he said. In this way, it is said, the cow's omasum, these piled up leaves, are the book of the Mru.

R8b) *Büi-ko chang-cia*

Büi-ko chang-cia mi m'yong: Rik po machi lök pe, rik po machi lök panö, köng khai t'ya, nami ka caa wöi pre-yaa tai. Taa lot-taa caa pre-yaa paka-caa-lö hanö, hai taa lot-taa caa pre-yaa mi, nami taa po machi wai-mi, u-ramma ngaa-mi ngaa-da-mi tuk a dö'böt pe. Uh, tong-nö en-ci tai-nau köi cang kön nö, pa-hanö apaa: Ang cam köi tak, ang cam köi tak, nami ngun pre-buk mi, en u köi en-ci hot yaa thum pat, hot thum yüa, ma-mi ngun cak hut-mi, ang wang khi cak hut döi, en-ci köi, ngen en-ci, en-ci tai-nau ka-e panö, apaa cam köi khai pe. Apaa cam köi khai panö, kayüa tai-nau-mi ni-

töng ni-töng pe cam ap cak, cam cak-leng, cam kom dong, cam cang chot panö kaa. Pai-mi ngun köi pe tö a, cak hut da, tö a, tö a, ngun pre buk pe tek, na khöt kayüa tai-nau hot kram-mi.

The story of the white bamboo rat

This is the story of the white bamboo rat. In the beginning he took a wife, it is said. He first took a wife, then she died. Of her he had two children. She left two children and under these circumstances the one with the two left children took the next wife who from this time on was their stepmother. He gave them a stepmother, and how it was that way, a stepmother as bad as she is not conceivable worse. "Oh, what shall I do with you two brothers?" the father says, and "I go away, I go away for some time. There are two baskets; you two can support your mother all the while, support her persistently; these baskets are for the game bones. When I come home and do not see any bones – I shall cut your neck, you two brothers." And then the father goes away. The father has gone away and the two brothers go daily to shoot game, to put loops, to bind spring poles, to set spear traps every day. And they get. The bones of the game, they put into the baskets, they always put in and fill the two baskets; so much it is what the two constantly get.

Apa, a-ni da hua wang mang böt paka-caa-lö hanö, cam run rui rep khai pe. Cam run rui rep khai, paka-caa-lö hanö, yu praa köi da, car ka da, tö pe rui rep, aluk ka da, tö pe büng, büng luk ka da, be pe wan, hot yüa pe, cung rau wang cace, na-khöt hot takhum taplang. A hua wang mang: Ali ö, en cu en caa, yung caa yung da caa, ang hut ang lang rau, khöt cam ti cam tep cam bop ang mi ö. Praa taplang rep-rep rep-rep pa. Ali ö, en cu en caa, yung caa yung da caa, khöt pat ce cam ti, cam bop ang, cam ti cam tep ang nak-na baa. Apaa wang pün, wang mü ce, ngun po yaa tek khai pe, apaa khaing-khya la mi, tek yong: uh, pai köi a u-ramma ti cang reng reng, maa praa taplang re-rep rep, yu praa taplang rep-rep rep, ang hut ang lang rau ö, en cu en caa, yung caa yung da caa. Uh, apaa kaa twak, kaa caa döböt.

"Today surely my husband will come home", she says and therefore goes to tear down lianas. She goes and tears lianas and puts the creepers down on the platform. Above them she puts a door and on the door she spreads a blanket. She lies down and waits, and since it just becomes winter she continually turns from the belly to the side. Her husband comes: "My dear, your child and grandchild, good children, such good children! My bones and limbs ache; how they have beaten, hit and thrashed me." She rolls herself on the side, rankle, rankle, rankle. "O dear, your child and grandchild, good children, such good children! How they have hit and beaten, hit and thrashed me all over!" The father climbs up, and when he looks, they even could fill the baskets, just so as had ordered them threateningly. But there, their stepmother, she moves herself constantly on the one side, moves herself on the other side: "My bones and limbs ache, oh your child and grandchild, good children, such good children!" Oh, the father does not know what to do any more.

Uh, paa, trau kh'e i cam kröi la pe. Tong i cam kröi kön. Ce, i cam thai büi la, pe Thai t'ya, büi tarap da, kat po khöt ang-cing cam ap cak, büi ang-cing khi döi, ang-cing kaa döi. Thai nö, khi döi tak, i cam thai nöm la. Paka-caa-lö hanö, apaa po la cang-mu-ma nai pä-kho pe, cam köi, cam köi nö, po la ngapok pre chu t'ya, kayüa ka ngapok wöi da, cam mang mang nö: Mi tong chüng pape, apaa tek pe. Ah mi da ang-cing wa-yak ap raa, tai tak le, wa-rik wa-ram ap raa le. Awai mang mang mang, Mi tong chüng cö pape. Ah mi da ang-cing mit nga-ki mi raa, tai tak le, yu-e yu ang-cing phyak nga-ki mi, ang-cing or-e ap a nö, wai mang mang mang nö. Mi tong chüng cö. Ah, mi o köi ce-böt ang-cing wang kom dong mi, ni köi kaneu utma chök. Wai mang mang mang nö, apaa: Mi tong chüng cö pe. Ah mi da ang-cing chot cang raa tai tak lamaa, or-e pak ap nö, yu-e ang-cing phyak ap a mi raa. Wai mang mang mang, Mi tong chüng cö, apaa nök a reng, mi tong chüng cö, pe. Ah, mi da ang-cing nga-hiu ap raa tai tak, or-e ang-cing ap nga-hiu, yu-e ang-cing yu yua la nö, yu-e ang-cing phyak nga-hiu mi. Wai mang mang mang nö: Mi tong chüng cö, apaa pe. Ah mi da, ang-cing cia-röng ap raa ce reng tak da, pai raa-e, cam tek, pai raa-e ang-cing cam ap cia-röng, ap raa raa-e nö, cia-röng tim raa-e, ang-cing hing phyak la nö, ki tai-böt, na-mi raa ce reng tak lamaa, mi chüng da, yaa mang a, mi chüng ca-reng tak kaa, pe. Uh yaa mang a, wai mang mang mang mang, Mi tong chüng cö, pe. Ah mi da, pe, pri ap raa ca-reng tak ka le, ang-cing pri ap raa nö, mang mang reng reng mang reng reng. Mi tong chüng cö, apaa pe. Ayö mi da ang-cing tuk döi, ang-cing tuk döböt, naa-ce-po, khi büi ko lüi döi tak da, löi-ma khi döi tak ce, mang reng reng, mang reng reng, mang reng reng, Mi tong chüng cö pe. Apaa, ang po tuk döböt mi da. Nö, en da. Ang po tuk döi.

"Hey sons, tomorrow we go to search." "What will we go to search?" "Well, we go to dig up bamboo rats." "We already dug; but how often we went to hunt game, bamboo rats we did not see, bamboo rats we did not catch." "You dug and saw none! We go to really dig them up." And then the father takes along a pot of birdlime, and they go away. They go away, and two guns they have taken along. The two have guns. They go and go, then: "What kind of mountain is this?", the father says to them. "Well, this still is our place to shoot birds spending the night, to shoot pheasant and jungle fowl." Further they go, go and go: "What kind of mountain is this?", he says. "Well, this still is our place to creep up to deer; we cut to pieces down there the deer which we have shot up there." And further they go, go and go. "What kind of mountain this is?" – "Well, at this brook now, we are putting our spring pole traps. One catches particularly many mice here." They go further, go and go, then the father says: "What kind of mountain this is?" – "Well, this in the meantime is our place to put spear traps. If we catch a pig here above, then the place where we cut it to pieces is there below." Further they go, go and go. The father asks them again: "What kind of mountain this is?" – "Well, this meanwhile is our place to shoot sambur. We shoot them above, we drag them down there and we take them to pieces there below, the sambur." The father goes further, goes and goes, then he says: "What kind of mountain this is?" – "Well, this is the place where we always shoot gaur. Let us say we shoot the gaur over there and where the gaur falls, in the place where we have shot it, we take it down to cut it to pieces and already are ready over there. We chased there on this mountain really again and again." They have passed it, that mountain where they chased again and again; they go on, go and go. (The father) says: "What kind of mountain is this?" "Oh here we shoot leopards again and again, we shoot leopards." And they go, always walk on, always walk on. He says "What kind of mountain is this?" "We do not know any more, we do not know this." Oh, but still they do not see earth thrown out by the white bamboo rats. Since they did not yet see thrown-out earth, they walk always further, always further. "What kind of mountain is this?", he says. "Father, I do not know this one." – "Well, and you?" – "I don't know it either."

Naa-ce-po mang reng reng, khi büi löi-rö raa ce mi, khi büi löi-rö raa hanö, krum khü ce: Mi thai chik, ma-mi thai bot chik, ma-mi büi ko ce-böt. I cang te tachik, ko i naa kom, i ni yak kön tai böt ma-e, cang te büa, cang te büa hanö, cang te büa nö ni: En-ci da, en-ci tai-nau pre-ya, en-ci da, en-ci rik thai yüa, ang müi tak, ang da long-tang, ko büi ton kh'e, krüng ang yüa, na kh'e ang thai lai. Apaa rik müi pe, apaa müi müi müi nö, kayüa tai-nau mi thai pe. A thai-mi thai-da-mi, thai reng reng, thai reng reng, hur-ti-ti, rek a pe ce, büi ton khök pe. Apaa, or-rü, ko da büi ton khök. Nöm nöm, apaa rü paka-caa-lö hanö, ko da müi yüa lai. Khöt kom mi ce da kayüa tai-nau da i-müi nöm t'ya. Nö, apaa thai ce, kaa da khai, pa-caa-lö hanö: Uh, kayüa tai-nau köi tong cö cang kön cö. Na-mi cang-mu-ma nai mia a phia, phia hanö, praa mik ka ben, praa mik ka ben, a mik tali klüng mia-taa pe klap, büi-ko mi dök a hanö: Phyöi, ma kayüa tai-nau mi ö, hang pakröi tuk chüng rap o rap la hang pakröi tuk tiama la, panö, ngen a nö, büi wi karo köi pe, lük pe lük pe panö, kat-chang apaa da mang-taa khai t'ya.

But they always go on to find a place where bamboo rats have thrown out earth. And when they see and find a place where bamboo rats have thrown out earth (they say): "Let us dig, we must dig here, these here finally are the white bamboo rats! Let us make a hut, soon we'll have to take a lot of trouble. Finally we shall have to stay here, make a hut!", and when they made a hut (the father says): "Now you two, you two brothers! You there, you two dig first. You two dig first, I sleep in the meantime, I am an old man. And as soon as you are close to the bamboo rats, wake me up, then I will dig for my part." At first the father sleeps, the father sleeps and sleeps, and they, the brothers, dig. And how they dig! They always dig further, dig always further, continually. When they poke, they are close to the bamboo rats. "Father get up, the bamboo rats are near now!" The father really gets up and as it is that way (he says): "Now you will sleep!" A little while later, the two brothers have already fallen firmly asleep. The father gets (a rich bag) and when he has it (he says): "Oh, what shall I do with the two brothers?" He spreads the sticky birdlime, spreads it and plasters it on the eyes here, on the eyes there. When he has completely sealed all four eyes, he grasps a white bamboo rat: "Phyöi! Oh, here the two brothers, do not let know them the mountains and rivers, do not let find them the way!", he says, cuts its neck and drips the blood of the bamboo rat on their throat. Hardly has he spread it on them, the father has already gone way in no time.

Apaa mang-taa hanö, kayüa müi-du kh'e da, or thör pace: Ing ö, ang-ka da mik tamia yan kabo, tong cö ön, ayüa le hng, ayüa le. Tai ö ang-ka po le ang po pe. Ataima rik thük, thük pa-hanö yaa klaa. Ayö le, ing ö en ka karo köi awima le, tong cö, pe. A nauma yaa thük ka yaa klaa ka. Ayö tai ö, en ka po karo köi awica le. Mü chüng mü o pace, kaa döböt amat acang. Cu, aci mang, apaa ti tö-taa aci hüa lamaa, apaa tö-taa aci hüa cace da, aci mang aci kröi apaa kung, mang yüa mang yüa, mang reng reng, mang reng reng, kaa döböt, kaa twak kaa caa döböt, chüng caa o caa ce-böt. Uh, ce ing ö, ang mü paci chüng tak, tong tong praa ce-ü, na-mi ching köi ang klo hanö, ang mü paci chüng kat tak. Nöm nöm or klo pace, wa-to bai wöi lök. Mi tong dui cü, hom chak-krai ce da, ang yon a le, or dök nö, or yon pe, or dök nö or yon nö rak hap-lap taa pe. Tong cü, naa-ce-po cöng-la wa-to mi cace da, akwak chak-wang döi pe, or mi da rak hap-lap nö, or köi plan yon lök. Wa-to dui le, ing ö. yu kwak ang lök, yu kwak ang lök pe. Or mi da yon pre-buk khök t'ya, apoi mur om khaiböt pe. Ayö, ing ö, ang da poi om khai le, yö kom ce-böt. Tai ö, yu kwak ang lök, kwak ang lök. Ni baa yu kwak en khai ba, pe, yu kwak pape, en yon nö, hang kakwar ba, ing ö ba, pe. Yon nö, ing da kuar taa yan yok pe, arak wang yayam ce. Tai ö, yu kwak ang lök dia. Hang kwar tham ba, pe. yu kwak pe, yon nö, kwar taa yan yok pe, arak wang ti, or mi da lap om yan böt pe, yö kön mi. Nö, ui po wöi döböt cace da, lök yaa-e pre-buk ci mi yon, takwak döböt. Or mi da yö kön ce-böt, yö kön ce-böt pa-hanö: Atai ö tai ö tai ö pe. Ing ö, ang da yö kön tai böt co, nö: Ang da ang-mat lök-ca le, krong köi le, ang lök-ca le. Hanö, ataima yö kön ce-böt t'ya. yu coi pe lap thür poi lök nö pacöng a charai, paka-caa-lö hanö, ni ba: Ing ö ang da yö kön tai böt co, panö yö. Tai ö, cong ang dia, cong ang dia. Anauma rep-pu rep-pa marong-heng-kai cada da, or mi da ateng-lengma kayö mi. Cam yö, yö nö, cong a panö: Ang wang-töm khök tai ö, kut plan yö. Tai ö, cong ang dia, cong ang dia pa-hanö, ataima cong a, ma hön'ya da rep-pu rep-pa marong-heng-kai ca, cam mang reng reng, mang reng reng, taima raa-e katöm. Katöm nö, ataima yaa ce-khai, pai waa chang-ching cace da, hing yö wing pe, taa döböt. Tai ö, cong ang dia, cong ang dia naa-ce-po ataima yö wing böt. Naa-ce-po, tai ö, pa-ni kh'e da, ang pa-ning tum tang mi khin kh'e da, ang dong-lu-kauma köi, en wang töi mön, ang cam pho löng-chur kam-bör mi, pho da ka-caa-lö hanö, ang pa-ning tum tang mi kh'e, en po wang töi. Na-yong pe, anauma kröi tek pe mi da lai-mak, na-ka, mang reng reng, mang reng reng, rep-pu rep-pa.

Away the father goes, and when the two have had a good sleep, when they wake up (the elder says): “Oh little brother, my eyes are quite stuck up, what is the matter with me? Oh my dear, my dear, oh!” – “Oh elder brother, mine also, me also!” The elder brother pulls first, he pulls and finally gets it off. “There is blood on your throat, oh little brother, what is this?” When the younger brother removes it, when he gets it off (he says): “Yes, oh elder brother, some blood is on your throat too!” When they look at the mountains, look at the rivers, they cannot remember any more, nor do they find the way. “We go, the father actually has left us back; father has left us, so we go, we follow the father.” The two go, go both, always walk on, always walk on: they do not find the way any more, do not know it any more, know neither mountain nor river. “Well, little brother, I'll take a look-out for us, where we actually are. I climb this tree and at first take a lookout for us.” When he climbs up right to the top, there is a bird's nest. “What kind of eggs are these? I am hungry, so I drink them up.” He above seizes (one) and drinks it. He seizes and drinks it above and it already itches him. “What is this?” But while he now changes into a drongo, the one up there, he has no desire to throw it away. It tickels and itches and he above again drinks one. “Drongo eggs, little brother!” “Throw one down for me, throw me down one!”, he calls. That one has already swallowed two pieces, a feather down already grows on him above. “I grow feathers, little brother, I soon will fly.” “Elder brother, throw one down for me, throw me one down!” “I now throw it down for you.” He throws it down to him and says: “Drink it and do not scratch yourself, little brother.” The younger brother also drinks it, but scrapes it off again, since the itching gets too strong. “Oh elder brother, still throw one down to me!” “But do not scratch”, he says, he throws him down one; he drinks it and scrapes the down off again when the itching comes over him. Wings have grown by now, however, the one up there will fly. There are also no more eggs, two pieces for each are drunk, thus he throws down none more. From this time onwards the one up there will fly. And as he will fly henceforth, the other calls "Oh elder brother, brother, brother!" "Little brother, from this time on I shall have to fly." – "And I, I am quite alone on the earth, I am alone, oh me!" And the elder brother definitely can already fly. He pulls himself out a wing feather and changes it to a hewing knife, and when this has happened, he says: "Well little brother, I will have to fly now", and flies. "Elder brother wait for me, wait for me!" The younger brother jumps over stick and stone like water rapids; the elder one flies straight on, flies and flies, waits for him when (he hears): "I have followed, elder brother", he flies again. "Elder brother, brother, wait for me, wait for me!", he calls, and the elder brother waits for him. From here, over stick and stone, like water rapids, he always runs on, always runs on, arrives where the elder brother was. When he arrives, the elder brother is away; since he has bird shape now, he has flown on and does not hear any more. "Brother wait for me, wait for me!" But the elder brother definitely has flown away. "But, then, elder brother, one day, when I blow the *tang*, that year, you will arrive, sit down on the bamboo mast at the head of my staircase; I will do it equal to the ants and the termites, and when it is then ready, when I will blow the *tang*, you also will arrive that year." These words the younger brother tries to tell him for memory. And then he walks always on, uphill, downhill.

Philu mang caa a mi, kim dong athong pe, caa pu caa pam kri-cham pung-tang, mong ning ka, ner ning ka, bök-bök caa-ta, philu mang mu lak köi, klap-klap na-mi, na-ka: Uh, amat-ci pu cu hot pat lok pe, o ka-e, pat lok pa-hanö: Uh, i mru chang-ching caa da malaa-wüa da, yu lün pe bon klu, pe. yu lün bon klu hanö, or mü ce khi tayan kacak-wüa ce: Ma-mi ang cang la hua, yung lamaa pa-mi, machi-wüa ce tuk tüng-khök, nau tüng-khök, apu köi tek pe mi: Apu ö, ko ang cam kaa mru e, tong cö en cang a da. Uh, khüm ci khüm caa, kwak ci kwak caa, ang cang a da baa, hu-hü, pa tak pe ren. Uh, lai döi, wöi chang-rum hanö. Uh, nöm da, apu ö, ang cam kaa mru, macu manar malök e, tong cö en cang a kön cö. Tong po ang cang döi, en khi baa, en khi baa. Machi-wüa da: Khi döi, daa-taa yan yan. Awai wöi nong nö: Apu ö, ang cam kaa mru, macu manar malök, tong cö en cang a tak ce la, pe. Uh, ang cang la mokca kom, en khi ba, en khi ü, ho yu ho. En dai. Dai döi. En dai. Dai döi nö, ho dia. Ho hanö, wang tatüng pace, wang tatüng naama ma-e. Uh, mokca ö, i pat lok la, i pat lok la, mokca ö, pe. Pat tanö, pe, pat lok pe. Ng, mokca ö, ni en rik mir, ang pat khai. Nö, machi le na-mi, nö, machi tek pa-mi, ma raa-e en mir hanö, yu hön'ya tök, yu raa-e en mir hanö, pai hön'ya tök, nayong cang a baa, pe, a machi tek pa-mi, nöm nöm a, en po yup yu mir, hanö, apu kröi pat roi-taa pe, caa kom ti, yu hön'ya tök pluk pe, kaa döi pe, töi döi, töi döi pe. Ni mir klen dia, mokca ö, mokca yup, pai hön'ya pluk tök, krau-krau pat a pe, kröi pat mi öi, pat da lok hai, küng küng, töi döi yok pe. Hom chak-krai khök, ni i röng kan dia chik, i röng kan dia chik, kan röng mokca katu. Ko ka i kan röng rö t'ya katu. Mi kan röng i lok dötham, apu köi pütma ching pe rai a ngot, cok a phau pe, bak-kaa da, a bak-kaa po raima ching pe rai a ngot, cang la bak-kaa, phük a klur tak, nö, röng kan büa. Röng kan büa pa-hanö, caa hom pe. Ki caa hom kh'e da, yu chet ang yüng yüa, hi-taa pe yüng hüa pe, marek cayang pe, pok, ching long cayang pe, chör, chet-müa da. Mat-ci chet yüng büa ka khök ko da, na khök, apu ö: Ko da i pat lok cace da, en mir lai, pe. Hng, naa döi mokca ö, en mir yok dia, ang pat yok tak, alai kh'e yong yok. Nö, rnok-caa mir. Mir nö, kröi pat, pai hön'ya tök-taa pe yan. Ng, acu chang ce-böt cace da: Ko da, pu ö, en mir lai, ang pat lai. Mokca da ma-e yup, nö pai-e krau pape pat, tadun a pup pe, apu, akat tarü bop yüa pace, bop reng reng, bop reng reng, achak tai pe hü dük-lük. Naa-ce-po cong dia, cong dia, achak da, hawang chai ang yüa, hawang chai ang dia, ang tek en-ci lai klö a nor tak, ma-lai pa nor tak ce da, naa döi ce-po en-ci ang achak chai kom ce-böt ce da, pe. Tong, pe. Ce, en-ci köng-taa ang khai pa-caa-lö hanö, ang ngaa mi kim dong athong mi, mün khöt cö kim wöi mi, pre dong-thong nö pre chek, chum dong-thong nö chum chek, ang ngaa mi chek hanö, kim töng kröi cong a, ang ngaa mi chek a leu hanö, kröi ret a, nö, ako en-ci kaa kim mi kim, üa mi üa, na pho kön tai.

What the cannibal has swallowed are one thousand houses, rice baskets, rice barns, brass and gold, old gongs and old plates; he has devoured everything, all this has disappeared in the belly of the cannibal. But now, uncle and niece, both throw the fish net on the river. When they throw the net (the younger brother sees her): "Oh, the one has the shape of us people, that girl there, I'll throw down to her a clod of earth", he says, throws down a clod of earth and when she looks up she suddenly sees a man (and thinks): "If I took this one there as a husband, wouldn't this be good?" As a woman she can prove it, she wants to prove it. She asks the uncle: "As soon as I get a man what then you do with him?" – "Oh, I grasp him a bit, take a mouthful and throw it a bit away, throw a mouthful away. This I do with him, hoho!", the uncle says and laughs. Oh, she speaks nothing, a time passes, then she says: "Oh uncle, really, if I get a man, rich or poor, somebody, what will you do with him?" "I do nothing. Do you see one, do you see one?" The girl denies: "I see nobody." But soon later on she says: "Uncle, if I should get a man, rich or poor, somebody, what still will you do with him?" "Oh, I will make him son-in-law. Do you see one? Well, you see one. Call, call (him) down!" – "You lie!" – "I don't lie" – "You lie!" – "I don't lie. So call him anyway!" She calls and when he lets himself be seen, really comes and lets himself be seen, (the uncle) says: "Oh, son-in-law, we just throw the net. We just throw the net, son-in-law." "Well then, throw", he says, "throw the net." "Hm, you first dive, son-in-law, dive first, I throw." And the one who wants to be his future wife tells him as a woman: "If you dive here, then appear from there again, if you dive there, then appear over there again, please do this", his wife tells him, "really jump and dive completely under water." And then the uncle quickly tries to catch him by the net, he will eat him. There he appears again from below, is not caught, is not in the net. "Now dive on, son-in-law." The son-in-law jumps, appears over there again; swish, he, the fisher, throws, throws the net, draws, draws, and again he is not in it. They got hungry. "Let us cook food in the quiver, let us cook food now in the quiver." When the son-in-law goes to cut a quiver, he does it in the form usual among us now. "This food quiver is not at all sufficient for us!" The uncle goes, grasps a whole *Pütma* tree, breaks it off and with his finger bores a hole into it. Now the supporting fork. Also for the fork he graps a *Raima* tree, breaks it off, makes a fork from this and bang!, beats it into the earth. Then they cook the food. After they have cooked the meal, they eat. He baring his teeth, he shows them to them. "Clean my teeth!" As toothpicks they take on the one hand an axe for getting the remains out, on the other hand they take a tree-trunk for cleaning. When finally they have finished tooth cleaning (the son-in-law) tells him: "When we throw the net now, then you dive, father-in-law." "Hm, no, son-in-law, you dive again, I throw once again, again like before." And the son-in-law dives. He dives and (the father-in-law) tries to catch him, however, he appears again, is not in the net. He dives over there. He throws after him, he, however, appears again. "Hm, shall the niece learn it." "Uncle, you dive, I now throw instead." He jumps here and the niece throws the net, swish, they catch him and wrap him up, the uncle, and at once they both start to beat, beat him up, beat always more. He still breathes weakly only: "But wait, wait", breathes he, telling them: "Do not kill me, do not kill me, I still have to tell you a word, another word for memory, particularly because otherwise you soon will bring my life to an end." "What?" – "Assuming that you have killed me, then put what is my meat, what is one thousand houses, on so many skewers as houses are available, two households then two skewers, three households then three skewers. Make skewers from my meat and insert them into the wall of each house. Cut up my meat, skewer it and distribute it. Then soon what is your house will beome a house, what is your field will become a field."

Paka-caa-lö hanö, uh, m'yong panö, kat-chang kayüa po tep-taa kat, chek taprot pe a ngaa. Kim töng long-haa pak-ko mi khöng a ranit rau nö, ko en-ci kaa mru kom wöi kön tai. Kröi cong, kröi tö hanö, long-haa pak-ko mi tek khök, rau töng, kua kom-mi-wüa lö, rü rü rü, long-haa pak-ko ang kaa, kri-cham pung-taang po ang kaa, kling-ui yam-baa po ang kaa, tham-pak wan-klai po ang kaa, mong ning ner ning po ang kaa, chik-chek po ang kaa, bök-bök mi ang po a. Kua kom-mi-wüa lö, rü rü rü, m'yong pe, taa döi, awai rau po plan khöng khai yok pe. Uh, kua kom-mi-wüa lö, rü rü rü, kling-ui yam-baa po ang kaa, kri-cham pung-tam po ang kaa, tham-pak wan-klai po ang kaa, ru, deng, caa pam po ang kaa, kua kom-mi ö, rü rü rü, naka, pre rau, chum rau khöng, taruk rau kh'e, röu tün yong mün yong, pe, taa. Taa klö dit-det-ca khök, taruk rau ka ranit rau kh'e, kua kom-wüa lö, rü rü rü, kri-cham pung-tang po ang kaa, mong ning ner ning, kling-ui yam-baa ang kaa, caa-pu caa-pam ang kaa, cia kom rat kom po ang kaa, rua kom-wüa, rü rü rü. Na khü ce, ling-kö krot klö da, waa kak klö da, cia kom naa kom klö da, a'ur yang pe, na khü ceyang pe, a'ur a'ur a, paka-caa-lö hanö: Uh, m'yong nö cöng khök, mru nak hai kayiia po caa-pu caa-pam t'ya, kim a kim, üa a üa, hai ka, m'yong paka-caa-lö hanö, uh, ko da nö khök lamaa, khöt ang cang ngare mi.

And as soon as he has spoken this, they beat him dead, make skewers from his meat and distribute it. "When the iron pig trough of each house is beaten for seven days, then many people will be around you." He inserts the skewers, puts up the said iron pig troughs and calls every morning: "Hey, I got iron pig troughs, you people of all villages, I also got brass and golden plates, everything, I got precious stones and carneols, I got woman clothes and skirts too, I got old gongs and old plates too, I also got odds and ends, get up, get up, I'll distribute these. Hey, you people of all villages, get up, up, up", he calls, they do not hear. Also the next morning he beats again: "Oh, I got precious stones and carneols, hey you people of all villages, I also got brass and golden disks, I also got woman clothes and rocks, I also have taro, bananas and rice barns, people of all villages, get up, up, up, up." He beats two mornings, three mornings and on. On the sixth morning one hears something like noise from foot rattles. One hears weak sounds on the seventh morning. When on the next morning he calls: "Hey, I also got brass and golden disks, you people of all villages, I got old gongs, old plates, precious stones and carneols, I got rice baskets and barns, I also got many goyal and many cattle, people of all villages, get up, up, up!", at the same time it moves and becomes noisy. One hears pepper pods being pounded, hears chickens crowing, hears many goyal, many buffalo, hears sounds and noises and, oh, as it is noisy and noisy, at the same time it is accomplished in this way. Like everybody, they too got baskets and rice barns, a house is their house and a field is their field. When in this way it had been accomplished (he said): "By now it is really finished; how much trouble I have endured. Are they still alive, mother and father? Is there any news?" The crow comes and brings him the news: "Your mother and father are still alive. From your parents' house I always seek food."

A'u apaa da, wöi reng tak ce ü, maa ü, wa-ak: En u en paa wöi reng tak, en u en paa kim raa-e ang kadök waa caa reng. Wang tek pe chatang pe, a'u apaa wöi reng tak hanö, nayong paka-caa-lö hanö, wa-ak köi öm döi ka, a'u apaa wöi reng tak ce ü, maa a. Wa-klingca li-liu: En u en paa kim raa ang tuk, en u en paa kim con ka-e, ang kacot kür-kha ui reng, pre-yaa khök, awai cam talop talop. Ah, a'u apaa wöi reng tak ce maa a, tüng-tüng tarüt-tarüt, en u en paa kim raa da ang tuk, en u en paa kim ka da ang kacot caa reng tai ka, caa-bai ka-e. Tong pa cö, wa-tur ka wa-tur ce da, ang pa-ning tum tang kh'e, ang po a pin phau kom, pape. Ang tuk, ang tuk. Nöm nöm a, cak yu cak hom, the yu the arak, cang taröng khai böt, ran cia ran naa. Uh, panö cöng khök. A'u apaa da, mia cö maa, köi po ang pin kom, wöi reng tak paka-caa-lö hanö maa a, wa-ak: Ang tarü chö, ang tarü chö, pe. En wa-ak da, en dai-lai tai mataa naa a. Dit po ang dai döi, en u en paa kim raa-e ang kadök waa caa reng kaa, ang da. Nö, ni: En wa-ak köi tarü maa dia. Wa-ak kayö thai-lai, hanö: Ak ak, en cu en caa pa-ni tum tang, wang po en pin phau. Alawü, en wa-ak, thui, ang cu ang caa pa-ning tum tang lak-da, ui-mukma ria-kwöng köi khom wöi döböt dök-ta, cang cöng löng-chur kam-bor böt dök-ta, ang cu ang caa, en wa-ak dök mai dia, nö, kröi kwak a kok pe nö, nor köi khot pape. M'yong kaa wa-ak nor numma-mi. Wa-ak ti cam hing yö, cam tahing yö, hing katöm döböt pai köi, kaphöng döböt chatang. Wa-kling wang kut yö, wa-kling phö, wang ein, wa-kling kakut tarü. Kling-ca li-Iiu, en u en paa kim raa ang tuk. Ni, wa-kling en tarü yok dia, pape. Ang tarü, ang da kacaa kür-kha ui reng ka ang tuk. Wa-kling phö, phö-löp phö-löp, mat-ni kim con köi kamön a klö, kling-ca: Li-liu, en cu en caa pa-ning tum tang. Ih, en wa-kling dai-lai, ang cu ang caa pa-ning tum tang lak da, löng-chur kam-bor khom wöi döböt dök-ta, ui-mukma cam e cam yan taprang böt dök-ta, en wa-kling dai-lai, hia po en mai-ring, hia kwak en dok-dak, nö, po pe mai-ring nö, kwak tadok pe. Wa-kling koi po, nayong kaa le ring ringca mi. Uh, wa-kling po cam hing yö, pai köi katek döböt. Wa-tur tüng-tüng tarüt-tarüt. Uh, mia po pin pok mi kacung kaman ang döi lamaa, a'u apaa wöi reng tak patek ang kh'e da, mia po katöm ang döi lamaa, pe. Wa-tur kham la khai yok: Tüng-tüng tarüt-tarüt, ang da katöm reng, ang kacot caa-bai reng. Nö, en wa-tur tarü teng-teng en wang kaa la dök-ta. Teng-teng ang wang kaa la. Wa-tur ma hön'ya phö phö, pa, yang-pang ka pe kamön, yang-pang kamön nö: en cu en caa pa-ning tum tang. Ih, ang cu ang caa pa-ning tum tang lak da, löng-chur kam-bor khom wöi döböt dök-ta. Kathom nga-ngam pe, caa-bai köi, wa-tur da, pot lak da, kathom nga-ngam, kacot caa nga-ngam pe: En cu en caa pa-ning tum tang, en öm döi ka-caa-lö hanö, trau pom la ang, taa pa-ning tum tang klö lot ang, pape, wa-tur takham. Nöm nöm, wa-tur taköi hanö, dap pe yu. Dap pe yu hanö, wa-tur tuk kham döböt yu krong köi, kham döi, paka caalo hanö, Ion pape, biaca hai wa-tur kham pe. Kham kham hanö, rau-e kh'e, ang köi pom la ang, en-ni öm döi hanö, pom la ang hanö, wa-tur köi pom la chüt pe mat-ci. Matci hua-yu töi che tarü. Mat-ci hua-yu töi tarü paka-caa-lö hanö, katöi tarü hanö, yu köi pa-nlng tum tang klö ce da longma-cangma hön'ya yu taa khai böt pe. Ko da lot ang, lot ang, en-ni po ṭaa klö taa kui khok ka pe. Lot-taa pe wa-tur köi, wa-tur phö yang-pang köi pe kamön: En u en paa wang ting khök, ang wang yaa ho la khök pe, wa-tur keng-yang pe pe a. Nayong ka wa-tur ngong preuma, kröi pe a, keng parüa tak pe yu hön'ya ka.

"Mother and father are still alive." But when he comes to know this, he does not believe the crow. "Announce it, if father and mother are still alive." "Where your parents' house is I know", (whistles) the red tail. "I always pick bitter berries on the edge of the house of your parents." After to the two, he talks to the next one: "If father and mother are still alive, announce it." The pigeon (coos): "Where your parents' house is I know. I always go there to the drying mat to pick grains at your parents' house." "How then, you pigeon of the pigeons, shall I invite them as guests of my this year's blowing the *tang*?", he says. "I know, I know." He indeed prepares rice, sets up beer pots, prepares beer, prepares spirits, all preparations are made, he buys goyals, buys buffaloes, oh, and then everything is completed. "Who will inform them? Who will go to fetch my guests, mother and father? If it is so, that they are still alive, then inform them." The crow says: "I start already, I already start." "You crow, you lie and pervert the facts." "I do not lie at all, I always seek food from your parents' house." "You crow go, start and inform them now!" The crow then flies like a plough through the field, and calls: "Krah-krah, your child and grandchild these days blow the *tang*. I come to invite you as guests!" "Shame on you, crow, ugh, our child and grandchild blow the *tang* this year! They do not live any more even in the appendix of the vampire, not turned into termites and ants. Our child and grandchild! Catch fire, you crow!" Then they throw a fire brand at it and hit it on the beak. Therefore the crow has a black beak. The crow flies up, flies up and does not arrive there any more, brings no more message. The red tail, the bulbul, comes to fly to them, the bulbul flies and takes over the work, the bulbul returns again (and whistles): "I know where your parental home is." – "Now bulbul, you start once more and tell it to them." – "I will start; since I always eat bitter berries, I know it." The bulbul flies away, wagging its tail up and down, sits down on the edge of the house and calls: "Liliu, your child and grandchild this year blow the *tang*. "Ugh, bulbul you lie, our child and grandchild this year blow the *tang*, you are kidding! Even as ants and termites they do not live any more, the vampire does not shit and piss them out any more. You lying bulbul, fetch embers, we throw them to you", and they take embers, throw them at it and scorch it. Therefore the bulbul has red feathers under its tail. Oh, also the bulbul flies away, it reports there nothing more. "Whoever invites the guests, achieves nothing. Oh pigeon, do it for me. Nobody who told me that mother and father are still alive really arrives", he says. The pigeon firmly promises again: "Tüng-tarüt, I always arrive; I always pick paddy on the drying mat." – "Then start, pigeon, for certain you will finally achieve it. "I quite certainly can manage it." The pigeon flies off, flies to them, and sits down on the platform banister. Sits down on the platform banister and (calls): "This year your child and grandchild blow the *tang*!" – "Ugh, our child and grandchild blow the *tang* this year, far from that! Even as ants and termites they do not live any more!" The pigeon hops cheekily down on the paddy mat. Far from clearing off it hops cheekily down and starts to pick the paddy. "Your child and grandchild this year blow the *tang*. If you do not believe me, shut me up in your basket, then tomorrow, when you hear the sounds of this year's blowing the *tang,* release me", the pigeon vouches. They tap a beer, but as the pigeon cannot drink at the drinking tube, as it does not drink, they suck the beer off and the pigeon drinks it from a little bowl. It drinks and promises: "When the morning comes, shut me up in the basket; when you do not believe me, lock me up." And then they shut the pigeon in the basket, they start, man and woman, join to visit. They, husband and wife, start with (the pigeon) and as they have set out together, they finally hear the sounds of this year's *tang* blowing there down from the ridge. "Release me, release me now, since you have heard the sounds, the babble of voices." They release the pigeon; and the pigeon flies and sits down on the platform banister. "Your mother and your father have arrived, I have been able to call them for you." He gives it a necklace for pigeons, and that's why the neck of the pigeon is spotted, this means the chain ribbon still is from that time.

Nö, kayüa hua-yu wang töm nö: En wang töm caa, pa pe, wang töm, ang da löng-chur kam-bor khom da cam wöi döböt, mi tai ang twak, pe, ang-cing cam wöi reng tak manaa, apaa pe. Wang töm ü töm pa kha-e taröng hot, taröng cuk-luk böt ce da, thai dong-lu-kauma kuam pe. Mia cö thai kom pe, a'u köi pathai a. Tong pa cö, paa, ang toi thai dong-lu-kauma kuam da-e, yung da baa, pe. Uh, pa-ning tum tang kh'e, en kük-lük pathai kom, döi nö pa-ning tum tang yung döi pe. Ce, u ö, en katöi thai dia, ce da, en ce-khai tak, ari wang khai. Ang thai döi ce-po yung dale, pa pe. Uh, machi-wüa thai mi döi, pe, döi nö, apyak aci wöi la, pe. Nö, amat u töi thai t'ya. A'u töi thai paka-caa-lö hanö, pa-yöng pa-yöng pa kh'e, a'u köi hing chür phük-taa klum pe, dong-lu-kauma po kröi phük-taa pe, a u köi hing chür tahüa, thia a cua. Hom khom kan khom caa döwöng tak, kut plan thai ling kwam pe. Kut plan thai ling kwam paka-caa-lö hanö, thai thai, thai na kh'e ho döwöng tak. Thai ngong mör khök pe, ko da apaa köi pathai a lai, ling kuam. Apaa köi pathai a lai alök kaho: Ni, apaa kathai ling kwam dia, pe, nö paa: Ling kwam thai tong cö, ang thai döi ce-po, mru bümma tak. Uh, en ce-khai tak, pe, apaa, nau pok po ang wang, rai pok raa pok po ang waug. Ang thai ling kwam mi naa döböt, ang da long-tang ceböt ce. Uh, thai döi nö yung döi, kua mi wüa po pruk-pruk, pa po pe tek. Ang naa döi, paa, ang töi thai döi ce-po, yung kön tai. Uh, en ce-khai tak, thai kom taklaa, en ce-khai tak, thai kom taklaa, apaa. Tom kar tom nguk pe kalot döi, töi kathai ling kwam. Apaa köi wat bon nö, köi taphang ö kh'e, ling hai phük-taa klum yok pe, na kh'e po thia tacua. Thia tacua khü ce, nami da plai ui.

Then they both, husband and wife, arrive, and he says: "You succeeded to make a feast. You got it, son. As for me, I had to believe that even as ants and termites you did not live any more. We two are still alive", says the father. During the "you succeeded", "you got it" the preparations continue.

All are just busy with the preparations and to dig the hole for the bamboo mast to be erected near the stair. "Who will dig?" is the question. "Let the mother dig." – "But how, can it be good, when I dig the hole for the mast?" – "Oh, for this year's blowing the *tang*, we let dig you alone. If not, this year's blowing the *tang* will not be good Therefore, mother go along and dig. As long as you are still there, we follow the rule." – "If I do not dig, it will be also good anyway, son", she says. "Oh, I do not have any women for digging and without this everything goes wrong and gets broken." Finally his mother helps digging. The mother digs and when she has dug a man's length deep, one knocks the mother down headfirst and rams her in, the bamboo mast is also pushed in. One leaves the mother stuck headfirst and fills up the earth. Before one starts to eat rice or side dish, one goes down again to dig the hole of the poles to tether the cows. When they dig the holes for the posts, they dig until they have dug neck-deep before they call again: "Let the father for his part dig the post hole." He goes to call the father, to let him alone dig for his part. "Now father go, dig the post hole", he said. "Well, son, me digging the post hole? Even if I do not dig, there still are sufficient people." – "Oh, you are still here, father, with small children I must manage, with bamboo roots and leaves I must get by." "I am no longer someone who digs post holes. I finally am an old man." "Oh, if you do not dig, it is not good." All village people also talk to him on and on. "Not I, son. It will become good, even if I do not dig with them." "Oh, since you are still here, it falls on you to dig. You are still here, you will have to dig, father." And as much as he cries and sighs, they do not let him out; and he goes to dig the post hole. The father accepts to dig out the earth, and when he descends into the hole they knock him down with the post again, ram him in and at once also fill up earth. While they are filling up the earth, they dance in a circle.

Nö, na kh'e da, ataima, cia tim ni kh'e, cia tim ni kh'e ataima wang khai böt pe, dong-lu-kauma köi wang mön klo pe: Ö ing ö, en pa-ning tum tang caa, ö ang pa-ning tum tang co tai ö. Naa le, deu ö, en pa-ning tum tang caa pe, ang pa-ning tum tang co. Kat-chang khong-töm tu pe or kwak pe, chang-cia köi dabaa, khong-töm tu pe or kwak panö, töi wöi la pa-ni pe, ma-mi ni hanö, cia-ngaa pe or pör pe. Nayong pe na wa-to i cang la wa-to mi dong-lu-kau ka-e, ataima dong-lu-kau ka-e wang töi mön. (Aröi Chüngprim, Waling-KP, Galengya-Mouza)

At that time, the elder brother, on the day of the cattle sacrifice, on the day of the cattle sacrifice also the elder brother finally comes. He comes and sits down on the bamboo mast at the head of the staircase. "Oh, little brother, your this year's *tang* festival shall also be mine this year's *tang* festival." – "To be sure, my dear, your this year's *tang* blowing shall also be my this year's blowing the *tang*." He quickly throws him up an old basket mat, the story goes. He throws up for him an old basket mat, and he remains one day. On this day one throws him up also beef. Therefore one says, the drongo (the drongo, which we make on the bamboo mast) is the elder brother, who comes to sit down on the bamboo mast at the head of the staircase.

S Folklore

S1 Preliminary remark

Mru folklore was (and still is) transmitted in two different forms: songs and stories. Though many young men could and did sing, what they sang (and are still singing today) were mainly love-songs, made up of traditional phrases and own inventions. Traditional songs, however, which are classed here as folklore, they rarely knew and rarely could come to know, in case there was no song master (*meng-charaa*) in their hamlet. And even when there was one, the young men would not go daily to his house and ask him to sing for them, so that they might finally learn the whole songs; instead, they could just imitate some phrases which they could use when singing their own love-songs. Married men normally would not sing love-songs anymore, though good singers could continue to invent songs of their own and might even try to note the text down, in case they knew a little bit of writing (in Marma or Bangla script). Song masters, on the other hand, would not do so: they knew all by heart, they remembered the contents of the songs and gave them their own form when singing them. Despite many assertitions to the contrary, I am convinced that, what the *meng-charaa* sang, was not immediately and fully understood by all who heard them singing. The main reason was that the words used by the song master were not commonly used in everyday language. All those whom I asked to tell me in detail the contents of the song or to explain difficult passages in it, which I had recorded on tape, finally admitted that only the song master would be able to do so. But I experienced that even he would not do this (see footnote 31), and I needed Menkröi and his son Dingte to help me in understanding the text line by line.

The best *meng-charaa* I had, undoubtedly was Menyom Tang. Unfortunately he visited me only rarely and I came to know more of him only during the last months of my stay. He had been an orphan and spent his youth with an old couple of *meng-charaa*, from whom he had learned all their songs. He had not been able to marry (he was too poor) and now tried to earn his living by walking around and singing his songs, at the same time plagued by an old infection of *tinea* (ring-worm), from which I finally could cure him. First (already in October 1956) I had recorded the *Nom-meng* (which I left for translating it later [but actually could not do]), then I recorded the *Khommi kua tuk meng*, and this long song will be given with translation under S2a. During the last days of my stay, Menyom Tang visited me once more and full of gratitude wanted to sing still other songs; my tapes, however, had been exhausted and I just took some small notes, intending to record these songs when I should have new tapes. But when I came back in 1964, Menyom was already dead, he had died from cholera. I tried to have more songs sung by other song-masters, but none of them had such a knowledge as Menyom had had.

Stories on the other hand are easier to be had, many people know them and can tell their own version of them. Therefore, they are easier to be translated.

S2 Old songs (*meng*)

S2a) Khommi kua tuk meng

Khumi village war song (sung by Menyom Tang 1957)[485]

ang kling-ui yun ö, u öi, kling-ui pöng-pör yun öi,
O my precious jewel[486], o you, precious beautiful jewel

ang yam-ui yun, yam-ui pöng-pör yun öi, paka,
about my precious gem, my precious beautiful gem.

u ö, kling-ui plong-rau, ang yam-ui yun ö,
o you beloved jewel, my precious gem,

yam-ui plong-rau da öi, ci chang-long le u ö,
beloved gem, alas, about our love,

adong tapong, kra tan dong pöng,
amidst the lines of stairs,

tan dong talüp, chum[487] thong[488] dong köi öi,

---

[485] Note on the song master (see also the preliminary remark): Unfortunately I could not ask Menyom about passages not really understood by my informant. Instead of giving an explanatory answer, he always immediately started to sing the passage anew, supposing that he had spoiled something. He indeed seemed unable to give any further information. Though I was told that only the song master (*meng-charaa*) did fully understand the text, I suppose that Menyom just repeated what he (and he alone) in his youth as an orphan had learned from his teachers who had taken him up. Still, he sung brilliantly, no other singer whom I came to know could surpass him. The detailed translation of several passages remains dubious, while most passages pose no problem. Still, the end of the song, containing a bit of old Mru philosophy, will remain difficult to interpret in our terms.

[486] *kling* in fact is (as will appear in the last lines) the name of a creeper (*Entada scandens*), with pods longer than 1 m, in English called "sword bean", *ui* its fruit, here probably not a whole pod, but only one bean (big like a small patella, used as "chips" in playing), *yun* is a numeral classsifier for little things of some worth like small coins or cowries. Hence a closer translation might be "sword bean coin", but my informant translated it as "jewel", and another passage in the end might also suggest "gold piece"; the parallel word *yam baa* than should correspond to the "silver" (mentioned in the same place), while in fact it might be "amber" or a carneol (a red-brown jewel).

[487] footnote on numbers: Nearly all numbers used in the song are to be understood symbolically; the following connotations result (English, other than German, lacks appropriate constructions):
2: countable single objects
3: multiples, for instance *chum pren* = countries, the earth; *chum chüng* = mountains; *bang chum krum* = all the branches
4: encircled, contained things, for instance *pat tali* = wrapping, *long um tali* = the house and its inhabitants, *ria tali* the interior of the body
6: unfathomable things, for instance *tui taruk* = (deeep) in the water, *war taruk* = deep in the night
7: big units, long spans of time, for instance *ranit mang plan* = again and again, *ranit ni kum döi-ron* = not before long, soon
8: uncountables, for instance *hia rayat kom thong* innumerable warriors
9: totalities, for instance *taku kui* = all meanders, *taku ning wa* = an "age-old" (fully grown) fowl
Man himself is conceived as a being with several layers (*lüp chong*), the higher units are: body (*ra chin* = ria, the guts) = 4, soul (*plong ram* = liver) = 5, and, as maintained in the

overtowering each other, between all the families,

ci mang bit[489] chari, nong ding lai ru khai ö,
there is gossip and talk.

mang bit chari, nong ding lai ru kha-nöm, u ö, ci chang long da
There is really slandering gossip about our love.

or ö ni lot preng pre, cer kom phak hönya, – kan plong chang neu, –
Up there the sun comes out in the east, love of my heart,

chum pren ni long po, ni ko da wang chot, paka,
the lord of all countries, the white sun pierces through,

chum pren ni long le, ni ko wang pün,
the lord of all countries, the white sun starts rising,

la[490] cung pö khaing, kra tan dong köi ö,
people squat down, between the lines of stairs,

pareng cüa-lak khung ko da, rin bang klim kwak nim rü rü, paka,
behind the broad store-room, and when the woven wall slowly throws its shade,

kimma tali[491] long um, klim wang kwak, kwak nim rü rü,
the encompassing house slowly throws its shade

ni kra tan dong köi, ci chang long ka da öi,
between the lines of stairs, of our love,

mang karbari, chang ri reng-ya ce öi,
the chief's headman, who always judges,

reng ce lak khung rin bang klim, kwak nim rü rü,
when broad behind the house the plaited wall slowly throws the image

mi kra tan dong köi, reng mong talang,
between the lines of stairs, the public place,

ku[492] tong prang yang, mi prang [pröng ?] nai con di ö, paka
on the raised round outside place,

---

song, undestroyable, spirit (*chak wang*, heart, the impulse of life) = 6 ?. Regarding the difference between *plong* and *chak* compare: *plong-rau* the (perishable) love between the sexes (passion), and *chak-rau* the (unperishable) general love among human beings, the compassion.

488 interpretation not certain, *chum thong* = 3,000 (but see footnote 33, on numbers); perhaps: rows of streets, but see also *dong thong* (household, family)

489 no literal translation; probably "there is much gossip and slandering" (because the intended marriage is against the rules)

490 interpretation not certain, according to informant: all gather for a meeting

491 whole phrase for: house

492 literal translation partly uncertain, sense: on the raised place in front of the house

kwak ting kum küa, wang phet phai cöm, kakhü e ce,
he throws bamboo halves to squat down, at that time

kwak ting kum küa, wang phet phai cöm pace,
when he throws bamboo halves to squat down

ang kling ui yun, kling ui pöng pör yun co,
my precious jewel, precious beautiful jewel,

mang karbari, chang ri reng-ya,
when the chief's official, who alway judges,

reng long wang po, po reng chua lai pace,
the headman comes to take, takes account,

mang long wang po, po reng chua lai pace,
when the king's man will take, takes account,

ang kling ui yun ö, kling ui or pe, pe reng chua lai mi,
my precious jewel, o, my jewel there gives, gives account,

di, ang yam-ui yun, yam-ui or pe, reng chua lai mi,
my precious gem, gem stone there gives her account,

u lai chong[493] da, u ö, teng man nöm,
what she says, o dear, is upright and correct.

u lai chong teng kla, kha-nöm di le pakö,
What she says to him must be upright and true indeed.

or ö, mang karbari, chang ri reng-ya, tong da ö,[494]
Up there, the king's official, the man to judge, how now,

ting[495] puk ram kar, kong tüama cün, hai ce öi,
with a bamboo piece, leaves removed, a kind of stalk,

pri krong then len, cang[496] ting a kür po, uh, ting kür ba pün khai ö,
sketches lines on the earth, talks big, oh, big words climb up,

tarua chuk-kri[497], reng long-ya, tarua chuk-kri mang long-ya, da öi,
the big man of the village, the chief's man, the deciding judge, the king's man

reng long wang po ang/[498], po reng chua lai,

[493] literally "plant", here: the developing words, the account
[494] literally "which?", probably: "oh, how now?"
[495] interpretation of phrase not certain
[496] whole phrase: to talk big
[497] perhaps instead of *taraa su-kri* (Burmese: judge), but cf. also: (Burmese) *rua su-kri* (headman of a village)

the chief's man, also will take from me, take account.

ang yong ing klang ka ce öi, yong ing or pe, pe reng chua lai,
from me, the black bird, up there I give account,

yong ing or tek, tek mang chua lai pace,
when black bird up there provides, provides account,

lai chong teng man, ang yong ing klang pace le di, paka,
his account is upright and correct. When thereafter, the black bird,

tarua chuk-kri reng long-ya, po ram kar kong tüama cün,
the village judge, the king's man, takes a kind of stalk,

ting puk ram kar, kong chingma cün, mi la[499] hai ce ba di ö,
a bamboo piece, leaves removed, a kind of stalk, and with this

pri krong then len, cang ting a kür, uh, ting kür ba pün yan, pace,
sketches lines on the earth, talks big, o, when mighty words climb up,

ang yong-ing klang le, u ö, ang chak charui, u ö,
I black bird, o you, my whole heart, o you,

ang kling-ui yun, kling-ui plong rau, ang chak charui ba u ö,
my precious jewel, beloved jewel, my whole heart,

ang yam-ui yun, yam-ui plong rau yun da ba,
my precious gem, beloved precious gem,

mang bök ru ton, kong tut ai krom, pace yong po,
like when the cultivated black taro dies off from the root,

ang yong ing klang, yong ing ai krom ka öi,
I, the black bird, I must wither away,

ang klik[500]-ui yun, klik-ui pöng pör yun u ö,
my precious jewel, precious beautiful jewel,

na le ngan po[501], ci chang long da ba öi,
nevertheless, we loving each other,

yu ö, reng chip mang kwaing[502], mang kwaing raca[503],
down there, on the border of the little place, where to draw water,

---

[498] The sign "/" is used here to indicate the rising tone.

[499] (Arakanese) to take (?)

[500] instead of *kling*

[501] perhaps: "however that may be", stereotyped phrase followed by consoling ideas

[502] arakanised form of Mru *kwam* (place to fetch water)?, see also footnote 50; *mang* probably not = raja, but "little place"

[503] despite *mang* (chief), *raca*: probably not = raja, but "little place"

kwaing[504]-ma wang po, po tui taruk[505], rüm ding con köi ba di ö,
where the guardian spirit's daughter comes to draw, draws the deep sweet forest water,

tar tang tui yia, cer chong pöng pör ui hai,
with the calabashes from the sunny side, the well-formed fruits to be put in the basket,

ting lung krök cak, krang[506] chi tapoing,
with the glugging calabashes to be put on the bamboo rack

yia dong[507] müm rüm, mi la hai ce öi,
with the glugging calabashs

chatma tui dir, no khoma cün pace,
after ladling the hot season water with a hollow halved gourd,

ting lung par pai, par di dör kür, u ö,
the openly plaited ting-bamboo basket powerfully carrying, o dear,

wang[508] lüng a rüm po, tui ding klam yia di ö,
the glugging sweet cold water gourds,

chai di non kwai da u ö, aplot taku wör, u ö,
drawing like an elephant's trunk, letting out many times, o dear,

u da, wang kwai ang/, kwai rüm tuima di a, pace,
when darling draws for me the water of the forest –

ang yong ing klang, thop thi pün lün,
I, the black bird, make up my mind,

thi thop kom cum po, mi klök oi wan daba,
think it up to the end, this song,

ang kling-ui yun, kling-ui plong rau ö,
my precious jewel, beloved jewel,

wang lüng a rüm po, tui ding klam yia,
fetches the glugging calabashes with sweet cold water,

---

504 interpretation not certain, possible *kwai* (to take out, excavate, draw water), but otherwise always *tui cot*, *kwaima* "the lady drawing water?"), or *kwaing* the guardian spirit of the place where to fetch water?

505 literally: six; but here "deep in" (see footnote 33 on numbers)

506 Arakanese?, refers to the calabash

507 uncertain interpretation; informant said: beautiful, more likely: the glugging (*müm-rüm*) of the water in the interior (*dong*) of the bottle-gourd

508 uncertain interpretation; *lüng* = to rotate; *rüm* probably not "forest", but like in the passage of the preceding footnote referring to the sounds of the water in the bottle-gourd; *wang ... po* = to fetch

chatma tui dir, no kho-ma cün, mi la hai ce, öi,
hot-season water ladled with the hollow half-gourd,

chai di non kwai po, kling ui wang kwai, kwai rüm tuima kha-nöm,
like an elephant's trunk draws, jewel fetches, draws forest water, indeed,

ang kling ui yun ö, u ö, ci chang long mi daba,
my precious jewel, o dear, we who love each other,

u ö, kan plong chang neu, ci chang long mi da,
o dear, my beloved heart, we who love each other,

püt löng ram thia, rüm ding con köi ö,
the gorjon tree's leaves lie in the sweet forest place,

mi rai löng ram thia, rüm ding con köi,
these gorjon leaves lying in the sweet forest place,

phung ram net khai, tui chu ram kong,
world leaf is pushing away a rotten water leaf's stalk,

ang yong ing klang di le, ko le tamat[509] nong kat-cang köi,
Me, black bird, now soon and at once,

phung ram wang cang ang/, tui chu ram kong le öi,
world leaf comes to make me a rotten water leaf's stalk, and so,

khunaik kri len, ni wang tacum cadöi[510] ri ron, le öi,
even before seven days will have passed, and so

ang yong ing klang, yong ing lang cöng,
I, the black bird, I on the other hand, will become

chüng ce nam pram po, kong tut ai dam, u, chi klam ce kon köm, ba di le,
a hill-spirit's foliage, the ginger at the tree's foot, o, the bulb of the cooling medicine,

ang kling ui-yun le u ö, kling-ui plong-rau, u ö,
my precious jewel, o darling, beloved jewel, o dear,

yong ing cam chök, chök Yam-ce rong köm co,
black bird is going to drive upwards, must drive up the Koladan region,

ang yong ing klang, yong ing cam chök,
I, the black bird, will detout to drive upwards,

chök Yam-ko rong köm ce öi, pakö, o le,
will drive up the Koladaing area to where

---

509 probably expanded from *ko mat* (now)
510 expanded from *döi ron* (before)

Yam-ce pre prai, kut tön hor lang,
the Koladan branches, splits again, flows differently.

macer pön cak, aröng ton ku, ton le mala,
The eastern side people, the females grown up in Arakan, grown-up girls,

mü klang thi yam khom ce, klang chong[511] rup lang
when seeing their males' appearance, the male figures are different,

mü tum[512] ching pung khom ce tum ching pung lang,
seeing the image of the female flower-trees, their figures look different,

mü cam[513] om pung khom ce, cam om pung lang,
seeing the image of the flowering sprouts, their figures are different,

mü klai ing pöng khom ce, klai ing pöng lang,
when seeing the tissue of the black skirts, the black tissue is different,

mü klai ing ret khom ce, klai ing ret lang,
seeing the border of the black skirts, the borders of the black skirts are different,

mü klai ing bong khom ce, ki-ce kong rik,
seeing the patterns of the back skirts, they are red-brown like Muntjac-hide,

kong dok wan klai ö, chek[514] ram chö kong mi or ö, paka,
pretty split skirts, split leaves like sticks[515], up there, and then

mi lö ka ce, ang yong ing klang, yong ing palö,[516]
of one of them, I, the black bird, I want to have

panöm u tang kön cö, nöm u chin rui kön cö,
as future real dear mother, a really affiliated own future mother,

öi, paka, ang yong ing klang, yong ing palö,
and then I, the black bird, I want to have

parüm[517] yan tang, rüm yan chin rui[518] u kön ce khai ö,
a real aunt, to be my future mother's closely related younger sister;

---

[511] Literally "plant", "stem", but here referring to humans. Similar comparisons in the following text.
[512] *Mesua ferrea* = grown-up female
[513] *Michelia champaca* = growing-up female, teeny
[514] interpretation not certain, descibes the pattern of the skirt
[515] The description may sound funny, but fortunately can be confirmed by the first photographs from the earliest 20th century, showing "Eastern Kuki" (Siyang) women with skirts made from split (palm) leaves.
[516] Refers, like the *lö* at the beginning of the line (*mi lö ka ce* = of these), to someone, but is never used as a full noun; the *pa-* is used here as a masculine prefix.
[517] *rüm*: (perhaps) same as *ram* = a younger relative of a parent, see footnote 68
[518] own, near, affiliated relative (*chin* = thread)

or ö, aröng ton klang, mi la hai ce öi,
up there, the grown-up Arakanese males, one of them

ang yong ing klang palö, panöm pa tang kön cö öi,
I, the black bird, want to be my true future dear father.

mü men[519] om pung khom ce, men om pung lang,
When seeing the figures of the sprouting orchids, their figures are different,

mü rang om pung khom ce, rang om pung lang,
when seeing the figures of the sprouting trees, their figures are different,

cam prö mani che ding woi loi[520] kön ba,
when they will discuss and exchange words,

tare lang wa[521] lai chong klak lang ya ce khai, paka,
the language, the valiant birds speak, sounds differently; and then

mü dong ing pöng khom ce, dong ing pöng lang,
when seeing the black tissue of the loin-cloth, the black loin-cloths are different,

mü dong ing ret khom ce öi, dong ing ret lang nöm mi ö,
when seeing the border of the black loin-cloth, the border is a really different one,

mü dong ing pöng khom ce, dong ing tali pat,
when seeing the tissue of the black loin-cloth, the black cloth slung all-around,

cam kwai lang ya, or ö, mi la ka ce öi,
the pulled-out end is different. Up there, of these men

ang yong ing klang, yong ing palö,
I, the black bird, I want to have

patai long[522] ken, tai long chin rui pa[523] kön, di ö,
a next elder brother, affiliated own elder brother in future,

yong ing palö, panöm pa tang, nöm pa chin rui pa kön di le, paka,

---

519 *men* = *Dendrobium fimbriatum* = male; parallel to this "rang" = (another flowering plant ?)

520 No literal translation, the line is concerned with the words to be exchanged.

521 name of a bird (?), *tare* (probably =) valiant, brave

522 parallel arrangement: *(pa)rüm yan* und *(pa)tai long*, *tai* = elder sibling, *long* (a length of bamboo) = numeral classifier for humans, therefore *yan* here probably "a piece (internode) of bamboo", *rüm* = younger sibling; prefixed *pa-* < *pe* (to give, to say) as causative (to let become)

523 *u* and *pa* suffixed to *rui* = "female and male (affiliated) relatives"

the black bird needs a true dear father, will have a truly affiliated own future father. And then

ari Lang-ke cir ching[524] charüt, khap ya na kum ka ce ö.
the eastern Kuki wooden comb, tucked in on the back-side,

cam kla rangeu po, cham dong prö rö[525], cin mat khom ce, ö,
letting the temples fall, while the hairs in the middle

macer cham kwai, cham dong pöng lang ya or,
are knotted on the top in the different eastern way,

mü chui pong mong, cin mat khom ce, chui don[g?] yolo ya or ö, paka,
when seeing the hill of the hair-knot turban, the topping hair-knot looks nice, and then

mi la ka ce, ang yong ing klang, yong ing palö, panöm u tang kön cö,
from them, I, the black bird, I want one who will be for me a real father
.

ang yong ing klang, yong ing palö, panöm pa tang kön ce khai. -
I, the black bird, I want one who in future will be for me a real father
.

ang kling ui yun, kling ui plong rau yun ö, pace,
When, my precious jewel, beloved precious jewel,

khunaik kri len, ni wang tacum, cadöi ri ron ö,
even before seven days will have passed,

ang yong ing klang pace da, po la ing-chom,
when I, the black bird, will take an axe,

bür dong pre[526] töng, po la ing chom, bür dong da pre töng,
with a handy shaft, take a well-shafted axe,

chong ce chiri, mang long o mör köi,
the river's brooklet goes to the stream with big boats,

chong ce chiri, tui öng mör köi ö,
the river's brooklet goes to the deep water,

ang yong ing klang, yong ing lang[527] pün, pün chüng bebat,
I, the black bird, I on the other hand climb, climb the sequence of hills,

ang yong ing klang, yong ing cam pün, pün nau[528] bebat, pace,
I, the black bird, I on the other hand start to climb, and having climbed

---

[524] no precise translation, name for the bamboo comb of the Khumi and "Kuki"
[525] onomatopoetic as also the following *yolo*
[526] perhaps = *prem töng* "well-formed, handy"
[527] proper meaning: "different", here in a rather restricted sense, "on the other hand; but"
[528] parallel word to "mountain" with unknown special meaning; the *be-bat* following it indicates the sequence of mountains following

hill after hill,

yong ing da u ö, cam tu ui kong[529] long le ö,
black bird am going to hew down a breadfruit-tree, and so

ang yong ing klang, yong ing cam tu, tu ui kong long,
I, the black bird, I start to cut, to hew down a breadfruit tree,

long kwaik chama[530] ya, kwaik long chö[531] lö po,
as a skilled worker can hollow out, also forcefully hollow out a boat,

long kwaik tangi, khü khü e ce, kan plong chang neu da ba,
at the time when I finally hollowed out the boat, my beloved heart,

ang yong ing klang le, chong ce chiri, mang long o mör köi,
I, the black bird, (take it to) the river's brooklet towards the stream with big boats,

chong ce chiri, tui öng mör köi, ang yong ing klang le,
the river's brooklet towards the deep water, I, the black bird,

yong ing long cam pün, pün chüng bebat,
I go to climb, climb hill after hill,

ang yong ing klang, yong ing plan tu,
I, the black bird, I return to cut a forest tree,

tu rüm pra naing[532], kong chüama tut,
hew down another forest tree, a yellow stem,

yong ing plan hui, hui kri tak ram po, tak bong wir wai, pace,
and when black bird returns to trim, to scrape also a big paddle blade with round paddle arm,

ranit ni kum, kum döi ri ron[533] po öi,
even before seven days have passed,

rayat war kum, kum wang döi-ron, pacö,
before eight nights have passed,

ang yong ing klang le, kan plong chang neu da ba,
I, the black bird, my beloved heart,

ang yong ing klang, kan bü[534] chang ram ce di le,

---

[529] according to the informant, a kind of jackfruit tree
[530] Arakanese: learned worker, specialist
[531] probably: *khyö lö*, an (Arakanese?) adverb relating to the work of excavating
[532] name of a tree
[533] *döi ri ron* = *dö-ron* (before) in common language

my, the black bird's, beloved little body leaf,

ang yong ing klang le, ari Lang-ke o-laing[535] long kot hai,
I, the black bird, shouldering the musket of the eastern Kuki,

ari Lang-ke o-laing long kot, mi la hai ce öi,
with the eastern Kuki musket shouldered,

macer rüm-ku, cin-dong[536] kok wai hai,
with the eastern sword, the sword in its sheath hung over the shoulder,

ang thwak reng klang wa, plo[537] ba thöng rüa
I set free the chief cock, release a volley,

uh, rui[538]-pyo tan yang klang, yong ing nök la, nök ni ko long kön ö,
o, Goldpleasure, the small cock, black bird's measure, measures day time,

yong ing nök la, nök war ko long kön,
black bird's measure, measures night time,

mong[539] ki kyaing wa cam tong kring po, tong ka khyöng long[540],
the small hill-cock will also crow, crow at the top of his voice,

nök war ko long, ka[541] war ko nök kön,

---

[534] Parallel word to *plong* (liver, the seat of emotions; but not really the same, as will appear at the end of the song); the following *ram* = leaf, but once more not to be taken literally, since relating to a human being.

[535] Arakanese pronunciation of "Holland", in the form of the Dutch East-Indian Company the first European trading power with mostly conflictive relations to the kingdom of Arakan until the middle of the 17th century (but unable to colonise it or to spread Christian belief), and as such importer of these old muzzle-loading guns, which subsequently were also acquired by the hill-tribes, among them (besides the Maraa [= Chamtu, Shendu, who had come into possession of some rare muskets at least at the beginning of the 19th century] and the Khumi) also some of the Arakanese Mru. In 1960 some Mru still knew how to manufacture from imported gas pipes their own (completely illegal) muzzle-loading guns. This does not imply that we should believe that Yonging did possess one before he had come into contact with the East: it is always called *ari Langke o-laing long* (mentioning it already here is – within the story of the song – premature). However, fact is that fire-arms were reported for the Burmese and Arakanese already in the beginning of the 15th century, and need not really have taken centuries to appear in the hills as well. Anyhow, Yonging first visited Bengal, so he must have lived in what is now Bangladesh south of Chittagong, which had been under Arakanese control until the middle of the 17th century and could not have been fully settled by Bangalis before the 18th century. Our song therefore, despite the *o-laing long*, need not have been composed much earlier than in the beginning the 19th century – but also not much later: the Mru of Bengal cannot be expected to know much of the rapids of the Koladan river (see footnote 94), and hence the song may be assumed to be roughly 200 years old.

[536] two kinds of swords, in daily language "cia-ran" and "dong"

[537] translation uncertain: informant said: "drive off"; but could be: "release". *ba* = bow; remainder: dubious

[538] Arakanese personal name: *hrwe-pyo* "gold-pleasure", here of the "*tan yang*" (Burmese: *tin ñang*.), a small kind of domestic fowl

[539] probably another Arakanese term for the cock mentioned in the previous footnote

[540] "crow at the top of his voice"

measuring night time, crows the measure of the night,

nök ni ko long ö, ka ni ko nök kön ba,
measuring day time, crows the measure of the day,

u, rui-pyo tan yang klang ce öi.
o, Goldpleasure, the small cock,

kan-kung leng toi po, toi muk chir chir, pace,
when joined and tethered by a cubit of jute string,

ang yong ing klang le, thwak reng klang wa, plo[542] ba töng rüa, pace,
I, the black bird, after setting free the chief cock, ...,

yong ing da u ö, cam pün, pün ui kong long,
I, the black bird, climb on, mount the breadfruit tree boat;

yong ing cam po, po kri tak ram, po tak bong wir wai,
black bird goes to take, takes the big paddle, the round paddle arm,

yong ing cam wui, wui kri tak ram po, tak bong wir wai ba,
black bird goes to pull, also pulls the big paddle, the round paddle arm,

ang yong ing klang le, kan plong chang neu, rui-pyo tan yang klang ö,
I, the black bird, my beloved heart, Goldpleasure, the small cock,

rui-pyo tan yang klang da, pün chang pho[543] long,
Goldpleasure, the small cock, climbs on the cucumber boat,

kar ko tong kring, tong ka khyöng löng,
the (white) cock crows, crows at the top of his voice,

pün chang-pho long ce, chang-pho long roi,
climbs on the cucumber boat, the cucumber boat sways,

wa cam tong kring, tong ka khyöng löng, pacö,
the cock starts to crow, crows at the top of his voice, and then

ang yong ing klang le, yong ing cam klen, klen kan kung leng
I, the black bird, I start to tie, to bind him with the jute string,

khanöm öi, pace, ang yong ing klang öi,
and then indeed, I, the black bird,

yong ing cam cun, cun ri chum kui,
I start to drive downward, down the meandering water,

yong ing cam pak, pak ling mang pren, po,

---

[541] to get the measure of night
[542] Could mean: to fire a volley; but this would not be useful in the line with footnote 146.
[543] *chang pho* = a long cucumber, in common language also used for ships and aeroplanes.

black bird tours, tours the Bangali country, and

yong ing cam pün, pün ling mang kyong ö,
black bird starts to visit, to visit the Bangali school,

ang yong ing klang, yong ing cam chang, chang ling mang lai,
I, the black bird, I start to learn, to learn Bangla standard language,

yong ing cam chang, chang kwar mang lai, pace,
black bird having started to learn, to learn Bangali local language,

kwar mang lai pi, yong ing cam chang,
speaking it brokenly, black bird starts to learn

chang ling mang lai po, ling mang lai cang khai ce,
also Bangla standard language; having done so,

ranit ni kum, kum wang döi-ron le,
before seven days have passed,

rayat war kum, kum wang döi-ron,
before eight nights have passed,

ang yong ing klang, yong ing cam chök, chök Yam-ce rong ba,
I, the black bird, I start to drive upward to the Koladan area,

yong ing cam pyaing, pyaing ok-cha pren öi,
black bird begins to visit, to travel through lower Burma,

yong ing or chang, chang ok-cha lai,
black bird there learns, learns lower Burmese,

yong ing or chang lai, chang ri-cha[544] lai, pace,
when black bird there learned the language, learned middle Burmese,

yong ing or pün, pün ok-cha kyong,
black bird there visits, visits the lower Burmese school

chang ok-cha lai, ok-cha lai cang,
to learn lower Burmese, lower Burmese done,

chang ri-cha lai, ri-cha lai cang, pace,
learns middle Burmese, having middle Burmese done,

ang yong ing klang le, yong ing or küng[545], küng ok-cha pung,
I, the black bird, there clothe like and take the appearance of a lower Burmese,

ok-cha pung küng yan nöm, yong ing or küng,

[544] literally "east-people", middle Burma
[545] literally "to draw (a hide over)", here: "to assume the appearance of ..."

looking like a real lower Burmese, black bird clothes

küng ri-cha pung, ri-cha pung küng khai öi,
like a middle Burmese and take the appearance of a middle Burmese.

ang yong ing klang le, yong ing or pün a öi,
I, the black bird, I there climbed up to Arakan,

pün röng-cha kywaing pace öi, kan plong chang neu da ba,
and when climbing up to Arakan, love my beloved heart,

ang yong ing chök po, chök Yam-ce rong,
I, the black bird, also drive up, up the Koladan area,

Yam-ce pre prai, kut tön hor lang, macer pön köi,
the Koladan branches, splits again, flows differently, in eastern direction,

ang yong ing klang, yong ing or preng, u ö, preng chang-pho long pace,
I, the black bird, I there untie, o dear, when I loosen the cucumber boat,

ang yong ing klang, yong ing or preng, preng kho-tu[546] long,
I, the black bird, I untie, loosen the pumpkin boat,

u, rui-pyo tan yang klang, wa pe tong kring, tong ka khyöng löng,
o, Goldpleasure, the small cock, the birds crow, crow at the top of their voice,

rui-pyo tan yang klang da, wa cam tong kring po,
Goldpleasure, the small cock, the bird starts to crow too,

tong ka khyöng nök war ko long, ka war ko nök,
crows loudly measuring night time, crows the measure of the night,

nök ni ko long, ka ni ko nök pace, u ö, kan plong chang neu da ba,
measuring day time, crows the measure of the day, o dear, love of my heart,

hung mang-tara[547], hua mang king kai ö,
the king of the mountain ridges, the lord of the rock-ledged gate,[548]

---

546 Bangla *kodu*, a kind of pumpkin, also used for a boat made of planks, without motor
547 *mang-taraa* (Burmese): supreme lord, king
548 This gate was also mentioned by Latter (1846:63) in a translation of a Khumi invocation ("Oh! spirit of the rock-ledged gate"). Latter adds the following explanation: "These are two very singular wall-like ridges of sandstone, running across the Kuladyne, about twenty miles the one above the other. They are not rocks like those of Colgong on the Ganges; but ridges perpendicular on each side, and only a few feet in width; the river has forced itself a passage through the centre. The tradition is that when the spirits found their domains invaded by a new faith from the plains, they endeavoured to raise a barrier; this was forced: a second attempt in like manner failed, and in despair they have given up the idea of a third." Correspondingly also Yonging cannot pass without Buddha's support (see below).

hung mang-tara, hua mang king kai, pace ö,
the king of the mountain ridges, the lord of the rock-ledged gate!

ang yong ing klang, yong ing or preng, preng chang-pho long pace,
When I, the black bird, untie, loosen the cucumber boat

hung mang tara, hua mang king kai,
the king of the mountain ridges, the lord of the rock-ledged gate,

chang-pho long mu kön co, uh, chang-pho long mu yan kön,
will surely tilt the cucumber boat, o, the cucumber boat will capsize,

ang yong ing klang le, yong ing cam plan, plan phek tangek,
so I, the black bird, try to turn, turn and move backwards,

plan kung thi rong, plan kung rü rü yan, ka öi,
return to a backside place, turn slowly fully backward, there,

yong ing cam khen, khen chang pho long,
black bird tries to secure, to fasten the cucumber boat,

yong ing cam khen, khen kri nan cün, pace,
having secured it and fastened to the big punting pole,

ang yong ing klang le, po kri poi-cha, che chong[549] pre yun,
I, the black bird, take two coins of copper money out of my girdle,

po ngui chin lap, khü khü e ce, ang yong ing klang di le,
and after I took a pure silver coin, I, the black bird, o yes,

yong ing cam ran, cam ku dong ran po,
black bird goes to buy, to buy on the market,

ling mang rua ton, rua di[550] yia yung,
a goat raised in Bengal, an unblemished good kid,

ang yong ing klang, hai pe ca pru, kri pok pre thar, khü khü e ce, da ba,
and after I, the black bird, I also took white rice, fully popped,

mi hung mang tara, hua mang king kai ka ce öi,
to this king of the mountain ridges, the lord of the rock-ledged gate,

ang yong ing klang nong po, nong rua di long[551], ka khü e po,
I, the black bird, also sacrifice, but though I sacrifice the goat

hung mang tara, hua mang king kai,
to the king of the mountain ridges, the lord of the rock-ledged gate,

---

[549] translation of the two syllables uncertain (informant said: hip)
[550] *di* = a term of endearment
[551] *long* = numeral classifier

hua cam nöng na pa[552] döi ya ce öi, ui höng[553] po nöng na döi cepo,
the rock is not really content, but also the local spirit is not contented.

ang yong ing klang le, yong ing lang chök, chök Yam-ce rong döi kabö[554],
I, the black bird, I, on the other hand, will drive upward, drive up the Koladan no more,

ang yong ing klang, yong ing long plai döi nöm, pace u ö,
since I, the black bird, I will not be saved, can surely not survive any more.

ang yong ing klang, ni ce taklök, mang nok ni khui,
I, the black bird, when the sun gets broken, goes down in the West,

yak döm khin wang, ka khü e ce,
at the time, when the evening comes,

chang pho phai luk, mi la ka ce öi, be phai chang li, phai chang li luk ka ce,
on the mats of the boat, one of the mats where to sleep, on the mats where to sleep,

ngek la chi-mi, pha-yong-toing cüng, mai di pre chek,[555]
I lighten candles, two sticks of Burmese candles,

ka khü e ce da, ang chup mani po, naik ing khyö lö,
and at that time, I now send my prayers to the spirit's house,

ang chup mani, naik mang khyö lö pace,
having now prayed to the lord of the spirits,

ang thop macin, ca chong krüm rüm[556].
I go to sleep, ...

taruk war mum, war mang rong köi ba,
Deep in the bud of the night, in the realm of the nightly dream,

taka chami, tui dong kreng pau[557],
Krengpau, the daughter of the water dragon in the middle of the water,

naik karek malaa ce, naik mang da u ö,
the spirit star maiden, the spirit lord,

[552] probably "he (*pa*), who (*ya*) not (*döi*) is (*ce*)"
[553] a local spirit ?
[554] exclamation of regret
[555] numeral classifier for long and thin things
[556] interpretation not certain
[557] name of the daughter (*chami*) of the waterspirit (= dragon, *dakaa*) in the middle of the water (*tui dong*)

wang pe ang/, mang plan thop ce,
comes and gives me back a dream,

taruk war mum, war mang rong köi,
deep in the bud of the night, in the realm of the nightly dream,

taruk war mum po, war mang ce lai:
a message in the night, deep in the bud of the night:

mi hung mang tara, hua mang king kai, mi la ka ce ö
to this king of the mountain range, the lord of the rock-ledged gate,

ni nong khan ka[558] ba di ö, u ö, mang kot-da-ma phura phwang wang,
once upon a time lord Gautama came to reveal,

mang nong khan ka ba di ö, krong reng[559] chang-kan, wan chüa li pai,
at that time to the lord of the earth, the yellow robe thrown over the shoulder,

krong reng ai dup, dup chüa klang kon mi ö,
the lord of the earth planted, planted turmeric bulbs.

krong reng ai dup, dup chüa klang kon,
Turmeric bulbs planted by the lord of the earth,

hung mang tara, hua mang king kai, hung rö long köi,
on the ridge of this king's, the lord of the rock-ledged gate's mountain range,

yap chüa klang kon, yap chüa re kon tanö,
looking for turmeric bulbs, after seeking turmeric bulbs,

ka chüa klang kon, ka chüa re kon, ka khü e ce,
you'll find turmeric bulbs. And then after getting turmeric bulbs,

mi hung mang tara, hua mang king kai ka da,
to this king of the mountain range, the lord of the rock-ledged gate,

kwak pe chüa re, pong ka wir wai, ka khü e ce öi,
in a small round basket, offering turmeric bulbs, at that time,

men[560] yong ing klang, yong ing long plai yang, men ö öi, pace,
you, black bird, will be saved, and then you, black bird, will survive.

taruk war mum, war mang rong köi,
Deep in the bud of the night, in the realm of the nightly dream,

taka chami, tui dong kreng pau, naik karek malaa ce,

---

558 (whole phrase:) once upon a time
559 declared to be the "younger brother" of Gautama; but here apparently he himself
560 see footnote 65, here term of address

Krengpau, the daughter of the water dragon, in the middle of the water, the spirit star maiden,

wang pe ang/, taruk war mum, war mang ce lai daba,
came and gave me the dream, in the bud of the night, the message in the dream of the night.

rau cin töm wang, mi hung mang tara, hua mang king kai,
The morning comes, for this king of the mountain range, lord of the rock-ledged gate,

ang yong ing klang, yong ing lang yap, yap chüa re kon,
I, the black bird, I on the other hand seek, seek turmeric bulbs.

yong ing cam yap, yap chüa klang kon, pace,
When I, the black bird, try to look for turmeric bulbs, seek turmeric bulbs,

ang yong ing klang ka, ka chüa re kon,
I, the black bird, find, get turmeric bulbs,

ang yong ing cam ka, ka chüa klang kon, da ba,
I, the black bird, start to find, to collect turmeric bulbs,

mi hung mang tara, hua mang king kai, mi lö köi ce,
to him, this king of the mountain range, lord of the rock-ledged gate,

ang yong ing klang, yong ing cam kwak,
I, the black bird, I go to offer,

kwak pe chüa re da, pong ka wir wai paka, ka khü e ce,
to throw a round basket with turmeric, and after this

ang yong ing klang, yong ing or preng, preng chang pho long,
I, the black bird, I there untie, untie the cucumber boat,

yong ing or preng, preng kho-tu long, pace,
and when I, the black bird, untie the pumpkin boat,

ang yong ing klang, kho-tu long plai khanöm,
the pumpkin boat is saved, survives indeed,

ang yong ing klang ö, chang-pho long plai ba di le kön,
my, black bird's, cucumber boat will safely survive,

ang yong ing klang le, yong ing long plai, yong töng tai nöm öi,
and I, the black bird, I am saved and really survive, in the same way too.

ang yong ing cam chök, chök Yam-ce rong,
I, the black bird, start to go upwards, up the Koladan area.

macer pön cak da, keng döm pön köi, paka,
To the people of the eastern side, downward on the other side, and there,

ang yong ing klang, yong ing cam khen po, khen chang pho long, daba,
I, the black bird, I also go to tie up, to fasten the cucumber boat.

or ö, aröng ton klang le, aröng ton ku,
Up there, the males grown up in Arakan, the females grown up in Arakan,

ton le mala ce nöm or,
and truly the girls growing up there,

mü tum ching pung khom ce, tum ching pung lang khai ö,
when seeing the image of the female flower-trees, their figures look different,

mü cam om pung khom ce, cam om pung lang,
when seeing the image of the flowering sprouts, their figures are different,

mü klai ing pöng, cin mat khom ce, klai ing pöng lang,
and also when seeing the tissue of the black skirts, the black tissue is different,

mü klai ing bong, klai ing bong lang, [paka, mi la ka ce,][561]
seeing the pattern of the black skirts, the pattern of black skirts is different,

ki ce kong rik, kong dok wan klai, chek ram chö kong mi,
they are red-brown like Muntjac-hide, pretty split skirts, split leaves like sticks,

mi lö ka ce öi, ang yong ing klang, yong ing palö,
and among them, I, the black bird, I want to have

panöm u tang le di, parüm yan tang, rüm yan chin rui u nöm, paka,
an own dear mother and an own aunt, an affiliated younger sister of my own mother;

u ö, or ö macer ton klang, aröng ton kong, ton le taklang,
o dear, up there, of the eastern grown-up males, of Arakanese stem, grown-up and male,

macer ton klang ka ce öi, ang yong ing klang, yong ing palö,
of the eastern grown up males, I, the black bird, I want to have

pahi[562] ron tang, hi ron chin rui pa nöm, pacö,

[561] Does not really belong here; and therefore is repeated two lines later.
[562] Father's younger brother, step-father. The parallel use of "pa-rüm" suggests an equation with "u-ram", mother's younger sister, step-mother (see footnotes 63 and 68).

an uncle, a younger brother affiliated to my own father,

mü men om pung khom ce, men om pung lang nöm öi,
when seeing the figures of the sprouting orchids, their figures are really different,

mü rang om pung khom ce, rang om pung lang,
when seeing the figures of the sprouting trees, their figures are different,

mü dong ing pöng khom ce, dong ing pöng lang khai ö,
when seeing the black tissue of the loin-cloth, the black tissue is different,

mü dong ing ret khom ce, dong ing tali, pat cam kwai lang,
when seeing the border of the black loin-cloth, the black cloth slung all-around differently,

mü dong ing pöng khom ce, dong ing tali, pat cam chua lang,
when seeing the black tissue of the loin-cloth, the black cloth all-around, will come out differently,

Cham-tu da, u ö, dong tom, dong ing tali, pat cam kwai lang le,
the Shendu, o dear, binding the loin-cloth pull out the black cloth, slung all-around, differently.

ari Lang-ke cir ching charüt, khap ya na kum, mi la hai ce,
The eastern Kuki wooden comb, tucked in on the back-side, with this,

cam kla rangeu, cham dong prö rö, cin mat khom da,
letting the temples fall, the hairs in the middle, however, are knotted on the top,

mü chui pöng mong, chui pöng rup lang,
seeing the arrangement of the hair-knot, the picture of the knot is different,

mü chui pöng mong khom ce, chui don[g?] yolo ya öi, paka,
When seeing the hill of the hair-knot, the topping hair-knot looks nice,

mi lö ka ce, ang yong ing klang, yöng ing palö,
from them, I, the black bird, I want to have

patai long ken, di le, yong ing palö,
an elder brother, and I want one to be black bird's

pa-hi ron tang, hi ron chin rui pa nöm khai ö,
dear uncle, a younger brother of my own father,

ang yong ing klang le, – kan plong chang neu, –
when I, the black bird, my beloved heart,

yong ing toi pün, pün chüng bebat khom ce öi,
I climb with them, climb hill after hill,

yong ing or tu, tu thüm palong,
black bird there cuts, cuts a length of thüm-bamboo,

büa pe pön[563] rin, pön bong ku klang,
to prepare small decorated tubes, tubes with male and female decoration,

yong ing toi kham, kham la pön rin, pön bong ku klang choi, paka,
black bird will drink with them, drink the oath of the male-female decorated tubes, by this,

ang yong ing klang le, yong ing toi doi a, doi la phura,
I, the black bird, I embrace with them, embrace their god,

am[564] puk mia mü choi, paka,
by this oath seeing the scraps from preparing bamboo lashing,

ö, wa-ca lu rai choi pe, choi cum röi man, pace öi,
called the oath of twisting off a chick's neck, an oath executed quickly and precisely, and when,

yong ing toi rai, wa ca lu rai choi pace,
when black bird twists off with them, the chick's neck twisting oath,

yong ing toi kham, wa ca lu rai choi pace le, paka,
black bird drinks with them the chick's neck twisting oath, and then after

choi cum röi man, ka khü e ce ö, rau cin töm wang di le,
this oath, executed quickly and precisely, when at that time the morning comes,

yong ing toi pün, pün reng bong cu, pace da ö,
black bird will climb with them, climb up to the chief's palace.

yong ing toi pün, pün reng bong cu,
After climbing up with them, up to the chief's palace,

ca reng hom hua, tom reng chak chia[565], aröng chak toi khai ö,
eating the chief's rice balls, I bind the chief's bad heart, have joined the Arakanese heart.

aröng chak toi, la khü e ce, aröng chak wang u ö,
When joining the Arakanese heart, the Arakanese mentality, o dear,

---

563 a small bamboo tube for pouring water during sacrifices, decorated (*rin*) with cut decorative patterns (*bong*).

564 name of the oath (literal translation not certain): *mü* = to see, *mia puk* = scraps from preparing bamboo lashes

565 Not certain. Informant said: (morally) bad heart; but *tom chak* = to knot, and *chia* also = iron arrow.

aröng ton klang, macer ton klang, ton le klang da ö,
the males grown up in Arakan, the eastern grown-up and growing up males,

kan[566] pa raneu po, kan plong chang ram da öi,
are also close to my heart, my beloved persons,

mang[567] bok kwö ru, kwö ing hök tök, nge dong prem töng,
the black king crow's, the rocket-tailed drongo's long beautiful tail feathers

cham chui ling cong, kau dong[568] prüng rüng li,
tucked upright into the hair-knot, over the middle of the forehead, dear!

di, mü men om pung khom ce, men om pung lang,
when seeing the figures of the sprouting orchids, their figures are different,

mü rang om pung khom ce, rang om pung lang mi,
when seeing the figures of the sprouting trees, their figures are different,

cam prö rö mani, ding ding woi loi, cin mat khom ce ba di ö,
when they will discuss and exchange words,

tare lang wa po, lai chong klak lang khai ö, mi lö ka ce,
the language, the valiant birds speak, sounds differently; with one of them,

ang yong ing klang, yong ing toi kham, kham tai long ken,
I, the black bird, I drink together, drink with a next elder brother,

yong ing toi kham, kham nau long kroi[569], nau long po chin rui pa nöm, khai ce,
black bird drinks together, drinks seekig a younger brother, a truly affiliated younger brother

u ö, reng[570] mong talang, ting tong prang lak,
On the chief's public place, ...

mi pröng nai con köi, rau cin töm wang, u ö,
on the raised flat round place, the morning arrives, o dear,

yong ing toi cang, cang kor ko kyang,
black bird joins mastering, getting accustomed to a horse,

---

[566] term of endearment, *ra-neu* "heart" in a metaphorical sense
[567] literal translation of the whole line not certain, name of a black bird (rocket-tailed drongo?, king crow?), the tail feathers of which are worn as a plumage in the hair
[568] informant: bamboo grove; but probably: "forehead centre", "over the forehead"
[569] word and meaning not certain; perhaps parallel to *ken* = next (younger), "next (elder)"
[570] text uncertain; similar to, but not identical with that to which footnote 38 refers

yong ing toi pün, pün kor ko luk pace,
when black bird joins mounting, riding on a horse,

ang yong ing klang, yong ing toi cang po,
I, the black bird, I also join them

cang kor ko kyang, ang kor ko kyang yan nöm, cö,
getting accustomed to a horse, being fully accustomed to my horse, then

yong ing toi pün, pün reng bong cu,
black bird climbs up with them, up to chief's palace,

ca reng hom hua, tom reng chak chia ce öi,
eat the chief's rice balls, binding the chief's bad heart.

ranit ni kum, kum wang döi-ron, rayat war kum, kum wang döi-ron, paka,
And then, before seven days have passed, before eight nights have passed,

ang phök ri-cha[571], hia kom pre töng,
I call together the eastern sons, the warriors ready to start.

pre töng ari Lang-ke, hia kom wang kla,
The prepared Kuki warriors of the east, will invade

ranit kom thong, kön co di le, pakö,
700,000 will it surely be,

ari Lang-ke da u ö, hia kom wang kla kön ö, rayat kom thong ba di ö,
the Kuki of the East, o dear, warriors, will invade, some 800,000,

chai rung[572] chai phe, chai kong höng tün khom ce öi,
with their shields made of elephant hide, when the elephant shields vibrate,

wai Yam taku kui nöm öi,
they truly round all meanders of the Koladan,

chai rung chai phe, chai kong höng tün khom ce,
when the elephant shields vibratingly resound,

höng krong dai rui, krong dai rui cü, ranit mang plan ba di le,
they let vibrate the navel-string of the earth, vibrate its navel-string again and again,

u ö, ari Lang-ke, hia kom wang kla ö, kla Yam taku kui khanöm,

---

[571] here not "Burmese", but literally: "eastern people"
[572] shields (*kong*) made of elephant (*chai*) hide

o dear, the eastern Kuki warriors will fall in, fall into all meanders of the Koladan, indeed,

ari Lang-ke, u ö, hia kom wang kla a, kla Yam ko kui,
the eastern Kuki warriors, o dear, invade, fall into the meanders of the Koladaing,

kla nok kom pren kön paco, u ö, aning ka ning, yu ö,
will surely invade the western countries. O dear, this very year, down there,

mang karbari, chang ri reng ya, reng long lang then, then[573] wa kom rui,
the king's officer, who always judges, the chief, however, who founded, founded the village,

rua chuk-kri öi, reng long ya mi ba, mi wa kom rui köi ka,
the big man of the village, who is the ruling one in this village,

lu kri mi pha[574], lu ra cün töm, tom kling lu ro[575], reng long ya mi ba,
the old man with brittle hairs coloured like those on the head of the babbler, who rules,

phung ram kwai[576] möt, pre chüng rong köi,
beyond the cultivated area of two hills

phung ram kwai möt, chum chüng rong köi, pace,
when beyond the cultivated area of three hills,

lu kri mi pha, lu ra cün töm, tom kling lu ro,
the old man with brittle hairs coloured like those on the head of the babbler,

reng long lang kir, kir ui[577] kom rong,
the chief, on the other hand, selected, selected the fields to be planted,

toi chöi thüm lung po, döm kau bön lön, ö,
also joined cutting the thüm-bamboo, cutting down bamboo row after row,

ni ui kom rong ba di ö, ni tin thi rong ba di ö,
that to be a paddy field, that to be a cotton field,

ret ui kom khai ce, ui kom phek ru, arai pröng töng,
the paddy fields divided up, the paddy grew nicely and equally,

---

[573] literal translation not certain, passage means: to found a village (*wa kom rui*)
[574] Arakanese: "father"
[575] *ro* properly "brittle", here: "in the colour of brittle bamboo", sense: (the hairs of) the upper sides of the (man's) head (*lu ra cün? töm?*) are coloured like the head-feathers of the bird (namely, the *wa-har*).
[576] literal translation uncertain, meaning: beyond the limits of the hitherto cultivated area
[577] literally: "fruit-many", i.e., "paddy"

ngak ce kre chot, kre ding yan püi, kre wang liu ru, pace,
bird cane is dibbled, the cane with sweet ears, when the cane seed grows,

la kui pren pon, pren chot[578] caloi,
when August maize shoots up, the cobs develop,

liu wang liu ru po, toi wang bong kla, pace ö ko,
the kernels come, the kernels also grow, the cobs are going to fall,

mi kra tan dong köi, ari Lang-ke,
then in between the lined-up stairs, the eastern Kuki

hia kom wang kla, kla Yam taku kui ce,
warriors fall in, fall into the meanders of the Koladan,

ari Lang-ke, hia kom wang kla, kla ui lim don,
the warriors of the eastern Kuki fall in, invade the fruit-bearing fields,

[a]ri lang-ke, hia kom klöt rui po, klöt ui lim don ka öi, pace,
the eastern Kuki warriors also devour all, eat up the fruit-bearing fields,

ari Lang-ke, hia kom wang cöng, cöng yuk ce lot[579] tai nöm mi,
the eastern Kuki warriors transform themselves, become really ravaging monkeys.

hia kom wang cang po, cang yuk ce lot,
The warriors also become, become destructive monkeys,

ching töm nga-pun, nga höng mang reng ce, lot ui kom rong po,
wild pigs of the forest, huge wild pigs to destroy the fruit-bearing fields.

rau cin töm wang köi, mün tu döi kabö,
The morning comes: Alas, how much evil? –

lu kri mi pha, lu ra cün töm, tom kling lu ro reng long,
The old ruling man with brittle hairs, coloured like those on the head of the babbler,

reng long mi da, kan plong chang neu, kan bü chang ram daba,
the headman, my beloved heart, my beloved little body leaf,

u ö, u ö, kan plong chang neu, kan bü chang ram,
o dear, o dear, my beloved heart, my beloved little body leaf,

la kui tim pör, thop lang kong cün hai ce,
the small well-formed August hoe, with the shaft to grasp,

578 to shoot up, sprout?, the state before ripening
579 to destroy the fruits of the field (by animals feeding there)

wang bot thüm krung, wang hoi leng leng po di ö,
goes weeding the thüm root clusters, to clean row by row:

mün lung[580] mün ya, mün ya ram kong kau[581] cö öi,
how many men, how many leaf stems, for heaven's sake,

reng long da wang cang ang, cang yuk ce lot, di,
headman, have made us the prey of monkey devastation.

yu a, mang karbari, chang ri reng ya ka ce öi, po reng töng-pang pace,
Down there when taking consultation with the king's officer, who always rules, discussing the case,

tara lang chen, chen pong ui yun, tara lang chen, chen chai ui yun, pacö,
discussing the Ficus fruits, discussing the fig fruits,

ho reng tara, ni töng lök rüa, pace ö,
when calling an official meeting, once every day:

mün lung, mün ya, mün ya ram kong kau pacö di ö,
how many men, how many leaf stems, for heaven's sake,

phung ram da wang cang en/, cang yuk ce lot,
world leaf, have made you the prey of monkey devastation,

phung ram da wang cang en/, cang dor ce lot ce ö,
world leaf, have made you the prey of gibbon devastation!

ang po twak chang pang[582] po, ka twak chang pang döi kabö,
Whatever I think, I find no solution,

ang po tuak chang kram po, ka tuak chang kram döi kabö,
how much I ponder, I find no answer.

men yong ing klang ce, yong ing cam chök, chök yam-ce rong,
Black bird, the man, black bird went to drive up, upward the Koladan region,

ang chak charui, kling-ui plong-rau da öi, pace,
my whole heart, beloved jewel,

ari Lang-ke, hia kom wang kla, kla Yam taku kui yan ce ü,
did the eastern Kuki warrior now invade, fall into all the meanders of the Koladan?

ari Lang-ke, hia wang kla, kla nok kom pren,
Will the eastern Kuki warriors invade, fall into the western countries,

---

[580] pejorative; *lung* = (numeral classifier for cattle), here: for persons; *ram kong:* "leaf-stalk" = person
[581] question particle, showing embitteredness (*kau-cä* = fate)
[582] instead of *chang krang?*

nok kom cüng[583] cü kön ce ü, rau cin töm wang da,
will they shake the west? The morning comes,

men ö, takeng tang rui[584], ram ra ce plum hön ce öi,
o you, from the foliage of the lianas,

kan plong chang neu, aru tarü, cam kreng pong mü tanö öi, pace,
my beloved heart, stretch yourself, try to hear and spy out,

lu kri mi pha, lu ra cün töm, reng long ya mi da ba öi,
and when the old man with brittle hairs, the headman,

takeng tang rui, ram ra ca plum hon ce öi, aru tarü, cam kreng pong mü,
from the foliage of the lianas, stretches himself and spies out,

mü men om pung khom ce, men om pung lang khai nöm ö,
when seeing the figures of the sprouting orchids, their figures are really different,

mü rang ching pung khom ce, rang ching pung lang khai nöm,
when seeing the figures of the sprouting trees, their figures are different indeed.

tarua chuk-kri reng long mi ö,
The big man of the village, who is headman,

mü men om pung, men om pung lang khai ö,
seeing the figures of the sprouting orchids, their figures are different,

mü dong ing pöng, ing dong tali pat cam kwai lang po, khai nöm ö, paka,
seeing the black tissue of the loin-cloth, the black cloth slung all-around is differently indeed,

Cham-tu da dong tom, dong ing tali, pat cam chua lang khai nöm ö, kö,
binding the loin-cloth, the Shendu pull out the black cloth, slung all-around, differently indeed,

mü men om pung, men om pung lang khai nöm,
seeing the figures of the sprouting orchids, their figures are different indeed.

ari Lang-ke hia kom wang kla,
The eastern Kuki warriors come to fall into

kla nok kom pren, kla Yam taku kui yan khai nöm,
the western countries, invaded all meanders of the Koladan indeed.

aning kaning, ning nöng long plai döi ö,

[583] uncertain: *cün cü* (?) = if perhaps; *ce ü:* question particle
[584] name of the creeper (*rui*)

This very year, this year we cannot safely survive,

hia kom wang kup, kup wa kom rui köm lama,
the warriors will enter, enter the village lines, alas,

hia kom wang kla, kla wa kom tan köm ce khai co pace,
the warriors will fall into, invade the village rows, to be sure.

tarua chuk-kri, reng long ya mi da, mang karbari, chang ri reng ya da,
The big man of the village, the headman, the king's officer, who always rules,

po reng töng-pang, kaa reng töng-pang döi di le,
takes counsel, gets no counsel,

po reng töng-pang mi, keng döm pön köi, tu ting palong, büa pe ting lung,
takes a council: on the other descending side, cut ting-bamboo, cut them into shape,

cher ya-ca mia, mia chong yop yop, da ba,
make nice lashings, good and hard lashings,

kom la ting lung, kha ra nen len, akri chum nöng dia, tanö,
bind the ting-bamboo pieces, make a surrounding fence of three big layers,

kom pe thüm lung, kha ra nen len po, akri chum nöng tai nöm,
bind the thüm-bamboo pieces, make a surrounding fence too, three durable big layers,

u ö, kom ting chur cam dia, u, kom teu[585] chur cam dia, tanö ö,
o dear, bind double-barbed ting-staples,

nam kom mi da ö, kom ting chur cam,
when all local men bind double-barbed ting-staples,

kom teu chur cam, kom la ting lung, kha ra nen len,
bind double-barbed teu-staples, bind with ting-bamboo the surrounding village fence,

kom la thüm pong, kha ra nen len, po, akri chum nöng pace,
when binding with additional thüm-bamboo the surrounding village fence, three big layers,

ang yong ing klang le, ang chak charui, kling ui plong rau,
I, the black bird – my whole heart, beloved jewel,

ang chak charui, ang yam ui yun, yam ui plong rau di ö,

[585] *thum, ting,* and *teu* are three kinds of bamboo

my whole heart, my precious gem, beloved gem,

ayak kayak, uh, nöng da long plai döi ö,
o tonight, this night, you cannot safely survive,

ayak kayak, yak nöng long le, uh, mit ce chüm chum dö öi,
tonight, this night you cannot be saved, and dying out are extinguished.

uh, reng ce khö kwak, ra ha[586] net nöng, uh, chur plon tom köi ba di le,
Where the offering baskets are thrown away, to push back the spirits, o, in the place of the barbed gate,

ang yong ing klang le, yong ing wang cang, cang ki pum nai,
I, the black bird, I will make, make a Muntjac bivouac,

yong ing cang po, cang ya pum nai khai ö,
black bird also made, made a Serow bivouac,

ari Lang-ke hia kom wang cang po, cang ki pum nai khai ö,
the eastern Kuki warriors were also going to sleep, made a Muntjac bivouac,

uh, hia kom wang cang po, cang ya pum nai khai ö,
uh, the Kuki warriors too were going to sleep, made a Serow bivouac.

yong ing klang le, kan pa raneu, kan plong chang neu daba,
The black bird, my dear heart, my beloved heart,

ang po pe ca pru, rai cang pre ma, da khü e ce,
I take two kernels of rice, and when

ang po pe ca pru, rai cang chum ma, da khü e ce,
I take three kernels of rice to see the omen, at this time

ang yong ing klang, yong ing lang mü, mü reng lung bang,
I, the black bird, I in my term will see, regard the omen

yong ing cam mü, mü mang lung bang, pace,
when I black bird, try to see, to regard the omen,

uh, reng ce ca pru, ca chin plang khoi, khoi tong[587] chang preng khai nöm öi,
o, the white rice, the rice which truly shows an answer, answers which way to take indeed,

---

[586] no literal translation, informant: "where the spirits live", but possibly: "where they can be pushed back (*net nöng*)"

[587] Informant: "1 Ari" (~ 10 kg in measuring paddy) = *atong-apreng;* more probably literally: "answer which teach take-off" (*preng*). For the omen paddy grains are placed on a bamboo lath and held over the fire; it is a good sign when they fly off in direction of the enemy.

uh, reng ce ca pru, ca chin plang khoi po, khoi tong chang preng,
o, the white rice, it also truly shows the answer, answers which way to take,

reng ce ca pru, uh, rai cang lung yung po, yung tong chang preng khai nöm,
the white rice, o, the rice provides a good omen, good which way to take indeed,

ang yong ing klang le, yong ing plan po, po kar ko dui po ö,
I, the black bird, I again take, take hen's eggs too,

yong ing cam mü po, mü kar ko lung,
black bird will also see, regard the egg omen,

yong ing cam mü, mü kar ing[588] lung,
black bird will see, regard the hen's egg omen,

ang yong ing klang, yong ing or köi le öi, po kar ko dui,
I, the black bird, up there I take a hen's egg,

yong ing cam mü po, mü kar ko lung,
also to see, regard the hen's omen,

kar ko lung yung po, yung thüm[589] dön lön da ba,
the hen's omen is good as well, all omens are propitious.

kar ko dui yung, yung thüm dön lön ce,
the hen's eggs are good, good one by one,

u, kan pa raneu po, kan plong chang ram da ba,
o you my heart, my beloved leaf.

reng ce khö kwak, ra-ha net nöng, uh, chur plon tom köi di da,
Where the offering baskets are thrown away, to push back the spirits, o, at the barbed gate[590],

---

[588] Literally: "black" as a play of words in relation to "ko" (white) used as the genetive particle (= *ka*). For the omen eggs are broken; it is propitious when the yoke is not mixed with the white.
[589] literal translation not certain, *thüm* perhaps = to develop, *lön* = one after the other, i.e., all omens are propitious
[590] Such barbed gates were still used by the Bawm warriors in 2005 (photo by courtesy of Céline Mouchet).

yong ing wang cang, cang ki pum nai,
black bird will make, make a Muntjac bivouac,

yong ing wang cang, cang ya da pum nai ce,
black bird will make, make a Serow bivouac,

ari Lang-ke, u ö, hia kom wang kla, kla Yam taku kui ba öi,
the eastern Kuki, o dear, warriors invaded, fell into all the meanders of the Koladan,

ari Lang-ke, hia kom wang cang po, cang ki pum nai, ranit mang plan, ca di le, paka,
the eastern Kuki warriors went to sleep, made a Muntjac bivouac, again and again, and then

cin dong war wang le u ö, war dong phong[591] long khanöm ö,
midnight arrives, o dear, midnight becomes high indeed,

cin dong war wang po, war dong phong long pace,
midnight arrived, and when midnight became high,

rau cin leng lot le u ö, leng wang chum chan, pün kwang rö rong,
the morning star appears, o dear, the star rises and climbs up to the sky,

rau cin leng lot po, leng wang chum chan da, pün kwang rö röng, pace,
when the morning star appears, and the star rises, climbs up to the sky,

rüm dong pring li, nor chong chi tak[592] wa ö,
the red forest bird with the bright beak,

rüm dong pring li, nor chong chi tak wa da lang mön po,
the red forest bird with the bright beak perches,

mön tang rui krök, rüm dong kheng ring, nor chong chi tak wa da,
perches on the liana network, and when the red Kheng, the forest bird with the bright beak,

[591] not certain: to swell?
[592] Informant: showing a strong colour; the whole line contains nothing but the poetical name of the bird, normally called *wa kheng* (mentioned 2 lines later).

lang mön ö, mön tang rui krök pace,
in his turn perches, perches on the lianas,

ta pe mani da, u ö, khok chong chör tün,
it now hears, o dear, the sound of shuffling legs,

ta pe mani po, khok chong da klök tün,
now also hears the noise of legs,

rüm dong pring li, wa da lang ta po,
the red forest bird with the bright beak, also hears,

ta pe mani, khok chong klök tün ce,
when it now hears the sound of legs,

rüm dong pring li, wa cam tong kring da,
the red bird in the forest, the bird crows,

klup pe mani po, lap hoi pün lün, pace da u ö,
and when it flaps and shakes its wings,

u rui-pyo tan yang klang ö, wa cam da kreng ta khanöm paka,
Goldpleasure, the small cock, listens and indeed then hears it,

ta pe mani da u ö, khok che chör tün ö,
hears now, o dear, the sound of shuffling legs.

rui-pyo tan yang, wa cam kreng ta ö,
Goldpleasure, the small cock, the bird listens and hears,

ta pe mani po, khok chong da klök tün, ka khü e ce öi,
now also hears the noise of legs, and at that time

u rui-pyo tan yang, mong ki kyaing,
Goldpleasure, the little cock

wa cam tong kring, tong ka khyöng löng, di ö,
starts to crow, to crow at the top of his voice,

mong ki kyaing wa da u ö, klup pe mani, lap hoi pün lün,
the small cock, o dear, now flaps and shakes its wings up and down.

klup pe mani, lap deu pün lün, pace, chöng lap di li,
When flapping now and shaking out its wings up and down, flapping its wings

klup pe mani, lap hoi pün lün ce,
now shaking them out, flapping the wings up and down

wa cam tong kring, tong ka khyöng löng,
the bird starts to crow, crows at the top of its voice,

nök war ko löng, kaa war ko nök,
measuring night time, crows the measure of the night,

nök ni ko löng, kaa ni ko nök, pace,
when measuring day time, crows the measure of the day,

u ö, yu ö, adong tapong, kra tan dong pong,
o dear, down there amidst the multiple, between the multitude of stairs,

wa kom talüp, chum thong[593] dong köi,
the village overtowering each other, in all families,

wa kom talüp po, pre thong da dong köi,
the village overtowering each other and in all the families,

ci ce[594] lu kut, lu lim chong reng, myö ya mi ö, rü tam tari, u ö,
those who wear their hair in a knot in the neck, get up and prepare themselves, o dear,

yu ö, khung[595] to do köi, reng ce cor cong[596],
down there, under the house, where the fowls are kept,

kar ko kreng ta da, ta pe mani, u ö, lap deu chöng tün, pace,
the cocks listen and when they now hear, o dear, the noise of flapping wings,

rüm dong pring li, lap deu chöng tün,
the flapping of wings of the red bird in the middle of the forest,

reng ce cor cong wa da, klup pe mani po, lap hoi pün lün,
also the cocks under the house now shake their wings up and down,

wa cam tong kring, tong ka khyöng löng, pace,
and when they crow at the top of their voice,

u ö, kar ko krüng thör, krüng wa kom rui,
o dear, the cocks wake up, wake the village lines,

kar ko krüng thör po, krüng tan dong lak,
the cocks wake also up, wake those in the lined-up stairs

ci ce lu kut, lu lim chong reng, myö ya mi,
those who wear their hair in a knot in the neck

rü tam tari, la khü e ce,
get up to prepare, and at that time,

---

[593] cf. footnote 34
[594] whole line: "those who wear their hair in a knot in the neck"
[595] the space under the *kim-mä*
[596] informant: the store-room, here: the place where the fowls are kept

po pe yong köng, u ö, chop chong pre ria,
they fetch, o dear, two pieces of dried and dead wood,

po pe yong köng po, chop chong chum ria,
they fetch three pieces of dry fuel,

po pe leng[597] köng, chop chong da pre ria,
they fetch two pieces of dried and dead wood,

ce phoi[598] chum krum, nam nai un köi,
put it crosswise on the hearth,

kre pe ding ring, mai di prön rön, ka khü e ce,
beat off red embers, the sparks of fire, and thereafter

ling mang chaing ö, chaing teng lak köi, te khong ko yun, ba di ö,
put white hulled rice kernels into the round Bengal iron pot, into the iron vessel,

ram löng tui thut, höng kröng rün rün, di ö,
fill in water and put it on the fire for cooking.

tui wöi ri luk, mai wöi ri dak po,
water being on top and fire below,

heu pe wa le, khong ko lüng lüng,
and turning the white hulled rice slowly round and round, they let it become dry,

mi kra tan da dong köi, ba di ö,
in between these lined-up stairs.

or ö, reng ce da, or ö, reng ce khö kwak,
Up there, where the round offering baskets are thrown away,

ra ha net nöng, uh chur plon da tom köi, öi,
to push back the spirits, o, in the place of the barbed gate,

rui-pyo tan yang klang, wa cam tong kring,
Goldpleasure, the small cock, starts to crow,

tong ka khyöng löng, ka khü e ce,
crows at the top of his voice, at that time

ang yong ing klang ba di ö, men yong ing klang ba di da,
to wake me, black bird up: you black bird man,

rü tam tari, la khü e ce, reng ce khö kwak,

[597] parallel word to *yong*; no own meaning?
[598] no literal translation; only part: at the hearth *un*(?) (the flat *nai* place *nam*)

get up and prepare yourself. Where the round offering baskets are thrown away,

ra ha net nöng, chur plon da tom köi,
to push back the spirits, o, in the place of the barbed gate,

chöng[599] ram prau rau tai ö, paka,
the warriors get up in the morning

ang yong ing klang, yong ing klang ö,
and I, the black bird,

macer rüm ku, cin dong kok wai,
the eastern sword in its sheath hung over the shoulder,

ari Lang-ke, o laing long kot,
the musket of the Kuki shouldered,

plo[600] reng klang wa, plo ba thöng rüa khai ö,
I release a large cock, ...

ang yong ing klang le, plo reng klang wa, plo ba thöng rüa, pace,
I, the black bird, release a large cock, ...

ang yong ing klang, ang rum ran thak ba di da,
and then I, black bird, I hew down,

ang thak ting chur cam, ang rum ran thak, thak teu chur cam,
cut off the ting-barbs, hew down, cut off the teu-barbs,

kha chur taku, büng nöng chüt cöi döi kabö öi,
of the fully barbed fence, alas, the closure cannot be pushed in,

kha chur taku, büng nöng chüt kla döi ö,
the closure of the fully barbed fence cannot be broken down,

ang yong ing klang ba di da, khöng chai di kong, chai di kong tün khom ce di ö,
I, the black bird, beat the elephant shield, and when the elephant shield sounds,

wai Yam ko kui öi nöm, wai Yam taku kui ö,
around all meanders of the Koladan, around all meanders of the Koladan,

ang yong ing klang da ba, u, kan pa raneu, kan plong chang ram,
I, the black bird, o, my father's heart, my beloved men,

ang yong ing klang ba di le, khöng chai di kong, chai di kong tün ö,
I, the black bird, beat the elephant shield, the elephant shield sounds,

---

599 literally: leaves (of an unidentified plant); here (as usually) "people"
600 for whole line see footnote 88

ang yong ing klang, chöng ram prau-rau, pace,
my, the black bird's, men get up in the morning;

ari Lang-ke, o-laing long kot hai,
the Kuki of the east, with the muskets shouldered,

ari Lang-ke, o-laing long kot hai ce di da,
the eastern Kuki, with shouldered muskets,

or ö, macer ton kong ö, macer ton klang reng long ya,
up there, the stems grown-up in the East, the men of eastern growth,

ang tai long ken, tai long chin rui ya ce,
my elder brethren, affiliated elder brethren,

kan pa raneu, kan plong chang neu da öi,
my father's heart, my beloved men,

ang tai long ken, tai long chin rui, reng long ya da öi,
my elder brethren, affiliated elder brethren, the chief's men

wang rum ran thak, wang thak ting pong,
go to hew down, will cut down the multiple ting-bamboo

kha ra nen len, akri chum nöng pace di da,
cut down quickly the three layers of the fence,

uh, thak ting chur cam, thak teu chur cam,
o, when hewing down, cutting down the teu-barbs

kha chur taku, büng wang da chüt cöi,
of the fully barbed fence, the closure will be pushed in,

kha chur taku, büng nöng chüt cöi da nöm ö,
the closure of the fully barbed fence can really be pushed in.

ang yong ing klang ba di le, u ö,
My, the black bird's, o dear,

ari Lang-ke, hia kom wang kup, kup wa kom rui ö,
eastern Kuki warriors, will enter, enter the village rows,

[a]ri Lang-ke, hia kom wang kup, kup wa da kom tan,
the eastern Kuki warriors enter, enter the village lines,

ng, kan pa raneu, kan plong chang ram, ba di ö,
ha, my father's heart, my beloved men,

a dong tapong, kra tan dong pong,
amidst the multiple, between the multitude of stairs,

wa kom talüp, pre thong da dong köi, di ö,
the overtowering village, in all the families,

ci ce lu kut, lu lim chong reng, pa tum[601] ching na mi ce,
those who wear their hair in a knot in the neck, the flowering daughters,

mi cam kröng thör: ang reng wang ya, pa deu plong rau,
they are waking up: My commander, my beloved father,

ang reng wang ya, pa deu plong rau, rü tam böt, rü tam böt,
my commander, my beloved father, please get up, please get up!

ö, or ö, reng ce khö kwak, ra ha net nöng,
Up there, where the offering baskets are thrown away to push the spirits back,

uh, chur plon da tom köi, di ö,
o, on the barbed village gate,

ari Lang-ke, uh, hia kom da wang kla, kla wa kom rui, ba di ö,
the eastern Kuki, o, the warriors are falling in, falling into the village.

rü tam böt, tanö[602] manö, rü tam böt,
Please get up as quick as possible, please get up!

arau krau rau, nöng da long plai döi dök nöm cö,
this morning, this very morning can we really not survive safely,

rau krau rau nöng long le, uh mit ce chüm kan döi dök, uh, men ö,
this very morning cannot be saved, o, pass over the extinguishment no more, o you?

wa kom lang cöng, cöng rüm kwai prü khök cö,
Has not the village, in its turn, become, has changed into a swarm of bees?

wa kom lang cöng po, cöng rüm kwai lot khai ö,
The village has transformed, changed into bees swarming out,

u ö, wa kom lang cöng, cöng tör[603] ui wi,
o dear, the village, however, becomes, changes into the blood of red fruits,

wi chin hor hök po, hor lau döm prang,
the own blood also gushes out, flows dug-out down the breast,

[601] cf. footnote 74, here: *pa ... na* "daughter"
[602] *tanö* is used to form superlatives, *manö* is adhortative, i.e. perhaps: "most urgently", "as quickly as possible".
[603] a creeper with blood-red fruits

kaa lau pön töng, ka[604] thüm mia pai khai cö,
getting dug out, starts to cross over, tinges all beyond.

rü tam böt tanö, pace, rü tam tari, la khü e ce,
When getting urgently up, getting prepared,

wa kom lang cöng po, cöng rüm kwai prü, cöng tör ui wi,
the village became a swarm of bees, became blood of read fruit,

wi chin hor hök, hor lau döm prang,
own blood gushed out, flowed dug-out down the breast,

kaa lau pön töng da, kaa thüm mia pai khai ö,
dug-out started to cross over, tinging all beyond,

nö, u ö, kan plong chang neu, u rau cin töm wang, keng döm pön köi ba di da,
and then, my beloved heart, the morning comes, on the other side of the slope,

ang tu ting lung po, kau chong ku klang,
I also cut a ting-bamboo, a male-female pair of bamboo,

ting lung cher hui, cher chong hui a,
shave bamboo thongs for drawing fire, a shaved male-female pair

mi la hai ce, ang yong ing klang ba di le,
with this, I, the black bird,

ang küng mareng[605], mai di ching liu po, u,
I draw the thong and the living fire seed too, o,

ting ko lak cak, mai ching liu rü, ka khü e ce öi,
living in the ting-bamboo, and when the fire seed grows,

ang can pe mai chong po, mai chong ku klang,
I also kindle a pair of torches, a male-female pair of torches

yong ing plan lot, lot mai di ching,
black bird then sets free, sets free the living fire,

ng, öi, ang yong ing klang le, yong ing or lot po, lot mai di ching,
ha, I, the black bird, up there set free, free the living fire,

mai ching lang len po, len wa kom rui da ba,
the living fire in its turn also runs, runs through the village lines,

rau cin töm wang, mai ching lang kam, kam wa kom rui ce öi,

---

[604] no literal translation
[605] the thong for drawing fire

as the morning comes, the living fire in its turn burns, burns the village lines.

ang yong ing klang da ba, wa kom lang cöng po, cöng rik khöm[606] ploi,
My, the black bird's village, however, also becomes, turns into red hot ashes,

tan dong lang cöng, cöng rik khöm ploi, pace,
and when the lines of stairs become, turn into red hot ashes,

ang yong ing klang le, kan plong chang neu da ba,
I, the black bird, my beloved heart,

yong ing or cam, cam rui chi yun,
black bird there seeks, looks for precious gold,

men yong ing klang, yong ing or cam di le,
I, the black bird, I there seek

cam rui chi yun, ang rui chi yun paka,
the precious gold, which is my precious gold,

yong ing or cam, cam kling ui yun, ang kling [r]ui plong rau ö,
black bird there seeks, the precious jewel, is my beloved jewel,

ang kling ui yun, kling ui plong rau da di le,
my precious jewel, beloved jewel;

yong ing or kaa, kaa rui chi yun,
black bird there finds, gets the precious gold,

yong ing cam nap, nap plong ko rau,
black bird seeks to fix, to fix his love,

yong ing cam nap, nap bü ko rau, pace,
when black bird tries to fix his love, to fix his pain,

yong ing klang le, nap plong ko rau, nap bü ko rau,
black bird fixes his love, fixes his pain.

nong neng nari, la khü e ce öi, ang yong ing klang di le,
Soon afterwards, however, when it is time for my, the black bird's,

Cham-tu da, u ö, hia leng le u ö, hia kom lang plan,
Shendu warrior lines, the warriors to return,

plan kung thi rong, plan long um lak, plan kung rü rü e ce ö,
return to the place where they came from, return home, by and by to return,

---

[606] meaning unknown, perhaps: "to bask (in the fire)"

ang yong ing klang, plan kung thi rong, plan kung rü rü da öi,
I, the black bird, return to the place where I came from, by and by go back,

reng ce khö kwak, ra ha net nöng, uh chur plon tom köi,
to where the round offering baskets are thrown away to push back the spirits, to the barbed village gate.

Cham-tu hia leng ba di ö, hia kom lang cang po, cang ki pum nai ce,
The Shendu lines, the warriors, however, make, make an other Muntjac bivouac,

kan pa raneu, kan plong chang ram,
my father's heart, my beloved men,

ang yong ing klang le, tu ling bang[607] ngua, ai wa kom lu,
I, the black bird, kill a tethered cow, to feed the heads of the village,

ang yong ing klang, cho ui kar[608] teng, tui tam prü[609] rü, paka,
I, the black bird, spew over the ginger vessel, spitting out ginger water,

ang yong ing klang, cho ui kar teng, tui tam prü rü, cin mat khom ce,
and when I spew over the ginger vessel, spitting out ginger water,

tarua chuk-kri, reng long ya ce khai ö, kan plong chang neu mi,
the big man of the village, the headman over the one I loved,

mang karbari, chang ri reng ya ce khai, tang tang nöm[610],
the kings official judge, who always ruled, that very one,

ang yong ing klang ka ce, cang ting et[611] re,
who made me, the black bird, discarded ting-bamboo,

yu ö, tarua chuk-kri, reng long ce wang cang ang/,
down there, the big man of the village, the headman, who made me

cang teu et re, paka, mila hai ce,
discarded teu-bamboo, when over him,

ang yong ing klang, cho ui kar teng, tui tam prü rü,
I, the black bird, spit ginger water,

---

[607] informant = tethering pole (= *ling pu*), but *bang* means "wall", and hence may indicate the Chin custom to hang the heads of sacrificed animals on the front wall of the house
[608] ginger?
[609] Also *prö rö*, onomatopoetic for the sound produced by the water spat. The sacrifice mentioned (and executed at the village gate) seems similar in execution to the still practised *cia mai lek choi* (cattle-tail-waving-oath).
[610] that very one (?)
[611] no literal translation, "to push to the side, expulse" (?)

cin mat khom ce, i long ö, i pan reng ya, i long chüi lüi,
to sleep, to be always tired, to fall deeply asleep,

men plong i nong döi ka, tu ling bang ngua le,
"your mind is not sleeping soon", I killed a tethered cow,

u ö, ai wa kom lu, ai tan dong lu khom ce ö,
o dear, to feed the heads of the village, when I fed the heads of the lined-up stairs,

men plong i nong, men bü i nong, mit ce chüm kan döi ka öi,
"your mind soon sleeps, your body soon sleeps", to those who did not escape extinguishment.

rau cin töm wang köi, ang po pe mareng, chön kong rot ya,
When the morning comes, I take a thong, a rattan without leaves,

ang po pe mareng po, chön lang rot ya ce,
having taken a thong, a rattan without leaves,

ang yong ing klang di le, yong ing plan cher, cher wa kom lu,
I, the black bird, I return to skewer, to skewer the village heads,

yong ing plan cher, cher tan dong lu da ba,
black bird returns to skewer, skewer the heads of the lined-up stairs,

ang yong ing klang le, kling-ce yö möt,
I, the black bird, beyond the mountain range where the songbirds fly,

chum chüng rong köi, har kom yö möt po,
also beyond the mountain ranges where the thrushes fly,

chum chüng rong köi da ba, pai ö tar pong long kwoi ka ce,
far off in the dale of the Ficus tree where to hang them up,

ang yong ing klang, tar pong long kwoi ka öi,
I, the black bird, in the Ficus tree dale where to hang them up,

pong bang chum krum, chai bang mia yör ka ce ba,
in the network of branches, in the drooping branches and twigs of the Ficus,

ang yong ing klang, yang ing cam bat, bat wa kom lu,
I, the black bird, I suspend, hang up the heads of the village.

yong ing cam bat, bat tan dong lu, la khü e ce,
When the black bird suspended, hung up the heads of the village,

ang yong ing klang, ang chak charui, kling-ui plong rau hai ce ö,
with my, the black bird's, whole heart, beloved jewel, my whole heart,

ang chak charui, yam-ui plong rau hai ce,
with my beloved gem,

ang thop macin, ca[612] chong rüm da ba,
I go to sleep, imbibing the forest.

yu ö, o tut tarö, phak[613] chong dong hön,
From down there, the shore of the deep water,

cha-ta li lan, krang li or kwak, kwak mau thüm rong ce,
the fresh wind of the ocean, the storm up there throws down the thüm-bamboo place,

cin dong da war wang, war dong phong long, le khü e ce,
the middle of the night comes, after midnight has passed,

ö, rau cin leng lot, leng wang chum chan, pün kwang rü rong, pace, da u ö,
the morning star appears, when it starts to rise, climbs up to the sky,

rüm dong pring li wa da, mön tang rui krök nöm le,
the red forest bird with the bright beak perches, perches on the liana network,

rüm dong pring li, wa da lang mön po, mön tang rui krök ce,
and the red forest bird perching, perching on the liana network,

ta pe mani po, khok che chör tün,
now also hears the sound of shuffling legs,

ta pe mani po, khok lang chör tün da öi,
now also hears the sound of other shuffling legs,

rüm dong pring li wa, klup pe mani, lap hoi pün lün ce,
the red forest bird, now flaps and shakes its wings up and down,

yu ö, khung to do köi, reng ce cor cong, kar ko da kreng ta,
down there, under the house, where the fowls are kept, the cocks listen and hear

rüm dong pring li, lap deu chong tün, kar ko kreng ta po,
the red bird in the middle of the forest, the noise of flapping wings, and when they listen and hear

rüm dong pring li, lap lang da khöng tün, da u ö,
the red bird in the middle of the forest, the sound of other wings quickly beating,

kan plong chang neu da ba, cin mat khom ce,

[612] see passage of footnote 102 ("translation uncertain")
[613] "to let return", i.e., break the waves; or: "direction (of compass)"??

my beloved heart,

kar ko tong kring, tong ka khyöng löng ce,
the cocks crow, crow at the top of their voice,

kar ko mi da, klup pe mani, lap hoi pün lün,
these cocks now shake out their wings up and down,

kar ko tong kring, tong ka khyöng löng ce,
the cocks crow, crow at the top of their voice,

kar ko krüng thör, krüng wa kom rui,
the cocks wake up, wake the lines of the village,

kar ko krüng thör, krüng tan dong lak da öi,
the cocks wake up, wake those in the lined-up stairs.

yu ö, o tut tarö, phak chong töm hön,
From down there, the shore of the deep water,

cha-ta li lan, krang li or kwak, kwak mau thüm rong khanöm a,
the fresh wind of the ocean, the storm up there throws down the thüm-bamboo place indeed,

krang li ö, u ö, krang li hür hek, hür tan dong lak,
the storm, o dear, the storm swirls, whirling to and fro in the lined-up stairs,

mi khü e ce, hür wa kom lu,
this time whirls the heads of the village

pong bang chum krum, chai bang mia yör ka ce ö,
within the branches of the Ficus, in the drooping branches and twigs of the Ficus,

kling ce yö möt, pre chüng rong köi,
beyond the mountain range where the songbirds fly,

har kom yö möt po, chum chüng rong köi,
also beyond the mountain ranges where the thrushes fly,

cop ku parui di ö, ku tün kling ling, khöt lang nak döi ö,
the hung-up golden chains, not different from tingling gold,

cop tai parui, tai tün kling ling, khöt lang nak döi,
the hung-up silver chains, not different from tingling silver.

cin mat khom ce, ang yong ing klang, chin ram kreng ta di le,
When at this time, I, the black bird, hear it with my own ears,

ang yong ing klang le, chin ram kreng ta, cin mat khom ce da ö,

I, the black bird, when hearing it with my own ears,

i long ö, i pan reng ya, i long chüi lüi,
I sleep, I remain tired, fall deeply asleep,

men plong i nong, i long chüi lüi,
my body sleeps, falls deeply asleep,

men bü i nong, uh, mit ce chum kan döi di le,
but my sleeping body does not escape extinguishment,

u ö, cop ku parui, ku tün kling ling,
o dear, hanging up golden chains, the gold tingles,

chin ram kreng ta, cin mat khom da,
when hearing it with my own ears,

ra chin tali, ra chin tali, chin yang kra köi, u,
in my interior, in my own guts,

ra chin tali, chin yang da dong köi,
in my interior, in the middle of my guts,

plong ram tanga[614], lüp chong kra köi,
deep in my mind, in between its layers,

ang po twak chang pang, kaa twak chang pang,
I take to pondering, get the reflection,

po twak chang kram, kaa twak chang kram mi da,
take the conclusion, come to this conclusion:

krong[615] dong chang kling, rui run da cheu,
tearing down the creeper of the sword bean and peeling it,

cheu reng wa[616] te, cheu mang liu hok, cin mat khom ce öi,
the seed comes out,

plong cheu la[617] leu po, bü cheu chak wang khai tak,
while peeling the mind and mincing it, the life impulse of the peeled body remains,

bü cheu la leu po, ang plong cheu chak wang khai tak.

---

[614] "five" to surpass the preceding *tali* (four), see footnote 33 (on numerals)
[615] name of the creeper, *Entada scandens, kling* = sword bean plant, *kling ui* = sword bean fruit = the beloved
[616] Meaning not certain, perhaps Burm. *wa*, to satisfy a need, *cheu* (to peel), subsequently also used in an enlarged meaning: judgement over life and death; the idea of "peeling" is transferred to the layers of the body.
[617] soul ?; probably only the usual undefined object

peeling and mincing the body, the life impulse of my peeled mind
remains,

ang chak charui, ang kling ui yun, kling ui plong rau yun le, paka.
my whole life, my precious (sword bean) jewel, beloved jewel you.

S1b) *Richi laama pung klai meng*
(Song of the Richi[618] casting a picture of the moon)
(sung by Changku Palau, 1964)

(Öi, u o öi,)[619] ci long tang da, khet rui[620] (u cading) nöm tai (le)
khet rui plong rau (nöm a u le), ci mang phung köi,
Oh dear, we two old people, truely caring for and loving each other,

pho khaa[621] ram tong[622], ci nöng tong rui (ang dötham)
po keng[623] ram tong, ya nöng tong[624] plong (ang döi böt tham)
like the ronkha tree and the kengma tree we can make love no longer.

Menleng ö yu, lai döm pren köi (le),
katoi khoi[625] paci, khoi ngui pre tong[626] (dia le),
toi khoi paci, khoi tai pre tong (dia öi)
O Menleng, go to the plains and borrow for us two baskets full of gold
and silver,

Menleng kakhoi, khoi ngui pre tong (kakhü le, u ö),
Menleng kakhoi, khoi tai pre tong (kakhü),
When (I) Menleng go to borrow two baskets full of gold and silver,

(ang kaa le u) richi tung ton[627], tung deng lak cak,
krang-chi (da) krang-laa, krang-ko kor-ngaa[628],
taruk mon[629] khin, khin ngok[630] ngolö (mi)
I get (for us) a pair of horses, braught up by Richi,
which easily can carry 6 Maunds,

[618] *richi* = Burmese word (written *ra'se')* for which Judson's Burmese-English Dictionary gives as translation "devotee, ascetic, hermit" plus a long explanation, which ends by a quote declaring that "their bodies become spiritualized to an extend which enables them to travel from place to place by following an aerial course".

[619] The words in brackets form part of the song. They were bracketed only in order to show the basic 4-syllable structure. In the following footnotes: *(M) = loaned from Marma (Burmese).

[620] khet rui = manau : to love each other

[621] ron-khaa tree, flowers in April

[622] tong = tacong ; able to marry (see footnote 36)

[623] keng-ma tree; unclear why mentioned (but see: keng tong rui : to meet with a lover; keng-ma : bead)

[624] nöng tong : to be able to stay together

[625] to borrow money (from a money-lender in the plains)

[626] tong : standing basket for 1 Ari; or: thong : thousand

[627] to raise (under the house)

[628] (whole line) : a pair of horses

[629] mon : Maund, t.m.kh. : to be able to bear a load 6 Maund

[630] to play (here: playfully, easily)

(ang kaa le u ö) ku chin mia toi[631] (mi), (ang kaa) tai chin (da) mia toi (mi),
krang-chi (da ce, da ba) krang-laa, krang-ko kor-nga,
taruk mon khin, kum döm luk[632] köi.
I get (for us), like gold and silver,
a pair of horses, which can carry 6 Maunds on their back,

Menleng or te[633], te ka-ri long le,
Chongpoi wang cöm (ang), cöm ka-ri long (böt chanaa)
Chongpoi or cöm, cöm ka-ri long,
Menleng drives a car: "Chongpoi come on and sit in the car"
and Chongpoi sits on the car,

Menleng yu küng[634], küng kri chang[635] leng (nöm),
richi tung ton, tung deng lak cak,
krang-chi (da ce, da ba) krang-laa, krang-ko kor-nga,
Menleng pulls the reins of the pair of horses elevated by the Richi,

(ang ka) kaa-ri mrang[636] pyaing, pyaing ri-chaa[637] rong (khaiböt),
pün kraa phung luk[638] (khaiböt), (u ö) pün kraa phung pren,
our car visits while traveling the Burmese place, to climb up to the sky country,

(or ö) cakraa wadi[639] (ö, u ö), pren lang kyoing[640] köi,
cakraa wang te[641], te kyong chüi-lüi (khai ö),
above the heavenly lord comes to lodge in the temple on an island (?)

ning ce (le) taria[642], chang kraing[643] wang klaa,
klaa dir thum[644] ru (khaiböt ce), kong-mu mang[645] nöm (oi),
the festival for the change of the year will take place and lord Buddha,

(oi, a u oi), kyokto[646] (da baa) phung-kri, phraa[647] ching ton-yaa,
kong-mu mang cöm (ang), cöm kyong chüi-lüi,
the priest of Kyokto, God's leader of life, lord Buddha sit in the temple,

(u ö) cham-tu lake[648], yap[649] bong chün töng,
ho reng taraa[650], paing-chaa[651] bön-lön (nöm kabö).

---

631 like gold
632 on the back
633 to drive
634 to pull, to draw
635 k.ch.l. : the reins
636 *(M) to see; m.p. : to vsit when traveling
637 *(M) a'hri'-sa : the eastern people, the Burmans
638 to climb up to the sky
639 *(M) ca'kra' wa'te : ? ?
640 l.ky.: ?? (kyoing < *(M) kywân :? island)
641 te < *(M) tañ : to stay for a little while, to lodge
642 he borderline of the year
643 *(M) sâng-kran : (Burmese) new year's festival
644 dir : ?, thum : ?
645 the lord of good deeds (Buddho?)
646 *(M) kyok-to bhûng-krî : the priest of Kyokto ?
647 *(M) bhurâ : god; ton-yaa : the leader ?
648 ch.t. l.k.: name of the pattern
649 fan; ch.t. : to hold
650 *(M) ta'ra: ho : to preach, expounding the religious law

holding a plaited fan, expounds religious law full of wisdom,

(ci le) ci or pün pang, pün kraa phung luk (nöm dia le)
ci or pün pang, (nö,) pün kraa phung pren (dia le)
cakraa dong pöng[652] (köi or a), cakraa dong pöng (köi ö)
let us climb up to the sky country by the heavenly ladder,

(u ö le) chang[653] khat marö, hung bü da khöp[654] (köi),
richi[655] chum töng, (le yu,) po kri phalaa[656], bia bong chang laa (kakhü e ce),
below the learn-1-town hill, 3 Richi take a brass bowl to make the moon,

richi klai ang, klai laa-ko pung (khaiböt), richi lang klai,
klai laa-ko mör (khai-böt, u ö u pa,) (u ö,) richi klai ang,
klai laa-ko pung (khai-ce), toi kreng tamü (manong dia),
the Richi cast an image of the white moon, let's go to have a look.

Chongpoi plong dön[657] (yayang), Menleng plong i (yayang nöm le),
ci le Menleng (ö), richi chum töng,
Chongpoi wishes, Menleng consents; the 3 Richi,

ni laa-ko long, k'rek cong[658] tui tang[659],
laa (ang da) mör ngi mi (da, u ö),
the white moon, a star in the water, the pleasing face of the moon,

cakraa dong pöng, kraa tan dong köi,
chang khat marö, hung bü khöp köi,
[ri]chi tangi, ni laa-ko long (da),
into the sky hamlet stairs, under the learn-1-town hill, the Richi are contend

mong pe kyang waa[660] (ka öi), wang pyak[661] ci pe,
(ci) chu tong[662] pün lün (kön ce camü),
the sendling comes to spoil us our wishes,

Menleng ö, küng kri chang leng (böt le),
(or a) kong-mu poi prö[663] (ba), nau mün döi-dök[664],
chang kraing poi prü, nau mün döi-dök
O Menleng pull the reins, up there the feast of good deeds will not be late,
the festival of the change of the year is near,

(yu le yu) chi laa[665] mang-phung (köi),

---

651 p-ch: (< paña : wisdom)
652 d.p. : to
653 ch.kh.m. : a mountain in the east (unknown to the informant)
654 below, at the foot of
655 *(M) ra'se' : hermit; here: wise man; ch.t. : three
656 *(M) krê pha'lâ : metal bowl
657 pl.d. : to wish, to want
658 karek cong : star
659 water in the bowl
660 m.p.k.wa = kraa-ma mu : sendling
661 comes to spoil (the "chu tong")
662 *(M) chu' tông : to wish, to pray
663 p.p. : festival
664 not very late, (that is): is near

ci yaa nöng tong rui (la) döi pace,
ci yaa nöng tong plong döi pace,
down here our duty in the world, when we cannot love each other,

ci or toi klai, klai caa bong[666] pöng (kön ka ce);
ci [or] toi klai, klai chu tong ram[667] (kön cekhai),
we up there will spoil changing our fate, spoil our demand to exchange our fate,

ci or wang pün, pün kraa phung luk (mi),
Menleng (ö) Chongpoi (u),
we are going to climb up to the sky, Menleng and Chongpoi,

richi kangi, ni laa-ko long (khaiböt), kreng mü böt dia (le),
richi (le yu) chang lot, lot laa-ko long (khaiböt), talöng[668] nim rua (paka),
the Richi are contend, the white moon must be finished, come and see,
the Richi set free a mirror image of the white moon,

richi or tong[669], tong laa-ko mör (ang khaiböt),
richi or tong, tong laa-ko yam (ang khaiböt ce),
the Richi up there produced (for us) the face of the white moon,

k'rek cong tui tang, laa lang mör ngi (ce),
richi (or) plang klaa (ce), klaa laa-ko long (ang khaiböt, tak dia[670],)
the star in the water, the perfect face of the white moon
the Richi have reproduced, thrown up (for us) the picture of the white moon.

(u ö) Menleng ö le, (or a le), chang thum (le) marö, li lang phung köi,
ca-kraa wang te, te kyong chüi-lüi (khaiböt),
O Menleng, up there, in learn-3-town, the other world,
cakraa comes to lodge in the temple

u wang pün ce, pün kraa phung luk,
phung wang pün, pün kyong chum bök (khaiböt),
phung wang pün, pün kyong rö lak (ang khaiböt, paka),
she (a wordly human) is going to climb up into the temple

küng kri chang leng (böt le), kong-mu poi prö, nau mün döi-dök
ci nöng toi klai, klai caa re ram (ang dötham),
ci or wang lü, lü chu tong ram (khom ce dia a),
pull the rains, the feast of good deeds will not be long,
we cannot spoil changing our fate, we will ask up there for a favour.

Menleng tong döi (lamaa), Menleng tong da[671] (lamaa),
ang töi plong i (döi nöm),
Menleng yu dong[672], dong klau[673] chum mur (khaiböt),
Menleng yu dok[674], dok klau chum bök (khaiböt),

---

[665] *(M) si la' : religious duty
[666] kl.c.b.p. : to exchange (the fate)
[667] *(M) chu' tông : to pray for, to entreat a blessing; tong ram = to ask for, demand
[668] mirror
[669] to produce an image in the sky (with the mirror)
[670] t.d. : having made
[671] <? *(M) : tong'ta' : to wish, to long for
[672] dong =? dok
[673] kl.ch. : ladder

Menleng does not wish, I am not pleased, Menleng seizes the ladder down

Menleng yu döm, döm klau chum mur (khai ö),
u ö, Menleng, toi kreng (ba) tang mü,
Menleng steps down the ladder, Menleng goes to see

richi lang klai, klai laa-ko pung (mi, ce),
richi tong da, (yu ö), awang pyak ang, pyak chu (le) tong ram (nöm khaiböt)
(u, ang) waing[675] ne krong wöi,
the Richi casting the white moon,
the Richis' wish has spoiled my asking for a favour, I feel sorry;

kakhü e ce, Chongpoi or pün, pün kraa phung luk,
chongpoi or pün, pün kraa phung pren,
at that time Chongpoi climbs up to the sky country,

cakraa wadi, chin kyong lak köi,
(u ö,) kong-mu poi prö (ang khaiböt), (nöm ö),
the heavenly lord, in his temple, the feast of good deeds,

(u le), kyok-to phung kri, caa kyang[676] reng yaa,
phraa ching ton yaa, pe praa chi lai (mi ö),
the priest of Kyokto, the learned man, God's leader of life, giving prayers,

kong mu mang da, ok-chaa cing yen[677], wan ching chüa pai, (hai a),
ok-chaa cing yen, wan ching chüa kup (hai),
the lord of good deeds, clad in his yellow robe, grown by the Burmese,

(u ö) be[678] chang khün li, chang khün li luk (köi, u,)
thop cin paca, nam nai klem-lem, (ang kharök oi),
makes his bed on a cloth
and goes to take a nap in his comfortable sleeping place,

(u le), alai[679] kaba (u ö), kraa-ma[680] mu leng, (u ö,)
kyong chu (da) kyong chang[681], kyong chang phung kri, tape[682] to ka
cakraa mu ba, mu ce lang cöm, cöm kyong (lak) chüi-lüi (ang khaiböt)
before that the messenger bird, the scholar of the temple person,
the teaching priest of the temple, the angel sits in the temple,

Chongpoi or le, Chongpoi or pün, pün kyong chum mur,
Chongpoi or pün, pün kyong rö lak, lak wöi[683] töng[684] long (da öi),
Chongpoi up there climbs the temple staircase, climbs up into the temple,

---

674 to put up (in a leaning position)
675 *(M) wâm (belly); wam ne : to be sorry, to take to heart; krong : cause, reason, circumstances; (whole line) : I am not happy in the mind, I feel sorry.
676 *M: câ kyang: versed in writing
677 to plant (like trees)
678 to make a bed with a cloth
679 before
680 *(M) kra: hma: to give information
681 (M) temple person, temple master
682 *(M) ta'pañ' to: disciple of an eminent teacher
683 left hand (?)
684 = ?

(u ö,) pau bong[685] tacum, ranit kom thong[686] (nöm dia),
lak yaa[687] töng löng, phyong toing[688] cün hai,
(u ö,) ph[u]raa tang pön[689], tang[690] yaa luk köi,
a bouquet of 70,000 flowers in her right hand and wax candle sticks,
the god sits crosslegged upon his buttocks

Chongpoi or ngek, ngek ch'mi[691] mai chong, pre toing po,
mai ching ang ang, ang kyong rö lak (ang döiböt),
Chongpoi kindles an oil lamp and also two candles.
The living fire (however) does not illuminate the interior of the temple.

Chongpoi or tang[692], tang paing chum tu[693] (döiböt),
(ang ka da) pau tu tang lang[694] (yan),
Chongpoi presents the flower bouquet no longer,
my flowers have become changed,

(ang ka le) mari[695] caing-kaa[696], krong khet dir pau,
yung rang ph[u]raa, lak chong[697] kau döi (mi ce), Chongpoi or tang (ang nöm),
my Michelia champaca are unfit as a present to god,
which Chongpoi up there presents

(u o,) kyong chu kyong chang, kyong chang phung kri, tape to ö,
kong-mu mang rü (tam böt le,)
persons of the temple, temple teaching priest, scholar, lord of good deeds, get up!

yu le yu ö, yong chüng taruk[698], yong o con moi (hön ce),
Chong tong (da, yu ö) wang cöm, cöm kyong rö lak (ang khaiböt),
From down there, from a far place,
Chong came with a wish to sit in the interior of the temple,

Chong wang ngek ang, ngek ch'mi mai chong, pre toing (ang) khaiböt (ce),
Chongpoi wang leng, leng chu tong ram (kom pe yu)
Chong came to light an oil lamp, two candles, Chongpoi came to ask for a favour.

(u ö, ang) kyok-to phung kri (tape to),
ang nöng pe pang[699] (ang le), pe reng chu yung (döidök)

---

685 flowers
686 70,000
687 *(M) lakya: right (hand)
688 *(M) bha'yong: töng : wax candle
689 *(M) tang pyañ: to sit crosslegged
690 ?< *(M) tang: buttocks, hip
691 *(M) chimî: (oil fire) light
692 *(M) tang : to put or place upon
693 three pieces (of flowers); but probably *cum tu* (see above)
694 flowers have become others (which one cannot present)
695 ?, probably = earth
696 *(M) cañ kâ : *Michelia champaca* = bong-rong-pau ~ krong khet dir pau, unfit for kyong
697 *(M) lak chong : present, gift
698 = alakma hönya : from a far place
699 can give

O you, I the scholar of the priest of Kyok-to, can give a good fate no longer,

(ang ka) naari rui ron[700] (khaiböt), alai khan ka (ba),
(en wang pe), pe chu[701] tong wan (manong dia ce),
kraa ma mu ce (lang pe ang), atu[702] tong ram (nöm pace),
my time has passed, when you had come before to confer a favour, soon the angel gave you together what you wanted,

(u ö khai ce), pho khaa ram tong, yaa nöng tong rui (ang döinöm tak),
(u), po keng ram tong, yaa nöng tong plong (ang döiböt)
like the ronkha tree and the kengma tree we can make love no longer,

(u), Chongpoi tong da(le), Chongpoi tong da, kaa chu tong ram (döi ce ö)
(mi nöng klai), klai chu tong ram (döi ce),
or mai makling[703], yam tui paprö (khü).
Chongpoi's wishes found no favour, could not exchange her fate;
wiping her tears away,

(u), cöng kut mang plan, mör chü na ka[704],
cöng kut mang plan, plan kung thi rong (ang nöm).
she returns with a trite face, returns back to where she came from.

As may be taken from the many loan words from Marma, this song probably is borrowed from them. But already Menyom Tang in August 1957 told me that it was an old Mru song. I'll add here the short note on the song, which I got from him. It will be seen that his version was somewhat different. Here, husband and wife always go together, and the end has a clearer statement.

Richi laama pung klai meng
(Song of the Richi casting a moon picture)
An old married couple sets out to visit Nari-Kyong (hour-temple) in the country of the heavenly god where the 3,000 wise men are, whom they want to ask for luck. When they go together, they meet on the way 3,000 Richi who cast the picture of the moon on a golden plate. They stop and watch the activities. When they come to the sky, they have missed the Changkran[705] day of the Nari-Kyong. They light candles, but the God turns his face away and does not want to see them. We have come too late, the man says. When the heavenly god turns his face away from us, the woman says, we go to the Praingma[706] lord to ask him for luck. When they arrive there, Praingma's woman is ill, he himself is away to look for the life medicine (*maching-chi*) in the East. When he comes back, he says to the old people: You are too greedy, first you watched those who cast the moon and now you also want to ask for luck. I can give you only one: if you have died, then you shall be reborn and get new possibilities, however, whether rich or poor, I cannot

[700] time has passed, is over
[701] *(M) chu' pê : to confer a favour
[702] ? < *(M) a'tû : together
[703] to wipe tears (from the eye)
[704] with a sinister, trite face (?)
[705] see footnote 189
[706] praingma < Old Burmese prañ-ña = paingchaa (wisdom), see footnote 197

determine. The couple returns to the earth and they remain poor just as before.

S1c) Two other old songs (in the short version told by Menyom Tang, 1957)

Pauring-malaa daing-ram klek meng

(Song in which Pauring-malaa, the red flower girl, picks *dal* leaves)

At the meal in the evening Pauring says to her parents: Tomorrow I go to the Bengali and fetch *dal* leaves. However, mother and father say: Do not go, you meet an elephant on the way. Pauring, however, says: I am not afraid of the elephant: wrap up rice for me. And after the morning meal she takes the wrapped-up rice and sets out. On the way in the forest she meets an elephant. She hastenes away, the elephant cannot follow so fast. She climbs on a Ficus tree, the fruits change into Areca nuts; where the bark is injured, the juice becomes lime and the leaves become betel leaves. The elephant waits under the tree, but Pauring waits on the tree and eats betel. The elephant goes back to the forest, worries about Pauring, and dies there. Pauring goes to the Bengali country and fetches the *dal* leaves. When she is at home again and eats the leaves, she says to her parents: Today I met the elephant, however, I escaped. It could not follow me. I climbed on a Ficus tree, its fruits became Areca nuts, its juice changed to lime and its leaves to betel leaves. The elephant waited below but I stayed on the tree and ate betel. The elephant went back to the forest again, worried much and died. Next morning the parents go to the Burmese country and call 6 princes. They come and fetch the elephant's teeth. The ivory is cut into strips and from these a bedstead is prepared. Pauring sleeps on it, sees the elephant in her dream and dies. The parents follow the custom and burn Pauring.

Ngachai prön-ding ui cak meng

(Song in which the elephant wants to eat mandarins)

An elefant comes from the east, rests below a tree and smells mandarins. From where the wind has brought the smell, there the elephant goes to look for the mandarins. When it reaches there, however, it gets thirsty and looks for water. While it is drinking at the river, a monkey comes and in the meantime eats up all mandarins. When the elephant comes back, it finds no more mandarins. It is very angry and stamps a mountain into a flat country and the flat country into a plain. In this plain a European prince sets up a Bazaar and in another place a Bengali makes a rape field. The rape grows and the Bengali fences in his field. When the elephant comes, it sees the flowering rape in the field and would like to eat it, however, cannot break the fence. In sorrow it orbits the fence, however, does not find any entrance. So it goes back to the big forest and dies. (The young man says to the girl: the sorrow is in me like the tooth of the elephant and since I cannot get you, I will die.)

S3 *plong rau meng* (a [modern] love song)

(Composed and sung by Menching Atwang, Tapwüa-Kua, 1956)

This love song was a really new composition by Menching Atwang, who even tried to note it down for himself (using Burmese letters with which he could write Marma and which he now had tried to adapt for Mru). For publishing it in 1980, I translated it into English.

Chong leng ö
ci chang long ba, ci chang long a
hom prek prai li, hom prek prai wöi döi ö
apre tarua, tan dong öi lang
hom tan döi le, u ö ba
chong wang tö ang, tö kung thi klai
ang plong i rui döi kabö, rau rök ö

O, beloved,
we two, ah, we two,
harmony rent asunder, separated, not together,
divided in villages, in different rows of houses,
in loneliness, ah, you,
beloved, you abandon me, leave me behind empty,
sorrow in my heart – o, young morning!

U ko ö le
kling chia chek hai, kar ko pom wöi kace
phung ram wui en, wui tum dong lu
kau phom cün yung, kön ce dönök nöm,
u öi

O, bright darling,
with the rod of the iron arrow in the chicken basket
a man pulls you down – you flower, down the staircase,
in order to bind you like bamboo canes bound together into a post,
o you.

U rau ö
yang bong tong teng, tong rö hung kada
kliu re ram min, re ram bön lön
manong wang ni, dang de la ram
ni nong e da

O, morning you,
on the mountains of the Yangbong, on the chain of mountains,
the leaves turn yellow, drop the leaves of the silk-cotton trees,
after those rainy days, the leaves of the cotton plants,
one day.

U rau ö
lung ko plong mon, palai döm con,
pawaa kom rui
kar ko plong mon, palai döm con,
pawaa kom rui yong da
chong om plong mon, khin wang li len
khin wang li len khök u ö, rau rök ö
en wang tö ang, tö kung thi klai

hom tan döi le

O, morning you,
as the cattle in the low plains
all together are pensive,
the chickens in the low plains
all together are pensive,
so will you become, beloved, pensive, when the time comes,
when the time has come, o you, o, young morning,
you abandon me, leave me behind empty
in loneliness.

Pai a ö
pröt cong tui tang, prang mom con hon
laa ko wang rung, ni nong khan ka
laa ko wang chot, chot thum phai phia
ni nong e ka, ci prö lö mani
kang ding wöi löi, rong nöm catöm
ce dia – rau rök a
en mang tö ang, tö kung thi klai
ang plong i rui dönöm, rau rök ö

Look there,
in the east, as clear as water, from the edge of the horizon,
the moon climbed up, on that day,
moonbeams fell upon the matting of the floor,
that day, at that time when we spoke to each other:
we two shall follow the rule, in the proper place,
ah, young morning,
you abandon me, leave me behind empty,
sorrow in my heart – o, young morning!

U ö le, u rau ö
en plong thüm rau, pangö e ka
lai dom cing yöng, cing mung krung yong
ang ce tanöm, u ö ba
ang ria cin daku, taklep pöng köi
ang ko cin tai rau, tut ce tang tang
en wang tö ang, tö kung thi klai da
ang plong i rui döi, u ö

Ah you, o, morning you,
when the pain of your love overpowered you,
like a lowland bean, like an entwining tendril,
surely I held you, o yes,
deep in my insides, in all the folds and coils,
in my innermost self I am aching,
you abandon me, leave me behind empty,
sorrow in my heart – o, you.

Dam li ö, dam li ö

chak klam ya ö, ang rui bong tanga
ang kri bu tang leng, ang bong chong tang töm
u ce dia le

O refreshing wind, o refreshing wind,
which can cool the heart, for the fingers of my searching hand
the brass vessel which I cover, which I embrace,
this should be you.

U rau ö
phung ram tuk lung padöi mi ce
paklik döng rui, cang klik döng naa
hom tan döi le, u ö ba
phung ram cang en, rui bang langa
ca ci cakang mi klai döm töm
hom tan döi nöm, u ö ba

O, morning you,
a man who does not know how to hold you,
seeks your embrace, presses you to his bosom,
without harmony, ah you!
A man forms you, with the fingers of a searching hand,
into a basket, to empty in the rice of the dried paddy,
without any harmony, ah, you!

U rau ö ba
en tum ching long alö mi da
nöm u rui en, rui rong nöm töm nöm
ang yan khök nöm le, u öi
ang ria cin daku, taklep pöng köi
ang ko cin tai rau, tut ce tang tang
en wang tö ang, tö kung thi klai da
ang plong i rui döi, u ö

O, morning you,
you flowering tree, which you are,
your mother seeks for you, seeks a proper place,
I was rejected, ah, you!
Deep in my insides, in all the folds and coils,
in my innermost self I am aching,
you abandon me, leave me behind empty,
sorrow in my heart – o, you.

U rau ö le
rang lö mi da, en kung thi rong köi
rang cam cöng a, pia tui nöm kar waa
kön ce dönök le
tur cam ui nguk, ui pia run run
da ce yong da, en kung thi köi
rang cam ui nguk, ui pia run run
rang ce dönök nöm, chang ngan a

O, morning you,
I, here at the place you left behind,
I shall turn into a chicken which pines for water,
into nothing else,
as a dove which sobs for berries, coos longingly for berries,
abandoned by you in this way,
I shall sob for berries, coo longingly for berries,
nothing else, remembering the beloved.

U rau ö
ci kling paing pyo dia chik baa ö
en nöm u tang, nöm u rui en
rui rong nöm töm, ce le ngan po
kar ki cau pün, pün kri bia pa
arong nöm töm döi lö ngan po
plong krek kröi ngön, ci keng tong rui daba
mi phung köi le, ci keng tong rui chik
mi phung köi baa öi

O, morning you,
let us yet be happy!
Your honourable mother, your mother seeks for you,
seeks a proper place – yet remember, where the fat chickens are piled up,
piled up on the plate of the brass dish,
is not the proper place, remember also
what the heart longs for – is it not that we two come together,
and in this world we should come together,
yes, in this world.

U rau ö le
ci lai thi wan, ci plong ko rau dale
u rau ö le, u rau ö baa
marüm klau poi, klau döng cahap
wang dang dia le, mang dang böt

O, morning you,
the content of our words, the pain of our love, ah,
oh, morning you, o you, morning you, – in the forest the leaves
of the *klau*, with the smooth top and the rough underside,
go and look for them, go and look.

S4 Stories

S4a) *Chongpau malaa cia*

The storey of Chongpau (told by Klangwai Atwang)

Nö, cük kh'e Chongpau malaa cia mi nayong pe.

Chongpau malaa lök pe, Wakek[707] *malaa lök pe. mat-ci au-yüa khetma. Matci auyüa khetma nö, Chongpau malaa ka rak mi chot cia pe. Chot cia nö, makh'e da, tö*[708] *khai chit krum-rum*[709] *pe. Chongpau malaa da hai Wakek malaa matci auyüa, cam po chop pe. Cam po chop hanö, matci chok pok raa hön'ya wang lan. Wang lan nö, tui cot ka-e, tui cot mör wang töm kh'e, lu kom-ka, cang taing-tik*[710] *mi. Wakek malaa tek da: uh, ang ka-e ayak da, ataima chot cia cace da, a-u apaa dia*[711] *ang wan-klai wan-ca char, wan-klai char-chör pa-e da, ang ka chop-töm thüp*[712] *tün tüng pa. Nö, dia ang döi paka-caalö hanö, prototo*[713] *pa patün a, chop töm thüp tün. Nayong pe, Wakek malaa thi thaing*[714] *ba. Wakek malaa thi thaing nö, nöm a nöm a, yu cöm pace, tüng pa, chop-töm thüp tün. Pai a taküi*[715] *döi, töm tathüp kh'e ba. Nö Chongpau malaa tek mi da: ang rak chot cia cace da, ayak a-u apaa nö, dia ang wan-klai char-chör pa-e da, ang ka chop-töm tün tüng pa. Dia döi paka-caalö hanö, prototo pa tün. Yu thüp pace, prototo pa tai pe, tün döi, dia pa döi.*

In the beginning the Chongpau story is as follows:

Chongpau is a girl, and Wakek is a girl, and the two are close friends. They are close friends, and Chongpau's brother makes a cattle feast. To kill the cattle he makes an enclosure with tassels. Chongau and Wakek, the two girl-friends, go to fetch firewood. They go to fetch firewood, and on the way back from the place where they fetched the firewood they come to the place where to fetch water. When they reach the place where to fetch water they want to take a bath, and they make an omen. Wakek says: This evening, because the elder brother makes a cattle feast, the parents will present me a new skirt. In case they will present me a new skirt, my firewood basket, when put down, will produce a loud sound. In case they will not, it will sound like brittle bamboo when put down. Saying this, Wakek casts lots. Wakek casts lots and truely, when she sits down, her firewood basket, when put down, produces the loud sound. The basket does not topple over when put down. Chongpau is saying: Because my brother makes a cattle feast, in case my parents in the evening will present me a new skirt, my firewood basket, when put down, will produce a loud sound. In case they will not, it will sound like brittle bamboo. When she puts it down, it sounds like brittle bamboo only, it makes no loud sound, they will not give her a present.

Nö, panö, wang nö, khai chit mün, taröng, yak-e taröng, kat-chang taröng. Taröng nö: wang döm plai. Wang döm plai nö, a-u mi: tong yong cö

---

707 parakeet

708 to make *chit-wai* (tassels)

709 enclosure

710 omen

711 to make a present, to present to

712 to put down on the earth

713 sound of broken bamboo thrown on the earth

714 to throw lots (= *thi than*)

715 to topple over

cang a. Nakh'e a-u hai apaa hai: mi deng klök[716] *mai-phum phum pe, awai tim-chük pe kabut a. Tim-chük kabut-but nö, mami tim-chük but po ngat pat. Ma deng klök mai-phum po tui*[717]*, tim-chük but po ngat khü ce, en töi kaplai köm cayang. But pe, u hai paa hai, but klok-lok. But klok-lok nö, yu köi Wakek malaa da döm plai. Nö, klung-lung*[718] *t'ia, plai tün, tum plung ca, tep mong ca, meng klö ca. Nö, Wakek malaa tek mi: o yüa, wang döm plai böt. Tim-chük but ngat dö'tak, pape. Nong nö: o yüa, wang döm plai böt tanö. Deng klök mai-phum tui dö'tak, pape, ayüa kabö. Nö, a-u paa takraa*[719] *köi pe tö a. Nö, apaa chüt tarüt*[720] *kh'e da, a-u et tarep*[721] *pe. A-u chüt tarüt kh'e da, apaa et tarep pe. Chongpau malaa tek mi: apaa chüt non*[722]*, chüt non, ang ka lang-che klüa*[723] *rau tang, pape. Apaa chüt non, chüt non, pe-e, nakh'e apaa chüt tarüt kh'e da, a-u et tarep pe, a-u chüt tarüt kh'e da, apaa et tarep pe. Yung le, war pathöi la dia*[724]*. War pathöi la dia, pape, Chongpau malaa mi tek. Wakek malaa kut plan ho a tatöng pe: Uh yüa, wang döm plai böt tanö le. Uh, ang ka tim-chük but ngat dö'tak, ang ka deng klök mai-phum tui dö'tak, kut pe yok pe. War dong klen*[725] *kha'böt pe. Cong ang reng, cong ang reng, yaa cong döi. A taklang lon chak rau lai böt ka, yu yong kim khung*[726] *bang ce, er a kwak*[727] *pe. Er a kwak nö, khong re tui*[728] *ca, kim bang kua hön'ya wang pe a pön pe, taklang. Nayong cang reng reng, plai reng reng. Uh, yüa wang döm plai böt tanö, deng klök mai phum tui dö'tak, tim chük but ngat dö'tak. Nayong nong nö, war thöi yan*[729] *pe. War thöi kha'böt, töi ka plai döi.*

After this, they come home: There are many tassels, and the first encircling dance. At night, the first encircling dance, immediately the first dance. The first encircling dance: Come down to dance. Come down to dance, and the mother: How to make it? At that time, her mother and her father: This banana trunk keeps the fire over night, and then give her a hone to cook. The hone is cooking on and on. Cook this hone until it is done. As soon as the fire-keeping banana trunk burns out and the hone is also completely cooked, you will be able to join the dance. It boils, with mother and father, boils bubbling. It boils bubbling. and down there Wakek comes down to dance. There is pleasant music, the sound of dancing, blowing the gourd-pipes, beating of the gong, singing of songs. and Wakek is saying: Oh friend, come down to dance. "The hone is not yet cooked soft." A little bit later: Oh friend, won't you come down to dance. "The banana trunk has not yet burned down", says her friend, "alas!", and she sits between her mother and father. When her father moves a little bit aside, her mother moves quite close. Her mother moves a little bit away, her father moves close. Chongpau

[716] trunk
[717] to burn out
[718] pleasant music
[719] in between
[720] to move a little bit to the side
[721] to move quite closely
[722] somewhat, more
[723] girdle pressure
[724] would
[725] midnight passed
[726] below the front of the private room
[727] pull apart to a hole
[728] = *yu lon tui,* beer drawn off from the beer pot
[729] definitely

says: Father, move a little bit, move a little bit, the pressure of my girdle gives me pain. On saying: father move a little bit, a little bit, when her father moves a little bit aside, her mother moves quite close; when her mother moves a little bit aside, her father moves quite close. Well, the next morning would come. The next morning will come, says Chongpau. Every time Wakek calls her again: Oh friend, won't you come down to dance? She says again: Oh my hone is not yet completely cooked, my fire-keeping banana trunk did not yet burn out. Midnight has passed. Wait for me, wait for me; they cannot wait any longer. Finally her most beloved young man from below draws apart the wall of the front of the private room. He draws apart the wall and the young man gives her through the hole a small bamboo tube of beer drawn off from the pot. Thus they do on and on, they dance on and on. Oh friend, won't you come down to dance. The fire-keeping banana trunk did not yet burn out. The hone is not yet cooked soft. Thus later, the night definitely passes. The night has passed, she did not go to join the dance.

War thöi kha'böt nö, tong yong cö. Chongpau malaa cang or rü puk[730] *nö, chi tök raa ka köi hanö, a yung a yung, yung klen ka yung mi pe karui*[731]*. Kaa taku hamüt cuk*[732] *pe, kaa yung nöm nöm mi. Yung nöm nöm mi hamüt cuk klen ka, rui rui, kaa tangaa cuk pe. Tangaa cuk mi da po la, klai la tui hanö, tatong la kaneu*[733]*. Klai la tui a par nö, po la mai-phum lök, po la chi tangaa cuk pe.*

The night has passed, how now? Chongpau gets up, and goes to the place where the tobacco leaves are kept, a good one, a good one, she selects the best of the good ones. Gets nine, ten strings, gets the really good ones. From ten strings of the really good ones, she selects, selects and gets 5 strings. She takes the 5 strings, pours out the water, and then dons all jewelry. Pours out a basket full of water, takes a fire-keeping trunk, takes 5 strings of tobacco leaves.

Chongpau malaa Wakek malaa matci pe katöng tui. Matci katöng tui nö, mang mang nö, kre[734] *la mai-kut köt*[735] *pe, kwak la chi a*[736] *ram pe. Cang pre kat chum kat nö, Wakek malaa lünta a pe: Uh yüa, tong cö en cang cale, nayong le. Nong nö en kre mai-kut nö, en kwak chi ram, nong nö, en kre mai-kut nö, en kwak chi ram, kabo le, yüa le, yüa le, tong cang cale. Mru kaa döi hang döi mi, kröi lüm, kröi kham, kröi caa la kön ka camü le pape, pe. Na "o" pe. Nö, matci katöng tui. Kre mai kut köt, kwak chi ram, cang la a, cang la a tiama rong*[737] *pe, malaa. Tui töng raa tum po, tim-chük rak luk köi cekhai pe. Tim-chük rak yu ka-e da, phyang phyang*[738] *tum. Tim-chük rak luk cekhai pe, tui töng raa. Matci töng tui. Töng tui nö, tim-chük rak luk hön'ya, Chongpau malaa yu yör*[739] *köm cang pe. Nö, kwai-che köi klen*[740]

[730] to get up (from sleep)
[731] to select
[732] string
[733] to don all jewelry
[734] to beat off (embers)
[735] to knock off
[736] one
[737] on the whole way
[738] very big
[739] to throw oneself down

patai pe, Wakek malaa dök ta khai. Nö, uh, yüa, en nau madök le, kwai-che. Panö, thua a nö, pe ta khai, pe, kwaiche po. Wai, töng tui töng tui, hanö, hüm[741] *yör kaköp pe, yör ta köm, rak köi. Ngong or köi klen pape kröi dök, hang plot ü, Wakek malaa. Uh, en nau madök, yüa. Panö, pe ta khai pe, ngong köi kaa kengma ten po. Nö, töng tui töng tui, cang cang nö, hüm yör yok pe. Na kh'e da, mat ke-lang köi kaa ke-kom*[742] *pe, kröi dök. Nami po en nau madök, yüa, pe ta khai yok, pe. Wai wai nö, hüm yör kaköp pe. Na kh'e da, harköi cekhai pe kröi dök. En nau madök, panö, pe la bökbök, Wakek malaa köi. Pe la bökbök ha'nö, wai ka katcang hüm yör kh'e da, cham-chüi köi pe kröi dök, kaa dök, dök taa, dök raa da. Cham-chüi ca kröi dök taklen pe. Khöt pö la chi ka kak mi ce da, plüi plüi plüi*[743] *cham tayut*[744]*. Phang patai pe dö khai, tim-chük rak luk, rak cöngma hön'ya. Yu tum köi Chongpau malaa dö khai, nö, ka mang, ka mang da, ka mang, rak luk hön'ya.*

Chongpau and Wakek, the two go to fetch water. The two go to fetch water, walk and walk, and (Chongpau) beats off the embers from the fire-trunk, knocking them off, throws away one tobacco leaf. She does this two times, three times, and Wakek asks her: Oh friend what are you doing this way? Now you beat off the fire-trunk and throw away one tobacco leaf, and then you beat off the fire-trunk and throw away one tobacco leaf. What for? Oh friend, friend, what are you doing? People who possess nothing will want to pick up, want to drink, want to eat it perhaps, says she, speaking in riddles. They go to fetch water. She knocks off the fire-trunk, throws away tobacco leaves, she does it, does it on the way, the girl. The pool to fetch water is above the hone cliff. Below the hone cliff is a very deep pool. Above the hone cliff is the place to fetch water. The two fetch water. They fetch water and from the hone cliff Chongpau wants to throw herself down, and quickly Wakek seizes her by the bracelet. Oh friend, you probably want to have the bracelet. With this she strips it off and gives her the bracelet. Soon after they draw water, draw water, and then she leans over again to let herself fall down the cliff. Wakek quickly tries to seize her by the neck, does not let her free. Oh you probably want this, friend. After this, she gives her all the bead strings which she got. Then they fetch water, fetch water, they do and do, and she leans herself over again. This time, (Wakek) tries to seize her by the chain she wears around her waist. You probably want this too, friend. She again gives it away to her. After a short time, she leans over again. This time, she tries to seize her on the upper arm ring. You probably want it, and after this she gives all to Wakek. She gives her all and then, after a while, when she suddenly leans over to let herself fall down, (Wakek) tries to seize her by the hair-knot. Tries to seize, seizes where to seize. Tries quickly to seize the little hair-knot. How much oil was it, which she became to seize, and the hair slips away. She falls down, from the top of the hone cliff, from the steep cliff, into the pool down. Chongpau falls down, and off she goes, off she goes, goes off from the top of the cliff.

---

[740] quickly

[741] to lean oneself over

[742] chain around the waist

[743] to be slippy (?)

[744] to slip away

Dö khai panö, na kh'e da, tanger cho pe, kua mi wüa. Na kh'e da, kut plan lai a, cia ngaa phom[745] *mün, plak hom chau mün, ka po la pe. Nö, ka lai a, u-paa. Chongpau u ö, wang mang, wang mang, wang caa plak hom chau, wang mang, pape. Ka ho a, nö, rung rung rung*[746] *pape, wang döng*[747]*. Yu tui dong hön'ya, or tök*[748] *köm. Ang caa le, pape yok pe. A-u, thui, en kai kwa war*[749] *tömma khong khöt, pape, hanö plan kup khai pe. Wet pe cia ngaa phom, wet pe plak hom chau. Yu ho a, yu ho a ha'nö, rung rung rung pe yok pe, yu hönya wang tök. A paa: yu mi da ang caa le, panö, en nia kwa war tömma khong khöt, palö hanö, plan kup khai yok pe, na kh'e po. Nö, taklang cang kau lung leng*[750] *ca, ker*[751] *a ker a hanö, a taklang thang*[752] *ta pe, hüa pe, tum köi. Thang ta pe hanö, pai a lup dingca*[753] *chau hai, plak hom chau hai, yu ho, yu ho, nö, rung rung rung, na kh'e da, wang or tök. Klök, pa, kau lung leng hai, a klang chök ta khai, a ngong köi. Taklang chök ta klök nö, che phum*[754] *köi yua ta hüa t'ia, chong prang*[755] *köi yua ta hüa.*

She fell down and then, after this, the villagers bewail her. At that time they call her to come back. They take some pounded cow meat, some sticky rice and call her, the parents. Chongpau, come, come! Come to eat sticky rice, come, they call her, and slowly, slowly, she rises, comes up from the water. From the water down there she will rise up. My child!, they say again. To her mother: Ugh, your vagina hole is broad like the hollow of a drum, she says, and then she dives again. They throw pounded cow meat, throw sticky rice. They call down to her, call down, and then, slowly, slowly, again she rises, comes up from down there. The father: that one down there is my child! Upon this: Ugh, your penis hole is wide like the hollow of a drum, she says, and then she dives again, also this time. Then the young man makes a thong from a new bamboo, twists it together, twists it, and then her young man makes a snare, puts it in the pond. He lays a snare and then with this sweet work, with the sticky rice, he calls down, calls down, and slowly, slowly, at this time, again she comes up. Crack!, with the new bamboo snare, her young man catches her by the neck. The young man catches her, pulls and keeps her on the pebble beach, drags and keeps her on the shore.

A köng pape, ka camü da, köng dö'le, acük kh'e mruca da, a chak da wöi reng madök, a twak wöi reng madök chö, nö, yua ta hüa. Nö: Uh, ang caa cace, ang kur[756] *la, mokca ö, apaa pe, mokca tai böt. Nakh'e da, pai a, taklang köi: Ang caa cace, ang kur la, ang kur la, pape, apaa. Kat-chang apaa kur pace, kwai-chek lök po yaa kan döi pe. Apaa tek pe mi: Uh mokca*

---

[745] dried or cooked tough meat wih chillies, salt, ginger, etc., pounded in the mortar until soft enough for eating

[746] to rise slowly

[747] to surface, to come up from the water

[748] to come up, to rise

[749] broad, wide

[750] new bamboo thong

[751] to twist together

[752] = thang leng, snare

[753] sweet

[754] pebbles

[755] river bank, shore

[756] to carry on the back

ö, ching wöi chak wöi panü[757] *da, cöng machi mi tai dia*[52] *cace, wang töi khöng kur lai*[758]*, en machi köi, pa pape, a mokca köi. A mokca kur la ce, ram pü klö ca, ram wer klö*[759] *ca, ting-teng*[760] *tün ca, prüi tün ca, plung tün ca, meng klö ca, nök ok-lok*[761]*. Nök ok-lok, pape, ta mang*[762]*. Nök ok-lok ta mang nö, kayüa hua-yu tek mi: Yu mi da, aci caa tün teu*[763] *mi kabö, aci caa ram pü ram wer tün klö yong, ting-teng taro tün yong kabö. Nayong pa yu tek, na kh'e mat mat*[764] *yu köi aklang luk köi chu*[765] *tui bot bot bot, bot bot bot pape cang. Chongpau malaa a-u tek mi da: Uh, ang caa cace, mokca ö, ang la lai thüp ang*[766]*, ang kur la lai. Ang mat caa cace da thüp ang, ang kur la lai yok chö, ang matya caa tang tang cace da, pape. Nö, a-u kur la, a di mi, di dami, ram chong lök po yaa kan döi pe, kwai-chek lök po yaa kan döi pe. Hing tacuk pret-ret*[767]*, tacuk pret-ret, yaa kur döi, a di mi. Di di di lai nö: Uh, ching wöi chak wöi panü da, en machi ön kön tai dia cace, mokca ö, en kur la yok tai, ang yaa kur döi. Nakh'e da, taklang yok pe kur la nö, ram pü, ram wer tün ca, ting-teng tün ca, taro tün ca, prüi tün ca, hiu klö ca, meng klö ca, nök ek lek, nök ek lek tayok pe. Panö: uh, yu mi da, aci caa, ram pü ram wer tün yong kabö, tingteng taro tün yong kabö, ci caa ren klö, lai klö yong kabö, pape tek yok. Yu köi, taklang luk köi, yu köi chu tui bot bot bot pape klok.*

She is dead, but she is not dead, in the beginning people said: her life probably still exists, her thinking probably still exists, for sure, then pull her along. And: Oh, this is my child, therefore load it on my back, son-in-law. (He has become a son-in-law.) At that time, that one to the young man: Because this is my child, I carry her, I carry her, says her father. As soon as the father carries her, he cannot step over even one tiny bamboo piece. Her father says: Oh, son-in-law, if life would be there, you would take her as a wife, therefore come and join, help to carry in your turn your wife, says the father to the son-in-law. When his son-in-law carries her, the sounds of whistling on leaves, of the bamboo zither, of the flute, of the gourd-pipes, of songs, are to be heard from afar. They sound from afar for those going behind. Going behind and hearing the sound from afar, husband and wife say: Alas, that down there is the broken sound of our child, like our daughter's sounds whistling on leaves, like sounds of playing on the bamboo zither and on the fiddle. While they are saying this, at the very same time down there on the young man drips putrid water, making drip, drip, drip. Chongpau's mother says: Oh, because this is my child, son-in-law, let her down for me, I carry her in turn. This is my own child, therefore let her down for me, I carry her in turn for sure, because she certainly is my own child. And her mother loads her up, heavy, too heavy, she cannot step over even a single leaf, cannot step over even a little piece of bamboo. She

---

[757] panü ... dia: if (contrafactual condition)
[758] in turn
[759] whistling on a leaf
[760] bamboo zither
[761] to sound together from afar
[762] to go behind
[763] broken sound
[764] at that very time
[765] putrid
[766] let down for me
[767] crouched down by the load

starts crouched under the load, crouched under the load, she cannot carry it, it is heavy, heavy, too heavy, and: Oh, if life would be there, you would take her as a wife, therefore, son-in-law, you carry her again, I cannot carry. At that time, when the young man carries her again sounds of whistling on leaves, sounds of bamboo zither, of fiddle, of flute, of whistling on the lips, of singing songs are to be heard from near, are again to be heard from near. At that: Oh, this down there is, alas, like our child's sound whistling on a leaf, like the sound of bamboo zither and fiddle, alas, like our child's laughing and speaking, alas, they say again. Down there, on the young man, down there putrid water bubbles, drip, drip, drip.

Mang mang ha'nö, töm la kim köi. Töm la ha'nö, cong a kim köi, cong a bök bök, ting-teng tün hai, taro tün hai, plung tün hai. Tuk meng mi tek pe meng, tuk tum plung mi tum plung, tuk tep ting-teng mi tep ting-teng, nök ok-lok, nök ok-lok, pe, cong a. A u-paa cong kh'e da chu tui bot bot bot, bot bot bot, pape, cang, chu nam, yaa wöi döi. Nö, mru hai, a taklang hai ce kham cong nö, a ren klö ca, lai klö ca, ting-teng tep klö ca, cho-cho le-le aklö yung, tün klung-lung khai pe. Nayong cang cang nö, a u-paa töi mü kh'e da, chu ca bot bot bot pe tayok pe. Alöwa, löwa, löwa, a u-paa kunma rau khai pe. Yong khöt ce wang töi mü en döi kh'e da, u ö, en, meng klö ca, lai prö ca, prüi tün ca, ting-teng tün ca, plung tün ca, meng klö ca, chöm döi[768]*. Wang töi mü en kh'e da, chu tui bot bot bot pa, en cang, en pahing chu, nayong en pahing köng, nö, tui pe dong taruk kat pe*[769]*. Tui pe dong taruk kat kh'e da, kat-chang ka a taklang ta' krüng*[770] *mi, plung hai, krüng reng reng, krüng reng reng, dit po rü dö'böt pe. Cang pe piati kha'böt mi, hing rü dö'böt pe. Rü dö'nö, köng yan böt.*

They go and go and then reach home. They reach home and then keep her in the house, keep her all, with bamboo zither sound, fiddle sound, gourd-pipe sound. Those knowing songs, sing songs, those knowing to blow gourd-pipes, blow gourd-pipes, those knowing to play the bamboo zither, play on the bamboo zither, they are to be heard from afar. To be heard from afar, they have guarded her. When the parents guard her, putrid water makes drip, drip, drip. Putrid smell, they cannot stay. And with people, with the young man when he achieves guarding, her laughter and speaking, playing the bamboo zither, all making noise, the sound is pleasant. That way, they do and do, when her parents join to see, putrid drip, drip, drip again. Oh my god, my god, her parents get angry. How many times coming to see you, when you are not there, darling. What an amount of singing, talking, playing the flute; when we come in order to see you, putrid water, drip, drip, drip, you make. You start to become putrid, that way you start to die and they kick the ladder 6 times with the foot. When they kick the ladder 6 times with the foot (cursing her), immediately her young man, who rose her up with the gourd-pipe, rises her, rises her up, she does not wake up anymore. They have made a curse, she does not start to wake up anymore. She does not wake up, she finally is dead.

---

[768] not small = quite an amount of

[769] to kick 6 times against the stair is a curse

[770] to let rise again

Ni, kaceng ta khai, panö, mat-ni ceng a pe. Kat kaceng a po, kam döi pe, maa po khöt tin[771] *a-e, mai loi khom wöi döi pe. Nom*[67] *pe chong chot*[772] *bong*[773] *pe, na kh'e, rut mi tatua che*[774] *pe. Nom pe ching pöng bong pe, na kh'e, khok mi tatua che pe. Nö, taklang kong*[775] *pe mi kho bong*[776] *nom pe, pe, nakh'e, a pik tapho*[777] *kröngca*[778] *pe. Wai, a taklang kong pe mi tum-pia bong*[779] *nom pe, pe, na khe cham chüi kam ditca pe. Nayong, cang reng reng, cang rengreng, dit po a taklang yaa mü döi pe. Mü mü nö, wui*[780] *kau lungca cem nö, chut*[781] *pape, töi tanom khai, a taklang po. Matci preya yem*[782] *pape, mai kam ta khai pe. Mai kam ta yem.*

Now she is to be burned, and they burn her. Once they try to burn her, she does not burn, and how often they throw her into the fire, the fire does not even blacken her. When they throw the embroidered sword of the weaving apparel into the fire, her fingers squirm. When they throw the decorated wooden bars, her feet squirm. And when they throw the little embroidered basket for keeping her little jewelry, plaited by the young man, into the fire, her skin blisters somewhat. Next when they throw the embroidered rainbow basket, plaited by the young man, for keeping her larger jewelry, into the fire, her hair-knot burns a little bit. That way they do on and on. Her young man cannot stand seeing this doing on and on any longer. He sees it, sees it and, pulling down a young bamboo top, he catapults himself into the fire as well. The two blaze up at once, the fire burns them. The fire burns them at once.

Mai kam ta yem nö, nami ceng loi plu[783] *ka-e chong-klang*[784] *ca wang ru a tut pe. Klang-yung tong da pe, chong-klang om. Nö, kalang lek*[785] *mi da au-yüa cace da, Wakek malaa pe, kalang lek reng. Wakek malaa kalek nö, cong ü, kalek a nö, cong a ü a, pe. Nayong cong reng reng, ü reng reng nö, Chongpau malaa u töi khi khai pe. Uh, tong pauca c'e, u ö le, en cam lek reng le. Pe la ang pre kongca dit, ni anam yungca, tong da. Nö, pe la pre kongca pe. Pe la pre kongca nö, tong yong cö cang a pa e da. Cham-chüi ka pe, or cong a. Cham-chüi köi or rik cong a pace, tom*[786] *cham-chüi pe dö khai. Alöwa löwa, mami pau da, pape. Yu kut plan lüm pe. Yu plan nö, param köi pe or ü a. Tom param pe dö khai yok. Alöwa löwa, nami pau da, plan pe yok pe. Yu plan lüm yok. Na kh'e, ke-lang köi plan kap a, pe, tom kelang yok pe, dö khai. Nö, kunma rau khai yok na-e po. Kat-chang ka a u*

---

[771] to throw into the fire
[772] sword of the weaving apparel
[773] decorative pattern (prepared by the young man)
[774] to roll oneself, to squirm, to double up
[775] to plait
[776] little embroidered basket for keeping girls' things
[777] to blister
[778] a little bit, somewhat
[779] emboidered basket
[780] to pull down
[781] to move aside
[782] to braze up, to burn right away
[783] heap of charcoal
[784] A plant with sweet-smelling leaves. The name actually represents a combination of the names of the lovers: Chong(pau) and (Ta)klang.
[785] to pick, to pluck
[786] together

kunma rau. Alöwa löwa, nayong cö nak naa[787]*, tong kabü cam kaa ü. Panö, na kh'e da, mit*[788] *a köi böt pe. Mit a ha'nö, pai a, Chongpau malaa ceng loi plu köi cekhai pe, ka khi mami. Na ka, den pe*[789] *tui chaa, ngur ngur*[790] *ka khü ka, kaci*[791] *ta pe plot*[792] *pe nami chong-klang tut köi. Nö, chong-klang po, hing köng tayok pe.*

Khöt nö, cur t'ia, Chongpau malaa cia da.

The fire burned them at once, and on the heap of charcoal grows a plant with sweet-smelling leaves. How nice, the sprouts of the sweet-smelling plant, Wakek says. And she picks them, because she is her friend, and she picks them always. Wakek picks them and plugs them into her hair, picks them and plugs them, wears them. That way she carries them always in her hair, wears them always, and Chongpau's mother joins her to see. Oh, what little flowers are this, darling, which you always go to pick. Please give me two stalks, how their smell is good. And she gives her two stalks. She gives her two stalks, and how to do it with them? In the hair-knot up there she plugs them. As she plugs them first up into the hair-knot, together with the hair-knot they fall down. Oh my god, my god, these flowers, she says, picking them up again. Up again, she puts them in her ears. Together with the ears they fall down again. Oh my god, my god, these flowers, she says again, picking them up again. This time she fastens them on her waist band. Together with the waist band they fall down again. Anger again, also here. At once her mother gets angry. Oh my god, my god, whatever way, wherever I can get her? Upon this, at that time, she goes to trace it. She traces it and then, there, where it is on Chongpau's charcoal heap, she finds it. For this, she heats water, when bubbling, she turns it over and empties it out on the Chong-klang plant, and the plant starts to die as well.

So much, and this was the end of the Chongpau malaa story.

S4b) *Karek malaa chang-cia*
The storey of the star maiden (told by Klangwai Atwang, Tapwüa-Kua, 1956)

U-paa wöi döi pe, na-mi ka. A u-paa wö'döi ha'nö, klöng bau[793] *khai pe. Caratca*[794] *caa, klöng bau khai paka-caa-lö ha'nö, karek malaa wang tu pe mo*[795]*, pe. Karek malaa wang tu pe mo ha'nö, na-mi ka mru tek pe: Tong cü, caratca lö, en kada üa chai lam-lam*[796] *ta da, en da klöng bau kh'e, pa pape. Tong cü en-ni tek ang, klöng bau klöng bau ang kada maa ka-e, klöng bau rau kar rön-rön tai, pape. Panö klöng bau ce-po, tong üa chai da, en kada, pa pape. Öm döi tai a, öm döi, öm döi panö, caratca ka klöng bau yung non kh'e, kamü üa pace, cakraa-mang camaa karek malaa mi ranit ram pe,*

787 whatever way
788 to trace
789 to make hot
790 to bubble up
791 to turn over
792 to empty out
793 swollen leg
794 orphan
795 *tu pe mo:* weed for him
796 tidy

wang tu pe mo. Uh, klang-yung khai, tong cö ang cang a dit kom. Ang kaa ta khai panö, ang cang la machi kom. Pa patek ha'nö, na-mi mi tu mo thüm[797] *kh'e, üa ria köi pe, a nge*[798] *da cong tarip, karek malaa baa, cong tarip ha'nö, lang klang-yung tanö tö mi, achoi tanö tö mi, nge kapo, ha'nö pet ta kha'pe. Pet ta khai, pa-caa-lö ha'nö, cang la machi kom pa-ha'nö, ci tacong la, ci tacong la, pape, khüm ta klen. A taima-wüa da, prung, payö. Cong la ang dia, cong la ang dia, ang kada nge wöi döi ka, pape. Cong la ang dia panö, or kwang dong köi ting payu chanaa kh'e, cong a pe, cong a panö, yaa cong döi, wang böt, wang böt, pape. Or hön'ya, yaa cong dö'nö, yö ta khai pe. Töi yaa yö dö'nö, ci tacong la, ci tacong la panö, tacong yüa khai pe.*

That one has no parents. His parents have died, and his leg is swollen. The orphan has a swollen leg and under these circumstances star maidens come to weed for him. The star maidens come to weed for him, and people say to him: What, orphan, your field is so clean and tidy, while you have a swollen leg? What do you tell me? I have a swollen leg, a swollen leg with wounds. The pain of the swollen leg makes me cry. – And so they say: Though you have a swollen leg, how is your field so clean? He just does not believe it, he does not believe, and since he does not believe, as soon as his swollen leg is a little better, when he goes to see the field, the heavenly lord's daughters, seven star maidens, weed for him. Oh beautiful, what shall I do? I get one, and then I'll take her as a wife, so he says. While they cut down the developing weeds, they deposited their wings on the edge of the field. The star maidens deposited their wings, and he goes to take the wings of the most beautiful, of the youngest one, and hides them. He hides them, in order to take her as a wife. And then: We will marry, we will marry, he says, grasping her quickly. Her elder sisters, at once, fly off. Wait for me, wait for me, my wings are not here, she says. Wait for me, and so, when they reach the middle of the sky, they show mercy and wait for her, wait for her and then can wait no longer, come on, come on, they say from above, we cannot wait any longer, and fly away. She cannot accompany them (and he says), "we will marry, we will marry", and then they marry.

Pet ta pe nge can[799]*, or ö klöng ka-e. Uk*[800] *pe klöng nö, a nge da, kapet ta pe khai. Tacong yüa, talang*[801] *yüa, nö, kaa caa lök pape*[802]*. Kaa caa lök panö, na-mi mi anmat-ya cong lai*[803]*, a u cong lai, a paa cong lai. A u cong kh'e da, a paa tu mo, a u tu mo kh'e da, a paa cong. A paa cong kh'e da, klöng köi, u ka nge kapo pe, nö lek pe köt-löt*[804]*. Lek pe köt-löt nö, ren, ni cut pe ren. Ren ren panö, ko da en cong lai, pe a u köi-e. Tuk ring-ria*[805] *khök pe, tuk lai dit-det khök, panö, a u cong pe. Na-kh'e da, kar mi, kar dami nöm-nöm pe kar. Kar pe, kar panö da, tong cü klöng praa köi cok*

[797] to develop (itself), to grow
[798] tail (of a bird), but here better to be translated by "wings"
[799] securely
[800] to make a hole into
[801] to conduct a married life
[802] Literally: "so one says, it is said". Here generally not translated.
[803] in turn
[804] flapping (?)
[805] to make the first walking attempts

lai[806] *da. En, tong cü en paa tö ta en khai, ha'nö, na-mi yak e nöm-nöm pe kar, na ka yak-e kh'e, tong ce ün cü, kar lai, ang hot cong kh'e da. En hot cong kh'e da, kar tak*[807] *baa, en hot cong kh'e po, pape. Ang hot cong kh'e da, ren kök-lök*[808] *ta'da, khöt ko yung mi. Kar döi da, tong en hot cong kh'e kar khai da, pape. Nö, trau da en cong yok, kar lai ang cong kh'e, pape. A u tek pe-mi, karek malaa. A paa da tüng pe a u nge ka da, panö, a paa cong kh'e da, lek pe nge ha'nö, ren kök-lök yok pe. Wai ni kh'e da, a u cong khai yok, kar kar nö, klöng köi cok khai yok. Tong cü wöi kar ka, ni-e katüng pe-mi, ang wöi döi kh'e tüng pe-mi, wöi pa cap da. Kamü pace, krek malaa nge, ayaa nge, kadök kha'böt.*

He hides the wings securely, upon the cross-beam. He makes a hole into the cross-beam and hides the wings there. They marry, conduct a married life, and they get a child. They get a child, and then they guard it in turns. When its mother guards it, the father weeds, when the mother weeds, its father guards it. When its father guards it, he goes to take the mother's wings from the cross-beam, and waves them flappingly. He waves them flappingly, it laughs, laughs the whole day. It laughs, laughs, now you guard it, he says to its mother. It has made the first walking attempts, knows to speak a little bit, and then the mother guards it. At that time, it weeps and cries, weeps really. It weeps and cries, and then: What does it point always to the cross-beam? You, what has your father put there for you? And then, when one evening it really cries and weeps, this evening: What is the matter? It cries too much when I guard it. When you guard it, does it still cry, when you guard it?, she says. – When I guard it, it laughs flappingly, how good it now is. It does not cry, why cries it when you guard it?, he says. – And tomorrow you guard it, it cries permanently when I guard it, says the mother, the star maiden. The father shows it the mother's wings, therefore when the father guards it, he waves the wings and it laughs again flappingly. The next day, its mother guards it again, it cries and cries and points again to the cross-beam. What is it crying for, shown to it yesterday, shown when I was not there, there is one thing. When she goes to look: the star maiden's wings, her wings, taken away from her!

Uh, a-ni tak, ang ka nge, ang paa pren köi kating la-mi, a-ni tak lamaa pape. A caa, mom pe hom üm[809]*, lu a lam*[810] *a, kaneu chim a mök*[811] *pe. A caa chim a mök paka-caa-lö nö, yö pe. Tatöng*[812] *yö, kwang praa köi, tatöng yö, paka-caa-lö nö, na-kh'e ka, a caa kar klö mi waa-mu ku-liu*[813] *lai yong taa pe, yu taa. Öh, ang caa le, ang chak rui panö, wang yu döm khai yok pe. Wang chim a üm, wang mom pe hom üm, wang lu a üm, wang chim ta a mök, na kh'e da, ma praa achet, na praa achet, pe, rung*[814] *la la-puk*[815]*, param ka e. Khök nö, yö yok kwang ka e, mai-chang ka or yö. Ma-mi da,*

---

[806] permanently, always
[807] still
[808] heartily
[809] optimal, best possible
[810] to tidy up
[811] to leave (behind)
[812] to depart
[813] to steal seeds (?)
[814] to plug up
[815] dressed cotton

ang hua wang kröi kh'e, pacöng a chop pren, pe, kwak la chop lung lök, pe. Ma-mi da, chön cu lang nom[816] *pren pacöng a, ang hua wang kröi kh'e, wang kröi kaa ang dö'pat, pe, kwak la chön cu lök pe. Mi da tui pren pacöng a, thok la kria pak*[817]*, pe, ang hua wang kröi kh'e, wang yaa kröi dö'pat. Nö, apaa cakraa-mang phung ka pe, ting kha'böt, ting kha'böt panö, uh, ani tak ang caa wang ting. Chok a chau*[818] *a tak, cang pe lak-chu maraa*[819] *poi tak, pape. Cang pe lak-chu maraa poi.*

Oh, still today, my wings will bring me in my father's country, still today, for sure. She feeds her child as good as posssible, bathes it and tidies it up, and leaves it fast asleep. She leaves her sleeping child, and under these conditions flies away. She departs flying into the sky, and while she is departing, at that time she hears her child crying with a voice like a robbing hawk. "Oh my child, my beloved baby!", and thus comes down again. Comes to lull it best posssible, comes to feed it best possible, comes to bath it best possible, comes to put it to sleep fully, and leaves it. At that time this side she makes deaf, that side she makes deaf, plugging dressed cotton into her ears. After that, she flies again into the sky, quickly she flies off. "This here, when my husband comes to search me, may become a firewood land", she says, throwing away one piece of firewood. "This here may become an overgrown thorny rattan country. When my husband comes to search me, he shall not find me, she says, throwing away one thorny rattan. This here may form a land of water – she spits out phlegm –, when my husband comes to search me, he shall not be able to find me. Then in her father's, the heavenly lord's, world she arrives. She arrived, and then: "Oh, today still, my daughter arrives. Still scrub it, cleanse it, I still make for her a welcome feast", he says. They make for her a welcome feast.

A paa, ma köi, a paa wang pace, wang mü üa[820] *ce, ange po wö'döi, a caa po kar reng-reng nö, mru wang po ta khai, pe, kim kho, üa klaima*[821]*, tatö tahüa. Uh, ang awui chömca, caratca caa, u po wöi döi, ming-ma*[822] *tua-ma po wöi döi, i-büi*[823] *nauca po wöi döi, khöt lamaa ngaa. Wang eng caa hom, wang eng caa hom, pe, uh a u wöi döi kh'e, machi wöi döi kh'e, mat-ya wang cang lü-lü. A prüa*[824] *nguk*[825] *khai ce, ang kakröi tak, pape, awai rau e kh'e, ang kakröi tak, ang ka machi pot khai, apu cakraa-mang tut, pape. Ha'nö, po la ngapok lök pe, kakröi, mang mang mang, paha'nö chop pren, chop pren kangö*[826]*, yaa mang döi pe. Uh, phyoi*[827] *a, ang*

---

[816] to float
[817] *kria pak:* phlegm
[818] *chok chau:* to scrub and cleanse
[819] *lak-chu maraa* = ?? probably from dubious Marma, in one instance [see footnote 414]: *lak-chuk maraa* [to not receive (with) a closed hand (?)], here translated as "welcome feast". Used here in a friendly context (*pe ... poi* : to give a feast to s.o.); afterwards mostly in a threatening way (*cang ... poi*: to make a feast of s.o. (probably by sacrificing him).
[820] = ?, to challange ?
[821] harvest failure
[822] (poet.) elder sister
[823] (poet.) elder brother
[824] to worry
[825] to have an oppressive feeling, to goan
[826] everywhere, all around
[827] (exlamation used when calling the spirita)

pacöng a along[828] *da ka caa pada, ma-mi chop pren ka e, ang pacöng a utma co*[829]*, chir-chir*[830] *chör-chör*[831]*, pa, ang payaa mang a, pape, ha'nö, chir-chir chör-chör panö, yaa mang khai. Mang mang mang, chön cu lang nom pren akraing mi, kraing da-mi, tong po tuk tek döi ce, pren da köi döi pe. Uh, tong yong cö ang yaa mang a köm, mi da ngaa lai khök kabö, pe, chön cu lang nom pren köi, ang pacöng a ngaklai ngakang*[832] *pada, ang along da ka caa pada, chir-chir chör-chör da, ang payaa mang a, a co pape. Panö, yaa khök pe.*

Its father, here, its father comes, and when he looks around, the wings are not there, his child also continually cries, people come to take it, the house is empty, the harvest fails, she has left him back. "Oh, my lot is small, I am an orphan, have no mother, have neither elder nor younger sisters, have neither elder not younger brothers, how bad indeed." Come cook and eat rice, come cook and eat rice, oh, when his mother is not there, when his wife is not there, he himself goes to do it alone. As he worries groaningly: I go to search her still, he says, tomorrow morning, I will go to search her, my wife has escaped to the grandfather, the heavenly lord, he says. And taking one gun he goes to search, walks on and on and on, and then the firewood country, firewood country everywhere, one cannot walk. Oh, he calls the spirits, let me become a small one to achieve what seems impossible, in this firewood country, let me become just a mouse, advancing slidingly I could go through it, and then advancing slidingly he can go. Going on and on and on, the country overgrown by thorny rattan, difficult to the outmost, one cannot tell, a country one cannot move into. Oh, how will I be able to go, alas, this is extremely bad. In this country overgrown with thorny rattan, let me become a small lizard, a small one to achieve what seems impossible, advancing slidingly, let me be able to just go through it, he says, and thus, he can go.

Mang mang mang a, chataa o köi ting, thok la kria mi. Ma-e war po iu kha'böt, chokka-tik chokka-tak[833] *ching ka pe, yak*[834] *klaa khai, ching-kri ngak-kri boima*[835] *waa po, chokka-tik chokka-tak ching plong ka-e, yak pre buk pe, mat-ci hua-yu. Panö, mat-ya po kayak ching tut ka e (along caa cakhai mi po baa), panö, müi müi müi, pe. Or ö, machi-wüa da tuk döi, kacak-wüa da tuk pe, aklang da ching tut köi, mru yung, mru ngak, wang yak pami, yak panö, wai rau-e, waa kaa*[836] *khök pe, wa-ram kaa khök, panö: Uh, kacak ö, trau tong praa cö ci cam to*[837] *caa kom, pape tek pe. Ah, hang popan reng le, trau da cam to ngak-tai*[838]*, pa pape. Uh, lai dö' machi, kraa nö, tek khai yok pe. Tek en po ki döi. Tek döi po ki döi kabö, ching tut ka e, mru yung mru ngak wöi ca. Cepo, ang tek en tai a, en chak i*[839] *döi cace. Or*

828 to accomplish (what seemes impossible)
829 just (with no other possibility)
830 slidingly ?
831 to advance
832 kinds of small lizards
833 (an imagined tree)
834 to stay overnight
835 literally: tree-big-bird-big-??-bird
836 to crow
837 ? (= jungle ?)
838 everything (what one pleases)
839 to be satisfied, quiet, to give peace

cakraa-mang camaa kapot khai, lu-chaa köi ka kacang dong-thong[840] *mi, caratca hot khüm ta' khai mi. Na-mi cakraa-mang chok a chau a pace, cang lak-chu maraa poi pe ma-mi, aci katöi caa kom, mi poi, chataa pai praa pön*[841] *köi caböt, pai cakraa-mang kua, pape. Ton khök lamaa, pai ö, (mi caratca twak mi) trau en-ci yö kön ce khök kh'e, ang pacöng a löng charin*[842] *natca waa pada, en-ci lap krüa*[843] *ka-e, en pata'ua*[844] *chüt*[845]*, chataa o köi payö la ang, ang along da ka caa pada, pape. Or tadan*[846] *khai pe, or kachöng*[847] *lap kh'e, yaa yö pace, yaa yö pace. War thoi khaböt, yaa yö pape. Lap adi mi, di dami, a machi ka, nöm-nöm pe lap di. Uh, kacak-wüa, ang kada lap di lai kabö, tong cang a kom. Ang thür*[848] *ta ü, pape. Uh, hang kathür, ang tek en dia, mru yung mru ngak wöi, ching tut ka e, pace ang tek en dia, en tuk taa döi. Pai ka ting pat hia*[849] *yö la, ako o pang köi phai*[850] *thür yang, pape. Yaa yö la, yaa yö la panö, o pang*[851] *köi ting kh'e da, thür ta bak*[852] *pe.*

He walks on and on and on, reaches the ocean, the one spit out as phlegm. Here it has become dark night too, and he must stay overnight under a huge tree. In the top of this tree two large birds, husband and wife, stay overnight. And so he also (who accomplished what seemed impossible) goes to stay overnight below the huge tree, and he sleeps on and on and on. Above there, the wife does not know, the man knows: the young man below the tree, a good man, a bad man, he who stays overnight, next morning the bird has crowed, the jungle-fowl has crowed. Oh, man, tomorrow which direction will we take to get jungle food?, she says. Ah, do not talk always. Tomorrow we do anything we like, he says. Oh, the wife does not say a word; after some time, she talks again. I speak to you, no end, I do not talk to you, alas no end too. Below the tree is a good man or a bad man. However, I talk to you, because you don't keep quiet. Above, the daughter of the heavenly lord escaped, she who contracted a marriage with a human, the orphan who seized her. For her the heavenly lord scrubbed it and cleansed it, to make a welcome feast; this we two will join to eat, this feast, it will be on the other side of the ocean, in that village of the heavenly lord, he says. I have come near to it, indeed, there (this is what the orphan thinks), tomorrow, when you two will fly, let me become a small red ant, you spirit birds, under your wings, let me hide in your feathers, let me fly over the ocean, a small one to accomplish what seems impossible, he says. He puts himself up in between the feathers when she flaps the wings, and then he can fly, then he can fly. The next morning comes, the wing of the wife is heavy, truly heavy. Oh, man, my wing is too heavy, what to do? I'll shake it out, she says. Oh, do not shake it out; didn't I tell you, there is a good man,

[840] to contract a marriage
[841] the other side of
[842] small red ant
[843] room under (the wings)
[844] to let hide oneself (in feathers or hair)
[845] to put into
[846] to put oneself in between
[847] to go to shake or to flap (the wings)
[848] to shake out
[849] from a distance
[850] however
[851] the other side
[852] to fall out

a bad man below the tree, when I told you, you did not know. They fly the whole distance to arrive there, now on the other side of the ocean however, she finally shakes him out. She could fly here, could fly here, and when she reached the other side of the ocean, she shakes him out and he falls down.

Or mang or mang nö, cakraa-mang tui cot köi ting khai pe. Cakraa-mang tui töng-mi narma[853]*, na-mi chok la chau la cakraa-mang camaa karek malaa kom-mi, tui pe wang töng. Na-mi ti hing yaa tatöng döi pe. Amia reng cu*[854] *mang cem-ma caa ce ü. Ang cin en tak, ang yaa tatöng döi lamaa, wang töng ang ditca, pape, panö, phyoi a, ang machi tawai rut tale*[855] *köi, ma-mi myak rut-büt, pe ta'krua*[856] *a chüt*[857] *co. Ako kathut*[858] *tui kong kh'e, pape, phang*[859] *pe chüt, nö, töng a pe, kathut pe, or kating nö, plot plot plot kathut nö, tawai rut tale köi chüt pape mi, na-mi myak rut-büt, matci tayüa ü rengmi. Ang hua wang ting kha'böt, ang hua wang ting kha'böt, ha'wang thut ang tham, pe. En krum la mru ne tak baa, tui chot köi, pe. Ta'krum lök, mia reng cu mang cem-ma caa ce ü.*

Up there he walks and walks and arrives at the heavenly lord's place where to draw water. The heavenly lord's female servants, who fetch water, those many scrubbing and cleansing heavenly lord's daughters, the star maidens, fetch water. They, however, cannot fetch for themselves. Which highest lord's child is this? I still send you, I cannot draw (water) for myself indeed, draw a little bit for me, she says, and then, he calls the spirits: on my wife's left hand let slip this ruby finger-ring, let it slide to be put just on her finger. Now when pouring out the water-pot, he puts it in, and they draw water to pour it out. Up there it arrives and when she pours it out and on the left hand slips what he put into, that ruby finger-ring, which they two always wore. My husband has arrived, my husband has arrived, don't go to pour out anymore. On drawing water you met a man too? We met with one, which highest lord's child was it?

Po la ngapok lök pape. Panö ni ka lai[860]*, cin pe mru tangaa yaa pe, mru a cakraa-mang töi cin tangaa yaa. Hng, en, en pu*[861] *cakraa-mang tek ce, töi bot pape. Hng, ang lu tak, ang lu ang lam tak, or rik mang büa. Ni ngapok phai or rik po la, pape, or kot la pace. Tangaa yaa mru cü*[862] *khom,*[863] *yaa cü döi pe. Di mi, di dami, na-mi ngapok pe. Uh, yaa döi, kamang khai yok pe, ang-ing tangaa yaa, ang-ing yaa kot döi. Khöt di ngapok, ce, lu lam tak pe, en mokca, pape, cakraa-mang köi e. Panö, kraa nong nö, kacin khai yok pe, wang ting döi reng-reng ka, ha-müt yaa pe kacin. En pu cakraa-mang tek ce, cu böt pe. En pu cakraa-mang kröi maa kha'böt. Ng ng ng, ang lu tak, ang lam tak, ma ngapok phai or kot la ang, ha-müt yaa cepo yaa kot döi*

---

[853] female servants

[854] *mia reng cu* = ? (*mia*: who, whose; *reng*: chief; *cu*: grandchild), here simply translated as "which", because the second part of the phrase will be sufficient.

[855] to slip (away), to dislocate (oneself)

[856] to put on (a ring), to slip into

[857] to put (s.th.) into

[858] *thut*: to pour out (water)

[859] to put (s.th.) into; to keep (s.th.) in

[860] lit.: that of word

[861] grandfather, mother's brother = father-in-law

[862] to move

[863] even

pe. Nö kabüa cong la, lu lam nö, matyaa da, tawai rutca hai tape, khwön[864] *la kö*[865] *khöt mör*[866] *pe, mat-ya da, kut plan yaa yok, mi ngapok köi-e. Alai kh'e or yaa mang kröngca tai tak, cakraa-mang, apu cakraa-mang tut ka-e. Ha'nö, ting khök pe.*

As told, he took along one gun (yet to be mentioned). The heavenly lord sent five men, five men to accompany him. Hm, you, your father-in-law, the heavenly lord, said "accompany him". Hm, I still take a bath, I take a bath and still tidy me up, you go ahead up there. That gun, however, take it up first, he says, carry it up. Even five men cannot lift it. Heavy as heavy can be, that gun. Oh, we cannot, we will go again, we five man, we cannot carry it. What a heavy gun. He still takes a bath and tidies up, your son-in-law, they say to the heavenly lord. A while later, he sends them again. As they continue not to arrive, he sends ten men. As your father-in-law, the heavenly lord, said, let's go. Your father-in-law, the heavenly lord, is looking for information. Ya, ya, ya, I am still taking a bath, still making me tidy, this gun, however, carry it up for me. But even ten men cannot carry it. And they wait for him, he takes a bath, and then he himself with his left hand hangs it (on his shoulder) and he himself can walk back with his gun. As before he easily can walk up to the heavenly lord, to his father-in-law, the heavenly lord. And then he has arrived.

Ting khök, apu cakraa-mang tut ka-e, ting khök paka-caa-lö ha'nö, na-mi war kayak pawar pe baa: mokca lö, awai rau-e kh'e, tek pe-mi, en mokca lö, en wang po ang camaa caböt cace da, arau da, yu a, chön cu lang nom pren köi ka, deng-rüa hap tawek[867] *döi yong köi po la pe ang. Dö'nö, cang la en lak-chuk*[868] *maraa poi, mokca lö, pa pape. Panö, kar la hi-li hü-lü*[869] *pe, po la ngapok lök. Ang kapo tai a, tong cang a cö, apu cakraa-mang lai cace yaa kwak döi. Kapo, paha'nö, cia-rua mang tiama rong ka e, cia-rua mang hot or phü-lü*[870]*, pe, paha'nö, kar la hü-lü, pe. Tong cü, mia reng cu mang cem-ma caa cö, mia macu manar caa cö, ni katöng yak kato kh'e da, taa anguk klö kar klö dami. Ani da, wang kar hü-lü kabö ce, nayong naa döi. Apu cakraa-mang tek, chön cu lang nom pren köi ka deng-rüa hap pe, tapo kom, tawek dö'yong. Tong pa cö ang yaa po ca ün ca ü. Nayong ka, ang taprüa ka, ang kar pape. Nayong naa döi, nayong paka-caa-lö ha'nö, ang dui phai hot cong ang. Ang naa*[871] *kapo en co, tawek dö'yong, pape. Köi po, kapo nö, kapo tai a, ha'nö, kachör yun-lun*[872] *nö, kakrem a ngot*[873]*, ha'nö yon la a kyok*[874]*, pe. Nö, wang kut plan po la yok ayaa tut e, ayaa bai*

---

[864] to hang up
[865] to hang down
[866] weight, heaviness
[867] worn out, ragged, torn
[868] < *Marma: *lak chup:* to close the hand to a fist; maraa < *Marma: ma'ra': to not receive; see also footnote 365.
[869] (onomatopoetic, here translated as "like a gibbon", because of *hü* = gibbon, *Hylobates huluk*, but could also be translated as "loudly", "heartrendingly", "heartbreakingly", etc.)
[870] = ?
[871] truly, really (?)
[872] (expressive) qickly wriggling
[873] *krem* (= *ngot*): to bite off
[874] = ? as a whole ?

raa e, wang ting kh'e da, wang ör[875] *pe plot pe. Wang ör pe plot, pa caa-lö ha'nö, na kh'e da, tawek dö'yong pe kapo la.*

He has arrived with his father-in-law, the heavenly lord. After he had arrived: Today you can stay for one night, son-in-law, tomorrow morning, he said, you son-in-law, because you came to take my daughter, in the morning, down there, go to the land overgrown by thorny rattan, go and fetch an untorn banana leaf, to give it to me. And if not, I'll make it to you a welcome feast, son-in-law, so he says. And then, he weeps like a gibbon, takes his gun. I am going to take it, what to do? Because he cannot throw away his father-in-law's, the heavenly lord's words, he takes it, and then the big snake king on the way, the big snake king rests looking out, and then he comes crying like a gibbon. What is?, whose highest king's child, whose slave's child, of whom every day, the whole night I hear sobbing and weeping? Today you weep like a gibbon, alas, not that way. The father-in-law, the heavenly lord, said: go to the land overgrown by thorny rattan to fetch an untorn banana leaf. How can it be that I fetch one? Therefore I worry, I weep, he says. Not that way. Under these circumstances, however, stay on my eggs and guard them for me. I truely just fetch it, untorn, he says, and goes to fetch it. Goes to fetch it, will fetch it and advances, quickly wriggling, bites it off, swallows it whole. Then he comes back again, bringing it back to him in his nest, after he arrives, brings it up from the stomach and spits it out. When he spits it out, what he brings is untorn.

Maing[876]*, ang mokca nöm en, pa pape. Paka-caa-lö ha'nö, awai ni kh'e da, mokca lö, pai a chüng hung co kap*[877] *köi ka, wa-kling*[878] *mru tui mi katöng dit, pape. Na kh'e po kar la hü-lü yok pe. Nö, wa-mu*[879] *mang hot or phü-lü, pe. Uh, mia reng cu mang cem-ma caa cö, mia macu manar caa cö, ni katöng yak kato kh'e da, taa nguk klö kar klö dami. Uh, ang le, apu cakraa-mang tek, pai a chüng hung co kap köi ka wakling mru tui, pe, ang katöng kom. Tong pa cö, taköp po hot ta nanong, tawö po hot ta nanong, tong pa cö ang kayaa töng kom. Nayong ka, ang kar, pa pape. Uh, nayong paka-caa-lö ha'nö, ang katöng en co, ang dui phai hot cong ang, pa pape. Pe a pön lök, paka-caa-lö nö, mai-chang ka kayö, tahou*[880] *kh'e, kakup chut pe, katöng nö, wang or lot yok pe, tahou kh'e. Taa taköp rap*[881] *pe, wang po la pe nö, wang pe a, pe. Panö, wang po a pe yok pe, apu cakraa-mang tut-e.*

Wow, you are really my son-in-law. Under these circumstances, tomorrow, son-in-law, there, in the mountain range folding up, fetch a little bit of the bird-man's water, he says. That time also he weeps again like a gibbon. And the eagle-hawk king rests looking out. Oh, which highest king's child, which slave's child, of whom every day, the whole night I hear sobbing and weeping? Oh, that's me. The father-in-law, the heavenly lord, said: There, in the mountain range folding up, I shall go to fetch a little bit of the bird-man's water. How, it closes itself for one moment, it opens itself for one moment, how can I go to fetch it? Therefore I weep, he says. Oh, under

[875] to bring up (from the stomach) and spit out

[876] < *M *hman*: true, correct, right

[877] to fold up, to flap

[878] song bird

[879] eagle hawk

[880] to open itself

[881] to clap

these circumstances, I just fetch it for you, my eggs, however, stay and guard them for me, he says. He takes a small bamboo tube, and under these circumstances, flies off at once, when it opens, he quickly enters, fetches (the water) and comes up and out again, when it opens. One hears the clapping, he brings it and gives it. And then, he brings and gives it again to his father-in-law, the heavenly lord.

Panö: Maing, ang mokca nöm, en tarap mace da, panö, awai ni kh'e da, yu a, kapo en deng-rüa hap mi, cia-rua mang ka ap köi, mokca lö, pa pape. Kar la hü-lü, pe, uh, alai kh'e da cang ang cen cu[882] *mi cadia, tong pa cö ang naa ap cö, pa-pe. Tong cö en wang kar hü-lü yok cö, caratca ö, pa pape. Uh, nayong baa, ang naa ap en pe, pu cakraa-mang tek, lai kh'e da, en cadia cang ang cen cu, pape. Ah, cakraa-mang tek kha'böt, ka-caa-lö ha'nö, ap dia. Tong pa cö ang yaa nga-rang*[883] *cö, pa pape. Mai-chang ap pace, tanaa mi tanaa dami, ching kau taröi*[884] *bon-lon*[885] *ta pe. Paka-caa-lö ha'nö, kayua la plöi-löi*[886] *yok pe. Uh mokca lö, en maing ang mokca nöm, pa yok pe. Nö, awai ni kh'e kut plan tek pe, khai yok pe. Paimi waa-mu mang ka ap, cang la lak-chu maraa poi la. Pu-chaa mi, nö, khöt pu-chaa mi wö'döi. Nöm-nöm madök, alongca nömca ü, en, pami ka ka. Paka-caa-lö ha'nö, ka ap yok pe. Kaköi carüt-rüt, kar la hü-lü yok. Tong cö en kar la hü-lü yok kabö, tong alö cö wöi khai yok, en ka. Uh, alai kh'e da, en cadia katöng ang wakling mru tui mi, en cadia cang ang cen cu mi, ko da pu cakraa-mang tek en ka pe, ang naa ap en kom. Ang chak prüa rau en ka ang kar lot ta pe. Tong cang a ang cö, cakraa-mang lai cekhai böt, kacaa lö ha'nö, yaa ngarang döi. Ap dia, panö, na-mi po ap yok pe. na-mi ti tanaa mi pütma raima ching taröi ban-lan da pe. nöm-nöm pe yuk. Na-mi po, kapo la yok pe.*

And then: Wow, my real son-in-law. But tomorrow, down there, shoot the big snake king, who brought you the banana leaf, son-in-law, he says. He weeps like a gibbon: Oh, he is the one who helped me last time, how shall I really shoot him?, he says. What do you come weeping like a gibbon again, orphan?, he says. Oh, that way I shall truely shoot you, father-in-law said, in former times, you were the one who helped me, he said. Ah, the heavenly lord has said it, under these conditions: shoot! How can I be not up to the expectation?, he says. When he at once shoots him, he wriggles and struggles, trees and bamboo he smashes row by row. Under these conditions, he pulls him away and drags him back. Oh, son-in-law, you truely are my real son-in-law, so he says again. Then, tomorrow go again and shoot that eagle-hawk, I make of him a welcome feast. Busy ones and how much busy one's aren't they. Truely, probably you really achieve what seems impossible. Under these conditions go and shoot again. He goes to get ready, weeping again like a gibbon. What do you again weep like a gibbon, alas, what do you want again? Oh, last time you were the one who went to fetch me the bird man's water, you were the one, who helped me, now father-in-law, the heavenly lord said: you, I shall really shoot you. My heart feels worrying pain for you, I weep forsakenly. What has he done to

[882] to help
[883] not up to the expectation, to not fulfill the expectation
[884] to break into pieces
[885] *bon-lon (= ban-lan = bön-lön)*: one after the other, row by row
[886] *yua ploi-loi:* to drag away (by pulling)

me, the heavenly lord's word is final. Under this condition, I must fulfill the expectation. Shoot! And then, this one too he shoots. That one too struggles and wriggles, all really big trees he smashes row by row. That one he brings back too.

Maing, mokca lö en, ang mokca nöm cekhai payok pe. Na khöt nö, wai rau e kh'e da, mokca lö en, en thung[887] *luk ka e, na cöm en arak bia, pang-kot ka e, na ap en ngapok. Arak bia taplek*[888] *ka-caa-lö ha'nö, en ka e cang la lak-chu maraa poi, pape. Cang nö, cang tai a, nayong cepo taa nö, a mokca thung luk cöm pe arak bia, pe, pang-kot ka e, apu ap pe ngapok pe, apu cakraa-mang, (pang-kot köi tang pe ngapok long). Cü khom cü döi pe, mokca, apu ap. Maing, ang mokca nöm en, pape. Apu ö, en po ang yong thung luk kacöm arak bia, en pang-kot köi ang ap en ngapok lai, pa pape. Mai-chang kahia ap. Tong po tuk tek a dö'ce, apu takrö*[889]*-mi, tatheu*[890] *a kun*[891] *pe. Paka caa-lö ha'nö: Maing, en yaa khai mokca lö, ang da yaa döi en yong, pape.*

Wow, you are a son-in-law, my real son–in-law are you, he says again. After that, tomorow, son-in-law, you, upon your thigh I place you a cup of *arak*, on your shoulder I truely shoot your gun. The cup of arak spills over, under this condition, I make of you a welcome feast, he says. He makes it, making it, but that way he hears it and upon his son-in-law's thigh he places a cup of *arak*, on the shoulder the father-in-law shoots the gun, his father-in-law, the heavenly lord. He not even trembles, he does not tremble, the son-in-law, his father-in-law shoots. Wow, you are really my son-in-law, he says. Father-in-law, you also in my way with a cup of *arak* placed on your thigh, on your shoulder I shoot your gun instead, he says. At once he shoots from a distance. He knows nothing to tell him, the startled father-in-law, falls down and remains lying. Under this condition: Wow, you could do it, I cannot do it your way, he says.

Panö, trau da a mang bot pa kh'e, apu or then a hom pe, kaneuma yung, or then a hom panö, wing wang töi lü pe hom maa reng-reng pe, mruca lü, caratca lü, pe ang hom maa lök, ang töi caa cekön nau, pape. Ko ang pe en praingya lai, pape. Pe lök pe, ko-mat, ma-mi hom maa ki caa khök nö, ko kh'e, en ki caa hom kh'e, cakraa-mang camaa ranit ram mi, ranit tök[892]*-mi, atök köi, a yang*[893]*, a yang caci, cham yang mi, wang hia lai en, cham yang lökca mi, na kh'e, en machi cham yang, en tuk mat dö'tham. Ako ang mön mi kaphai, mi co ang machi cham yang, pami en tek a baa, pa pape. Nö, ki caa hom kan pröng*[894]*, nö, ki caa hom kan pröng kh'e da, pe tök köi, lök lök, lök lök, ci wöi a ha'nö, a yang a yang caci pe, wang hia lai pe, cham yang. Paka caa-lö ha'nö, ko da, mü ma dia, mokca lö, en machi cham yang mi, en tuk mat tak ca ü. En tuk mat döi ha'nö, cang la en lak-chu maraa poi, pape khai yok pe, apu cakraa-mang. Panö, wing-mi klö pape, kamön klö pape*

887 thigh
888 to spill over
889 to be startled
890 to fall down (on the earth)
891 to lie (fallen down to earth)
892 to appear
893 to be alike
894 to finish, to bring to an end

(wing-mi hia mön klö pe), machi cham yang köi. Mi co ang machi cham yang, pape. Maing, en tuk mat dia.

And then, next morning, when he has to go, his father-in-law up there offers him cooked rice, very good cooked rice he offers him up there, and then a fly continually comes to ask for a grain of cooked rice, it begs the man, it begs the orphan: give me one grain of cooked rice, I want to join eating, it says. Now I give you a word of knowledge, it says. He gives it: Now that grain of cooked rice eaten, now, when you finished eating, seven appearing daughters of the heavenly lord come to appear, alike, alike they all, hairs alike, come here they call you, hairs being alike, a single one you will not know anymore. Now, I sit on one, however. "This one is my wife, to be sure", you should say, so it says. And he finshes his meal, and when he finished his meal, they appear, one one, one one, here are they and all are alike, come here they call him, hairs alike. Under these conditions: Now, look! son-in-law, your wife with hairs alike, you still know her yourself? When you don't know yourself, I make of you a welcome feast, he says again, his father-in-law, the heavenly lord. And then, a fly's sound, it comes to settle down on the wife's alike hair. "This is like my wife's hair, for sure", he says. "Wow, you know it yourself!"

Ko-mat da, mang ci-yüa böt. Mi wakling mru tui po, kapo la ci-yüa, pape. Nö, yu mang, yu mang, yu mang, nö, na-mi tui kan po, cakraa-mang camaa lai klö hai, na-mi pia-ti hia pe ta mi chataa o po hang po-wöi a, tiama köi ka, chön cu lang nom pren po, hang po-wöi, chop pren po hang po-wöi, paha'nö, tong po wöi döi, tiama yung tang, nong nö kating khai pe. Ting khai pa-caa-lö ha'nö, kakröi cet[895] *acaa pace, acaa da long-tang prang*[896] *khök pe. Long-tang prang khök panö, na-mi wakling mru tui hai yok pe kalu a plek a caa, na kh'e da, ngia-wüa por, amat-ci tatö ta yong talek teng*[897] *pe. Khöt khü ce*[898]*, amat-ci wang te*[899] *dong-thong, wang cang dong-thong, alai ka yong tu-tu, a lai ka yong thaing-thaing*[900] *yok pe, wang kut plan cang dong-thong, yüa yok.*

Cur kha'böt.

Now, you two will go off. Also this bird-man's water, take it along, you two, he says. And they walk down and down, and also that water pond, which the heavenly lord's daughter with the sound of her voice, which she left by that curse, the ocean also is not there; on the way, the land overgrown by thorny rattan also is not there, the firewood country also is not there, and then, nothing is there, the way is really good and shortly after they have arrived. Under these conditions when they search and call for their child, the child has completely become an old man. As it has become a completely old man, with that bird-man's water when they wash and pour it over their child, it becomes a tender baby, the two keep it anew as before. In short, they set up a family, make a family as before, as before they do their duties again, renewedly they form a family, they again.

---

[895] to search, to ask
[896] truly, fully
[897] *lek teng:* afresh, anew
[898] (?) sooner or later
[899] to stay, to lodge, to set up
[900] redublication word for *tu-tu*: alike, or (?) <*M. *thâm*: to shoulder, to do one's duty

Selected Bibliography

Ba Myaing, U. 1934, The Northern Hills of Ponnagyun Township. Jl. of the Burma Research Society 24:127–148.

Ba Thin, U. 1931, The Awa Khamis, Ahreng Khamis, and Mros in the Chin Hill Area (Saingdin) Buthidaung Township. Census of India 1931, Vol 11,1:249–256.

Belitz, H. 1987, Im Schatten des Waldes. Naturvölker in Bangladesh. Delta 1:32–43. München.

Bernot, L. 1967, Les Paysans Arakanais du Pakistan Oriental. L'histoire, le monde végétal et l'organistion sociale des réfugiés Marma (Mog). Paris/La Haye.

Bessaignet, P. 1958, Tribesmen of the Chittagong Hill Tracts (Asiatic Society of Pakistan Publications 1). Dacca.

Grierson. G. A. (ed.) 1904, Linguistic Survey of India III,3. Calcutta.

Hughes. W. H. 1881, The Hill Tracts of Arakan. Rangoon.

Huq, M. M. 1982, Government Institutions and Underdevelopment. A study of the Tribal Peoples of Chittagong Hill Tracts, Bangladesh. Birmingham.

Hutchinson, R. H. S. 1906, An account of the Chittagong Hill Tracts. Calcutta.

Hutchinson, R. H. S. 1909, Chittagong Hill Tracts (Eastern Bengal and Assam Districts Gazetteers). Allahabad.

Kauffmann, H. E. 1962, Observation on the Agriculture of the Chittagong Hill Tribes. In: J. E. Owen (ed.) Sociology in East Pakistan (Occasional Studies of the Asiatic Society of Pakistan 1). Dacca.

Latter, T. 1846, Note on some hill tribes on the Kuladyne river, Arracan. Jl. of the Royal Asiatic Society of Bengal 15:60–78.

Lewin, T. H. 1869, The Hill Tracts of Chittagong and the Dwellers therein. Calcutta.

Lewin, T. H. 1870, Wild Races of Southeastern India. London.

Löffler, L. G. 1960, Khami/Khumi-Vokabulare. Vorstudie zu einer sprachwissenschaftlichen Untersuchung. Anthropos 55:505–557.

Löffler, L. G. 1966, L'alliance asymétrique chez les Mru. L'Homme 6,3:68–80.

Löffler, L. G. 1975, Mru Tu Long. In: H. Berger (ed.) Mündliche Überlieferungen in Südasien. Schriftenreihe des Südasieninstituts der Universität Heidelberg 17:8–28.

Mills, J. P. 1931, Notes on a tour in the Chittagong Hill Tracts in 1926. Census of India, 1931, Vol. 5:514–521.

Mahmud. 2007, Ang ing Mru Rung. Life and Struggles of the Mro People of Bangladesh. Matri production.

Ohn Pe, U. 1931, The Awa Khamis, Ahraing Khamis and Mros in the Ponnagyun Chin Hills, Ponnagyun Township. Census of India Vol 11,1:257–264.

Phayre, S. P. 1941, Account of Arakan. Jl. of the Royal Asiatic Society of Bengal 10:679–712.

Rajput, A. B. 1964, Les Murung de la forêt de Banderban (Pakistan oriental). Objets et Mondes 4,2:119–148.

Riebeck, E. 1885, Die Hügelstämme von Chittagong, Ergebnisse einer Reise im Jahr 1882. Berlin.

Shafer, R. 1941, The linguistic relationship of Mru. Traces of a lost Tibeto-Burmic language? Journal of he Burma Research Society 31:58–79.

Spielmann, H. J. 1968, Die Bawm-Zo. Eine Chin-Gruppe in den Chittagong Hill Tracts (Ostpakistan). Heidelberg.

St. Andrew St. John, R. F. 1972, A short account of the Hill Tribes of North Arakan. Jl. of the Royal Anthropological Institute 2:233–247.

Van Schendel, W., Mey, W., A. K. Dewan. 2000, The Chittagong Hill Tracts. Living in a Borderland.-Bangkok.

ENDNOTES

i Menyang kindly provides both a Bangla and an English transcription of the Kramma names. My transliteration into „Latin" is based on the Bangla version (adjusted to the phonemes of Mru). Menyang's „English" version reads: ta, gne, u, mim, ba, da, a, fe, kai, how, di, cho, khaw, all, mom, nin, pa, o, aw, ro, c [!], thaw, oa, wa, ea, co, lan, la, hi, re, take.

ii In the following text, // will mark phonemes; | | English and transliterated Kramma letters (graphs); „" their pronunciation.; all written in „simplified Latin."

NOTE about the footnotes and endnotes in this book

Due to a formatting mistake, the two notes on this page have been formatted as endnotes; they refer to pages 475 and 476.

More importantly, due to the same formatting mistake, *all* footnotes have been numbered from 447 up (page 152), while this should have been note 1. Thus, there are no footnotes are missing before no. 447 on that page.